Advances in Science, Technology & Innovation

IEREK Interdisciplinary Series for Sustainable Development

Advances in Science, Technology & Innovation (ASTI) is a series of peer-reviewed books based on important emerging research that redefines the current disciplinary boundaries in science, technology and innovation (STI) in order to develop integrated concepts for sustainable development. It not only discusses the progress made towards securing more resources, allocating smarter solutions, and rebalancing the relationship between nature and people, but also provides in-depth insights from comprehensive research that addresses the **17 sustainable development goals (SDGs)** as set out by the UN for 2030.

The series draws on the best research papers from various IEREK and other international conferences to promote the creation and development of viable solutions for a **sustainable future and a positive societal** transformation with the help of integrated and innovative science-based approaches. Including interdisciplinary contributions, it presents innovative approaches and highlights how they can best support both economic and sustainable development, through better use of data, more effective institutions, and global, local and individual action, for the welfare of all societies.

The series particularly features conceptual and empirical contributions from various interrelated fields of science, technology and innovation, with an emphasis on digital transformation, that focus on providing practical solutions to **ensure food, water and energy security to achieve the SDGs.** It also presents new case studies offering concrete examples of how to resolve sustainable urbanization and environmental issues in different regions of the world.

The series is intended for professionals in research and teaching, consultancies and industry, and government and international organizations. Published in collaboration with IEREK, the Springer ASTI series will acquaint readers with essential new studies in STI for sustainable development.

ASTI series has now been accepted for Scopus (September 2020). All content published in this series will start appearing on the Scopus site in early 2021.

More information about this series at https://link.springer.com/bookseries/15883

Antonella Versaci • Hocine Bougdah •
Natsuko Akagawa • Nicola Cavalagli
Editors

Conservation of Architectural Heritage

Second Edition

Springer

Editors
Antonella Versaci
Faculty of Engineering and Architecture
University of Enna "KORE"
Enna, Italy

Hocine Bougdah
School of Architecture and Cities
University of Westminster
London, UK

Natsuko Akagawa
Faculty of Humanities and Social Sciences
School of Languages and Cultures
University of Queensland
Brisbane, QLD, Australia

Nicola Cavalagli
Department of Civil and Environmental Engineering
University of Perugia
Perugia, Italy

ISSN 2522-8714 ISSN 2522-8722 (electronic)
Advances in Science, Technology & Innovation
IEREK Interdisciplinary Series for Sustainable Development
ISBN 978-3-030-74481-6 ISBN 978-3-030-74482-3 (eBook)
https://doi.org/10.1007/978-3-030-74482-3

This Springer imprint is published by the registered company Springer Nature Switzerland AG
The registered company address is: Gewerbestrasse 11, 6330 Cham, Switzerland

Scientific Committee

Alessio Cardaci, University of Bergamo, Bergamo, Italy
Antonella Curulli, Sapienza University of Rome, Rome, Italy
Antonella Versaci, University of Enna "KORE", Enna, Italy
Chiara Pepi, University of Perugia, Perugia, Italy
Hocine Bougdah, University of Westminster, London, UK
Joost J. Cote, Monash University, Melbourne, Australia
Marco Filippucci, University of Perugia, Perugia, Italy
Maria Margarita Segarra Lagunes, Roma Tre University, Rome, Italy
Maria Panta, German University in Cairo, Cairo, Egypt
Michela Moretti, University of Florence, Florence, Italy
Natsuko Akagawa, University of Queensland, Brisbane, Australia
Nicola Cavalagli, University of Perugia, Perugia, Italy
Romain Iliou, Laboratoire HiCSA, Université Paris 1 Panthéon Sorbonne, France
Rosella Rossi, University of Florence, Florence, Italy
Valeria Menchetelli, University of Perugia, Perugia, Italy

Preface

Our built heritage has a wide-ranging impact on creating a sense of place and how we feel about the world we live in. In this, the physical representation of architectural heritage can play a central role in sustaining and, at times, reassuring our identity in a fast-changing world, but it also has significant economic implications.

The idea of conserving architectural heritage has seen a paradigm shift in past decades. The notion of authenticity and acknowledgement of cultural diversity has deepened and enriched our understanding of the places we aim to safeguard. Increasingly, new techniques are becoming available for undertaking this important task. However, what has remained central to the idea of heritage has been the recognition of its intrinsic value in taking forward what we cherish to the next generation. We need, then, to think profoundly about how we balance extrinsic and intrinsic values related to how we manage and conserve architectural heritage.

This book, based on key papers selected from a recent IEREK International conference, explores the broader social and financial benefits that can accrue to property and business owners by the restoration and adaptive reuse of restored buildings. In doing so, however, it also pays attention to the critical questions involved in such processes of managing and conserving the built environment. As increasingly new technologies and new materials are available to conduct conservation work, it asks how do we use these new technologies in a sensible manner? How do we embrace the use of traditional skills and materials in these processes of conserving? In exploring these and related questions, this book offers fresh insights into this important area of heritage safeguarding practice.

Natsuko Akagawa
Ph.D., Master of Cultural Heritage & Museum Studies
MBA, Expert Member ICOMOS, ICOM
The University of Queensland
Australia

Acknowledgments

We would like to thank the authors of the research papers that were selected for addition in this book. We would also like to thank the reviewers who contributed with their knowledge and constructive feedback in hopes of ensuring the manuscript is of the best quality possible. A special thanks goes to the editors of this book for their foresight in organizing this volume and diligence in doing a professional job in editing it. Finally, we would like to express our appreciation to the IEREK team for supporting the publication of the best research papers submitted to the conference.

Conservation of Architectural Heritage

In the world's constant fight for a global technological utopia, history became exactly just that. Efforts to preserve what already exists gave way to the world's vision of the future and its hovering cars. However, one does not need to be sacrificed for the other. Today's generation must stay aware of the fact that heritage preservation is still one of the most effective investments in a community, it rewards societies tenfold, as the impact of this movement is evident in even the smallest scales, not only in the most renowned projects.

This book focuses on the management and conservation of architectural heritage, with the aim of supporting the tourism industry and generating economic return, while respecting the effect on our environment. The development objective, therefore, must be to harness the source, while safeguarding the future of the natural resource, the historic town and its community.

Of the greater benefits of preserving heritage is that it provides a sense of identity and continuity in a fast-changing world for future generations. Moreover, management and conservation of the architectural heritage provide benefits to property owners, businesses and to the community as a whole by increasing the property values of both the restored buildings and surrounding properties. In addition, heritage conservation uses less than half the energy of new construction and reduces the construction waste as well. Our authors broaden the conversation on these matters honing in on this conservation's identification, importance, management and challenges.

The book also discusses various vital topics and strategies that should be considered during the process of protecting the heritage landscape such as governance, partnerships and the communities' contribution.

In conclusion, this book explains the importance of architectural heritage conservation and its potential in acting as a catalyst in areas such as tourism and economic development. It also discusses how heritage conservation inspires new development of imaginative and high-quality design that pushes the boundaries of development.

Cairo, Egypt

Mourad Amer

Contents

Identification of Architectural Heritage Conservation

Influence of Language on the Built Environment: The Forgotten Vernacular of Traditional Indian Architecture 3
Noor Fatima Zaidi, Zaira Butool Zaidi, and Gaurav Shorey

Characterization of Earthen Building Materials in Gölcük Vernacular Houses 13
Üftade Muşkara and Seda Karayünlü Bozbaş

Conservation of Gravitational Architecture 27
Ramiro Sofronie, Radu Cazamir, Liviu Gruia, and Sebastian Lungu

Luxor Temple as a Reactivated Holy Site. Sacred Architecture Between Cosmology and Authority 41
Rasha Said Abd el-Lateef, Yasser Mansour, Shaimaa Kamel, and Ruby Morcos

Partnerships for Sustainable Tourism Development in the Cultural Heritage Sites 57
Reham Kamel El-Khadrawy, Amr Abdalla Attia, and Rowaida Rashed

The Inner Courtyard as an Important Element in Architectural Heritage and Interior Spaces 71
Sarah Fathy Ahmed Fahmy

Save the Troglodytic Heritage of Beni Zelten 79
Houda Driss and Fakher Kharrat

Metrology in Egyptian Architecture of the XVIII Dynasty, in Thebes 91
Mónica M. Marcos González

Conserving Symbolism in Architectural Heritage—Case Study Eloquence in Depicting Philosophical Ideas Inspired by the Principles of Islam on Islamic Architecture Through Ages 109
Khaled El-Daghar

Cultural and Architectural Heritage Values of the Qift-Quseir "Myos Hormos" Road 125
Ahmed Ibrahim Othman

Management and Conservation of Architectural Heritage

Architectural Heritage and Tourism Development in Urban Neighborhoods: The Case of Upper City, Thessaloniki, Greece 139
Stella Kostopoulou

A Methodological Framework for the Conservation and Planning of Urban Spaces in Historical Centers Around Riverfronts. The Case of Arequipa, Peru 153
Carlos Zeballos-Velarde

Conserving the Past, Ameliorating the Future—Situating Conservation of Architectural Heritage in the Urban Context—A Case of India 163
Prathyaksha Krishna Prasad

Cultural Heritage and Contemporary Armed Conflicts in the Arab World 181
Mayas Nadim Ahmad Taha

Rehabilitation and Exploitation of Heritage Buildings. An Investment Approach 199
Mohamed M. Shawky Abou Leila and Magdy M. ElBastawisy

Towards a Sustainable Use of Architectural Heritage: How to Assess Accessibility and the User's Role 213
Maria Luisa Germanà and Elvira Nicolini

Heritage Conservation and Community Development Through a Sustainable Management Approach: The Case of Abu Mena 227
Hossam Ragab Mahmoud Ahmed

From Conservation to Revitalization. Ambiguous and Critical Conditions in the Heritage Conservation Experiences of Erice 239
Giuseppe Abbate, Maria Luisa Germanà, Raffaele Savarese, and Ferdinando Trapani

Importance of Architectural Heritage Conservation

A Review of Existing Literature in Saudi Domestic Architecture 261
Mohammed ALghafis, Magda Sibley, and Eshrar Latif

Syrian Cultural Heritage Conservation and Management for Future Generations: Issues and Challenges 273
Mahmoud Abdulkader Alghafri and Mohannad Ali Mohammad

Urban Conservation of Heritage-Sensitive Zones in India: A Methodological Approach 281
Shahim Abdurahiman and A. K. Kasthurba

Historic Preservation as a Tool for City Branding Case Study: Khedivial Cairo 291
Mona Y. Shedid and Noha H. Hefnawy

Conserving Oral Heritage Through Architecture 305
Ar. Renu Khanna and Er. Nakul Khanna

An Inquiry and Documentation of Architectural Styles of Residential Building Facades in Old Rajshahi, Bangladesh 317
Md. Asaduzzaman, Z. H. M. Monjur Murshed, and Md. Nazmul Hoda

Vjosa River Valley; Strategies for Sustainable Tourism 331
Irina Branko

Investment of Heritage Villages in Saudi Arabia—Case Study of Al-Khubara Village in Qassim 345
Mohamed Ibrahim Elbelkasy and Mohsen Mohamed Ibrahim Mustafa

Protection and Regeneration of Urban Historical and Cultural Area in New Period—A Case Study of Pingshijie Area in Nanjing 357
Yifan Cheng

The Impact of Symbolic Values in Regaining Lost Heritage Buildings After War: A Case Study from the Modernist Architectural Heritage in Mosul 367
Aseel Fatahullah Abdullah Al-Ta'ai and Saba Sami Mehdi Al-Ali

Inspiring Traditional African Art & Architecture in the Modern African Hotel Design Field 379
Alshaimaa Mortada Abozaid Ali

Advancing Technologies and Innovation to Conserve the Architectural Heritage

Towards Developing a Heritage-Based Architectural Methodology to Design New Buildings in Historic Areas, the Case of Beith in Scotland 401
Maryam Al-Irhayim and Adrian Friend

Importance of Digital Techniques of Documentation for the Conservation of Cultural Heritage 417
Mohammad Moein Danesh and Ahmad Rajabi

Strategic Architectural Interior Design Methodology for Recruitment of the Hieroglyphs Line in Developing Ancient Areas by Hi-Tech Technology to Conserve the Egyptian Heritage 429
Hala Hassanein

Preserving the Italian Heritage in Alexandria: The Don Bosco School as an Example 459
Heba Mahmoud Saad Abdelnaby

Architectural Innovative Technologies for Improving the Built Environment: A Technical Framework 475
Magdy M. El-Bastawisy and Mohamed M. Shawky Abou Leila

Building Virtual Environment for Safeguarding Archeological Ruins Case Study—El-Sheikh Ibada (Antinoopolis) 491
Noha Abbas Hamed, Mayada Belal, and Mona Raafat

Decision-Making Processes in Landscaping the Garden of the Museum at the Lowest Place on Earth, Dead Sea, Safi, Jordan 503
Alexandra-Aikaterini Andrianou, Georgios Papaioannou, and Konstantinos D. Politis

Heritage and Sustainability: Motives Behind the Use of Colours Derived from Natural Materials in Al-Qatt 519
Arwa A. Alghamdi and Abeer A. Alawad

Identification of Architectural Heritage Conservation

Part one opens this volume with a series of chapters that address issues of architectural heritage conservation in urban contexts in Africa, the Middle East and more. They explore a range of issues, values and approaches which are influencing architectural heritage practice, specifically those of which the authors argue, should be approached. In this part, the book deals with ways of breaking the barriers to heritage conservation by looking at a number of issues relating to contemporary heritage conservation initiatives from across the world.

For example, in the chapter titled "The author of the chapter titled "Cultural and Architectural Heritage Values of The Qift—Quseir 'Myos Hormos' Road", the author seeks to explore the idea of regeneration through cultural tourism by inserting the road in the Red Sea tourist program, in Egypt, as a new cultural tourism destination targeting a possible reconstruction that usually starts with the visible remains using virtual technologies. Dealing with a rare and endangered heritage is also a topic of discussion in this part, specifically in the chapter with the title of "Save the Troglodytic Heritage of Beni Zelten" as it looks at ways of saving the world-famous Troglodytic settlement in Tunisia. The authors, who adopt field surveys and literature analysis argue that the conservation of this endangered cultural heritage requires, on the one hand, raising the awareness of local people of the heritage value of their ancient built environment. The heritage authorities, on the other hand, have a role and a duty to intervene and form a multidisciplinary team that is working on the development of an action plan to save and protect this specific heritage before it is too late. Any conservation measures to be adopted must not be seen as an end in itself but as a means that enables the sustainable conservation of buildings for present and future generations. The third chapter in Part I is a moment where the topic of heritage conservation becomes a destination of a study into gravitational architecture that is bridging human endeavor in terms of buildings, from ancient times to the modern era. In "Conservation of Gravitational Architecture", the authors explain the role of gravity in the act of architectural creation the authors adopted an indirect investigation consisting of reprocessing the information acquired from the past. Only facts directly involving gravity are considered. Such a package of selected data is analyzed from the perspective of the knowledge reached by the 21st century and then accordingly interpreted into an original vision. Investigations are carried out in three successive steps, and each step is based on five different sources. The conclusion of the research is overwhelming. By the act of creation, the gravitational architecture praises the earthen life and its achievements.

The chapter titled "Luxor Temple as a Reactivated Holy Site. Sacred Architecture Between Cosmology and Authority", takes an unusual view of heritage as the act of reactivating ancient temples to be used as religious buildings of different religious denominations representing the power and religion of the day. The link between ancient festivals and modern celebratory traditions indicates the temple transformation as a reactivated holy site over the time. The research is using ethnographic, iconographic, archaeological and astronomical methods for the interpretation of findings, within a methodology of hermeneutic phenomenology. Other chapters in this part deal with strategies for heritage conservation with a view to promote sustainable tourism. In Chapter five, "Partnerships for Sustainable Tourism Development in The Cultural Heritage Sites", the paper evaluates the roles of different bodies that are directly related to the conservation of cultural heritage sites. Using a case study from Aswan, Egypt, a field study at Anakato ecolodge was conducted. The major result that has emerged from the study showed that achieving coordination and partnership across these parties is challenging, but it's a key to achieve sustainability. In the chapter titled "Characterization of Earthen Building Materials in Gölcük Rural Vernacular Houses" the authors take a different approach to architectural conservation. They carried out an archaeometric investigation to determine the physical and chemical characteristics of the materials that make up the buildings to be restored in order to develop appropriate restoration strategies and materials compatible with the principles of sustainable architecture. On a separate yet related note, an indispensable tool, in the field of conservation and restoration of architectural heritage, is metrology. It is applicable in the composition of architecture of past times, for archaeological excavations, and for reconstitution of constructive elements. That being said, the chapter entitled "Metrology in Egyptian Architecture of the XVIII Dynasty, in Thebes (Egypt)" has the objective to determine the geometrical and metrological relationships in the Theban architectural models of the XVIII dynasty, Thebes, Egypt.

While the final chapters of this part share, in one way or another, the same message as they explore ways of better understanding the heritage site targeted for conservation, they take and present different routes to making their analyses. Each of the chapters is ultimately concerned with highlighting the importance of recognising inherent local values when considering questions of the preservation or reconstruction of architectural heritage. The section argues that the local significance of historical Monuments and buildings in cities warrants the preservation of their architectural heritage, and suggests ways in which their legacy can be preserved and celebrated. In the chapter "Conserving Symbolism in Architectural Heritage—Case Study Eloquence in Depicting Philosophical Ideas Inspired by the Principles of Islam on Islamic Architecture through Ages", the author presents the reader with a clear exposition of the crucial Islamic religious beliefs that underpin key features of traditional Islamic architecture. Addressing a quite different cultural context, that of India, is the the chapter entitled "Influence of Language on the Built Environment: The Forgotten Vernacular of Traditional Indian Architecture". The authors present a powerful argument for the importance of recognising the principles of the vernacular in architectural heritage that underpins traditional Indian architecture, if not the architecture of the world, against what they show are the pressures of globalisation embedded in a linguistic hegemony. The next two chapters deal with questions of urban conservation and reconstruction. The final parts of this section considers further examples of architectural heritage in distinct cultural contexts challenged by foreign intervention. While.... Altogether, this part introduces the reader to several distinct cultural contexts in which issues of architectural heritage conservation are currently being addressed.

Influence of Language on the Built Environment: The Forgotten Vernacular of Traditional Indian Architecture

Noor Fatima Zaidi, Zaira Butool Zaidi, and Gaurav Shorey

Abstract

This study explores language and architecture, two domains that take shape from their ambient social structures, and attempts to examine the link between them. It examines the impact of changing patterns of the spoken language of urban India on its building philosophy, to draw conclusions for the future impact of the hegemony of Western languages on local architectural identity. Throwing light on the expanding practice of English-medium instruction and documentation of knowledge in the Indian subcontinent, the study argues that the grammar of Western languages does not imbibe the essence of traditional Eastern concepts, thus losing the philosophy behind them in the process of translation. Moreover, it argues for the proliferation of native languages in India, as a counter-hegemonic tool towards the increasing proclivity of urban India for Western style of architecture. It makes the case for bringing the focus of Indian architecture back towards notions of sustainability that are inherent to the traditional Indian vocabulary.

1 Traditional Architecture; Linguistic Hegemony; Vernacular Language

1.1 Introduction

Urban India of the twenty-first century exhibits a dichotomous built landscape. On one side, an influx of tall, innovative, technologically proficient structures is fast becoming the architectural reality of the present; while on the other side, standing in stark contrast and almost overshadowed by the former, stand the ancient surviving monuments of Indian cities, testament of a forgotten past. These monuments do not remain as much more than tourist attractions anymore, much less as having a utilitarian value of any sort. And yet, much has been written about the extraordinary knowledge base and architectural expertise that drove the genesis of these structures, not only because of their survival over centuries and the simplicity of the philosophy behind their designs, but also due to their inherent ability to incite a reminiscence of the period and region they were erected in. Conversely, the new-age, chrome and glass specimens of contemporary architecture are much sought-after, if the rising trend of *modern* houses and corporate hubs in Indian cities is any indication of the trajectory that development is taking. What is the reason behind this expeditious transition in the discipline of architecture in urban India? Is there an external influence that is driving it? The answer may lie in the field of linguistics. But what does a material science like architecture have to do with the study of language and meaning?

The spoken word is as old as human consciousness. It is an expression of human needs, creative perception and experience. The purpose of language is to enable humans to have an expression for conscious thought. It thus may be a legitimate claim that language, in its most fundamental form, is as old as man. The same can be said for shelter, providing relief from the wrath of nature. Architecture and linguistics are two vastly different domains and may overtly not seem to be related. But on closer scrutiny, it is evident that they share certain fundamental similarities. For instance, it has previously been argued that both systems are the progeny of the need of the people of the time and so evolve with changing needs over time. Just like the language of a place, its architecture can only be as proficient as the expertise of its people. Thus, an examination of these two aspects of culture paves the way for further inquiry into all cultural phenomena

N. F. Zaidi (✉)
Indian Institute for Human Settlements, Bangalore, India

Z. B. Zaidi
Secure Engineering, Abu Dhabi, United Arab Emirates

G. Shorey
Swaraj, New Delhi, India

A. Versaci et al. (eds.), *Conservation of Architectural Heritage*, Advances in Science, Technology & Innovation,
https://doi.org/10.1007/978-3-030-74482-3_1

of a community, material or otherwise, that evolve along with each other due to continuous mutual interaction. Both these domains originate from two fundamental needs of man. The former arises from his need to communicate with his natural environment in order to survive and the latter from the need of the social animal to communicate with other humans. Both can be and historically have been used as tools by hegemonic powers to establish their political rule. Incidentally, there have been proposed theories of both architectural determinism as well as linguistic determinism of society, and hence at different points in time, both domains were thought to be the dictators of characteristic features of the human culture they originate in. Each of the similarities stated here will be elaborated as the paper progresses.

This paper examines the convergent and divergent aspects of linguistics and architecture as two components of culture and further presents a deconstruction of the impact of changes in the former, on the dynamics of the latter.

2 Language and Its Evolution

It is suggested by Alfred Korzybski, the man considered to be the father of semantics, that semantics, the study of meaning, is arguably an integral component of virtually every aspect of human endeavour. He claims that the arts, natural sciences and humanities are so deeply influenced and affected by semantics, that no inquiry can be objectively conducted and no understanding completely reached, without taking the role of language habits of the individual or community under scrutiny into account (Korzybski, 1995). It was thus inevitable that a scrutiny into the genesis and evolution of architecture opened up avenues of influence from the domain of linguistics.

Language is the foundation of culture and knowledge creation. Wolfgang Köhler, a German psychologist, conducted a study of chimpanzees to understand their patterns of communication and how their ways of communication impact their social relationships. He eventually discovered that their speech impediment constitutes the primary cause that prevents them from attaining any cultural development whatsoever (Köhler, 1973). Speech is, thus, a superlative advantage that man enjoys over all other animals. Moreover, his ability to express himself, narrate his experiences and learn from the experiences of his fellows is the main reason behind his ability to understand and partake in the mysteries of the world. In his book, Manhood of Humanity, Korzybski accredits the evolution of language as the driver of civilization. "Man improves, animals do not; man progresses, animals do not; man invents more and more complicated tools, animals do not… man is a builder of civilization, animals are not" (Korzybski, 1921). Human beings are not the only animals that engage in labour; but they are undoubtedly the only beings who improve their workmanship based on their past experiences. And these experiences take birth from the culture that gives direction to human lives.

A theory that has dominated the arena of linguistics is the idea of linguistic relativity. Given by Benjamin Lee Whorf, it suggests, "The background linguistic system…of each language, is not merely a reproducing instrument for voicing ideas, but rather, is itself the shaper of ideas." (Carroll et al., 2012). Following a more radical and less believable theory of linguistic determinism which claimed that language substantially shapes and limits the thought process and perception of its users, linguistic relativity theorises that language has a distinguished influence on thought. Whorf describes the significance of this principle in his book, Language, Thought and Reality as follows, "…users of markedly different grammars are pointed by their grammars toward different types of observations and different evaluations of externally similar acts of observation, and hence are not equivalent as observers but must arrive at somewhat different views of the world" (Carroll et al., 2012).

Thus, if language influences the thought process of its speaker, different languages give different directions to the philosophy behind the practices of each culture, be it eating habits, choice of occupation, leisure or the construction of dwellings.

Drawing parallels in the domains of language and architecture, Bill Hillier, Professor of Architecture and Urban morphology at University of London articulates that, "In vernacular buildings, the non-discursive aspects of space and form are handled exactly like the grammar of language, that is, as an implication of the manipulation of surface elements, or words and groups of words in the language case, building elements and geometrical coordination in building. In the vernacular, the act of building reproduces culturally defined spatial and formal patterns" (Hillier, 1996).

In the view of Roger Scruton, as he writes in The Aesthetics of Architecture, "…all that we ever find in architecture is found, at least in embryonic form, in the everyday vernacular in which most of us participate through our everyday lives". This leads him to a definition: "Architecture is primarily a vernacular art, it exists first and foremost as a process of arrangement in which every man may participate" (Scruton, 1979).

3 The Legacy of Vernacular Architecture

Vernacular architecture can be defined as "any regional system of domestic building design and construction that is passed on through generations as local tradition, through practice more than formal academic study" (Dharmalaya, 2020). The term *vernacular* in architecture implies the same as in the field of linguistics. Vernacular language is the way ordinary people talk in a given region, with all its idiosyncrasies and intricacies. Similarly, vernacular architecture is the manner in which the common folk designs and builds their residences and other structures of functional significance as per their culture and customs.

Henry Glassie defines vernacular buildings as *material culture*, the physical evidence of the ubiquitous cultural norms of a region, of which individual buildings are merely a part. He claims that all structures are *profoundly matters of social order*. Since social order cannot be isolated from the economic aspirations, and religious and political ideologies of any time, thus it is self-evident that the built environment cannot be understood outside of a cultural dimension (Glassie, 1990).

Another term that has lately become a zeitgeist of its own accord is that of *traditional buildings*. Without getting into the (valid, in author's opinion) argument that they should be simply called buildings, and all the innovations must be called *non-traditional buildings* or *modern buildings*, it is better to run through the definition of a traditional building. Traditional buildings are those built with solid walls made of mud, brick and/or stone, with timber or metal windows (frameless, in early days) and a mud, thatch or timber-framed roof; clad with slate or tiles (Dharmalaya, 2000). This has been the predominant process of building construction from ancient times until the nineteenth century across the globe. Throughout the Indian terrain as well, since times immemorial, architecture has been seen to be holistically responsive to the ambient climate and local setting. Of late, arguably since post-colonial times, this tradition has lost its significance.

4 The Hegemony of Language—English as the Lingua Franca

According to David Crystal, the reasons behind a language becoming a global language is dictated, not by the number of people speaking it, but who is speaking it (Crystal, 2002). Latin became an international language first because of the power of the Roman Empire and later continued to spread even after the downfall of this empire due to Roman Catholicism. He emphasises that a language may become a global language if it is a vehicle of great literature, or due to an association with a great culture or religion. But these factors singularly, or even in combination, do not guarantee its popularity. Crystal further states that, "a language has traditionally become an international language for one chief reason: the power of its people". For instance, even as early as two millennia ago, the armies of Alexander were the reason behind the international spread of Greek, the militancy of the Roman Empire was behind the popularity of Latin. On the contrary, a language is seen to become obsolete with time, when its speaker population is consensually convinced by foreign elements to abandon its usage and adapt to the foreign dialect. (Suarez, 2002).

One of the defined methods of moral and intellectual leadership, as described by Italian political writer Antonio Gramsci, is *leadership without force* (Gramsci et al., 1971). It is a form of rule where the dominant group exercises leadership over subordinate groups by a gradual development of a consciousness, and remodelling their ideas, rather than an exertion of overt strength. He coined the term hegemony, defining it as "intellectual and moral leadership through consent and persuasion". His concept of hegemony provides a philosophical framework to examine the power relations between dominant and minority groups, and the means by which the former secures its power and position over the latter. *Linguistic hegemony* or *linguistic imperialism* is a subset of this.

The term linguistic hegemony has been identified as what is achieved when dominant groups convince others to accept their culture, language, norms and usage as standard or paradigmatic and create such a consensus with no inclusion of or discussion with minorities or fringe groups. According to Wiley, linguistic hegemony is said to be ensured when some people, or their agents and supporters, can convince those who fail to meet their standards, to view this failure as being the result of the ineptitude of their own language (Wiley, 2008).

An example of linguistic hegemony is the spread of the English language across the globe. In the case of the English language, the evident link between language dominance and economic, technological and cultural power in the story of its spread throws light on its phenomenal success. From the beginning of the nineteenth century, English found itself at the centre of an explosion of international trade, scientific research and technology, most importantly communication technology and all these factors together inevitably gave it a global status. During the twentieth century, this status continued due to the economic development of the American superpower. It is for the same reason that English slowly became the lingua franca, i.e. the bridging language for interactions between linguistically diverse populations, which is again not dictated by geographical extents, but mostly by political factors. And now, with the advent of the United Nations and related forums since 1945, up to 190 countries have come to interact on the same platform. Thus, the need for a lingua franca is evident to avoid a large number of unnecessary translation exercises. But in this race of nations towards learning and mastering the English language to achieve an international standing, the use and proliferation of native languages seem to be coming to a sorry end.

The use of English has now become ubiquitous in international communication. Once confined to a handful number of tribes in British Isles, English barely escaped extinction itself, at the hands of the French language. Today, it has emerged as the spoken language of the second-largest number of people in the world, following Chinese. In terms of territorial spread though, English comes second to none. Ulrich Ammon comments on the dominance of English, claiming that it has the greatest number of speakers—close to 1.5 billion. It is the official language of 62 nations across the globe, the most dominant language in scientific publications with 70–80% of academic journals publishing in English, as well as the de facto working language in most international organisations. It thus comes as no surprise that English is the most widely taught foreign language in the world (Ammon and Hellinger, 1992).

Fishman and colleagues claim that, "English is used internally for official purposes almost exclusively by all non-English speaking countries that currently are, or have formerly been, under the political or economic hegemony of English-speaking powers." (Fishman et al., 1979). Phillipson goes on to outline that English qualifies people to operate the technology to which English itself provides access and brings *modern* ideas with it, serving as an inlet for social and cultural values. In some countries, English is the only permitted language in legal processions and authoritarian work (Phillipson, 1992).

Many academicians, who are opposed to the linguistic hegemony of English, are not fundamentally against the usage of the language itself. However, they see it as a vehicle using which the English-speaking world is engaging in a neo-colonialism of its own design. Wherever the language goes, the cultural perceptions and ideologies from English-speaking countries do not remain far behind, having reshaped the cultural landscape of various countries in the past, by way of marginalisation of their native cultures. The Western economic and diplomatic muscle has also been accused of introducing commodities that usurp or replace existing, indigenous products.

One of the main areas in which this is a recognisable issue is academia. The ever-increasing use of English language in documentation and proliferation of all scientific knowledge has led to a retraction from and discouragement of the knowledge base of native cultures, even in their own countries. A consequence of the hegemony of English language in Eastern cultures can be seen in the curriculum taught to their students of architecture. As noted by Udo Kultermann, in his book, architecture in the twentieth century, "There is an inexcusable gap between the research devoted to European architectural tradition and the traditions of India, Africa, China, Southeast Asia, and Latin America. The neglect has reigned to such an extent as if the old architectural monuments in the Southern Hemisphere are of no value, as much of its architectural heritage has been and continues to be destroyed" (Kultermann, 1993).

Pennycook has described the spread of English in terms of Galtung's concept of imperialism in which there exists a dominant centre, the powerful countries of the west, and the dominated peripheries, or the developing and least-developed countries (Galtung, 1971). The English language enjoys an elitist perception by non-English speakers, especially in peripheral countries. In institutions of higher education, writing theses in English is widely encouraged, even enforced as a rule, in non-English-speaking countries. Its position across the globe is also reinforced by the fact that international conferences are inevitably in English. Any ensuing publications are documented primarily in English as well, along with translations in several languages.

He further points out how English media from the centre have penetrated the media of peripheral countries, a linear flow of information that has led to persistent diminishing of cultural identity and political independence of the peripheral countries, meanwhile creating a dependency suffered by the periphery. He maintains that every language carries the essence of a civilization. Hence, a shift in language implies a shift in the importance and acceptance enjoyed by the new language, a change in its cultural matrix (Pennycook, 1998).

The problem arises when the prevalence of a lingua franca creates an elite section that has a proclivity for dismissing other languages, deeming them unnecessary or of a lower standing than the one that serves their political and financial interests. Knowledge generated in other languages has been largely ignored by the international community until very recently, but the solution was not what we hoped for. To be able to access treasures of native languages, they are being translated into, again, English. We have seen how different languages convey the same things differently, hence the term, lost in translation. Any bilingual person can easily think of at least a few terms in each language that cannot be translated into the other with the exact same meaning and intention easily, or at all, a restriction felt most strongly in the translation of philosophy and art.

In some countries, the dominance of the English language on education has taken a turn for the worse. For instance, schools in Kenya strictly profess only European literature. The students are thus only able to look at history and science and every other domain of knowledge as created and developed by Europeans (Ngũgĩ wa Thiong'o, 1997). Similarly, in most countries that are former colonies of the Western powers, the English language is seen to enjoy a higher status than their native languages, which are consciously sidelined in their education systems and not given primary importance, even during the initial years of schooling. English is also taught as a foreign language in more than 100 countries across the globe.

English is one of the several official languages recognised by the Government of India, along with major native languages across the Indian landscape. However, the formal recognition of the language in the Indian Constitution is not a problem. The problem is that English is a priority in not just education at all levels in schools, colleges and universities in all major cities and most towns, but also in India's foreign language teaching. It is the language that all students are taught and the one most available to adults.

5 The Impact of Language on Architecture

In a text called Space is the Machine, Bill Hillier defines architecture as "the application of speculative and abstract thought to the non-discursive aspects of building, as well as its application to the social and cultural contents of building" (Hillier, 1996). He further writes that, "…through the idiosyncrasies of style, building form becomes one of the primary expressions of culture." He describes architectural theories as not just analytic, depending on an understanding of what the human species is like, but also as normative, leading the way to how the world should be, more strongly than saying what it conventionally is.

Going by this definition, does the mimicry of Western architecture that can be seen more and more in the urban Indian landscape, go on to say that we are eager to abandon our traditional norms of building and adopt the Western way of design, an imitation that is aesthetically innovative on the face of it, but one that does not suit the climate we live in? The answer seems to be a resounding yes. If the intellectual process behind the manifestation of our building changes to a mere mockery of Western style of building, no integrity remains in the process of design. Inevitably, the result is an urban landscape filled with structures that the common people cannot identify with anymore; much less see them as a manifestation of the culture that shapes the society around them.

Following centuries of colonisation under European powers, and the more recent economic and technological dominance of the United States of America, over the course of the past few decades, architecture all over the world has come to follow a monotonous standard, a phenomenon popularly termed as neo-colonialism. This mimicry in architecture has been the subject of severe criticism across the globe, yet it continues to spread in practice, due to the convenience of mass production brought about by technological advancement, another area dictated and dominated by Western powers.

With no taught and practised literary tradition, the experience of any space remains confined to the present. It is only visible in its concrete form, without an imagination of the journey of its past. For instance, in an article titled A City without Time, Anand Vivek Taneja laments the loss of Urdu as the language of North India, as he writes, "In the violence of the partition, Delhi lost its literature. Urdu, rather than remaining the literary language of the *sheher* as it was, became the ghettoized language of Muslims alone, unavailable to others to read and learn. Without its literature, Delhi lost its ability to remember what it once was" (Taneja, 2015). He further goes on to beautifully capture the importance of the language of the capital, once upon a time: "Without Urdu, Delhi is a city eloquent only in its violence. Without Urdu, the monuments of the city are just dead congregations of stone, without ever having been a part of the city's life."

Western style of architecture has slowly crept up and encroached upon the ideology of Indian architecture. So much so that we are blindly following a technology and philosophy, that comes from neither a similar climate, nor culture. The Government of India has tried to preserve Indian heritage in its policies, both in terms of promotion of native languages and ancient monuments, yet, in practice, the official heritage protection and management are grossly redundant and inadequate (Thakur, 1990). Historical architecture seems to be at the mercy of the trajectory that development takes in due course; meanwhile languages dwindle at the onset of *well-meaning* globalisation. Criticising Western technology, Glassie writes that it is, "an exaggeration of… the wish to free will from environmental conditions, the intention to set the human being righteously in the role of exploiter" (Glassie, 1990).

It is not as if Eastern societies, especially the professionals, do not realise that the knowledge of their cultural heritage is slipping from their fingers. In the last couple of decades, there has been a growing disillusionment with market-driven architecture and Western-influenced concepts of design, which have given birth to a vernacular renaissance. However, a large-scale drive of translating ancient texts and documentation of construction methods and technologies into English has anti-climactically defeated the purpose of the self-realisation.

All of this is not to say that tradition must be perceived as a form of stagnation, or worse, as a noose around the neck of architectural innovation. It does not even mean an isolation of all cultures to preserve them as they have been since their genesis, something which is not even possible in this age of *one global village*. It is healthy to learn from other cultures, to trade technologies and to even indulge in the juxtaposition of different cultures, as long as one does not overpower another. As Joshua Fishman articulates in his book Reversing Language Shift, "Persistence no more means equilibrium, than change means chaos" (Fishman, 1991).

As touched upon earlier, the theory of architectural determinism given by Maurice Broady in an article titled *Social Theory in Architectural Design*, suggests that the built environment of a region chiefly determines social behaviour (Broady, 1966). Although this theory is not entirely convincing, since design factors cannot solely be expected to determine social interaction, it is understood that the built environment is a significant part of the effective environment, which includes the social structure and the cultural attributes of the people that inhabit it. Architecture is the largest, most impactful form of art, resulting from the knowledge and technology prevalent in an era. As Freeman describes in his book, Mental Health and the Environment, architecture can bring about anti-social behaviour, induce stress and depression in individuals, as well as create vulnerability to crime (Freeman, 1978). It has the largest environmental impact of all human endeavours. It is also seen to be capable of causing social malaise. This is an insight that most architects fail to apply in their method of practice. It is for this reason that Broady advocated rigorous instruction of the social sciences in schools of architecture and calls for the formation of joint sociologist–architect design teams.

Brian Ladd calls buildings, the *symbols and the repositories of memory* (Ladd, 1998). It is for this reason that throughout history there have been instances where the built environment has been used as a political weapon to impose hegemony of a ruling power on a region. Architecture creates the visual identity of the physical landscape, which lasts longer than any most professional practices, spanning over several generations at a time, and, if constructed with meticulous thought, can introduce and legitimise a new regime in the eyes of the public. An instance of this was seen in Nazi Germany in the 1930s and '40s, where Hitler took great care to create visual representations of his regime. Meticulously planned structures of neoclassical and art deco styles of architecture were erected to visually enhance and impose the presence of Hitler's power. He constructed urban monuments, buildings and concentration camps in this manner to modify the image ability of Berlin and other major cities in his favour. The converse of this has also proven to be truthful in history, that is, a destruction of important specimens of architecture can delegitimise a fallen regime. In a lecture on Rare and Endangered languages, Professor Tim Connell, an honorary life fellow of Gresham College, London elaborates saying, "With any regime change, a shift in architectural representation also takes place in order to redirect memory to shine new or revitalized light on a particular era that is in line with the ideals and discourse of the new ruling regime. At the heart of power is the idea of what is deemed preservable, what should be demolished, and who or what should be silenced" (Connell, 2011).

Is then the arrival and popularity of high-rise, glass buildings the harbinger of changing power structures in urban India?

6 Vernacular Architecture in Ancient India

Emperor Babur showed considerable interest in constructing buildings to establish the Mughal Dynasty in India in the 1500s, a passion that was followed by each of his successors as well. The calligraphic inscriptions on Mughal monuments were a political and religious proclamation of each ruler, and visually proclaimed where they come from, and the culture they want to propagate. Mughal structures in the Indian subcontinent are revered as the most beautiful surviving structures in the world, yet the architects behind them are little known personalities. Why was it so that the masterminds behind these marvels of architecture were merely called as *amla wa fa'la-i imarat* (building staff) and *sarkardaran-i-imarat* (supervisors of building construction) (Qandhari and Ahmad, 1993)? The answer is more philosophical than scientific. These structures, considered to be the epitome of architectural knowledge and expertise, were least concerned with who is making them, but made entirely as a tribute, out of love and reverence for a higher being, or a lover, or a kin. The intention was not to achieve fame or monetary elevation, or to impress an employer, or even to win a competition, but to be able to experience being a part of something greater than oneself, a spoke in a giant wheel of collective human effort. Such a practice of building meant that at the end of the process of construction, it did not matter who made it, only who it was made for. A significant factor that contributed to the phenomenal architecture of this period was that there was a hierarchical dissemination of skills and knowledge, often also through a system of apprenticeship. Such a system meant that the same skill was learnt, polished and excelled by craftsmen over generations, be it a *sang-taraash* (stone-cutter), *naqqaash* (carver), *durudgaraan* (carpenter), *khisht-mataan* (brick layer), *parchinkaar* (engraver), *gul-taraash* (floral design carver), *mambatkaar* (embosser) or *tabdan-taraash* (glass window maker) (Rezavi, 2013).

Vernacular architecture in erstwhile India was so entirely based on skill and experience that there was no need for multitudes of plans and drawings and construction details. The art was shown by hand, learned by doing. Has then, the modern practices of drawing and modelling and 3D imaging paved the way for a weakening of building traditions? Having worked extensively on folklore, vernacular architecture and material culture, Glassie provides valuable insight into the old ways of building and construction as he writes, "The key to vernacular technology is engagement, direct involvement in the manipulation of materials, and active participation in the process of design, construction and use (even if the social organization of architectural creation includes high levels of specialization). The product of engagement is knowledge" (Glassie, 1990).

A lack of engagement between architect and artisan in the contemporary age, in the opinion of Glassie, is the reason behind a lethal social transition, "…from communal to individualistic enterprise, from self-sufficiency to dependence; the gain in control over nature, accompanied by a loss of personal involvement in creation; (and) the gain of convenience, accompanied by the loss of pleasure in work…" (Glassie, 1990).

A study of the architectural practices of a period alone can be used to understand the collective identity of the society at the time. What will the collective identity of urban India in the twenty-first century look like, if mapped in this manner? What will it say about the cultural attributes of this time?

In this more secular century, subtle public policies are the tools that are used by modern-day nation states, instead of the brute force of redundant monarchies, to consciously or ignorantly eliminate from view the marginalised. The visual narrative created in today's cities, through the relatively nascent (in formal education) domains of city planning and architecture speaks volumes about who has a voice and who is silenced in an urban landscape (Dorroll, 2012). Today's glass corporate houses seen in the economic hubs of all major and most minor Indian cities thus go on to not just display the hegemony of the West-inspired education that is being taught to scholars of architecture in the classroom, but also point the way towards a future where our own culture will be completely lost at the hands of this neo-imperialism. As if to promote this McDonaldisation of Architecture, a term inspired by George Ritzer's book, The McDonaldisation of Society (Ritzer, 1993), but equally appropriate in this context, elements of regional architecture are occasionally added on the façade of structures for ornamentation, resulting in the ubiquitous production of tasteless and unoriginal kitsch.

Along similar lines, Glassie describes the consequences of cultural detachment on the dynamics of society, "Loss of vernacular tradition is usually associated with the creation of barriers to direct social interaction, compartmentalization of functions within a building, and the imposition of an external mask of symmetricity. These changes usually correspond to changes in the nature of a society, from one that is based on trust to one that is based on exploitative socioeconomic relations" (Glassie, 1990).

Connell suggests that if a region sees a shift in language, its speakers may experience a significant change in the perception of their needs as well (Connell, 2011). There needs to be a resistance towards English linguistic imperialism, where we increase public awareness of the pragmatic rewards of English language acquisition and use, but at the same time, take a stance of negation and denial of the cultural hegemony of English, or any foreign language for that matter. Sir Syed Ahmad Khan, an Islamic modernist and social activist of nineteenth century India, was a firm propagator of Western and Modern education, but for 16 years he fought against the instruction of this education being English, saying that higher education especially must be in *maadari zubaan* or the vernacular language of the people. Subsequently, he suggested that the medium of instruction in each of the four presidencies of the Bharat of his time should be their official languages—Bengali in Bengal Presidency, Gujarati in Bombay Presidency, Tamil and Telugu in Madras Presidency and Urdu in the United Provinces of Agra and Oudh (Youtube, 2016).

It is not merely a hypothesis that native languages like Urdu, and possibly even the more widely spoken Hindi, may not sustain over the coming decades. If this happens, it would not come as a surprise when the cultures and building philosophies that they shaped die a natural death with them. This is not even a problem of India alone. Connell emphasises that, "It is perhaps time that we feel a sense of urgency in the realization that half the world's 6000–7000 languages are so endangered that they are not likely to survive into the next century" (Connell, 2011). It has even been claimed that one language is being lost on average every two weeks (Dalby, 2003). Although Crystal argues that it is not possible to generalise the rate at which languages of the world are dying out, as the definition of threatened languages varies greatly, defined by *less than 100* to *less than 100,000* speakers (Crystal, 2002). Regardless of the definition, a sharp downward trend is evident in the diversity of native languages on all continents. The Foundation for Endangered Languages, established in 1995, emphasises that, "..a majority of languages are about to vanish", and that, "Now there is a fear that along with the gain in ease of communication there may be serious loss in the long run, not only for the people most closely affected, but even for monolingual English-speakers. The potential for danger lies in loss of diversity, on the analogy of genetic biology, where it can be shown that increasing standardization and uniformity holds dangers for the long-term survival of the population as a whole" (Foundation for Endangered Languages, 1995).

Since the start of the twentieth century, there have been a few Indian architects that have tried to mobilise the contemporary trend of architecture back towards its roots and discourage the visibly rising disregard for culture in urban societies. Laurie Baker for one, said in an interview, "My observation is that vernacular architecture almost always has good answers to all our problems. In every district, wherever you go, the people themselves take an active part in making their houses. Now, for whatever reasons, they have lost their skills, and need to look outside for help" (Bhatia, 1991).

7 Independent India—The Search for a New Vernacular

Post-colonial Indian vernacular architecture does not have a solid, singular identity. In hindsight, temple architecture of the south and indo-Islamic architecture of the north and north-east have served as the roots of independent India's architectural identity. One of the first major projects after 1947 was the Capitol Complex in Chandigarh, the seat of power in the upcoming capital city of Punjab. It remains ironic that the task of building an iconic complex showcasing India's newly independent identity was allocated to a European architect, in versatile forms of concrete, a material foreign to the Indian building culture. In the name of modernity, the complex emerged as an isolated structure, inciting awe from a distance, instead of an intimate connection with its admirers.

Analysing Corbusier's formative processes in designing the Capitol Complex, authors Puteri Shireen Jahnkassim and Norwina Mohd Nawawi theorise that his ideas may have Mughal roots, "To disassociate the new capital from its Colonial past and to create a new sense of spatial drama symbolizing the nation's hopes for the future, Fatehpur Sikri's renowned orthogonal and gridded urban plazas with its interconnected courtyards and cloisters, became part of Corbusier's arsenal of precedents, and these were abstracted and reworked into a new orchestration of urban spaces; and integrated with Modernised concrete architectural forms. The garden archetype and recurring traditional Mughal devices such as the *chattri* and the trabeated terraces allusions were simplified and synthesised with overlapping *spaces between buildings* such as bodies of water, platforms and a series of roofscapes" (Jahnkassim and Nawawi, 2016).

On the evolution of Corbusier's work in India, William Curtis comments in his book *Modern Architecture since 1900*, "Like Lutyens, he learned his lessons from the Moghul tradition, with its generous provisions of deep loggias, romantic roofscapes and water" (Curtis, 1996). Kenneth Frampton also observed similar inspirations in Corbusier's work: "Unlike Lutyens, who had exploited only the secondary elements of Moghul architecture, Le Corbusier appropriated the traditional parasol concept of Fatehpur Sikri as a monumental coding device to be varied from one structure to the next… The evident intent was to represent a modern Indian society that would be free from any association with its colonial past" (Frampton, 1985). This analysis suggests an effort on Corbusier's part to commemorate the Mughal roots of India, but the Capitol Complex itself has been criticised for not having aged well, only symbolising the idealism held by India after its independence from Britain (Curtis, 1996).

The vision of Lutyens' Delhi began with the decision of moving the capital of the Government of India from Calcutta to New Delhi in December 1911. Though started before independence in the 1930s, Lutyens' Delhi has an Anglo-Mughal character. Lutyens' Bungalow Zone until today continues to house the most prestigious real estate in the country. In their book *After the Masters: Contemporary Indian Architecture*, authors Vikram Bhatt and Peter Scriver summarise the works of the *masters*—Le Corbusier and Louis Kahn, whose works have blazed the trail that Indian architecture has tread post-independence (Bhatt and Scriver, 1991). Since then, everyone connected to India's built environment—architects, engineers, sociologists, academicians—have struggled to put a label to its identity. "Some of the tallest political leaders in the land lent their support to the revivalists, who sought to reach back a thousand years for architectural forms and details which symbolised various classical eras and golden ages of Indian culture. On the other side were a handful of intellectuals and architects who argued that monuments should be viewed in context of their times, that they were not be imitated and modern India required modern architectural symbols and forms to express the dynamism of a free people on their march to economic development" (Sebastian and Ravishankar, 2018).

The veteran league of Indian architects, including the likes of BV Doshi, Charels Correa, Chitra Vishwanathan, Raj Rewal, Achyut Kanvinde, etc., have explored indigenous built forms and materials in their work to a significant extent and continue to do so. These essentially took root from rural traditions of building. But the spread and knowledge of such forms are not nearly satisfactory in the country. Contemporary students of architecture are often taken on a tourism of the rare adobe building or isolated monument in their vicinity during the 5 years of their training, but it is hard for faculty to find actual sites engaging directly in these building customs to ground the understanding in their minds. Laurie Baker's building solutions based in the local climate and resource availability have a *rural* reputation.

8 Discussion

Supporting and encouraging localised modernity may be the answer to some of the questions this paper attempts to ask and embedding modernity in evolving traditions across the globe in the India of today's hyper-connected age. It is pertinent to mention the traps of hybridity as we think of a way forward. A hybrid of our colonial past and modern future with the country's predominantly rural identity can result in kitsch-like elements borrowed with a haphazard understanding. This may well lead to another seventy-five years of an identity crisis. It is important to give breathing space to local traditions under the inevitable onset of modernity and educate the young entrants into the field of architecture about the journey of the century that has gone by, in a language that they understand well. This is the challenge that India's developing society faces.

9 The Way Forward

This paper is an attempt to urge academics, professionals and civilians alike, to not let the persistent draft of Western culture and languages inundate the heritage of knowledge of their own unique cultures. In a book titled Reversing Language Shift- Theoretical and Empirical Foundations of Assistance to Threatened Languages, Joshua Fishman, who has written and worked extensively on revitalisation of threatened languages, claims that a plebeian beginning is necessary for the success of language revival, instead of formulating and adopting newer and more pleonastic heritage conservation laws that would merely be *on the books* (Fishman, 1991).

The damage that has already been done may not be easily reversible, but it is imperative that a proactive approach now be taken, especially in the education system of nations that have been colonised in the past, to curb an insurgence of foreign culture on their society. An effective method of this may be to introduce subtle, but persistent modalities in their education system, to expedite the intellectual independence of the coming generations, so as to ensure future protection from hegemony of any form.

In the profession of architecture, the solution may not be to find another new approach towards building design, but to go back to old systems of building and relearn native languages to understand the vocabulary and sentiments that were involved in the building philosophy of the past. No foreign language can help retain and secure a culture, or validate the inefficiency of the *innovative* structures to be seen in contemporary Indian cities.

A look at the old systems of knowledge dissemination concludes that a self-serving, performance-based regime, driven by a constant fear of being outdated, in the field of architecture today, may very well end up annihilating the diversity of our built environment. Every scholar of architecture today is taught to experiment, to innovate and to bring up yet another ground-breaking design. The old practice of polishing the same skill over generations through apprenticeship and hands-on training is, most unfortunately, lost on the contemporary 'formal' education system. It is time to revise the way we are teaching our students, and this does not apply merely to the field of architecture or engineering, but every domain of knowledge. Teaching the limited paradigm of knowledge documented by the English people, in the English medium of instruction, with the crutches of English books and English research to our children, will not take us far.

This study also shows that the answer does not lie in any one knowledge domain. An isolated study of the past and present practices of architecture cannot expedite an understanding of what the future may look like. And if we do not see where we are headed, we cannot infer if it is good or bad for us, thus we cannot remodel it for the better. Similarly, simply studying language and its evolution may tell us about our changing perceptions, but not the impact it will have on lifestyle, heritage and the environment. The study does not exhaustively conclude that language is impacting only architecture, or vice versa. The solution to good architecture thus, we can conclude, lies in a multidisciplinary exploration of linguistics, anthropology, ecology, philosophy, sociology and theosophy as well.

References

Ammon, U., & Hellinger, M. (1992). Status change of languages. *De Gruyter*. https://doi.org/10.1515/9783110851625

Bhatia, G. (1991). *Laurie Baker: Life*. Works & Writings: Penguin Books India.

Bhatt, V., & Scriver, P. (1991). *After the masters: Contemporary Indian architecture*. University of Washington Press.

Broady, M. (1966). Social theory in architectural design. *Arena: The Architectural Association Journal, 81*, 149–154.

Carroll, J., Levinson, S., & Lee, P. (2012). *Language, thought, and reality: Selected writings of Benjamin Lee Whorf* (2nd ed.). The MIT Press.

Connell, T. (2011). *An introduction to rare and endangered languages*. Gresham College, London. https://www.gresham.ac.uk/lectures-and-events/an-introduction-to-rare-and-endangered-languages

Crystal, D. (2002). Language death. *Cambridge University Press*. https://doi.org/10.1017/CBO9781139871549

Curtis, W. (1996). *Modern architecture since 1900* (3rd ed.). Phaidon.

Dalby, A. (2003). *Language in danger: The loss of linguistic diversity and the threat to our future*. Columbia University Press.

Dharmalaya. (2000). *Society and network, dharmalaya institute for compassionate living*. http://dharmalaya.in

Dharmalaya. (2020). *Eco-architecture: What is it & why does it matter? —Dharmalaya*. http://dharmalaya.in/eco-architecture-what-and-why/

Dorroll, C. (2012). On the visual apartheid in Western Europe. *American Journal of Islamic Social Sciences, 29*(3), 85–105.

Fishman, J., Cooper, R., & Conrad, A. (1979). *The spread of English: The sociology of english as an additional language*. Newbury House.

Fishman, J. (1991). Reversing language shift: Theoretical and empirical foundations of assistance to threatened languages. *Multilingual matters*.

Frampton, K. (1985). *Modern architecture: A critical history*. Thames & Hudson.

Freeman, H. (1978). Mental health and the environment. *British Journal of Psychiatry, 132*(2), 113–124. https://doi.org/10.1192/bjp.132.2.113

Foundation for endangered languages. Home. Ogmios.org. (1995). http://www.ogmios.org/ogmios_files/22.htm

Galtung, J. (1971). A structural theory of imperialism. *Journal Of Peace Research, 8*(2). https://doi.org/10.1177/002234337100800201

Glassie, H. (1990). Architects, vernacular traditions and society. *Traditional Dwellings and Settlements Review, 1*(2), 9–21. https://doi.org/10.2307/23566248

Gramsci, A., Hoare, Q., & Nowell-Smith, G. (1971). *Selections from the prison notebooks of Antonio Gramsci*. Lawrence & Wishart.

Hillier, B. (1996). *Space is the machine*. Cambridge University Press.

Jahnkassim, P., & Nawawi, N. (2016). Allusions to mughal urban forms in the monumentality of Chandigarh's capitol complex. *Journal of Architecture and Urbanism, 40*(3), 177–190. https://doi.org/10.3846/20297955.2016.1210050

Köhler, W. (1973). *Dynamics in psychology*. WW Norton and Co.

Korzybski, A. (1921). *Manhood of humanity*. E. P. Dutton & Company.

Korzybski, A. (1995). *Science and Sanity* (5th ed.). Institute of General Semantics.

Kultermann, U. (1993). *Architecture in the 20th century*. Van Nostrand Reinhold.

Ladd, B. (1998). *The ghosts of Berlin: Confronting German history in the urban landscape*. University Of Chicago Press.

Ngũgĩ wa Thiong'o. (1997). Writers in politics. *Heinemann*. https://doi.org/10.2307/524619

Pennycook, A. (1998).*English and the discourses of colonialism*. Routledge.

Phillipson, R. (1992). *Linguistic imperialism*. Oxford University Press.

Qandhari, M., & Ahmad, T. (1993). *Tarikh-i-Akbari*. Pragati Publications.

Rezavi, S. (2013). *Fathpur Sikri Revisited*. Oxford University Press.

Ritzer, G. (1993). *The McDonaldisation of society*. Pine Forge Press.

Scruton, R. (1979). *The aesthetics of architecture*. Princeton University Press.

Sebastian, S., & Ravishankar, K. (2018). Post-independence architecture in India: A search for identity in modernism. *Journal of Emerging Technologies And Innovative Research (jetir.org), 5*(7), 1467–1473. https://doi.org/10.6084/m9.jetir.JETIRC006253

Suarez, D. (2002). The paradox of linguistic hegemony and the maintenance of Spanish as a heritage language in the United States. *Journal of Multilingual and Multicultural Development, 23*(6), 512–530. https://doi.org/10.1080/01434630208666483

Taneja, A. (2015). A city without time. *The Indian Quarterly*. https://indianquarterly.com/?p=2386.

Thakur, N. (1990). A holistic framework for architectural conservation education. *Third World Planning Review, 12*(4), 344. https://doi.org/10.3828/twpr.12.4.7696403438812g18

Wiley, T. (2008). Language policy and teacher education. *Springer*. https://doi.org/10.1007/978-0-387-30424-3_17

Youtube. (2016). *"Sir Syed Ahmad Khan"—views of Prof. Iftikhar Alam Khan* [Video]. https://www.youtube.com/watch?v=vcSm5BPOcY4.

Characterization of Earthen Building Materials in Gölcük Vernacular Houses

Üftade Muşkara and Seda Karayünlü Bozbaş

Abstract

Gölcük in Kocaeli city is located in the northwestern region of Turkey on the Anatolian fault line. It has been subjected to various civilizations according to archeological evidence and historical sources. Among 23 villages located in Gölcük district, Saraylı, Örcün, and Selimiye that have a long history dated back to around 1326 when the region was conquered by Ottomans are known for their well-preserved rural architectural heritage. The rural houses show architectural design properties that increase the structural resistance for earthquakes due to their various characteristics such as construction materials' properties. The study involves the determination of raw material characteristics of earthen construction materials of vernacular houses of Saraylı, Örcün, and Selimiye villages.

Archaeometric investigation carried out on samples from earthen bricks, mud plasters, and mortars of representative houses in the mentioned villages. The earthen materials were investigated from the mineralogical and chemical point of view including particle size distribution, aggregate/binder ratio, total soluble salt content, X-ray diffraction (XRD) analysis, Fourier transform infrared analyses (FTIR), and X-ray fluorescence (XRF) analyses.

The results of the study allowed an understanding of the building technologies and the properties of raw materials employed in the houses in different locations. The determination of the original construction materials would help to develop appropriate restoration strategies and restoration materials compatible with the principles of sustainable architecture.

Keywords

Vernacular architecture • Raw material characteristics • Hımış-type houses • Earthen materials • Sustainable architecture

1 Introduction

Vernacular architecture is broadly defined as the traditional buildings constructed by non-experts of a specific place through shared knowledge and experience of this place formed in time (Brown & Maudlin, 2012, 19). This knowledge is developed by the environmental factors, the local traditions of lifestyles, space-dwelling perceptions, religion, and the economic structure of society (Gür & Batur, 2005, 165). Accordingly, the vernacular architecture is described as "the architectural language of the people" and the product of non-experts (Oliver, 2006). The environment is the prime agent that affects the selection of construction materials. For instance, Kuban has grouped Anatolian traditional civil architecture into seven regions according to local materials and construction practice (1995). By definition, it could be concluded that the essential components of the concept of vernacular are the original construction material and building technique of a specific location. Merely with the presence of these, a dwelling could obtain its vernacular contexture either in urban or rural areas and gain authentic identity. This inference is in accordance with Nara Document on authenticity.

Nara Document mentions that authenticity appears as the essential qualifying factor concerning heritage values. According to text, understanding authenticity plays a fundamental role in all scientific studies of cultural heritage. The

Ü. Muşkara (✉)
Faculty of Fine Arts, Department of Conservation and Restoration of Cultural Properties, Kocaeli University, Kocaeli, Turkey
e-mail: uftade.muskara@kocaeli.edu.tr

S. K. Bozbaş
Faculty of Science and Letters, Department of Analytical Chemistry, Kocaeli University, Kocaeli, Turkey

A. Versaci et al. (eds.), *Conservation of Architectural Heritage*, Advances in Science, Technology & Innovation,
https://doi.org/10.1007/978-3-030-74482-3_2

authenticity of cultural heritage components could be recognized through analytical evaluation of the aspects given in the Nara Document as form and design, materials and substance, use and function, traditions and techniques, location and setting, spirit and feeling, and tangible and intangible values involved. In Turkey, Eres indicates five key issues in the conservation of vernacular architecture in rural areas (2016), including social, economic, environmental, technical, and legal issues.

All these issues restrain the preparation of restoration materials according to original building materials. Because of this, some researchers suggest that contemporary construction materials and designs should be developed that would be compatible with urban texture (Eres, 2013). Sey describes "the restored" civil architectural examples in Turkey as "new oldies," and the restoration applications are carried out as demolishing and rebuilding with concrete (2003, 4–6). However, Turkish restoration regulation officially registers these houses as cultural heritage property that is under protection.

Especially in rural areas, however, vernacular architecture is the sum of multiple components such as local lifestyle, which depends on the relation between human and nature, and moral values of a society, which provides retrospection of its identity and the shared history. Preserving rural heritage in the context of a globalizing world through developing technology and communication has the priority for preserving cultural diversity and richness.

Meantime, rural architecture has gained much more attention over the past decades with respect to sustainable architectural design principles. "Sustainable architecture is defined as the body of activities which minimize the given harm to the environment, care about ecological balance, and use materials, water, and energy effectively during making, using and destroying process of needed construction." (Yılmaz & Bakış, 2015, 2261). Nine major sustainability evaluation categories include energy, ecology, pollution, and materials (Yoon & Kim, 2017, 43). Among them, energy is the highest weighted criteria, and materials are the third. Earthen materials are known to be low-energy and cost-efficient construction materials (Costa et al., 2019, 6). The researchers indicate that when the thermal behavior of the houses is measured, the utilization of local materials decreased the embodied energy used in building dramatically compared to a typical concrete house (Morel et al., 2001). Literature also mentions that cement-based construction is the reason for 40% of global energy consumption and 50% of the total waste load of the planet (Papayianni & Pachta, 2017). Cement is associated with high CO_2 emissions and requires a high temperature for production. On the contrary, earthen materials are also renewable and recyclable resources that reduce energy consumption. Therefore, analyzing the construction materials used in a specific area is substantial to maintain sustainability there.

Historic building materials could have very different mechanical, physical, and chemical properties. The physical and chemical properties of the raw materials and their interaction with each other would define the behavior of earthen construction materials. Earthen plaster, mortar, and adobe blocks are traditionally prepared by mixing soil, which contains the proper amount of clay, aggregates, and binder. They are generally named after the main binding material used in the preparation. Natural aggregates in the soil range in size in a wide range. Boggs indicates that because of this wide range of sizes, the most useful grade scales are logarithmic or geometric that have a fixed ratio between successive elements of the series (2009). Among these scales, the most widely used is the Udden–Wentworth scale (Wentworth, 1922). Sand is known as the stable fraction of the soil; however, it has a high degree of internal friction force, and it gives the appearance of cohesion when slightly moist (Joel, 2016). Meantime, silt has a lower frictional force than sand, and when wet, it displays good cohesion. The term consistency denotes the degree of plasticity of soil (Reddi et al., 2016, 155–171). Plasticity index, consistency index, liquidity index, shrinkage ratio, volumetric shrinkage, and linear shrinkage are important parameters to define the consistency of the soil. The plasticity characteristics of soil mainly depend on mineral contents. Therefore, the chemical and mineralogical compositions of the soil, the type of aggregate, and binder and the aggregate/binder ratios affect the behavior of mud plaster (Özgünler, 2017, 33–41; Meriç et al., 465–473). High aggregate/binder ratio, the addition of lime or gypsum, and organic additives such as dung or straw, cannabis, pine needles, and other fibrous materials were preferred to stabilize the mud mixture and reduce shrinkage tension of the drying process and increase cohesion and durability (Davey, 1961; Singh et al. 2016a, 522–535). Inclusions such as charcoal particles, shell fragments, brick dust, and crushed brick were also used as aggregate. For instance, Stefanidou and Papayianni mentioned the good cohesion between the binder paste and the crushed brick aggregates (2005, 848).

Gölcük District, which is located on the Anatolian fault line, has been subjected to various civilizations according to archeological and sources (Galitekin, 2005). Saraylı has a long history back to the fourteenth century AD, and they were founded during the Ottoman conquest (Köroğlu, 2007). Örcün was known from the Ottoman archives since the sixteenth century. Vernacular houses in the rural areas of Saraylı, Örcün, and Selimiye show the civil architectural characteristics of the nineteenth century in Anatolia. Among the 51 houses in Saraylı that were registered as samples of rural architectural heritage, four were collapsed down

(Köksal, 2013). Registered houses have undergone various modifications, but some still maintain their original facades. In Örcün, only 25 houses were preserved in the original construction, while in Selimiye, only ten houses remained with original architectural characteristics.

The construction system in Saraylı, Örcün, and Selimiye vernacular houses is the hybrid construction technique called "hımış" (Güçhan, 2007, 1–20). The sampled houses in Saraylı, Örcün, and Selimiye were built using this system with walls are infilled by mud bricks and plasters with mud and straw coverings on outer walls (Fig. 1). Exterior mud plasters were coated with a colored lime wash to grant protection and durability to walls in addition to aesthetic reasons. It is determined that along with their architectural heritage values, Gölcük vernacular houses are compatible with the principles of sustainable architecture. That is especially accurate because building materials employed for the houses comprise timber and adobe, used as plaster or filling material (Bayraktar & Kishali, 2013). The houses in Saraylı also show architectural design properties that increase the structural resistance to earthquakes due to their various characteristics, such as building materials. After the earthquake in the Kocaeli region in 1999, a study was conducted to measure the damage assessment of the timber-framed buildings of settlements in the area (Güçhan, 2007). According to this study, in most of the "hımış"-type houses investigated, only some plaster cracks and plaster loss were detected. It was concluded that besides the location, application of the lath and plaster technique (bağdadi) and nails increased the earthquake resistance of timber-framed houses. However, the use of mud brick as infill material was evaluated as "Wall materials such as stone, brick, and mud brick usually cause damage to the buildings during earthquakes as they add extra load to the structure or in some cases because they have weak binding mortars." (Güçhan, 2007, 848).

Fig. 1 Hımış-type house located in Saraylı

Another study was on the evaluation of seismic resistance of hımış-type houses by means of reverse cyclic frame tests conducted on different frames with various infills or claddings (Aktaş et al., 2014). It was observed that the measured capacity to demand ratios decreased when the infill or cladding was applied, while the capacities consistently remained more significant than the demand. Meantime, the performance of the lath and plaster technique (bağdadi) was found to be superior to the others. These studies on "hımış"-type houses mainly focus on infill material and frame geometry. The infill mud materials in the latter study were prepared based on traditional recipes. Nevertheless, the recipes are defined in a general sense.

"Mortar preparation differs from each other based on where to be used. In the brick masonry infill walls, mortar was prepared by mixing two units of sand, one unit of lime and water. Water was added slowly and its amount was adjusted according to the consistency of the final compost. When mortar was to be used in adobe infill walls, on the other hand, sand, water and straw were mixed, in undefined proportions, until the mixture has the desired consistency" (Aktaş et al., 2014, 1719).

This assumption is commonly accepted in many studies, and the historic earthen materials used in rural areas have not been gaining much academic attention. Therefore, a more detailed analysis should be applied for raw material characteristics of earthen materials of "hımış"-type houses. The consequences would further contribute to understanding the seismically resistant property of building materials of houses in the Gölcük rural area, where the earthquake in 1999 occurred.

Regarding the state of decay of the vernacular houses in Saraylı, Örcün, or Selimiye, any diagnostic study or in situ mapping concerning the decay factors, the process of decay, or the urgency of preservation has been not applied yet. Some studies mention the observations of the current state of the houses in Saraylı (Köksal, 2013; Usanmaz, 2007). According to this, the buildings in good condition refer to the houses, which maintain the original construction materials and facades.

The objective of this study is to define the mineralogical and chemical characteristics of earthen materials in Saraylı, Örcün, and Selimiye vernacular houses. Raw material compositions of earthen plasters and complementary materials, mortars and bricks, were determined by chemical analysis using XRF, the particle size distribution of aggregates, SEM-EDX analyses, and FTIR analyses. It is known that using new construction materials, such as concrete, would be incompatible with the original materials and induce further deterioration problems such as salt decay. Therefore, understanding the raw material characteristics of the original construction materials will help develop appropriate restoration strategies for vernacular houses in Gölcük

District. The results of the study enlighten the ingredient of the admixtures of plasters, mortars, and blocks; however, further experiments should be applied to obtain a standardized formula. It will also be important to define the main construction materials of the "hımış" type to interpret the earthquake behavior of the houses in the region.

2 Materials and Methods

The mud plasters, mortars, and adobe blocks were sampled from five houses, of which original construction techniques and construction materials are preserved. Two houses are located in Saraylı while others are in Örcün and the Selimiye village (Fig. 2). The sampled houses are generally dated to the nineteenth century; however, more specific dates have not been given even in registration records (Usanmaz, 2007, 22–29). The houses are in good condition with partly damaged timber load-bearing systems. The samples that were collected are listed in Table 1. Visual examination under a magnifying lens showed that more than one layer of lime-wash was applied on the mud plasters.

2.1 Particle Size Distribution of Aggregates

Except for the lime mortar from house 01 in Selimiye, all the samples were dissolved in 10% hydrochloric acid for overnight. The aggregates, then, were sieved after drying and separated as per grain size to obtain particle size distribution and to determine the binder/aggregate ratio of earthen materials. The particle size distribution of aggregates was determined by standard sieve analysis using 1000–500–250-125 and 63 µm sieves were performed with Retcsh Vibratory Sieve Shaker AS 200 basic. The particle size distribution of aggregates was evaluated according to the Udden and Wentworth scale (Caner, 2003; Sharma & Singh, 2019).

2.2 Soluble Salt Components

The amount of salt in the sample, as percent by weight, was determined by measurement of electrical conductivity. Powdered mud samples were dried for 24 h. 1000 mg was collected from each sample and mixed with 50 ml of distilled water and filtered (Caner, 2003). The amount of soluble salt in the samples was determined by the conductivity measurements using Hanna Instrument EC 215 conductivity meter. The results were calculated with conductivity equations and expressed in percentage (Black, 1965, 999–1010).

2.3 XRD Analysis

X-ray diffraction analyses were carried out on powdered samples of brick, mortar, and plaster the identify binders and of aggregate. The powdered samples were obtained by grinding. The unoriented powdered samples were examined by a Philips type PW 1800/00 X-ray diffractometer with CuKα radiation, at 40 kV and 40 mA.

a b

Fig. 2 **a** House 01 in Örcün (OR01). **b** Degradation of the case OR01

Table 1 Description of material samples

Sample name	Location	Sample description
SR01T	Saraylı	Adobe
SR02T		
OR01T	Örcün	
OR02T		
SE01T	Selimiye	Brick
SR01H	Saraylı	Mud mortar
SR02H		
OR01H	Örcün	
OR02H		
SE01H	Selimiye	Lime mortar
SR01S	Saraylı	Mud plaster
SR02S		
OR01S	Örcün	
OR02S		

The fine fraction with the diameters below 45 μm, which is also called binder-enriched fraction in earthen samples, was examined to identify crystalline phases. For this purpose, the oriented samples of the clay constituents were prepared by wetting the powders below 63 μm with distilled water and then keeping them until they were dried out at room temperature (Meriç et al., 2013).

2.4 XRF Analysis

XRF analysis was applied using Thermo Fisher Scientific ARL Perform to determine the element composition of earthen materails using pressed powder pellets.

2.5 FTIR Analysis

FTIR spectra were obtained in the region of 4000–500 cm^{-1} using powders directly with ATR Bruker-Tensor 27 (Bruker, Bremen, Germany) spectrometer. FTIR analysis was applied for the identification of organic components present in mud plasters, mortars, and brick.

2.6 SEM–EDX Analysis of Mud Plasters

The instrument for SEM–EDX analysis was JEOL JSM-6060LV scanning electron microscope operated at 20 kV.

3 Results

3.1 Particle Size Distribution of Aggregates

Particle size analysis which characterizes the raw material properties of soil used for plaster, mortar, and adobe blocks provided a distribution graph given in Figs. 3 and 4. Some of the components of aggregates separated by the sieve analysis are shown in Fig. 5. In most of the samples, charcoal particles were observed (Fig. 6a), and in some of them crushed bricked inclusions, cloth pieces, nutshells, or even a fly was also detected (Fig. 6b).

Using Udden and Wentworth scale, aggregates are classified as fine aggregate if the highest weight percentage of aggregates has the size ≤ 250 μ. When the highest weight percent of aggregates has the size ≥ 500 μ, then the aggregate is called coarse aggregate. According to this, all the samples that were analyzed are having coarse aggregates. The silt clay content of samples (below 0.063 mm) was found to be between 4 and 6% by mass for SR01S, SR02S, SR01H, OR02H, and OR02T and between 7 and 12% by mass for the rest of them. These values are not within the acceptable limits mentioned in Delgado and Guerrero for adobe blocks which were between 25 and 45%.

Except for the house 02 in Örcün (OR02), the aggregates above 2 mm (gravel) had the largest content with the ratio of 45–58%, while the portion of very fine particles (below 0.125 mm) was between 6 and 14%. The aggregate particles in OR02 were found to be evenly distributed in size. Sand

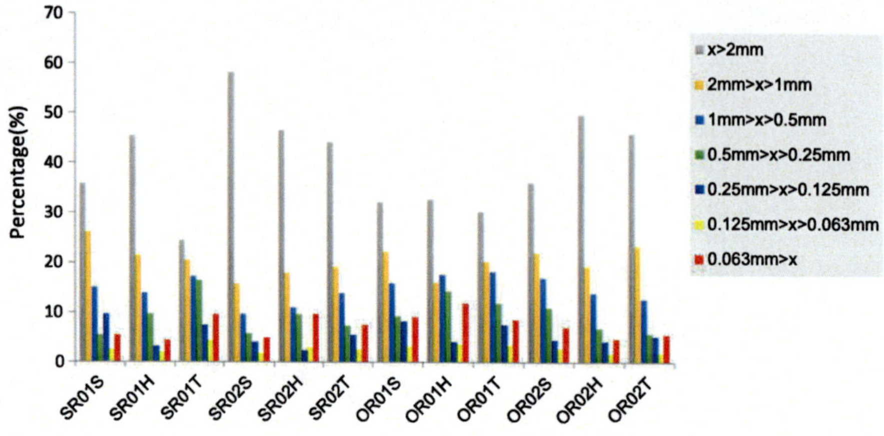

Fig. 3 Particle size distribution of aggregates for mud plasters, mortars, and adobe blocks in percentage

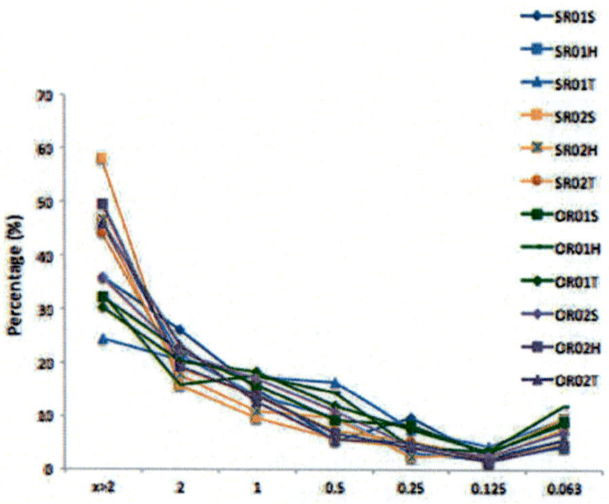

Fig. 4 Curves showing cumulative passing percent of aggregates for mud plasters, mortars, and adobe blocks

aggregates (2–0.063 mm) improve cohesion between particles and reduce the shrinkage and swelling as providing higher density (Bahobail, 2012, 23). Meriç et al. suggested that using very coarse aggregates in mud materials was expected to contribute to volume stability and, therefore, to the long-term durability of earthen structures (2013). The sand-to-silt ratio, which determines the quality of earthen materials,, was calculated between 6.5 and 11.5%, except for the mortar samples from house 02 in Saraylı and house 01 in Örcün (4.5%). The clay content and type of clay minerals are important because it will determine the strength and the cohesion of earthen material against shrinkage. The recommended clay content is usually around 10–15%, but values outside this range could also be suitable depending on the type of clay.

3.2 Soluble Salt Contents

Salt weathering is considered among the causes of the decay process of historic building materials. The growth of salt crystals within pores can create sufficient stresses to overcome the tensile strength of mud plasters, mortars, and bricks. Hydration and deposition of soluble salts are often associated with the capillary movement of water through pores in adobe materials (Matero & Moss, 2004, 2012). Therefore, pore size and porosity and nature of salt are important factors that affect salt decay (Matero & Moss, 2004, 2012; Moussa et al., 2009). Air pollution is the major source of sulfates and nitrates resulting in salt deposition in the form of gypsum ($CaSO_4$) that causes deterioration. Gypsum is the product of the reaction of calcareous inclusions in mud materials with the regional sulfate.

The total soluble salt contents of the mud plasters, mortars, and bricks were in the range of 0.81–1.73% by weight (Fig. 7). Brick samples SR02T and OR02T along with plaster samples SR02S and OR01S contained a relatively high amount of soluble salt. Except for these four samples, the soluble salt content of other samples was between 0.80 and 1.17%, the average being 0.96%. A recent study on the determination of the fertility and current status of soil nutrient content conducted in Marmara Region reported the soluble salt content of the soil in the Gölcük region to be less

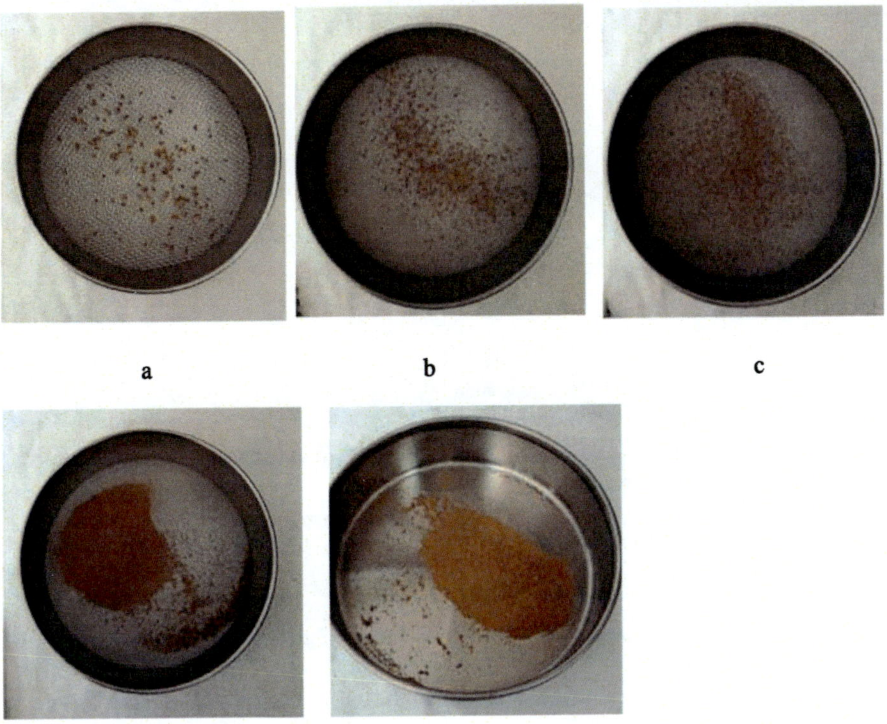

a b c

d e

Fig. 5 Sieve analysis of plaster sample from house 01 in Saraylı **a** x > 2 mm, **b** 2 mm > x > 1 mm, **c** 1 mm > x > 0.5 mm, **d** 0.125 mm > x > 0.063 mm, **e** x < 0.063 mm

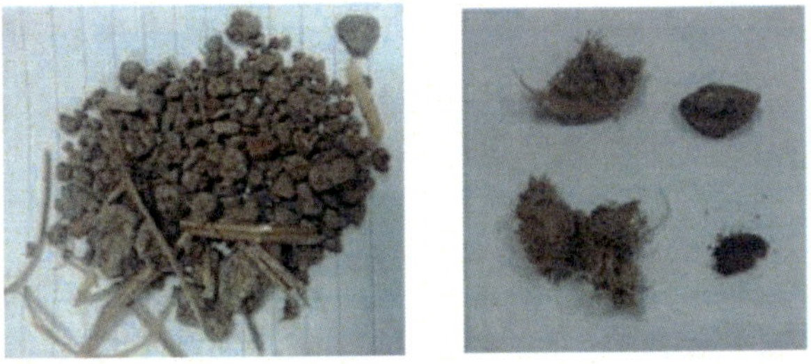

a b

Fig. 6 Aggregates of plaster and mortar samples from house 02 in Saraylı **a** crushed brick particles, **b** cloth pieces and nut shells

than 0,15% (Taşova and Akın 2013). According to this result, the percentage of total soluble salts in Saraylı samples are much more than the reports for the region.

3.3 XRD Analysis

The mineralogical compositions of earthen samples obtained by XRD analysis are presented in Table 2. Quartz is the main component of the bricks, mortars, and plasters, associated with plagioclase, mica in different amounts, and potassium feldspars (K-feldspars) as accessory minerals. XRD results of samples indicated the presence of clay minerals, which is mainly illite together with kaolinite, chlorite, mixed-layered chlorite–vermiculite, and sepiolite. Montmorillonite was also detected in two brick samples (SR01T and SE01T). Meantime, gypsum was observed in some samples in small and trace amounts (OR02T, ST02T, SE01T, SR02S, OR01S, OR02S). Three samples, SR01T, SE01H, and SR02H, and two samples SR01S and OR02H revealed the presence of amphibole and pyrite, respectively. Zeolite mineral was detected in the lime mortar from Selimeye house 01.

3.4 Chemical Chracterization

The chemical composition of the earthen materials is shown in Table 3 in the form of major oxides obtained by the XRF technique. The results indicated that the earthen materials, except for the mortar sample from Selimiye, have higher amounts of aluminosilicate minerals, as Al_2O_3 the

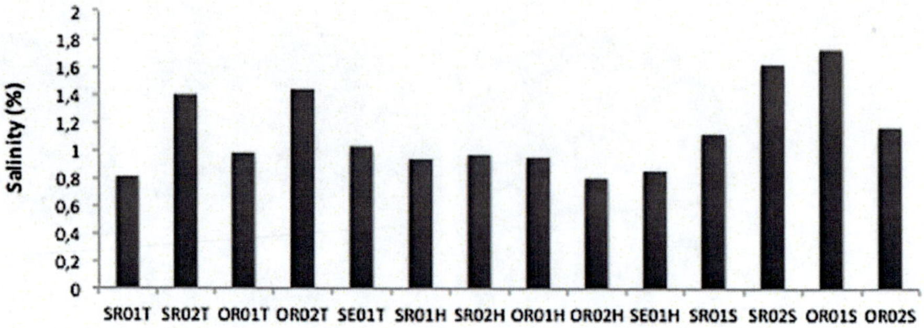

Fig. 7 Total soluble salt content of samples in percentage

Table 2 Mineralogical compositions of earthen samples (XRD)

Sample	Mineral composition
Bricks	
SR01T	Quartz (++ +), Plagioclase (+++), Mica (+++), Illite (++), Amphibole (++), K-feldspars (++), Montmorillonit (+)
SR02T	Quartz (++ +), Plagioclase (++ +), Mica (+++), Illite (+++), K-feldspars (++), C-V* (++), Sepiolite (++), Gypsum (+), Chlorite (+), Kaolinite (+)
OR01T	Quartz (+++), Mica (+++), Illite (+++), Plagioclase (++), K-feldspars (++), Kaolinite (+)
OR02T	Quartz (+++), Mica (+++), Illite (+++), Plagioclase (+++), Chlorite (++), Gypsum (++), K-feldspars (++), Kaolinite (+), C-V (+), Sepiolite (+),
SE01T	Quartz (+++), Calcite (+++), Mica (+++), Illite (+++), Plagioclase (++), K-feldspars (++), Chlorite (++), Gypsum (+), Kaolinite (+), Montmorillonit (+)
Mortars	
SR01H	Quartz (+++), Mica (+++), Illite (+++), Plagioclase (++), K-feldspars (++), Chlorite (++), C-V (+), Sepiolite (+), Kaolinite (+)
SR02H	Quartz (+++), Mica (+++), Illite (+++), Plagioclase (+++), Chlorite (++), Kaolinite (++), K-feldspars (++), Amphibole (+), C-V (+), Sepiolite (+),
OR01H	Quartz (+++), Mica (+++), Illite (+++), Plagioclase (++), Chlorite (++), C-V (++), Sepiolite (+), K-feldspars (++), Kaolinite (+)
OR02H	Quartz (+++), Mica (+++), Illite (+++), Plagioclase (+++), K-feldspars (+++), Chlorite (++), C-V (++), Sepiolite (+), Kaolinite (+), Pyrite (+)
SE01H	Quartz (+++), Calcite (+++), Plagioclase (+++), Muscovite (++), Illite (++), Chlorite (++), K-feldspars (++), Zeolite (+), Amphibole (+)
Plasters	
SR01S	Quartz (+++), Mica (+++), Illite (+++), Plagioclase (++), Chlorite (++), K-feldspars (++), C-V (+), Sepiolite (+), Pyrite (+)
SR02S	Quartz (+++), Illite (+++), Plagioclase (++), K-feldspars (++), Chlorite (++), Kaolinite (+), Gypsum (+)
OR01S	Quartz (+++), Mica (+++), Illite (+++), Plagioclase (++), K-feldspars (++), Kaolinite (+), Gypsum (+)
OR02S	Quartz (+++), Mica (+++), Illite (+++), Chlorite (++), K-feldspars (++), C-V (++), Sepiolite (+ +), Gypsum (+), Calcite (+), Kaolinite (+)

*mixed-layered chlorite–vermiculite

percentage varied between 14.70 and 18.00% by weight and SiO_2 percentage between 55.3 and 61.9%. SiO_2 and Al_2O_3 contents of the Selimiye mortar sample were 7.1 and 6.76%, while CaO was 47.32%.

The combined percentage of CaO and MgO in mud plasters, mortars, and adobe blocks vary between 2.80 and 5.40%, which could be due to slaked lime added to local soil in order to increase bonding between clay, silt, and sand in earthen paste (Singh et al. 2016b, 436).

The earthen materials are also marked by the presence of Fe_2O_3, detected between 5.70 and 7.20%. K_2O content is related to the K-feldspars and mica minerals. Due to the lower amount of aluminosilicate minerals, the Selimiye mortar sample contained 1.55% percent of Fe_2O_3.

Table 3 Chemical compositions of earthen samples % by weight (XRF)

Sample	LOI*	Al_2O_3	CaO	Fe_2O_3	K_2O	MgO	MnO	Na_2O	P_2O_5	SO_3
Saraylı House 01										
SR01T	4.15	14.70	3.40	6.90	2.10	2.00	0.10	2.70	0.20	0.33
SR01H	6.80	16.90	1.90	6.40	3.30	2.30	0.10	1.70	0.50	0.23
SR01S	6.30	17.60	1.50	5.70	2.80	2.40	0.10	1.60	0.20	0.21
Saraylı House 02										
SR02T	6.40	15.10	1.60	6.50	3.20	1.60	0.10	1.40	0.20	0.25
SR02H	6.60	16.00	2.30	6.60	3.20	2.00	0.20	1.40	0.90	0.23
SR02S	7.60	18.00	1.30	6.70	3.00	1.50	0.10	0.90	0.20	0.28
Örcün House 01										
OR01T	6.15	17.60	1.20	7.20	3.20	1.80	0.20	1.20	0.20	0.21
OR01H	8.35	16.10	2.00	6.80	3.00	1.80	0.20	1.20	0.70	0.27
OR01S	8.10	14.90	1.60	6.50	2.30	1.40	0.20	1.20	0.20	0.65
Örcün House 02										
OR02T	6.45	16.20	1.60	6.70	2.80	1.60	0.20	1.10	0.30	0.56
OR02H	6.80	17.40	1.80	7.20	3.00	2.20	0.10	1.30	0.40	0.33
OR02S	9.20	15.30	3.10	7.00	3.00	2.10	0.20	1.10	0.90	0.53
Selimiye House 01										
SE01T	8.00	15.90	9.10	6.60	2.50	2.50	0.20	1.00	0.20	0.46
SE01H	30.65	6.76	47.32	1.55	0.08	5.40	0.04	0.03	0.03	1.04

*Loss on ignition

The P_2O_5 content of the samples is particularly interesting, since the mortar samples contained almost two or three times more P_2O_5, except for the house 2 in Örcün. Here the most P_2O_5 content was in plaster samples. This phenomenon is considered to be related to the dung additive in the earthen samples. The difference visible in the color of filtered samples after the acid treatment could be due to P_2O_5 content (Fig. 8).

XRF results indicated that SO_3 concentration in the samples is found between 0.23 and 1.04%. Proportion of SO_3/CaO in the composition of samples was similar, between 0.10 and 0.22, despite of sample OR01S and OR02T (0.41 and 0.35, respectively). Figure 9 shows the scatter plot of SO_3 versus CaO in the composition of all samples. According to this plot, OR01, OR02S, OR02T, SR01T, and SE01T appeared different from the rest. As observed in XRD analyses, OR01, OR02S, OR02T, and SE01T contained significant amounts of gypsum. It is known that gypsum ($CaSO_4$) is the product of regional sulfates related to air pollution with calcite content in the building materials, especially in humid environments (Airborn particles). Since gypsum exists in only six samples in lower amounts, the origins of it should be due to air pollution rather than intentionally added to the mixture, especially for exterior plasters.

The amount of loss on ignition (LOI) observed in the earthen samples is due to CaO contents and the plant-derived additives in the form of straws to the plaster except for one sample. The higher loss on ignition for the mortar sample from Selime (SE01H) is related to CaO amount and calcination loss. The chemical composition of samples supports the data obtained through XRD, SEM-EDX, and FTIR analysis.

Fig. 8 Filtered solution of samples

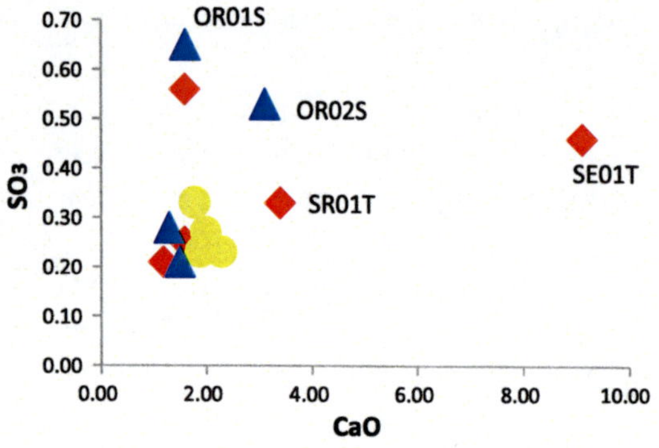

Fig. 9 Scatter plot of SO_3 versus CaO in composition of all samples (blue plasters, red bricks, yellow mortars)

3.5 FTIR Analysis

The earthen samples were observed under FTIR spectroscopy. Except for the mortar sample from Selimiye House 01, all the samples exhibited similar characteristics. Therefore, only The FTIR spectra of mud plaster samples from Saraylı and Örcün, SR01S, SR02S, and ORC01S, ORC02S are given in Fig. 10a. The characteristic band of silicate as SiO_2 tetrahedra is seen at the spectral region at around 1000 cm^{-1} corresponding to Al-Al-OH which is observed at 914 cm^{-1}. The symmetric and asymmetric Si–O bending modes at 690 cm^{-1} show the quartz mineral. Meantime, a small peak at 748 cm^{-1} indicates the presence of magnesium carbonate.

The IR spectra also showed an absorption band around 1740 cm^{-1} could be due to the presence of ester (Derrick et al., 1999, 94). Various literatures on restoration and conservation science indicate vegetable and animal oils consist of glycerol esters of higher fatty acids which lies in the type and composition of the fatty acids (Mills & White, 1994; Derrick et al., 1999, 102). The intensity of this band is known to be depending on the state of dryness of the oil, where in well-dried oils, it would be very small. The presence of esters could be the evidence for the addition of adhesive materials such as vegetable or animal oil in the mixture during the preparation. The small at 1366 cm^{-1} due to methyl group intensified by its direct bond to a carbonyl group confirms the presence of an acetate ester (Singh & Arbad, 2014, 384–395).

In the case of the mortar sample from Selimiye House 01, SE01H, the stretching band at 1400 cm^{-1} and angle bending at 874 and 712 cm^{-1} may show the presence of calcite (Fig. 10b). The band at 1034 cm^{-1} is the vibrations of in-plane Si-O stretching. Beside calcite, gypsum could be identified from the peaks of gypsum ($CaSO_4.2H_2O$) at 669, 1113, 3404 and 3524 cm^{-1}, and the presence of calcium oxalate is evident mainly from the band at 1323 cm^{-1} for sample (Akyuz et al., 2015, 744–750). Glycerol esters oils are also detected from the band around 1740 cm^{-1} for each lime wash sample. Besides esters, IR spectra of lime wash layers also showed a small peak around 1622 cm^{-1}. The spectra of primary and secondary amides contain a strong carbonyl band in the region of 1650 cm^{-1}. Therefore, this band could be an indication of the presence of primary and secondary amides in the samples (Derrick et al., 1999, 108). Since proteins are characterized by amide I and amide II bands, it could be concluded that proteins-based materials such as animal hair or egg could have been added to the mixture along with glycerol esters oils.

Fig. 10 FTIR spectra of plasters samples (left) and mortar samples (right)

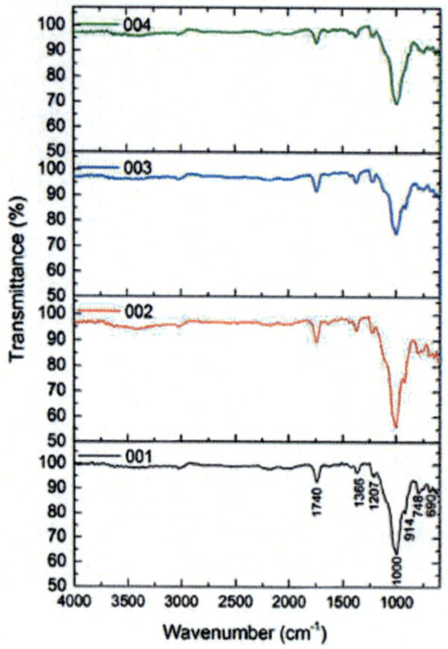

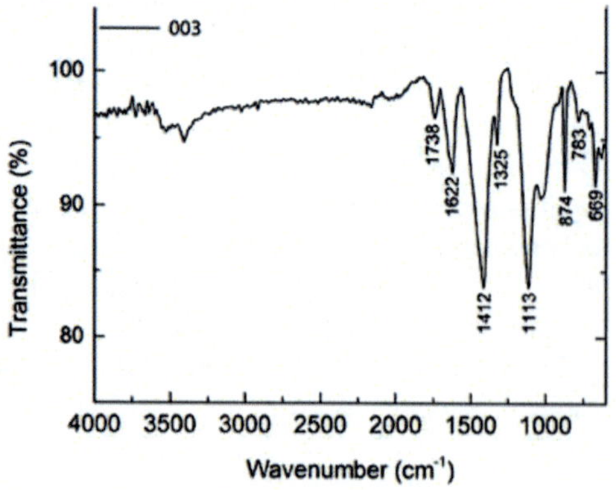

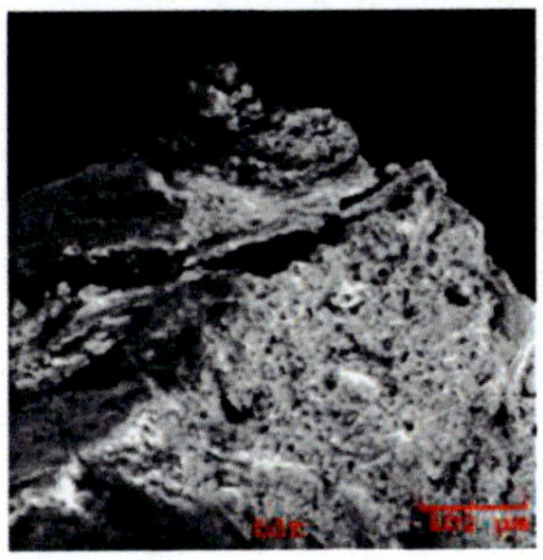

Fig. 11 SEM images of plaster sample SRO2

3.6 SEM–EDX Analysis of Mud Plasters

SEM analysis was applied to understand the morphology and microstructure of the binder and the aggregates of the samples. As seen in SEM image of SR02S shown in Fig. 11, the colored lime wash layer could be easily distinguished from the mud plaster layer from its width and porosity differences. The lime wash layer was also slightly detached from the mud plaster layer. The interconnected pores providing permeability for plaster were observed in each sample, except for OR01S, which has a more compact network structure (Fig. 12). Interconnected pore type of texture is known to reduce permeability, without affecting porosity.

Elemental analysis (EDX) of the matrix showed that besides main element silicium, varying amounts of Al, Si, Fe, Cu, Na, Mg, and K are also present. These elements may be related to minerals used as aggregates (Fig. 12). EDX analysis of OR02S also indicates the presence of iron due to iron oxide, clay minerals, and calcite inclusion used as the binder (Fig. 13).

4 Discussion

XRD analysis revealed that the main binders of historic mud plasters, mortars, and adobe blocks are the clay minerals illite and kaolinite. The presence of montmorillonite belongs to the smectite group, mixed-layer smectites, and chlorite were also observed, however, in trace amounts. Illite and kaolinite are known to be stable clay minerals due to their lower swelling characteristics. Therefore, the illite content seems to contribute to the dimensional stability of mud materials and their durability.

Mixed layer clays are common and consist of clays with a stacking sequence of two or more types of single layers (Al-Ani & Sarapää, 2008, 28). Chlorite–vermiculite mixed-layer clay minerals are detected in most of the samples. Vermiculite is characterized by its high-water holding capacity due to its larger surface area (Matera & Moss, 2004, 222). It is also among the lightweight aggregates, which have lower specific gravity than traditional aggregate sources because of their network of internal pores (Jones, 2013, 16). Vermiculite was chosen as a protective material for earthen architecture at the Çatalhöyük site regarding its water-holding capacities (Matera & Moss, 2004).

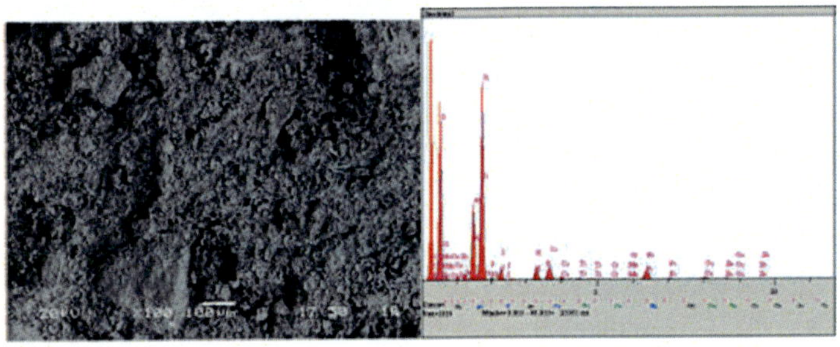

Fig. 12 SEM images of plaster sample SRO2

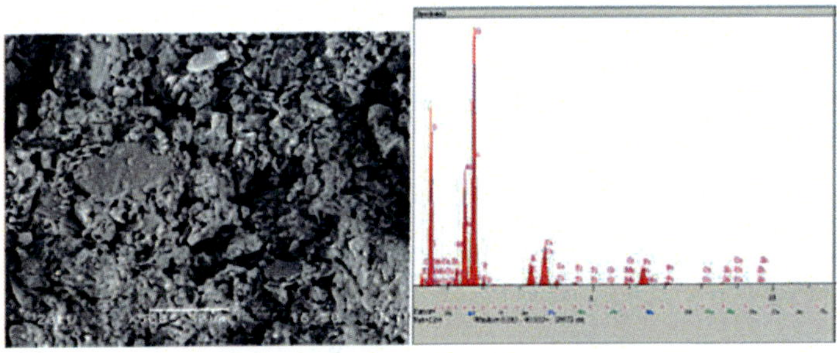

Fig. 13 SEM images of plaster sample ORO2

Sepiolite mineral was found in samples except for the samples from Selimeye house 01. Sepiolite, which has a fibrous nature, was elongated. Recent studies evaluated the effects of sepiolite minerals in historical materials regarding the researches on the properties of sepiolite in cement mixture and repair mortars (Kavas et al., 2004; Andrejkovičová 2012; Singh and Ardab 2014; Sharma & Singh, 2019). Sepiolite acting as a natural pozzolanic additive was found to increase the compressive and bending strength of cement mortar when added up to 10% (Kavas et al., 2004). As described by ACI, pozzolan is "a siliceous or silico-aluminous material that will, in finely divided form and in the presence of moisture, chemically react with calcium hydroxide at ordinary temperatures to form compounds having cementitious properties (there are both natural and artificial pozzolans)." It could be concluded that when the CaO amount in the samples is considered, the proper amount of sepiolite is present in the mixture. Instead of sepiolite, zeolite mineral detected in the lime mortar from Selimiye house 01 was probably added for its pozzolanic properties. The study on the effects of zeolite used as a pozzolanic admixture to lime mortar showed that in humid environments, lime mortars containing zeolite obtained higher compressive strength (Pavlik et al., 2015). It was suggested that lime–zeolite mixtures could be effectively used with specific knowledge of conditions of use.

Meantime, fired mud brick sample SE01T did not show the presence of kaolinite, while illite was dtected for the same sample. Kaolinte tends to disappear at firing temperatures of 550 °C and above; however, illite disappears at temperatures of around 800 °C. Therefore, it suggests that firing temperatures of mud bricks of Selimeye house 01 occurred within this range.

Besides straws, organic additives of protein-binding material were used in the preparation of earthen materials, both plaster, mortars, and adobe blocks. XRF results and FTIR results indicated that the quantity of protein material used in mortars for better cohesion was almost three times more than plasters and adobe blocks. This is reasonable when we consider the function of mortars. The nature of protein material could be dung; however, further analysis should be appealed to reveal its origins.

The particle size distribution patterns for mud materials seemed to vary depending on the region and the type of the building material. Mortar samples contained coarser aggregates in the mixture, which also had much more protein-binding material. The aggregates above 2 mm (gravel) had the largest content with a ratio of 45–58%, while the portion of very fine particles (below 0.125 mm) was between 6 and 14%. Sand aggregates (2–0.063 mm) improve cohesion between particles and reduce shrinkage and swelling by providing higher density. Meantime, the clay content of earthen materials appeared to be mainly illite and kaolinite, which seem to contribute to the dimensional stability of mud materials and their durability. The local natural sources of silt clay and aggregate of Saraylı, Örcün, and Selimiye Districts need to be investigated to explain their preparation techniques.

5 Conclusion

The archeometric investigation of historic building materials provides important contributions to understand traditional construction techniques of the region. Determining particle size distribution and soluble salt content of plasters, mortars, and bricks can provide information on the performance of historic structures. XRD, FTIR, SEM-EDX, and XRF analyses contribute to identify the mineralogical composition, morphology, microstructure, and elemental composition of the component materials. It is concluded that the earthen materials of the houses in the rural areas in Gölcük were prepared primarily from the local soil containing stable clay minerals, although the mud mixture was rich in coarser aggregates. The mixture was further qualified by combining of calcite and dolomite and organic adhesives for their cementing properties. The ratio of additives seemed to depend on the type of construction material. Due to the aggregate/binder ratio, the binder and additive types, and ratios in the mixture, the mud plasters, mortars, and adobe blocks show strong cohesion.

This study, including the characterization of historic building materials, provides an understanding of vernacular architecture in the area and traditions of local craftsmanship in the region. Thus, these specifications obtained by the analysis can be used as guidelines in preparing intervention for houses to compete for historic fabric. The results will help to form conservation decisions for the region which is under a high earthquake risk and provide a basis for the integration of local craftsmen in heritage efforts.

Acknowledgement This project was funded by Kocaeli University, Scientific Research Project Coordination Unit (BAP) in 2019, Project Number: 2018/088.

References

Aktaş, Y. D., Akyüz, U., Türer, A., Erdil, B., & Güçhan, N. Ş. (2014). Seismic resistance evaluation of traditional Ottoman timber-frame Hımış houses: Frame loadings and material tests. *Earthquake Spectra, 30*(4), 1711–1732.

Akyuz, T., Akyuz, S., & Gulec, A. (2015). Elemental and spectroscopic characterization of plasters from Fatih Mosque-Istanbul (Turkey) by combined micro-raman FTIR and EDXRF Techniques. *Spectrochimica Acta Part a: Molecular and Biomolecular Spectroscopy, 149*, 744–750.

Al-Ani, T., & Sarapää, O. (2008). Clay and clay mineralogy. *Physical-chemical Properties and Industrial Uses*. Geological Survey of Finland.
Andrejkovičová, S., Ferraz, E., Velosa, A. L., Silva, A. S., & Rocha, F. (2012). Air lime mortars with incorporation of sepiolite and synthetic zeolite pellets. *Acta Geodynamics Et Geomaterialia, 9* (1), 79–91.
Bahobail, M. A. (2012) The mud additivesand their effect on thermal conductivity of adobe bricks. *Journal of Engineering Sciences, 40* (1), 21–3.4.
Bayraktar, N. T., & Kishali, E. (2013). Evaluation of thermal behaviour of traditional Hımış type buildings in Kocaeli. In *Kerpiç 13-New Generation Earthen Architecture- Learning from Heritage- International Conference*.
Black, C. A. (1965). *Methods of soil analysis, part 2*. American Society of Agronomy.
Boggs, S. (2009). *Petrology of sedimentary rocks*. Cambridge University Press.
Brown, R., & Maudlin, D. (2012). Brown, R., & Maudlin, D. (2012). Concepts of Vernacular Architecture. The SAGE Handbook of Architectural Theory, 340–368.
Caner, E. (2003). *Archaeometrical investigation of some Seljuk plasters* (Doctoral dissertation, METU).
Costa, C., Cerqueira, Â., Rocha, F., & Velosa, A. (2019). The sustainability of adobe construction: Past to future. *International Journal of Architectural Heritage, 13*(5), 639–647.
Davey, N. (1961). *A history of building materials*. Phoenix House.
Derrick, M. R., Stulik, D., & Landry, J. M. (1999). *Infrared spectroscopy in conservation science, scientific tools for conservation*. Los Angeles: The Getty Conservation Institute.
Eres, Z. (2013). "Türkiye'de Geleneksel Kırsal Mimarinin Korunması: Tarihsel Süreç, Yasal Boyut", *Mimari ve Kentsel Koruma, Prof.Dr. Nur Akın'a Armağan*. ed. K.K. Eyüpgiller - Z. Eres. İstanbul: Yem Yayınları. 457–469
Galitekin, A. N. (2005). *Gölcük Tarihçe ve Kültür Mirası Eserleri* (1. Baskı), İstanbul: Gölcük Belediyesi Kültür Yayınları.
Güçhan, N. Ş. (2007). Observations on earthquake resistance of traditional timber-framed houses in Turkey. *Building and Environment, 42*(2), 840–851.
Gür, Ş. Ö., & Batur, A. (2005). Rural Architecture in the Eastern Black Sea Region/*Doğu Karadenizde Kırsal Mimari*, İstanbul: Milli Reasürans TAŞ.
Joel, O. A. (2016). Conservation of traditional earth building in Nigeria: Case study of origbo in IFE North Osun State. *International Journal of African Society, Cultures and Traditions, 4*(2), 20–31.
Jones, C. (2013). Jones, C. (2013). Effect of Lightweight Aggregate Moisture Content on Internally Cured Concrete. Graduate Theses and Dissertations Retrieved from https://scholarworks.uark.edu/etd/83.
Kavas, T., Sabah, E., & Çelik, M. S. (2004). Structural properties of sepiolite-reinforced cement composite. *Cement and Concrete Research, 34*(11), 2135–2139.
Köksal, G. (2013). *Gölcük'teki Mimari Miras Kitabı* (1. Baskı), Kocaeli: Çınar Ofset.
Köroğlu, J. (2007). *17 Ağustos 1999 Marmara Depremi'nin Gölcük ve Civarı Çevresel Etkileri*, Yayınlanmamış Yüksek Lisans Tezi, İstanbul Üniversitesi, Sosyal Bilimler Enstitüsü, İstanbul.
Kuban, D. (1995). "Türk Evi Geleneği Üzerine Gözlemler", *Türk İslam Sanatı Üzerine Denemeler*. İstanbul: Arkeoloji ve Sanat Yayınları.
Matera, F., & Moss, E. (2004). Temporary site protection for earthen walls and murals at Çatalhöyük, Turkey. *Conservation and Management of Archaeological Sites, 6*(3–4), 213–227.
Matero, F., & Moss, E. (2012). Temporary site protection for earthen walls and murals in Çatalhöyük Turkey (2004). In S. Sullivan & R. Mackay (Eds.), *Archaeological sites: Conservation and management* (Vol. 5, pp. 392–406). Getty Publications.
Matero, F., & Moss, E. (2004). Temporary site protection for earthen walls and murals at Çatalhöyük, Turkey. *Conservation and Management of Archaeological Sites, 6*(3 & 4), 213–227.
Meriç, I., Erdil, M., Madani, N., Alam, B., Tavukcuoglu, A., & Caner Saltik, E. N. (2013). Technological properties of earthen building materials in traditional timber frame structure. *Proc Kerpic, 13*, 465–473.
Mills, J., & White, R. (1994). *Organic chemistry of museum objects*. Routledge.
Morel, J. C., Mesbah, A., Oggero, M., & Walker, P. (2001). Building houses with local materials: Means to drastically reduce the environmental impact of construction. *Building and Environment, 36*(10), 1119–1126.
Moussa, A. M. A., Kantiranis, N., Voudouris, K. S., Stratis, J. A., Ali, M. F., & Christaras, V. (2009). The impact of soluble salts on the deterioration of pharaonic and Coptic wall paintings at Al Qurna, Egypt: Mineralogy and chemistry. *Archaeometry, 51*(2), 292–308.
Oliver, P. (2006). *Built to meet needs: Cultural issues in vernacular architecture*. Elsevier.
Özgünler, M. (2017). Kırsal Sürdürülebilirlik Bağlamında Geleneksel Köy Evlerinde Kullanılan Toprak Esaslı Yapı Malzemelerinin İncelenmesi. *Süleyman Demiral Üniversitesi Mimarlık Bilimleri Ve Uygulamaları Dergisi, 2*(2), 33–41.
Papayianni, I., & Pachta, V. (2017). Earth block houses of historic centers. A sustainable upgrading with compatible repair materials. *Procedia Environmental Sciences, 38*, 274–282.
Pavlík, Z., Fořt, J., Pavlíková, M., Kulovaná, T., & Černý, R. (2015). Characterization of cement pastes containing natural zeolite as a pozzolanic admixture. In *Applied mechanics and materials* (Vol. 719, pp. 206–209). Trans Tech Publications.
Reddi, L. N., Jain, A. K., & Yun, H. B. (2016). Soil materials for earth construction: Properties, classification and suitability testing. In M. R. Hall, R. Lindsay, & M. Krayenhoff (Eds.), *Modern Earth Buildings* (pp. 155–171). Woodhead Publishing.
Sey, Y. (2003). Vizyon 2023 Öngörü Panelleri: Sürdürülebilir Kalkınma / Tarihi Kültürel Mirasın Korunması. https://www.tubitak.gov.tr/tubitak_content_files/vizyon2023/csk/EK-13.pdf .
Sharma, A., & Singh, M. R. (2019). Multi-analytical investigation of the composition and binders used in the earthen support layer of fifth–fourteenth century CE painted fragments from Bezeklik, China. *Studies in Conservation*, 1–17.
Singh, M., & Arbad, B. R. (2014). Characterization of traditional mud mortar of the decorated wall surfaces of Ellora caves. *Construction and Building Materials, 65*, 384–395.
Singh, M., Kumar, S. V., & Waghmare, S. (2016). Mud plaster wall paintings of Bhaja Caves: Composition and performance characteristics. *Indian Journal of History of Science, 51*, 431–442.
Singh, M., Kumar, S. V., & Waghmare, S. (2016). An analytical and geoarchaeological approach on earthen plaster of 2nd century BCE Karle Caves of Western Deccan, India. *Journal of Archaeological Science: Reports, 9*, 522–535.
Stefanidou, M., & Papayianni, I. (2005). The role of aggregates on the structure and properties of lime mortars. *Cement and Concrete Composites, 27*(9–10), 914–919.
Taşova, H., & Alper, A. (2013). Marmara Bölgesi topraklarının bitki besin maddesi kapsamlarının belirlenmesi, veri tabanının oluşturulması ve haritalanması. *Toprak Su Dergisi, 2*(2), 83–95.
Teutonico, J. M. (1988). *A laboratory manual for architectural conservators*. ICCROM.

Usanmaz, B. (2007). *Gölcük-Saraylı Köyü Sit Koruma Projesi: Fiziksel Düzenleme İçin Ön Koşul ve İlkelerinin Araştırılması* (Yayımlanmamış Yüksek Lisans Tezi). İstanbul Teknik Üniversitesi, Fen Bilimleri Enstitüsü Mimarlık Anabilim Dalı, Restorasyon Programı, İstanbul.

Wentworth, C. K. (1922). A scale of grade and class terms for clastic sediments. *The Journal of Geology, 30*(5), 377–392.

Yılmaz, M., & Bakış, A. (2015). Sustainability in construction sector. *Procedia-Social and Behavioral Sciences, 195*, 2253–2262.

Yoon, J., & Kim, K. H. (2017). Critical review of the material criteria of building sustainability assessment tools. *Sustainability, 9*(2), 186.

Conservation of Gravitational Architecture

Ramiro Sofronie, Radu Cazamir, Liviu Gruia, and Sebastian Lungu

Abstract

Architecture devoted to gravity was ingeniously used in the past for most of the buildings erected along the Silk Road, for instance. This art of building is defined by (1) massiveness, (2) vertical balancing, (3) exuberant decorations, (4) chromatic effects, and (5) environmental integration. Along those famous routes, even the roads were gravitationally shaped by appropriate slopes and serpentines. From the beginning, gravitational architecture assumed two functions: (1) expressing gratitude to gravity for the chance of living, and (2) insistent seeking a lasting identity instead of the ephemeral one that is limited to the life of the physical body, and still in use. The reasons for the two functions assumed then by the primitive architecture have never been disclosed to people. The available data about the early life of humans are scarce and confusing. Existing legends, stories, habits, beliefs, and traditions suggest an ancient history dynamic and rich in cryptic symbols. Unfortunately, the doors of that history remain closed. For instance, the year 11,451 B.C.E, advanced by Jonathan Black for the consciousness of homo erectus is of great interest. However, it has not been proved yet, even if then the Milky Way became parallel with the Nile again (Black, The secret history of the world, Quercus Publishing, 2007). To explain the role of gravity in the act of architectural creation, the authors adopted the indirect investigation. This method consists of reprocessing the information acquired from the past. Only facts, and not words, which directly involve gravity, are considered. Such a package of selected data is analyzed from the perspective of the knowledge reached by the twenty-first century and then accordingly interpreted into an original vision. Investigations are carried out in three successive steps, and each step is based on five different sources. The conclusion of the research is overwhelming. By the act of creation, the gravitational architecture praises the earthen life and its achievements.

Keywords

Addiction • Aeroelasticity • Anthropomorphism • Artificial intelligence • Automorphism • Awakening • Balance • Bipolarity • Black-hole • Bridging • Conception • Congruence • Consciousness • Cryptic • Disconnection • Gradient • Identity • Idiomorphic • Isomorphism • Meditation • Mirroring • Shaping • Topology

1 Introduction

The paper deals with the involvement of gravity in the architectural act of creation since ancient times. By disclosing some less known conceptual details from the past, historic civilizations are better understood. The twenty-first century reached a high level of technological development that now allows cultural heritage to be regarded from different specialized perspectives. Since gravity offers the opportunity of bridging the civilization of present, conventionally called advanced, with the still enigmatic civilization of the past, gravitational perspective was selected with particular interest. This is why shaping the buildings and structures according to gravity requirements, like vertical symmetry, for instance, was called gravitational architecture in this paper.

For the civilized world, it seems the twenty-first century started under the sign of gravity. Indeed, on 9/11, 2001 the Twin Towers of the World Center in New York were submitted to the terrorist attack. It was considered then that the potential energy of the towers was lower than the kinetic energy of pirate planes. Otherwise, they would not have fallen. Potential energy means the energy of a position that directly depends on gravity. The question is whether the

R. Sofronie (✉) · R. Cazamir · L. Gruia · S. Lungu
UNESCO Chair #177, Bucharest University, Bucharest, Romania

A. Versaci et al. (eds.), *Conservation of Architectural Heritage*, Advances in Science, Technology & Innovation,
https://doi.org/10.1007/978-3-030-74482-3_3

terrorists were aware of Clapeyron's conservation theorem? Anyhow, it was a true lesson for those with an open mind. By coincidence or fate, on the same day, the UNESCO House of Paris hosted the International Millennium Congress "More than two thousand years in the history of Architecture. Safeguarding the structures of our architectural heritage." At the very time, when in New York, the Twin Towers were still burning; in Paris, two papers were presented to a large audience, one on the Leaning Tower of Pisa and another on Brancusi and his obsession with gravity (Sofronie, 2001).

In the same year, 2001 Stephen Hawking resumed the Grand Unification Theory, shortly GUT, created five years before (Hawking, 2001). To the Einstein objection that "God does not play dice" Hawking answered that all the evidence showed that "God is quite a gambler." It was supposed then to be a quantum theory that unifies the electromagnetic, strong, and weak forces with the gravitational force, the weakest of the four fundamental forces of nature. Unexpectedly, four years later, in 2005, Hawking published the book "The Theory of Everything. The Origin and Fate of the Universe" (Hawking, 2005). After reminding the old paradox: "Can God make a stone so heavy that He can't lift it?", later clarified by St. Augustin, Hawking admitted that, for the time being, the nature of gravity remains unknown. The same idea was written in the Bible long before but from a different perspective. For many scientists, the news produced confusion, coming in the centenary year of Einstein's "The Principle of Relativity." Despite it, Hawking continued his works on the black holes that he discovered in 1974. However, he never saw one during his life. It was supposed that a black hole was a region of space–time from which nothing, nor even light, can escape because gravity was so strong. Shortly after his death, on March 14, 2018, the first image of a black hole was captured by the Event Horizon Telescope (EHT) in the Virgo constellation on April 10, 2019. Six months later, some physicists disputed the authenticity of that image. And thus, the contest for capturing true images with black holes is continuing.

All buildings are material bodies with their masses. This is why, without any exception, they are submitted to the action of gravity and therefore are designed based on code requirements as such. However, the architecture of buildings is independent of their weight. Building aesthetics is established by functional and location criteria. Only when the shaping criteria are explicitly based on gravity actions, architecture becomes gravitational. For instance, any geometrical composition with vertical symmetry is gravitationally balanced. This kind of design is usually chosen for monuments. The ordinary buildings are shaped according to the severe urbanistic restrictions imposed by the available or existing spaces as well as by the plastic coherence of each zone (Robertson, 2017). Therefore, architecture and gravity have their distinct fields. However, architecture means aesthetics, and beauty is mathematically translated by equilibrium or balance. On the other hand, in mechanics, gravity is the force of equilibrium. The whole motion of all cosmic bodies is governed by Newton's law of universal attraction. Therefore, by targeting the same idea of equilibrium, architecture and gravity are a perfect match.

As a force, gravity is the cause of motion. It was defined by Newton's laws of motion since 1687. Gravity is a latent force physically identified only by its effects. Even so, it is suitable to be rigorously used in all analyses requested by practice. However, gravity essentially differs from the abovementioned three fundamental forces of nature. Gravity is a bipolar force of attraction only, and one of its two pols is the Earth, much larger than the other, opposite pole. The Earth as a pole is not accessible to humans for any kind of intervention. In these conditions, gravity appears as a conservative or potential force that depends only on its vertical position. As Hawking stated in 2005, the nature of gravity is still unknown. That is not a tragedy. Probably in the process of evolution, its time has not come yet. What is however intriguing remains the memory of the time when the knowledge of gravity was prohibitive. Fortunately, that time has passed, but gravity is still puzzling. Prohibition is against human nature. It seems that physics has consumed its resources of knowledge, while metaphysical digressions are not acceptable. Only by facts, true reality, and open-minded research, the honor of science could be saved, at least for the time being. This is why this paper intends to use the favorable circumstances created by the twenty-first century to understand now what seemed inconceivable in the past. The existing data coming from the past should be regarded with new eyes accordingly endowed with advanced concepts and tools. All the former societies did the same thing. The obsession of gravity comes from ancient times like an endless flow of new ideas (Sofronie, 2018a). Therefore, this paper proposes a step forward in the direction of gravity. It is expected that the twenty-first century will bring radical changes in human society as André Malraux and Stephen Hawking predicted some time ago.

2 Legends

2.1 The Legend of Oedipus

In ancient Greece, educated people displayed the results of their searching by metaphoric oral short stories which modern times called legends. Such was the well-known story of the young Oedipus. At Thebe's doors, he was stopped by the winged monster. Under the death threat, his consciousness was checked with a riddle. Oedipus answered correctly. The one who at the beginning of life walks in four legs, then, at maturity, in two and ends life in three, with a walking stick, for instance, is the human being. The whole support for passing from the quadruple condition to the bipedal one comes directly from gravity. It is gravity that provides the standing position of humans.

Gravity was inserted into the genetic program of humans. It is not the case of other beings. The vertical equilibrium of the birds, for instance, is supported by their wings. The ostrich lacks wings, but it was endowed with a long neck and reaches its vertical stability by inserting its head in the sand. In the case of humans, gravity does not act constantly throughout the whole life span. Only during maturity, gravity acts on humans at its highest intensity. At the beginning of life, the support of gravity is low but gradually increases. Then the support of gravity gradually decreases after maturity, toward the end of life. The rate of gravity variation is not constant and the same intensity throughout life, but is personalized by individuals. In these conditions, it looks like life span is extended between two poles, i.e., the point of connection to gravity by conception and the point of disconnection by death. This is an accurate and very suggestive definition of life. Moreover, death is not a disease or illness, but strictly the end of earthen duty. Starting two years ago, the archetypal model of the human satisfying such an algorithm was called *homo gravitas* (Sofronie, 2017). That model expresses the bipolar Newtonian attraction between individual humans and the Earth. It is a very special bipolar, with a high gradient. The human's pole is very small compared with the pole of the Earth. Oedipus was indeed the first *homo gravitas*, and the appellative is still of actuality (Fig. 1).

Fig. 1 Oedipus with the winged monster

2.2 The Legend of Narcissus

In a day with quiet weather, Narcissus, a young and handsome hunter, observed his image reflected in the waters of a pool. It was like a natural and gravitational mirror. Suddenly, he was surprised to observe that he imagine appeared as free of any weight while at the same time he felt the pressure of his own body like a burden. Around that instant, the legend developed a long story with details devoted to impressing readers. In fact, at that very moment, Narcissus confirmed that the Legend of Oedipus was correct. Gravity is inoculated in human blood, and the idea of choosing *homo gravitas* as an existential model seems at least inspired and realistic for a while (Fig. 2).

2.3 The Legend of Knowledge Tree

From the gravitational perspective, the metaphorical tree of knowledge is an apple tree. When its fruits are ripe enough, they may fall. Therefore, the existence of gravitational force in full action is recognized. The other details of the legend are not relevant for the paper. The reason why that tree is forbidden does not change the bipolarity attraction between the apple and Earth. And something more, the apple was not

Fig. 2 Narcissus

Fig. 3 Knowledge tree

chosen by chance in this legend. Owing to its nutritive qualities, the apple is the fruit of life like gravity is the force of life on Earth (Fig. 3).

2.4 The Legend of Diogenes

As a Greek philosopher, Diogenes of Sinope (412–323 BC) was wrongly called cynical. On the contrary, he understood the essence of life. Like Buddha, the spiritual teacher who was born in Nepal as Siddhartha Gautama in 563 BC and passed away in 483 BC in India, who was called a *tathagata* or *tathata-ist*.

Diogenes was also a clairvoyant witness of life. For instance, he used to walk with a lit lamp in the daytime, ironically claiming he was searching for light. It was a paradox. Light cannot be discovered and defined in itself, it is something that already does exist and is currently used. It is also the case of gravity, a well-known force of bipolar attraction, but with a nature that is still unknown. According to Diogenes' philosophy, *homo gravitas* is unable to disclose the secret of gravity, and in 2005, Stephen Hawking agreed with that idea (Fig. 4).

2.5 The Legend of Babel Tower

All biblical legends are cryptic, and that of Babel Tower seems more difficult than the others. For their boldness to erect such a tower, the believers were punished with amnesia and then scattered in all cardinal directions. So much the more that verdict is surprising as long as the tower did exist physically, in 3D, for more than 1400 years. Indeed, it was erected during Hammurabi (1792–1750 BC) and then successively consolidated during Nabucodonosor (1125–1103 and Nabucodonosor II (605–562). Herodotus (485–425) visited Babylon in 460 BC when the tower was a citadel of astrology. Alexander the Great first saw the tower in 331 BC and was fascinated by its beauty. Seven years later, in 324 BC, considering tower's bad condition, Alexander ordered its demolition and reconstruction. Unfortunately, next year, in 323 BC, a mosquito decided otherwise. Both fates, that of the tower and that of Alexander were simultaneously closed. The tower was never demolished but brutally vandalized during centuries. About 85 million bricks with Nabucodonosor's signet were spread around over a large area, e.g., including Cyprus. After copiously bombing Baghdad by US Army, many original bricks of the Babel Tower were easily recovered from the ruined villas. Finally, the mystery of Babel Tower was disclosed after 1898 when the German architect Robert Koldeway (1855–1925) started a 19-years-long searching program in the village Babil, on the left bank of the Euphrates River (Schneider, 1966). He was surprised to find out that paradoxically, the massive foundation of the tower was made by adobe masonry. Adobe is a brick made of clay mixed with straw and dried in the sun. Since the underground soil was sandy, the infiltrated waters of Euphrates easily softened the massive foundation of adobe masonry extended over 90 × 90 m^2 which slowly generated unequal settlements. As a gravitational force, the weight of the tower promptly acted. As concerns the upper ground structure of the tower, it was made normally by masonry with solid bricks carefully backed and lime mortars. It was unbelievable how for one long millennium, the builders could not understand the real cause of tower's damages. Therefore, the builders of Babel Tower were punished not for their adoration of God, but their unpardonable professional incompetence. Unfortunately, things have not changed much since then. Nowadays, in the twenty-first century, the philosophy of UNESCO for preserving cultural heritage is well defined. The recommendations of both the Venice Charter in 1964 and ICOMOS-Iscarsah in 2001 are also easily available. Despite that, by traveling along the Silk Road, for instance, one randomly finds out that some of the old masonry buildings or monuments, of obvious patrimonial value, are still retrofitted with

Fig. 4 Diogenes of Sinope

cement mortars or reinforced concrete members, which is sad. Is the lesson of Babel Tower already forgotten? Or was it not learned, which means lack of culture (Sofronie, 2018b) (Fig. 5).

3 Proofs Regarding Practical Connections of Humans with Gravity

3.1 Dance

Mircea Eliade, the philosopher in the history of religions, wrote in his 1932 *Soliloquies* when he was 25 y.o., that dance has an astral meaning. Indeed, all people of any cultural level, young or old, joyful or sad, single or committed, with or without music, at any time of the day or night, are endowed with a natural predisposition to dance. Humans proved to be non-fatigable dancers. By dancing around some imaginary centers, they display their appurtenance to the universe where all cosmic bodies similarly move according to their programs of revolving (Fig. 6).

3.2 Adornments

There are two types of jewelry with a high content of gravity: (1) earrings and (2) chains. Gravitational earrings are heavy pieces of solid materials that are freely hung by earlobes. In a resting state, earrings display the standing position of the body, while in motion they slowly oscillate. Their small oscillations are harmonic. Gravitational chains are flexible pieces of heavy materials that freely hang around the necks (Fig. 7).

The shape assumed by the balancing chain under its weight is called a catenary. In the Cartesian coordinates, this perfect shape is described by a hyperbolic cosine function. The aesthetic values of both, earrings and chains, are bestowed by the weight of the materials which they are made of. The specific gravity of usual materials varies along with a large range of values as follows: oak 0.74, glass 2.4–2.8, marble 2.6–2.8, iron 7.85. silver 10.50, lead 11.35, gold 19.29, platinum 21.50, and iridium 22.50. Thus, golden earrings and chains are 19.29:0.74 = 26 times more beautiful than the similar oak jewels. The rate of costs is not

Fig. 5 Supposed model of Babel Tower

Fig. 6 Dance

Fig. 7 Catenary chain

identical to this one, but it remains high enough. Additionally, in the long clothes for ladies and gentlemen devoted to some ceremonies, small heavy pieces are inserted at their bottoms, usually made by lead. In this way, the large vestments, e.g., like clerical robes in standing positions, assume firm vertical lines reminding of the gravitational verticality of the field in which they live. In this way, their authority is much enhanced. It is like secretly they would keep the control of gravity.

3.3 Hydrology

Water is one of the four Aristotelian elements that support life. It has been collected, conducted, and stored with the aid of gravity. Since ancient times, advanced knowledge of mathematics and physics was necessary for water management. It was accordingly combined with soil investigations, climatic information, and environmental statistical data. On all the five continents, there are vestiges of water engineering (Fig. 8).

In the Kingdom of Morocco, for instance, the former underground system Khattara, which provided water for the former Imperial Capital Marrakech and surrounded lands, at about 10 m in depth, is well preserved and still in service.

Fig. 8 Machu Picchu and the Inco trail

3.4 Artesian

The joy of natural cascades was brought in public places of towns in the form of artesian fountains. They represent gravity in motion. The shapes of the water fascicles are parables and together with the circle belong to the conics family. However, parables essentially differ from the catenaries of heavy chains. The motion of water in artesian

Fig. 9 Artesian in Unirii Square, Bucharest

fascicles is in dynamic equilibrium. The spontaneous affection of humans for such creations is explained by their natural content of gravity. Gravity is the force of life (Fig. 9).

3.5 Roads

If gravity invisibly connects bodies by bipolarity, the roads are visible and physical connections between places. Since they are completed with the aid of gravity, they are called engineering works of art. The art of roads consists of decomposing the vertical force of gravity according to its two components: one normal on the road surface and the other tangential on it. The normal force provides pressure, while the tangential force opposes or superposes with friction. This is the gravitational secret of roads. The rest are geometrical cosmetics, although they are not negligible at all. For instance, the structure and depth of foundations are essential. Also, the lateral canals for collecting and evacuating waters are important. For their function, the cross sections of the roads should be accordingly shaped with appropriate slopes. The famous Silk Road, extended from Venice to Xi'an in China, made numerous physical and spiritual connections (Fig. 10).

Fig. 10 Tibetan road toward Himalaya (Adi Gligor)

Corollary: Artificial Intelligence

The five case studies presented above show that the bipolar connection of humans with gravity undoubtedly seems strong enough. However, one cannot avoid the question: If the bipolar connections of *homo gravitas* are, one way or other, vulnerable or not? History reminds us that at the beginning of the twentieth century, professor Theodore von Kármán was questioned by a reporter in Berlin if robots present any threat to humans. No, was his prompt reply. As long as the robots are not able to reproduce by themselves, like humans, there is no peril at all. In subsequent decades, Aldous Huxley wrote that overpopulation and over-organization expose humans to mind manipulation (Huxley, 1959). No later than two years ago, in 2017 Stephen Hawking warned that humans are subordinated by the advanced technologies to artificial bipolar connections. Alternatively, in the same period, Osho's followers suggested the practice of meditation beyond gravity. The gravity was inserted to *homo erectus* by consciousness and lately disclosed by Oedipus. It seems that, according to Narcissus, an irreversible process is hard, but not impossible to reach. Today, only one year after Hawking passed away, one speaks about human's addiction to artificial intelligence beginning with 2030, by psychological bipolar connections with super-intelligent machines. The success obtained nowadays by mobile telephony supports that target as verisimilar. (Fig. 11).

The consequences of such mixed bipolarity for *homo gravitas* are not yet predictable, but according to Murphy's law worth of raising concern.

4 Case Studies: Outstanding Gravitational Creations of *Homo Gravitas*

4.1 Pyramid of Cheops About 2500 BC in Giza, Egypt

Ancient people were extremely afraid of death. They imagined death as a continuation of the earthen life. This is why they carefully buried their dead and secured their graves. In the early times, corpses were not buried underground but kept sitting at the ground level, in gravity direction, and covered with local soil. During those periods, in the sandy soil of Egypt, the corpses were covered with sand and shaped as pyramidal figures, often aligned according to cardinal points. In the same periods, in the Far East similar sitting corpses were covered by clayey soil, shaped like round heaps or piles, and located behind the shelter or house of family to which the corpse belonged. In the middle of the twentieth century, this tradition was still preserved in some isolated villages of the Far East. In old Egypt, the graves of sand figures were gradually replaced by

Fig. 11 Cloud

pyramidal graves made of stones. Archeologists have discovered on Egyptian territory about 70 such pyramidal graves shaped like isosceles pentahedra. Most of these funeral monuments were small, accordingly adapted to human size. Pyramids of Giza, in Cairo, are the only exceptions by their larger sizes. Most importantly, the shapes of all pyramidal monuments were identical. They were achieved manually by laying down stone by stone without using any binder materials. The stones supported each other by the friction between them. To shape them, a gravitational condition was used: the vertical pressure overall horizontal sections of the pyramid to assume the same constant value. Ancient builders used for this purpose their senses. In 1930, S. P. Timoshenko, Professor at Stanford University, published in 1930 a mathematical solution for this engineering problem (Timoshenko, 1930). Thus, the Pyramid of Cheops in Giza, Egypt, became the first Memorial erected and devoted to gravity by *homo gravitas* (Fig. 12).

4.2 Eiffel Tower, 1889 in Paris, France

For about 4,500 years, Cheops Pyramid in Giza, Egypt, was the highest construction in the world. After the industrial revolution, which according to Hawking started in England around 1750, the challenge of the Cheops Pyramid with its height of about 150 m became unbearable. To their honor 130 years ago, in 1889 the French succeeded in erecting the Eiffel Tower with a height of 300 m. Regarded from a gravitational perspective, the Eiffel Tower has the same conceptual base of shaped like the Cheops Pyramid. Indeed, if in Timoshenko's abovementioned equation changes the values of only two parameters, e.g., of the strength to centric compression in *kPa* and the specific weight in kN/m^3, corresponding to stone and iron, one finds the unique and elegant shape of the Eiffel Tower. Both constructions are provided with squared bases and four planes of symmetry. Guy de Maupassant praised the Tower in Paris with his subtle and intelligent humor. However, as Napoleon Bonaparte once admitted the glory is transient. Only 41 years later, in 1930, the height of the Eiffel Tower was overpassed by Chrysler Tower at 318.9 m, and one year later, in 1931 by Empire State Building at 381 m, both located in New York as simply towers, nothing more. A long competition of skyscrapers has started then. Despite it, Eiffel Tower remains the second Memorial devoted to gravity and construction of art, an important reference for gravitational architecture (Fig. 13).

4.3 Endless Column, 1937 in Targu Jiu, Romania

Brancusi lived in the shadow of Eiffel Tower since 1904 when he was 28. In 1907, he spent two months practicing in the studio of the famous sculptor August Rodin. Then he started to work on his resources. In 1916, in some unexpected circumstances, Picasso recognized his true value. As a consecrated sculptor in 1935 Brancusi was invited at Targu Jiu, in Romania, a town nearby his birthplace, to create a memorial devoted to the local heroes of the First World War. He started his work in the summer of 1937 and finished it three months later. The funeral monument and the surrounding sanctuary were inaugurated on October 27, 1938, when Brancusi called his masterpiece Endless Column, without any other explanations. A long period of radical political and economic changes occurred then in Romania. That is why Brancusi's concern for gravity was discovered very late, and it was disclosed for the public beginning with 2001.

With an inspiration of genius, Brancusi chose coffins not statues for his funeral monument, actual coffins 1.80 m in length. That was his anthropomorphic concept (Sofronie, 2012). For the other two dimensions, he assumed that the coffin was composed of two hexahedral models of Babel Tower reduced by the ratio of 1:100. That was his great idiomorphic concept (Sofronie, 2003). Finally, since the two halves of each coffin were identical in the middle of the

Fig. 12 Pyramid of Cheops in Giza, Egypt

Fig. 13 Eiffel Tower by night

decahedral module such formed, a transverse plane of symmetry resulted. He composed the funeral monument by successively laying one decahedral module over the other, in the vertical direction of gravity. That was his automorphic or isomorphic concept based on the congruence of modules (Sofronie, 2004). Topologically, the resulted height was endless (Sofronie, 2005). The process of mirroring was not limited (Sofronie, 2006).

Physically, Brancusi has chosen for his funeral monument 15 full modules and two halves together with a supporting base of 0.55 m, so that the total height is $15 \times 1.80 + 2 \times 0.90 + 0.55 = 29.35$ m. The depth of the foundation is 5.00 m so that the ratio depth/height assumes an optimum value 1:6. After the year 2001, the models of Column were successfully tested for wind and seismic stability (Sofronie, 2013). In the wind tunnel of the Yokohama National University of Japan, the phenomenon of *aero-elastic indifference* was identified on the 2D model and then confirmed by the 3D one (Yamada, 2005).

Thus, although endless, physically the Column is five times smaller than the Pyramid of Cheops in Giza, Egypt, and ten times than Eiffel Tower in Paris, France. Amazingly, the height of the Endless Column is almost identical with the diameters of two ancient circular sanctuaries. One is that of Stonehenge, built between 3,100 and 2,800 BC in England, and the other is a Dacian Temple built during the first century BC, at Sarmisegetuza Regia, in the Transylvanian

Fig. 14 Endless column

Carpathians, as discovered by Prof. Constantin Daicoviciu from Cluj-Napoca. The same height of Column divided by the length of the decahedral module was almost ten times the value of golden mean $\varphi = 1.618$, i.e., $29.35{:}1.80 = 16.31 \approx 10\varphi$. It seems that both the Babel Tower and Endless Column still contain undisclosed secrets (Sofronie, 2015). What is certain now is that the Endless Column in Targu Jiu, Romania, is worth to be classified as the third Memorial devoted to gravity, after Cheops Pyramid and Eiffel Tower (Fig. 14).

4.4 Brancusi Sanctuary, 1938 in Targu Jiu, Romania

As mentioned in a former paper, the late architect Silvia Paun ascribed a megalithic stylization to that Memorial of modern art (Paun, 2001). She identified the Table of Silence and its twelve Stools with a *Cromlech* or a Circular Temple of *Menhirs*, the Alignment of Stools with a horizontally extended Linear Temple of *Menhirs*, The Gate of the Kiss with a *Dolmen*, and The Endless Column with a vertically extended Linear Temple of *Menhirs*. Once these functions were recognized as such, the Targu Jiu Memorial was integrated with other vestiges of World Megalithic Cultures like Carnac in France, Stonehenge in Great Britain, and Sarmizegetusa in Romania. It was for the first time in history when modern sculpture at such a large scale reflected so faithfully the ancient art expressed by a stone not only in shape but mainly in the meaning of their messages (Fig. 15). It may be interesting to add that The Table of Silence was alternatively interpreted as representing a Solar System and its twelve satellites, while the Gate of the Kiss the indestructible unity between a man and a woman (Fig. 15).

4.5 National Orthodox Cathedral, 2019 in Bucharest, Romania

Based on the design elaborated by the architect Constantin Amaiei and selected by competition in 2010, the Cathedral was inaugurated on November 25, 2018. It is an ecclesiastic building gravitationally shaped. Cathedral's imprint conserves the traditional three-lobed plan of the Orthodox Church on Romanian territory. Its slender body confers dignity and elegance while Pantokrator's prominence

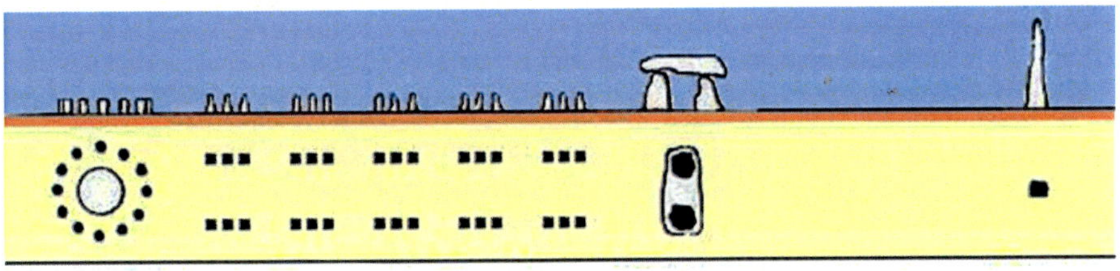

Fig. 15 Megalithic stylization of Brancusi Sanctuary in Targu Jiu

Fig. 16 Southeast view of the Cathedral

authority and much more. By its central and dominant position, the Pantokrator Steeple assumes the function of *Menhir* from the Megalithic Cultures. Its true height is four times higher than that of the Endless Column. All four facades are well marked with vertical lines in open and plastic compositions. There is nothing redundant and ostentatious in them. By the chosen proportions, the building seems engaged in a bipolar attraction with everyone and everything around. As a whole, the Cathedral bears the distinctive mark of the twenty-first century from which there are great expectations, some not yet confessed. Owing to the outstanding height of Pantokrator Steeple, the Cathedral can be seen from anywhere in the city, day and night. Together with the Endless Column and Brancusi Sanctuary in Targu Jiu, this monumental Cathedral is a creation of *homo gravitas.* Their common message consists of the confidence that human life starts and ends with gravitation. (Fig. 16).

5 Conclusion

Gravitational architecture is the creation of *homo gravitas*. It was a personal initiative, nobody asked for it to be done. *Homo gravitas* has been obsessively searching for its unknown identity, which has not been disclosed yet. By doing such outstanding works, probably *homo gravitas* believed they would replace its lack of identity. But that did not happen. Fame is ephemeral, while identity is intrinsic. Cheops, Eiffel, and Brancusi acted spontaneously, each of them, by themselves, perhaps under some atavist feelings or unusual aptitudes. *Homo gravitas* still does not know where it comes from and to whom it belongs. It is probably a transitory state that occurs in a natural process of evolution. Currently, gravitational architecture is the art of building with priority for life by using for this purpose in designing the gravity forces. That means satisfying some well-defined requirements such as safety, equilibrium, beauty, comfort, and/or conformity at affordable costs. It is a traditional architecture, originally born in ancient times and based on creative freedom. It is not submitted to functional or technological restrictions. Once fulfilled, gravitational architecture connects people and brings confidence into the earthen life. Awakening comes through gravity, as the force of life. This is why gravitational architecture is worth to be carefully conserved. It is not an easy task due to the hard concurrence of business companies. The first steps should be taken by appropriate education in the Universities of Architecture and Civil Engineering. Then, by acting with determination through the network of UNESCO Offices, then the 12 Scientific Committees of ICOMOS, the International Council for Monuments and Sites, and finally by local offices of planning and design. Gravitational architecture was also mentioned in a recent report of the authors to ICOMOS Paris entitled "The Point of Trifurcation" as an inevaluable and irrecuperable treasure.

Acknowledgements The unconditional support of the UNESCO Chair #177 in Bucharest, Romania, granted to all authors as expert members of ICOMOS-Iscarsah, for writing and presenting the paper is gratefully acknowledged. The authors are also much obliged to the two official reviewers, designed probably by IEREK, who carefully read the paper and made pertinent remarks together with valuable suggestions, helping thus to improve its quality.

References

Black, J. (2007). *The secret history of the world.* Quercus Publishing.

Hawking, S. (2001). *The universe in a nutshell.* Transworld Publishers Ltd.

Hawking, S. (2005). *The theory of everything: The origin and fate of the universe.* Phoenix Books.

Huxley, A. (1959). *Brave new world revisited.* Chatto and Windus.

Paun, S. (2001). *Constantin Brancusi et la mémoire des héros de la Grande Union de la Roumanie - Proc. Int. Colloquium Brancusi - 125.* Bucharest, Romania: Romanian Academy.

Robertson, E. L. (2017). *The structure of design.* In J.A. Strong (Ed.), *An engineer's extraordinary life in architecture.* New York: Monacelli Press.

Schneider, W. (1966). *Überall ist Babylon.* ECON-Verlag. G.m.b.H.

Sofronie, R. (2012). The endless column. *ISCARSAH Newsletter, 1*, 6–8.

Sofronie, R. (2001). Brancusi and the obsession of gravity. In *Proceedings of the International Millennium Congress "More than two thousand years in the history of Architecture. Safeguarding the structures of our architectural heritage"* (pp. 307–313). Paris: Maison de l'UNESCO.

Sofronie, R. (2003). Idiomorphic concept of the endless column. In *Proceedings of the 28th Congress of the American-Romanian Academy.* Targu-Jiu: HAFA#4.

Sofronie, R. (2004). Why the column is endless? In *Proceedings of Romanian-Japanese UNESCO Workshop "Wind Engineering and Cultural Heritage"*. Yokohama: National University of Yokohama: #1.

Sofronie, R. (2005). The topology of the Endless Column. In: *Proceedings of the International Symposium on Shells and Space Structures* (pp. 871–876). Bucharest: IASS.

Sofronie, R. (2006). The challenge of the endless column. In *Proceedings of the 10th International Conference on Inspection, Appraisal, Repairs & Maintenance of Structures*, (pp. 119–126). Hong Kong.

Sofronie, R. (2013). Physical evaluation of the endless column. In *Proceedings of the Second International Conference on Structures and Architecture. Structures and Architecture: Concepts, Applications, and Challenges* (pp. 171–173). Guimaraes: Cruz (ed) Taylor & Francis Group.

Sofronie, R. (2014–2015). The wisdom of the earth—La Sagesse de la Terre. *ISCARSAH Newsletter, 8–9*, 27–31.

Sofronie, R. (2017). Homo gravitas. In: *Conference Proceedings of the 2017 ICOMOS CIPA-ICORP_ISCARSAH Joint Meeting on December 9th in New Delhi.* New Delhi.

Sofronie, R. (2018). *Homo astralis.* Elixir Social Studies 118C.

Sofronie, R. (2018). Intervention methods. In *Proceedings of the First International Workshop on Restoration and Strengthening of Historic Structures.* Tehran: Iranian Cultural Heritage Handicraft and Tourism Organisation, ISCARSAH.

Sofronie, R. (2018c). *The probabilistic approach to intervention durability.* International Scientific Seminars. The University of Isfahan.

Timoshenko, S. (1930). *Strength of materials. Part 1. Elementary theory and problems.* Michigan: D. Van Nostrand Company, Incorporated.

Yamada, H. (2005). The response of the "Endless Column" under wind action. In: *Proceedings of the International Symposium on Shells and Space Structures* (pp. 229–235). Bucharest: IASS 2005.

Luxor Temple as a Reactivated Holy Site. Sacred Architecture Between Cosmology and Authority

Rasha Said Abd el-Lateef, Yasser Mansour, Shaimaa Kamel, and Ruby Morcos

Abstract

The ancient Egyptian temples were the axis of the cosmos and the source of power, with their lingering charisma in the ancient sacred landscape. After the religious paradigm shift from polytheism to monotheism since late Antiquity, their transformation started. It had so many phases or faces: abandonment, destruction, being a source for retrieving building materials, the reuse for a religious or secular purposes. The fate of these temples in late Antiquity cannot be understood without approaching the most broader meaning of the landscape and the metaphysical position of temples in it, as a part of how the traditional man saw his cosmos and the complex system of his ontological understanding. This paper will focus on Luxor *Amun-Re* Temple in the heart of Ancient Thebes, following its reuse as a military fortress in the third century AD, with the divine king chamber reused for an imperial cult, then reaching the impact of the domination of Christianity, and then reaching the link between the ancient "*Opet* Festival" and the modern celebratory tradition "Mulid of Sidi Abu'l Hajjaj al-Uqsori". All indicates the temple transformation as a reactivated holy site over time, within the public ontological understanding of the sacred power and its points of access, in spite of changes in the perceptions of religion. This assumption is based on applying Mircea Eliade's phenomenological analysis of the religious experience. The research is using the tools of ethnographic descriptions, iconographic representations, archaeological evidences and archaeoastronomical studies for the interpretation, with the hermeneutic phenomenology as a methodology.

Keywords

Luxor temple • Cosmology • Christianizing • Late Antiquity

1 Introduction

Luxor Temple in the heart of the Ancient Thebes has a long story to tell, a story that starts with its significance as a sacred space and the theological doctrines that rooted this monumental construction into the ancient sacred traditional landscape, which gave it a metaphysical position. A story following the experience of its development and transformation, which cannot be understood without removing the alienation of its far world, a world as it was lived by the individuals, regarding their traditional spirituality that shaped how they saw their cosmos. This will be done by examining the cultural history, the relation between society and religion, the interplay between power and religion, and how politicized theology and public cosmic vision were entangled and reflected architecturally.

Considering architecture as a cultural embodiment, the paper will be following themes to help the recognition of meanings of the iconographic representations, the symbolism of the temple's transformation over the time in parallel to the change of the political power, the impact of the sense of national identity and the various religious perceptions.

2 Cosmology, Mythology and the Popular Consciousness

For the Ancient Egyptians, their complex theological system fulfilled their ontological needs within their cosmos, and it provided them with the sense of stability which was defined by a divine order. In the same time, its doctrine's validity was linked to the political power that served the ideology of

R. S. Abd el-Lateef (✉)
Faculty of Energy & Environmental Engineering, The British University in Egypt, Cairo, Egypt

Y. Mansour · S. Kamel · R. Morcos
Faculty of Engineering, Ain Shams University, Cairo, Egypt

A. Versaci et al. (eds.), *Conservation of Architectural Heritage*, Advances in Science, Technology & Innovation,
https://doi.org/10.1007/978-3-030-74482-3_4

the divine order. Such system was characterized with constancy and unity in the domain of sprit until the rise of Christianity.

The traditional (archaic) Egyptian understood his universe as an eternal cosmic cycle reflected in nature by the movement of day and night and rotation of seasons, etc., the appearance and reappearance formed a sacred rhymed order that was kept in cyclic motion defending life against chaos (Spencer, 2007, p. 70). Maintaining this divine order, "*Maat*" in harmony was the ultimate satisfaction for him, as this was the heart of the traditional *Kemetic* spirituality (Frankfort, 1961, pp. 36–80). The rhymed and sacred order is transmitted through myths that served as paradigmatic models, and by their virtue the cosmos and society were periodically regenerated (Eliade, 1991, p. 3). The symbols and rites played essential roles making those myths as expressions of the theological and cosmogonic doctrines (Lubicz, 1985, pp. 31–32). One of the oldest myths are related to "the *Ennead*" or the Nine of Heliopolis, their myths dominated the Egyptian religious literature until the Roman times (Spencer, 2007, p. 74), and they were manifestations to the natural powers and the cycle of regeneration and fertility (Fig. 1).

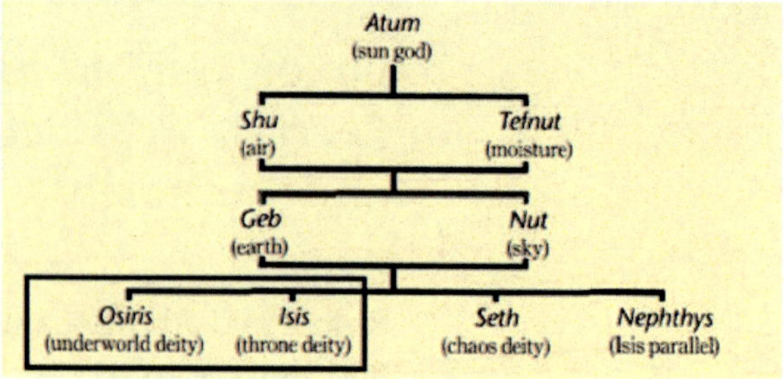

Fig. 1 Ennead/the Nine of Heliopolis (Wilkinson, 2003, p. 17)

Fig. 2 Double representation of *Hapi* "the spirit of the Nile" representing Upper and Lower Egypt—The colossal statue of Ramesses II—Luxor Temple (photo taken by Rasha Said, 2020)

The popular consciousness of the traditional society is complex, with the mechanism of the mythicization of historical figures, a repeated process to serve the recollection with the mythical models "Architypes", within the conformity of the traditional ontology of the spiritual horizon of the general populace (Eliade, 1991, pp. 39–44). This is a main concept interpreting the transformation of the sacred space over time.

2.1 The Nile: The Cosmos and the Sacred Time

The Nile was vital to the Egyptian and affected all the aspects of their lives, and the inundation water was worshipped and feared in the same time. The Nile was in the core of the cosmic cycle that shaped their universe or how they understood it (Fig. 2). Mythically, the Nile came from the place where the universe began in the south (Watterson, 1984, p. 30).The myths of creation in the Egyptian theology were derived from the annual events in the Nile cycle starting by the emergence of land after each flood, resembling the moment of creation (Spencer, 2007, p. 72).The creation myth was recited in connection with the agricultural activities, as every practice was part of the sacred and had a mythical meaning (Eliade, 1991, pp. 22–23) symbolizing stability and continuity.

The Nile was an essential part of the Egyptian traditional comprehension of time, the time was sacred as it was cyclic and reflected in the cycle of the universe, and the Egyptian calendars combined the annual cycle of flooding of the Nile with the astronomical observations (Fig. 3). The Egyptian astronomy was a religious science dominated by the priestly elite and guided the orientation of the temples, tombs and all ancient Egyptian monuments. (Canhão, 2013), (Belmonte, 2010).

Later, the Nile cult was merged into the Hellenistic idiom by the Ptolemies, and the Romans elevated the Nile cult to an official status (Frankfurtar, 1998, p. 64). After the Christian dominance, the Nile represented the Syncretism of Coptic Christianity (Cannuyer, 2001, p. 11). Frankfurtar argued that the Nile shaped the Egyptian Christianity, transforming it into a religious cosmology (Frankfurtar, 1998, p. 46), the Nile was Christianized as it rose through the power of Christ or by the intercession of the saints (Kakosy, 1982, p. 297). One of the vivid transformations serving the cosmic position of the Nile within the spiritual horizon of the Egyptian populace was the Coptic feast of the Epiphany, it took place after the winter solstice when the Nile water was the purest, and it included a procession from the church with a tour of the villages before ending at the Nile with throwing the cross into the river (Oestigaard, 2009, pp. 155–160). Another one is the recollection of Virgin Mary with the mythical model of *Isis* iconographically

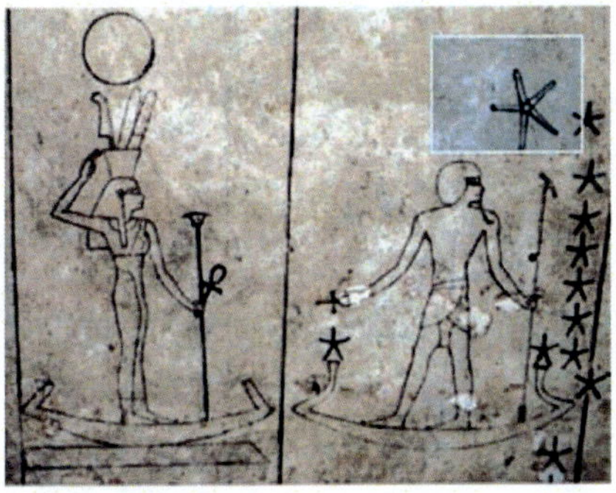

Fig. 3 Goddess *Sopdet* as a Celestial manifestation for Sirius that rose over the horizon in the north, announcing the upcoming of the flood (Belmonte & Shaltout, 2009, p. 217)

(Kamil, 2002, p. 18), and *Isis* was the manifestation of the flood or its herald (Ancient Egypt: Myth and History, 1997, p. 20), with the change in the perceptions of religion; a Christian archetype occurred and Virgin Mary's power was believed to control the flow of the Nile and for centuries to come (Six, 1999, p. 53).

2.2 *Amun-Re* as the Universal God

The Egyptians believed in one supreme god "the creator" (Watterson, 1984, p. 27), under this embodiment of power came a group of gods and goddesses providing the magic or power for maintaining the order for the Egyptians' lives (Harris, 1998, pp. 10–11). There were numerous theological myths about the creator, like *Ptah* of Memphis (Spencer, 2007, p. 75) and *Khnum* of Elephantine who formed people from clay on a potter and controlled the flood at the First Cataract (Wilkinson, 2003, p. 32). But in general, the creator was the sun, which "*Atum* and *Re*" were among its names, the sun was worshipped throughout Egypt, and it embraced authority and political weight (Watterson, 1984, p. 31), with its triumphal character, rebirthing every day, defeating the power of darkness and sailing through the boat by day and through underworld by night (Frankfort, 1961, p. 18).

Amun (the hidden one) was worshipped in Thebes, he became the supreme god *Amun-Re* by merging with the sun god in the middle kingdom and was recognized as the creator (Spencer, 2007, p. 74). Even though he could be considered the nationwide-foundation of advanced theological thought, and the closest approach to a universal god within the Egyptian polytheism (Frankfort, 1961, p. 22), his rise cannot be separated from the political rise of Thebes. Its power expanded during the eighteenth and nineteenth dynasties, and Thebes was claimed to be the birthplace of the universe, where *Amun* the never sleeping god ruled time (Fig. 4). Later, even the Greeks worshipped *Amun* as

Fig. 4 Theban Triad (Amin, 2015)

Zeus-Amun, and the Romans worshipped him as *Jupiter-Amun*. (Watterson, 1984, pp. 140–141). His importance was architecturally expressed in the massive and huge size of the Karnak complex (Frankfort, 1961, p. 155).The figures (Figs. 5 and 6) demonstrate the topographical and astronomical alignments of the Karnak, the main gate is perpendicular to the Nile, and in the same time the main axis is aligned with the sunrise at the winter solstice (Belmonte, 2010). The Karnak also has another axis (north–south) parallel to the Nile providing the passage to Luxor Temple for the *Opet* Festival events (Bell, 1997, p. 158).

2.3 The Divine Kingship

The divine kingship is an integration of religion and politics; the Egyptian estate was part of the universal order (Frankfort, 1961, pp. 30–50). Mythically, the creator produces a son to rule earth; they were in partnership fighting chaos (Spencer, 2007, pp. 85–87). Bowman considered the portray of Roman emperors in the traditional style on Egyptian temples relieves, as relating the "Roman imperial cult" to the traditional believe in divine kingship (Bowman, 1986, p. 166). This concept will be discussed further in this paper, analysing Luxor Temple transformation.

3 The Egyptian Temple: Space, Myth and Populace

Earth, sky, netherworld, life and death, water and the sun all are reals of the Egyptian universe that were turned into symbols and were represented in the Egyptian temple

Fig. 5 Karnak complex (Templo De Karnak-Vista Aerea, 2010)

Fig. 6 Sunrise at the winter solstice in the main axis (Belmonte, 2010)

(Shafer, 1997, p. 18). The temple was the soul of the countryside and the beginning of the settlement, and it was the local centre for the sacred power, as the populace would go to access that power and had it integrated into the daily life (Frankfurtar, 1998, pp. 39–41). The temple was the source of the phenomenon of life with its mysterious core, as it housed the intangible "The god". This gave the temple a sacred character with a dominating infinite respect (Lubicz, 1985, pp. 13–17). Each temple was situated in a sacred place and sacred time (Bell, 1997, p. 133), and it had its own significant features that integrated it with the sacred landscape. The landscape housing the temple represented mythical and theological believes of the populace; these believes borrowed historical and geographical facts to give concrete meanings for what were considered realities (Lubicz, 1985, p. 31). Such realities were inherited through the traditional cultures with the recollection with the mythical models "archetypes".

As mentioned before, the temple significant features required topographical and astronomical alignments, and the solstitial orientation was a dominating astronomical custom with respecting the Nile in Thebes (Belmonte, 2010). But these features were not the only link to the Nile. The Nile cult represented the interplay between the temple religion and the popular ritual tradition, rhyming the agricultural cycle, the appeal for the surge was a duty of the temple's god, and there was a Nilometer in each assigned temple (Frankfurtar, 1998, p. 42).

3.1 The Temple Festival: The Interaction with the Divine

By the New Kingdom, there was a shift in the lifestyle of temples, from cult to festivals, which received great architectural expressions like the lined sphinxes or ram-headed avenues (Spencer, 2007, p. 95). The yearly agricultural cycle determined the religious festivals (Bell, 1997, p. 158), the festivals connected several Theban Temples, like Karnak and Deir el-Bahri, Karnak and Luxor, and Luxor and Madinet Habu (Bell, 1997, p. 134).

The festival represented the interaction between the temple and the social environment, the divine image of the god which was confined by the priests inside the temple, appeared outside and proceeded a traditional route, accompanied by rituals and hymns, Terra-cotta lamps and incense burners, and later by Hellenistic performances, allowing the opportunity for people to access the deities' power for their needs, through the interaction between priesthood and populace. It is a phenomenon symbolizing the social articulation of the cosmic order (Frankfurtar, 1998, pp. 52–53). The cult of festival became a necessary aspect of the Egyptian social and theological world, and it was the time when the temples were accessed by commoners, reaching the outer courtyards. Those festivals lasted from the New Kingdom until the Coptic period (Spencer, 2007, p. 96). There was decline by the third century AD with the financial drop of temples, but as papyri documented, some festivals survived until the fourth and fifth centuries AD. But in general, the cult of festival transformed into regional cult processions as the Egyptian religion became a local phenomenon (Frankfurtar, 1998, pp. 57–64).

4 Luxor Temple: The Origin

The original Egyptian name of Luxor Temple was *Ipt rsyt* "the Southern Sanctuary", with "holy of holies" at the southern end (no. 18)—(Fig. 7) (Redford, 2001, p. 309), and it was founded probably under Ahmose, but its earliest architectural remains are related to Hatshepsut reign (Magli, 2014, p. 35). Architecturally, Luxor Temple is much simpler than Karnak complex, which is a city of temples. Both temples were devoted to *Amun-Re* and flourished during the New Kingdom, but while Karnak was the "residence" of the god; Luxor Temple cult remained mysterious for a long time. Bell defined it as a representation of a divine cultic temple, and it had the main role in the process of the rejuvenation of the royal divine power of the Pharaoh, or renovating his *Ka* "The divine aspect of the king" (Bell, 1985, pp. 251–256). Luxor Temple was the mythological power base of the living divine king, and the national shrine for his cultus, where the divine spirit was handled from one king to another, and his right to rule was confirmed as *Amun-Re* powers were transformed to him. Then, he would come out to the Egyptians in the public areas of the temple (no. 3)—(Fig. 7) (Bell, 1997, p. 157). The enlargements of the temple can be understood within the Egyptian conception of sacred landscape (Magli, 2014).

The main axis of the temple is oriented almost north–south, which is considered parallel to the Nile in general. The sanctuary was the first part to be built (Bell, 1997, p. 148), and its orientation is according to a cardinal custom, which would be achieved through observations of stars configuration in the north according to how Ancient Egyptians organized the cosmos since the Old Kingdom (Belmonte, 2010, pp. 537–538). Beside the main axis, there is a deviation in the Ramesside Court (no. 3)—(Fig. 7), it offered a symbolic feature to the temple from outside, it is aligned with the sacred path to the Karnak "the Avenue of the Sphinxes" (no. 5)—(Fig. 7) to serve the yearly *Opet* Festival, Magli interpreted this as "*Luxor Temple became every year for a short period a sort of double-faced temple*", and so the architectural design of the Luxor Temple developed over the time, emphasizing the significance of the divine authority, its relation with the creator and their roles in the regularity of the cosmic life cycle maintaining the *Maat*. That was celebrated by the public during the "*Opet* Festival" within the monumental sacred space (Magli, 2014, p. 42).

In the Ramesside Court (no. 3)—(Fig. 7), Ramesses II statues stand before the Pylon and at the entrance to the Colonnade (Fig. 8). These statues were interpreted by Labib Habachi as *ka* statues, or cult statues of the king, as embodiment of the royal *Ka* (Bell, 1985, p. 259). This is another empowerment to the divine authority within a space, which was accessed by the common populace during the festival, which will be elaborated upon.

The Barque Sanctuary (no. 15)—(Fig. 7) was built by Amenhotep III, and the columns were removed for a shrine rebuilt by Alexander as the son of "*Zeus-Amun*". This sanctuary was the end of the temple, followed by a separate temple in the south—which is not the case today, the temple was called the Southern *Opet*, which is the sanctuary dedicated to *Amun* of Luxor Temple "*Amenemopet*" (no. 18)—(Fig. 7), where the god's cult image resided on a mythic mound of creation. The entrance was through the southern end, while the axis of Luxor Temple that leads to the barque sanctuary was north–south; the axis of the Southern *Opet* was solar (east–west); both sanctuaries were separated physically but related ritually (Bell, 1997, p. 156).

4.1 The *Opet* Festival: The Sacred Landscape of *Amun-Re* Dominance

The *Opet* Festival was the longest and the most important event in the Theban calendar, the festival began around the middle of the second month of the first season "the inundation", when the Nile water reached its expected level, it lasted for eleven to twenty seven days (Bell, 1997, p. 158). Later, the Hellenistic *Semasia* festival was celebrated as anther format (Kalimi, 2016, p. 214). It was first celebrated during the reign of Hatshepsut; its name was probably related to an earlier "Heliopolitan prototype" (Darnell, 2010) (Fig. 1). This assumption was confirmed by Fukaya, who indicated that the festival could have been part of an echo of earlier seasonal festivals celebrating the Nile flood, which were embedded in the Theban theologies. Among them, the *Ssp-ftrw* "Taking of the River", which was a festival related to the Osirian myth since the Old Kingdom (Fukaya, 2015, p. 119). The festivals transformations could be interpreted as a strategy of Theban theology, rooting the Theban divine power within the public cosmic vision, for political reasons.

During the *Opet* Festival, the temples of Karnak and Luxor functioned together for empowering the expansion of the religious, sacred or mythical landscape of *Amun-Re* cult centre. The veiled Theban triad statues "*Amun, Mut, and*

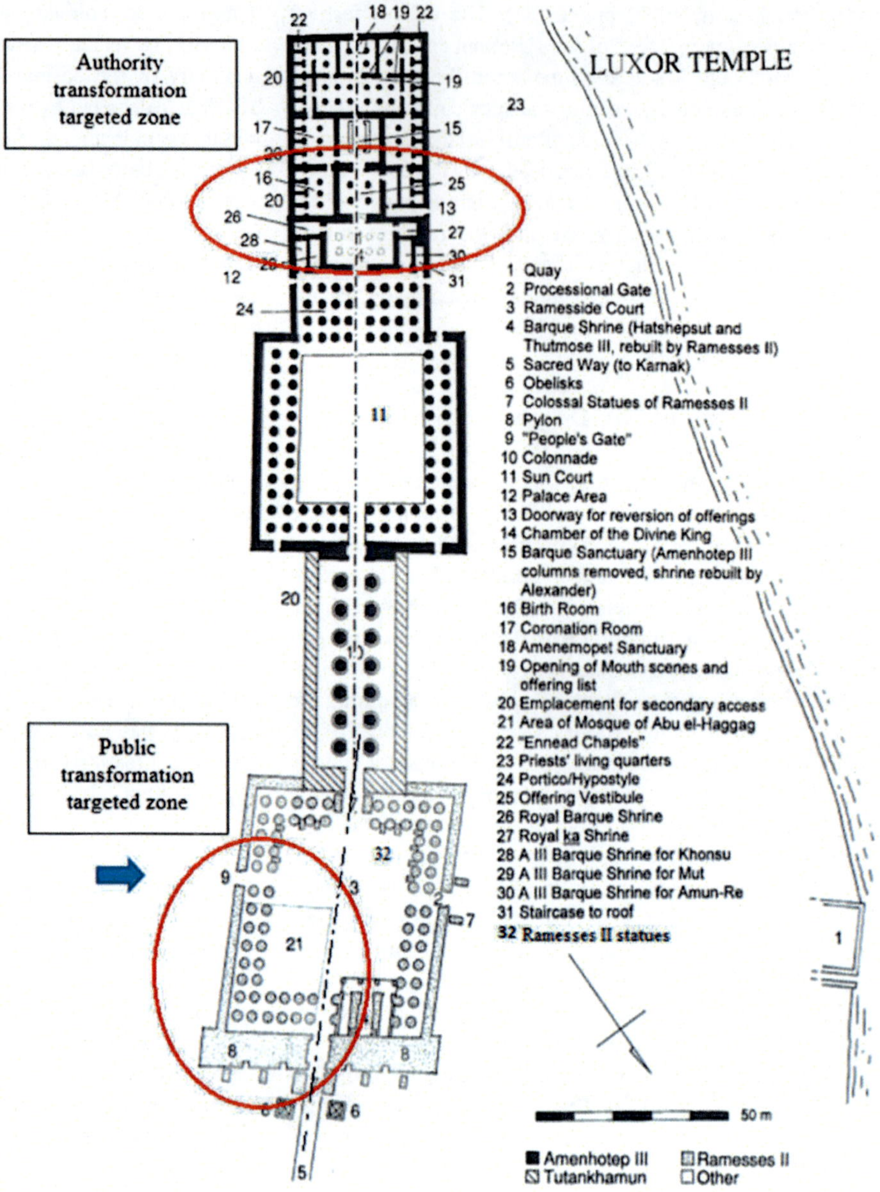

Fig. 7 Luxor Temple—Drawn by Carol Meyer (Bell, 1997, p. 149)

Khonsu" (Fig. 4) travelled from their sanctuaries in Karnak complex to Luxor Temple, at the triple-barque shrine station (no. 4)—(Figs. 7 and 9) to renew the life force (Wickett, 2009, p. 408). The *Opet* Festival had a cosmic significance, and it secured the regeneration of the creator *Amun* of Luxor, the recreation of the cosmos and the beginning of a new agricultural cycle (Shafer, 1997, p. 157). This was a repetition of the mythical moment manifesting the passage from chaos to cosmos (Eliade, 1991, p. 56), such operation was seen by the traditional Egyptians as dangerous, for the risk of unleashing disorder, it had to be done within careful rituals during the *Opet* Festival, and presided by the divine, so the most prominent part of the celebration was the procession of the king from the Karnak to Luxor Temple, to meet his father *Amun-Re* and back (Shafer, 1997, p. 158). Darnell interpreted the journey by land as a symbolism of the dry period before the union of the god and the return to the north by river as emphasizing the returning flood (Darnell, 2010, p. 10)—(Figs. 10 and 11).

4.2 The Water Symbolism: The Nile as Part of the Political Power

The cosmic water symbolism between *Amun-Re* and the divine king can be understood through the research of

Fig. 8 Ramesses II *ka* statues (photo taken by Rasha Said, 2020)

Fig. 9 "Barque Stations" inside Ramesses II's court (photo taken by Rasha Said, 2020)

Gulyás, he interpreted the repeated reliefs in the southern wall (Fig. 12), which is part of the Southern *Opet*, it is located symbolically towards where the Nile comes from in the Barque Chamber, it is portraying the king rendering back the divine gift "the water in the land", and the land is portrayed in dual representation, expressing the totality of Egypt, to assure the continuation of the inundation (Gulyás, 2007). An extended interpretation could be concluded, as the enhancement of the dramatic ruler with the ultimate god, the Nile, and the cosmos as Egypt.

Since Amenhotep III, the *Opet* Festival portrayed a transformation of an individual ruler to the personification of the royal *ka* and its transmission, mingling the ruler identity with that of an earlier incarnation of the royal *ka* (Darnell, 2010, pp. 4–5) and giving more strength to the theological and mythical position of the living pharaoh. In the same time, the association of the *Opet* Festival with the inundation was a manifestation of "the primeval water *Nun* flowing from Thebes" (Fukaya, 2015, p. 119), this was a comic empowering for the ruler, reflecting how the sociopolitical system was working. The chamber of the divine king (no. 14)—(Fig. 7) was where the real power point in Luxor Temple existed, where the coronation rites were repeated (Bell, 1997, p. 173), and this chamber was where the *principia* stood later. The *Opet* Festival and the temple rituals were confirmation of the reflection of the king's prominent

Fig. 10 Sacred path between Karnak and Luxor, known today as avenue of the Sphinxes (photo taken by Rasha Said, 2020)

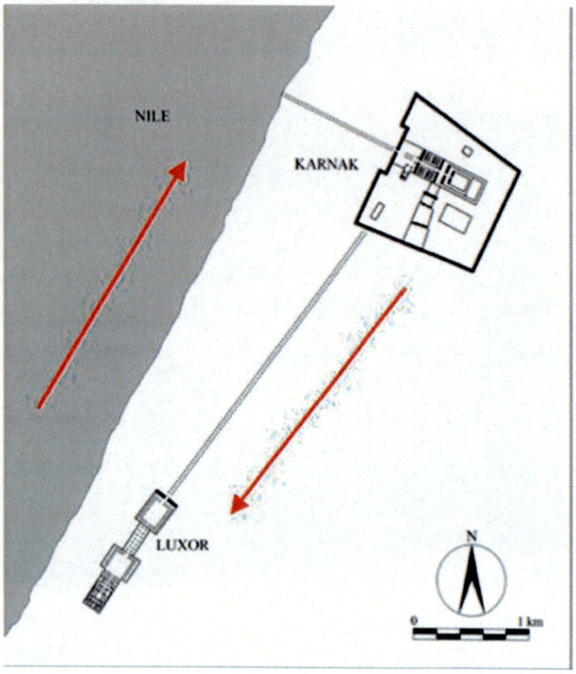

Fig. 11 Sacred path between Karnak and Luxor Temple, land and water (Magli, 2014, p. 37)

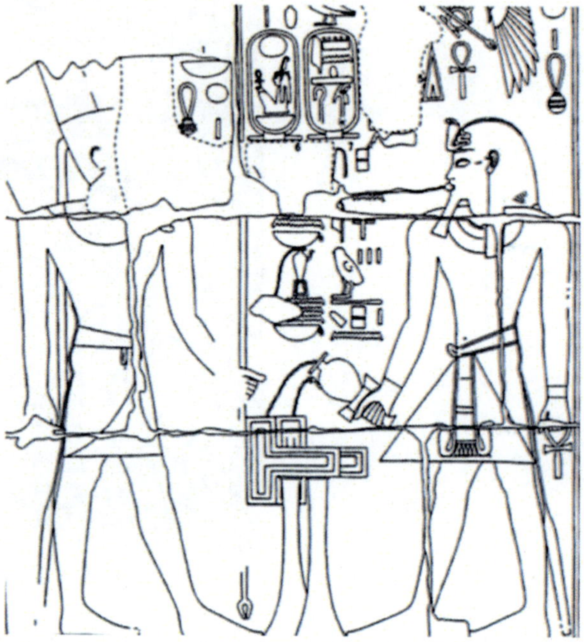

Fig. 12 Libation on the southern wall the Southern *Opet* (Gulyás, 2007)

role in the traditional Egyptian society, confirmed as a living incarnation of a deity (Bell, 1997, p. 127).

The public would enter Luxor Temple from the east gate (no. 9)—(Fig. 7), and they used to watch the celebration from the eastern part of the court, shearing for the rebirth of the world with the exit of the king. Today, it is where the mosque of Abu'l Hajjaj was built and before it a church (no. 21)—(Fig. 7), in front of the Barque Shrine (no. 4)—(Fig. 7).

5 Luxor Temple: The Transformation

The last repairs for Luxor Temple were done in the reign of *Tiberius* in the first century AD (Atiya, 1991, p. 331). In late third century, Luxor Temple was transformed into military fortress (Fig. 13), and constructions were added such as circuit wall, streets, barracks and other buildings. Bagnall concluded the transformation as secular or civil of a temple and that was not a religious construction anymore, especially that a Greek graffiti inside the temple showed clearly that the ancient Egyptian cult was abandoned before the reuse (Bagnall, 1993, p. 263). This is what would be argued in the following points.

5.1 The *Principia* at Luxor Temple

The transformation of Luxor Temple into a Roman fortress cannot be separated from being a political act, it was done by Diocletian after the suppression of the Upper Egyptian revolt in 297 AD (Atiya, 1991, p. 331), and the transformation could be understood as a part of his imperial propaganda, or as an expression of his glory and triumph (Kalavrezou-Maxeiner, 1975, p. 240). The fortress contained a *principia* which is an imperial cult temple, dedicated to a based civic religion centred on the emperor (McIntyre, 2019). Barbagli pointed out that this was an act of the continuity of the previous established ruler cult, through the worship of the emperors (Barbagli, 2019). This assumption was also brought up by Frankfurtar, interpreting the act of converting Luxor *Amun* Temple into an imperial audience hall to assimilate the emperor divine person to the Egyptian high god (Frankfurtar, 1998, p. 36).

The columns of the chamber of the divine king were removed (no. 14)—(Figs. 7 and 14), and the Romans also raised the floor converting it into a worshiping place for the divine emperor (Bell, 1997, p. 156) (Fig. 15). The hall was painted in monumental art of the emperor and his greeters or other emperors, but most of it is now gone (Fig. 16). Maxeiner did a detailed study visualizing those scenes, she suggested that part of the paintings was a special celebration of *Adventus* which was a Roman ceremony in which an emperor was formally welcomed into a city, the scene was probably for the Egyptian gods' sacred image visit to the temple, as a celebration that continued until the Roman rule, and a statue of *Jupiter* "the equivalent to *Amun*" could be carried to there in the scene (Kalavrezou-Maxeiner, 1975, pp. 242–243). Definitely, this brings to the minds the "*Opet* Festival".

Architecturally, Karelin indicates that there was monumentalization of the principia's architecture, increasing the significance of the Roman imperial cult was clearly obvious in the presence of an apse (Fig. 15), which was a feature of the late Roman religious architecture (Karelin, 2015, p. 360). In the 3D reconstruction of the Roman Imperial cult temple at Luxor which was constructed by Kulikova & Karelin (Fig. 18), they distinguished some architectural features indicating the Egyptian culture influence, such as the placement of the four statues of the tetrarchs in front of the hypostyle hall (Fig. 17), which was interpreted as an iconographic emphasizing of the connection between the Roman rulers and the great kings of the New Kingdom (Kulikova & Karelin, 2017). These statues show a resemblance to Ramesses II *ka* statues (Fig. 8). These architectural evidences support Bell's insistence, that the choice of this part of the temple for the imperial cult was deliberate, according to the Roman awareness of the long Egyptian tradition of the divine kingship (Bell, 1985, p. 274).

5.2 The Christianization

After the transformation into a Roman fortress, a civilian settlement remained around the camp, which kept working until the Persian occupation in the seventh century AD, with remaining churches until today (Dawud & Fakhri, 2008). The archaeological investigations reached the existence of five churches outside the temple, they were dated to the sixth and seventh centuries AD, and the oldest one is in front of the temple pylon (Dawud & Fakhri, 2008) (Fig. 19). Inside the temple, the *principia* was often mistaken for the reuse as a church (Atiya, 1991, p. 331), which is unlikely as the *principia* was oriented to the south (Syriac & George, 2002), while there are remains of an early Christian church which lies beneath the mosque and tomb of Abu el-Haggag (Fig. 20). The church could be accessed from the court, and the windows are clear under the mosque (Weeks, 2005, p. 124). It is one of the few churches from the early Christian period that is still standing to the height of the side windows (Atiya, 1991, p. 331).

It is an interesting phenomenon that the church inside the temple (no. 21)—(Fig. 7) was in the space that used to house the public interaction with the holy, during the sacred events of the *Opet* Festival, in front of the "Barque Stations" inside Ramesses II's court. There are no evidences of Christian establishments inside the more sacred parts of the temple, where the Romans made their modifications, even though the landscape became clearly Christian and that the public continued holding its sacredness as part of their cosmic view. The Christianization process was obvious in the numerous churches around the temple and the Christian consecration signs in the temple (Fig. 21). In comparison with other temples that had Christian transformation for obvious cosmic value related to the Nile, such as *Isis* Temple on Philea Island and *Khnum* Temple on Elephantine Island, as sacred locations for controlling the flood at the First Cataract, it is documented that Christian sanctuaries were established inside the *Pronaos* (Dijkstra, 2011) of each

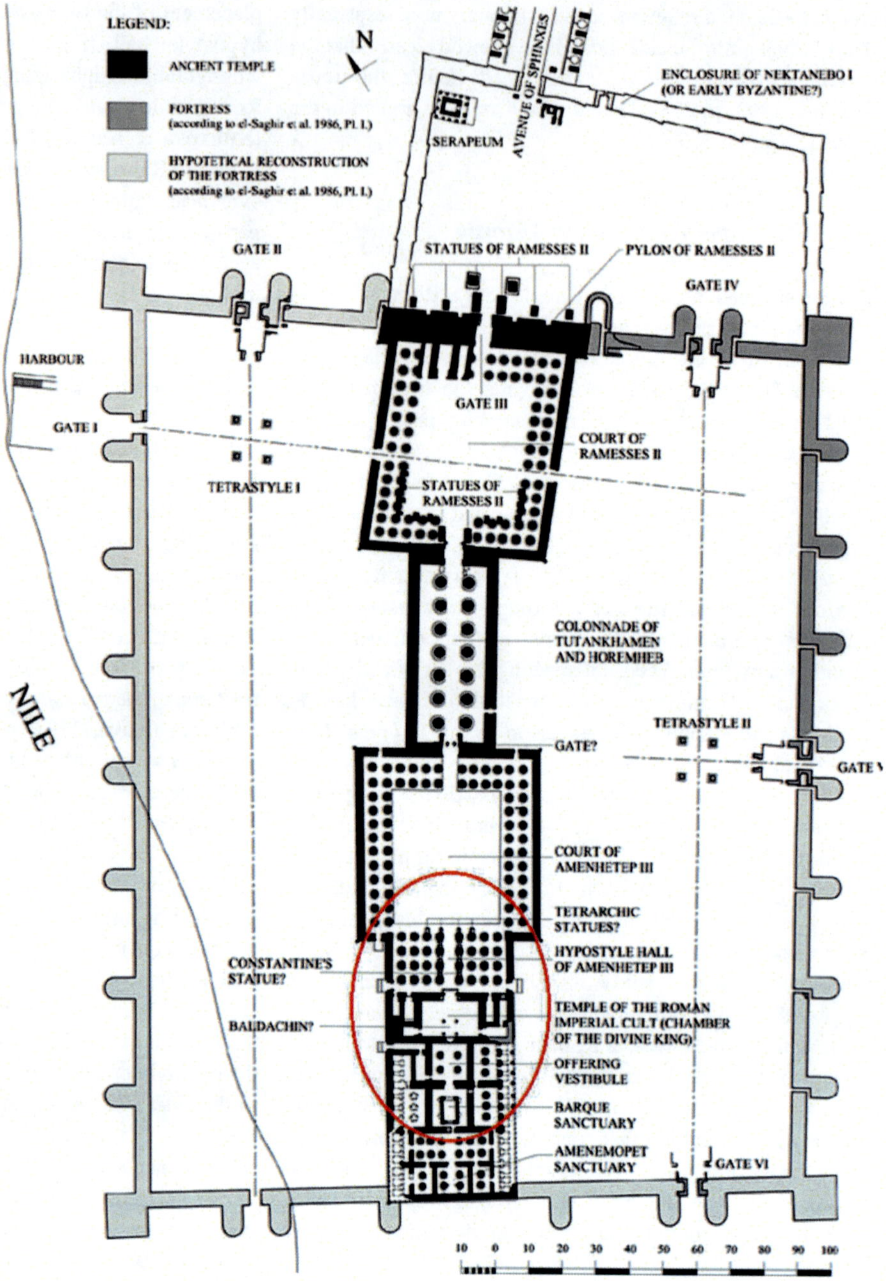

Fig. 13 Plan of the Luxor Fortress (Karelin, 2015, p. 361)

temple, closer to where the god lived, or Holy of Holies, which is not the case in Luxor Temple. This brings up the *Opet* Festival and the Coptic interaction with the sacred space. Logically, the Egyptian Christians continued to celebrate the festival in some format, as its sacred significance must be Christianized. Yet, it is not clear how the *Opet* Festival took its Coptic version, only the Islamic Mulid of Sidi Abu'l Hajjaj remains as a phenomenon to relate to.

5.3 Mosque and Tomb of Abu Al Haggag "Sidi Abu'l Hajjaj Al-Uqsori"

Abu'l Hajjaj (died 642 hejra–1243 AD) became a legend or a patron saint of Luxor, as the father of pilgrims who had power on water, and he had different myths forming his memorial image, which is clearly a transformation of myths supporting the architype theory, serving the concept of the

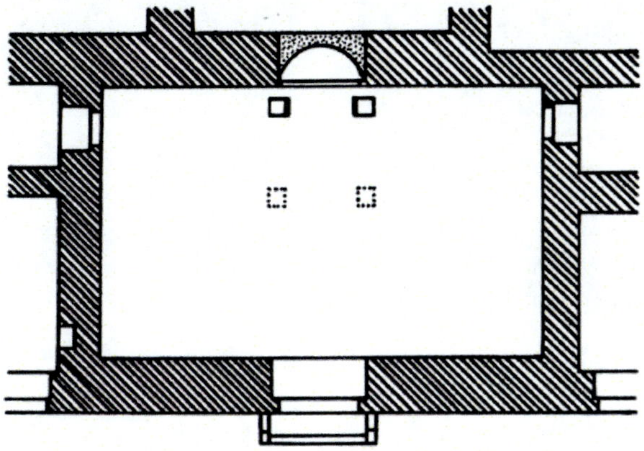

Fig. 14 Architectural modification of the chamber of the divine king (Kalavrezou-Maxeiner, 1975)

Fig. 15 Roman apse (photos taken by Author, 2020)

power controlling water or the Nile over time. For example, there were links to his sacred boat and the Nile, some myths explained the construction of the mosque "Over the church inside the temple" (Fig. 22) because angles carried his dead body and put it in this specific location to be buried (Wickett, 2009, p. 405). An interpretation could be reached through understanding the significance of this part of the temple and its position within the populace memory, or the social memory.

Murray described a local celebration of the Mulid by the beginning of the twentieth century, she indicated the existence of an ordinary Nile boat on the roof of the mosque, which was freshly painted in midsummer before the rise of the Nile[1] (Collecting & Documenting Egyptian Festivals, 2008, 2009), then it was placed on a carriage and driven round the fields of the town, followed by the locals, and this march was accompanied by police officers, horseback riders, camels, religious men and members of Abu'l Hajjaj family "Al-Hajajiyya". She interpreted this celebration as the continuity of the ancient cult, which was strongly attached to the temple under a new name (Murray, 2004, p. 115). Wickett made a deep analysis of the similarities in rituals between the ancient festival and the mulled, and she concluded the statement "*A fusion ritual re-enactments and belief in the interrelationship of the divine and the human in the cosmic order—Mulid festivals for saints and sheikhs are not anachronistic or archaic celebrations, therefore. They embody the tradition of millennia in an ancient space*" (Wickett, 2009, p. 406) (Figs. 23 and 24).

5.4 Luxor Temple Today

Focusing on the archaeological evidences that were referred to and analysed in this research paper, some of them do not exist anymore, as it was pointed out in multiple parts. Luxor Temple as the rest of monuments of Eastern Thebes are under danger. The deterioration accelerated in the last decades because of the rising groundwater levels and climatic conditions. Sadly, it resulted in exfoliation of stones and the decrease of their durability, and the loss of paintings and texts (Ahmed & Fogg, 2014). This priceless heritage should be maintained for future generations.

6 Conclusion

Luxor Temple was an architectural embodiment of the political authority's power. Yet, it was not a secular one, but a theological power base for the divine king. This base was strongly tied with the cosmic Egyptian mythology, which the Nile was in the core of it. The temple's features: location, design, the topographical, and astronomical alignments, all reflected the respect to the Nile, and the rhyming of the agricultural cycle, controlling the Egyptian cosmos. The main purpose of the temple and its strong link to the Karana; were expressed urbanely and ritually; it was a magnification of the sacred space, representing the universal god, accompanied by the enhancement of the divine ruler and the Nile cult. That concept was inherited by the foreigner ruler "the

[1] Later it was modified and aligned to the lunar Islamic calendar on the eve of the half of Shaban month. (Collecting & Documenting Egyptian Festivals 2008, 2009).

Fig. 16 Reliefs of Amenhotep III covered with Roman painting (photo taken by Rasha Said, 2020)

Fig. 17 Entrance to the Roman imperial cult temple

Roman" who used the symbolism of the temple to reflect his power, especially that the Romans elevated the Nile cult to an official status. The holiest parts of the temple were inaccessible by the public, housing the sacred process of the divine cosmic regeneration and its link to the ruler; this part of the temple was modified by the Romans for the imperial cult, reflecting the mixed cultural heritage.

The outer court has another story as interpretation for the transformation of the temple, as it was a sacred space where the divine authority could be reached by the common populace during the festival, it was a phenomenon symbolizing the social articulation of the cosmic order. The repeated mythicization of the historical figures served the recollection with the mythical models, within the conformity of the

Fig. 18 3D reconstruction of Roman imperial cult temple—by D. Karelin (Kulikova & Karelin, 2017)

Fig. 19 Church on the west of Luxor Temple

cosmic vision of common populace. The new Archetypes were linked to the cosmos whether Christians or Muslims, both used the same architectural space to confirm the connection with the holy power controlling the Nile. May be the Egyptians lost the true understanding of the *Opet* Festival over time, but its cosmic implication was not lost, represented in the archetype and the tie with water. Accordingly, the significant phenomenon of the "Ramesside court" as a sacred space was reflected by the public yearly celebration, and by housing a church then a mosque in the same location,

Fig. 20 Church beneath the mosque of Abu el-Haggag (photos taken by Rasha Said, 2020)

Fig. 21 Consecration engraved crosses in The Hypostyle Hall (Photos taken by Maha Abu Bakr)

Fig. 22 Relation between the mosque, the temple and the Nile (Rovielo, 2018)

Fig. 23 Procession of mouled (Ficalora, 2013)

Fig. 24 Painting of the *Opet* Festival by Mark Millmore (Millmore, 2019)

which could be interpreted as reactivated of a public holy site.

References

Ahmed, A., & Fogg, G. (2014). The impact of groundwater and agricultural expansion on the archaeological sites at Luxor, Egypt. *Journal of African Earth Sciences*, 93–104.

Amin, O. S. (2015). *Amun, Mut, and Khonsu*. https://www.ancient.eu/image/3463/amun-mut-and-khonsu/. Accessed 23 Oct 2019.

Ancient Egypt: Myth and History. (1997). New Lanark: Geddes & Grosset, David Dale House.

Atiya, A. S. (1991). *The coptic encyclopaedia*. Macmillan Library Reference.

Bagnall, R. S. (1993). *Egypt in late antiquity*. Princeton University Press.

Barbagli, N. (2019). *From pharaohs to tetrarchs? Continuity and change in ruler cult within the temple of Luxor*. https://www.academia.edu/39717474/From_Pharaohs_to_Tetrarchs_Continuity_and_Change_in_Ruler_Cult_within_the_Temple_of_Luxor. Accessed 25 Nov 2019.

Bell, L. (1985). Luxor temple and the cult of the royal Ka. *Journal of near Eastern Studies, 44*(4), 251–294.

Bell, L. (1997). The New Kingdom "Divine temple": The Example of Luxor. In B. E. Shafer (Ed.), *Temples of ancient Egypt* (pp. 127–180). Cornell University Press.

Belmonte, J. A., & Shaltout, M. (2009). Astronomy, landscape and symbolism: A study of the orientation of ancient Egyptian temples. *In Search of Cosmic Order: Selected Essays on Egyptian Archaeoastronomy* (pp. 215–250). Supreme Council of Antiquities Press.

Belmonte, J. A. (2010). Keeping Ma'at: An astronomical approach to the orientation of the temples in ancient Egypt. *ScienceDirect*, 532–539.

Bowman, A. K. (1986). *Egypt after the pharaohs 332BC-AD642 from Alexander to the arab conquest*. The University of California Press.

Canhão, T. (2013). A timeless legacy: The calendars of ancient Egypt. In R. Sousa (Ed.), *Alexandrea ad Aegyptum: The legacy of multiculturalism in antiquity* (pp. 283–301). Coimbra: CITCEM.

Cannuyer, C. (2001). *Coptic Egypt. The christians of the Nile*. Thames & Hudson.

Collecting & Documenting Egyptian Festivals. (2008). *2009*. UNESCO.

Darnell, J. (2010). *Opet festival*. In UCLA Encyclopaedia of Egyptology: http://digital2.library.ucla.edu/viewItem.do?ark=21198/zz0025n765.

Dawud, N. K., & Fakhri, A. (2008). *History of christianity and its monuments in the Parish of Naqada, Esna, Luxor, and Armant*. St. Mark's Foundation for the Study of Coptic History.

Dijkstra, J. H. (2011). The fate of the temples in late antique Egypt. In L. Lavan (Ed.), *The archaeology of late antique 'paganism'* (pp. 389–463). Leiden. Boston: Brill.

Eliade, M. (1991). *The myth of the eternal return: Or, cosmos and history*. Princeton University Press.

Ficalora, S. (2013). *Abu el-Haggag festival: The modern-day opet*. https://nilescribes.org/2018/05/12/abu-el-haggag-festival/. Accessed 14 Oct 2019.

Frankfort, H. (1961). *Ancient Egyptian religion, an interpretation*. Harper Torchbooks.

Frankfurtar, D. (1998). *Religion in Roman Egypt, assimilation and resistance*. Princeton University Press.

Fukaya, M. (2015). Dates and precursors of the opet festival. 哲学•思想論叢 *(Tulips-R)-University of Tsukuba Repository*, 128–100.

Gulyás, A. (2007). A cosmic libation. Researches on the Theology of Luxor Temple. In *The ninth International Congress of Egyptologists—Grenoble 2004* (pp. 895–905). Leuven: Uitgeverij Peeters en departement oosterse studies.

Harris, E. L. (1998). *Ancient Egyptian divination and magic*. York Beach, Maine: Weiser.

Kakosy, L. (1982). The Nile, Euthenia, and the Nymphs. *The Journal of Egyptian Archaeology*, 290–298.

Kalavrezou-Maxeiner, I. (1975). The Imperial chamber at luxor. *Dumbarton Oaks Papers, 29*, 225–251.

Kalimi, I. (Ed.). (2016). *Bridging between sister religions: Studies of jewish and christian scriptures offered in honor of Prof. John T. Townsend.* Boston: Brill.

Kamil, J. (2002). *Christianity in the land of the pharaohs. The coptic orthodox church*. The American University in Cairo Press.

Karelin, D. (2015). Reflection on some architectural peculiarities of the late Roman principia in Egypt by the example of Luxor. In *Limes XXII—The 22nd International Congress of Roman Frontier Studies Ruse,Bulgaria, 2012* (pp. 359–368). Sofia: National Archaeological Institute with Museum Bulgarian Academy of Sciences.

Kulikova, I., & Karelin, D. (2017). The Roman imperial cult temple at Luxor: Its architecture and possible connection between Roman and Egyptian cultures. In *The XI International Congress of Egyptologists, Florence, 23–30 August 2015* (pp. 720–721). Oxford: Archaeopress and the Authors.

Lubicz, R. A. (1985). *The Egyptian miracle: An introduction to the wisdom of the temple*. Inner Traditions International.

Magli, G. (2014). Some cognitive aspects of the Luxor-Karnak relationship, time and mind. *The Journal of Archaeology, Consciousness and Culture*, 33–45.

McIntyre, G. (2019). *Imperial cult*. Leiden: Brill.

Millmore, M. (2019). *Opet festival at Karnak*. http://markmillmore.com/opet-festival-at-karnak/. Accessed 20 Oct 2019.

Murray, M. A. (2004). *Egyptian temples*. Routledge.

Oestigaard, T. (2009). Christianity and islam as Nile Religions in Egypt:Syncretism and continuity. In T. Oestigaard (Ed.), *Water, culture and identity. Comparing past and present traditions in the Nile basin region* (pp. 141–164). Bergen: BRIC Press.

Redford, D. B. (2001). *The oxford encyclopaedia of ancient Egypt* (Vol. 2). Oxford University Press.

Rovielo, E. (2018). *The mosque of Abu al-Haggag, the Luxor temple*. https://www.flickr.com/photos/eliasroviello/30622172827. Accessed 20 Oct 2019.

Shafer, B. E. (1997). Temples, priests and rituals: An overview. In B.E. Shafer (Ed.), *Temples of ancient Egypt*. Newyork: Cornell University Press.

Six, V. (1999). Water, the Nile, and the Tä'amrä Maryam: Miracles of Virgin Mary. *Aethiopica, 2*(1999), 53–68.

Spencer, A. J. (2007). *The British museum book of ancient Egypt* In:. A. Spencer (Ed.). Cairo: The American university in Cairo press.

Syriac, S., & Georgi, B. H. (2002). *A guide to churches and monasteries in Egypt*. Al-Naam Printing and Supplies Company.

Templo De Karnak-Vista Aerea. (2010). https://historizarte.wordpress.com/historia-del-arte/karnak2/.

Watterson, B. (1984). *The gods of ancient Egypt*. Facts On File Inc. Publications.

Weeks, K. R. (2005). *Illustrated guide to luxor. Tombs, temples and museums.* Cairo: The American University Press.

Wickett, E. (2009). Archaeological memory, the leitmotifs of ancient egyptian festival tradition, andcultural legacy in the festival tradition of Luxor: The mulid of Sidi Abu'l Hajjaj al-Uqsoriand the Ancient Egyptian "Feast of Opet" . *Journal of the American Research Center in Egypt, 45*, 403–426.

Wilkinson, R. H. (2003). *The complete gods and goddesses of ancient Egypt*. Thames & Hudson.

Partnerships for Sustainable Tourism Development in the Cultural Heritage Sites

Reham Kamel El-Khadrawy, Amr Abdalla Attia, and Rowaida Rashed

Abstract

Cultural heritage sites are among the most important tourist attractions in any country. The importance of "Sustainability" has emerged as a result of the direct relationship between the environment and development, and thus "Sustainable Tourism" must be environmentally sensitive and culturally appropriate, in addition to achieving economic benefits. And to develop cultural heritage sites, we should consider the intertwined relationship of many relevant parties. This research paper evaluates the roles of different bodies that are directly related to the conservation of cultural heritage sites. Hence, to establish guidelines for a sustainable development approach, the research relied on the descriptive-analytical approach through conducting an extensive literature review followed by a comparative analytical study of some global experiences. This paper thrived to exchange the gained knowledge to the local case study of Anakato. Anakato has unique and important elements of cultural heritage, and this ecolodge is an individual initiative to achieve sustainability. The research proved that achieving coordination and partnership is challenging, but it is a key factor to achieve sustainability, and the paper also concluded that coordination and partnership are essential for achieving sustainable tourism development in cultural heritage sites.

Keywords

Cultural heritage • Sustainable tourism development (STD) • Conservation • Partnerships • Aswan

Abbreviations

ADF	African Development Fund
AWF	African Wildlife Foundation
EQI	International Environmental Quality Company
CH	Cultural Heritage
CT	Cultural Tourism
IFC	International Finance Corporation.
MMO	Mundo Maya Organization
NGO	Non-government Organization
STD	Sustainable Tourism Development
UNDP	United Nations Development Program
UNWTO	World Tourism

1 Introduction

Cultural heritage sites in many countries were subjected to many cultural and social changes that have led to their deterioration. However, as of the last century, many intellectual streams have called for the preservation of cultural heritage sites have emerged (Ibrahim, 2013). With the increasing trend of the tourism industry and its revenues, there is a need to strike a balance between the protection of cultural heritage and tourism development (Al-Naiem, 2005). International organizations such as UNESCO and UNTWO emphasize the importance of cooperation and the participation between different parties in the process of cultural heritage site conservation. The contribution of this research is to address and analyze this interrelation, justify the importance of this issue, and explore the role of various parties and bodies to achieve STD in the heritage sites. This research highlighted three main sectors that have a direct impact on the conservation process (government sector, private sector, and NGOs). The research methodology is based on the descriptive-analytical approach. Accordingly, a literature review for references dealing with the topic of the

R. K. El-Khadrawy (✉) · A. A. Attia · R. Rashed
Department of Urban Planning, Ain Shams University, Cairo, Egypt

A. Versaci et al. (eds.), *Conservation of Architectural Heritage*, Advances in Science, Technology & Innovation,
https://doi.org/10.1007/978-3-030-74482-3_5

research was conducted, selecting global experiments for the analytical study, and Anakato Lodge in Aswan was chosen to conduct the field study.

2 Cultural Heritage and Sustainable Tourism

CH in the past meant only tangible heritage. Now the definition of CH is much broader and includes *"All creative expressions of people's existence in the past, near past and present that have been passed on to the present generation by past generations"* (UNESCAP, 2008). According to the previous definition issued by UNESCO, cultural heritage is divided into tangible and intangible heritage as the following:

- Tangible heritage: "*Physical manifestation or symbol of cultural expressions or traditions of the societies that are living or lived in the area. Monuments, traditional buildings, archeological sites, temples, historic cities,* etc. *are examples of tangible heritage.*" Tangible cultural heritage is divided into the tangible cultural heritage that can be transferred and cannot be transferred (UNESCAP, 2008).
- Intangible heritage: "*Non-physical manifestation of cultural expressions and traditions of a society that has its roots in the cultural values and practices of the previous generations. Traditional ways of life, social practices, festivals, music, crafts,* etc., *are examples of intangible heritage*" (UNESCAP, 2008).

There is a dynamic relationship between tourism and heritage. Tourism has been seen as a tool for bringing visitors to cultural sites, as cultural heritage and its events are sources of attraction for tourists to do cultural experiences (EENCA, 2017). Tourism can generate funding and educate societies, and when managed well, it can be an effective factor for sustainable development. Meanwhile, mass tourism or poor tourism management can threaten the nature and the characteristics of heritage, in addition to the deterioration of the environment, culture, and lifestyles of the hosting communities. We must emphasize that tourism should achieve economic benefits to the host communities and provide important means and motivations to preserve their cultural heritage and practices (UNESCAP, 2008). Also, CT is an incentive to preserve the elements of CH in tourist destinations, where it helps to preserve these elements as important tourist attractions (Bahaa-El-Din, 2005). The definition of CT according to UNWTO is *"The movements of persons, essentially for cultural motivations such as study tours, performing arts and cultural tours, travel to festivals and other events, visits to sites and monuments, travel to study nature, folklore or art"* (UNESCAP, 2008). According to this definition, the concept of CT is no longer limited to visiting monuments and temples only but has become an interactive experience with the local community. Hence, social values, customs, and traditions of the local community have become one of the most important tourist attractions, by attending exhibitions, festivals, events, and purchasing traditional local products.

As a result of the rapid expansion of the tourism sector, a lot of tourist destinations are facing pressure on their natural, cultural, social, and economic environment. Besides, unbalanced tourism growth aiming to achieve short-term benefits often leads to negative impacts that harm the environment and societies (Mahmoud, 2011). The importance of "Sustainability" has emerged as a result of the direct relationship between the environment and development, thus "Sustainable Tourism" must be environmentally sensitive and culturally appropriate, in addition to achieving economic benefits, especially to the host communities (EENCA, 2017). Furthermore, the involvement and cooperation of local community representatives, environmental protection workers, tourism operators, and policymakers are essential to achieving sustainable tourism and strengthening the protection of heritage resources for future generations (UNESCAP, 2008).

3 Challenges of Sustainable Tourism Development in Cultural Heritage Sites:

The following part is illustrating the expected challenges facing sustainable tourism development in cultural heritage sites:

- Protection of cultural heritage: CH resources are including both tangible and intangible heritage. These resources are usually scarce and cannot be recreated, so protecting and managing these resources is the biggest challenge for STD. As the main goal of the preservation is to protect the cultural significance of CH (UNESCAP, 2008).
- Lack of information: To have an integrated approach to sustainable development in cultural heritage sites, it is important that the following relevant information is collected: cultural resources, tourism needs, and available infrastructure. This information helps to determine what is available and what needs to be done (UNESCAP, 2008).
- Technical training: Many countries are lacking trained and qualified personnel to manage CT resources and

high-quality tourism products. Providing training for local people to participate in cultural tourism activities leads to achieving economic benefits for the local community, which leads to STD (NWHO, 1999).

- Management of cultural heritage sites: The lack of plans to manage tourism in cultural heritage sites leads to many problems such as pollution, overcrowding, and other problems. Good management in CT sites can identify and prevent threats and conduct corrective actions in time (Ruoss & Alfarè, 2013).
- Local community behavior: Local communities play an important role in tourists' reception at tourist destinations. The host communities should therefore be empowered to take decisions concerning tourism and to set tourism guidelines if they desired to. Accordingly, the challenge, however, is to facilitate the integration of residents into the tourism economy (NWHO, 1999).
- Laws and legislations: It means having a framework of the legislative act for construction works and changing in the cultural heritage sites, and the purpose of these laws is to control and direct processes to serve the overall development (Al-Saleh, 2002).
- Funding: One of the biggest obstacles facing tourism projects in cultural heritage sites is securing the funds needed for sustainable development. The funding has many forms such as funds from the government, funds from the private sector through investment (BTB, 2008), donations of goods and services, funds from corporate foundations, funds from a festival or other large special event, and funds from international institutions (THC, 2010).
- Marketing: It is considered a major challenge in CH sites. These challenges are represented in increasing visitors' numbers, increasing length of stay, increasing daily spending, and ensuring tourists return and/or pass good recommendations to others. There is a strategy meant to attract "big spenders." And this strategy can make a major contribution to sustainability, by increasing benefits (revenue and jobs) without increasing visitor numbers (NWHO, 1999).
- Coordination and partnership: Cultural heritage sites should be where all stakeholders cooperate through effective partnerships to maximize conservation and presentation outcomes, while minimizing threats and adverse impacts from tourism (Ibrahim, 2013).
- The negative impact on the environment: Environmental impact assessments (EIA) should be carried out before any new facility such as accommodations, visitor centers, roads, waste treatment and disposal, heritage trails, tourist shops, and other constructed facilities, and then identify the necessary procedures to reduce the negative impact on the environment, to achieve the sustainable development (Ibrahim, 2013).

4 The Role of Relevant Parties in Facing the Challenges of Sustainable Tourism Development in Cultural Heritage Sites:

There are many parties whose relations are intertwined with the development of cultural heritage sites, including the government sector, the private sector, local community, civil society (NGOs), international organizations and bodies, local and international financial institutions, in addition to experts (engineers, project designers and managers, and environmental experts) (Ibrahim, 2013). However, the focus of this research will only be on three sectors, which are the main parties that have direct contact with the cultural heritage sites and the tourism process therein. These sectors are the government sector, the private sector, and civil society sector (NGOs):

4.1 Governmental Sector

The responsibility of the government is the following: developing plans and schedules for sustainable tourism development; provides an integrated system of legislation and laws; creating a comprehensive database of cultural heritage sites; establishing heritage and natural reserves; granting licenses after studying the environmental impacts of projects; and conducting environmental awareness for society (El-Khasawna, 2008).

4.2 Private Sector

The role of the private sector is to take responsibility for basic infrastructure, ownership, and/or management of the natural and cultural heritage that serves as tourism attractions (NWHO, 1999), to rely on local laborers after providing them with the required training; to develop competitive tourism products suitable for sustainable tourism development; and to promote sustainable tourism projects (El-Khasawna, 2008).

4.3 Civil Society Sector (NGOs)

The importance of NGOs is that they have the ability and flexibility to work and interact with the local community and the different levels of the population (Ghawy, 2010). They also can coordinate partnerships between stakeholders (NWHO, 1999). Their role is in providing cultural awareness and technical training for the local community to emphasize the economic benefits that can be obtained by implementing the principles of sustainable tourism development and protection of cultural and natural resources (Ghawy, 2010).

A key goal of sustainable tourism is to provide benefits—especially economic benefits—to local people. These benefits can be achieved through the participation of residents in the tourism or auxiliary industries. Therefore, when local communities can connect to their personal, local, regional, or national heritage, they are often eager to protect their cultural resources and achieve STD (NWHO, 1999). The following Table 1 gives a summary of the three sectors' role in addressing the challenges of achieving sustainability.

Table 1 Role of the three sectors in addressing the challenges of STD in cultural heritage sites (Adapted by the researcher)

Challenges	The role of the three sectors		
	Government sector	Private sector	NGOs
Protection of the cultural heritage	Establishment of natural and heritage reserves with special provisions to preserve cultural heritage sites and natural areas	Commitment to laws and strategies for the protection of cultural heritage	Connecting the sense of ownership to community members, so that the tourism and heritage resources are taken care of and preserved
Lack of Information	Conduct a comprehensive database on cultural heritage sites, internal and international tourist movement	Participation in the database compilation	Providing the necessary information for all participants in the tourism development process
Technical training	Conduct training courses for the local community to participate in preservation projects and tourism services	Reliance on national employment after their training and rehabilitation	Conduct technical training for the local community needed for the tourism development process
Management of cultural heritage sites	Identify tourist sites, working on the construction and support of infrastructure and support services. In addition to developing programs for monitoring and evaluation	Distributing tourism investments in different regions and taking advantage of tourist potentials and components	Training local community cadres to participate in the management of tourism projects in cultural heritage sites
Local community behavior	Conduct environmental awareness for all segments of society through various types of media	Utilizing local community capacity in employment and tourism project management	Contribute to raising the community awareness of the economic benefits that can be obtained as a result of the application of the principles of sustainable tourism development
Laws and legislations	Provide an integrated system of legislation, laws, and guidelines for the protection of environmental resources	Compliance with laws and regulations to protect the architectural heritage approved by the government sector for the investment projects	Raise the local community awareness of the importance of compliance with the laws and legislation to preserve the architectural heritage
Funding	To attract and encourage investments in the field of ecotourism by providing incentives and facilities to investors	Provide necessary infrastructure for the development of ecotourism represented in the establishment of hotels, restaurants, and amusement parks	Coordinate with international donors to obtain the necessary funding for the preservation process

(continued)

Table 1 (continued)

Challenges	The role of the three sectors		
	Government sector	Private sector	NGOs
Marketing	Put cultural heritage sites in the country plans and strategies and promote them through different media	Marketing tourism projects to reach the largest number of the target audience	–
Coordination and partnership	Coordination and partnership between the three sectors, achieve the expected economic benefits of tourism activity while facilitating the process of preserving cultural heritage sites, and then achieving sustainable tourism development		
The negative impact on the environment	Study the environmental impacts of projects before granting licenses	Conduct the study of environmental impact assessment of projects before their implementation	Raise the awareness of the local community about the importance of preserving the natural and cultural heritage

5 Global Experiences in Preserving Cultural Heritage

The following part is a presentation of global projects aimed at achieving sustainable tourism development for cultural heritage sites. All projects share several points, such as the existence of a distinctive urban heritage stemming from the local environment, the existence of natural heritage, and the existence of a community with its distinctive characteristics, culture, and products, while agreeing on a single goal of "conservation of natural and cultural heritage and achieving sustainable tourism development."

5.1 Dana Biosphere Reserve Project—Jordan

This reserve is located in the south of Jordan, Tafilah Governorate, and it was founded in 1989. The project received an international award from the Hanover exhibition in 1998, and it was described by World Bank President James Wolfensohn as "One of the most successful projects that the bank has contributed to financing for sustainable development" (Jordan, 2009). The project aims at preserving the natural and cultural heritage legacy of the region, exploiting natural and heritage resources in the region to maintain its sustainability, taking advantage of ecotourism to encourage and develop and promoting cooperative action in the community. In this project, the NGO (The Royal Society for the Conservation of Nature) has played a key role in sustainable development, with limited roles of the government and the private sector. The Dana project has succeeded in involving the local community from the first phase of planning, and by taking advantage of the skills of the local community, and the economic return has been achieved for this community, and the urban heritage (98 archeological sites) has been preserved and traditional crafts have been revived, as seen in Figs. 1 and 2.

The project has succeeded and achieved the expected results of sustainable development, through prepare a comprehensive study on the region, reconstruction of the village with its original stones and the traditional way, training the local population to work as tour guides, encouraging ecotourism to the reserve through the Rummana Camp, the funding problem was resolved through community contributions, then funding from the World Bank for investment project, and the contributions by the international organizations (RSCN, 2011). The conclusion is this project involved many relevant parties, where the NGO has played the key role, the local community has participated in the execution, in addition to the participation of several parties in the planning, funding, and execution of some programs that needed funding or expertise, especially investment projects (governmental and non-governmental organizations, private sector, local initiatives, international financial institutions, and agencies).

5.2 Santawani Lodge Project—Okavango Delta —Botswana—Africa

The Santawani Lodge is located in Botswana, Okavango Delta, with an area of 8000 hectares. It is a model of Botswana's sustainable tourism development. It is located in a nature reserve, constructed from local building materials and using local labor. It is the first ecotourism destination to be fully managed by local community members (Kalikawe, 2001). The project aims at poverty alleviation while achieving the objectives of preserving urban, natural, and cultural heritage, by providing financial incentives to the community to encourage participation in sustainable development and natural resource management practices (AWF,

Fig. 1 Dana cultural heritage (RSCN, 2011)

Fig. 2 Dana traditional crafts by the local community (RSCN, 2011)

2005). In this project, it was noted that the NGO played a key role in the conservation process, while the government's role was limited to renting the lodge, with the absence of the role of the private sector.

The NGO has developed a conservation strategy within the overall framework of Botswana's conservation strategy and following its principles. The urban and cultural heritage has been preserved, the traditional crafts have been revived, and an economic return has been achieved for the local community. The lodge has succeeded in achieving the results of sustainable development, through a commitment to international conventions signed by Botswana, which support conservation of biodiversity and sustainable tourism, a comprehensive study of the project was carried out before its execution, using local building materials and local labors only (see Fig. 3), the revival of traditional local crafts, engaging the local population in the tourism activities (see Fig. 4), using part of the lodge's profits to develop local

Fig. 3 Santawani Lodge (AWF, 2005)

Fig. 4 Santawani local community (AWF, 2005)

facilities, funding has been collected with several ways: firstly, African funding institutions (AWF) and (ADF), then from the local community, and then from the international grants (Sampéré, 2004). The conclusion is this project is the result of a partnership between NGO and community members who managed the lodge, and African and international funding institutions, with the absence of governmental & private sectors' roles.

5.3 Yaxunah Camp Project, Yucatán, Mexico

Yaxunah village is located in the central part of Yucatán, Mexico. It is characterized by the distinctive cultural and urban heritage of the local community. It is a traditional village of the ancient Mayan civilization in Mexico, in addition to the nature reserve and urban heritage in the village. A sustainable tourism development project has been carried out in this village, preserving the distinctive natural and cultural heritage of the local community and achieving economic returns (MRP, 2007).

This project aims to develop community-based tourism in Yaxunah Village, by taking advantage of the distinctive natural and cultural heritage of the Mayan civilization, in partnership between researchers and academics, and the village's local leaders (Magnoni et al., 2007). It was noted that the camp was developed by NGO Mundo Maya Organization (MMO) who has played the leading role in the

Fig. 5 Yaxunah Camp (MRP, 2007)

conservation process; the government's role was limited to renting the lodge, with the absence of the private sector's role. The camp was originally established by American scientists to carry out archeological research on Yaxunah Village, and Fig. 5 illustrates Yaxunah Camp. After the completion of "Yaxunah Archeological Project," the camp was developed with tourism services provided by the village's members (voluntarily) under the umbrella of (MMO), to achieve economic return while preserving the natural and cultural heritage of the local community. The project has succeeded in achieving sustainable development, through engaging the local population to conduct a comprehensive study on the village, building the camp of local environment materials and the traditional construction way, the NGO (MMO) trained and educated the local population to work in tourism field (see Fig. 6), and funded the rehabilitation projects of archeological and historical sites (Magnoni et al., 2007). The conclusion is this project included partnership between many relevant parties, where the Mexico Government provided the land for project construction, a US academic institution has built the camp for archeological research, the NGO prepared a strategy to develop the camp and provide the materials needed for camp restoration, and the local community provided tourist services in the camp voluntarily.

Fig. 6 Mayan local community (MRP, 2007)

5.4 Al-Dariyah Development Project—Riyadh—Saudi Arabia

Al-Dariyah is located in the northwestern part of the Arabian Peninsula, 20 km from Riyadh city, and it is one of the important projects to preserve the urban and cultural heritage and achieve tourism sustainability in KSA, Fig. 7 illustrates Al-Dariyah cultural and natural heritage. The Tarif District in Al-Dariyah was listed on the UNESCO World Heritage List in 2010 AD (Riyadh, 2009). The project aims to develop the historic Dariyah through its reconstruction and transform it into a cultural center at the national level according to its historical, cultural, urban, and environmental characteristics, as well as making it a tourist attraction by preserving its historical cultural heritage, for example, Fig. 8 illustrates Salwa Palace in Tarif District. In this project, the government body "The High Commission for the Development of Riyadh" has taken the leading role in the development process alongside the experts, with the involvement of the private sector by investing in certain projects. (Riyadh, 2009).

The development program has been developed following the laws and regulations for the protection of natural and cultural heritage and was based on conducting a comprehensive study of the current situation, developing a plan to raise the awareness of the local population, preparing

Fig. 7 Al-Dariyah cultural and natural heritage (Riyadh, 2009)

Fig. 8 Salwa Palace in Tarif District in Al-Dariyah (Tatweer Magazine, 2010)

specialized studies and training courses for the workforce in the field of heritage protection and ensuring the full application of the determinants of heritage preservation in the restoration, rehabilitation, and utilization projects, the project was funded by Saudi Arabia, with specific projects being assigned to the private sector to invest and execute (Tatweer Magazine, 2010).

The conclusion is this project involved many relevant parties, where the government body has played the key role in planning, construction, management, and operation works, the private sector has invested in some development programs according to the plan in a manner that ensures the preservation of the natural and urban heritage of the historical Dariyah, the NGO organized voluntary participation by individuals and institutions to raise environmental awareness and culture of the local population. A group of experts from outside the kingdom was engaged to study the current situation, prepare the development idea, and to develop an operational program, and estimate its cost (Tatweer Magazine, 2010). Although the three sectors were involved in the development project, the success in achieving sustainability is not clear yet. As in order to sustain a project, a part of the economic benefits must accrue to the local community, which motivates them to help to maintain the project, and this is not yet clear in the development project.

Fig. 10 Nubian local community

6 Case Study: Anakato in Aswan—Egypt

"Anakato" is an ecolodge hotel located in Gharb Soheil Nubian Village, on the west bank of The River Nile in Aswan, and it was built in 2008. Anakato means "Our Home" in the Nubian language (Anakato, 2017). This ecolodge started with one exclusive Nubian House (see Fig. 9) and expanded to encompass five complexes along Gharb Soheil's Nile bank which all maintain the same traditional architecture (El-Beih, 2017). These small enterprises are reflecting the Nubian culture in the designs of their homes with colorfully painted houses and furniture (Talaat, 2017). Anakato ecolodge is owned by a local businessman who has built the house as a private residence before turn it into a lodge (Shampoo & Ghalib, 2019). This project aimed to put Nubian Cultural Heritage on the tourism map by attracting tourists interested in Nubian culture. Besides, this project achieved economic benefits for the local community, by providing job opportunities to the residents through tourism activities, such as renting camels, or taking guests for felucca rides across the riverbank, or buying handicrafts from the local market, which led to bringing further income to the village (Sonbol, 2012). Anakato ecolodge has taken a unique step in partnering with key Nubian families in managing their guesthouses and boutique concept hotels. In this way, the tourist can get the experience of the families and lifestyles of Nubian culture, as seen in Fig. 10 (Anakato, 2017).

Fig. 9 Anakato ecolodge

Anakato has taken a unique step in partnering with Nubian families in managing the guesthouses. In this way, the tourist can get the experience of coexisting with the real Nubian culture. Besides, Anakato managers are paying attention to train their employees continuously to ensure the quality of hospitality (El-Beih, 2017). And according to Sonbol (2012), Anakato's owner has built a mosque in the village from the hotel's profits (see Fig. 12). Anakato's success led to the spread of the idea of ecohotels in the village, which encouraged the local villagers to transform their houses into ecological hotels, as seen in Fig. 11

The idea of staying in the Nubian guesthouse spread through social media, until it became a well-known destination on many online booking Web sites (Sonbol, 2012). The major problem is the bazaars and the camel ride trips in the village are disorganized, and a lot of souvenirs in bazaars are not Nubian-made (El-Beih, 2017), Figs. 13 and 14 illustrate boats, camels, and bazaars in Gharb Sohail village (Site Visit). There is no sewage network in the village, and with the increase in the homes' transformation into ecohotels, the increase in unplanned construction of hotels, and the increase of the visitors, the wastes had been increased, which will affect the natural environment in the village (Shampoo & Ghalib, 2019).

In this project, the private sector played a key role in the ecolodge initiative, by building the hotels in the traditional architecture, train the residents and employ them in the hotel, and involve the local community directly in tourism projects. However, the absence of the governmental role especially in setting guidelines and strategic plans could lead to destroying the nature and culture of the village.

Fig. 11 Gharb Sohiel villagers transform their houses into hotels

Fig. 12 Gharb Sohiel mosque

Fig. 13 Boats and camels in Gharb Sohiel

Fig. 14 Bazaars and camels ride

7 Conclusion

From the literature review, the global experiments, and the work field conducted in Anakato, the following results were achieved:

- The main objective of STD in cultural heritage sites is to develop these areas in a sustainable economic, urban, and cultural manner that preserves the continuity of the heritage identity and makes it an economic resource for the local population, by benefiting from tourism in creating job opportunities for the local community through the revival of handicrafts and traditional industries which were prevalent in the cultural heritage sites.
- Engaging the local community in the rehabilitation and development of cultural heritage sites, leading to feeling more responsibility toward cultural sites.
- Coordination and partnership between the three main sectors (Government, Private, and NGO's) will lead to achieving tourism sustainability, and the neglect of any sector's role is leading to the deterioration of the cultural and heritage sites.
- Although Anakato is a good initiative, it is missing key steps to achieve sustainability. The major and important step is by activating the government role.
- The absence of planning or guidelines from government side for achieving sustainable development in Gharb Sohail could destroy the nature and the culture of the village.
- The key steps needed in Anakato to achieve sustainability are: conducting a comprehensive database on Gharb Sohiel cultural heritage and internal and international tourist movement; providing an integrated system of legislation, laws, and guidelines for protecting cultural and environmental resources; studying the environmental impacts of projects before granting licenses; constructing and supporting the infrastructure and transportation services; and finally marketing Nubian cultural heritage through different media.
- The basic roles of the main three sectors to achieve tourism sustainability in cultural heritage sites are:
  - Government Sector: Its key role is to develop conservation laws and policies.
  - Private Sector: Its key role is to invest and support tourism projects.
  - Civil Sector (NGOs): Its main role is to raise community awareness and to train and engage the local community in tourism projects.

References

Al-Naiem, A. A. (2005). The tourism development of heritage areas in tourist sites [in Arabic]. In *International Conference: Development and Tourism in Coastal Areas*. Sharm El Sheikh: Arab Urban Development Institute. Retrieved December 22, 2016, from http://www.araburban.org/AWF.

Al-Saleh, E. (2002). Organizational policies to deal with urban heritage—Upgrading policies in the old city of aleppo as a goal for comprehensive development [In Arabic]. *Research Paper*. Syria: Faculty of Architecture, Aleppo University.

Anakato. (2017). https://www.anakato.com. Accessed 25 Jan 2019.

AWF. (2005). *Community owned and run: A case study of Santawani Lodge, Ngamiland, Botswana, Working Paper*. African Wildlife Foundation (AWF): https://www.awf.org. Accessed 10 Sep 2017

Bahaa-El-Din, R. M. (2005). The study of economic and social effects of tourism development in Aswan Governorate [In Arabic], pp.259 . *Unpublished MSc Thesis*. Faculty of Tourism and Hotel Management, Tourism Studies Department, Helwan University.

BTB, B. T. (2008). *Botswana ecotourism best practices guidelines manual*. http://www.botswanatourism.co.bw. Accessed 22 Feb 2017.

EENCA—European Expert Network on Culture and Audiovisual. (2017). *Sustainable cultural tourism - A mapping document for the OMC*. Report. http://www.eenca.com/index.cfm/publications/sustainable-cultural-tourism-amapping-document-for-the-omc/. Accessed 16 July 2019

El-Beih, Y. (2017). Dreamy Anakato Guesthouse Transforms Nubian Tourism. *Invest-Gate*. https://invest-gate.me. Accessed 23 Jan 2019.

El-Khasawna, M. S. (2008). The role of local community in sustainable tourism development: Applying on Petra city in Kingdom of Jordan [In Arabic]. *Unpublished PhD Thesis*. Faculty of Tourism and Hotel Management, Tourism Studies Department, Helwan University.

Ghawy, A. (2010, May 18). *The urban heritage of Saudi Arabia Kingdom fortifies the cultural identity in front of challenges [in Arabic]*. http://www.alriyadh.com. Accessed 14 Feb 2017.

Ibrahim, M. A. (2013). *Tourism and cultural heritage in tourism anthropology [in Arabic]*. Alexandria: Dar El-Maarefa.

Jordan. (2009). *Dana Reserve*. http://www.natcom.gov.jo. Accessed 10 Mar 2017.

Kalikawe, M. C. (2001). Botswana: Integrating biodiversity into the tourism sector. In *A presentation made to the UNEP International Workshop on Best practices and Country Case Studies*. Mexico City: UNEP—Biodiversity Planning Support Programme. https://www.cbd.int/. Accessed 23 Aug 2016.

Magnoni, A., Ardren, T., & Hutson, S. (2007). Tourism in the Mundo Maya: Inventions and (Mis) representations of Maya identities and heritage. *Archaeologies: Journal of the World Archaeological Congress*, 353–383. https://www.researchgate.net. Accessed 11 Apr 2018.

Mahmoud, A. A. (2011). The role of specialized NGOs in sustainable tourism development in Egypt, applying on Hurghada and Aswan [In Arabic]. *Unpublished MSc Thesis*. Faculty of Tourism and Hotel Management, Tourism Studies Department, Helwan University.

MRP, M. R. (2007, October 25). *Volunteer opportunities in a Yucatán Maya community*. http://www.mayaresearchprogram.com. Accessed 10 Feb 2016.

NWHO. (1999). *Sustainable tourism and cultural heritage, A review of development assistance and its potential to*. http://documents.worldbank.org. Accessed 23 Aug 2018.

Riyadh, H. C. (2009). Addiriyah—The originality of the past and the present brightness [In Arabic], Historical Addiriyah Development Program. (1st). Riyadh, Kingdom of Saudi Arabia. http://rda.gov.sa/res/ada/ar/Publications/historical_addiriyah/index.html#19. Accessed 24 Mar 2017.

RSCN. (2011). *Environmental tourism with a social basis: Dana nature reserve initiative [in Arabic]*. https://www.rscn.org.jo/. Accessed 15 July 2018.

Ruoss, E., & Alfarè, L. (2013). *Sustainable tourism as driving force for cultural heritage sites development, planning, managing and monitoring cultural heritage sites in South East Europe*. CHERPLAN Report, National Research Council of Italy. https://www.researchgate.net. Accessed 14 Nov 2018.

Sampéré, E. (2004). Botswanan communities cash in conservation is good business at new Botswana Lodge. *Botswana: The African Journal*: https://www.ajol.info/. Accessed 10 Feb 2016.

Shampoo, A., & Ghalib, S. (2019). Site Survey & Interview With Anakato Managers. (R. K. El-Khadrawy, Interviewer).

Sonbol, S. (2012). My home is yours. *Al-Ahram Weekly*. http://weekly.ahram.org.eg. Accessed 25 Jan 2019.

Talaat, N. (2017). Keeping up with the nubian dream. *Invest-Gate*. https://invest-gate.me. Accessed 23 Jan 2019.

Tatweer Magazine, T. (2010). Historical Addiriyah registration in the world heritage list [In Arabic], High Commission for The Development of Riyadh (pp. 39–43). http://www.ada.gov.sa/ada_a/Tatweer_ADA_A. Accessed 15 May 2017.

THC, T. H. (2010). *Heritage tourism guidebook*. https://www.thc.texas.gov/. Accessed 05 Sep 2017.

UNESCAP. (2008). *Cultural tourism sites management, a training manual for trainers in the greater Mekong Subregion*. https://www.unescap.org/publications. Accessed 05 Sep 2018.

The Inner Courtyard as an Important Element in Architectural Heritage and Interior Spaces

Sarah Fathy Ahmed Fahmy

Abstract

The inner courtyard is considered one of the most important elements that characterized Arab architecture in general and Islamic in particular. It is the product of the human interaction with the surrounding environment and adapting it to achieve the approach to the interior of human being to feel privacy and protection from strangers. The inner courtyard of the Islamic society has gained a new dimension, as the society related it to Prophet Muhammed when he built his mosque and home after his departure from Mecca to Medina with an inner courtyard. In addition, the inner courtyard contains many values, which give new dimensions and different visions, due to the integration of spaces and fit with their environmental conditions to which most activities have moved and achieved privacy and security, and the historical dimension gained by the emotional value that attracts us towards it automatically. The problem of research is the scarcity of the inner courtyard in modern housing, with its various philosophical values in modern housing. The aim of research is to shed light on the importance of internal courtyard as an influential element in the creation of internal spaces and the development of most important values it contains.

Keywords

Inner courtyard • Functional values • Aesthetic values • Interior spaces • Architecture • Heritage

S. F. A. Fahmy (✉)
Associate professor of Interior Design and Furniture Department, Faculty of Applied Arts, Damietta University, Damietta, Egypt

1 Introduction

In spite of the presence of the inner courtyard in the ancient civilizations from prehistoric times through the Pharaonic, Roman and Greek civilizations, it is attributed to the Arab Islamic architecture because of the evolution of it, and the resulting harmony between the natural and social conditions of the Arab nation and its compatibility with the principles and the teachings of Islam which linked both functional and ideological aspects (Yahiawe 2006).

Since the design of the Islamic house came from the inside to the outside, and not the other way around. Therefore, the emphasis was placed on the interior design of the courtyard through the attention of architectural and decorative treatments, also the different forms of the cells around the openings, doors and windows.

2 The Philosophy of the Inner Courtyard

The inner courtyard overlooking the clear sky was not a casual gap in the Arab dwelling, but it carried a philosophical dimension even before the advent of Islam. The Arab man was closely associated to the sky, which he considered to be the face of nature. The land around him represented nothing but a barren desert that was disturbed by its dry sand and winds. In an attempt to bring down the purity and sanctity of the sky into the house, the internal courtyard was the way to achieve that concept. We will find that the house has turned out walls without windows, while each room inside overlooks a courtyard where only the sky can be seen, and therefore the courtyard for the Arab man is more than just a means of obtaining privacy and protection. The four sides of the courtyard were the columns that held the dome of the sky and even the sky itself the roof of the courtyard, so the courtyard is considered the source of peace inside the house (Abdul-Rahman 2000).

A. Versaci et al. (eds.), *Conservation of Architectural Heritage*, Advances in Science, Technology & Innovation,
https://doi.org/10.1007/978-3-030-74482-3_6

Fig. 6.1 The inner courtyard as a major design element across historical Ages. *Source* http://www.touregypt.net/featurestories/manzil.htm. Accessed 2–4-2013. Designed and translated by: author

The relationship between Arab man and the courtyard is a relationship of affection and contact with the sky. The architect "Hassan Fathi" showed that the inner courtyard is the special part of the sky owned by the owner gives him comfort and security and it attracts this sky down to connect in affection to pass with him to the rooms (Fathi 2000).

3 A Brief History of the Origin of the Inner Courtyard

The courtyard was built from ancient times. It took many forms, treatments and names from one era to another. It was found in Mesopotamia in the Sumerian and Assyrian times, as well as in the architecture of ancient Egypt, Greek architecture and Roman architecture until it reached the Islamic architecture. In which the internal courtyard gained a new dimension and special attention, which he attributed to the Arab-Islamic civilization as shown in Fig. 6.1.

(a) In ancient Egyptian architecture: The interior courtyard of the Old Kingdom era appeared in the houses of the villagers of symmetrical form. In the Middle Kingdom period, it was found in the large dwellings. It was surrounded by four suites (the master suite, the living room, the servants' wing, the kitchen suite and the stores) on a small courtyard inside. The inner courtyard of the modern state continued to be seen in the house of Nab Amoun and a house (carving). It was decorated with floral and human ornamentation units of pleasant colors.

(b) In Greek and Roman architecture: The inner courtyards of the Greek civilization emerged as essential elements in the planning and architecture of their cities. It was of paramount importance that man put statues of his gods and practiced sports to build his harmonious body in addition to its climatic role. He found the inner courtyard plan in "OLYNTH" and the availability of two types of inner courtyards.

The first inner courtyard with the courtyard surrounded by columns, and the second is the distinctive feature of a rectangular reception hall preceded by a two-column entrance, opening the entrance to the inner courtyard (Abdel-Gawad 1997).

The inner courtyard, surrounded by columns in Roman architecture in Pompeii, was in the center of the house. The rooms were ideally organized around it. The courtyard became the main center for all family activities and was coordinated with gardens and water fountains.

(c) In Islamic architecture in Egypt: The House and Mosque of the Prophet (PBUH) is the first of the Islamic residential designs with the inner courtyard, and the courtyard was determined by the walls and rooms that were built on one side and open it (Azab 2003).

After the Islamic conquest of Egypt in the seventh century AD, Amr Ebn Al-Aas took the city of Fustat and the building was open on the inner courtyard as shown in Fig. 6.2.

Fig. 6.2 The inner courtyard in the city of Fustat. *Source* Azab (2003) pg.68-Translated by: author

The interior courtyard was mainly in the Fatimi and Mamluk buildings, but it was distinguished in the Ottoman era. Some of the new architectural elements that were completely opened to the inner courtyard were introduced. To benefit from the cold air of the northern air in addition to protecting it from the sun and the view of the courtyard garden, which is full of shadows and one of the most important examples of houses with internal courtyard in Islamic architecture in Egypt as shown in Fig. 6.3.

4 Design Elements for Inner Courtyard

The inner courtyard was not merely a casual gap within the dwelling, but an affirmation of the privacy of family life, which is a social and religious value. The architectural design emphasized this peculiarity by unifying the heights and not being able to reveal the inner courtyard of adjacent surfaces. In the Holy Quran, the descriptions of Paradise were mentioned about one hundred and forty times, from which we derive the vocabulary of the inner courtyard which lies in the following:

- **Fountain**: Often mediated by the inner courtyard, and is a symbol of the inevitability of the presence of a component of water in the inner courtyard, and it is the most important elements, which works to cool the temperature of the atmosphere through the connection between cold water and dry air, leading to evaporation of water and low temperature And to soften the atmosphere in the surrounding areas.
- **Plants and trees**: Plants and fruit trees in the inner courtyard were a quote from the trees of Paradise and their fruits. They also shaded and also minimized the impact of dusty winds and played the role of the natural candidate for the dwelling. The plants appeared in the form of rows and in a free image within regular basins (Mustafa 1998).
- **Wooden furniture units**: Some wooden seats are made of wood type, as is the case so far in the house of Al-Sihemi.
- **Murals and written tapes**: The murals of the inner courtyard were characterized by a kind of plastic art that merged organically into the house. It was not detachable from it, and we notice this through the murals surrounded by epigraphs and written ribbons which varied between the Qur'anic and poetic verses, Geometric patterns are replicable in identical units around one or two axes. The tapes and decorations are an alternative to the arts of sculpture and photography in the non-Arabic arts.

5 Structural Features of the Inner Courtyard

Despite of the diversity of the interior courtyards of the Islamic house, the most important characteristic of the quadratic quadrilateral is the quadrilateral shape. Sometimes the courtyard takes a trapezoidal shape. From the climatic point of view, the rectangular courtyard and the axis with the movement of the sun are better than the square courtyard with similar characteristics. The smaller the size of the yard,

Fig. 6.3 Some models of the internal courtyard of the Islamic house in Egypt. *Source* (Fekry 2017) pg. 5-Translated by: author

the higher its climatic performance, where the light of the sun entering the courtyard is reduced and therefore the percentage of solar radiation are reduced (Khatabi 1997).

6 The Various Courtyard Values

The design of the inner courtyard was not just an independent space in the center of the house, but the interior spaces were a vision of design that included a set of principles and values that formed this type of construction and made it special in the interior design and exterior of the building. What is the product of the thought and essence of the Islamic content reflected in all the work at the time including the inner courtyard, when the analysis and analysis of the inner courtyard we find contains a set of key values and falls under a set of sub-values and these values are interrelated and mutually influential and integrated each other as they form a unique combination of body, mind and spirit requirements as shown in Fig. 6.4.

First: The emotional values of the inner courtyard

Emotional values express the interaction between man and the inner courtyard in terms of perception and meditation, so that it comes from the mere recognition of the external form to the contemplative entity to live with it and its feelings. The historical aspect directly affects the extent of our interaction with the inner courtyard, Here is not the past but the temporal dimension of the consciousness of society in a past time period and we generate empathy for the inner courtyard through three basic aspects:

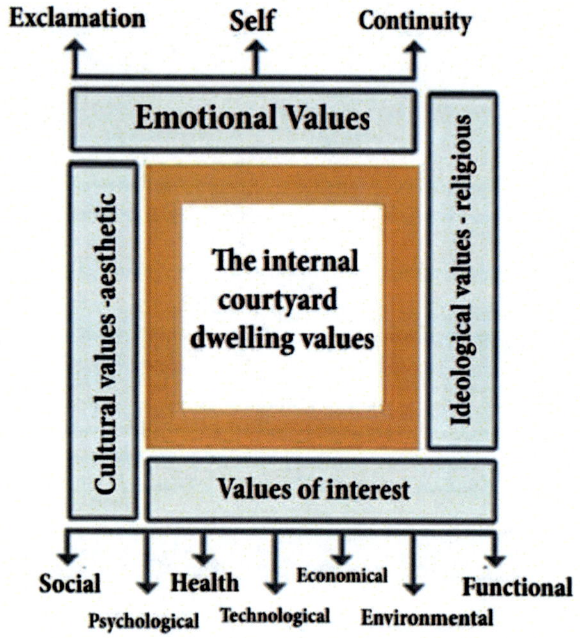

Fig. 6.4 A diagram showing the inner courtyard values of the Islamic house. *Source* by author

- Exclamation: Excitement is a direct and irresistible emotion and is the product of an unconscious process. When we coexist with the inner courtyard of the dwelling, we find that the exclamation result from the cosmetic beauty crystallized in the form of internal courtyard and the engineering relations of the decorative units and the wonder of content and functional performance. Social, environmental and other, as well as the surprise of moving between dark spaces of the house and lighting the sky.
- Personalization (Self): When we realize that the internal courtyard is only the product of the thinking of our ancestors, that is, our historical dimension confirms our identity as it contains our historical roots and gives us hope that the identity of our heritage is not lost. The sense of identity is the sense of belonging and familiarity in our heritage, to increase sympathy with him (Bayoumy 2002).
- Continuity: It is the product of the sense of self-identity. When we see how the models of housing in the Islamic era, which contain the internal courtyard such as "House of Sihami and House of Creedalism" and other historic houses, we see that continuity earns the place character, which makes us sympathize with him.

Second: The benefits of internal courtyard

Is a physical value of physical requirements and is concerned with human life as it is a social organism living in societies that have many social, economic, functional and health aspects, etc. These values are in general the values of interest that give a clear indication of the design thought created in The Islamic era and how to accommodate it to the conditions that work to create a suitable environment for the human and include the values of interest of the inner courtyard many aspects as follows:

- **The functional aspect**: The functional values of the inner courtyard are illustrated by the appropriate shape of the function, which is called the truth in expression. The inner courtyard achieves the efficiency of the various activities that are performed within it. To realize the functional aspect in the inner courtyard:
  - Safety: The internal courtyard provides stability and safety, which is one of the most important requirements of the dwelling.

- Privacy: The internal courtyard provides privacy by going inside so as not to compromise the privacy of the dwelling.
- Efficiency: In terms of the compatibility of the courtyard and the treatment of all external conditions, it creates the environment suitable for human to perform all its activities efficiently.

- **The social aspect**: The work of the internal courtyard on the transfer of most of the external activities of the person to the inside and the sense of social cohesion among individuals through the following:
  - Provide a suitable and safe space for children to play under the eyes of the family.
  - Accommodating the patio to perform various household chores and welcoming guests in periods of mild weather.
  - Achieving the rights of the neighbor through the inward closeness and personal freedom of the Muslim in his home (Yahiawe 2006).
- **The health aspect**: The courtyard contributed to the entry of the sun in the daytime, which helped to purify and purify the atmosphere of bacteria and moisture and bacteria, where doctors advise the need to expose the rooms to the sun daily, and contributed to sunlight play an important role in the formation of vitamin D, Play an important role in the calcium deposition in the bones, where the lack of growth and weakness of bone.
- **The psychological aspect**: It allows the internal courtyard of the family to interact in harmony with natural elements such as the sun and the moon and the sky and green areas and these factors affect positively on the mental health of the human, as it is scientifically proven that colors affect the mental health of man and his behavior and behavior, Looking at the color of the blue sky, it relaxes the self as well as looking at the green color of the green areas, and the internal courtyard plays a positive role on women, allowing them to exit from the space of the rooms to the spacious space, the loss of the inner courtyard of the house leads to the frequent exit Pain Ah housing and psychologists explain it in the loss of their right to look to the universe. Psychologists believe that man aims to achieve a clear balance between the environments, and when this balance is exposed to imbalance, it leads to tension, which leads to move in order to restore balance.
- **The environmental aspect**: The courtyard plays an important role in creating the external conditions of the human being from the environmental point of view. The existence of the courtyard was not spontaneous, but the Muslim architectural understanding of the extreme conditions of heat and wind and protection of the dwelling from noise, 20.5 to 26.5 °C), adequate humidity range (30% to 60%) and adequate air movement of 00.35 m / s. All these conditions are usually not available outside the dwelling. The internal courtyard contributes to the creation of these conditions through (Waziri 1987) as shown in Fig. 6.5:

(a) Improving the temperature: The inner courtyard to regulate the temperature where it has a large area of shadows, because of the height of the courtyard is larger than the longest ribs and therefore the presence of shadows throughout the year, and also shows the mechanism of thermal regulation of the courtyard through the day At night, during the night, different parts of the yard lose part of its temperature and cool its surface and cool it with the air that touches it. It descends to the bottom and is replaced by the warmest air which rises up to the top (cold air is heavier than hot air). The night turns into a storehouse for the air Ward.

By providing the inner courtyard with water and plants, they help to cool the air. When the warm air passes over the water, it evaporates. This process absorbs a large amount of air heat. It is more useful if the spray fountain is used. Purify the air from the dust particles suspended by it and note the difference in the amount of water evaporation depending on the type of fountain as shown in Fig. 6.6.

The inner courtyard also plays an important role in the movement of air inside the building. The exposure of the inner courtyard to the sun works on the hot air, leading to a decrease in weight. Recent studies of houses with a courtyard found that the temperature is less than 5 degrees

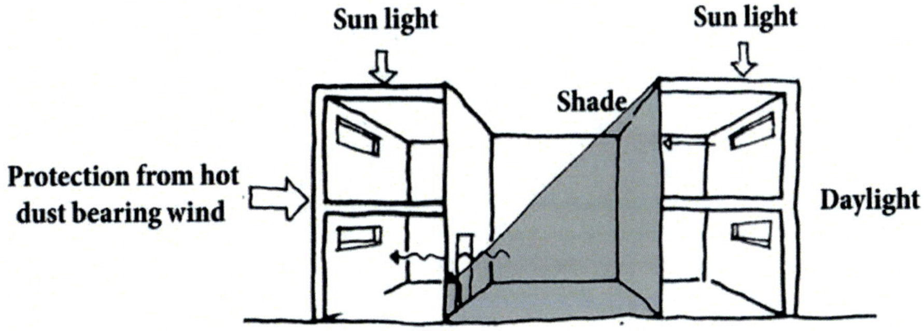

Fig. 6.5 Demonstrates a large area of shadows inside the yard of the dwelling and thus improve the temperature. *Source* (Fekry 2017) pg. 6-Translated by: author

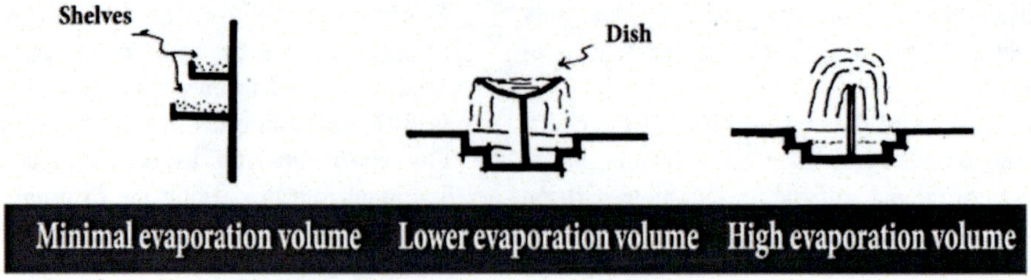

Fig. 6.6 Different amount of water evaporation depending on the type of fountain. *Source* By author

Fig. 6.7 The role of the yard in protecting the home against sandstorms. *Source* (Fekry 2017) pg. 7-Translated by: author

compared to houses that do not have an internal courtyard, and increases in the winter by five degrees.

(b) Resistance to sand storms and dust: The courtyard plays an important role in protecting the dwelling from sandstorms and dust, and protection from the wind and its effects of the fundamentals of design in areas with a hot dry climate where deserts predominate and sand abound, and formed the pavements semi-secure spaces of the impact of wind and pollution, and this was reflected on the spaces of the house in general, and in a study to determine the dimensions and proportions required during the design to protect it from the wind shows that the yard square does not exceed the half height of the building is given good protection from wind and pollution either If the courtyard is a rectangle it should not be decorated A depth of about three times the height of the building as shown in Fig. 6.7.

(c) Resist of noise: Resistant of noise is considered one of the most important features of the inner courtyard protection from noise and thus ensure the tranquility of the inhabitants of these housing, by studying the sound waves found that they move from their sources in the long continuous spherical waves and the speed and intensity of the greater the distance from the source, Noise is 5 dB. Thus, the best defense against noise is to increase the distance as far as possible between the source of light and the building to be protected, and note that the overlapping planning in the Islamic cities was able to control the spread of noise through the end of the closed streets, the building has been developed The rooms that are not affected by noise in the dwelling in the parts close to the noise source and vice versa.

Three Islamic houses were chosen in Cairo to measure the effect of noise on neighboring streets on different sites within these dwellings, the result was as follows:

- The first house (Al-Sihami House): The noise level in the street is about 68 dB, and the noise level in the reception hall near the inner courtyard is 36 decibels, ie, the reception noise level is about half the noise level in the street.
- The second house (The House of Creed): The street noise level was about 56 decibels, the noise level in the harem hall was 38 decibels, the internal noise level was 42 decibels.
- The third house (the traveler's house Khana): - The men's reception hall was found to be at a noise level of 38 decibels from the street.

From the above, it is clear that the Islamic designer and his knowledge of the sounds are well known and that he was adopted in planning and arranging the horizontal projection of the dwelling on the requirements of the inner space and its degree of noise.

- **The economic aspect**: The economic side of the internal courtyard has been achieved through the optimum use of materials and materials and the ease of dealing with them environmentally to achieve the required function. The internal courtyard also reduces the use of energy in ventilation and cooling, as well as reduces the use of industrial lighting during daylight hours where lighting is available naturally.
- **The technological aspect**: The internal courtyard contains many technological values are in the following points:

- Internal courtyard floors: The interior courtyard floors are paved with burnt clay bricks, which are characterized by thermal insulation properties resulting from the presence of fine pores in its mass. The air permeates them to achieve low surface heat. In the summer, these surfaces are washed in the middle of the day. Evaporation of water on the ceramic surfaces in the cooling of this surface.
- The surrounding walls: The walls surrounding the courtyard are thick, and are built of mud bricks or burnt bricks. These walls delay the heat transfer to the interior from sunrise to sunset. Over time, the air temperature is much lower than the temperature of the wall. The wall loses a large amount of heat into space.
- The furniture used: Some wooden seating units were used. We note the Islamic man's skill in inventing various means to implement the different types of wooden carving. He also presented some technological ideas for the cohesion of these units with each other and also for fixing them inside the furniture units.

Third: The Religious values of the inner courtyard

The religious values of different ages are the most important factors influencing the human mind in general and therefore reflected in the planning of his home, and those religious values carry the meanings of symbolism we find them achieved in the internal courtyard as follows:

- Make the person in the highest degree of privacy where to move away from the roads where the embodiment of the right way to say the Prophet peace be upon him "you and sit in the streets" and what it means to go inside, and see "Dr. Ali Raafat" Islamic Islam is directed to the invisible inside, full of calm and peace away from the systems of engineering and rationality (Yahiawe 2006).
- Helping to realize the concept of Paradise on Earth through the inner garden and the need to have the element of water that represents the rivers in Paradise, as well as the abundance of fruit plants and a quote to describe the Koran to the fruits of Paradise such as grapes.
- Work to link prayer times through the movement of sun and sunrise and sunset, as well as the vision of the moon in the month of Ramadan.
- Establishing the connection of humans to the sky.

Fourth: The aesthetic values the inner courtyard

The design of each design is the birth of its age, and the age is often the original source of our individual emotions. Hence, each cultural stage produces a design product to which it belongs. The aesthetic values on the side represent one of the main determinants of the value of the building. The interior courtyard has its various elements of plants, trees, water, and the coordination of its floors and walls, and the treatments surrounding it aesthetic dimension of the vacuum and reinforced by the element of aesthetic surprise felt by the human after a piece of narrow streets free of plants and when coming to the courtyard surprised to see the garden and it contains aesthetic values.

We also notice the beauty in the windows of the wooden windows, the contrast in the types and forms of wooden carving used and narrow distances under the windows and the wideness at the top, creating a kind of harmonious beauty that enjoy the human when looking at it.

7 Conclusion

We can summarize what has been analyzed in this research, the need to return to the elements of Arab architecture and to use them; also the philosophy of the use of the inner courtyard includes various values.

The building laws must be reviewed through the establishment of many requirements that emphasize the need for an internal courtyard, which helps to maximize the efficiency of the use of spaces according to the environmental situation.

8 Results

Research and analysis resulted to the following points:

1. The inner courtyard spread in most ancient civilizations. It was found in the ancient Egyptian architecture, through the Roman and Greek architecture until it reached the Islamic civilization in which it acquired a new dimension and special attention attributed to it.
2. The inner courtyard is an important element in the house that cannot be dispensed, as it contains many values utilitarian and aesthetic, emotional and religious, etc., which are integrated with each other to give the requirements of a decent life for man.

References

Abdel-Gawad, A. T. (1997). *The History of Architecture and Art in the Early Ages*. The Anglo Egyptian Library Press.

Abdul-Rahman, M. M. (2000). Sahn Eldar in Islamic Architecture, article. *Ammar Magazine, 42*, 62.

Azab, K. (2003). *Heritage of Islamic Architecture*. Dar Al Ma'arif Press.

Bayoumy, A. (2002). *Architectural Value and Art*. Dar El Ratib Press.
Fekry, D. (2017). Islamic content and its effect in crystalizing the design vision of contemporary housing. *Journal of Architecture. Arts and humanistic science*. *2*(5), 80–93. doi:https://doi.org/10.12816/0036595.
Fathi, H. (2000). *Architecture of the Poor*. Egyptian General Book Organization Press.
https://mjaf.journals.ekb.eg/article_20636_7fe410ca3274eb8b490b91dc2c6805c1.pdf
Khatabi, S. M. (1997). *Designing the Inner Courtyard of the Residential Units in the Kingdom of Saudi Arabia and the Impact on its Inhabitants* [Doctoral dissertation, Faculty of Fine Arts. Helwan University].
Mustafa, A. F. (1998). Islamic values in contemporary architecture. *World of Construction Journal, 41*, 130.
Selqini, M. K. (1994). *Architecture*. Dar Gabes Press.
Waziri, Y. H. (1987). Islamic architecture-a modern look. *World of Construction Journal, 81*, 8–10.
Yahiawe, F. (2006). Internal Courtyard in Contemporary Architecture between Originality and Modernity. published article: The First International conference of in Architecture and Urban Planning- Ain Shams University (pp. 342–343).

Save the Troglodytic Heritage of Beni Zelten

Houda Driss and Fakher Kharrat

Abstract

Tunisia contains diversified architectural heritage spread over the entire territory. Several institutions have taken care of certain sites and monuments; however, many others have not had the same fate. They threaten to ruin, such is the case of the ancient troglodytic village of Beni Zelten. This latter houses semi-troglodytic dwellings as well as vertico-lateral cave dwellings. In the following research, we shed light on the study area through its natural environment, its sociocultural context, and its architectural setting. We approach this part essentially by the historical method. The second part focuses on the identification of the architectural specificities of semi-troglodytic dwellings and vertico-lateral cave dwellings. For this, we used the in situ observation method (architectural survey technique), as well as the typo-morphological method. The third part tries to identify the pathologies and threats that affect both types of housing, based on the in situ observation method and the historical method. In fact, the region of Beni Zelten is located north of the Matmata range; it belongs to the semi-arid bioclimatic stage. A Berber community coming from Libya founded the village in the late fifteenth century. The interaction of the natural and the human environments has given rise to a rich and specific architectural product. Concerning the architectural specificities, the semi-troglodytic dwellings are located in the east flank of the mountain. We enter to these habitations through a vestibule that leads to the animal shelter and the patio on which rooms, kitchens, and granaries open. Workers use limestone to build walls and plant and mineral components for flat and vaulted roofs. As for the vertico-lateral cave dwellings, they are located at the foot of the mountain. They have, in turn, a vestibule that articulates service spaces and the patio that lead to the rooms, kitchens, and granaries. Laborers dig from top to bottom and then follow a horizontal direction in loess. The deterioration of this heritage is essentially due to two factors. The first is abandonment and lack of maintenance, while the second is the result of inadequate interventions that harm the authenticity of this architecture. The abandoned semi-troglodytic dwellings have cracks in their walls that are bare of their plaster. Flat roofs lost their protective layers, and wooden beams became apparent and in some cases collapsed. As for still occupied houses, the inhabitants use industrialized materials (cement, brick, and steel) for maintenance and extension. Concerning the neglected vertico-lateral cave dwellings, the patio walls fall down and this court become even a rubbish dump. Some roofs of the underground rooms collapsed. Some owners have reinforced the walls of the patio with masonry, and others have introduced built parts in these houses that were originally dug. The conservation of this endangered cultural heritage requires, on the one hand, the Zeltnis' awareness of the heritage value of their ancient built environment. On the other hand, the heritage authorities must absolutely intervene and form a multidisciplinary team that is working on the development of an action plan to save and protect this specific heritage before it is too late.

Keywords

Troglodytic heritage • Conservation • Beni Zelten • Architectural specificities • Threats • Southeast of Tunisia

H. Driss (✉) · F. Kharrat
PAE3C, National School of Architecture and Urbanism of Tunis, Carthage, Tunisia

A. Versaci et al. (eds.), *Conservation of Architectural Heritage*, Advances in Science, Technology & Innovation,
https://doi.org/10.1007/978-3-030-74482-3_7

1 Introduction

Unlike the built architecture, the troglodytic architecture is the result of subtraction of material in a site that is suitable for digging. "In the human collective imagination, all civilizations and religions combined, the underground sojourn seems to be reserved for exceptional beings or situations" (El-Wakil 2012). For a long time, the built architecture is better valued than the excavated one. However, the latter is experiencing a renewed interest in several countries of the world for its environmental qualities.

Tunisia is home to several troglodytic sites; the funerary chambers of Djebel Sidi Latrech in Hammamet, the catacombs of Hadrumete in Sousse, the hollowed houses of Djebel Sened in Gafsa, and troglodytic dwellings and equipments in Matmata, Douiret, Chenini… However, this heritage, halfway between nature and culture, has not been the subject of a well-conducted restoration and valorization project.

In the absence of an explicit heritage policy toward this increasingly degraded architecture, we have chosen to work on the troglodytic village of Beni Zelten. It is one of the few sites that have, at the same time, two forms of troglodytism; semi-troglodytic dwellings and vertico-lateral[1] cave dwellings.

Being aware of the need to intervene on this architecture to preserve it, our research has three objectives. It aims, first, to have a knowledge of the study site by identifying its natural and anthropogenic environments as well as its ancestral architectural setting. These environments consist of all the agents that contributed to producing an architectural result in a precise space–time framework.

Then, our work aims to reveal the architectural specificities of the studied dwellings through three aspects (implantation mode, the principle of the space organization, and the constructive mode). In fact, the implantation mode concerns the way in which people organize a rural or urban fabric in a natural environment. The principle of space organization is concerned with the functionality of spaces, as well as the relationships between them. The constructive mode deals with the technical aspect of an architecture including the materials and techniques of the construction or production of spaces.

Finally, our research tries to identify the threats and pathologies that endanger this troglodytic heritage, which deserves a lot of protection and preservation. The threats and pathologies group the alterations that have affected this architecture due to direct human action or because of time.

[1] « Vertico-lateral» Expression borrowed from Boukhchim et al. (« Vertico-Lateral» Cave Dwellings in Haddej and Guermessa: Characteristic Geocultural Heritage of Southeast Tunisia Geoheritage, 2017, 10, 4, p. 575–590).

2 Methodology

In order to achieve the above objectives, we have adopted a multidisciplinary methodological approach that combines tools specific to the humanities and architectural techniques.

For the data collection relating to the presentation of the study site, we used, on the one hand, the historical method based on documentary research and the use of the literature review (books, articles, theses). On the other hand, we relied on the survey method by using the direct interview technique used with isolated or small groups of inhabitants. This technique enables to interrogate the target population, by means of a question application, in a semi-directive way in which the informant answers questions freely without deviating from the subject.

In order to define architectural specificities, we adopted the historical method, the architectural survey method as well as the photographic report technique. We treated the collected data by the typo-morphological method that combines the architectural typology with the urban morphology. This approach is useful in detecting the material reality and the morphological characteristics of the concerned dwellings. It allows conducting an analysis through three scales: the urban scale, the architectural scale, and the technical one.

As for the last part of this work, which consists in identifying the threats and pathologies that affect semi-troglodytic and vertico-lateral cave dwellings, we have used the historical method, the photographic report technique, and the method of in situ inspection. The latter aims to recognize and classify the intensity and type of apparent damage that the different parts of the building suffer from. The on-site visit allows the collection of the information necessary to understand the condition of the dwellings (location, orientation, environment, materials …).

3 Results

3.1 Presentation of the Village of Beni Zelten

3.1.1 The Natural Environment

The village is located in the southeast of Tunisia forty kilometers southwest of Gabes. It belongs to the delegation of New Matmata. It is part of "The monoclinal chain of Dahar […] which is more than two hundred kilometers long […]. The eastern end of the Dahar chain, significantly stretched westward, bearing the name of Matmata, was the seat of a lœssic filling during the Pleistocene" (Sghari, 2009). The geological bedrock contains limestone benches in the mountainside while the flat part of the site is made up of argil-sandy sediments.

The region belongs to the semi-arid bioclimatic stage. At the end of spring and in summer, the site receives warm winds from the southwest while in winter it receives sand winds from the north side. The temperature knows large daily and seasonal variations. The region has almost two seasons; the summer season starts from May to October (from 16 °C to 50 °C), while the winter season runs from November to March (from 10 °C to 20 °C). The precipitations are rare and irregular; they do not exceed the average of 250 mm/year.

We find an adapted flora to drought like palm, olive, and figs trees. The agricultural sector is developing cereals and vegetable crops. Several wild varieties are present on the site such as rosemary, alfa… The inhabitants used available flora for craft activities and for construction.

3.1.2 The Human Environment

The first group of the inhabitants has occupied the site toward the end of the fifteenth century and the beginning of the sixteenth one. They came from Libya. According to the local population stories, this group constituted a family clan that settled on the site of Z'hiriya (fertile agricultural land) where they became active in the agricultural sector. However, with the arrival of new waves of migrants in times of war and conflict, the villagers have turned back to the mountain to escape the invaders. An Arabized Berber community forms the social group in Beni Zelten. People still respect ancestral values such as sincerity, trust, solidarity…

Beni Zelten's inhabitants worked the land (arboriculture and cereal crops) to supply their food needs. They were also active in the goat-rearing sector. The wool work ensured the need for clothing and blankets. The Zeltnis used, also, plant components such as alfa and palm branches to make mats, baskets, hats… The migration phenomenon affected the region for a long time, Nasr speaks of "a very old dynamics since more than four centuries (organized emigration)" (Nasr, 2004). These movements peaked in the mid-twentieth century after the Second World War.

3.1.3 Architectural Framework:

The architectural product in the traditional village of Beni Zelten is quite diverse. We find equipment (places of worship and oil mills) as well as residential architecture that includes built houses, semi-troglodytic houses, and vertico-lateral cave dwellings.

The old village houses a mosque and several mausoleums. This first one occupies a strategic location on the southeastern slope of the mountain, which makes it possible to control the whole village and the surrounding land. The mosque has a prayer hall formed by three galleries; Boukhchim estimates that "the third gallery was expanded around 1819" (Boukhchim, 2011). Barrel vaults cover the three galleries and a dome surmounts the mihrab, located in the middle of the first gallery. The stone walls are approximately 70 cm thick and receive an interior coating consisting of clay and plaster, and an exterior coating based on lime.

As for the mausoleums, these small spaces shelter the tomb of the Saint. The zeltnis visit three holy personages: Sheikh Thameur, Lella El Khadhra; and sidi Baâgoub. Sheikh Thameur's shrine is at the top of the mountain. It is a small dome supported by a circular wall. However, the m'zar of Lella El Khadhra is at the top of the mound east of the first mountain. It has the form of a circular wall surmounted by a conical dome. Sidi Ali Baâgoub's mausoleum is more important in terms of dimension. This monument has a rectangular gallery covered by a barrel vault supported on the side of the exterior façade by a series of arches. The gallery gives access to a square room surmounted by a dome; it shelters the tomb of sidi Ali Baâgoub. The monument contains a rectangular prayer hall covered by two-barrel vaults.

The agricultural activity is very important for the economic life in the village of Beni Zelten, especially cereals and olive growing. To extract olive oil, the villagers used oil mills that alternate between built, dug, or mixed spaces. The underground spaces allow having a constant temperature, in winter as well as in summer, which favors a better thermal environment for the extraction of olive oil. The oil mills consist of four main areas: a storage area for olives, a crushing area, a press, and an oil collector.

The ancient village of Beni Zelten has three forms of domestic architecture. It contains built houses situated at the top of the mountain, the semi-troglodytic dwellings on the mountainside, and the vertico-lateral cave dwellings (fully dug) at the foot of the mountain. The built houses constitute the first nucleus of the village that appeared toward the sixteenth century. They are now in a state of ruin. This study focuses just on the study of the other two forms.

Houses located on the mountainside constitute the second level of this human establishment; it is a "stepped transitional site" (Prost, 1954). Unlike other troglodytic villages of the Matmatas that followed a direct descent from their perch at the top of the mountain to the bottoms of the valleys, the inhabitants of Beni Zelten have come down gradually by realizing their houses at half-height. This implantation always allows for a defensive site while getting closer to farmland and water sources. Prost estimates "some 250 years ago; like the Beni Zelten, the Toujane descended gradually" (Prost, 1954).

As for the realization of vertico-lateral cave dwellings, it began after the semi-troglodytic dwellings and ended at the beginning of the twentieth century. This underground location allows, on the one hand, the protection from invaders.

On the other hand, it constitutes a perfect adaptation to the climatic conditions of the region. We define the architectural features of the semi-troglodytic houses and the buried houses in the following paragraph.

3.2 Determination of the Traditional Domestic Architecture Specifities

3.2.1 Semi-Troglodytic Dwellings

These dwellings are located on the eastern side of the mountain. This opening on the east/southeast orientation allows enjoying the good sunshine as well as the humid and cool southeast wind coming from the Gulf of Gabès. Semi-troglodytic houses define a moderately dense fabric, distributed according to the level curves and according to tribal membership.

Semi-troglodytic dwellings house extended families who share the available resources and harvest. We enter to the houses through a vestibule that constitutes a transition space between the private interior and the public exterior. The vestibule serves a stable, a storage space for agricultural equipment as well as open patio, the unique source of air and light for the dwelling. This central courtyard articulates the other spaces, housing units, kitchen(s), and granaries.

Each small family appropriates a housing unit and shares the other spaces with the other members of the large family. This chamber houses several activities, including the parents' sleeping space on one side of the room. The inhabitants use the central part, usually in front of the door for weaving, living and a space where children can sleep. The cooking space is a small cell opening onto the patio. The houses contain granaries located always upstairs. With the demographic growth, the dwellers transform some granaries into rooms accessible by stairs.

The building materials used are of local origin. The exterior walls do not have foundations; the workers build them directly on the geological bedrock. These walls are 50 to 60 cm thick and are made of carefully cut stone, limestone travertine. The inner walls are also made of travertine and have an average thickness of 30 to 40 cm. They use gypsum, obtained by calcination of limestone travertine, as a binder. The walls receive a coating of gypsum subsequently brushed with white lime.

The roofs of the different spaces are flat or vaulted. Dried olive wood beams, almost 2 m long and having a diameter of 20 to 40 cm, support the flat roofs. Then, vegetal layers coated by aggregates and local sand surmount the wooden beams. Finally, they apply a layer of gypsum and lime to the various components, which provides a waterproof and heat resistant roof (35 to 40 cm thick).

Inhabitants use unidirectional vaulted roofs to cover a long and little wide space. The construction requires the presence of a formwork based on half-trunks of palm trees, arranged in the form of a semi-circle, supported in turn by other wooden boards. The work begins with the application of a clay layer topped by gypsum layer. Then, they place small stones (travertine), simultaneously, from the sidewalls that support this roof and they are linked by a quick-acting gypsum mortar. Finally, the entire cover receives a lime mortar layer.

The occupants obtain the excavated spaces by digging in the rock using pickaxe. The laborers recover the extracted stone to realize the built parts of the house.

3.2.2 Vertico-Lateral Cave Dwellings

Vertico-lateral cave dwellings are located at the foot of the mountain; they are imperceptible at first sight. They are dug vertically into the earth thanks to the presence of the "Lœss; soft clay-sandy soil, soft to the touch, powdery and yet cohesive enough to maintain itself in vertical walls" (Leplat, 1968). Some houses belonging to the same family clan are communicating in the underground (Fig. 7.1).

The houses home multi-generational families. We enter through an outdoor corridor. This space is a descending slope toward the entrance of the dwelling, which allows making up the level's difference between the interior and the exterior. The corridor leads to the vestibule, which is equipped on both sides with farm equipment storage and a barn. The vestibule serves also the patio that articulates rooms, cooking spaces, and granaries. This central open-air courtyard is the unique source of air and light for the dwelling. It houses several female household activities.

Each small family occupies a room and shares the other spaces with the other members of the large family. The room is multifunctional; the inhabitants reserve the deepest space for the couple, while the closest part of the door is for children, for weaving, for the stay...

The cooking space is a shallow excavation situated on one of the patio walls. If the dwelling has a granary, it is

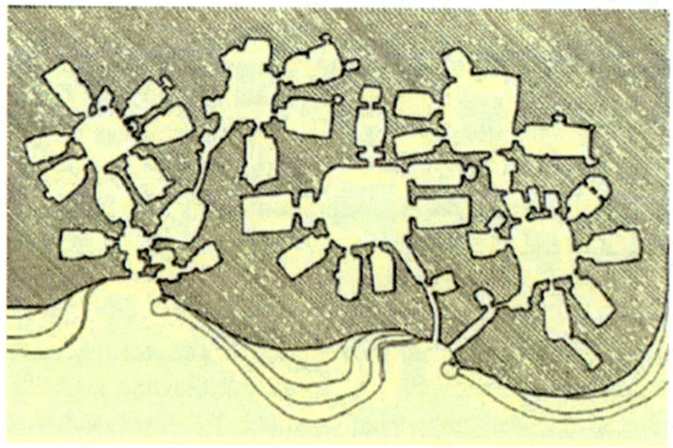

Fig. 7.1 Interconnecting underground dwellings (Labs, 1975)

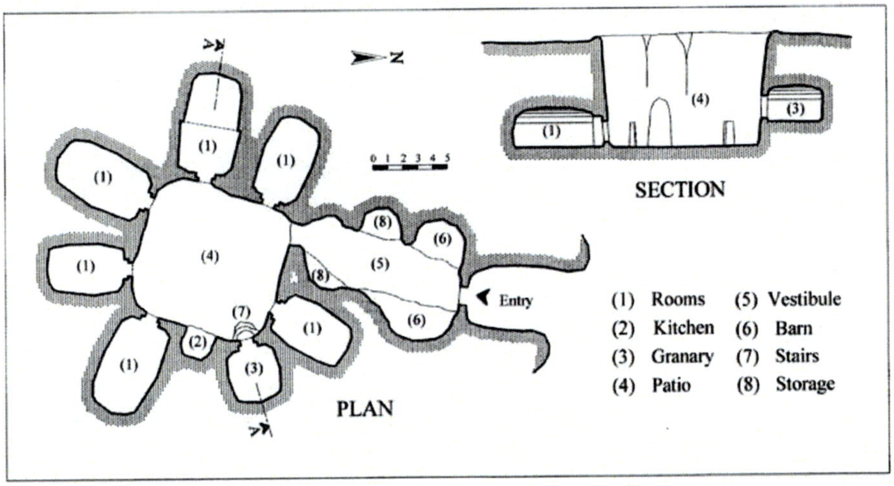

Fig. 7.2 Plan and section of vertico-lateral cave dwelling (Author's drawing)

raised by half a level or an entire level compared to the other spaces. The granary serves for the storage of food supplies and especially olives. It is equipped at its roof with a pipe that allows the passage of olives from the outside to the inside without crossing the courtyard (Fig. 7.2).

We obtain this architecture by the subtraction of material. Louis explains "the cave dwelling practically does not require building materials: no woodwork is necessary, except that of the door and its frame; no bricks, no rubble stones to mount the walls; no beams or joists to support the ceilings"(Louis, 1968). People can excavate a new dwelling at the top of a small hill whose ground allows the excavation (lœss).

The realization of the house begins with the digging of the patio deep of 6 to 10 m. The workers proceed, then, to the digging of the access tunnel. They excavate simultaneously from the inside to the outside and from the outside to the inside. Once they clear the tunnel, they proceed to the digging of the rooms and the kitchen. Laborers dig the chambers gradually from top to bottom, in the form of a step up to the desired depth. They have an arched or triangle roof; forms that allow loads decomposition and transmission without risk of collapse (Fig. 7.3).

To avoid flooding the house, they dig a sump in the center of the patio. Deep from 3 to 4 m, this well allows the collection of runoff. To increase its performance, they place bags of sea salt at its bottom, which accelerates the absorption of water. At the end of the work, the chambers and the walls of the central courtyard receive a layer of lime mortar.

3.3 Identification of Pathologies and Threats

Semi-troglodytic dwellings and vertico-lateral cave dwellings are at risk and are in an alarming state of degradation. This architectural heritage is in danger and can disappear because of two main factors. On the one hand, the original population quitted the majority of these habitations and no longer maintain them. On the other hand, occupied ones have undergone inappropriate transformations with their original style.

3.3.1 Pathologies and Threats Due to Abandonment and Lack of Maintenance

For the semi-troglodytic dwellings, no one maintain the walls made of double limestone siding. They lost, then, their coating and finishing layers, which favors the infiltration of the water that deteriorates the gypsum mortar. Gharib confirms, "mortar degradation problems are mainly caused by climate change and the presence of humidity"(Gharib, 2015). The deterioration of the mortar leads to the cracking of the walls and even to their collapse.

Cracks also affect vaulted roofs in the longitudinal or transverse direction (Fig. 7.4). These fractures may be due to instability of the walls or eccentricity of the loads. Raftani and Hassouni add "the pathologies affecting vaulted structures are usually the result of the combination of several stresses, starting with the proper weight of the vault which creates a state of initial constraint and deformation, to which is added the application of extra actions such as concentrated or not loads or movement of supports (walls, pillars, etc.) under the effect of the thrusts generated by the vault itself" (Raftani and Hassouni 2018).

With the departure of the inhabitants, no one maintain the flat roofs of the semi-troglodytic houses and no one renew the sealing layers based on gypsum and lime. Water then seeps in and attacks the mineral and plant components leading to the degradation of the roof until it completely collapses (Fig. 7.5). According to Zanaz, "the common denominator in the occurrence of most alterations is the presence of water. Water mobilizes soluble salts, freezes,

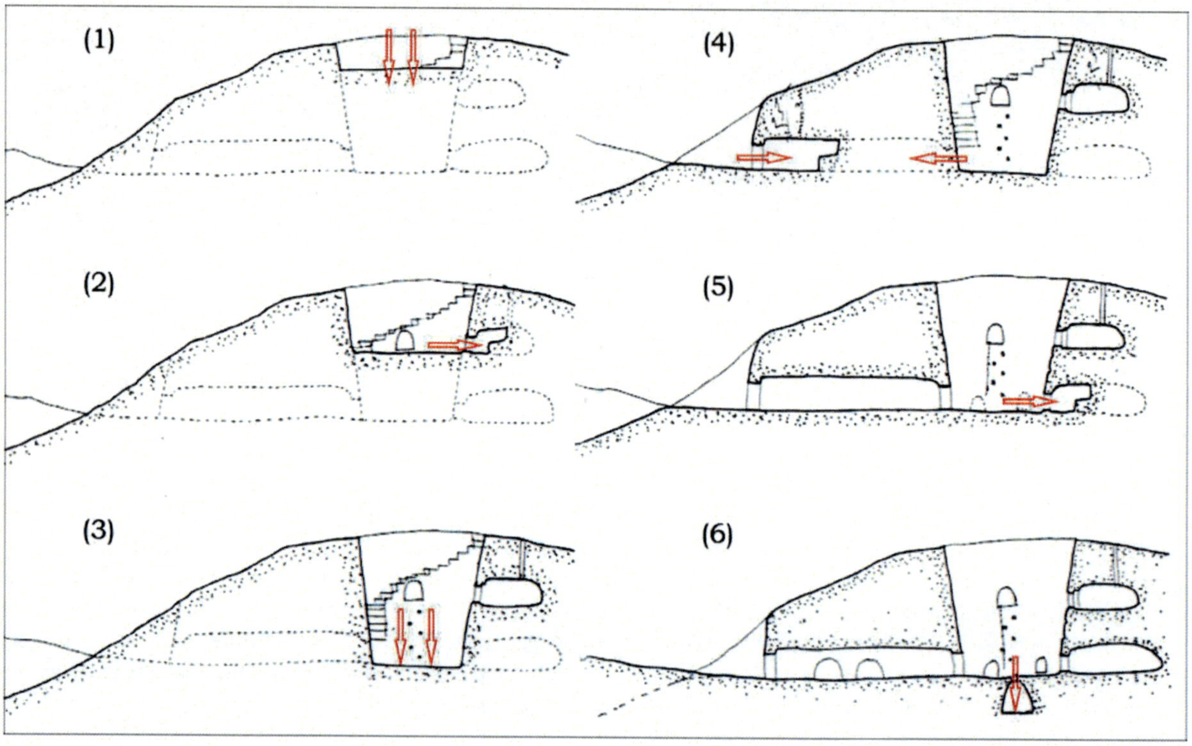

Fig. 7.3 House digging stages: 1—Patio digging 2—Granary(ies) digging 3—Patio depth: 6 to 10 m 4—Vestibule digging 5—Rooms digging 6—Sump digging (Aouadi 1984, rectified by the author)

Fig. 7.4 Cracked vault (Author's photo)

Fig. 7.5 Degraded flat roof (Author's photo)

dissolves minerals, promotes plant growth, alters mechanical properties, etc. It is therefore important to ensure a perfect seal associated with a good drainage system while avoiding trapping water in the core of the materials" (Zanaz 2016).

The patios of the vertico-lateral cave dwellings, whose owners (heirs) left the village, became a kind of public dump (Fig. 7.6). The inhabitants allow themselves to throw away their household waste and the objects they no longer need. It is a way, in fact, to devalue this heritage which some consider, on the one hand, being a threat to the safety of their children and, on the other hand, a symbol of poverty and indecent living.

Some other abandoned dwellings suffer from the collapse of the patio walls due to lack of maintenance and the rain-water action. This leads, over time, to the backfilling of these inner courtyards (Fig. 7.7). The filling of these big

Fig. 7.6 Patio becoming a public dump (Author's photo)

Fig. 7.7 Backfilled patio (Author's photo)

«hollows» causes the burial of an architectural tradition specific to the region.

No one cares for the sumps dug in the center of the patios to absorb rainwater, and they no longer play their role. As a result, in case of heavy rains, the houses flood, the lower part of the patio walls and those separating the different spaces soak in water. This last erodes the walls and the sandy clay dissolves in the water that leads to the loss of several support points and threatens the stability of the structure (Fig. 7.8).

Concerning the characteristics of the lœss, Krencker and Cruz-Mermy cite "the granular structure is consistent can collapse when the water conditions change under a static charge (saturation) or liquefy when cyclic stresses cause excessive pore pressures in the saturated lœss"(Krencker and Cruz-Mermy 2009). The loss of matter is an irreversible disorder that we cannot remedy.

Fig. 7.8 Disappearance of a part of the walls (Author's photo)

3.3.2 Inadequate Interventions

Inhabitants or public institutions (INP: National Heritage Institute) have restored some semi-troglodytic dwellings, even in part. However, these interventions are inappropriate; they harm the authenticity of the construction and constitute a disfiguration of the local architectural vocabulary (Fig. 7.9).

Some other occupants have extended their habitations so that the old and new spaces open onto the same patio. However, they build the new ones with industrialized materials (brick, cinder block, concrete, etc.) (Figs. 7.10 and 7.11), a symbol of modernity.

The major transformation that some vertico-lateral cave dwellings have undergone is the introduction of the built element in this architecture initially dug and obtained by subtraction of material. The addition of built parts appears in three forms. The first is the consolidation of access to the dwelling and the walls of the patio by a stone or brick

Fig. 7.9 Change in architectural vocabulary (Author's photo)

Fig. 7.10 Ancient and new materials in ancestral dwelling (Author's photos)

Fig. 7.11 Only new ones in the same dwelling (Author's photos)

apparatus at half-height or over the whole height of the wall (Figs. 7.12 and 7.13). This reinforcement makes the patio space safer and keeps the walls that are in a state of progressive collapse.

The second form of addition of built elements consists in demolishing certain underground parts and replacing them with fully constructed spaces. The dwelling, then, acquires a "hybrid" character; it comprises both visible and concealed spaces, all of which open onto the same central courtyard. This configuration proves that the owner is low or medium income. He cannot take over the construction of a total built house on a new terrain. "Hybrid" habitation provides access to modern housing while recovering the terrain and taking advantage of the thermal comfort qualities of underground spaces.

The third form of intervention is to give the dwelling a built appearance. On the one hand, the owner reinforces the walls of the patio and the contours of the access to the rooms

Fig. 7.12 The reinforcement of an entry by a stone apparatus (Author's photos)

by a stone apparatus. On the other hand, he is building a partial floor (Figs. 7.14 and 7.15).

4 Discussion

In order to preserve the troglodytic architectural heritage of the village of Beni Zelten, we tried, first, to know more about the study site. Then, we turned our attention to defining the characteristics of traditional troglodytic dwellings. Finally, we looked at the identification of the various pathologies and threats that may lead to the disappearance of this regional specific heritage.

The studied houses have similarities as well as differences in their architectural specificities. Semi-troglodytic dwellings and vertico-lateral cave dwellings have two different implantation modes. The former are located halfway up and have a built and dug part or they are completely built and leaned against the mountain. However, the second ones are fully dug at the foot of it.

In both cases, inhabitants took defensive measures in a territory where insecurity and conflict prevailed for a long time. During the Hilalian invasions, the Berbers took refuge

Fig. 13 The reinforcement of patio walls by a stone and brick apparatus (Author's photos)

Fig. 7.14 Built spaces in a cave dwelling (Author's photos)

Fig. 7.15 Built spaces in a cave dwelling (Author's photos)

in the mountains in order to create villages difficult to reach by the enemies. With the establishment of a relative peace, the inhabitants descended to occupy the plain while sinking into the earth. This allows them to move closer to farmland and water sources without their homes being perceptible by invaders.

In terms of spatial organization, the two forms of studied habitat have many similarities. They have the same spaces and a practically identical organizational logic. We enter to these dwellings through a transition space (vestibule) between the interior (private) and the exterior (public) space. They define a concentric organizational pattern where the different spaces revolve around a central courtyard. The inhabitants separate service spaces from living ones, but all components of the dwelling are within the same enclosed perimeter for security reasons.

The semi-troglodytic dwellings and the vertico-lateral cave dwellings house extended families. This reflects, on the one hand, "the union, solidarity and mutual aid between members of the same family and who strive to consolidate their dignity by defending all their interests" (Zaïed 2006). On the other hand, it reflects the austerity of this society living in a territory known by scarcity and shortage of resources.

Concerning materials and techniques for the production of spaces, workers built both forms of housing with local materials available on site and transformed by knowledge transmitted from generation to generation. The realization of these two forms of housing was possible thanks to the simultaneous presence of two types of geological formations. The local population have built the semi-troglodytic dwellings of limestone that forms the base of the mountain. While she have dug the vertico-lateral cave dwellings into the lœss (clay-sand), the major constituent of the flat part of the village.

It is an architecture adapted to its climate. The more the walls are thick, the more the thermic ambience of the interior is softened especially during the long warm days of summer and during the cold winds of winter. The subterranean architecture offers an even softer interior thermal atmosphere thanks to the strong earth thermic inertia. On cold winter days (6.6 °C outdoors), underground rooms are at 15 °C and during summer heat events (41.1 °C outdoors), they are at 25.5 °C. (Krarti 1998).

The troglodytic architectural product of Beni Zelten is the result of the interaction of the natural environment and the imagination of the people. This architecture respects the environment and has satisfied the needs of its occupants for centuries. To evoke the troglodytic architecture qualities, Loubes says, "Integration in landscapes, material and energy economies, interesting generated spaces, light effects, are all dimensions of an architecture that is described and lived from the interior…" (Loubes 1984). The troglodytic habitat is thus a lesson of architecture and a source of inspiration for designers.

Nevertheless, this cultural heritage is in a dilapidated state and may even disappear. The studied homes are at risk; they are abandoned or they have undergone inadequate interventions.

The walls and roofs of the abandoned semi-troglodytic dwellings cracked due to lack of maintenance and exposure to various weather conditions. Flat roofs lost their different layers and in some cases collapsed. As for the vertico-lateral cave dwellings, some patios became a public dump, and others are backfilled due to the collapse of their walls.

The original population have abandoned these ancestral homes for several reasons. In fact, from the second half of the twentieth century, the majority of Tunisia's rural areas have known massive migratory movements of their populations moving toward cities in search of better living conditions. The village of Beni Zelten, like all repulsive rural areas, has not retained the active labor that sought income support elsewhere. The father leaves his native region, and if he finds gainful employment, he brings his wife and children to settle permanently in the host city.

In addition, after the independence, the President Habib Bourguiba have devalued vernacular architecture by the country's modernization policy. He considered that the troglodytic villages are the sign of animal life. For this reason, he decided to relocate the populations of certain mountain sites (Matmata, Douiret, Zeraoua, etc.) to new "modern" villages. In other sites such as Beni Zelten, the local population quitted ancestral dwellings. They prefer settle in new houses built on the same family terrain. These recent constructions acquire the various networks and give access to new communication and information means. "Several ancestral traditional and troglodyte housing formulae were sacrificed on the altar of technological progress and modern comfort" (Golany-Gideon 1988).

In terms of inadequate intervention, some of the semi-troglodytic houses have been restored but without respecting their original style. Others have known extensions but using industrial materials. As regards underground dwellings, the aggression appears in the introduction of the built element into initially excavated spaces. The changes in the studied dwellings show that the inhabitants are seeking to adapt their homes to the contemporary life standards. Traditional homes were adapted to the traditional lifestyle, but they do not ensure the current population needs.

5 Conclusion

The knowledge, understanding, and identification of troglodytic heritage evils of the village of Beni Zelten are the first steps toward its conservation and valorization. However, the rehabilitation and restoration of this heritage are not an end in itself; it is rather a question of giving it a new life to ensure its sustainability.

The village of Beni Zelten and many others is located in marginalized regions that have not known the same economic and social development as coastal cities. This exclusion dates back centuries and it became more pronounced during the colonial period. As for the contemporary period, the governments base development policies on exclusion and inequality principles. For this reason, these policies have failed to harmonize between the exploitation of available natural resources and the establishment of prosperity and the creation of new balances.

As such, the architectural heritage of each region can be its driving force for development. An intervention based on a well-thought-out scientific approach would allow its conservation and its participation in the development process. In Matmata El Guedima, several individuals have taken the initiative to transform certain troglodytic dwellings into tourist accommodation units (hotels). Others have opened their underground homes to visitors and tourists "Although some of the exploited houses were recently dug (after 2000)" (Hammami 2018). This form of exploitation valorizes, to a certain extent, the troglodytic heritage and creates a new source of income for both the inhabitants and the micro-enterprises. However, it is always a punctual operation managed on a small scale.

It is time for the different actors in the heritage field, the concerned authorities as well as the citizens to mobilize to enforce an action plan that takes charge of the conservation of the Tunisian troglodytic heritage. In our opinion, it is necessary, first, to intervene on all the troglodytic villages (or at least the most representative ones) in order to restore and valorize them. Second, the competent authorities must integrate this cultural heritage into a global tourism circuit that contributes to regional and even national socioeconomic development.

Acknowledgements My deep thanks go first to the inhabitants of the village of Beni Zelten for their welcome and availability (especially Mrs Hédia). I would like to express my thanks to Dr Nouri BOUKHCHIM and Dr Zayed HAMMAMI, who have kindly made available to me valuable books and their previous publications, which have facilitated my work. I thank also Mr Anis AYADI and all those who have contributed in any way to this work.

References

Aouadi, Z. (1984). *La construction et les constructions dans les monts troglodytes*. [Thèse de 3ème cycle, Institut Technologique d'Art, d'Architecture et d'Urbanisme de Tunis]. Dépôt de recherche de l'Ecole Nationale d'Architecture et d'Urbanisme de Tunis.

Boukhchim, N. Ben Fraj, T. & Reynard, E. (2017). Lateral and « Vertico-Lateral » Cave Dwellings in Haddej and Guermessa : Characteristic Geocultural Heritage of Southeast Tunisia. *Geoheritage*, 1–16.

Boukhchim, N. (2011). *Population and architecture in Djebel Matmata :case of Beni Zelten* [Doctoral dissertation, Faculty of Human and Social Sciences of Tunis]. Faculty of Human and Social Sciences of Tunis Research Repository.

El-Wakil, L. (2012). Architecture troglodyte: théories, réalités et sauvegarde. *First Conference on Troglodytic Architecture*. https://archive-ouverte.unige.ch/unige:24214

El-Wakil, L. (2015). Buried or Half-buried Contemporary Settlements Between Fantasy, Genius, Loci and Fiction. *Second International Troglodytic Architecture Conference*. https://archive-ouverte.unige.ch/unige:88390

Gharib, T. (2015). *Renforcement des structures historiques en maçonnerie par matériaux composites : Application aux murs en pierres calcaires, Matériaux*. [Thèse de Doctorat, Université Claude Bernard de Lyon]. Université Claude Bernard de Lyon. https://tel.archives-ouvertes.fr/tel-01186203/file/TH2015GharibTurath.pdf

Golany-Gideon, S. (1988). *Earth Sheltered dwellings in Tunisia: Ancient lessons for modern design*. University of Deleware Press.

Hammami, Z. (2018). L'architecture troglodytique verticale et la mise en valeur touristique aux Matmata (Sud-Est tunisien). *Isnar, C., Mus-Jelidi, Ch. et Zytnicki, C. (dir.). Fabrique du tourisme et expériences patrimoniales au Maghreb (XIXe-XXIe siècles)*, 204–220. https://books.openedition.org/cjb/1568?lang=fr

Krarti, M. (1998). Thermal Performance of Ancient Underground Dwellings in Tunisia. *ASME Solar Energy Engineering Conference*, 65–72.

Krencker, M. & Cruz-Mermy, D. (2008). *Cartographie de l'aléa retrait-gonflement des sols argileux dans le département de Bas-Rhin*. BRGM.

Labs, K. (1975). *The architectural use of underground space : issues and applications*. [Master's Thesis, Washington University]. Washington University. http://www.parameterpub.com/KenLabs/KenLabsThesis-HiRes.pdf

Leplat, J. (1968). Limons et loess. *Bull. Liaison Labo. Routiers P. et Ch., 29*, 2–27.

Loubes, J. P. (1984). *Archi troglo*. Parenthèses.

Louis, A. (1968). L'habitation troglodyte dans un village des Matmata. *Cahiers Des Arts Et Traditions Populaires, 2*, 33–60.

Nasr, N. (2004). Agriculture et migration dans les stratégies productives des jbalia du Sud-Est tunisien. *Picouet, M.(dir), Environnement et sociétés rurales en mutation. Approches alternatives*, 247–257. https://horizon.documentation.ird.fr/exl-doc/pleins_textes/divers10-07/010037102.pdf

Prost, G. (1954). Habitat et habitation chez les Ouderna et les Matmata. *Les Cahiers De Tunisie, 2*, 239–253.

Raftani, K. & Omar, H. (2018). *Pathologies des bâtiments traditionnels dans le contexte spécifique des médinas marocaines*. Coopération Municipale – CoMun Gouvernance locale et participative au Maghreb.

Sghari, A. (2009). Les loess subdésertiques de Matmata, un patrimoine géomorphologique à revaloriser. *Géomorphosites 2009 : imagerie, inventaire, mise en valeur et vulgarisation du patrimoine géomorphologique, Vol. des résumés*, 69–70.

Zaïed, A. (2006). *Le monde des Ksours du sud tunisien*. Centre de Publication Universitaire.

Zanaz, A. (2016). *Approche probabiliste d'aide au diagnostic des voûtes en maçonnerie*. [Thèse de Doctorat, Université de Limoges]. Université de Limoges. https://tel.archives-ouvertes.fr/tel-01382448/file/2016LIMO0052.pdf

Metrology in Egyptian Architecture of the XVIII Dynasty, in Thebes

Mónica M. Marcos González

Abstract

Within the framework of the Conservation and Restoration of the Architectural Heritage, the purpose of the present study is the applied metrology in the composition of the religious architecture realized throughout the 18th Dynasty in the ancient Thebes (Egypt). The objective will be the determination of the geometrical and metrological relationships supposedly used in the Theban architectural models and the module used in the initial project of the buildings. The study and data collection of significant religious buildings, tombs, and temples in the ancient Theban necropolis is completed with the analysis and drawing of their plans. The systematization of the measurements and the modulation of the buildings allow establishing common compositional parameters, the module marked by the measuring unit used. The measurement system corresponding to that time and period (18th Dynasty) was the Egyptian royal cubit. The analysis and study of the measurements in this unit, used in its architectural design, provide exact numbers in the building spaces dimensions. It allows establishing proportional relationships between them, and finding a module of geometric composition, on which the original project was based. This responds to a philosophical and functional concept of the projected spaces. The study of architectonic typologies, and the drawing of plans in the correct unit of measurement (the Egyptian cubit), help to understand the process by which the geometric composition of the project is generated. Being a result of this design process, we can see an evolution in the distributions of the spaces, dimensions, measurements, as well as ritual and functional justification. In the field of heritage rehabilitation and restoration, the knowledge of these patterns and modules help in excavation work, reconstruction, and restoration of construction elements. The correct use of metrology contributes to the identification of possible work areas, by helping locate where the damaged or disappeared areas are. Furthermore in restoration projects, metrology is useful to rearrange and locate the scattered pieces or parts of the buildings decontextualized. The translation of the measurements taken in the current International System of Units, into the measurements in which the original project was conceived, allows understanding of its conceptual purpose and functionality, which makes any interventions easy to undertake.

Keywords

Metrology • Architecture • XVIII dynasty • Luxor • Measuring unit • Geometry • Egyptian Cubit • Mathematics

1 Introduction

In the field of Conservation and Restoration of Architectural Heritage, an indispensable tool is metrology.

It has direct utility in works of reconstruction of ancient architecture, in archeological excavations, and for reconstitution of constructive elements.

The geometric and metrological relationships, supposedly used in the architectural models, and the module used in the initial project of the buildings, define the basis of the investigation, and of the possible prospections in the land for its later analysis and understanding. The starting point is the data collected, both historical and of the work carried out in the last century, and of the measurements carried out in situ.

The work developed stems from the metrological, construction, and architectural data collected throughout projects of excavation, reconstruction, consolidation, study, and

M. M. Marcos González (✉)
Architect, Egyptologist, Conservation and Restoration of Architectural Heritage, Polytechnic University of Madrid, Madrid, Spain
e-mail: mmarcovich@telefonica.net

A. Versaci et al. (eds.), *Conservation of Architectural Heritage*, Advances in Science, Technology & Innovation,
https://doi.org/10.1007/978-3-030-74482-3_8

repair of religious architecture, located in Luxor. Considering the existing constructions, corresponding to different periods, it focuses on the historical period of the New Kingdom, during the 18th Dynasty. The analysis of the models and typologies used both in composition, as in measures and dimensions, can establish a conceptual relationship between the projected and constructed architectural ensembles, with the philosophical and religious beliefs that governed this historical and cultural period.

The measurement study, as well as the geometry and proportionality that guided the spatial thinking of the old builders, give guidelines for understanding the building, its genesis and knowledge for possible actions, by having the modular pattern that created them.

Within the framework of the architectural heritage, the understanding of the work is very important, in order to approach it and intervene as faithfully as possible to its true historical and conceptual identity.

The analysis includes:

- Definition and characterization of the measurement system used in architecture.
- A compilation of references, regarding the proportions and measurement units used in the historical and cultural period corresponding to the 18th Dynasty in Egypt.
- Establish an analytical methodology based on geometry and proportion, according to the units of measurement used during that period.

1.1 Historic Context

In the historical context, a chronology could be established to place the studies. The Egyptian civilization began approximately in 3000 BC with the first prehistoric cultures of the predynastic and protodynastic periods and ends in 31 BC with the conquest of the Roman Empire.[1] The dates are approximate, due to different criteria and dating techniques, the chronological inaccuracy of ancient historians, and new archeological discoveries, which evolve each year. The main sources used have been chronicles of antiquity, according Egyptian historian Manetho,[2] and the royal lists as Royal Papyrus of Turin.[3] Some authors agree, not exactly, but roughly on dates estimation; Redford (1967) and Parker (1971) in their studied of 1967 and 1971, respectively, and more recently Kemp (1991) and Shaw (2000) in 1991 and 2000.

It is structured in three large periods; Old Empire (3200–2680 BC), Middle Empire (2055–1650 BC), and New Empire (1550–1069 BC), among which there are times of transition: First, Second, and Third Intermediate Period, and Late Period (664–30 BC). These intermediate periods are characterized by political instability and coincide with the main war events. The First Intermediate Period was a crisis time and economic decline that leads to trade being stopped.

Chronologically, it goes from 2160 to 2055 BC (Dynasties 9–11th), with great social and political instability, documented, among others, in the Ipuwer Papyrus, and Instruction for Merikare (12th Dynasty). In the Second Intermediate Period, there was the invasion of the Hyksos. It goes from approximately 1650 to 1550 BC (Dynasties 13–17th). The Third Intermediate Period was characterized by the invasion of the Hittites and the Aegean people, from 1069 to 664 BC (Dynasties 21–25th).

The location of the sites defines its purpose. At a general level, the settlements were determined by the landscape and the orography. Locally, the function of the building being constructed determined its location, East or West of the main axis, the Nile River.

As a reference example for the development of work, certain temples and funerary constructions are chosen, and these are located in ancient Thebes, Luxor, in Egypt. The period in which these constructions were developed extends, approximately, from the year 1550 to 1352 BC. This buildings belongs to the 18th Dynasty, whose chronology is referenced, among others, by Shaw and Bryan in The Oxford History of Ancient Egypt. (Chronology, pages 481–489) (Shaw, 2000).

Luxor is located around a bend of the Nile River, in the central area of the country (Fig. 1). The climate and geography determined the development of the settlements on the riverbank, to take advantage of the floods and the fertile area of Egypt.

The Nile was a physical and geographical border (material), and conceptual one (immaterial), that separates two worlds, that of the living and that of dead. The Eastern area, where the sun rises, the one that belongs to this world. The Western zone, where the sun sets, the one that belongs to the other world, or to the one beyond. The Eastern bank of the river is where the ancient Thebes located, as a city. Still on this shore, ancient Thebes is now the current city of Luxor. The western bank was the sacred area, where tombs and temples were located, according to the solar orientation that characterized the Egyptian culture. In the 18th Dynasty, Thebes became the capital of Egypt, reaching its greatest

[1] In 30 a. C., after the death of Cleopatra VII, the Roman Empire declared to Egypt a province (Aegyptus) that was to be governed by a prefect chosen by the roman Emperor.

[2] Manetho was an Egyptian priest and historian born in the third century. He wrote the *Aegyptiaka* (*'History of Egypt'*), where he organized the chronology of his history in the form of dynasties from the beginning until the conquest of Alexander the Great. This division has been generally accepted by modern Egyptology.

[3] Papyrus of hieratic writing, from the time of Ramses II, includes the kings of Egypt list. It was discovered by Bernardino Drovetti in 1822.

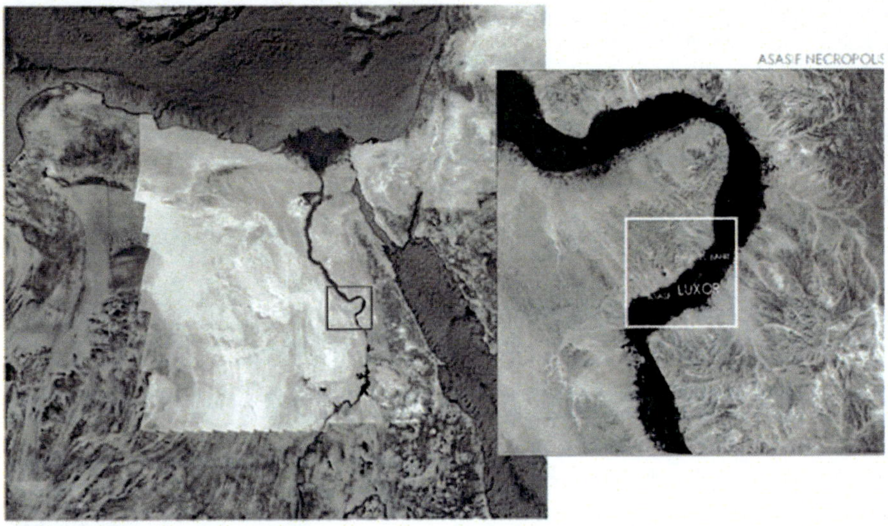

Fig. 1 Location of Luxor, Egypt

splendor. The most important temples were built, and its necropolis included most of the luxurious royal and nobles tombs.

The funerary buildings corresponding to this area and period are mostly composed of a processional exterior area and an interior cult area. The first used to be made of an exterior porticoed courtyard, and the interior by a hypogeum carved into the rock, with a columnar room. But the evolution of beliefs and rituals throughout the centuries evolved types and compositional schemes. As for temples, they vary in typology according to their function and location. In the nineteenth century, Mariette (1877) made here the first archeological surveys, but not documented. The area began to be excavated in early 1881, with Gaston Maspero (1846–1916), as director of the Egyptian Antiquities Service. Excavations in tombs and temples began in the 1890s, directed by the French archeologist Naville (1907) (1844–1926), among others, and relevant archeology figures, such as Winlock (1942), who performed excavation and restoration work. Since the mid-twentieth century, numerous excavation, reconstruction, and restoration works have been conducted by various international missions in collaboration with the Supreme Council of Antiquities. The image in Fig. 2 shows the current locations of the necropolis and the urban center on opposite banks of the river.

Fig. 2 Temples and Necropolis location in Luxor

1.2 Geographical Context

In the geographical context, location is important. At a general level, the settlements were determined by the landscape and the orography. Locally, the function of the building being constructed determined its location, East or West of the main axis, the Nile River. The geographical location of Egypt, surrounded by desert and with an autochthonous economy articulated on both banks of the river, made it a country focused on agriculture. This being its main form of subsistence, and the army was considered a limited need in times of political instability or with a defensive purpose.

We are facing the process of forming a state with a structuring axis, the Nile, and a territory limited in both margins by deserts. The presence of this desert in almost all its borders constituted in itself an important barrier against its enemies. The delta served as the basis for the control of trade routes. The war spirit was limited to certain epochs throughout history, and associated to leaders who undertook expansionist policies. However, Egypt did not really stand out in military exploits until late times, after the expulsion of the Hyksos in 1527 BC. It could be considered that it was here when a military expansionism was originated, fueled by the opportunity to conquer new territories (Fig. 3). During the New Empire, Egypt reached its greatest extension (Davies, 2001), due to military policies and being the period of greatest splendor.[4]

Theban Necropolis is located in the valleys that make up the funerary area (Fig. 4), located on the west bank of the river, very close to the monumental complex such as Deir el Bahari. It extends from North to South, along about 3 km between the hills that delimit the desert and the Nile riverbank. It is divided into several sectors: El Tarif, Dra Abu El-Naga, Asassif, El-Khokha, Sheikh Abd el Gurnah, South Asassif, and Gurnet Murrai. The area of Deir el Bahari, the object of study is located in the small valleys that make up the funerary area on the west bank of the river.

In this area, there are many funerary constructions, overlapping each other physically and chronologically. They are currently the object of work and study. It is currently a study and excavation area for various archeological missions, both Egyptian and international. The works have been carried out since the nineteenth century, with discoveries and findings that are collected in editions by renowned authors. Many Egyptologist documented the necropolis, with old photographs, plans, and reports that have been disseminated in specialized magazines and publications. The works of Nims (1965), Porter & Moss (1972), Davies, and Winlock, among others, were important, and their references are still considered.

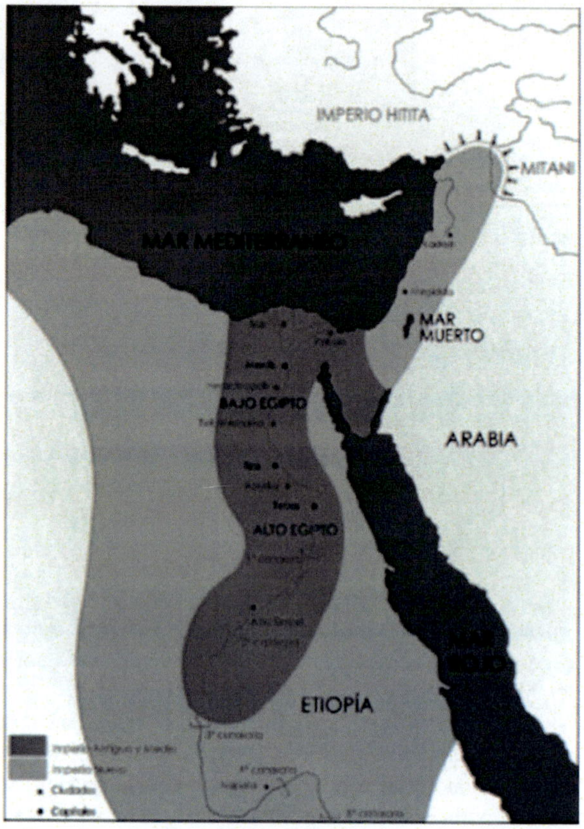

Fig. 3 Military expansion in different periods

2 Methodology

2.1 Case Studies and Field Work

The first part of the study will examine the metrological references, and the concepts of determining geometry and proportion of the research. Next, the architectural models and typologies will be analyzed; its origin, utility, evolution, materials, and construction systems. Finally, and through the significant examples, relations with religion, thought, and the meaning of concepts that were genesis of this architecture will be concluded. The work is developed based on the metrological, construction, and architectural data collected in several excavation, reconstruction, consolidation, study, and repair projects of religious architecture, located in the west bank of Luxor, corresponding to the indicated period.

For the temples analysis, Deir el Bahari Temple has been chosen, due to its state of conservation, by its adaptation to the historical period and the corresponding architectural typology. Wilkinson (2000) provides a broad summary of these typologies, specifically analyzing the area of Western Thebes. For the tombs, as they are smaller constructions, several examples have been chosen in order to homogenize the study sample. Most of the examples analyzed belong to the Sheikh Abd el-Qurna, El-Khokha, and Asassif necropolis. A large part of the nobles and high dignitary's tombs are located there, in a burial site close to the Kings Valley as a burial place. The necropolis was reused after the 18th Dynasty, and sometimes constructions of different times and periods are superimposed. The reports were compiled by authors as Kampp (1996). Many of the tombs were documented by the first archeological missions; others very deteriorated did not have this luck. The working methods were not as rigorous as the current ones; so much information has been lost. Conservation has not been a priority either, as part of them are closed and neglected, so data collection is complicated. The measurements and drawing of the corresponding plans have been carried out in intermittent seasons, from 2010 to the present. The field work was alternated with bibliographic documentation in archives and libraries, as well as specialized publications. Table 1 shows the examples analyzed.

The basis of study are the data collected, both historical and bibliographic, the work and the measurements carried out in situ. The drawing of plans for restoration and

[4] Pharaoh Thutmose III began a reign in which the Egyptian Empire reached its greatest extent by reinforcing the Egyptian presence in the East. With decisive battles, such as Meggido, and later Mitanni on the Euphrates, he brought Egypt to the limits of the empire.

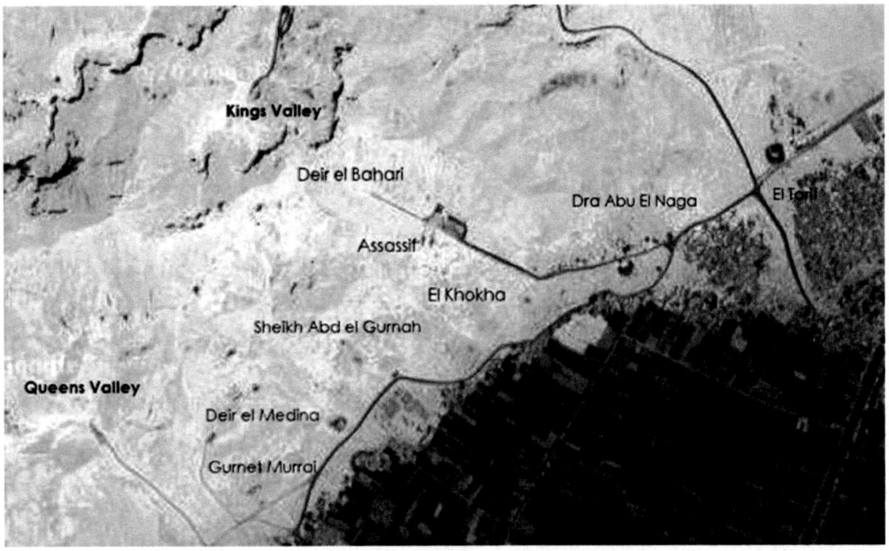

Fig. 4 Theban Necropolis

Table 1 Tombs featuring in the analysis of the Theban Necropolis

Necropolis		No.	Owner	Dynasty	King
Sheikh Abd el-Qurna	TT	31	Khonsu	19	Ramses II
Sheikh Abd el-Qurna	TT	41	Amenemopet,Ipy	21	Seti I
Sheikh Abd el-Qurna	TT	51	Userhat	19	Ramses I
Sheikh Abd el-Qurna	TT	52	Nakht	18	Thutmosis IV
Sheikh Abd el-Qurna	TT	55	Ramose	18	Amenhotep IV
Sheikh Abd el-Qurna	TT	56	Userhat	18	Amenhotep II
Sheikh Abd el-Qurna	TT	57	Khaemhat	19	Amenhotep III
Sheikh Abd el-Qurna	TT	69	Menna	18	Thutmosis IV
Sheikh Abd el-Qurna	TT	96	Sennefer	18	Thutmosis III
Sheikh Abd el-Qurna	TT	100	Rekhmire	18	Thutmosis III
Sheikh Abd el-Qurna	TT	110	Djehuty	18	Hatshepsut
Dra Abu el-Naga	TT	148	Amenemope	20	Ramses III
El-Khokha	TT	178	Neferrenpet, Kenro	19	Ramses II
Assaif	TT	192	Kharuef, Sesh	18	Amenhotep III
Deir el-Medina	TT	218	Amenakhte y Iymway	19	Ramses II
Deir el-Medina	TT	219	Nebenmaat	19	Ramses II
Deir el-Medina	TT	220	Khaemteri	19	Ramses II
El-Khokha	TT	255	Roy	19	Horemheb
El-Khokha	TT	295	Djehutymes, Paroy	18	Thutmosis IV
El-Khokha	TT	296	Nefersekheru	19	Ramses II
Sheikh Abd el-Qurna	TT	343	Bernia, Pahekamen	18	Thutmosis III
Deir el-Medina	TT	359	Inherkhau	20	Ramses III

excavation work has resulted in the identification of a metric module recognizable in the examples analyzed.

The conversion of the current International System of Units (meters) to the original creating unit (Egyptian cubit) allows identifying patterns and geometrical modules that are common to constructions. The apparently inaccurate measurements when making such a conversion are exact numbers and multiples of the project module.

2.2 Measurements and Metrology

All artistic representations in ancient Egypt have a religious, ritual, or magical purpose from sculpture to painting or architecture. Depending on the period, the reigning pharaoh, the liturgical programs, or the prevailing deities, the iconography changes, as well as relative dimensions. The culture of ancient Egypt is not easy to understand any artistic discipline without first analyzing the people´s ideas in those times and maximize from the architectural point of view. Architecture cannot be well understood without a prior study of Egyptian philosophy, language, and thought. Every architectural ensemble responds to ritual purposes, so the measuring unit that responds to these intended purposes is decisive in its conception. Everything is projected under some rules, modules, and functional schemes that link the construction to its ritual and magical end, so that in the analysis of all examples the measuring unit is repeated to make sense of the regulatory and geometric paths.

We start simply from the error in the measuring unit. The numbering systems are the decimal and the sexagesimal. The decimal numbering system is based on a set consisting of ten symbols, (Arabic numbers), where the position of the digit is what gives it its value. It seems that its origin goes back to the ten fingers of the human being to count. The Egyptian system was decimal, since it used the 10 digits, but not positional, since until late times the number had its value by itself, represented by a hieroglyphic symbol or ideogram, but not by the position it occupied. Hieroglyphic writing was multidirectional, being drawn indistinctly vertically or horizontally depending on the esthetic composition in which it was framed. That is why, the position of the digit to give it value was complicated. Initially, the numbers were written in hieroglyph, and gradually, the hieratic notation, used in administrative and accounting texts, was reached, being faster than hieroglyphic writing.

The Egyptian numbering system was a decimal system by juxtaposition. The birth of the hieroglyphs was not immediate and was the result of a linguistic process that evolved over the centuries. Initially, the numbers were written with basic symbols, simply the annotation of as many strokes as numbers were counted, and later the numbers evolved into hieroglyphic symbols as described below in Fig. 5.

The numbers from 1 to 10 were represented with simple strokes; however, the most significant numbers had their own ideogram or hieroglyph. The number 10 represented as cattle, 100 a rolled rope, 1000 a lotus flower, 10,000 a finger, 100,000 is for a frog or a duck, and 1 million for a god with raised hands in worship.

This last symbol, stopped being used with numerical meaning, and for this, other hieroglyphs appeared to indicate higher numbers. Considering one of the authors consecrated

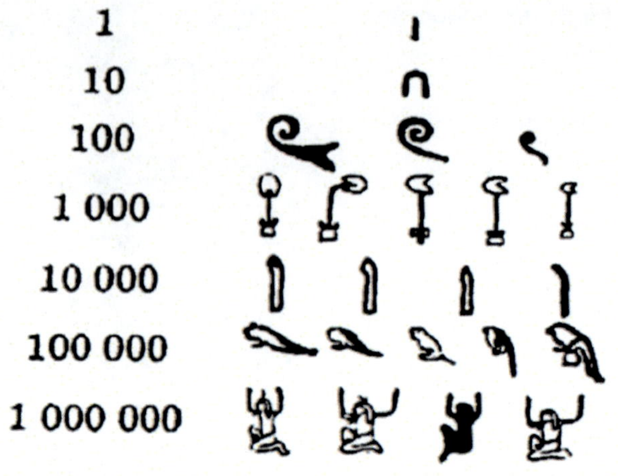

Fig. 5 Egyptian numbers. Representation in hieroglyphic writing, from 1. *Source* https://www.simboloteca.com/numeros-egipcios/

in hieroglyphic writing, Gardiner, the product was used for larger figures, as indicated in Fig. 6.

The zero appears as a symbol in the 18th Dynasty, in the Boulaq Papyrus through the symbol nfr (Lumpkin, 2002). Hieratic writing, equivalent to the cursive representation of hieroglyphs, was born as a result of speeding up the writing of pictographic signs, many centuries later. In both hieratic and demotic, the number system was no longer additive but rather simplified with symbols for each number.

The hieratic was used for commercial communications, while the demotic appeared in the late period, around 650 BC, as the shorthand language of the scribes when transcribing ancient documents.

Units, weights, and measures have related different civilizations with the appearance of commerce. With the beginnings of archeology and the search for the ancient, interest in treasures was the engine of the first approaches to the remains of history. However, many other data were overlooked and give us a clear idea of the relationships between different cultures and countries. The meaning of weights and measures used in ancient times is a clear clue to follow the direction of commercial transactions and the spread of culture geographically.

Ancient civilizations had a system of weights and measures that allowed them to organize their exchanges and construction. It seems likely that the first measures established were those of length (delimitation of the land), and subsequently those of capacity (accounted for what was produced by the land) and those of weight (with the appearance of metals). The oldest scale is Egyptian with limestone, dating back to 3800 BC. The first measures of length dated, also belong to Egypt, deriving from the parts of the human body; elbow (distance from the ring finger to the

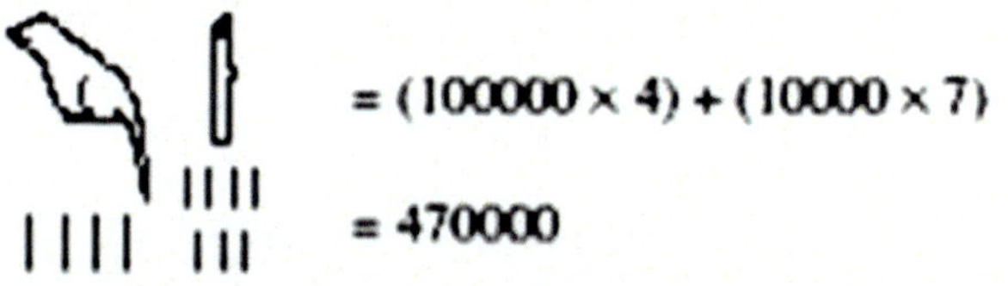

Fig. 6 Egyptian numbers. Mathematical operations. *Source* https://www.simboloteca.com/numeros-egipcios/

elbow), arm (distance from the ring finger to the chest), and the arm (distance between the two ring fingers). Palm and digits arise as fractions of the cubit.

The cultural connection between most of the excavated buildings is given simply according to the size or type of elbow, in the case of Egyptian culture, with which they were planned. The Sumerians had the foot as a unit of measurement in architecture, which was approximately half a cubit. The Greeks also used it as a linear measurement (according to the legend it corresponded to the measurement of the foot of Hercules), but it varied slightly from one city to another. In greater distances, the step (18.3 cm) and the stadium (100 steps) were used.

The measurement system in Rome was adopted according to the conquered countries, but it was duodecimal when dividing the foot into 12 parts. The pass was equivalent to 5 feet, and the 1,000 steps to the Roman mile. The Roman numbering system persisted until the appearance of the metric system, unifying the criteria.

In Egypt, the measurement of the lands testifies to the initial count of time, since it is the length of a decimal pendulum dividing the day; this passed to the prehistoric Bronze Age in England and was the base of the Roman foot. From here, the Egyptian elbow was formed, a similar length appeared in Babylon and New Mexico. With the end of the Roman Empire, unit systems were lost, turning to primitive natural measures. However, the Arab culture preserved the ancient systems during the seventh and eighth centuries, studying the previous measures and sciences such as the Greek ones. In Europe, the multitude and heterogeneity of weights and measures during the Middle Ages, remained until the sixteenth and seventeenth centuries when the development of the sciences began, and their need for an exact system. Finally, in the eighteenth century, the metric system was created. We could affirm that weights and measures were reliable indicators of human history in terms of trade and population. This highlights the constant concern of man to observe, measure, control, exchange, and ultimately grow, both culturally and geographically.

In the Egyptian civilization, we can find a series of units of measurement whose use is evidenced in numerous documents that have remained to this day. The unit of length is the decisive one in architecture.

Egyptian cubit dated approximately in 3000 BC, it was the main unit of measure, divided into palms, and in turn into 4 digits. The minor or short cubit was divided into 6 palms or 24 digits. In architecture, the unit of measure adopted was generally the real cubit, corresponding to 7 palms or 28 digits. It was standardized by the real master cubit, of which some example is preserved.

Figure 7, above, shows fractions of a certain unit of length. On the right, we see the hieroglyph formed by a folded fabric, which represents 1/2. And just below, a unit divided into two equal parts. To the left of the 1/2 sign are the hieroglyphs that denotes 1/3, and then, a unit is divided into three equal segments. This continues with 1/4, 1/5. …

The original cubit measured about 457 mm, which was measured from the elbow to the end of the middle finger with the fingers extended. From the 3rd Dynasty, the real cubit was taken as a unit of measurement, which is the cubit plus a palm and was equivalent to 523–524 mm, and was subdivided into 28 digits.

This was the side of a square, the diagonal that was the ground measurement base, 29,161. The cubit, as a unit of measurement, was used from 2700 BC. Many specimens of cubit have endured; some of them were made as ceremonial measurement units that were kept in temples. Toward the year 600 BC, during 26th Dynasty, an important reform was introduced to unify several measures. The short cubit disappeared and the so-called reformed cubit, equivalent to the old royal cubit, was installed. Almost all of these magnitudes were related to body measurements. For example, the short cubit was a distance from the elbow to the fingertips, equivalent to 6 spans of 4 fingers each, then one more span was extended, which is when it was called the real cubit.

Later, during the Greco-Roman period, the Greek cubit (~462.5 mm) and the Roman cubit (~443.5 mm) were used. The cubit was divided into seven spans or palms. There were also other fractional units of the cubit, such as the digit, which represented 1/28 of the cubit, or a quarter of a palm.

It linked measures of the sides and diagonals of a square, enabling areas that doubled or halved, without altering the proportion. Half of this diagonal is the actual cubit called Remen. Thus, a square cubit is twice that of a square Remen. This is where the measurement digits are derived, the basis of many other things. The double or diagonal row of the cubit 29,161 (740, 69) is almost the natural length of a pendulum that oscillates 100 times a day; at the latitude of Memphis which would be 29,157 (740.57). The close relationship of 29,161 to 29,157 seems highly unlikely to be accidental. Table 2 show the equivalences between ancient measurement units.

In Egypt, the cubit was divided into 7 spans, 28 fingers of 0.737, but the usual measurement figure was 0.729, formed

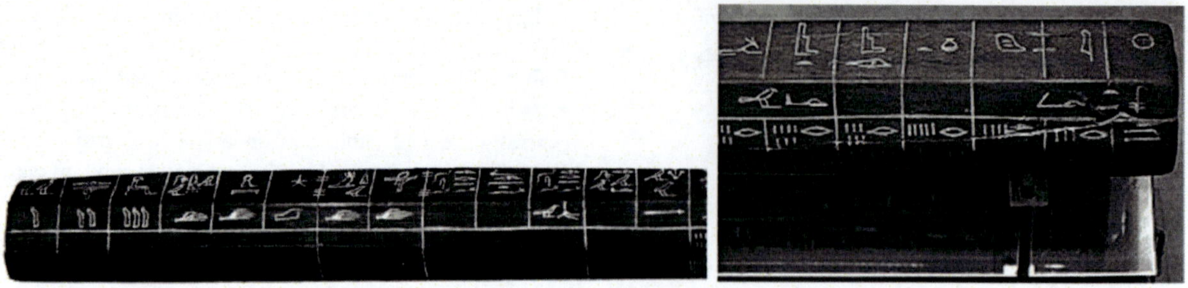

Fig. 7 Egyptian numbers measuring rod of Maya (Tutankkamun Treasurer), Saqqara. Louvre Museum, Paris

Table 2 Equivalences between ancient measurement units

Name	Egyptian name	Equivalence
Cubit	Meh (sherer/nesu)	59.29 cm
Palm	Shesep	7.471 cm
Digit	Dyeba	1.87 cm
Rod	Khet	52.5 m
River	Iteru	10.5 km
Relationship between ancient measurement units		
1 Royal cubit =	7 Palms =	28 Digits
1 Short cubit =	6 Palms =	24 Digits
1 Remen =	5 Palms =	20 Digits
1 Palm =	4 Digits	
1 Khet =	100 Royal cubits	

as a fraction of the Remen. In Babylon, the cubit was divided into multiples of 6 (Fig. 8).

Numerous authors have carried out research on the Egyptian canon of measurement, especially in art (Iversen, 1955, Giedion, …), based on the unit of measurement of the hand, in a closed fist or span, and making the proportions grids in function of this.

Similarly, the cubit were introduced, which the exact measurement has been the subject of numerous dissertations, and the foot. This preset module fitted the layouts for the models in painting, sculpture, and architecture. Thus, man was related to the arts, which were a projection of human proportions, but on a larger scale, preserving references. With this practice, technical utility was combined with meaning in composition.

The knowledge of geometry by the Egyptians is based on documentary mathematical papyri that have been preserved to this day, such as the Rhind Papyrus. The calculation methods differed from the current ones, since their arithmetic reduced division and multiplication to addition and subtraction (Gillings, 1972). Until the nineteenth century, data concerning Egyptian architecture were barely non-existent. Napoleon's expedition was the first relatively reliable documentation that reproduced plans and measurements of the archeological remains. Until then, the lack of information, together with the slow decipherment of the Egyptian texts, did not make its study possible. It was during the Renaissance, when there was a manifest interest in Europe for the ancient, and in particular for the proportions in architecture following the discovery in Montecasino of the Treaty of Roman Architecture, signed by Vitrubio. Thus, the concept of proportion and harmony is rescued.

In Egypt, the origin of geometry was functional. The taxes corresponding to the lands that were cultivated were proportional to the surface, and due to the periodic flooding of the Nile River, these delimitations disappeared. After each flood, the boundaries had to be marked again, so a precise calculation system was developed that favored the speed of the process. Egyptian knowledge of geometry made possible achievements such as building pyramids or measuring land. Egyptian geometry along with the Babylonian were the precursors of the powerful Greek geometry. The dominance of the triangles was possible by the knots spaced in ropes, which were used to measure and check the right angles. Pythagoras collected all this geometric experience for his theorem. In Egypt, the relationship between the hypotenuse and the legs in a right triangle was already known. They had formulas to measure areas, such as the square, rectangle,

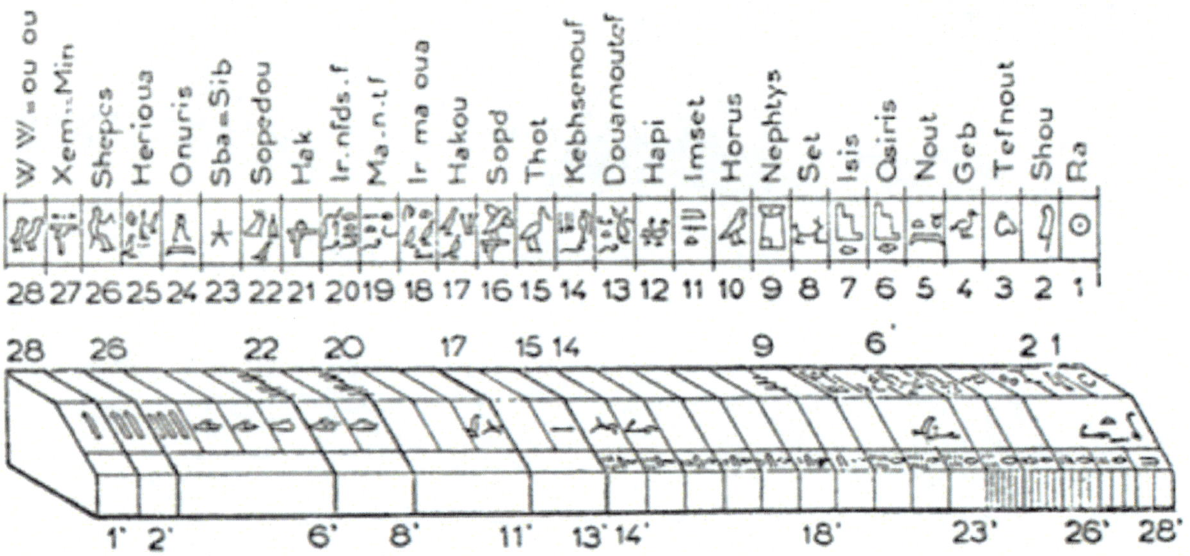

Fig. 8 Moon associations "Royal Cubit stick" Egyptian Museum, Turín[5]

rhombus, and trapezoid. The most used geometric figure in Egyptian architecture was the triangle (Figs. 9 and 10). According to leading authors (such as Legon (1996), Viollet-le-Duc and Auguste Choisy), the most recurrent typology could be included in three different triangles.

Viollet-le-Duc

- Triangle 3-4-5: Corresponds to the group of right triangles, in which the dimensions of its three sides correspond to whole numbers.
- Equilateral triangle: With three equal sides.
- Egyptian triangle: Isosceles triangle, with two equal sides and a different one, in which the radius between the base and the height is 8:5.

Auguste Choisy

- Equilateral triangle: With three equal sides.
- Egyptian triangle: Isosceles triangle, with two equal sides and a different one, in which the radius between the base and the height is 8:5.
- Golden Triangle: Built according to the golden section.

It seems likely that the architects used the Fibonacci series or the 8:5 isosceles triangles in the realization of the planes, then traced an axis of symmetry using triangles 1:2, 1:4 and 1:8 to determine specific points, and subsequently, the construction diagram was based on the combination of squares and triangles that defined the general profile of the building. Finally, the use of triangles 8: 5 on the main axis, marked at the intersection with it, the transverse stays. Similarly, angles, columns, and architectural elements were projected.

Religion and beliefs were decisive in architecture. Everything was governed by order or Maat, and Pharaoh was in charge of maintaining that order in this world. The intention to reflect the sky on earth was translated in the location and orientation of its constructions, which responded to an intention, the universal order in which this Maát balance reigned.

The evidences can be found documented in Papyri, the Pyramid Texts and the Coffin Texts, religious books, star clocks, festive calendars, and temples and religious buildings (Fig. 11).

Egyptian architecture came marked by factors linked directly to its genesis, for being inherent to the place where it was based. The growth of the Nile, the solar cycle, and the geography of the place were decisive for the location and projection of the buildings. The direct link of nature with the vital and religious cycles linked the constructions to the end for which they were hatched. That end was the continuation of these cycles but in relation to man as a continuation of nature itself. Most of the temples of Ancient Egypt were located according to the orientation of the east–west axis, which in most cases determined the Nile River, but in some

[5] Reading from right to left in the upper register, the digit XIV on a cubit stick was marked in 16 equal parts. The next digit was divided into 15 parts and so on, to the figure of 28, which was divided into two equal parts. Thus, measurements could be made to fractions of digits with any denominator from 2 to 16. The smallest division, 1/16 of a digit, was equal to 1/448 part of a real cubit. The precision of the cubit lever is attested by the dimensions of the great pyramid of Giza; although thousands were employed in construction, its sides vary no more than 0.05 percent of the average length of 230,364 m.

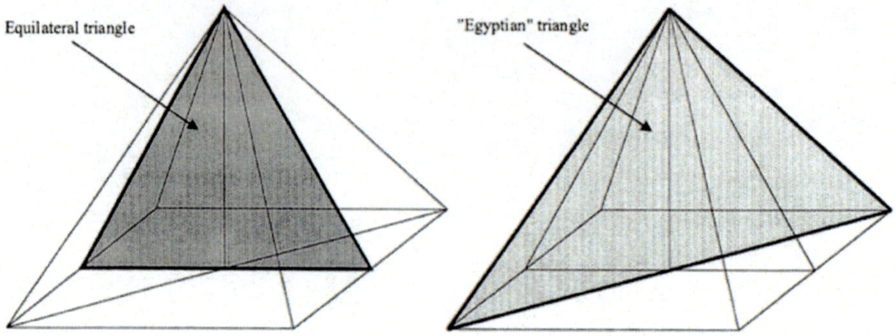

Fig. 9 Equilateral and Egyptian triangles used in Architecture "Entretiens sur l´architecure" Viollet-le-Duc[6] (1863)

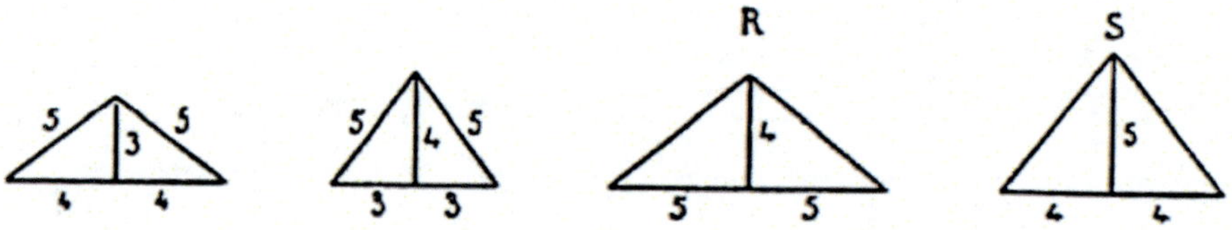

Fig. 10 Triangles used in Egyptian architecture "Historie de l´architecure" A. Choisy[7] (1899)

Fig. 11 Representation of Seshat, in the Red Chapel of Karnak (**a**). Representation in the Temple of Seti I in Abydos (**b**)

sections that the course of the Nile varied, the orientations also did (the delta or the Thebes area). Orientation variations according to various authors have different justifications. They could be due to the temple being oriented to canals perpendicular to the river for access, or simply to its perpendicular, so that the orientation would be merely topographic. However, the knowledge of the star map, as well as the organization of constellations, doubts that astronomy was not used for the alignment of temples. The winter solstice and other stellar alignments were clearly used. Probably, the topographic choice for the foundation of Thebes was due to the fact that in this area the river, its perpendicular coincides with the east–west axis, linking the sunrise in the winter solstice with the sunset in the summer solstice, contributing to its identification as a sacred place. The orientation was done according to the sun or the stars. Poles were used in the "*founding ceremony,*" and according

[6] Eugène, Viollet-le-Duc (1814–1879), French architect, archeologist and writer. The Neb (or gold) triangle, better known as the isosceles triangle, is the most widely used in the construction and harmonic diagram of Egyptian art and architecture. Viollet-le-Duc called it the "Egyptian Triangle".

[7] François Auguste Choisy (1841–1909), French engineer, historian, and theorist of architecture. Good student of Egyptian architecture, he defended that if the different parts of a building are compared, between its dimensions there are simple relationships; 3/5, 1/2, …

to the intersection of the shadow that was thrown with a circle marked the alignment of the main axis.

The knowledge of these orientations and alignments in archeology are fundamental, both when planning excavations in specific places, as well as to know in which area or area of each building to continue doing surveys, for knowing the structure, axes, and distributions of the deposits.

The religious and philosophical origin of the genesis of the project justifies the use of a regulatory module. According to ancient tradition and the Egyptian conception of sacred geometry, Egyptian temples were designed to operate as interactive energy devices that could activate physical, mental, and spiritual bodies. They were called houses of the soul or spirit, designed to awaken the senses. They were built according to lines of law, with a specific combination of stones, using the energetic properties of water, aligned with specific solar and stellar events, according to patterns of sacred geometries that are reflected in our natural world, and the upper cosmos. The Egyptians employed numerology, sacred geometry, the measures found in nature, and mathematical models based on celestial movement to design sacred spaces, with references in Badawy (1965) and Schwaller de Lubicz (1998). A complex approach to the construction of its temples, tombs, and pyramids including the use of symbolic art and building materials.

The basis of Egyptian culture is religious. Everything is for a spiritual purpose that contributes and maintains the cosmic order. The creation of the world through god's enneads justifies the genesis and the end of humanity in his thinking, so that the creative nature of a god will be embodied on earth by the pharaoh (king and god at the same time). That is why what was important for the Egyptians, originated life, and the pharaoh was responsible for maintaining this balance and universal harmony. The divine order is what guarantees that all things must be found in their corresponding place. That is why the artist or architect is simply the means for the work to be part of eternity, or the universal law established in the creation of the world. According to John Wilson; "*The same balance comes from Egyptian cosmology and theology, which sought a counterweight to each observed phenomenon or to each supernatural element*" (Wilson, 1988).

2.3 Architecture. Constructive Typologies

The multitude of funerary constructions surrounding the area, as well as the twinned position between them, created a complex set of burials, varied both in the level of the land on which they settle, and in the time in which they were built. From earlier times, and massively during the reign of Amenhotep III (1350 BC), nobles and officials tombs were excavated and built in this area. The structure and distribution of the hypogea varied according to the owner, the reign, and the time. It was in this period when the hypogea construction in the area was generalized, and when greater architectural and artistic representation levels were reached. The necropolis was also reused later with constructions belonging to the 20th and 22th Dynasties. It is a tombs set of great dimensions of the Egyptian nobility characters, located in the Theban necropolis, consisting of a hypostyle hall under the solar patio. The periods range from 21 to 22th Dynasties (2061–1080 BC).

The types and morphology vary according to the time they were built, and the dynasty prevailing at the time. They evolve in distribution, proportion, and dimensions, according to the funeral rites, and the prevailing ideology at each stage. The types of tomb, as well as the architectural plan changes, evolving over time, according to the beliefs of each era, and the funeral ritual. The functional scheme is very similar, only that more importance is given to certain parts of it, as well as to the decorative program. The typologies varied according to the construction factors of each burial, but the religious conception that determined the architecture was the same in the area (Fig. 12).

Their sizes vary, according to the importance of the nobility character which they belong, and are composed of a columnar room and the solar courtyard. The architectural structure is defined by religious beliefs, and each space is destined to a different part of the cult or ritual. The plan varies, following a T-shaped scheme, with a room transverse to the entrance, and a longitudinal room that leads to the chapel. In general, scenes from the life of the deceased and his family are depicted in the cross section, and in the longitudinal section, scenes of the funeral procession.

They are funerary sets in which architectural concepts were set by religious beliefs, especially structurally speaking. The conception of the tombs is directly related to the cult. In this way, the components of the spaces are linked to the intention of the building, being able to differentiate three modules, the space to the solar cult, the space for the cult of the deceased, and the Osirian space, collected in the following scheme (Fig. 13).

The orientation of the funerary constructions follows a longitudinal main axis in the east–west direction, on which the solar patio is located, a room with columns and the closed part that contain the funerary area. The courtyard was accessed from a ramp, and it is common for this access to be shared for several tombs. This is because the funerary space was subsequently reused in other dynasties.

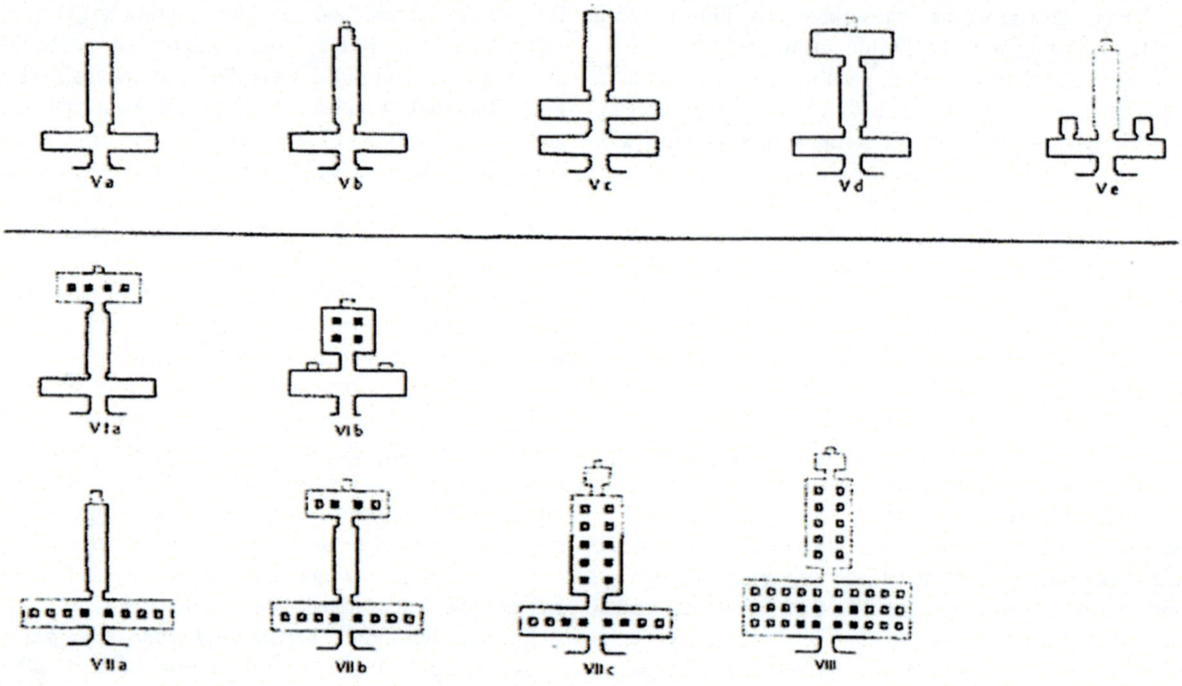

Fig. 12 Typological classification of the Theban tombs (F. Kampp) (Kampp-SeyFried, 2003)

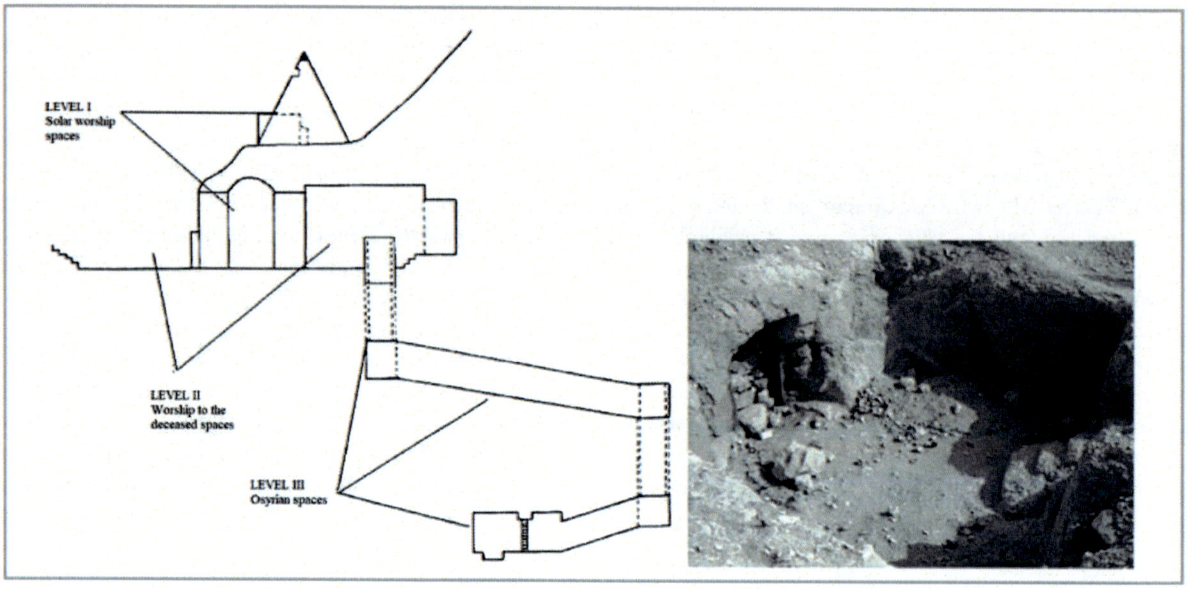

Fig. 13 Ritual levels of the Theban tomb (K.J. Seyfried, 1987)

The Egyptian tomb was divided into three parts: the outer chapel, the pit, and the sepulcher chamber (Fig. 14).

- The outer chapel: With several rooms, accessible to parents of the deceased and priests. There the family went to make offerings. It was also used to celebrate religious rites. In it, the deceased showed how his life had been, his achievements, and the wealth obtained
- The vertical well: Located next to the chapel, to introduce the sarcophagi in the sepulchral chamber.
- The sepulchral chamber: It kept the mummy and was closed to preserve it.

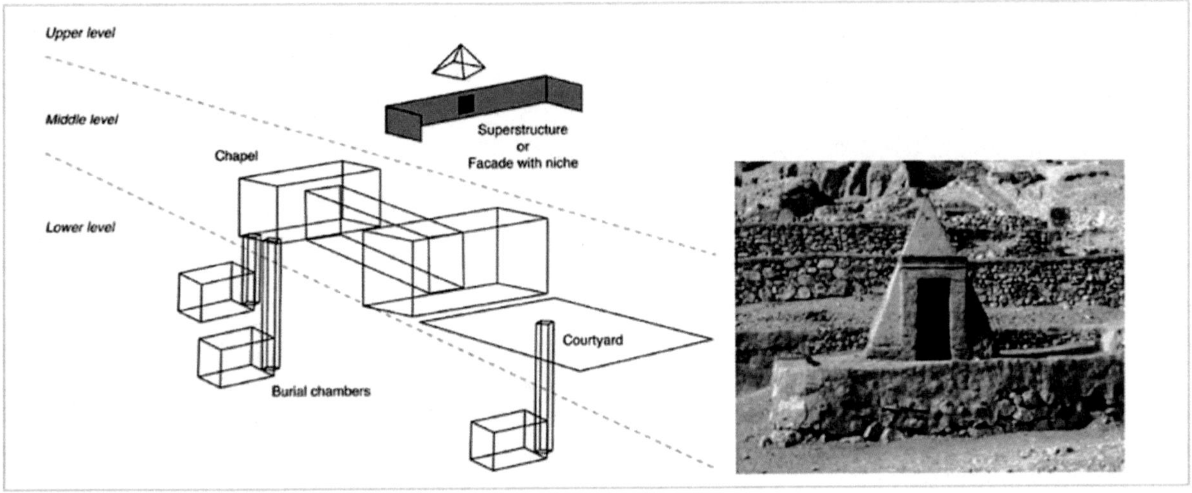

Fig. 14 Parts of the Theban tomb

The orientation of the constructions follows the directionality of the tombs of the time, distributed along a longitudinal main axis in the east–west direction. On it, a solar patio is aligned, a porticoed enclosure, and then the enclosed enclosure that contains the funerary area. The longitudinal axis of the tombs is oriented from east to west, finding a descending ramp first. These ramps usually constitute the access to the patio, having found several more tombs, due to the reuse of the funerary enclosure in later times to that of its original owner. Currently, both the ramps and the courtyards are buried under several meters of sand and rubble with archeological remains that are being recovered as the surveys progress.

In Thebes, a great necropolis was developed and occupied a considerable space, in the left margin of the river. It had tombs intended for the great characters of the kingdom. Here, we can find the three essential parts of the tombs belonging to the Old Kingdom and the Middle Kingdom, but instead of being all together, or close, they are separated at considerable distance. The chapels were located in the area accessible to the population, in order to facilitate the worship of families and religious rituals. With the passage of time, many became temples (Goornah, Rameseum, Medinet Habu, and Deir el Bahari). The temples are located on the banks of the Nile River, and are large, with the most significant being partially rebuilt. We find two types of temples according to their location: those of the east bank and those of the west bank.

- East Bank Temples: In the case of Deir el Bahari, it is worth noting the peculiar character of the temple on the left bank of the river in Thebes. Here only certain types of special buildings could be found, and all examples date from the comparative period of the 18th to the 20th Dynasty. The great temples on the right bank of the river (Karnak, Luxor, ...) are first buildings erected for the worship of the local deity and the work of many generations. Since the time of the 12th Dynasty, of the Amenemehats and Usertesens, almost every sovereign or reigning family had the duty to add or repair the structure of Karnak. Each generation insisted on being represented here, which is why the walls of the great temple of Amun became the annals of the Egyptian monarchy.
- West Bank Temples: It was very different in the temples on the left bank. Here, the pharaoh started a temple with the intention of finishing it himself. The plan was his and he tried to carry it out from beginning to end, in fact, these monuments were built by the king for his own glory and memory, and the inscriptions with which he was decorated often took the form of chapters of his own autobiography. To understand the purpose of these temples, they must be compared with the Egyptian tombs of the Old Kingdom.

It is Mariette (Mariette, Auguste, 1821–1881) who we should give the credit of having fully recognized the nature and function of these temples on the left bank. They are large funeral chapels directly connected to the royal tombs whose existence they imply. Its special character is determined, and we must proceed to divide these funerary temples into two categories; those erected for a single sovereign, and those that served for several royal tombs.

- Temples for a single sovereign: To the first category belong the Rameseum and Medinet Habu. Ramses II, the greatest of the Egyptian pharaohs, and then who seemed to imitate his father in everything, Ramses III. He built his own funeral chapels, monuments designed to perpetuate his exploits, and to bring to posterity what we estimate as his main titles to fame. Therefore, each of these temples was connected only with a grave.

- Temples for several royal tombs: It was the opposite with those of Goornah and Deir el Bahari. Seti I began the construction of Goornah, and here, he erected the funeral chapel of his father Ramses I. The inscriptions on the walls represent Ramses I dead and sitting in his sanctuary bearing the emblem of Osiris. In his honor, the temple was built and the ceremonies celebrated. Seti I did not finish the building, and his son continued by adding a long relief where his father, returning from his grave, invoked Ammon in favor of Ramses II, represented before him offering perfumes to the sacred boat of the god. The temple of Goornah is in this way, the great chapel of two tombs, and it is as in the Mastabas of the Old Kingdom, or as in the tombs of Beni Hassan, there were two open wells from the same room.

3 Results

After the studies and examples analyzed, the systematization of the geometric and architectural composition is evidenced, from the actual measurement and proportion, in the corresponding metrology. The module in architecture and its use defines the proportions used in the layout of floors, elevations, and sections of the buildings. The measurements of the constructions in units that do not correspond to the measurement unit that originated the original project, creates discrepancies. However, if we convert the current measurement units to the module used, the Egyptian cubit, both in temples and tombs, the proportions fit into perfect compositions. The regulatory geometric model of the plans confirm the use of this modulation and systematization of the proportions.

The results of measurements can be complicated when expressed in palms or fingers, but they are relatively simple, depending on the geometric subdivision of the cubit. This introduces an important theme, the adoption of a module in architecture in the design, not only of architectural details, but of the entire building. Modules and whole dimensions are proven in numerous monuments, by several authors Badawy (1965), Wisocky (1986), Carlotti (2005), and Arnold (1991).

The results have been obtained from many examples of both tombs and temples, in which measurements have been made, and the plans have been drawn. The data collection in situ, carried out in the current measurement system (meters), when they were converted to the original unit (cubits), mean exact numbers, which have enabled established significant compositional relationships. These are based on the spaces modulation in relatively rigorous grids, as well as on the compositional use of the triangle typologies previously analyzed. As significant studies, two have been chosen, which respond to the typology of temple, and of tomb.

As a possible impact of the geometry of the project on the construction of the building, the Deir el Bahari Temple and the Tomb of Nakht TT52 in Luxor have been analyzed as examples. There are detailed drawings, made by Davies Garis, (1917), that reproduce the decoration of the tomb, but the interest of his works was more pictorial than architectural.

Nakht, was an officer in Ancient Egypt. He held numerous posts within the government, including astronomer of Amun, scribe, priest, head of the granaries, and keeper of the king's vineyards, during the reign of Tuthmosis IV, in the 18th Dynasty. Follow the outline of the 18th Dynasty tombs, with an open courtyard, two internal chambers with a T-structure, and the underground area. It is raised in a north-northwest-west direction, although symbolically the courtyard should be oriented to the east (sunrise and life), and the inner chambers to the west (sunset and death). This variation of the stakeout could be due to the morphology of the mountain, or the state of the rock where it was excavated.

In the Tombed Nakht (Fig. 15), hypogeum excavated in the mountain, we can observe the modulated compositional scheme, and the distribution of spaces for ritual purposes. Considering that the walls are excavated in rock, and the difficulty of measurements and staking, the proportions are practically exact. The measurements, in Egyptian cubits, are modulated by series of 3×9 in the transverse room, and 7×7 in the courtyard, identifying the regulatory triangles in the composition of the spaces.

It is analyzed the case of the Temple of Deir el Bahari, as representative of the 18th Dynasty Architecture in Thebes. We could draw some conclusions from the studies carried out, as far as the system of composition and measurement is concerned. Previously studied by named authors, Wysocki (1986), leading the Polish mission, edited detailed information of their work seasons, confirming the existence of a pattern, modulated in cubits.

The construction of the temple was done, between the reigns of Hatshepsut, V century BC. Excavations of the present century have identified numerous changes in the project throughout the construction process. It was structured with three terraces connected by ramps, on a rear facade that is the Theban Mountain.

Considering location on site, the temple has a solar orientation as a sacred construction, had to be in communion both with the earth, through the sacred topography, and with the sky, through the astronomical orientations. In the Thebes area, the orientations coincide, as well as its perpendicularity to the axis of the Nile. The orientation of the Karnak Temple, prior to Deir el Bahari was to the west, focused on the sunset over the Nile. When Hatshepsut built the temple

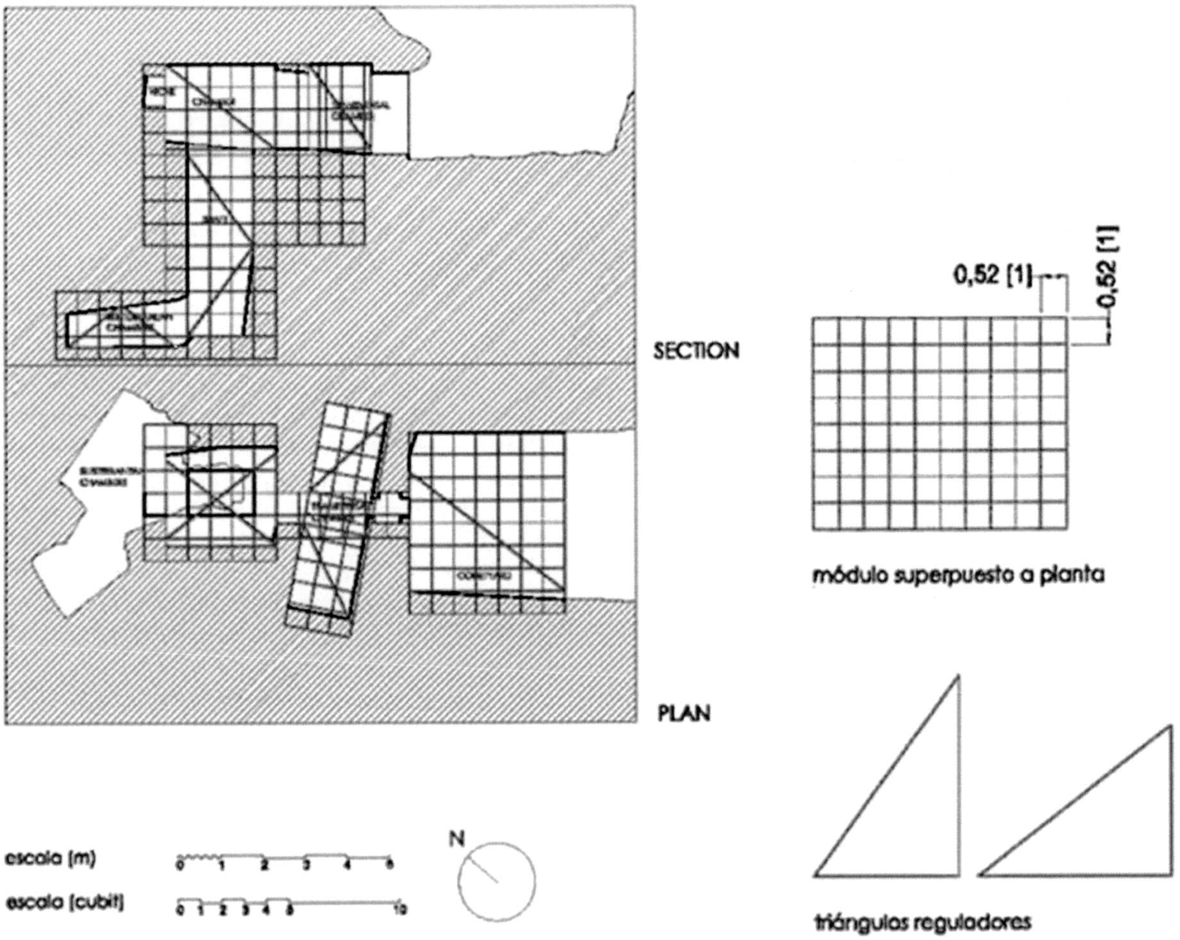

Fig. 15 Tomb of Nakht, TT-52, plant and section (left), and modules of composition in elbows (right)

Fig. 16 Alignments temples of Deir el Bahari and Karnak, Luxor

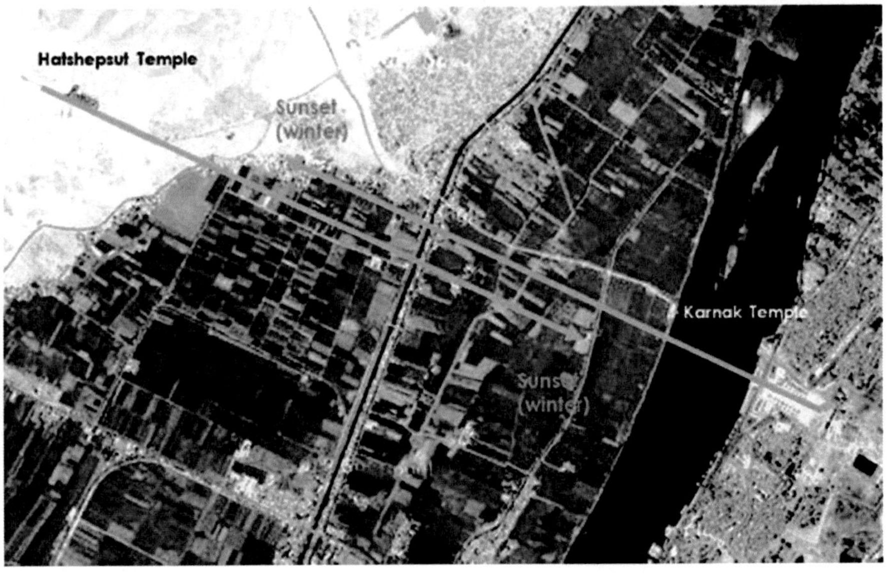

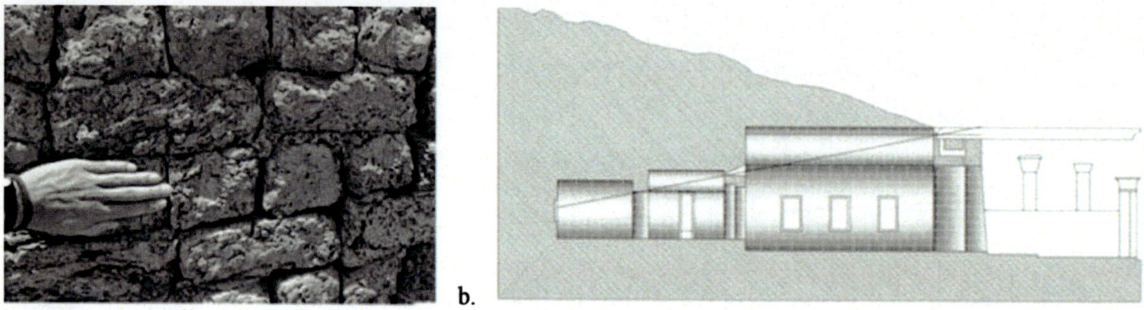

a. b.

Fig. 17 Temple of Deir el Bahari, module (**a**). Temple of Deir el Bahari, section (**b**)

intended for Amun-Ra within the Karnak Temple, oriented to the sunrise on the winter solstice, on the west bank built on the same axis the temple of Deir el Bahari, in such a way that every morning the sun illuminated the Amun-Ra chapel in Karnak and simultaneously her chapel in Deir el Bahari. In the following image (Fig. 16), you can see the orientation of both on opposite banks of the Nile.

The control of the dimensions, measures, and orientations, together with a perfect geometric conception, made it possible that on certain days of the year the sun would illuminate the Sancta Sanctorum of the temple (Figs. 17 and 18).

The temple fits in a module of 1.5 Egyptian royal cubits (0.523 m). All alignment of structural axes and pillars coincides with the measurement. The geometric composition is mainly based on the application of this module, and the conformation of rectangles. It can see the correspondence in the layout, with triangles 3-4-5, mentioned above. The modulation of the different levels of the terraces, as well as the compositional part of the porticoed parts, follow an axial scheme. The plans respond to exact regulatory layouts, dominated by the geometry of both the module used and the triangle (Figs. 19 and 20). The knowledge of the modulating measurements has allowed in the reconstruction of the temple, the location of the main constructive axes, both columns and pillars, in order to establish correct restitution hypotheses.

Numerous authors have maintained theories about geometry and numerology applied to Egyptian architecture. However, few have been the definitive results, due to the lack of certain archeological evidence. Considering the geometric analyzes of Siegler (1970), Viollet-le-Duc (1863), Lauffray (1979), Hinkel (2000), Lauer (1936), it should be noted that some mysticism has always surrounded the possible interpretations of the results, but the connection between geometric figures has been repeated in most of the plans.

After the analysis of numerous examples corresponding to the period studied, and considering the measurement errors in staking, as well as the possible light deviations of the numbers, due to the material where they were built, the results are very similar. The existence of this creative module seems confirmed, and the proportions established in the genesis of the architectural project. These modules fixed by ropes, marking the measurements and construction elements (adobe bricks of a section, to be easily handled by a worker), were repeated in the constructions. All this modulation and measurement seem to respond to the intention of reflecting a cosmic order in the buildings. The intent of respecting that symmetry and geometry in the composition in some way wants to transmit the balance marked by the prevailing religion of the moment, the Maat, or the divine order. Egyptian architecture seems be the result of mathematical, geometric,

Fig. 18 Temple of Deir el Bahari, Sancta Santorum of the temple

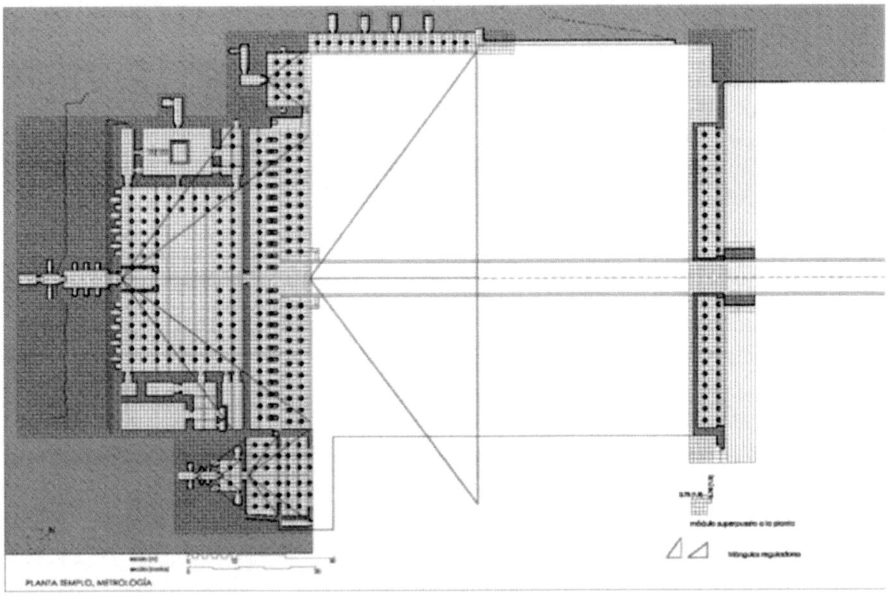

Fig. 19 Temple of Deir el Bahari, floor plan, and regulating triangles

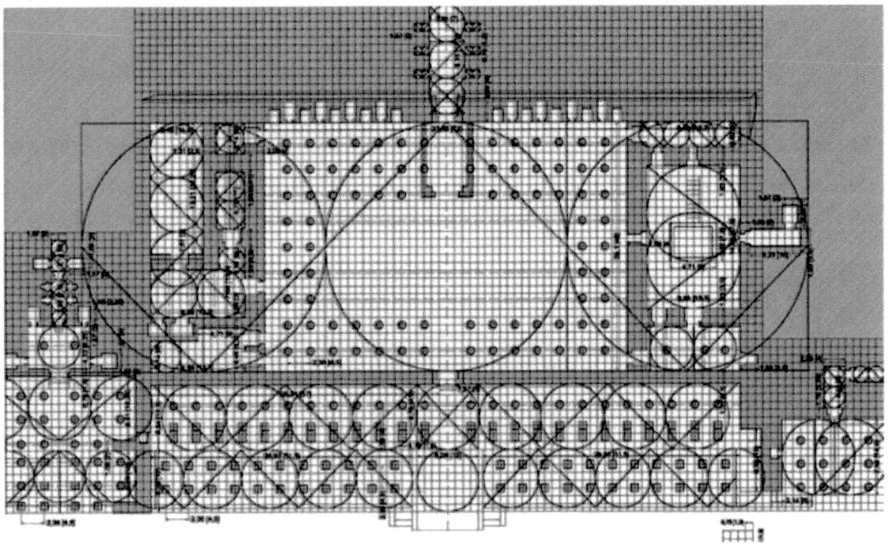

Fig. 20 Temple of Deir el Bahari, floor plan, regulatory geometry

and astronomical concepts, linked to religious and philosophical thought of the time.

The measurement of the buildings in the appropriate unit, Egyptian cubit, allows to discover the modulation that existed when they were projected. Understanding the geometric modules that created them helps to systematize the typologies and measurements of the spaces. In this way, we will be able to deduce approximately the dimensions of a courtyard, a hypostyle room, or a temple chapel if we know the compositional models used.

The appropriate metrological use in drawing plans makes possible to reach conclusions of the original design of the building. The knowledge of the construction system establishes premises for the possible repercussion in pathological states (structural, functional, and material), making possible to intuit the distribution of loads, based on the supposed measurements of the buildings and their distribution in space.

4 Conclusions

After taking data and drawing plans of numerous architectural examples, corresponding to the 18th Dynasty in Thebes, we can draw several conclusions.

Each construction was conceived according to a canon numerically, geometrically, and spatially.

The measurements make no sense unless they are taken in the original unit, the Egyptian cubit, which was used in its architectural design. The translation of any measurement to this unit (Egyptian cubit) gives exact numbers and proportional geometric relationships. This measurement in the correct unit is essential to prepare graphic documentation, plans, and coherent studies. Making drawings leads to obtaining geometric regulatory paths that are repeated in all the analyzed examples.

The mathematical units that govern this original approach to the project respond to a philosophical and religious conception of the spaces, which materializes the ritual purposes for which they were conceived.

Investigation of the measurement units of original projects, as well as geometric regulatory layouts, contained in the different spaces that make up the building, constitute evidence of the use of a numerical module that was the origin of the architectural project. The use of this module allows systematizing the structure of the buildings and the distribution of the spaces, according to their design typologies.

Applied metrology in the analysis of religious architecture, realized throughout the 18th Dynasty in the ancient Thebes, allows discovering the compositional modules of the projected architecture, the numerical canon that designed the building, and the relationship between the measurements of different parts of the building. The use of metrology implies the proper use of measurement units. The analysis of the buildings in the original measurement unit enables us to understand their compositional model. It also helps to establish typologies and standardize size of the spaces repeated between examples from the same period. The application of metrology in archeology and restoration facilitates both prospecting on the ground and composition rebuilding monuments.

The knowledge of the constructive typologies, and the measurements of the buildings, is an essential help in archeological excavation works. Being able to intuit or guess the architectural footprint of tombs and temples allows to organize the diggings, and to delineate efficiently the possible work areas.

References

Arnold, D. (1991). *Building in Egypt*. Oxford University.

Badawy, A. (1965). *Ancient Egyptian architectural design: A study of the harmonic system*. University of California Press.

Davies Garis, N. (1917). *The tomb of nakht at Thebes*. Metropolitan Museum of Art.

Gillings, R. J. (1972). *Mathematics in the time of the pharaohs*. MIT Press.

Hinkel, W. F. (2000). *The royal pyramids of meroe architecture* (p. 4). The Sudan Archaeological Research Society Bulletin.

Iversen, E. (1955). *Canon and proportions in Egyptian art*. London: Sidgwick and Jackson.

Kampp, F. (1996). *Die Thebanische Nekropole, zum Wandel des Grabgedankens, von der XVIII. bis zur XX. Dynastie. Theben XIII*. Mainz: Philipp von abern.

Kemp, B. J. (1991). *Ancient Egypt: Anatomy of a civilisation*. Routledge.

Lauer, J. P. (1936). *La pyramide à degrés, I et II, l'architecture*. Fouilles à Saqqarah.

Lauffray, J. (1979). *Karnak d'Egypte: Domaine du Divin*.

Legon, J. A. R. (1996). *The cubit and the Egyptian canon of art. Discussions in Egyptology (N.35)*. Oxford.

Mariette, A. (1877). *Deir El-Bahari. Documents Topographique, Historiques et Etnographiques*. Leipzig J.C Hinrichs; Cairo.

Naville, E. (1907). *The XIth dynasty temple at Deir El-Bahari*. Egypt Exploration Fund.

Nims, C. F. (1965). *Thebes of the pharaohs pattern for every city*. Stein and Day.

Parker, R. A. (1971). *The calendars and chronology, the legacy of Egypt*. J. R. Harris (Ed.), Oxford.

Porter, B., & Moss, R. (1972). *Topographical bibliography ancient Egyptian hieroglyphic text, reliefs, and paintings*. Clarendon Press.

Redford, D., & B. (1967). *History and chronology of the 18th dynasty of Egypt: Seven studies*. University Press.

Schwaller de Lubicz, R. A. (1998). *The Temple of Man*. Bd. II, New York.

Shaw, I. (2000). *The Oxford history of ancient* (Egypt). Oxford University Press.

Siegler, K. G. (1970). *Kalabasha, Architektur und Baugeschichte des Tempels*. Berlín.

Viollet-le-Duc. (1863). *Entretiens sur l'Architecture*. Paris.

Wilkinson, R. A. (2000). *The complete temples of ancient Egypt*. The American University in Cairo Press.

Winlock, H. E. (1942). *Excavations at Deir el Bahari, 1911–1931*. New York: The MacMillan Company.

Wysocki, Z. (1986). *The temple of queen hatshepsut at Deir el-Bahari: Its original form*. MDAIK 42.

Conserving Symbolism in Architectural Heritage—Case Study Eloquence in Depicting Philosophical Ideas Inspired by the Principles of Islam on Islamic Architecture Through Ages

Khaled El-Daghar

Abstract

Symbol includes a meaning not just a sign, and a meaning can be read whenever it is understood since the idea of communication is available in the significance of this meaning. This explains why human existence is closely related to symbol and symbolism. The architectural heritage attaches great importance to the idea of symbolism and constantly uses different symbols and distinctive vocabulary. The general geometric shape of the building or outline, the formation of spaces, the degree of simplicity or complexity, the mass of the building, the quality of the materials used, and the colors are architectural elements rich in symbols. In addition to the architectural vocabulary, such as columns/pillars, arches and domes, ornaments, and circular openings borrowed by some architects from different periods of time to enrich the symbolic meaning of their design approaches. Though, this metaphor remains superficial if the vocabulary is not in harmony with the rest of the space and functional components of the building attached. The study aims to illustrate the meaning of symbol and symbolism in architecture and then to discuss the symbolic reflection on architectural heritage, where some architectural paradigms were taken expressing various historical periods. Finally, review and compare selected examples from different chronological periods of Islamic architecture. Moreover, the objective behind this is to highlight a set of ideas and architectural elements and vocabulary contained in these examples, which can be adapted nowadays in the design concepts to bear the same moral significance. This paper is clarifying an attempt to highlight the eloquence in depicting philosophical ideas that inspired by the principles of Islam and their relationship to Islamic architecture. The purpose of this study is to analyze and understand these ideas and display their importance to conserve symbolism in architectural heritage. In general, in order to reach a set of recommendations that can be used to enrich contemporary Islamic architecture in particular, hoping to raise the level of the spiritual environment and confirm a distinctive architectural identity.

Keywords

Conservation • Symbolism • Architectural heritage • Philosophical ideas • Islamic architecture

1 Introduction

Symbolism and symbol have been known for long. In Greece, the name "Symbolon" is considered a source of the verb "Symballein" and it means to develop with something that may be different. In Latin, "Symbolum" or French "Symbole" means the symbol by which a person can read something or something that expresses something else, especially something that is distorted, and in the last general connotation, it means to use something to denote something else. The success of the symbol in this case reflects the arrival of the information or expression to the recipient of this information or the person dealing with the symbol, for example: using the crescent to express Islam, or some musical, or mathematical symbols (Kassem, 2011).

However, the special meaning of the word symbolism lies in the spread of an approach or style that reflects or translates the trends and visions of the owners of these works. As a movement of Romantics in art and literature that sought to establish the symbolic mental images, "unrealistic" or as some like to call it unscientific to address others based on their convictions on the need to reverse the trend started in the period of the so-called Renaissance, which is famous for emphasizing the academic curriculum and realistic

K. El-Daghar (✉)
Faculty of Architecture–Design and Built Environment, Beirut Arab University, Tripoli, Lebanon
e-mail: k.eldaghar@bau.edu.lb

A. Versaci et al. (eds.), *Conservation of Architectural Heritage*, Advances in Science, Technology & Innovation,
https://doi.org/10.1007/978-3-030-74482-3_10

expressions "imitator of nature" (Lapunzina, 2005; Thuwainy, 2017).

This definition is similar to what the Ikhwan Al-Safaa explained in Islamic history, where the symbol here means to suggest any indirect expression of hidden psychological aspects that language is not able in its significance positivist to clarify. The symbol is the link between self and things so that emotions are generated by psychological illumination rather than by naming and direct statement. Thus, it can be concluded that the symbol is associated with the meaning. The meaning includes the idea of communication, which is clearly traceable in architecture (Lapunzina, 2005; Kassem, 2011).

2 Symbolism in Architecture

2.1 Meaning and Expression of Symbolism

Symbolism formed a cornerstone in the development of thought and architectural work. It is known that architecture is an expression of human reaction to its surroundings. Architecture is also a means of expression with a variety of symbols and terminology that has multiple meanings. Architect since antiquity knew this and tried to adapt various architectural designs to let them have a specific meaning and identity through designs and symbolic indicators, and it is confirmed by the theorists of architecture. The very well-known Roman architect Vitruvius, the author of the first written theory of architecture, stresses the importance of three fundamental aspects of determining successful architectural work (as shown as Fig. 1), which are: function, durability and aesthetic. "Architecture must be built with due reference to durability, convenience and beauty" (Hodge, 1981; Kassem, 2011; Thuwainy, 2017).

The observation of many contemporary works can deduce the lack of symbolic connotations; it is not easy to distinguish between an architectural work in the East and another that appeared in a different geographical spot. As a consequence of the fact that symbolism in architecture is a result rather than a goal or an aim, there is laconic or jargon for the symbolism of forms in the explanations on arts and aesthetics philosophy, which are contained in architecture via the study of plans and vertical elements. Thus, the explanations of forms received from the ancient architecture need to decipher their talismans through analytical research comparative conscious and realistic (Lapunzina, 2005; Thuwainy, 2017).

2.2 Symbolism and Architectural Heritage

There are three major models of civilizations according to their artistic orientations. First, civilizations of image; including European and west civilization; second, civilizations of symbolism, such as Islam, East Indian and Chinese civilization; and third, civilizations of rhythm, like African civilization that differentiate through the myths and perceptions of the world and is not free of inventive formation, which inspired creative artists, such as Picasso and Giacometti (Osborne, 2019; Thuwainy, 2017).

Vitruvius (70 BC/23 CE), "Roman author, architect, civil engineer and military engineer," stipulated that there should be three origins in architecture, that are: strength and sobriety, practical utility and aesthetic influence, and was concerned about the feasibility of the symbolic dimension through architectural beauty, and the realization that there was sensual beauty and symbolic beauty. Good architectural work is in incorporates foundations that evoke the direct senses of proportions, form, rhythm, pivotal and phenotypic properties (as shown as Figs. 2 and 3), as well as symbolic objects, such as "frankness, ambiguity, decoration, texture and color," which moves in different emotional reservoirs within the human soul (Reiner-Roth, 2015; Hodge, 1981).

Westerners persisted on symbolism in post-Christian and medieval architecture, and it escalated during the Renaissance, especially at Alberti (1404–1472 CE) "Italian Renaissance humanist author, artist, architect, poet, priest, linguist, philosopher and cryptographer." People were passionate about the symbolic dimension in architecture (as shown as Figs. 4 and 5): the balance of plan, the simplicity of decoration and artistic processing, especially in religious buildings (Beniscelli & Furlan, 2005; Grafton, 2003).

For example, the domination of the dome in Byzantine architecture has symbolic and spiritual connotations derived from oriental philosophies in the Levant and Iraq (as shown as Fig. 6) and also also linked the shape of the construction dome with the sky dome (i.e., mixing the element with the sacred); meantime, it is apparently a roofing element for rounded spaces, and the most ancient is found in Jermo civilization (located south of Kirkuk in Iraq) around 5500 BC, perhaps even older from birds that nested in a vaulted form (Beniscelli & Furlan, 2005; Grafton, 2003; Paoli, 2004).

Gothic architecture affirmed the symbolism and prestige of the church in the plot of the urban fabric of the site. Its elements, especially in the facade with its dynamic play of forms, such as rose or wheel windows, colored-glass windows and flying columns (as shown as Figs. 7 and 8), came

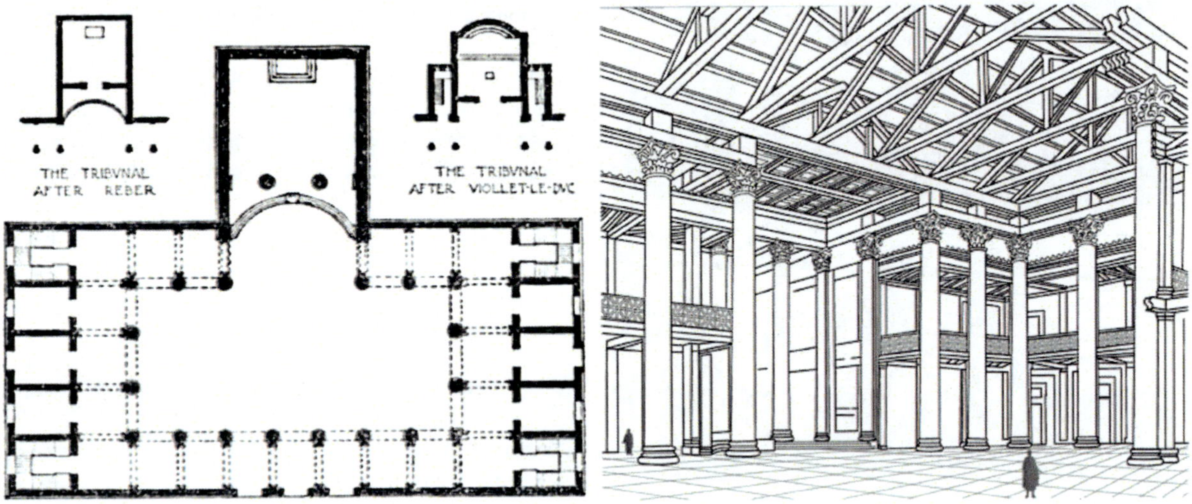

Fig. 1 Basilica in Fano, designed and supervised by Vitruvius (Hodge, 1981)

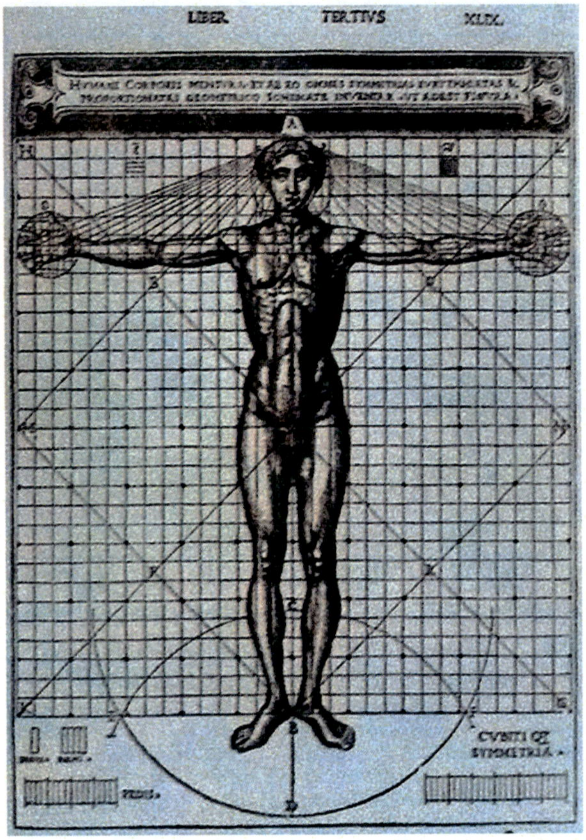

Fig. 2 "Vitruvian Man," illustration in the edition of De architectura by Vitruvius (Hodge, 1981)

Fig. 3 Measurement of height and angle by Vitruvius (Hodge, 1981)

to give a sense of frank awe. Despite the emergence of this style in the Islamic Andalusian architecture, which was not desirable for the Christians of the South and was applied in central and Northern Europe, it carries a high prudence in combining construction, architecture and artistic treatments. Later on, this trend has found its loss in the architectural

Fig. 4 Palazzo Rucellai in Florence, Italy by Leon Battista Alberti 1446–1451 (Beniscelli & Furlan, 2005)

Fig. 5 Façade of Santa Maria Novella, a church in Florence, Italy, by Leon Battista Alberti 1420–1470 (Beniscelli & Furlan, 2005)

product of the Arab and the Islamic countries, which have been searching for symbolic features emanating from their architectural heritage (Lapunzina, 2005; Paoli, 2004).

A group of orientalists wanted to prove that Greek symbolism had influence and dominance over Islamic art and architecture, and that the thought of Euclid, Plato and Aristotle penetrated the geometric shapes and symbolism in the beginning of the mid-eighth century. Also, the Greece built its projects around the seventh century BC, which is about 3000 years late compared to the pioneer architects found in Iraq and Egypt (Lapunzina, 2005; Osborne, 2019).

The Islamic architecture did not drift extreme for the obsession of image and sign, however approached the core of architecture as a feasibility art intended to achieve functional ends, without scrambling to give images and signs that strip them out of prudence and interference with an iconic frame. Nonetheless, the Islamic architecture is plentiful with symbols "numerous meanings and semantics hidden in shapes." Whereas, a sign or image refers to something or a particular impression, the symbol defines and implies the object, but does not fully describe it like the sign or image (Lapunzina, 2005; Thuwainy, 2017).

3 Symbolism in Islamic Architecture

3.1 Principles of Islam and Islamic Architecture

There are various principles of Islam that can be symbolized in Islamic architecture and convey a sense of peace and existence of Allah/Creator. The interaction of illumination done by attaining the benefit of normal light of the day as well as false illumination to generate a particular affection, which moreover do an enormous role as it is up to generate a strong bond with Allah/Creator and arouse tangible practice, where humans are to be expected. The strategic orientation and location as well unite humans to proceed and represent prayers owing to its simple access. Furthermore, the unique included carvings on doors, windows, and partitions, as well as the surrounding water features such as fountains and the existence of a courtyard that serves an environmental function generates a strong bond with Allah/Creator. So, the unifying principles of Islam can be summarized as follows (Vermeulen & Smedet, 1998; Omer, 2009; Taib & Rasdi, 2012):

- Architecture as Tawhid: Unity & Ubiquity of the Creator
- Architecture of Ihtiram: Respect
- Architecture with Ikhlas: Sincerity
- Architecture as Pursuit of 'Elm: Knowledge
- Architecture for Iqtisad: Balance
- Architecture of Haya': Modesty
- Architecture as Dikr: Remembrance.

3.2 Philosophical Ideas Inspired by the Principles of Islam and Islamic Architecture

Architecture as Tawhid: means believing in the unity and ubiquity of Allah/God. Thus, Muslims should have faith in the shahada: "There is no god but God/Allah and Mohammad is his Messenger/Prophet," and the Ka'aba in Mecca is the point where all Muslims face for their prayers. Architecture as Tawhid also signifies the essence of humanity; unified, border less community, transcend race, rank and wealth. Everyone is equal as a human being; this can be seen

Fig. 6 Sixth-century church of Santa Irene in Istanbul, as a superb sample of the domination of the dome in the early Byzantine architecture (Hodge, 1981)

Fig. 7 Piazza Pio II in Pienza, in Italy 1459 (Grafton, 2003)

Fig. 8 Tempio Malatestiano in Rimini, Italy 1447–1453/60 (Grafton, 2003)

through the performance of Hajj. Tawhid also mentions the notion for: place, shape, order and orientation that defined by the unity of the Ka'aba; therefore, the selection for location and orientation of mihrab, niche and qibla wall in mosques worldwide is influenced by the architecture as Tawhid (Hamouche, 2012; Omer, 2009).

Architecture of Ihtiram: translates into the appearance of showing honor, integrity and good manners to Allah/Creator, and toward all his creations (living and nonliving creatures), and also in front of self. As a guideline, Islamists obey the 5 pillars that are: shahada, salat, zakat, fasting and haji that encourage and show respect to faith and God. Islamic artwork can recreate natural patterns in religious messages form. The idea of these patterns is one of creating unity within art, by generating a unity across artworks that shows a connection to a divine creation as well as a unity among people. Although patterns may interweave in and out of each other, and may change throughout an artwork, in the end, they work together to form a unified piece of art as part of a grand design. Therefore, architecture of Ihtiram is mainly seen by symbolism, harmony and the purity of geometric shapes as the Ka'aba. These are set in all holy Islamic architecture throughout the Islamic world. Allah/Creator made the Ka'aba "Sacred House" Cube-shaped, as a sign for humanity. The Cube geometry linked the concept of "centralism"; thus, the concept is the essential of classical architecture (Vermeulen & Smedet, 1998; Omer, 2009).

Architecture with Ikhlas: expression of submissions in the way of Allah/God with outward physical actions of the body from prayer inward action of the heart. A sign of honesty and sincerity is also demonstrated through the art of architecture within the conformity to God's will by removing the existential forms (human, animals, Angels, etc.) to avoid using figurative images in ornamentation. This spiritual ornamentation is introduced by integrating a branch of mathematics concerned with questions of shape, size, relative position of figures, and the properties of space to come out with and apply the geometrical ornamentation. Therefore, architecture with Ikhlas can be defined as the architecture, whereby buildings and other architectural aspects are constructed in a sincere and pure manner respecting some definite rules and regulations in Islam. Thus, Islamic building should follow a certain degree of balance through geometric subtraction or addition and to reflect some characteristics of heaven, such as Gardens (Ahmed, 2014; Omer, 2009).

Architecture as Pursuit of 'Elm: following up with knowledge is detected in the plentiful framework which Islamic architect constructed for their societies to realize and scope Islam instructions. Decoration and patterns acquire a careful and clever master in the geometrical rules. The mosque rendered as a pedagogical study point, where people collect and encounter/commerce information. The knowledge of decorations and patterns has been propagated and exercised in the construction. The knowledge of geometrical regulation, architectural technology and the Islamic calligraphy has been afflicted and as yet being exceedingly applied in the new ages. Therefore, architecture as pursuit of 'Elm can be defined as: architecture in which knowledge is mainly emphasized. Learning and sharing knowledge is the main principle in this architecture where it is attracting scholars from around the world under one roof. Thus, its characteristic can be summarized in: architecture of inscriptions (calligraphy is found in Mosques mostly at the entrance) and architecture of light as expression (illuminating effects of sunrays/moonlight inside Islamic buildings) (Omer, 2009).

Architecture for Iqtisad: the act of achieving balance within spaces; functional and spiritual balance, and by using factors, such as spatial proportions/scale, textures, materials and lighting. It is also an understanding of nature and its portrayal within spaces and so forth, like floral motifs. Thus, architecture for Iqtisad is a building style that reflects the balance, symmetry and harmony in nature through the shape and form of the Islamic buildings. In terms of design, these Islamic principles refer to the rule of the Golden Ratio. It can be found in abundance all around nature that created by Allah/God. The golden ratio when used in design creates a perfect balance, which is naturally pleasing to the eye. As such, the key of the integration of the architecture of "Iqtisad" is to implement the golden ratio when designing a building (Khaki & Abad, 2015; Omer, 2009).

Architecture of Haya': Muslims preserve their dignity by having a sense of modesty. From an architectural perspective, this is best shown by shielding activities happening within from public view. This refers to spaces that offer a sense of privacy whether through screening or separation of spaces; this may also be linked to climatic control where screens are used for shading from sun and hence a sense of coverage. Therefore, architecture for Haya is described as a building style, whereby dignity, decency and modesty are given great importance, and all Islamic buildings should showcase these principles through their architectural design. As preservation of dignity, we have screen, sanctuary and privacy as articulation of Islamic life in space adoption, such as wall of Mosque, Salamlek, Haramlek "woman quarter" and Veil/Hijab as a guarding, covering or screening to gain respect, privacy and identity. Thus, characteristics for architecture of Haya are based around design elements, such as open courtyard and oriel window/mashrabiya which not only avail an ecological aim but also directs the several positions and duties for men and women where it permits the rightness "to see" and not to "be seen" (Omer, 2009).

Architecture as Dikr: remembrance within architecture may be represented through the harmony of monotonous spaces repetition of various features that offer rhythmic precision. This factor signifies a sense of "chanting" like a continuous representation of God's innumerable attributes. This can be seen through the use of rows of arch's, which is a technique that quite normally seen within Islamic architecture. Therefore, architecture for Dikr can be defined as architecture of monotony which concentrates mainly on the remembrance of Allah/God through rhythmic precision of various Islamic motifs (pillar, arcade, etc.), carvings and calligraphies. Most of the Arabic calligraphy is placed at the entrance of Mosque to remind Muslims of their real master. In addition to the repetitive use of ornaments, pillars, arches for the remembrance of Allah/God, however, most of the Islamic buildings are usually covered with repeated geometric shape ornaments that include nature forms. The rhythmic regulation reflected in the meditative intonation of Creator's untold characteristics / Dikr, and the exhibit of observation preserves a receptor of combination by harmony and is visible as the everlasting and boundless nature of Creator's core (Ahmed, 2014; Omer, 2009). "It has a portico of surpassing beauty; like the soul of a believer, immersed in the remembrance of God."

4 Methodology

Symbolism in architecture takes a different approach from other arts. The shape of the building is not shaped by the desire to express meanings; it needs a favorable benefit and it engages with functional, environmental and economic influences. It is not a symbol of emotion, as much as an expression of its truth. Need and feasibility have become a symbol, and they include the reasons and principles that found its course by architectural forms and possible interpretation according to layers of consciousness. The study will handle the great importance to the idea of symbolism and distinctive vocabulary in architectural heritage. The research depends on a systematic methodology via reviewing the meaning of symbol and symbolism in architecture, which is: meaning and expression of symbolism, and symbolism and architectural heritage. Moreover, it will demonstrate symbolism in Islamic architecture, the principles of Islam and Islamic architecture, the philosophical ideas inspired by the principles of Islam, and its relation with Islamic architecture. In addition to these, the case study for conserving symbolism in architectural heritage will explore and analyze the eloquence in depicting those philosophical ideas on Islamic architecture through its different chronological periods in order to achieve special guidelines and recommendations as bases and rules to conserve symbolism in architectural heritage in general. Furthermore, to enhance modern Islamic architecture in particular in order to raise the level of the spiritualistic atmosphere and stress this distinct architectural character.

5 Case Study Conserving Symbolism in Architectural Heritage

5.1 Islamic Architecture Through Ages

Islamic architecture can be defined as building traditions of Muslim in the Middle East and any countries, where Islam has been dominant from the seventh century. Thus, it is not concerned about the form of buildings only; however, it signifies a process where all the phases and aspects are equally important, which is almost impossible to identify a phase or an aspect in that process and consider it more important than the others. Islamic architecture process starts with having a proper understanding and vision. Then, it continues with planning, designing and constructing stages and ends with how people make use and benefit from them. Thence, it is a fine blend of all these factors, which are interwoven with the treads of principles, teachings and values of Islam (Bloom & Blair, 2009; Petersen, 2002).

Islamic architecture is often referred that it is stemmed from Christian, Persian and Indian roots. More specifically, the architectural forms and styles came largely from the Byzantine and Sasanian Empires. It includes religious buildings intended for worship by Muslims (Mosque, Mausoleum, Palace, Madrassa, etc.), and secular buildings built in a predominately Islamic region or used by Muslims (public baths, fountains, domestic architecture, etc.). Also, it indicates to specific architectural features (domes, arches, mihrab, minarets, courtyard, etc.), elements (decorative shapes, geometric pattern, calligraphy, etc.) and values "character" (balance and symmetry, formal landscape, centered upon God, etc.). Interestingly, as Muslims conquered different regions once controlled by other cultures, their early buildings were either renovations of existing buildings or new buildings constructed from the ruins of older buildings. In general, Islamic buildings have limited architectural expressions on the exterior; instead, the focus is on the interior, where most of the decorative elements are inside of the building (Bloom & Blair, 2009; Huq & Haque, 2017; Omer, 2009).

5.2 Eloquence in Depicting Philosophical Ideas Inspired by Principles of Islam on Islamic Architecture

Islamic architecture is diverse, because of the large geographic and chronological scope of Islam. It has been a major religion and cultural force for over fourteen centuries and continues till our days. Islamic architecture is organized into three chronological periods: Early, Medieval and Late, which helps to understand how architecture works relate to each other in time and space. There were dynasties and empires whose periods of ruling stretched across these chronological divisions: Early (Umayyad, Abbasid, Tulunid and Aghlabid), Medieval (Fatimid, Seljuk, Ayyubid and Mamluk) and Late (Ottoman) (Macaulay-Lewis, 2011; Omer, 2009).

First: Umayyad period

Umayyad architecture, for instance as an early chronological period, developed in the Umayyad Caliphate between 661 and 750 A.D, primarily in its heartlands of Syria and Palestine. It drew extensively on the architecture of other Middle Eastern civilizations and the Byzantine Empire, but also it introduced innovations in decoration and new types of building (Omer, 2009; Talgam, 2004). The following example will illustrate and analyze the eloquence in depicting philosophical ideas that are inspired by the principles of Islam on Islamic architecture, through Umayyad period represented in Dome of the Rock (as shown as Table 1), which was built between 687 and 692 A.D by "Abdel Malek Eben Marwan":

Second: Mamluk period

Mamluk architecture, for instance as a medieval chronological period, was a breathtaking flowering of Islamic art during the reign of the Mamluk Sultanate (1250–1517 A.D),

Table 1 Eloquence in depicting philosophical ideas through Umayyad period represented in Dome of the Rock (Author)

<table>
<tr><td>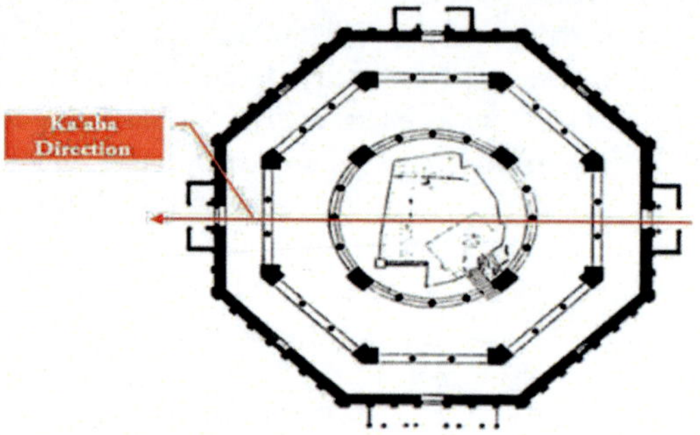
</td><td>
</td></tr>
<tr><td colspan="2">Architecture as Tawhid: uniting all mosques throughout the world, their entire mihrab and semicircular niche set into the middle of "qibla wall" in order to indicate a single direction of Mecca (Ka'aba)</td></tr>
<tr><td>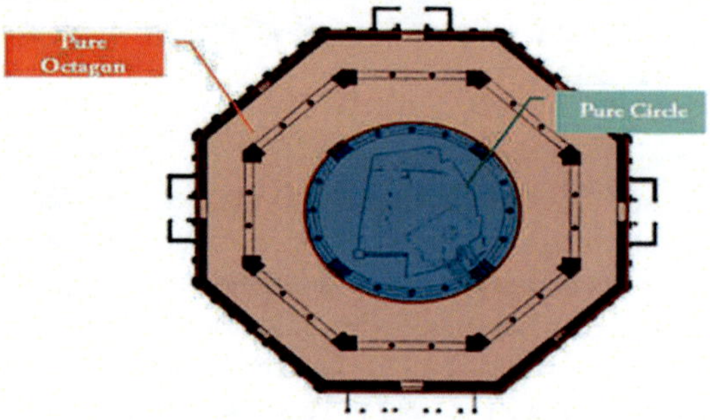
</td><td>
</td></tr>
<tr><td colspan="2">Architecture of Ihtiram: the design concept of harmony and purity for geometric shapes, which contains natural geometrical patterns to create unity within art showing a connection to divine creation and respect to Allah/God</td></tr>
<tr><td>
</td><td>
</td></tr>
<tr><td colspan="2">Architecture with Ikhlas: integration of mathematics with ornamentation, where the geometric proportions embodied through subtraction or addition and balance or harmony of shapes, with the presence of greenery garden to symbolize heaven</td></tr>
</table>

(continued)

Table 1 (continued)

Architecture as Pursuit of 'Elm: calligraphy is a sacred form of visual expression portraying the message of Quran. Also, praying hall with its minber facilitate education, where people gather and receive/trade knowledge

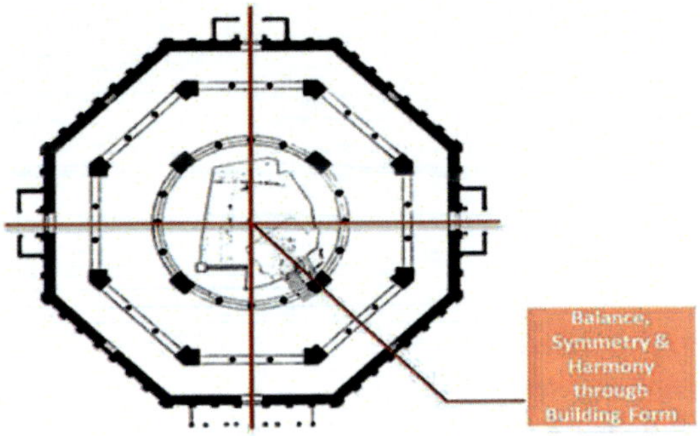

Architecture for Iqtisad: incorporate balance, symmetry and harmony in nature within space, and through the building form using spatial proportions/scale, textures, materials and lighting

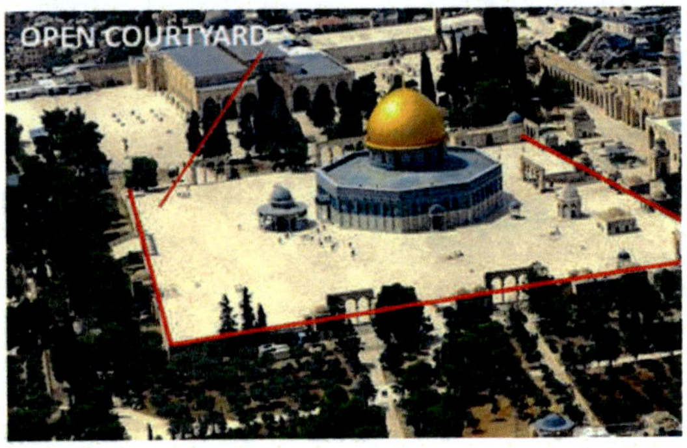

Architecture of Haya': the mashrabiya or partitions in praying hall serve as Veil/Hijab among men and women; also, the open courtyard represents modesty and provides a filter and privacy inside and outside the building

(continued)

Table 1 (continued)

Architecture as Dikr: repetitive of ornaments, pillars and arches that are used as remembrance of Allah/God where most of the buildings are covered with repeated calligraphy and geometric shape ornaments that includes nature forms

which is most visible in old Cairo. Religious zeal made them generous patrons of architecture and art. Trade and agriculture flourished under Mamluk rule where Cairo, their capital, became one of the wealthiest cities in the Near East and the center of artistic and intellectual activity. This made Cairo, in the words of Ibn Khaldun, "the center of the universe and the garden of the world," with majestic domes, courtyards and soaring minarets spread across the city (Ismail, 2010; Omer, 2009). The following example will illustrate and analyze the eloquence in depicting philosophical ideas that are inspired by the principles of Islam on Islamic architecture, through Mamluk period represented in Mosque, Madrassa and Mausoleum of Sultan Hassan (as shown as Table 2), which was built between 1356 and 1363 A.D by "Hassan Al Nasser Mohamed Ibn Qalawun".

Third: Ottoman period

Ottoman architecture, for instance as a late chronological period, is the architecture of the Ottoman Empire, which emerged in Bursa and Edirne in fourteenth and fifteenth centuries. The architecture of the empire developed from the earlier Seljuk architecture and was influenced by the Byzantine architecture, Iranian as well as Islamic Mamluk traditions after the conquest of Constantinople by the Ottomans. The Ottomans achieved the highest-level architecture in their lands since that period. They mastered the technique of building vast inner spaces, confined by seemingly weightless yet massive domes, and achieving perfect harmony between inner and outer spaces, through a dynamic architectural vocabulary of vaults, domes, semi domes and

Table 2 Eloquence in depicting philosophical ideas through Mamluk period represented in mosque, Madrassa and Mausoleum of Sultan Hassan (Author)

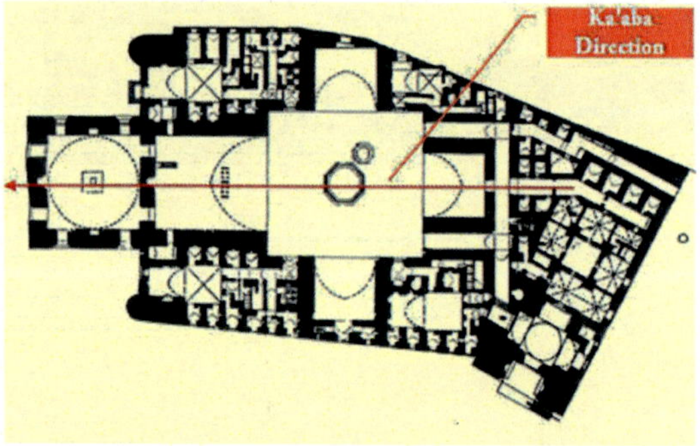

Architecture as Tawhid: uniting all mosques throughout the world, their entire mihrab and semicircular niche set into the middle of "qibla wall" in order to indicate a single direction of Mecca (Ka'aba)

(continued)

Table 2 (continued)

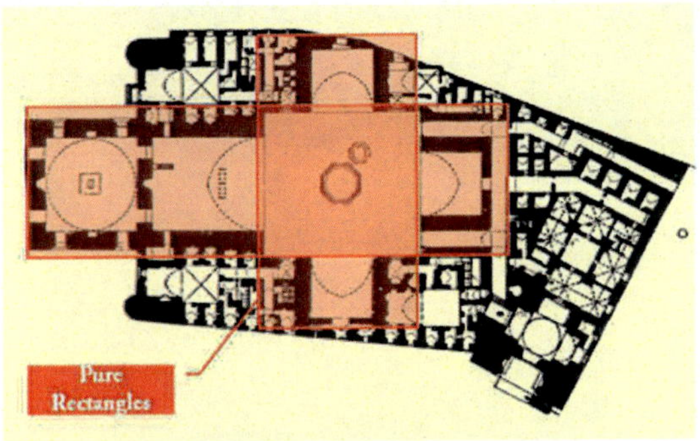	
Architecture of Ihtiram: the design concept of harmony and purity for geometric shapes, which contains natural geometrical patterns to create unity within art showing a connection to divine creation and respect to Allah/God	
	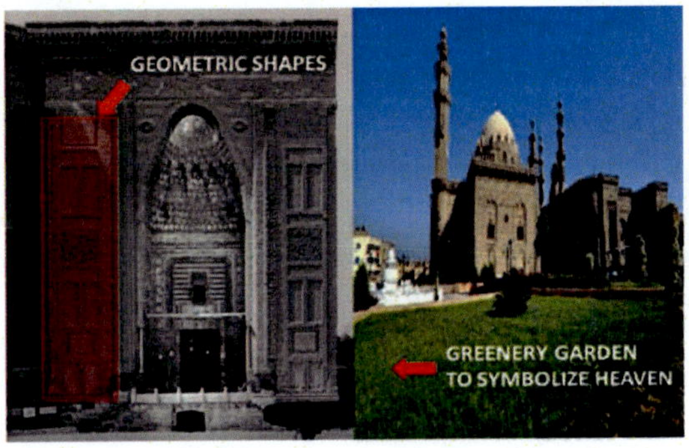
Architecture with Ikhlas: integration of mathematics in ornamentation, where the geometric proportions embodied through subtraction or addition and balance or harmony of shapes, with the presence of greenery garden to symbolize heaven	
	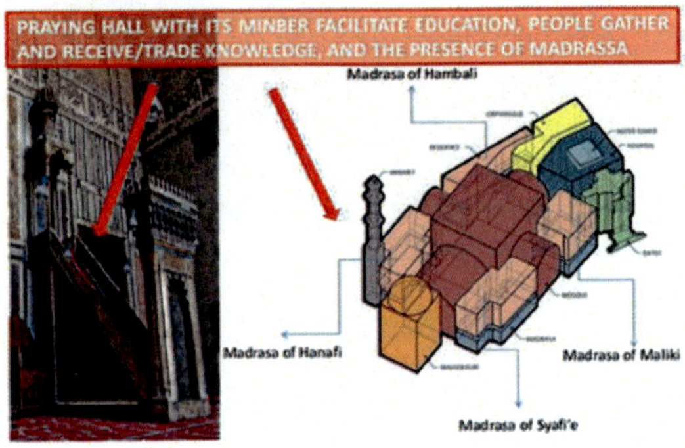
Architecture as Pursuit of 'Elm: calligraphy is a sacred form of visual expression portraying the message of Quran. Also, praying hall with its minber facilitate education, where people collect and extradite/commerce information, in addition to presence of Madrassa	

(continued)

Table 2 (continued)

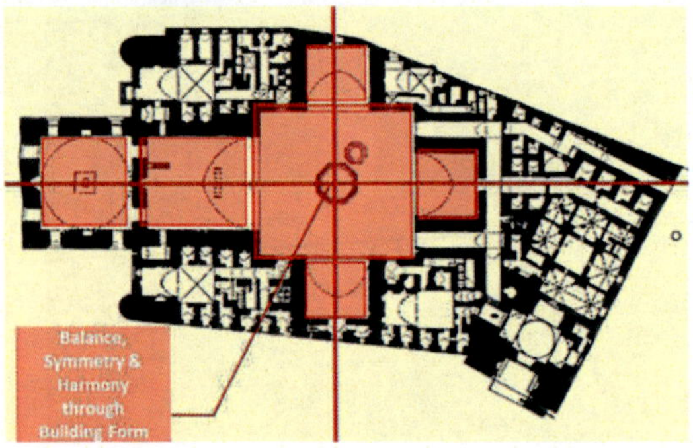

Architecture for Iqtisad: incorporate balance, symmetry and harmony in nature within space, and through the building form using spatial proportions/scale, textures, materials and lighting

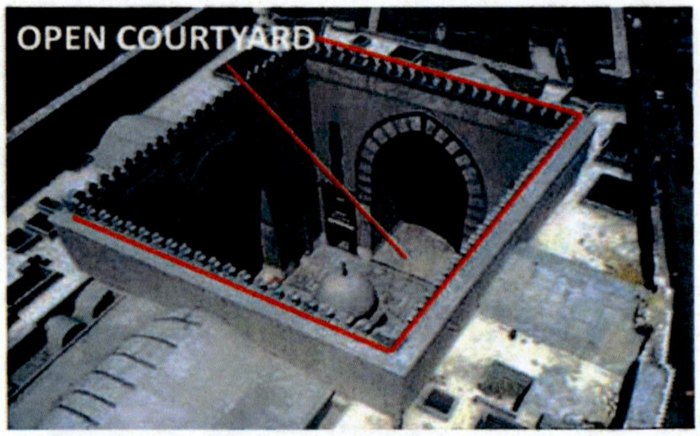

Architecture of Haya': the mashrabiya or partitions in praying hall serve as Veil/Hijab among men and women; also, the open courtyard represents modesty and provides a filter and privacy inside and outside the building

Architecture as Dikr: repetitive of ornaments, pillars and arches that are used as remembrance of Allah/God where most of the buildings are covered with repeated calligraphy and geometric shape ornaments that includes nature forms

columns (Freely, 2011; Omer, 2009; Saoud, 2004). The following example will illustrate and analyze the eloquence in depicting philosophical ideas that are inspired by the principles of Islam on Islamic architecture through Ottoman period represented in Sultan Ahmed Mosque or Blue Mosque (as shown as Table 3), which was built between 1609 and 1616 A.D by "Ahmed I".

Table 3 Eloquence in depicting philosophical ideas through Ottoman period represented in Sultan Ahmed Mosque or Blue Mosque (author)

<table>
<tr><td>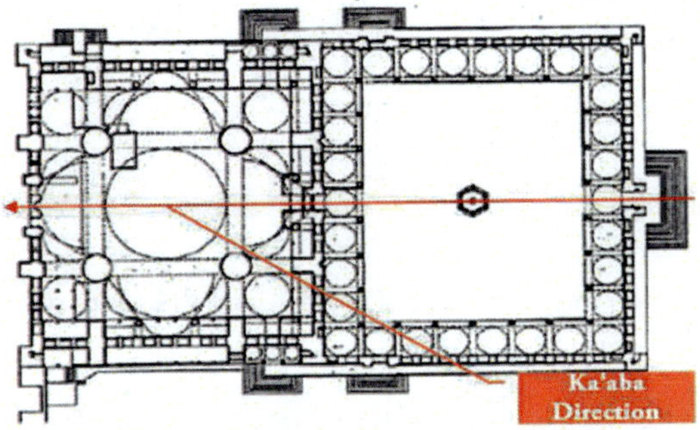
</td><td>
</td></tr>
<tr><td colspan="2">Architecture as Tawhid: uniting all mosques throughout the world, their entire mihrab and semicircular niche set into the middle of "qibla wall" in order to indicate a single direction of Mecca (Ka'aba)</td></tr>
<tr><td>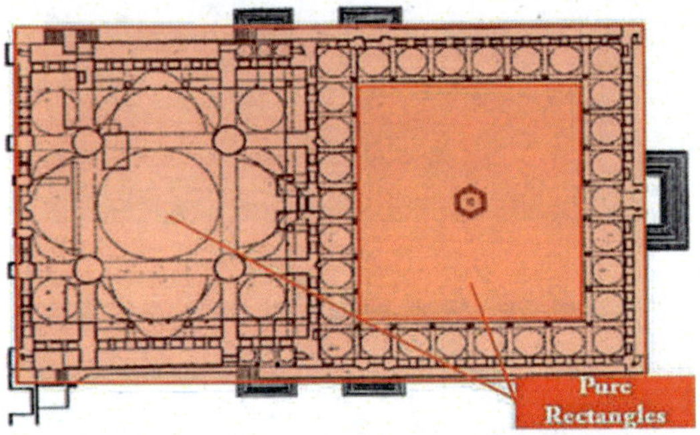
</td><td>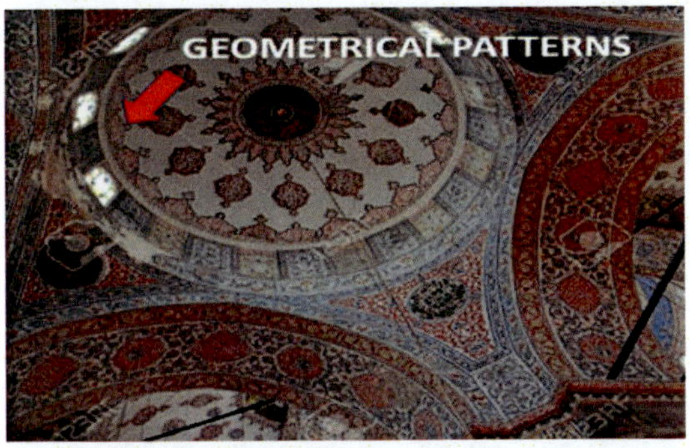
</td></tr>
<tr><td colspan="2">Architecture of Ihtiram: the design concept of harmony and purity for geometric shapes, which contains natural geometrical patterns to create unity within art showing a connection to divine creation and respect to Allah/God</td></tr>
<tr><td>
</td><td>
</td></tr>
<tr><td colspan="2">Architecture with Ikhlas: integration of mathematics in ornamentation, where the geometric proportions embodied through subtraction or addition and balance or harmony of shapes, with the presence of greenery garden to symbolize heaven</td></tr>
</table>

(continued)

Table 3 (continued)

Architecture as Pursuit of 'Elm: calligraphy is a sacred form of visual expression portraying the message of Quran. Also, praying hall with its minber facilitate education, where people gather and receive/trade knowledge

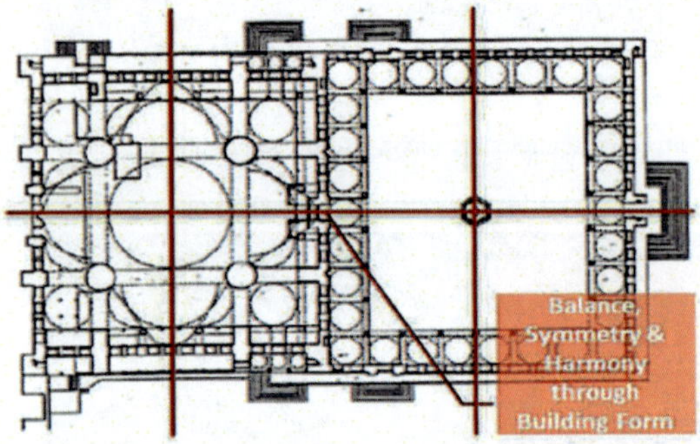

Architecture for Iqtisad: incorporate balance, symmetry and harmony in nature within space, and through the building form using spatial proportions/scale, textures, materials and lighting

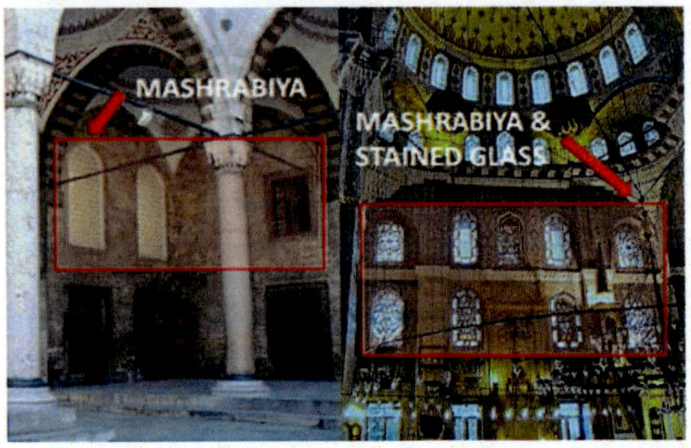

Architecture of Haya': the mashrabiya or partitions in praying hall serve as Veil/Hijab among men and women; also, the open courtyard represents modesty and provides a filter and privacy inside and outside the building

(continued)

Table 3 (continued)

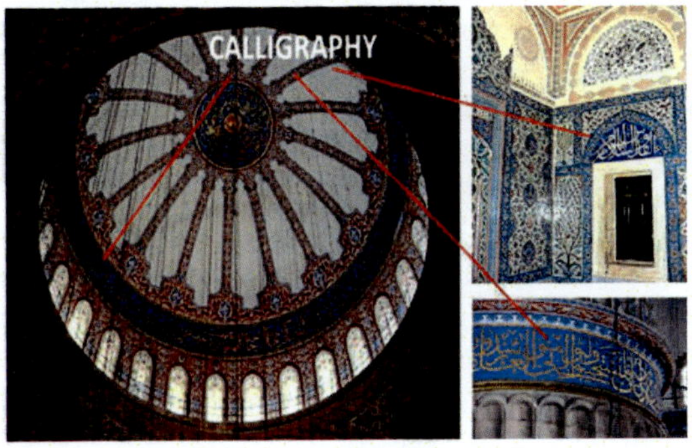

Architecture as Dikr: repetitive of ornaments, pillars and arches that are used as remembrance of Allah/God, where most of the buildings are covered with repeated calligraphy and geometric shape ornaments that include nature forms

6 Conclusion

Symbolism has a social role; it is a means of communication between human beings, where the reflection of job, social status and culture in a person's home are symbols that are used to send information to others. Therefore, architecture has been subjugated to what these symbols can carry in providing many of needs and confirming identity.

Cities variation, especially traditional ones, resulted from the different symbols embraced by these communities. The building with its outer envelope and interior spaces is an important way to satisfy the human need for communicating and transferring information to others.

There is still a difficulty in understanding the importance of symbolism. Thence, appreciating its effective use to enrich the living environment where in many newly designed buildings, "strange" elements appear on the facades and interiors without a regard to the building user, perimeter or even meaning.

There is a need to strive toward proposing environmentally friendly designs with distinctive identity that use legible architectural vocabulary, to ensure building with expressive symbolic contents that is connected to the surroundings.

It is difficult to see architectural heritage as a reflection of an individual thought, while design approach is a collective social output, which could be seen in the richness of the architectural values of local/historical cities.

The philosophical ideas inspired by principles of Islam are the fundamental characteristics that make the Islamic architecture similar to one another. However, the basis of what makes the buildings unique and diverse between them is the way the styling responds to the local context, where the design language from a culture is adopted into the architecture.

Furthermore, the following guidelines can contribute to conserve symbolism in architectural heritage, through using features, elements and values of Islamic buildings, to preserve the eloquence in depicting these philosophical ideas, which could stimulate the rebirth of an authentic Islamic architecture nowadays and at all the degrees of the Islamic existence, as following:

1. Architecture as Tawhid: the principle of unity of Allah/God could be portraying by using a large open courtyard as the gathering point of all those who are coming from the surroundings for praying or uniting all mosques worldwide to indicate a single direction of Mecca (Ka'aba) by using the mihrab, niche and qibla.
2. Architecture of Ihtiram: harmony and purity of geometric shapes or wall detail that contains natural geometrical patterns creating unity and representing a symbolic meaning of respect to Allah/God.
3. Architecture with Ikhlas: the absence of figurative images and statues in buildings to prevent idolatrous worship, where the ornamentation consists only verses of Quran and geometric shapes that embodied through subtraction/addition and balance/harmony, or portraying heaven by using greenery garden.
4. Architecture as Pursuit of 'Elm: a wisdom word of Quran inscribed by precise and sacred calligraphy, which can be found on the entrances and interiors, as well as the presence of Minber in praying hall, or Madrassa where people gather and receive knowledge.
5. Architecture for Iqtisad: balance and symmetry could be achieved through ornamentation or geometric forms, as Islamic buildings are proportionate, harmonious and moderate.
6. Architecture of Haya': mashrabiya maintains the respect of Islamic women. She gives the right "to see" or not to

"be seen." The women within the structure has the independence perform their actions within the feeling of particularity or modesty that could be portrayed through open courtyard that provides privacy between the building and the outside.

7. Architecture as Dikr: duplicate elements to invoke humankind of Allah/Creator. As elements are invoked: calligraphy and geometric shape, the reminder is invoked. The use of ornaments, pillars, arches and many more is often repeated in Muslim's buildings, which portraying a symbolic meaning of remembrance.

Acknowledgements My profuse thanks and gratefulness go to all my colleagues at Beirut Arab University, specially Prof. Mohamed Assem Hanafi and Ms. Maya Safar for their support, in addition to, all my BAU third-year students (class 2018–2019) for their participation and continuous hard work.

References

Ahmed, S. (2014). The spiritual search of art over islamic architecture with non-figurative representations. *Journal of Islamic Architecture, 3*(1), 1–13. Retrieved from https://www.academia.edu/8365972/THE_SPIRITUAL_SEARCH_OF_ART_OVER_ISLAMIC_ARCHITECTURE_WITH_NON-FIGURATIVE_REPRESENTATIONS. https://doi.org/10.18860/jia.v3i1.2538.

Beniscelli, A., & Furlan, F. (2005). Leon Battista Alberti (1404–72): Tra Scienze e Lettere. *Accademia Ligure di Scienze e Lettere* (pp. 287–308). Genève: Switzerland.

Bloom, J. M., & Blair, S. S. (2009). The grove encyclopedia of islamic art and architecture. *Oxford University Press* (ISBN 9780195309911, 9780195373042). https://doi.org/10.1093/acref/9780195309911.001.0001.

Freely, J. (2011). A history of Ottoman architecture. In Boston:WIT Press (ISBN 9781845645069). Retrieved from https://www.scribd.com/document/350434810/J-Freely-A-History-of-Ottoman-Architecture-pdf.

Grafton, A. (2003). Leon Battista Alberti: Un Genio Universale. In *Laterza,* Roma/Bari: Italy (ISBN 10: 8842069108/ISBN 13: 9788842069102).

Hamouche, M. B. (2012). Islamization of contemporary architecture: shifting the paradigm of islamic architecture. *Journal of Islamic Architecture, 1*(2), 70–88. Retrieved from https://pdfs.semanticscholar.org/61e7/607456c380276fec0824b76c75062d94ab29.pdf?.

Hodge, A. T. (1981). Vitruvius, lead pipes and lead poisoning. *American Journal of Archaeology, Archaeological Institute of America, 85*(4), 486–491. Retrieved from https://www.jstor.org/stable/504874. https://doi.org/10.2307/504874

Huq, M.U., & Haque, A. (2017). P-0263 "Does islamic architecture reflect islam (faith) ?". In *UIA 2017 World architects congress, forum 3*. Seoul: Korea. Retrieved from https://pdfs.semanticscholar.org/8f3d/f17c384e9a5b8cceac8b210751345068f8ec.pdf?_ga=2.216932243.405452782.1576503696-541634702.1576503696.

Ismail, K. (2010). Sultan Hassan mosque: An islamic architectural wonder analytical study of design and its effect on islamic Cairo. *Journal of Islamic Architecture, 1*(2), 94–105. Retrieved from https://pdfs.semanticscholar.org/2598/435227246d15f856f4eaeea7b2c7ba7fd877.pdf?.

Kassem, R. (2011). Symbolism and architecture. In *Inheritance blog*. Tripoli: Libya. Retrieved from http://mirathlibya.blogspot.com/2011/02/blog-post_13.html.

Khaki, A., & Abad, A. S. H. (2015). The essence of islamic architecture of mosques: a reflection on the spiritual identity and sacred suppositions. *Cumhuriyet University, Faculty of Science, Science Journal (CSJ), 36*(4), 2314–2321. (Special Issue, ISSN 1300–1949; 5000138834–5000231993–1-PB). Retrieved from https://www.researchgate.net/publication/308888082_The_Essence_of_Islamic_Architecture_of_Mosques_a_Reflection_on_the_Spiritual_Identity_and_Sacred_Suppositions.

Lapunzina, A. (2005). Architecture of Spain; reference guides to national architecture. In London: Greenwood Publishing Group (ISSN 1550–8315–ISBN 0313319634).

Macaulay-Lewis, E. (2011). About chronological periods in the islamic world. In *Khan academy*. CA. Retrieved from https://www.khanacademy.org/humanities/ap-art-history/cultures-religions-ap-arthistory/a/about-chronological-periods-in-the-islamic-world.

Omer, S. (2009). Islamic architecture: Its philosophy, spiritual significance, and some early developments. In *A.S. Noordeen baltimore MD*. USA (ISBN: 9789830652863).

Osborne, R. (2019). *Renaissance colour symbolism*. London: Thylesius Books (ISBN 9780244454760).

Paoli, M. (2004). Leon Battista Alberti (1404–1472). In *Éditions de l'Imprimeur*. Paris/Besançon: France.

Petersen, A. (2002). Dictionary of islamic architecture. In *Routledge*. London. (eBook ISBN 9780203203873). https://doi.org/10.4324/9780203203873.

Reiner-Roth, S. (2015). Vitruvius architecture's golden rules: The vitruvian triad. In *Kukun home intelligence magazine, Kukun academy*. U.K. Retrieved from https://blog.mykukun.com/vitruvious-architecture-golden-rules-vitruvian-triad/.

Saoud, R. (2004). Muslim architecture under Ottoman patronage (1326–1924). In *Foundation for science, technology and civilisation; FSTC Limited* (vol. 4064, pp. 2–26). Retrieved from https://pdfs.semanticscholar.org/5dee/81f841b9ca496459273e897f30c02e3ce01f.pdf?.

Taib, M. Z. M., & Rasdi, M. T. (2012). Islamic architecture evolution: Perception and behaviour. *Procedia—Social and Behavioral Sciences, 49*, 293–303. https://doi.org/10.1016/j.sbspro.2012.07.027.

Talgam, R. (2004). The stylistic origins of Umayyad sculpture and architectural decoration, part 1. Wiesbaden, Germany: Otto Harrassowitz Verlag (ISBN 3447047380, 9783447047388).

Thuwainy, A. (2017). Symbolism in architecture between the means and the end. In *Al roya journal of Oman—al roya, Muscat*. Oman. Retrieved from https://alroya.om/post/201563.

Vermeulen, U., & Smedet, D. (1998). Philosophy and Arts in the Islamic World: In *Proceedings of the Eighteenth Congress of the Union Européenne des Arabisants Et Islamisants*. Uitgeverij Peeters. (ISBN 9068319779).

Cultural and Architectural Heritage Values of the Qift-Quseir "Myos Hormos" Road

Ahmed Ibrahim Othman

Abstract

Wadi Hammamat enjoyed valuable architectural and cultural heritage potentials that can be promoted through the sustainable tourism. Stone quarrying and gold mining were the main social and economic activities at Wadi Hammamat (Quseir-Qift road), known also as Myos Hormos, during the course of time from the pre-dynastic period until the twentieth century. These activities envisaged both identity and heritage values through materials, tools, inscriptions and settlements of the quarry society. Moreover, the site represents a comprehensive image of the Egyptian Eastern Desert civilization. The study investigates both architectural and cultural heritage aspects of Myos Hormos in terms of

- Pharaonic and Greco-Roman traces.
- Architectural elements.
- Transport and movements.
- Religious beliefs.

The paper includes a project that targets promoting the surveyed area as a new cultural tourism destination through the virtual reconstruction. Some sites can no longer be visited; therefore, a facsimile should virtually replace the reality through scenic reconstructions, visual interfaces and holograms in a virtual museum.

Keywords

Eastern Desert • Hydreumata • Water stations-desert roads • Myos Hormos • Wadi Hammamat • Greywacke • Bir Fawakhir • Quarries • Quarry sites • Mines • Sustainable tourism • Archaeological park • Sustainable tourism development

A. I. Othman (✉)
Hotels and Restoration, Abu Qir, Alexandria, Egypt

Dipartimento Dei Beni Culturali: Archeologia, storia dell'arte, del cinema e della musica, Università Degli Studi di Padova, Padua, Italy

1 Introduction

The Eastern Desert was a reservoir of natural resources such as the greywacke of Wadi Hammamat, a stone that was used in producing bowls, palettes, statues and sarcophagi. Moreover, other raw materials came from the Eastern Desert such as lead, copper, galena and gold from Bir Umm Fawakhir (Klemm et al., 2002; Harrell, 2002).

Qift-Quseir road, known also as "Myos Hormos", was an important road, equipped with watchtowers "*Skopeloi*" and watering stations "*hydreumata*" that once sheltered the trade convoys going to and from the Red Sea.

These *hydreumata* with their additional constructional elements indicated that the modern Qift-Quseir road witnessed the traffic of pack-animal caravans and heavy-wheeled transport (Zitterkopf & Sidebotham, 1989). Petroglyphs along the surveyed road were used as route maps leading to the quarries and mines. These inscriptions included references to "gold" as a raw material, in addition to the names, titles and epithets of officials in charge of gold mining and stone quarrying expeditions (Harrell, 2002; Klemm et al., 2002; Morrow et al., 2002).

Inscriptions representing cargo transport, trade activities and imported forest animals were previously concentrated along the road leading to Berenice, and then they moved to the north towards Myos Hormos perhaps due to the great revolt of Thebes during the Roman Period (Casson, 1993; Redon, 2018).

Excavations at Quseir brought to light archaeological findings that date back to the I century BC (Whitcomb, & Johnson, 1979, 1982), Ptolemaic material (Weigall, 1913) and Claudian inscriptions from Laqueita (Bernard, 1972).

A. Versaci et al. (eds.), *Conservation of Architectural Heritage*, Advances in Science, Technology & Innovation,
https://doi.org/10.1007/978-3-030-74482-3_11

2 Wadi Hammamat

Wadi Hammamat was a famous trade road, a rich source of greywacke, gold and granite in addition to the easy access to the subterranean water (Arkell, 1975; Debono, 1951; Vermeersch et al., 1989).

Greywacke quarries (25° 59′ 25″ N; 33° 34′ 05″ E) (Fig. 1) are examples of a recently approached discipline that deals with the desert life and the archaeology of mobility (Riemer & Förster, 2013).

Quarry sites, such as greywacke quarries of Wadi Hammamat, were social, technological and cultural experiments associated with stone extracting activities. Therefore, the importance of documenting the value of ancient quarries and the transformation of landscape over time (Bloxam, 2005: 23–27; Bloxam et al., 2014). The term "quarry complex" refers to the more and less visible materials scattered in different layers that might belong to each other in time, space and function (Abu-Jaber et al., 2009; Bloxam & Heldal, 2008; Fairclough et al., 2008; Mason, 2008).

Quarries share some common features such as

- Stone heaps.
- Production remains (heaps and instruments).
- Life traces such as houses, petroglyphs and wares.
- Logistics and infrastructure (Bloxam & Heldal, 2008; Judd, 2003).

Wadi Hammamat represents a mixture of these elements extending on about 10 km^2 and indicating a long history since the Early Bronze age until The Roman Period (Bloxam, 2015).

Qift-Quseir road was prepared and equipped to facilitate mining and quarrying processes and the consequent transport. It also linked the Nile Valley with the Red Sea ports for commercial purposes with the Horn of Africa and Arabia that required an outstanding road network and infrastructure maintenance ready for the traffic intensity (Shaw, 2013).

3 Cultural Heritage Aspects

Ancient road heritage values can be examined through manuscripts, foreign relics, exchange of languages, religious beliefs, technologies, in addition to the archaeological remains of trading centres and depots, caravanserais, check points along the more important road lines, the ancient tracks and the barren landscapes (Friedman, 2002; Riemer & Förster, 2013).

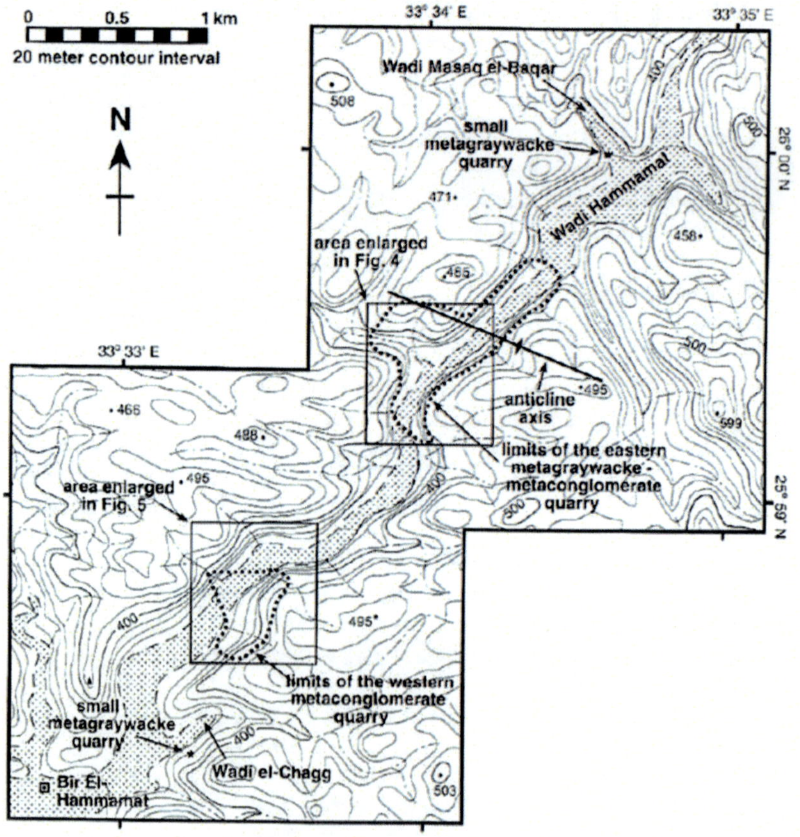

Fig. 1 Greywacke quarries of Wadi Hammamat. After, Harrell et al. (2002, 208, Fig. 2)

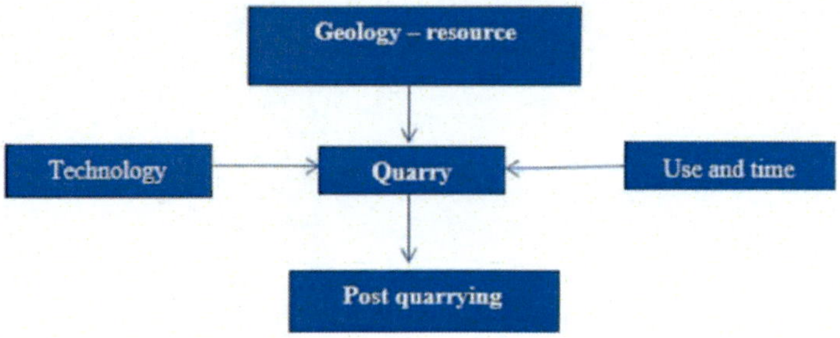

Fig. 2 Characteristic elements of the quarry site complex. After, Tom & Gurli (2015)

Physical remains of ancient stone quarries, despite loss of materials, are represented in paths, houses, ports, petroglyphs and quarrying materials and instruments (Fig. 2).

Quarry sites are the result of stone extraction over time. Therefore, these layers and findings should be identified as a baseline to assign values emphasized through the relationship with relics and landscape features (Bloxam, 2008; Cleere, 1996; Ingold, 1993; Mason, 2008; Thomas, 2001).

3.1 Transport and Movement

Trade convoys travelled between Qift and Quseir till the inauguration of "Canal of Trajan" around 112 AD.

Along this road:

- Wagons, conductors and camels were used by the military units for supply purposes.
- Donkey convoyed provisions and private transport.
- Horsebacks transported more official correspondences.

These means of transportation were used for commercial purposes (fish and vegetables brought by merchants, traders and peddlers), military purposes and also private individual movements exploiting any free capacity (Bülow-Jacobsen, 2006; Lankester, 2012).

3.2 Communications

Communications between the *hydreumata* occurred in two different means.

3.2.1 Official Correspondences

Private letters and exchange of information, payments and light goods, such as mullet and parrot-fish, from "Persou-Bir Fawakhir", up to 20 kg on a donkey back between the people of the *hydreumata* or from the coast to Qift, helped by the post-riders and horsemen who used to calculate the carrying capacity for both the delivery/return journey. These riders escorted important people and military affairs as the road was sometimes dangerous due to the presence of Bedouin gangs (Bulow-Jacobsen 2013; Cuvigny, 2005).

3.3 Watch Towers "*Skopeloi*"

Romans constructed about 65 roughly squared watch towers along the surveyed road, on the slopes of the mountains, in visual contact with each other in order to safeguard the trade caravans, the gold mines and the stone quarries. However, no paths or ceramic findings were discovered to prove the function of these constructions.

These towers were referred to as "*Skopeloi*", and they are square in shape, with maximum 2–2.5 m height to permit controlling the road (Fig. 3). The visibility facilitated messaging through optical telegraph, mirrors and flags. They were constructed of stones collected from the neighbourhood and stacked without mortar (granites, schists and metabasites) (Riemer & Förster, 2013; Sidebotham et al., 2008).

These watch towers were probably used to

- Alert the nearby *hydreumata* to prepare either reception or defence as soldiers should have been organized in rotation only when a signal had to be sent (Bagnall, 1982).
- Inform Qift with the commercial convoys coming from the Red Sea that could be suggested by the inter-visibility of the towers.

By the Arab period, the road might have been used for pilgrimage to Mecca, and these watch towers provided no traces of this period (Bülow-Jacobsen, 2013; Bernand, 1972).

3.4 Water Stations (*Hydreumata-Praesidia*)

Nine square *praesidia,* similar in design and construction style, were distributed along the Qift-Quseir road "from the Red Sea towards the Nile: Bir Nakhil, Dawwi, Bir Sayyala, Bir Al-Hamra, Al-Zarqa (Maximianon(, Bir Umm Fawakhir,

Fig. 3 *Skopeloi* of the Myos Hormos Road. Left. Photo by the Author. Right. After, Sidebotham (2008, 92, Fig. 4.17, Pl. 4.18–4.19)

Bir Hammamat, AL-Muwayh (Krokodilo), Qusur al-Banat" (Fig. 4). They are small, roughly squared constructions with a main entrance, surrounded by enclosures reinforced with projecting outer corners and gate towers. The inner space usually included a well. They were nearly 24 h distant with about 15–20 km from each other (Cuvigny, 2005; Harrel and Brown 1992). These constructions are simple and "self-contained" entities where travellers were hosted together with their animals as archaeological surveys referred to numerous elements that supported a large animal traffic (Zitterkopf & Sidebotham, 1989).

By the II century A.D. when Roman trade was steady, these stations helped in safeguarding the conveyance of goods. When ships arrived to the Red Sea port, merchants accompanied by their commerce did not stay at Myos Hormos (Quseir) port due to the lack of suitable infrastructure in addition to the threats that existed at least since the III century A.D. (Sidebotham, 1991). Petroglyphs attested Roman military activities along the surveyed road since the Augustan period onwards. The duration and intensity of occupation in the *hydreumata* is a determining factor concerning the number of military units as each *praesidium* should have hosted about 50–70 auxiliaries. Moreover, some soldiers had their own dromedary camels as well as equites. In addition to the military presence, miners and quarrymen used to reside temporarily along the Myos Hormos road for specific projects and time-limited expeditions (Gueraud, 1942; Maxfield, 1993). *Hydreumata* were mostly inhabited between the I and the III centuries A.D. depending on the supplies coming from Coptos or the Red Sea ports. These supplies were brought by the civilians as evidenced through the archive of Nicanor who operated with his transport company during the I century AD (Adams, 1995; Othman 2017a; Gundlach, 1982: col. 1101–1102; Bülow-Jacobsen 2013: 560, Fig. 3).

Wine indicated the site chronology through the amphora, of which also size determined the variety of goods either produced in the *hydreumata–praesidia* or transported from the adjacent Red Sea ports. Inhabitants consumed different types of wine such as the Italian, Rhodian, Cretan, Laodicean, wine of Gaulish and Ephesus, Aegean Islands and African wine, and these types were mostly consumed between the I and the III century A.D. (Brun, 2007; Fournet, 2000; Tchernia, 1986; Tomber, 2002).

3.5 Water Supply

Drinking water used to be secured through several means of procurement such as

- Wells and cisterns inside the *hydreumata* (Fig. 5).
- Intense rain and powerful floods (Seil).

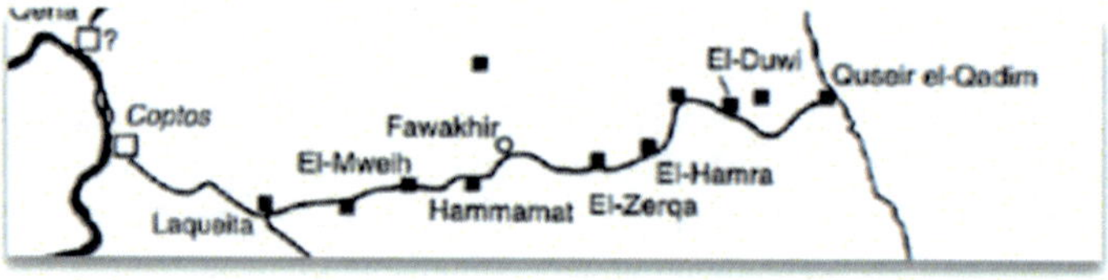

Fig. 4 Water stations, *hydreumata*, along the Qift-Quseir road. After, Harrel & Brown (1992, 4, Fig. 1): Maxfield (1993, 10, Fig. 1)

Fig. 5 Above: The source of water in Bir Sayyala. After, Photo by the author. Below left: The cistern of Krokodilo (Al-Muwayeh). After. Brun (2006), Othman (2017b, 91). Below right: The bath and basin of Maximianon. After. Adam et al. (2006), Othman (2017b: 87)

- Catchment basins (qalts) of surface water which is an unreliable source due to the infrequent rains.
- Subterranean water as the main source, despite being mostly saline or brackish.

Pharaonic inscriptions at Wadi Hammamat refer to the digging of wells during the mining or quarrying expeditions. However, during the Roman period, water stations were easily reachable in the desert road shelters (Zitterkopf & Sidebotham, 1989).

Today, some Roman wells and cisterns disappeared as they are erased due to the climatic effects; moreover, secondary road wells should have existed, these wells were large, and sometimes, they reached the diameter of 30 m (Othman, 2017b).

Water, necessary for the inhabitants of the *hydreumata,* should have been carried by voyagers arriving from Coptos. However, people coming from the Red Sea used the existing water sources due to the lack of drinking water along the coast.

Water in the Myos Hormos road was required for the large mining and quarrying settlements (Fig. 7 left), the small gardens nearby the wells and the needs of animals that varied according to the season and the food moisture (Demougeot, 1960; Zitterkopf & Sidebotham, 1989).

3.5.1 Water in the Gold Mining Process

Bir Fawakhir is the most complete settlement in the Central Eastern Desert, and the site was referred to in the Turin Papyrus (Ramesside Period) as a gold mine provided with a source of water probably a cistern (Fig. 6). Then, the site became the largest town in the Eastern Desert, inhabited by more than 1000 miners during the Greco-Roman Period. This continuity should have necessitated the existence of water sources either for human daily use or for the mining process (Meyer, 1995). Water was used as well in separating gold from the pulverized vein quartz through the gravity separation process. Gold ore was crushed and washed in a laboratory that was probably located nearby the above-mentioned cistern (Harrel and Brown 1992; Othman 2017b).

Inclined gold washing tables were constructed of stone fragments, consolidated by primitive clay/sand mortar. The surface used to be covered with a layer of the same material. They can be observed nearby the N.K. gold mines of Bir Umm Fawakhir. These washing tables varied between 2.2 and 4 m length, 40–60 cm width and 80–100 cm height. They were constructed with an inclination angle of about 15–20 grades. Washing the gold on board of a washing table appeared in the scenes of The Ramesside tomb of "Kha'y" at Saqqara (Martin, 1991; Ogden, 2000).

Diodorus described the use of water in the gold mining process during the Greco-Roman Period as follow:

- Miners used to heat and crack the quartz veins; they crushed the smashed pieces with the help of hammers and sledges to reduce the fragments into smaller pieces.
- Women participated in transferring the resulting dust during the final phase, and then, they poured water in order to wash the gold veins on board of inclined washing tables.
- Gold coming out of this process used to be kept afterwards in terracotta jars.[1]

[1] http://penelope.uchicago.edu/Thayer/E/Roman/Texts/Diodorus_Siculus/3A.html.

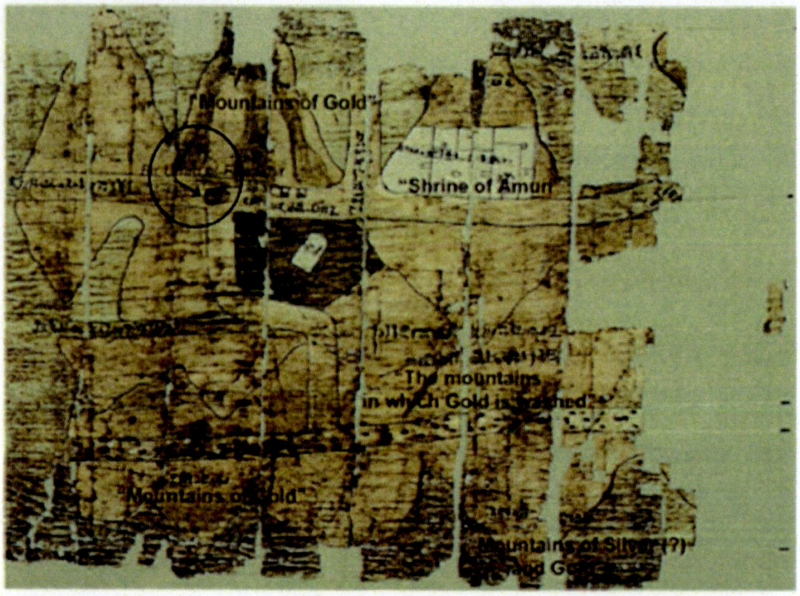

Fig. 6 Turin Papyrus: position of the water cistern. After, Othman (2017a, 208, Fig. 134)

Fig. 7 Left: settlement in the greywacke quarry. After, Bloxam (2015: 802, Fig. 10). Right: Wadi Hammamat quarries (Harrell et al. 2002, 211)

3.5.2 Water in the Quarrying Process

Quarrymen depended mainly on the road wells and cisterns during their long voyage of about 85 km in the desert. The inscriptions of "Henu" during the reign of Montuhotep III referred to 12 water sources "ir.n (.i) Xnmt 12" "I have dug 12 wells" (Couyat & Montet, 1912; Goyon, 1957; Gonzalez-Tablas, 2014).

In addition to the use for drinking and human needs, water was used in the quarrying process especially during the transport of extracted blocks of greywacke. After successful extraction, stone blocks had to be brought down the hill on ramps, constructed from spoil heaps (Fig. 7), until they reach the wadi level, then comes the horizontal overland transport performed usually with the help of wooden sledges[2] on lubricated paths until the expedition arrives to the city of Qift, and from there, blocks of greywacke were shipped and transported to their final destination.

Workers used to throw water in order to lubricate the road paths, while others carried water containers, dragging instruments and ropes. This phase included the transport of quarrying tools, supplies and the extracted "Bekhen" stone blocks. Contractors should have calculated weight, land inclination, climate, number of labourers and carrying animals in addition to the suitable infrastructure (Arnold, 1991; Barbotin, 2008).

These paths created the new aspect of transformed landscape still visible in Wadi Hammamat (Bloxam, 2015).

3.6 Petroglyphs

Since the pre-dynastic period, suitable surfaces where vegetation can be found until today attracted more inscriptions, probably because animals were gathered to feed, and

[2] An example can be found in the New York Metropolitan Museum of Art (MMA 24.1.84); Arnold (1991, 276, Fig. 6.36).

Fig. 8 Left inscription zone (circled left) and the Sanctuary of Min, "Paneion" (circled right). After, Bloxam, (2015, Fig. 8); Hart (2005). Right inscriptions of Sety I. After, Photo by the author

consequently, they were followed by hunters who left their traces (Lankester, 2012).

These inscriptions functioned as an illustrative photographic dossier to guide the future missions in Wadi Hammamat indicating the quarrying level (Fig. 8). Moreover, they offer a clear knowledge about the administrative hierarchy and site logistics during the quarrying period as these inscriptions include names, titles and roles of certain participants in the expedition (Sweeney, 2014).

Inscriptions of the Eastern Desert witnessed numerous threats during the last decade such as the illegal Bedouin quarrying activities, treasure seekers, the repeated trials to copy the existing scenes and texts through the impressions and the erasing of certain social and religious significant symbols (Kmally, 2011).

4 Religious Aspects

4.1 Pharaonic Gods and Goddesses

4.1.1 Min

Min appeared along the "Myos Hormos" road as the lord of Coptos (Pantalacci, 2018) and the protector of the Eastern Desert until the Greco-Roman Period. He was referred to in the M.K. texts as "Lord of the Eastern Desert", while quarries were "The lands of Min" where his miracles occurred.

Middle Kingdom inscriptions of the vizier (Imn im hAt) who was sent to the "Bekhen" hill under "Montuhotep II" stated

"wD Hm .f saHa wD pn n it (.f) mnw nb xAswt m Dw pn"

"This stela was constructed by the king to his father Min, god of the desert lands in this mountain" (Couyat & Montet, 1912).

In a successive point, the text indicated

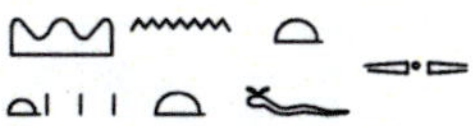

"*xAswt nt it (.f) Mnw*" referring to "The desert that belongs to his father Min".

Min was associated with the east in the text of the 13th Dynasty funerary stela preserved in Parma Museum (Italy).

"inD Hr.k Mnw—Imn nb sHnt mi

"Naty .k (r) sxwt .k ksbwt .k"

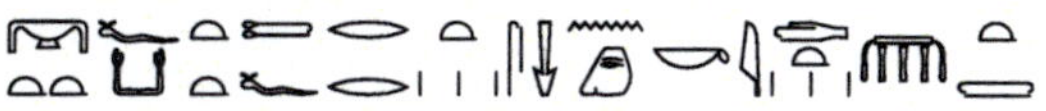

"Nbwt mfkttf rrwt sn.k idwt Ssmt".

"Greetings to your face, O, Min-Imn, Lord of (sHnt) when you walk through your fields, your golden, blue and green fragranced trees and when you smell the fragrance of the east"(Lange, 1927).

The tent of "Min" that probably functioned as a chapel is similar to the modern Bedouin tents. This tent used to be raised in a special ceremony since the New Kingdom to indicate his role as "Lord of the Eastern Desert". Min was also associated with the moon, guide of the voyagers and the inhabitants of the desert (Helck, 1980; Vernus, 1998).

The shrine of Min in Wadi Hammamat (Fig. 9) includes Hieroglyphic, Hieratic, Demotic and Greek inscriptions in addition to the pictorial reliefs that date back to the reign of Nectanebo II (Late Period). The scenes of the shrine refer to the pantheon of Wadi Hammamat in three registers (Fig. 10). Afterwards, the site was used as a space for inscriptions until the Roman Period (Cruz-Uribe, 2001; Lichtheim, 1973).

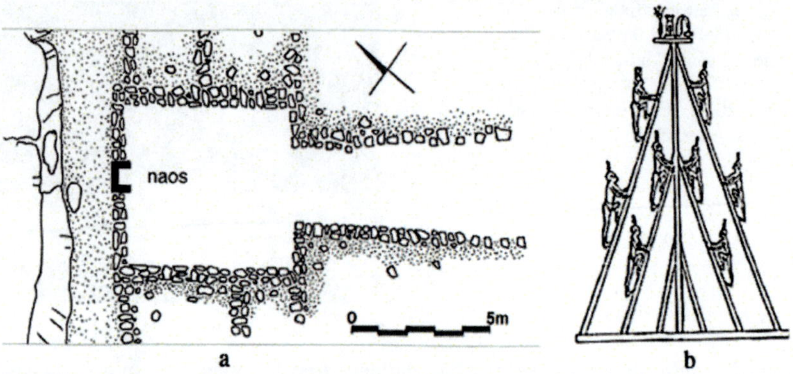

Fig. 9 a The shrine of Min at Wadi Hammamat. b The tent of Min. After, Othman (2014, 48, Fig. 16)

Fig. 10 Gods and Goddesses of Wadi Hammamat. After, Traunecker (2002, 379, Figs. 1–3)

I—Upper right side:
Amun-Re (King of gods), Horus-Ra (The Great God), Min, Group composed of Ptah and Hapy, Osiris and Harpocrates, Geb, Horus, Harpocrates and Thot.

II—Upper left side:
Isis, Nephthy, Horus (great of powers), Montu, Min of Panopolis (Akhmim).

III—Lower register:
A—First Naos (A Hippopotamus goddess accompanied by the sign (sA) and described as (Hwt nTrw).

B—Second Naos (Another Hippopotamus goddess "an evil form").

C—The god Shed (The protector) (Traunecker, 2002).

4.1.2 Bekhenew or "PA—Bekhenew"

Bekhenw appeared in the "book of the dead (Chapter CLXV—arrival to the port)" and in the book of the dead of "Iuefankh" preserved in the Egyptian Museum of Turin. The vignette refers to a beetle-bodied ithyphallic divinity with double-plumed crown and a raised right hand and arm (Fig. 11). This divinity was usually depicted in a blue colour composed of "Lapis-Lazuli" and "qamai" water (Budge, 1901).

Recitations tell: "O, Bekhenw, Bekhenw! O, Great one Great one! O Amun, Amun! O Lion of Yewkasa[3]! O God, Great one between the gods of the eastern part of heaven!".

Bekhenew was associated with "Min" in his external aspect as an ithyphallic divinity with an uplifted arm. Therefore, "Bekhenew" should have been invoked as a deity of Wadi Hammamat, "Bekhen" region.

4.2 Greek and Roman Gods and Goddesses

Excavations brought to light about 2400 Greek ostraca along the Myos Hormos road. These textual evidences illustrated certain aspects of life inside the *hydreumata*.

[3] "Iky" means "Stone Hewer", Wb, I: 139.

Fig. 11 Bekhenew, god of the greywacke quarries in a papyrus preserved in the Egyptian Museum of Turin. After, Left: Photo by the author. Right: Lucas and Rowe (1938)

The ostraca referred in more occasions to the "*proskynema*" that means (bowing) in front of the tutelary god or goddess including the names of Apollo (in Krokodilo), Athena / Athena in Persou (in Persou), Dioskouroi (in Krokodilo) (Dioscuri, n.d.), Pan (rarely mentioned in the letters), Sarapis (Maximianon), Philotera (sister of Ptolemy II, in Maximianon and the Red Sea region) and Tyche Simiou (mentioned in the commercial exchange letters) (Bülow-Jacobsen 2006).

5 Development Project

The study proposes some ideas in order to evaluate both cultural and landscape heritage values of the surveyed road as follows.

5.1 Archaeological Park that Aims to Encourage Tourism and Preserve the Remaining Sites in Order to Develop the Archaeological Research and the Consequent Reconstruction

This park targets

- Highlighting the previously excavated sites along the Myos Hormos road.
- Creating visit itineraries between the *hydreuma* and the quarrying sites.
- Technical support and plan—segmentation to realize a comprehensive management of the archaeological resources.

5.2 Open Air Museum

The long history of Myos Hormos road lasted from the pre-dynastic period until the III century AD, leaving behind further archaeological traces in an open museum that recounts the Pharaonic stone extraction, the gold mining activities, the imperial Roman commerce and finally the Arabic Hajj (pilgrimage) (Thorpe, 2009).

5.3 Virtual Museum

Materials, tools, inscriptions and settlements identify the quarry site complex (Fig. 2). These elements can be digitally reconstructed as several archaeological sites around the world were brought back to life through the remaining traces using "virtual reality" in order to preserve the standing site or a facsimile when the ruins are completely demolished (Bloxam, 2007; Djindijian, 2007).

A virtual museum can be a true commercial and promotional instrument for the site marketing. Therefore, a catalogue of greywacke is a step towards an illustrating digital database for both researchers and non-specialized curious visitors. This catalogue can be inserted in the virtual reproduction contents so as to follow the various extraction steps starting from quarrying, transport and then sculpture of the final product that reached us in the form of plates, sarcophagi, statues and minor arts.

The exhibition can function as an educational and cognitive centre of stone culture and technology, where the existing and the imaginary could form a new learning and entertainment method, probably hosting a site museum that proposes a virtual tour in the past (Niccolucci, 2007).

5.4 Site Panels

The Quseir-Qift road is lacking any explanation panel, indications and road directions, and therefore, the researcher has prepared some explanation panels during the site survey in order to promote the heritage values of each archaeological stop (Fig. 12).

6 Recommendations

- Authorities should pay attention to enhance more control and inspection along the road including toll booths with more attention to the dangerous slopes at about 70 km from Quseir, described by the locals as "The death road" due to the repeated mortal accidents.
- Tourist and hospitality services (restaurants, cafeterias, outlets and bazars) should be increased in order to promote the surveyed area and to create more job vacancies.
- Satellite coverage and phone networks are urgently needed.
- Guided visits and archaeological restoration work between the ancient port of Quseir (Myos Hormos), the

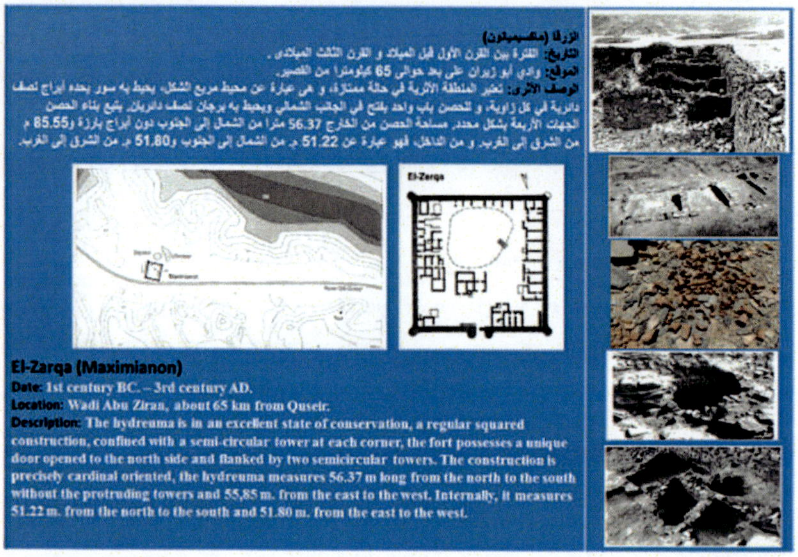

Fig. 12 Site panel of El Zarqa (Maximianon). After, The author during his Ph.D. survey

greywacke quarries, *hydreumata* (especially Maximianon "El Zarqa" as the most complete site), gold mines (still containing the mining instruments used during the last twentieth century investment) and Qift (Coptos).

- More studies, conferences and seminars will help in illustrating the cultural heritage values of the surveyed region.

Final Remarks

Cultural aspects of the Myos Hormos (Qift-Quseir road) were clearly intensive and well-equipped during the surveyed period, with stops regularly prepared to facilitate travel between the Nile Valley and the Red Sea. The road represents the role of commerce in the economy, society and culture during the Greco-Roman Period. However, the *hydreumata*, stone quarries and gold mines distributed along the surveyed road represent an exceptional image that flourished despite these unfavourable circumstances for centuries.

The lack of drinking water and water supply on Myos Hormos road are two important issues dealing with a need that was fulfilled through temporary cisterns and wells during the Pharaonic Period to supply the quantity of water needed for miners and quarrymen. However, *hydreumata* were constructed during the Greco-Roman period to function as watering stations, usually equipped with water cisterns that used to store a quantity of water for the eventual needs of large caravans.

Archaeological findings in Coptos and Quseir indicated that the road reached its peak during the I century AD, then it continued to be used until the IV century AD afterwards, and the descending curve of the Roman empire, probably starting from the III century AD, was reflected in the Roman trade and consequently the change of some aspects in certain *hydreumata* such as Maximianon "Al-Zarqa".

The study represents a theoretical approach not only to analyse but also to reconstruct the social aspects of the quarrymen community of Wadi Hammamat.

Promoting the sites along the Qift-Quseir (Myos Hormos) road through a virtual digital reconstruction aims to

1. Improve access and awareness about the Eastern Desert heritage sites in general and the surveyed area in particular.
2. Facilitate future archaeological and tourism visits, even virtually, for public and private entities.
3. Preserve the existing materials and constructions from different dangers and risks.
4. Petroglyphs and rock art spots are usually situated in shaded sites along the route leading to the coast, and therefore, more inscriptions were added upon the course of time including modern graffiti. Hence, the need for action is urgent to prevent the destruction of the central Eastern Desert's rock art in addition to the threats represented in the form of illegal quarrying, mining, treasure-hunting and vandalism.
5. Elaborate the importance of Wadi Hammamat and the greywacke masterpieces distributed in different museums around the world, offering a chance for further studies in order to reconstruct a chronological sequence of the site's long history.

The development of a heritage site can be achieved through

A. **Community development**
Creating a quality environment for the Bedouin inhabitants of the Eastern Desert, in terms of education, increased security forces and sufficient health facilities.
B. **Regional layout**
Administrative authorities should apply new designs concerning the road layouts, open spaces and traffic flow in addition to the cleanliness and environmental quality.
C. **Urban planning**
Rules and regulations in terms of land use and archaeological zoning, density and circulation.
D. **Economic development**
Introducing and promoting tourism industry will enhance the competitiveness and productivity.

Watering stations, quarrying and mining sites, trade routes and ports represented a sketch of the Egyptian Eastern Desert civilization that should be evaluated on a broader scale to guarantee an eventual site development. Therefore, comprehensive archaeological strategies should be applied not only on the surveyed region but the whole Eastern Desert heritage sites to start a non-destructive approach considering the deteriorated status of the surveyed sites.

References

Abu-Jaber, N., et al. (2009). QuarryScapes: Ancient stone quarry landscapes in the Eastern Mediterranean. *Geological Survey of Norway Special Publication 12*, Trondheim.

Adam, J. P. et al. (2006). Al Zrqa (Maximianon), in Cuvigny, H. *La route de Myos Hormos—L'armée romaine dans le désert oriental d'Égypte*, Vol. I, II Edition, IFAO.

Adams, C. (1995). Supplying the Roman army: O Petr. 251. *Zeitschrift Für Papyrologie Und Epigraphik, 109*, 119–124.

Arkell, A. J. (1975). *The prehistory of the Nile valley*. Leiden, Brill.

Arnold, P. (1991). *Building in Egypt; pharaonic stone masonry*. New York and Oxford.

Bagnall, R. S. (1982). Upper and lower guard posts. *CDE*, LVII, 125–128

Barbotin, C. (2008). *Ahmosis et le début de la XVIII e dynastie*. Pygmalion.

Bernard, A. (1972). *De Koptos a Kosseir*, Leiden.

Bloxam, E. (2005). Who were the Pharaohs' quarrymen? *Archaeology International, 9*, 23–27.

Bloxam, E. (2015). A place full of whispers: Socializing the quarry landscape of the Wadi Hammamat. *CAJ, 25*, 789–814.

Bloxam, E., et al. (2014). Investigating the Predynastic origins of greywacke working in the Wadi Hammamat. *Archeonil, 24*, 11–30.

Bloxam, E., & Heldal, T. (2008). *Identifying heritage values and character defining elements of ancient quarry landscapes report*.

Bloxam, E. (2007). *From ancient quarry site to ancient quarry landscape, in QuarryScapes report: Risk assessment and monitoring of Ancient Egyptian Quarry Landscapes*, Work Package 5, Deliverable no. 6: 45–61.

Bloxam, E. (2008). New directions in identifying the significance of ancient quarry landscapes: four concepts of landscape. In Abu-Jaber, N. et al. (Eds.), *QuarryScapes: Ancient stone quarry landscapes in the Eastern Mediterranean* (Vol. 12, pp. 165–183). Geological survey of Norway, special publication.

Brun, J. P. (2006). Al-Muwayh (Krokodilo). In Cuvigny, H. (Ed.), *La route de Myos Hormos—L'armée romaine dans le désert oriental d'Égypte* (Vol. I, II Edn.). IFAO.

Brun, J. P. (2007). Amphores égyptiennes et importées dans les praesidia romains des routes de Myos Hormos et de Bérénice. In S. Marchand & A. Marangou (Eds.), *Cahiers de la Céramique Egyptienne 8, Vol. II: Amphores d'Egypte de la basse époque a' l'époque arabe*, Le Caire, IFAO.

Budge, W. (1901). *The book of the dead*. Chicago, London.

Bülow-Jacobsen, A. (2013). Communication, travel and transportation in Egypt's Eastern Desert during Roman times (1st to 3rd century AD). In F. Förster & H. Riemer (Eds.), *Desert road archaeology in ancient Egypt and Beyond*, Köln.

Bülow-Jacobsen, A. (2006). Toponyms and Proskynemata. In H. Cuvigny (Ed.), *La route de Myos Hormos—L'armée romaine dans le désert oriental d'Égypte* (Vol. I, II edn.). IFAO.

Casson, L. (1993). Ptolomy II and the hunting of African elephants. *Transactions of the American Philological Association*, 247–260.

Cleere, H. (1996). The concept of outstanding universal value in the world heritage convention. *Conservation and Management of Archaeological Sites, I, 4*, 227–233.

Couyat, J., & Montet, P. (1912). *Les Inscriptions Hiéroglyphiques et Hiératiques du Ouâdi Hammâmât*, Le Caire.

Cruz-Uribe, E. (2001). Demotic graffiti from the Wadi Hammamat. *Journal of the Society for the Study of Egyptian Antiquities, 28*, 26–54.

Cuvigny, H. (2005). Ostraca de Krokodilo. La correspondance militaire et sa circulation. Praesidia du Desert de Berenice 2, *Fouilles de IFAO*, 51, Le Caire.

Debono, F. (1951). Expédition archéologique royale au désert oriental (Keft-Kosseir). *Annales Du Service Des Antiquites De L'egypte, 51*, 60–62.

Demougeot, E. (1960). Le Chameau et l'Afrique du Nord romaine. *AESC, 15*(2), 209–247.

DIOSCURI (Dioskouroi). Greek Gods of Horsemanship & Protectors of Sailors. (n.d.). http://www.theoi.com/Ouranios/Dioskouroi.html

Djindijian, F. (2007). The virtual museum: An introduction. In P. Moscati (Ed.) *Virtual museums and archaeology: The contribution of the Italian National Research Council*, Archeologia e calcolatori, suppl. 1: 9–14. http://soi.cnr.it/archcalc/images/VM.pdf. Accessed 15 July 2015.

Fairclough, G., et al. (2008). *The heritage reader*. Routledge.

Fournet, J. L. (2000). Coptos dans l'antiquité chrétienne, in *Coptos, l'Egypte antique aux ports du désert, Cat. Expo. Lyon, musee des beaux arts*.

Friedman, R. (2002). Foreward. In R. Friedman (Ed.), *Egypt and Nubia: Gifts of the desert*. London: British Museum Press.

Gonzalez – Tablas, J. . (2014). Quarrying beautiful Bekhen stone for the Pharaoh: The exploitation of Wadi Hammamat in the reign of Amenemhat III. *Journal of Egyptian History, 7*(1), 55.

Goyon, G. (1957). *Nouvelles Inscriptions rupestres du Wadi Hammamat*.

Gueraud, O. (1942). Ostraca Grecs et Latins de Wadi Fawakhir. *BIFAO, 41*, 141–196.

Gundlach, R. (1982). Wadi Hammamat, *LÄ*, VI: col. 1101–1102.

Harrell, J. A., & Brown, V. M. (1992). The oldest surviving topographical map from ancient Egypt (Turin Papyri 1879, 1899 and 1969). *Journal of the American Research CenterCentre in Egypt, 29*.

Harrell, J. A. (2002). Pharaonic stone quarries in the Egyptian deserts. In: R. Friedman (Ed.), *Egypt and Nubia: Gifts of the desert*. London: British Museum Press.

Hart, G. (2005). *The Routledge dictionary of Egyptian gods and goddesses* (II). Routledge.

Helck, W. (1980). Klettern für Min, *LÄ*, III, col. 454.

Ingold, T. (1993). The temporality of the landscape. *World Archaeology, 25*, 152–174.

Judd, A. (2003). *The trade with India through the Eastern Desert of Egypt*, (revised 2007). http://archaeology-easterndesert.com/assets/applets/LYSA_Judd.pdf. Accessed Jan 2016.

Klemm, D., et al. (2002). Ancient gold mining in the Eastern Desert of Egypt and the Nubian desert of Sudan. In R. Friedman (Ed.). *Egypt and Nubia: Gifts of the desert*. British museum Press.

Kmally, H. A. (2011). Deterioration and rates of weathering of the monumental rock inscriptions at Wadi Hammamat, Egypt. *E-Conservation*, 21, 66–79.

Lange, H. O. (1927). Ein Liturgisches Lied an Min, *SPAW*, S.S. 331–338.

Lankester, F. D. (2012). *Pre-dynastic and Pharaonic era rock-art in Egypt's central Eastern Desert: Distribution, dating and interpretation*. Durham thesis, Durham University.

Lichtheim, M. (1973). *Ancient Egyptian literature, Vol. 1: The old and middle kingdoms*. Los Angeles (CA): University of California Press.

Lucas, A., & Rowe, A. (1938). The Ancient Egyptian Bekhen Stone. *ASAE, 38*, 127–156.

Martin, G. T. (1991). *The hidden tombs of memphis*. Thames and Hudson.

Mason, R. (2008). Assessing values in conservation planning: Methodological issues and choices. In G. Fairclough, R. Harrison, J.H., Jameson, & J. Schofield, Jr. (Eds.), *The heritage reader* (pp. 99–124). London: Routledge.

Maxfield, V. A. (1993). The Eastern desert forts and the army in Egypt during the principate. In D. M. Bailey (Ed.), *Archaeological research in Roman Egypt: The proceedings of the seventeenth classical colloquium* of the department of Greek and Roman antiquities, British Museum, held on 1–4 December, 9–19.

Meyer, C. (1995). A byzantine gold-mining town in the Eastern Desert of Egypt: Bir Umm Fawakhir, 1992–1993. *Journal of Roman Archaeology*, (8), 211 ff.

Morrow, M., Morrow, M., & Cherry, P. (Eds.). (2002). *Rock art topographical survey in Egypt's Eastern Desert*, Bloomsbury.

Niccolucci, F. (2007). Virtual museums and archaeology: An international perspective. In P. Moscati (Ed.), *Virtual museums and archaeology: The contribution of the Italian National Research Council*, Archeologia e calcolatori, suppl. 1: 15–30. http://soi.cnr.it/archcalc/images/VM.pdf. Accessed 15 July 2015. The MAV, Virtual archaeological museum of Ercolano. http://www.museomav.it/museo.

Ogden, J. (2000). Metals. In P.T. Nicholson & I. Shaw (Eds.), *Ancient Egyptian materials and technology*. Cambridge University Press.

Othman, A. (2014). IAbtet (The East). and Imntet (The West). In Ancient Egyptian civilization: A linguistic, artistic and religious study, Ph.D. Thesis, Egypt.

Othman, A. (2017a). *Evaluating the cultural heritage and landscape of Quseir-Qift road: With a special focus on the gold mines and greywacke quarries*, Ph.D. Thesis, Italy.

Othman, A. (2017b). Cisterns, wells and use of water in the mining and quarrying sites of the Egyptian Eastern desert: A special focus on the central Myos Hormos road. *JFTH, 14*(2), 79–101.

Pantalacci, L. (2018). Coptos, gate to the Eastern Desert. In J.P. Brun et al. (Eds.), *The Eastern desert of Egypt during the Greco-Roman Period: Archaeological reports*, Paris.

Redon, B. (2018). The control of the Eastern Desert by The Ptolomies: New archaeological data. In J.P. Brun et al. (Eds.), *The Eastern Desert of Egypt during the Greco-Roman period: Archaeological reports*, Paris.

Riemer, H., & Förster, F. (2013). Ancient desert roads: Towards establishing a new field of archaeological research. In F. Förster & H. Riemer (Eds.) *Desert road archaeology in Ancient Egypt and Beyond*, Köln.

Shaw, I. (2013). We went forth to the desert land: Retracing the routes between the Nile Valley and the Hatnub travertine quarries. In F. Förster & H. Riemer (Eds.), *Desert road archaeology in Ancient Egypt and Beyond*, Köln.

Sidebotham, S. E. (1991). Ports of the Red Sea and the Arabia – India trade, in Begley, V. et. al. *Rome and India: The ancient sea trade*, Madison.

Sidebotham, S. E. et al. (2008). *The red land: The illustrated archaeology of Egypt's Eastern Desert*, The American University in Cairo.

Sweeney, D. (2014). Self-representation in Old Kingdom quarrying inscriptions at Wadi Hammamat. *JEA, 100*, 289–290.

Tchernia, A. (1986). *Le Vin de L'Italie Romaine*. Roma.

Thomas, J. (2001). Archaeologies of place and landscape. In I. Hodder (Ed.), *Archaeological theory today* (pp. 156–184). Political Press.

Thorpe, K. (2009). Protection of cultural heritage as social, political and economic issues. In H. Arlt & D.G. Daviau (Eds.), *Culture, civilization and human society: Encyclopedia of life support systems* (Vol. I, pp. 175–185). UNESCO.

Tom, H., & Gurli, M. (2015). Assessment of ancient stone quarry landscapes as heritage sites. In G. Lollino et al. (Ed.), *Engineering geology for society and territory* (Vol. 8) (Preservation of Cultural Heritage). Switzerland: Springer International Publishing.

Tomber, R. (2002). Amphorae from the Red Sea and their contribution to the interpretation of late Roman trade beyond the Empire. In J. Eiring & J. Lund (Eds.), *Transport amphorae and trade in the Eastern Mediterranean: Acts of the international colloquium at the Danish Institute of Athens* (Vol.5, pp. 393–402).

Traunecker, C. (2002). Le Pantheon du Ouadi Hammamat (Inscription no. 58). *Autour Du Coptos, Topoi, 3*, 357–369.

Vermeersch, P. M., Paulissen, E., & Van Neer, W. (1989). The Late Palaeolithic Makhadma sites (Egypt): Environment and subsistence. In L. Krzyzaniak, M. Kobusiewicz (Eds.), *Late Prehistory of the Nile Basin and the Sahara*. Poznan: Polish Academy of Sciences.

Vernus, P. (1998). *The Gods of ancient Egypt*. Cairo, Egypt: AUC.

Weigall, A. E. P. (1913). *Travels in The Upper Egyptian Deserts*. Edinburgh and London.

Whitcomb, D. S., & Johnson, J. H. (1979). *Quseir al-Qadim 1978*. Preliminary Report, American Research Center in Egypt, Cairo.

Whitcomb, D. S., & Johnson, J. H. (1982). *Quseir al-Qadim 1980: Preliminary report*.

Zitterkopf, R. E., & Sidebotham, S. E. (1989). Stations and towers on the Quseir-Nile road. *JEA of Egyptian Archaeology, 75*, 155–189.

Management and Conservation of Architectural Heritage

This part of the book specifically focuses on the conservation of heritage of the modern era and the growing interests in identifying the recent past as important and bringing together, as a result, sectors of the architectural and conservation community to conserve the legacy of modernism by bringing together contemporary architects, critics, historians and conservationists. Considering the many years and achievements since the emergence of a distinct area of practice in the 1990s known as the conservation of modern heritage, it would be easy to assume that modern heritage is well taken care of and preserved. However, there still seems to be a gap in research and a lack of dedicated training opportunities.

Consequently, in earlier chapters of this part, specifically that entitled "Rehabilitation and Exploitation of Heritage Buildings. An Investment Approach", the author investigates the validify of former international heritage renovation attempts and employs a descriptive and analytical methodology to make sense of architectural and urban heritage typologies to finally raise awareness for heritage areas discussed and consider their preservation, rehabilitation and development, which would encourage the private sector to launch major investment projects. "Conserving the Past, Ameliorating the Future—Situating Conservation of Architectural Heritage in the Urban context—A Case of India" turns the reader's focus to a different site, India, arguing that the conservation of architectural heritage in the context of the modernising Indian city will be significant in making an important contribution to the well-being of its inhabitants in the future.

Moreover, the role of urban planners and decision-makers is questioned in the chapter titled "A Methodological Framework for the Conservation and Planning of Urban Spaces in Historical Centers Around Riverfronts. The Case of Arequipa, Peru" where issues of displacement, the lack of physical and social accessibility to rivers, and the disappearance of important heritage around these natural habitats lower the quality of public life, as phrased by the author. The chapter explores a methodology to preserve and upgrade a public space of a historic center as well as proposes improvements to the riverfronts. The research examines one main case study which is the historic center of Arequipa in Peru and reviews other case studies of the continent to propose 4 axes of action for the management of public spaces around the river. The importance of Accessibility to Architectural Heritage in "Towards a Sustainable use of Architectural Heritage: How to Assess Accessibility and the User's Role" is discussed as a main contributor to a conscious and sustainable use of architectural heritage and conservation. The role of the users is also highlighted in this chapter in order to reach and increase sustainable processes in the field of architectural heritage. The authors begin by summarizing the evolution of the controversial relationships between use and conservation of the architectural heritage and outline a performance-based approach as a design methodology and, finally, addresses several points of view after yielding results of multidisciplinary and multiscale studies on the topic. In the chapter entitled, "Architectural Heritage and Tourism Development in Urban Neighborhoods: The Case of Upper City, Thessaloniki, Greece", the link between architectural heritage preservation and tourism development in traditional urban neighbourhoods is explored through a series of heritage implementation measures for the Upper City of Thessaloniki. The chapter concluded that there is a need to examine the perceptions of the residents so as to form a holistic research approach to the topic, to also include "touristification" issues awareness.

This part of the book presents where future and past come together and takes into account existing challenges to achieve widespread recognition for the conservation of 20th century heritage. It represents a stepping stone towards a shared vision that should guide future efforts and help identify where new paradigms are needed.

Architectural Heritage and Tourism Development in Urban Neighborhoods: The Case of Upper City, Thessaloniki, Greece

Stella Kostopoulou

Abstract

Inherited from past generations, cultural heritage, the legacy of physical objects and intangible qualities of culture, comprises a range of terms, such as architectural heritage, archaeological sites and cultural landscapes, providing opportunities for leisure and tourism, when properly maintained and wisely handled. On the urban level, the dynamic relationship between cultural heritage and tourism is even more apparent. Urban cultural tourism, which also focuses on traditional communities, is a growing trend worldwide, attracting attention from different fields such as tourism management, architecture, urban planning and environmental sciences. This chapter focuses on the subject of traditional urban communities' architectural heritage and tourism growth based on a holistic and integrated understanding of the dimensions embedded in urban heritage ensembles, e.g., historical, spatial economic and social. The case study makes reference to the Upper City (Ano Poli) of Thessaloniki, a historic neighborhood in one of Greece's most popular cities for cultural tourism, due to its continuous urban presence for more than twenty-three centuries and its multicultural history. Ano Poli is among the areas of Thessaloniki's urban fabric endowed with historic architectural and spatial planning elements, attracting many visitors. The chapter investigates the perceptions of Erasmus students–tourists toward architectural heritage tourism in general and specifically for the Upper City and the city of Thessaloniki, so as to explore a most promising segment of the rising cultural tourism market. The study analysis and research results lead to a better understanding of the significance of architectural heritage in urban tourism.

S. Kostopoulou (✉)
Aristotle University of Thessaloniki, Thessaloniki, Greece
e-mail: kostos@econ.auth.gr

Keywords

Architectural heritage • Urban neighborhoods • Tourism • Thessaloniki

1 Introduction

Cultural heritage, as the result of accumulated human activity articulated and projected in a material or immaterial sense (historic city cores, built complexes, iconic monuments, landscapes or music, dance and literature) (Nijkamp, 2012:76), denotes an invaluable asset in historical, cultural and socioeconomic terms. It should therefore be adequately maintained and handled, so as to prevent loss of place identity and cultural deterioration, while providing leisure and tourism opportunities (Kunzmann, 2004; Geppert, 2004 cited in Papandreou & Papageorgiou, 2019: 2039). Tourism is used to demonstrate the economic benefits of cultural heritage, often intrinsically linked with cultural heritage restoration and management projects (Alonso & Meurs, 2012: 6).

In the urban arena, the dynamic relationship between cultural heritage and tourism is much more apparent, generating opportunities and challenges (Hmood et al., 2018: 210). Urban tourism has become one of the fastest growing segments of the global tourism market in modern societies, primarily due to the rising interest in the cultural heritage resources of cities (Castela, 2018; Cazes & Potier, 1996). Urban cultural tourism is a growing socioeconomic phenomenon worldwide, attracting research interest from various scientific fields, such as tourism management, urban planning, archeology, environmental sciences and computer science (Barrera-Fernandez et al., 2016).

Cities are some of the most popular tourism destinations in the world today, attracting an ever increasing number of visitors who tend to concentrate in urban historic centers, getting substantially inspired or motivated by cultural

A. Versaci et al. (eds.), *Conservation of Architectural Heritage*, Advances in Science, Technology & Innovation,
https://doi.org/10.1007/978-3-030-74482-3_12

heritage. Urban cultural tourism is also focused on traditional neighborhoods, where culture and authenticity are stronger (Castela, 2018). The growing interest of visitors in the urban architectural heritage and the revitalized historic city cores is also related to the concept of nostalgia, for "what has gone" (Lengkeek, 1995 cited in Gospodini, 2001: 928). This leads to a renovated interest in visiting historical urban complexes (Richards, 2001), while also being interpreted as an economic resource (Barrera-Fernandez et al. 2016). Since the 1990s, architectural preservation projects in historic urban districts have been widely used to achieve urban regeneration, and several cities have committed to developing urban cultural tourism as a remedy for decaying urban settlements (Barrera-Ferrandez, 2016, Barrera-Fernandez et al., 2016).

Furthermore, it is important for urban policies to take into consideration the scale of tourism activities' impact within the city, especially in historic city cores. Positive impacts often spread out all over the destination, whereas negative impacts tend to concentrate on the most attractive, crowded localities of the destination (García-Hernández et al., 2017). Moreover, the problems associated with tourism policy that each city faces vary considerably, and as a result, initiatives and planning tools differ and need to be appropriately adapted, based on an interdisciplinary approach to cultural heritage conservation, tourism exploitation and local community engagement (Papandreou & Papageorgiou, 2019). Thus, it is important to analyze and evaluate the situation on a case-by-case basis and adapt, focused on the specific characteristics of each case and its changing social and economic environment, instead of bringing standardized models of policy recommendations (Papandreou & Papageorgiou, 2019; Barrera-Fernandez et al., 2016).

This chapter focuses on the topic of urban architectural heritage preservation and tourism development in traditional neighborhoods, articulating the need for a holistic and interdisciplinary approach to cultural heritage tourism development, to include historical, spatial, economic and social aspects (Gospodini et al., 2007a, b). The case study is the Upper City (Ano Poli) of Thessaloniki, a historic neighborhood in one of Greece's most popular cities for cultural tourism. For more than twenty-three centuries, Thessaloniki, situated in northern Greece, at the crossroads of the main land and sea routes connecting east and west, has had a continuous urban presence (Basteaand & Hastaoglou-Martinidis, 2013). The city was founded in the fourth century BCE and holds the marks of its historical trajectories through Hellenistic, Roman, Byzantine and Ottoman periods imprinted in the contemporary urban landscape. Today Thessaloniki, the second largest city in Greece, is an important business, administrative and cultural center, with three major universities, an extended industrial sector, an active port and a population of more than one million in its metropolitan area. As a post-industrial port city that lost the trade character that its port once had, Thessaloniki in recent times faced the negative results caused by the decline of manufacturing and trade, while, since 2009, all the fundamental economic and social facets of the country have been impacted by an ongoing unprecedented debt crisis (Mavridis, 2018). Greece experienced a severe loss in real per capita income and high unemployment that forced young high-skilled Greeks to massively emigrate abroad (brain drain). The country managed to exit from the bailout program on August 20, 2018, entering a new phase of national development, with its economy maintaining a good momentum and the international community increasingly optimistic about Greece's development prospects (World Economic Forum 2018).

The city retains a significant part of its centuries-old heritage amid natural disasters, war devastations, social changes and comprehensive post-war urban modernization, particularly in the historic center and the preserved traditional neighborhoods, e.g., Ladadika, Upper City, which could be mobilized further to improve the reputation of cultural tourism in the city (Hastaoglou-Martinidis & Christodoulou, 2010). Upper City or Upper Town (Ano Poli in Greek) is probably the only part of Thessaloniki's urban fabric that has collected so many characteristic elements of the city's history, elements of architectural, morphological, urban and spatial planning, as well as of particular social and demographic character, attracting many visitors (Kefala & Samaras, 2019).

The chapter is organized, including the introduction, into six parts. The second segment provides a literature review on the protection of urban cultural heritage and tourism development. An outline of the urban history of the city and the Upper City of Thessaloniki is presented in the third section, preceded in the fourth section by the presentation of the initiatives for the preservation of architectural heritage introduced in the Upper City. The challenges for the Upper City to be designated as an architectural heritage tourism destination is then explored in the fifth section, and the survey research findings are analyzed. Finally, in the sixth section, concluding remarks are discussed.

2 Urban Cultural Heritage Conservation and Tourism Development

Cultural heritage is recognized as the exposé of historical and architectural assets from the past (Yaldiz et al., 2014), the legacy of physical artifacts and intangible community (or society at large) characteristics inherited from past generations (Nijkamp, 2012: 76). Therefore, efforts to safeguard and maintain architectural heritage in the present are indispensable so as to be bestowed to the generations to come that is through the concept of sustainability (Gholitabar et al.,

2018: 2). The global debate on sustainable cities considers built heritage not only as a cultural value, but also as a crucial economic developmental factor and an important material and energy resource (Haspel, 2011).

However, due to the complex combination of use and non-use principles and public and private benefits, the quantitative evaluation of the economic benefits of conservation activities remains difficult (Mason, 2005 cited in Alonso & Meurs, 2012: 7). Use and non-use values express the tangible and non-tangible dimensions of the built heritage that in economic terms are distinguished by the heritage marketable and non-marketable elements. Use values, widely represented in historic city cores, can be identified and often measured with great accuracy, being related to the use of heritage, e.g., to tourism directly (visits to museums, monuments) or indirectly (accommodation, catering, shops and services) (Ost, 2012). Therefore, benefits are related to the direct use value (e.g., tourism revenue), as well as to the wider spillover effects and externalities on the whole "urban ensemble," considering cultural heritage as a basically "club good," that is, shared by many individuals in a large group (Buchanan, 1965 cited in Nijkamp, 2012: 88). Non-use values are not marketable and cannot be directly measured in monetary terms because they do not involve direct economic benefits, but instead reflect, for instance, the values of recognizing that specific goods exist and can therefore be identified in relation to historical districts, architectural complexes and public spaces (Alonso & Meurs, 2012; Ost, 2012).

Recent economic studies have shown that state architectural heritage funding has a strong economic multiplier impact where all stakeholders, landlords, developers and traders, social security, health care and job agencies can profit (Haspel, 2011). In Germany, the government's public subsidies to private parties contribute, on average, to an investment four or five times as high. Similar findings have been shown by research elsewhere in Europe, where government support for built heritage has also been appraised, e.g., in the UK (Positively Plymouth, 2010), Norway (Bowitz & Ibenholt, 2009) and Spain (Government of Spain 2009 cited in Barrera-Fernandez et al. 2016). Government support for heritage also has a major impact on employment, its direct and indirect effects leading to keep existing jobs and create new ones (Behr, 2000). In addition, urban cultural heritage assets are important locational factors for companies moving into an area, particularly among highly qualified employees, as smart businesses focus on cities as attractive locations and highly regard urban architectural heritage as an element of housing supply and leisure activities (Haspel, 2011).

Architecture is also considered as an important element of cultural tourism that includes visits to heritage and contemporary sites (Hughes, 2000). Historically, most of the definitions of cultural tourism suggest learning about other people and their culture, emphasizing humans' insatiable curiosity as an essential motivator to conduct travel. Cultural tourism can be defined in modern times as the commercialized expression of the desire of people to see how others live in their "authentic" environment and to experience local culture, such as arts and crafts, music, literature, dance and gastronomy. (Dewar, 2000). The cultural framework of tourism and recreation in modern societies has changed in recent times, with significant alterations in people's experience and aesthetic appreciation of nature, urban landscapes and local communities (Dietvorst & Ashworth, 1995; Lash & Urry, 1994). Consequently, leisure and recreational activities in cities tend to require more space within the urban fabric, e.g., historic urban cores, urban waterfronts. Among the urban structures that attract tourists, historic cores and traditional neighborhoods, as cultural tourism resources, have been the most popular since the last decades (Ashworth & Tunbridge, 1990; Gospodini, 2001). Most European cities (Athens, Rome, Berlin and Madrid) have many world cultural heritage sites and therefore offer a rich choice of diverse sights (BBSR, 2007), with historical buildings and architectural ensembles attracting continuously growing interest as tourist attractions. City neighborhoods, with their own traditional architectural complexes and listed buildings, are growing in popularity, since tourists increasingly designate their holiday to be also an educational or cultural tour. For example, in Germany, increasing tourist incoming flows are based mostly on the historic city cores and sites (Boberg, Fechner & Feist, 2009; Rescher, 2003), while almost one-third of all foreign travelers make classical city tours (Haspel, 2011).

Cultural tourism, being a labor-intensive sector, allows for the greatest impact on employment, with the least investment in resources, exploiting the cultural potential of the existing historical stock, which is particularly important especially in times of economic difficulties. Investing in the architectural heritage conservation is particularly important during periods of economic crisis, since it largely benefits the building industry and the regional labor market. In addition, cultural tourism focused on urban cultural heritage can help to enhance the image of the city and build value (Haspel, 2011). Law (1996) refers to tourism's important role in enhancing the image of the city, while also strengthening the sense of identity of the town and well-being within the local community. ICOMOS acknowledges that tourism can make a positive contribution to heritage conservation funding, boost the local economy, encourage the maintenance of urban services and raise awareness of heritage conservation (Barrera-Fernandez et al., 2014). Furthermore, the creation of new cultural infrastructures and the introduction of new cultural events are invigorated, for the benefit of both tourists and citizens (Barrera-Fernandez et al., 2016).

However, the relationship between cultural heritage and tourism may also generate potential conflicts that impose the need for sustainable management (Hmood et al., 2018: 210). During the last few decades, the sustainability of the historic urban landscape is being threatened by a rapid "touristification" process (García-Hernández et al., 2017). Worldwide, hustling urban tourism growth is often construed in strained infrastructure and overcrowding in many destinations, i.e., Barcelona, Venice, Amsterdam, be worrying for environmentalists and generating tensions between locals and visitors, whereas "overtourism" has become a buzzword (UNWTO, 2018). Historic neighborhoods are transformed into exclusive leisure districts with no residents, where, though the architectural aspects of the urban fabric are being maintained, the conditions as living urban space are being downgraded (García-Hernández et al., 2017). The transformation of cultural components into tourism products risks leading to loss of authenticity, which is an indispensable element of the cultural significance of urban landscapes, as expressed in the physical environment, collected memory and local traditions (André, Cortés & López, 2003; Barrera-Fernandez et al., 2016: 353). A "museification" effect (Lusiani & Panozzo, 2016) in historic urban districts may thus result in monofunctional areas, where new commercial activities and increased land values threaten to drive away the local population and traditional activities. Appropriate policies should be therefore planned in order to preserve and promote the authenticity of places and cultural experiences to boost the understanding and appreciation of cultural heritage (Pedersen, 2002 cited in Hmood et al., 2018: 210).

3 The Urban History of the City and the Upper City of Thessaloniki

Thessaloniki, historically a key port of the South-East Mediterranean and the most important gateway of the Balkan Peninsula to the Mediterranean, is one of the few cities in Europe that has been operating as an urban center for more than two thousand years (Vlami, 2009; Livadioti, 2013; Kostopoulou, 1996; Kefala & Samaras, 2019). Thessaloniki, Salonica or Selânik, endowed with a long lasting multicultural history, had one of the most varied communities in Europe (Jews, Latins, Armenians and others), fostering its cosmopolitan character (Hastaoglou-Martinidis, 1997; Kostopoulou, 2019; Svoronos, 1996; Vacalopoulos, 1972). Mazower (2004) portrays the distinctive multicultural features of the city through the centuries, as a Byzantine capital city, an administrative and commercial center during the Ottoman period, a Greek port, a Christian, Muslim and Jewish city, with a Balkan and Mediterranean mentality at the same time (Vlami, 2009), a city aptly described by Moscof (1978) to be sensed as a "feeling of the other, the love of the neighbor".

Thessaloniki was founded around 315 BC by king Cassander of Macedon, who named the city in honor of his spouse, half-sister of Alexander the Great (Vakalopoulos, 1972, 1985). Cassander gathered together in a single joint town the people of 26 small agricultural settlements scattered around the Thermaic Gulf (Hastaoglou-Martinidis, 1997). The plan for the first phase of Thessaloniki was typical of the early Hellenistic city (Lagopoulos, 2005), where the agora and other significant buildings occupied the central area, while the second agora was possibly near the harbor. The town was enclosed by walls on three sides, and the fourth was not walled on the seafront, with an acropolis containing a fort (Lagopoulos, 2005; Vakalopoulos, 1985; Vickers, 1985). Thessaloniki, situated at the crossroads between Europe and Asia, on the main routes connecting Europe with the East and the Northern Balkans with the maritime routes of the Eastern Mediterranean, became a significant economic, administrative and cultural hub, one of the most important cities of that period, increasing in importance both during the Roman Empire and in Byzantium (Kostopoulou, 2013). In Roman times, the city followed the Hellenistic grid plan (Lagopoulos, 2005: 64). The Byzantine period has been of great importance to the city that developed to become bigger, endowed with churches, sea walls fortifications and other buildings. However, the richness and prosperity of the Byzantine city enticed the conquerors who tried to and conquer it occasionally (Kefala & Samaras, 2019). Finally, in 1430, Thessaloniki was occupied by the Ottomans (1430–1912).

After the invasion of the Turks in Thessaloniki, those who escaped from the massacres left the city, and only 7.000 people stayed. In order to give life to the city again, Turkish families were brought from surrounding areas and settled in the houses of the fugitives. At the close of the fifteenth century, a large community of Jewish refugees from Spain settled in the city, fostering its multicultural and multi-religious profile (Vacalopoulos, 1972; Hastaoglou-Martinidis, 1997; Lagopoulos, 2005; Basteaand & Hastaoglou-Martinidis, 2013). With the arrival of the Jews Sephardim, the spatial separation of ethnicities began. The Jews settled near the seaside wall, the Christians gathered around their churches, and the Turks moved and built their houses up the hill, in today's Upper City "Ano Poli", a neighborhood noted for its mild climate, open spaces with trees and fountains and also a panoramic view of the city and the sea (Kefala & Samaras, 2019).

The city underwent a gradual modernization from 1870 onwards when the Ottoman Empire introduced westernizing reforms: the municipality accelerated the remodeling of urban space with a Roads and Buildings Regulation; the seaside walls were demolished, allowing the city to extend beyond its traditional limits and a new central business district (banks, offices, factories, modern shops, hotels) was developed; a modern quay was built that gave the city a new western character and new port facilities were built, while at the end of the century, a new rail connection linked the city with Europe and Constantinople (Yerolympos, 1992; Karadimou-Yerolympos, 1995; Gounaris, 1993; Hastaoglou-Martinidis, 1997). Since the 1890s, new suburban areas have been developed beyond the conventional center, sharpening the socioeconomic stratification of the city (Hastaoglou-Martinidis, 1997, Basteaand & Hastaoglou-Martinidis, 2013). The ethnic composition of the city's population was about 30,000 Turks, 16,000 Greeks, 12,000 Jews and 2000 Western Europeans in the early nineteenth century (Lagopoulos, 2005: 65). Until the beginning of the 19th century, Thessaloniki was strictly defined within the walls. The residential districts were assembled based on the religion and ethnicity of their residents, and the social hierarchy was more of an internal process for each ethnic group separately rather than a collective matter (Karadimou-Yerolympos, 1995 cited in Papadam n.d.).

Thessaloniki turned twentieth century into a multiethnic city, both oriental and occidental, with a population of 150,000 inhabitants. The spatial segregation of the population was clear, Christians, Jews and Muslims living into separate ethnic-religious quarters (Hastaoglou-Martinidis, 1997; Basteaand & Hastaoglou-Martinidis, 2013). The Turkish population was settled in the upper part of the city (Ano Poli), a relatively new quarter dating from the nineteenth century, away from the noise and polluted conditions of the congested lower urban areas, enjoying fresh air and views to the sea underneath and the hills behind the town. The Jewish population lived in the lower quarter next to the sea walls and the harbor, and Greek neighborhoods occupied part of the central area and the south-eastern part of the city, inside the walls, and along the western edge (Mazower, 2004; Morin, 2009; Stein, 2013). The trade sector and the European quarter were located in the western part, Western Europeans, mostly French, living in the south-western part of the city, near the port. (Lagopoulos, 2005). This mixture of cultures, characteristic of multiethnic East Mediterranean cities, persisted until the early twentieth century, when the political map of the region was altered by the disintegration of the Ottoman Empire and the 1912–13 Balkan Wars. Thessaloniki and the surrounding regions were effectively conquered by the Greek army from the Ottomans in 1912, absorbing them into the Greek state, thus drowning a new period in the history of the city. In 1913, of the city's 157,889 inhabitants, 61,439 (38.91%) were Jewish, 45,867 (29.05%) Turkish, 39,956 (23.31%) Greek, 6263 Bulgarian and 4364 belonged to other ethnic groups (French, English and Italian) (Karadimou-Yerolympou, 1985; Vickers, 1985; Lagopoulos, 2005). Thessaloniki became an important base for the Entente Powers after the outbreak of World War I. The increased demand for goods, created by the presence of 200,000 Allied troops, dramatically increased imports and boosted the economic life of the city (Hastaoglou-Martinidis, 1997).

Thessaloniki witnessed two consecutive major events within ten years of its liberation, which turned the cosmopolitan Balkan city into a modern regional metropolis: the destruction of its historic center by the great fire of 1917 and the arrival of 117,000 refugees after 1922, following the exchange of population between Turkey and Greece at the end of the Asia Minor campaign (Hastaoglou-Martinidis, 1997).

In August 1917, a major fire leveled 128 ha (316 acres) of the tightly built historic center, including the commercial sector, leaving some 70,000 people homeless. The Jewish community was hardest hit, the fire destroying three-quarters of the Jewish neighborhoods, having some 50,000 Jews massively resettled (Papastathis, 1978, Hastaoglou-Martinidis, 1997, Lagopoulos, 2005, Papastathis & Hekimoglou, 2010, Basteaand & Hastaoglou-Martinidis, 2013). Immediately after the devastating fire, the Ministry of Communications responsible for town planning set in motion the process of replanning the city according to the new modern urban planning methods (Basteaand & Hastaoglou-Martinidis, 2013). The homeless were moved out of the walls into temporary shelter, and new building was absolutely prohibited in the devastated zone (Hastaoglou-Martinidis, 1997; Yerolympos, 1996). An International Commission of specialists was set up for the New Plan of Thessaloniki, guided by the French architect Ernest Hébrard (1881–1933). Hébrard's plans (Fig. 1) aimed to transform Thessaloniki into a national urban center of monumental scale, balancing between East and West.

The commission completely rebuilt the downtown area under Hébrard's leadership, initiating a central north–south civic axis, Aristotelous Avenue, still visible today, linking the Upper Town with the sea (Volkgenannt 1995, Lewkowicz, 1999, Basteaand & Hastaoglou-Martinidis, 2013). Except for the Upper Town, which was preserved because of its picturesque features and beauty, remnants of the architecture of the old city were swept away (Lagopoulos, 2005). Lavedan () describes the "exotic" Upper City, indicating that Hébrard chose to "fully respect its appeal to the imagination" (Karadimou-Yerolympou, 1985, Lagopoulos, 2005).

In 1922, at the end of the war between Greece and Turkey and the Smyrna tragedy, Thessaloniki was overrun by refugees from Asia Minor, thus becoming the "refugee capital". The compulsory exchange of minority communities between the two countries under the terms of the Treaty of Lausanne in 1923 followed this initial forced migration. Between 1920 and 1923, 1.2 million Greeks left Asia Minor, and 355,000 Muslims emigrated to Turkey (Hastaoglou-Martinidis, 1997: 498).

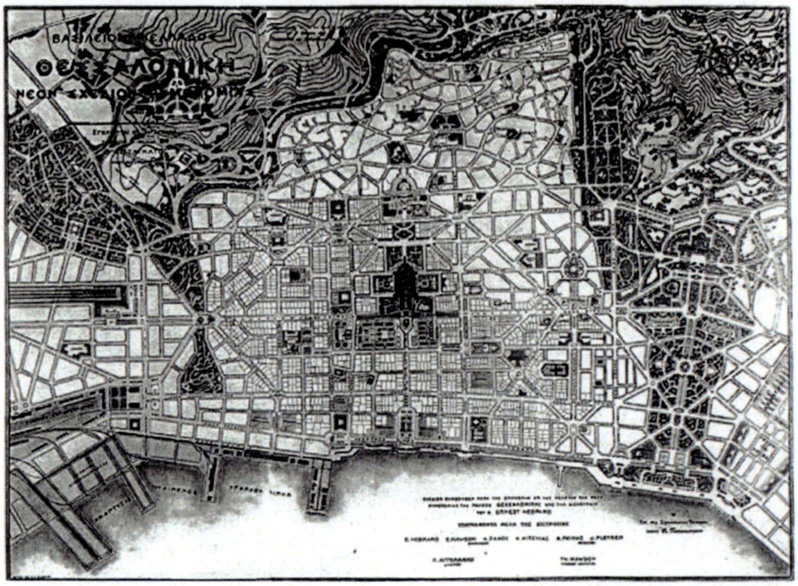

Fig. 1 Plan for the intra-muros section of Thessaloniki by Hébrard, 1918. *Source* Yerolympos (1996)

Thessaloniki was the largest destination point for the urban refugees: 117,000 of them settled in the city between 1920 and 1928 (as against the 25,000 exchanged Muslims), significantly raising the population and causing an acute housing shortage (Hastaoglou-Martinidis, 1997; Lewkowicz, 1999). The area of the city was doubled in size, and new neighborhoods developed on the outskirts eastwards and westwards of the city changing the urban landscape, while the traditional multicultural character of the city faded away (Kefala & Samaras, 2019). The more affluent refugees moved to the houses left behind by the Turks in Ano Poli, while the poor filled every open space with minuscule squatter houses in the spacious quarters of the Upper Town, the last manifestation of real vernacular architecture (Hastaoglou-Martinidis, 1997; Hastaoglou-Martinidis & Yerolympos, 1986).

With the start of World War II, all architectural and planning operations in the city ceased. In 1943, Thessaloniki experienced the mass expulsion of virtually the entire Jewish community (50,000 people) to the German camps, where most were executed, a loss of a minority that had been a significant part of the city's character for centuries. In the post-war years, a large number of internal migrants moved to the city, increasing the population to over 300,000. Lacking any effective planning control to respond to the demanding issue for habitation of the great wave of internal migration in the 1950s and 1960s and in an effort to reinvigorate the local economy, the government decided to introduce a considerable increase in building coefficients that permitted an extensive exploitation of building plots and triggered unleashed intense construction operation in the 1960s and 1970s. Concurrently, the large wave of internal migration increased both the population (557,000 inhabitants by 1971 and 706,000 by 1981) (Greek Statistical Authority, 2011) and the size of the city, establishing squatter settlements on the western outskirts (Hastaoglou-Martinidis, 1997; Basteaand & Hastaoglou-Martinidis, 2013). Thessaloniki was reconstructed quite fast although in a fragmented way; large-scale infrastructure construction combined with the (re)industrialization of its economy from the 1950s until early 1980s formed the city's modern character.

4 Upper City's Architectural Heritage Preservation

Ano Poli is one of the most characteristic areas of Thessaloniki. It occupies the north and highest part of the old city within the Walls, delineated by the Byzantine City Walls on three sides and by Olympiados Street to the south. It is built amphitheatrically on a hill with steep slopes, overlooking the Thermaic Gulf and the city. According to historical sources, the systematic inhabitation of the area begun after the conquest of Thessaloniki by the Turks in 1430 and especially after the seventeenth century. In the beginning, the Turkish residences in Ano Poli were not close to each other, but with the installation of Turkish refugees from Bosnia in 1876, the empty plots got filled with additional residences, and new neighborhoods were created. The area was densely built in the late nineteenth century, when the city was liberated in 1912. The great fire of 1917 did not hit this zone, and thus, many homeless people moved there. After the Asia Minor Catastrophe in 1922, Ano Poli was inhabited by many refugees that came in Thessaloniki to find a shelter (Kefala & Samaras, 2019).

During the Ottoman domination, the residences of the area were similar to those in the wider Balkan region with elements of folk Macedonian and Turkish architecture. The Turkish residences were decorated with timber pilasters and frames between the storeys, had large openings on the external walls and a triangular pediment with carved decorations on their facade. The vernacular Macedonian residences did not have carving decoration, rather small openings on the external walls, while their characteristic morphological elements were the "sachnisi", the closed protrusions on the floor, and the "hagiatia", covered open space on the first or ground floor (Kefala & Samaras, 2019). There were also many Jewish residences with strong influence on typology and decoration from the neo-classicism of Western Europe. The residences that were constructed by Greek refugees after 1923 had mostly only one storey, and they were made without any trace of luxury. However, they were distinguished for their functional structure and the bright colors with which their facades were painted. A street plan was imposed in the area in 1931, however, limited to the bounded widening of the old roads and the construction of few new ones. A free zone around the Byzantine Monuments and a green belt on both sides alongside the City Walls were introduced by a Presidential Decree (Gov. Gazette 406A/04.12.1931). Upon implementation of the Presidential Decree, about 900 houses were partially or fully demolished, the buildings of this zone ever since called "kastroplikton" (Kefala & Samaras, 2019).

By the late 1950s, the Ano Poli settlement, built around early Christian, Byzantine and Ottoman monuments, was a cohesive community. The residential growth of Thessaloniki in the early sixties right up until the seventies, with land compensation, had detrimental effects on the settlement of Ano Poli. Commercial apartment blocks were built, incompatible with the existing historic buildings in size and form, while many prominent buildings were demolished, and traditional neighborhoods were eliminated, marking the disappearance of much of the historic architectural heritage (Hastaoglou-Martinidis & Christodoulou, 2010; Vyzantiadou & Selevista, 2019).

At the beginning of the 1970s, the city undertook efforts to restore fragments of past periods into the urban fabric, reactivate the architectural heritage and strengthen its rich historical stratification (Hastaoglou-Martinidis & Christodoulou, 2010; Basteaand & Hastaoglou-Martinidis, 2013). According to the Greek Constitution, the State is primarily responsible for the strategic spatial planning, including the building policy which is practically used as a strategy of spatial organization (Aravantinos, 2007). Nationwide, the protection and enhancement of cultural heritage is a responsibility of the Greek State pursuant to Article 24 of the Constitution, as a consequence of the country's international obligations, legally recognized by the Greek State (the 1964 Venice Charter, the Declaration of Amsterdam of 1975, Paris Convention of 1972, the World Heritage Convention of 1981, the Granada Convention of 1985) and relevant laws. After the 1975 Constitution was signed, a series of laws defined in greater detail the terms of natural and cultural environment, following the new concepts of the protection and management of cultural environment. Within 1975, the first law was suggested for the urban planning and the protection of the environment, which included, among others, the protection of the cultural environment and of historic spaces. A while later, steps were taken to establish a policy of protection and preservation of the historic centers of the greater cities, like Ano Poli in Thessaloniki, while a Presidential Decree (GG 594/Δ/13-11-1978) recognized and recorded 400 traditional settlements (Avgerinou-Kolonia, 2015: 330). Since 1986, the Ministry of Interior, sector of Macedonia and Thrace, conducted a significant part of this task for Thessaloniki (G.G.I., 157/A/10-10-1986). Within the institutional framework, architectural heritage is defined as "an irreplaceable expression of the richness of the national cultural heritage and an invaluable testimony of the past". The definition of architectural heritage includes traditional architectural structures, residential complexes, traditional villages, general elements of the human environment and sites, landscapes or natural formations. (G.G.I. 61/A/13–4-1992).

In the late '70s and over the '80s, the cityscape of Thessaloniki began to change through renovation of old buildings and gentrification of areas that were considered as most representative of the city's history. The targeted areas were mainly Ano Poli and "Ladadika", the port district, both areas not destroyed by the "Great Fire" of 1917, thus preserving the "old flavor" of the city's past, with old buildings, narrow cobblestone streets and traditional urban activities. These areas were characterized as "preserved historical areas", and it was therefore not allowed to alter the buildings without the license of the Ministry of Culture (Kokot, 2003). The traditional Upper Town quarter became the site for the introduction of a controversial project for restoration and renovation, which allowed two- or three-storey apartment buildings to be constructed (Hastaoglou-Martinidis & Christodoulou, 2010).

The first systematic effort to restore the area's identity was after the major earthquake in 1978, when a series of administrative operations were carried out (1979–1985) on the recommendation of the Minister for the Environment, Physical Planning, and Public Works (Vyzantiadou & Selevista, 2019). These actions contributed to the designation of Ano Poli as a traditional settlement according to the published Decree of 12.03.1979 (G.G.I. 197/D/2-4-1979), the definition of special terms for building and land limitations under the urban plan (G.G.I. 313/D/31-5-1979) and the amendments and updates to the Thessaloniki Urban Plan

(1980 and 1985) (Mavromatis, 1997). Only 48 out of the 132 buildings that were proposed were designated by the Ministry of Culture as pieces of art requiring special state protection, 40 residences in Ano Poli and eight residences in the area along the walls (G.G.I. 680/B/14-8-1979). Anastasiadis (1989), based on an on-site survey carried out in 1979, reports that the houses of the Ano Poli were morphologically divided into five categories: buildings of distinct architectural and historical interest with clear typological features of popular Macedonian, Balkan or neoclassical architecture and a significant number of decorations (merely 1.51%); buildings with less significant morphological features that do not fully specify the time of their construction, have been damaged and have lost the authenticity that once distinguished them, and their decoration is limited (4.49%); buildings without essential architectural and morphological features, mainly refugee residences that have been altered and modified since the time of their construction (dominate the area, 76.77%); new buildings (8.95%); blocks of flats (8.28%) (Kefala & Samaras, 2019).

The qualitative improvement and revitalization of the Upper Town was included in the Thessaloniki Master Plan, approved in 1985 (G.G.I. A· 420/27.4 1993), together with other traditional districts of the city (Hastaoglou-Martinidis & Christodoulou, 2010; Basteaand & Hastaoglou-Martinidis, 2013). To preserve the architectural and urbanistic character of Ano Poli, successive presidential decrees and decisions were adopted in 1999, 2000, 2002 and 2007 (Mavromatis, 2009) which set out both the town planning rules and the rules for reconstruction plots after the demolition of old buildings that were not designated as listed (Kefala & Samaras, 2019).

The recent spatial planning system at the municipality level endeavors to meet spatial development weaknesses, by providing two main planning tools: the "General Urban Plan" (GUP) and the "Town Plan Study" (TPS) (Serraos et al., 2005). The GUP constitutes the first step of the local spatial planning procedure for the whole municipality, providing the general guidelines for spatial development. In April 2019, Thessaloniki's Municipality Council approved the final stage of the General Urban Plan Study (Urban Masterplan) (Ministry of Environment and Energy, 2015). The Plan's main strategic goals focus on "Thessaloniki's emergence into a modern European metropolis and a tourist, transit, exhibition and innovation center of the broader area, capable of hosting international events, promoting quality of life for the inhabitants and providing extended services by the Local Authorities". The main development axes of the proposed urban plan include the promotion of Thessaloniki as a tourism destination, e.g., religious, cultural and conference tourism. Under the former General Urban Plan of Thessaloniki (1993), the city was organized into 101 urban areas (neighborhoods), while the General Urban Plan of 1988 distinguished only three urban areas. The recent spatial plan proposes an operational restructuring into 24 neighborhoods, grouped into two wider domains, the Central West with ten neighborhoods (S1–S10) and the Northeastern with 14 neighborhoods (S11–S24) (Fig. 2). The contemporary land use model for Thessaloniki Conurbation Area refers to a compact city center with the traditional settlement of Ano Poli and Eptapyrgio (S5) as a capstone, with linear expansions to the east and west.

Located within the Central West domain, over a surface of 749 thousand m^2, Upper City, with the Neighborhood Code

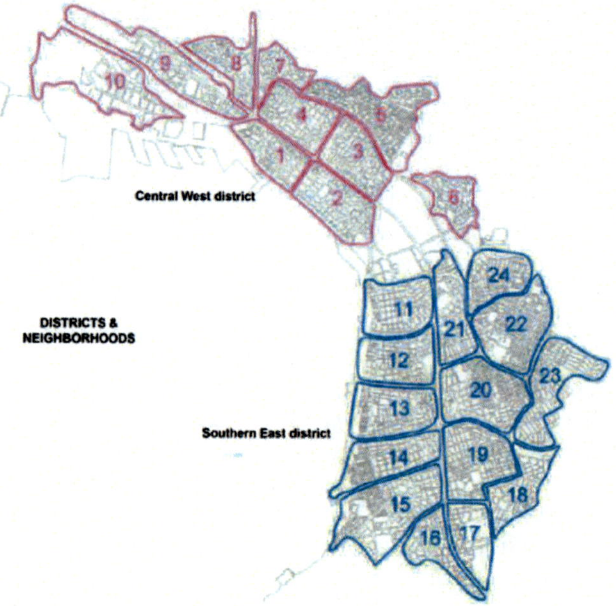

Fig. 2 Proposed spatial organization of Thessaloniki Municipality into neighborhoods (GUP). *Source* Ministry of Environment and Energy, Thessaloniki Municipality (2015)

S5 Ano Poli—Eptapyrgio, is organized into two urban areas (UA 5a and UA 5b) with the internal acropolis wall as a limit. The traditional settlement of Ano Poli is characterized as a residential area, where the following land uses are allowed (Government Gazzette 904Δ'/21.12.1999): residence, hostels and buildings of social infrastructure and public service. By way of exception, the following uses are allowed, under specific conditions along the streets (Article 4): retail stores; trade shops (leather and fur tanning, coppersmiths, etc.); pastry shops, coffee shops, restaurants and exhibition rooms; offices along Theofilou str (Amendment of Article 4 PD 17.5.79 by Article 4: subparagraph c (3) after Committee opinion). The establishment of disturbing industries and other similar uses with environmental impacts is strictly prohibited (Ministry of Environment and Energy, 2015).

Even though protected with specific regulation for land uses, plot ratios and building restrictions (Government Gazzette AAΠ 396/03.09.2007), Ano Poli has to confront a number of significant problems: (i) There are numerous listed buildings and buildings of architectural interest which are abandoned or not restored by their owners, resulting in a decay image of the wider area; (ii) The narrow, labyrinthine streets result in lack of appropriate accessibility. During the last years, the area is served by public transportation, however, without the expected results in reference to the connection of the area with the city center; (iii) The narrow streets network combined with restricted private and public parking areas for the residents, and even more for visitors, result in a severe mobility constraint; (iv) The concentration of land uses incompatible for the area, i.e., small restaurants, tavernas, bars and other recreational uses in specific locations, creates problems to the residents due to overcrowding during tourist seasons/hours especially in the most attractive sites, severely aggravated by traffic congestion and parking problems. More specifically, for the Eptapyrgio neighborhood, there is a tendency for the establishment of recreational uses and restaurants that disturb the residents (Ministry of Environment and Energy, 2015). Thus, the revision of the urban plan study of Ano Poli is recommended, with the inclusion of Eptapyrgio, to better serve the inhabitants.

5 Architectural Heritage Tourism Destination: Upper City Thessaloniki

A city with an abundance of monuments, religious buildings (e.g., Rotunda, Galerius Arch, White Tower, Mustafa Kemal Ataturk Residence, etc., and numerous museums (e.g., Archaeological Museum, Byzantine Culture Museum, Jewish Museum, etc.), Thessaloniki is considered a privileged cultural heritage tourism destination. The most emblematic features of urban heritage are clustered in the historic center of the city, covering an area of 330 ha. However, as Hastaoglou-Martinidis and Christodoulou (2010: 122–124) aptly indicate, the city's historical character and cultural identity can not only be remembered by its historical core, but should also include the historical elements dispersed across the metropolitan area. In addition to the main archaeological and historical complexes of the historic center (e.g., Aristotle Square, Forum's Ancient Greek and Roman Ruins), there are also several interesting traditional districts for tourists to explore, such as the Kapani and Modiano street markets, Ladadika and Upper Town.

New initiatives have been put forward since 1985 to boost urban heritage in order to draw more tourists, including hosting international cultural activities and promoting the cultural identity of the region. In 1993, the cultural activities of the Mediterranean crossroads were set in widely unknown or inaccessible historical sites of the city (the Roman forum, the old port market, the surroundings of Eptapyrgio in Ano Poli), while in 1996, the Biennale of new artists took place in extraordinary, yet abandoned heritage buildings and sites (Yeni Cami, Alaca Imaret), thus incorporating heritage into contemporary cultural life of the city (Hastaoglou-Martinidis & Christodoulou, 2010). Thessaloniki's designation as the Cultural Capital of Europe for 1997 served as a turning point for increasing cultural tourism, while numerous urban and architectural projects were advanced, e.g., the integration of the old harbor into the urban fabric, restoration works in the Galerius and the Roman Forum complexes, landscaping works of the Byzantine heritage like the Eptapyrgion in the Upper City, restoration works on Ottoman monuments, e.g., the Bedesten, the Aladja Imaret and the Yeni Cami (Hastaoglou-Martinidis & Christodoulou, 2010). Furthermore, the fact that several of Thessaloniki's monuments were included in the UNESCO World Heritage list significantly upgraded the city as a tourist destination.

However, the city has yet to capitalize on the cultural heritage of the peripheral urban landscape. Ano Poli, built on a higher ground than the modern city, partially surrounded by the old citadel walls, with numerous Byzantine monuments, churches and monasteries, exudes the atmosphere of an earlier era. One of the city's most important sights is the Byzantine fortification walls built by Theodosios the Great in the fourth century, while the walls end at Trigonio Tower offer a unique overview of the city, attracting locals and visitors: the forest on the left and the Thermaikos Bay stretching out beneath with a majestic view to Mt Olympus. Ano Poli is also hosting a major part of the city's metabyzantine urban fabric, as this was transformed after the refugees' establishment in the district during the 1920s. Besides the impressive Byzantine and Ottoman monuments, several parts of the conventional city grid, narrow stone-paved alleys, traditional buildings with the unique elegance of mixed Macedonian and Ottoman vernacular architecture, while also neoclassical and eclectic style, compose Thessaloniki's most cohesive and colorful neighborhood, a unique area of interest for cultural tourists. Despite the construction of newer buildings and the opening up of new streets in recent decades, the district has kept the old traditional urban fabric with the features of medieval cities, the labyrinthine narrow cobblestone streets and scant squares, the charm of architectural heritage preserved, capturing the history of the area and the city (Kefala & Samaras, 2019).

The particular atmosphere of this picturesque landscape has lured intellectuals, artists, architects, also traditional bourgeois and middlemen, who chose it for settlement in recent years. Furthermore, the Upper Town, with the traditional neighborhoods and a strong multiethnic urban character, played a central role in the literary portrayal of the city since the 1950s, most of which has concentrated on the physical deterioration of the built environment. Many artists, such as the poets George Themelis, George Vafopoulos and Ntinos Christianopoulos, the writer Vassilis Vassilikos, best known for his political novel "Z", wrote about the disappearance of historic parts of the city and its medieval and Ottoman street pattern (Kefala & Samaras, 2019; Basteaand & Hastaoglou-Martinidis, 2013).

Nowadays, the Upper City attracts daily a considerable number of visitors, all year round. According to the Trip Advisor evaluation of the area and also the ratings through international and national/local webpages of the tourism industry magazines and professionals, Ano Poli is on average included in the lists of the "10 best places to visit in Thessaloniki". However, a thorough academic research about the tourism development of the area resulted in the lack of relevant literature. On the opposite, documentation about the architectural heritage of Ano Poli is very rich, in both Greek and international studies and publications. The lack of research on the cultural heritage tourism topic in the Upper City of Thessaloniki initiated the research undergone within this study, as a first approach to the perceptions of visitors about the architectural heritage as a major attraction for the area and the city. As a first step, a survey was conducted among young visitors, to investigate the perceptions of the most promising segment of the rising cultural tourism market, about the architectural heritage tourism in general, and specifically for the Upper City and the city of Thessaloniki.

In this exploratory research, a self-administered questionnaire was designed, to examine the relationship between architectural conservation and tourism development within an urban environment. The survey questionnaire was addressed to Erasmus+ and Erasmus International students at the Department of Economics, Aristotle University of Thessaloniki (AUTh). The students were invited, by e-mail, to participate in the online questionnaire, while no incentives were used to encourage participation. The questionnaire was administered from December 4, 2019, to January 4, 2020. At the top of the questionnaire, a brief explanatory note introduced the respondents to the research survey topic and aim that is to profile their perceptions about the architectural heritage and its role in urban cultural tourism development, with specific reference to the Upper City (Ano Poli) district of the city of Thessaloniki. It was also confirmed that the questionnaire is part of an ongoing academic research and their answers would remain anonymous.

The survey included two sections in order to answer the research questions: The first section contains questions on the sociodemographic characteristics of students-visitors and the second section on their perceptions of the research subject, architectural conservation and tourism growth. The second section of the study, which was divided into three parts, aimed at understanding the views of students-visitors about the value of architectural conservation in urban tourism. The first part of this section dealt with the impressions of respondents of the elements that contribute to the understanding of the authenticity of a location (architectural heritage (historical buildings, traditional neighborhoods), traditional culture (music, dance, etc.), traditional food and drink, interacting with local residents, traditional souvenirs and then the significance they assign to the architectural heritage during their visit. The second part tested how the participants rate the main tourist attractions and their general impression of architectural heritage within the city of Thessaloniki and more specifically within the research area of Upper City (Ano Poli) in Thessaloniki. The final part of this section asked participants to rate the effectiveness of promoting Ano Poli as a tourist attraction, their willingness to explore the area's architectural heritage, for example, by engaging in cultural tours, as well as their intention to recommend Ano Poli as a worth visiting architectural heritage attraction.

A sample of 103 students were invited, practically all the Erasmus students that visited the Department of Economics, Aristotle University of Thessaloniki, within the academic years 2017–2018, 2018–2019 and 2019–2020. A total of 42 completed questionnaires were obtained, with a response rate of approximately 41%. The majority of the participants for the study were female (52.6%), within the age range of 20–27 years old, an age profile that was expected since they were University students, hence considered to be well-educated visitors. Most of the respondents had been or still are at Thessaloniki for one academic semester that is 5–6 months. The main countries of origin of the participants are Russia (16%), Romania (15%), Italy (11%), and Germany (10%), while at lower rates, there were also students from France, Spain, Belgium, Cyprus, Moldova, Ukraine, Georgia, Uzbekistan, and Lebanon. The multinational synthesis of the sample gives the opportunity to collect perceptions from a variety of countries, socioeconomic and educational systems and cultures.

The research asked which elements contribute most to the visitors' perception of the authenticity at a destination. Visitors rate the contribution of five elements to their authenticity perceptions at a destination using a Likert scale on which 1 was "does not contribute at all and 5 was" contributes very much to authenticity perceptions. With a ranking of 4.1, architectural heritage (historical structures, typical neighborhoods) added the most to visitor prospects. Traditional food and beverage (3.9) had the same score as interacting with local residents (3.9), indicating that both material (food) and immaterial (interaction) aspects of the tourism experience are equally important, followed by traditional culture (music, dancing, etc) (score 3.5). By comparison, the purchasing of traditional souvenirs (score 2.4) does not seem to contribute to the comprehension of authenticity by visitors compared to the previous four elements, thereby increasing the "End of Tourism as We Know It perception by the young generation of tourists (Wonderful Copenhagen 2017)". In reference to the general perception of the importance of architectural heritage while visiting a place, on a Likert scale on which 1 was "not at all" and 5 was "very much", the vast majority of the respondents (42%) rated it as very important, a considerable percentage (32%) rated it important enough, while only 5% rated it as slightly important.

With regard to the rating of the main tourist attractions of architectural heritage in Thessaloniki, the Ano Poli (Upper City) district received the highest score (4.3), followed by the historic district of Ladadika (3.8), near the port area. It is worth noting that students gave priority to Ano Poli as a peripheral traditional neighborhood with mixed uses and not to Ladadika, an area easily accessible in the city center, that turned out to become an exclusively leisure district. Respondents rated their general impression of architectural heritage in Thessaloniki and in the Ano Poli (Upper City) district of Thessaloniki, with a total score of 3.9 and 4.4, respectively. On a Lickert-scale from 1 "very bad impression" to 5 "highly rated impression", most respondents rated their general impression of Thessaloniki's architectural heritage as "highly rated" (26.3 percent) and "well rated" (52.6 percent), and respectively for the Ano Poli's (Upper City) district architectural heritage as "highly rated" (42.1 percent) and "well rated" (52.6 percent). Results confirm the wealth of architectural heritage of the city and even more of the Ano Poli district, both very much appreciated by young travelers.

Consistent with the previous result, almost 2 out of 3 of the respondents expressed their strong will to explore the architectural heritage in the Ano Poli (Upper City) district of Thessaloniki (not professional understanding, rather educational and cultural experiences, e.g., tours, workshops). On a Likert scale on which 1 was "not at all" and 5 was "very much", the vast majority of students are willing to explore the architecture of the area "very much" (36.8%) and "too much" (31.6%), thus designating the potential of the area to attract the next generation of mature cultural tourists. The responses about recommending Ano Poli as an architectural heritage tourism attraction were even more supportive, since roughly more than 80% of respondents were in favor. On a Likert scale on which 1 was "absolutely not" and 5 was "yes, without doubt", the vast majority of students rated "very much" (63.2%) and "too much" (26.3%) willing to promote the area. Given the experience of students with communication tools, such a "promotion campaign" could go much further than the "world of mouth" impact, to the exponential viral promotion effect of social media (Facebook, Instagram, Twitter, etc.). The multinational, multiethnic, multicultural Erasmus networks of the universities all over the globe may act as a powerful tool for such a goal. This proves to be of major importance for the Upper City area and the city of Thessaloniki, considering the perceptions of the respondents about the effectiveness of the actual promotion of Ano Poli as an architectural heritage tourism attraction. On a Likert scale on which 1 was "not at all" and 5 was "very much", a considerable percentage of respondents rated the promotion as medium (36.8%) and rather poor (10.5%), while only (26.3%) rated it "very much". This result confirms that, although the Upper City architectural heritage stimulates the visitors' interest, it remains a potential tourist destination still untapped. The lack of an organized policy to promote the area as an architectural heritage tourism attraction underlines the necessity to organize a well-documented research and tourism development project, exclusively prepared for the promotion of Ano Poli as an architectural heritage destination of the city.

6 Concluding Remarks

Tourism exploitation of traditional neighborhoods in cities should fit to their unique historic character and traditional settlements, exploring their histories over time in search of the *genius loci*, the spirit of the place, the tangible built environment, together with the intangible cultural heritage and "invisible" unique memories of the neighborhood, so as to form a distinctive cultural tourism milieu profile.

The case study introduced to highlight this approach, the historic neighborhood of Upper City (Ano Poli) in Thessaloniki, is considered as a typical example of a traditional architecture urban ensemble. Even though well documented regarding architectural conservation policies, the analysis indicated the lack of research on architectural heritage tourism development in the area and the city.

The findings of the report highlight the review of the field study, resulting in a deeper understanding of the significance of architectural heritage in urban tourism. Analysis limitations should be noted, however. First only visitors to Ano

Poli were examined in the report, so the findings could not be generalized to visitors to other historical districts in Thessaloniki or other cities in Greece or elsewhere. Second, the sample plan, which was centered on a specific set of Erasmus students, did not meet the criteria for random sampling and should be regarded as experimental. In addition, the duration of sampling was limited to just one month, but included frequent academic holidays, when students had more free time. Finally, the survey questions were chosen on the basis of a review of the related literature, so important topics that could contribute to the study might be overlooked.

Architectural heritage research and its role in cultural tourism development in urban settlements, such as the Upper City (Ano Poli) in Thessaloniki, are at a rather early stage, so there are great opportunities to further explore the issue. This research examined only the perceptions of a particular category of visitors, so it would be instructive to analyze the perceptions of other groups of visitors, e.g., in terms of age, occupation and income. Additional research to examine the perceptions of the residents is also required, so as to form a holistic research approach to the topic, to include also "touristification" issues awareness. It is widely acknowledged that architectural heritage conservation generates opportunities for socioeconomic and urban tourist development, while also tourism contributes to the preservation and reintegration of traditional settlements into the urban fabric, still challenging sustainability principles have to be met.

References

Alonso, V.I., & Meurs, V. M. (2012). Assessing the performance of conservation activities. In S.M. Zancheti & K. Similä (Eds.), *Measuring heritage conservation performance* (pp. 1–14). Rome: ICCROM.

Anastasiadis A. (1989). Thessaloniki: Upper Town, *Greek Traditional Architecture, 7*, 25-58. Athens: Melissa.

André, M., Cortés, I., & López, J. (2003). Turismo cultural: Cuando el recurso cultural supera al destino turístico. El caso de Figueres. *XII Simposio Internacional de Turismo y Ocio*, Barcelona, ESADE-Fira, pp. 5–22.

Aravantinos, A. J. (2007). *Urban planning: Towards a sustainable development of the urban space (in Greek)* (2nd ed.). Symmetria Publishing.

Ashworth, G. J., & Tunbridge, J. E. (1990). *The tourist-historic City*. Belhaven.

Barrera-Fernandez, D., Castillo, L., & Azevedo Salomao, E. M. (2014). Tourist use of historic cities. Review of international agreements and literature. *International Humanities Studies, 1*, 1–11.

Avgerinou-Kolonia, S. (2015). Greece and the year of European Architectural Heritge 1975. *Monumenta*, Vol. *3*, pp. 322–331.

Barrera-Fernández, D., Hernández-Escampa, M., & Balbuena-Vázquez, A. (2016). Tourism management in the historic city. The impact of urban planning policies. *International Journal of Scientific Management and Tourism, 2*, 349–367.

Barrera-Ferrandez, D. (2016). *Attracting visitors to ancient neighborhoods: Creation and Management of the historic city of Plymouth, UK*. Groningen: InPlanning. https://doi.org/10.17418/PHD.2015.9789491937194.

Basteaand, E., & Hastaoglou-Martinidis, V. (2013). Modernization and its Discontents in Post 1950s Thessaloniki: Urban change and urban narratives. In P. Panayota (Ed.), *Landscapes of development. The impact of modernization discourses on the physical environment of the Eastern Mediterranean* (pp. 91–117). Harvard Graduate School of Design, Aga Khan Program of the Graduate School of Design.

BBSR—Federal Ministry of Transport, Building and Urban Development and Federal Office for Building and Regional Planning (2007). *Urban preservation and tourism development—With special attention to the UNESCO World Heritage Sites*. Berlin-Bonn. http://www.bbsr.bund.de/cln_032/nn_22414/BBSR/DE/.

Behr, A. (2000). Economic and structural effects of the conservation of urban heritage. In *Heritage conservation and employment, international conference within the framework of the German EU-presidency, 15–16 April 1999. Publications of the German Cultural Heritage Committee, 62*, 44–65. Bonn/ Bühl.

Boberg, K., Fechner, O., & Feist, R. (Eds.) (2009). *Cultural tourism. Future for the historic city. Sustainable and economical city management through interdisciplinary*. Act. Weimar.

Bowitz, E., & Ibenholt, K. (2009). Economic impacts of cultural heritage-research and perspectives. *Journal of Cultural Heritage, 10*, 1–8. https://doi.org/10.1016/j.culher.2008.09.002

Buchanan, J. M. (1965). An economic theory of clubs. *Economica, 32* (125), 1–14. https://doi.org/10.2307/2552442

Castela, A. (2018). Impacts of tourism in an urban community: The case of Alfama. *Athens Journal of Tourism, 5*(2), 133–148. https://doi.org/10.30958/ajt.5-2-4.

Cazes, G., Potier, F. (1996). *Le Tourisme Urbain (Urban Tourism)*. Paris: Presses Universitaires de France, coll. "Que sais-je?"

Dewar, K. (2000). Cultural tourism. In J. Jafar (Ed.), *Encyclopedia of tourism* (pp. 125–6). London and New York: London Routledge.

Dietvorst, A. G. J., & Ashworth, G. J. (1995). Tourism Transformations: An Introduction. In G. J. Ashworth & A. G. J. Dietvorst (Eds.), *Tourism and spatial transformations—Implications for policy and planning* (pp. 1–12). CAB International.

García-Hernández, M., de la Calle-Vaquero, M., & Yubero, C. (2017). Cultural heritage and urban tourism: Historic city centres under pressure. *Sustainability, 9*(8), 1346. https://doi.org/10.3390/su9081346

Geppert, A. (2004). Planning systems facing heritage issues in Europe: From protection to management, in the plural interpretations of the values of the past. *European Spatial Research and Policy, 21*(2), 9–12. https://doi.org/10.1515/esrp-2015-0001.

Gholitabar, S., Alipour, H., & Costa, C. M. M. D (2018). An empirical investigation of architectural heritage management implications for tourism: The case of Portugal. *Sustainability, 10*(1), 93. https://doi.org/10.3390/su10010093.

Gospodini, A. (2001). Urban design, urban space morphology, urban tourism: An emerging new paradigm concerning their relationship. *European Planning Studies, 9*(7), 925–934. https://doi.org/10.1080/09654310120079841

Gospodini, A. (2007). Cultural and leisure clusters in Greek cities: Spontaneous formation and 'laissez-faire' development. *International Journal of Sustainable Development and Planning, 2*(2), 119–133. https://doi.org/10.2495/SDP-V2-N2-119-133

Gospodini, A. (2007). Cultural clusters in the post-industrial city: The Greek experience. *Sustainable Development and Planning III, WIT Transactions on Ecology and the Environment, 102*, 769–780. https://doi.org/10.2495/SDP070732

Gospodini, A., Beriatos, E., & Raskou, E. (2007). Architectural heritage conservation: Progress in politics in Europe and new challenges for Greece (in Greek). *Aeihoros, 6*, 146–173.

Gounaris, B. (1993). Salonica, *Review*, *XVI*(4): 499–518.

Government Gazette Issue (G.G.I.). (1979). *Designation of Ano Poli Area of Thessaloniki as Traditional Settlement. 197/D/2-4-1979* (in Greek) (pp. 2323–2325). https://www.et.gr. Accessed 11 Dec 2018.

Government Gazette Issue (G.G.I.). (1979). *Definition of special terms of construction and restrictions on the land areas of the urban plan of Ano Poli of Thessaloniki 313/D/31-5-1979* (in Greek) (pp. 3489–3495). https://www.et.gr. Accessed 11 Dec 2018.

Government Gazette Issue (G.G.I.). (1979). *Designation of (40) Houses in Ano Poli of Thessaloniki and (8) at protected area of ByzantineWalls as pieces of art that need the state protection. 680/B/14-8-1979* (in Greek) (pp. 5904–5906). https://www.et.gr. Accessed 11 Dec 2018.

Government Gazette Issue (G.G.I.). (1992). *Convention for the protection of the architectural heritage in Europe. 61/A/13-4-1992* (pp. 1340–1352). https://www.et.gr. Accessed 11 Dec 2018.

Government Gazette Issue (G.G.I.). (1993). *Law 1561, Government Gazette A·148/6.9.198S, and the general urban plan for Thessaloniki, approved early in 1993, Government Gazette A· 420/27.4 1993*.

Greek Statistical Authority. (2011). *Population of Thessaloniki* (in Greek). http://www.statistics.gr/portal/page/portal/ESYE/BUCKET/General/A1602_SAM01_DT_DC_00_2011_02_F_GR.pdf.

Haspel, J. (2011). Built heritage as a positive location factor—Economic potentials of listed properties. ICOMOS Atellers—débats, *Session 1: Le patrimoine, moteur de development* (pp. 905–916).

Hastaoglou-Martinidis, V., & Christodoulou, Ch. (2010). Preservation of in Thessaloniki. *Rivista di Scienze del Turismo*, *1*(2) 121–147. ISSN:2037-7916.

Hastaoglou-Martinidis, V., Yerolympos, A. (1986). Thessaloniki 1900–1940: From the contradictions of cosmopolitism to the homogeneity of a neohellenic city. History Centre of Thessaloniki, *Thessaloniki after 1912* (in Greek). Municipality of Thessaloniki.

Hastaoglou-Martinidis, V. (1997). A Mediterranean City in transition: Thessaloniki between the two world wars. *Facta Universitatis, Ser. Architecture and Civil Engineering*, *1*(4), 493–507.

Hmood, K., Aljumaily, H., & Melnik, V. (2018). Urban architectural heritage and sustainable tourism. *WIT Transactions on Ecology and the Environment, 227*, 209–220. https://doi.org/10.2495/ST180201

Hughes, H. (2000). *Arts, entertainment and tourism*. Oxford: Butterworth-Heinemman. 0-7506-4533-4.

Karadimou-Yerolympou, A. (1995). *The reconstruction of Thessaloniki after the Fire of 1917: A landmark in the history of the city and the development of Greek urban planning, (in Greek)* (2nd ed.). University Studio Press.

Kefala, Ch.-Z., & Samaras, N. (2019). The 'Kastroplikton' Residences in Ano Poli (Upper Town) of Thessaloniki, Greece. Status and prospects of preservation. *Sustainable Development, Culture, Traditions-Journal, 1a*, 35–55. https://doi.org/10.26341/issn.2241-4002-2019-1a-3.

Kokot, W. (2003). European port cities: Disadvantaged urban areas in transition. A collaborative project of the community action programme to combat social exclusion 2002–2006 VP/2003/023. Transnational exchange programme interim report: Phase II, Year 1, Head of Project Prof. Dr. Waltraud Kokot. https://www.europeanportcities.de.

Kostopoulou, S. (1996). City-port development and cultural heritage: The case of Thessaloniki. In: Hoyle, B. (Ed.), *City-ports, coastal zones and regional change: International perspectives on planning and management* (pp. 137–158). Chichester: J. Wiley. ISBN: 978-0-471-96277-9.

Kostopoulou, S. (2013), On the revitalized waterfront: Creative Milieu for creative tourism. *Sustainability, 5*, 4578–4593. https://doi.org/10.3390/su5114578.

Kostopoulou, S. (2019). Maritime silk road heritage and tourism development: The case of Thessaloniki, Greece. In: World Tourism Organization, *The 21st Century Maritime Silk Road—Tourism opportunities and impacts* (pp. 137–139). UNWTO, Madrid. https://doi.org/10.18111/9789284418749.

Kunzmann, K. R. (2004). Culture, creativity and spatial planning. *Town Planning Review, 75*(4), 383–404. https://doi.org/10.3828/tpr.75.4.2

Lagopoulos, APh. (2005). Monumental urban space and national identity: The early twentieth century new plan of Thessaloniki. *Journal of Historical Geography, 31*, 61–77. https://doi.org/10.1016/j.jhg.2003.08.023

Lash, S., & Urry, J. (1994). *Economies of sign and space*. Sage.

Lavedan, P. (1920). Un probleme d'urbanisme: La reconstruction de Salonique. *Gazette Des Beaux-Arts Series, 64*(5), 238–240.

Lavedan, P. (1933). L'oeuvre d'Ernest Hebrard en Grece *Urbanisme. Mai, 1933*, 149–151.

Law, C. M. (1996). *Urban tourism. Attracting visitors to large cities*. Mansell Publishing Limited.

Lengkeek, J. (1995). Materializing the imagined: on the dynamics and assessment of tourist-recreational transformation processes. In: G. J. Ashworth, & A.G.J. Dietvorst (Eds.), *Tourism and spatial transformations-implications for policy and planning* (pp. 17–36). Wallingford, UK: CAB International.

Lewkowicz, B. (1999). *The Jewish community of Thessaloniki: An exploration of memory and identity in a Mediterranean City*. [Doctoral dissertation, London School of Economics, University of London]. https://www.etheses.lse.ac.uk/id/eprint/2910.

Livadioti, M. (2013). The harbor of Thessaloniki in the Hellenistic, Roman and early Christian era. In *Medieval Ports in North Aegean and the Black Sea. Links to the Maritime Routes of the East*, Conference Proceedings. International Symposium, Thessalonike: European Centre for Byzantine and Post-Byzantine Monuments, 4–6 December 2013.

Lusiani, M., & Panozzo, F. (2016). *Culture on top: Beyond museification and culture-led regeneration of industrial heritage*. Università Ca' Foscari Venezia, Italia, Working Paper n. 2/2016.

Mason, R. (2005). *Economics and historic preservation: A guide and review of the literature. A discussion paper prepared for the Brookings Institution Metropolitan Policy Program*. University of Pennsylvania.

Mavridis, S. (2018). Greece's economic and social transformation 2008–2017. *Social Science, 7*(1), 9. https://doi.org/10.3390/socsci7010009

Mavromatis, M. (2009). Upper town 1978–2009 evolution of Interventions in the traditional settlement of the historic center of Thessaloniki. *Technografima, 373*, 12–13.

Mavromatis, M. (1997). Ano Poli (1978–1997). In Moutsopoulos, N. (Ed.), *The Ano Poli of Thessaloniki (1978–1997), The Revival of a Degraded Settlement* (in Greek) (pp. 77–85). Thessaloniki: Organisation of the European Capital of Culture' 97.

Mazower, M. (2004). *Salonica, city of ghosts: Christians, Muslims and Jews, 1430–1950*. Alfred A. Knopf.

Morin, E. (2009). *Vidal and his family: From Salonica to Paris*. Sussex Academic Press.

Moscof, K. (1978). *Thessaloniki, a compradorial city (in Greek)*. Stoxastis.

Nijkamp, P. (2012). Economic valuation of cultural heritage. In: G. Licciardi & A. Rana (Eds.), *The economics of uniqueness: Historic cities and cultural heritage assets for sustainable development*

(pp. 75–106). Urban Development, Washington D.C: World Bank. https://openknowledge.worldbank.org/handle/10986/12286.
Ost, Ch. (2012). Mapping heritage economics for spatial analysis in historic city cores. In G. Licciardi, & A. Rana (Eds.), *The economics of uniqueness: Historic cities and cultural heritage assets for sustainable development* (pp. 245–282). Urban Development, Washington D.C: World Bank. https://openknowledge.worldbank.org/handle/10986/12286.
Papadam, M. (n.d.). Thessaloniki. The strategic spatial planning after the great fire of 1917. *Regional Strategies and Territorial Governance 2015–2016,* Prof. Vincent Nadin.
Papandreou, A., & Papageorgiou, M. (2019). Planning for the enhancement of the modern built heritage in Thessaly region: The case of the "Konakia" monuments. *Heritage, 2,* 2039–2052. https://doi.org/10.3390/heritage2030123
Papastathis, K., Hekimoglou, E. A. (2010). *The great fire of Thessaloniki (1917).* Thessaloniki: EN Manos Ltd.
Papastathis, C. (1978). A memoir on the fire in Thessaloniki in 1917 and the relief of the victims (in Greek). *Makedonika,* 18.
Pedersen, A. (2002). *Managing tourism at world heritage sites: A practical manual for world heritage site managers.* UNESCO World Heritage Centre.
Plymouth, P. (2010). Presentation at the culture means business summit, celebrated at the National Marine Aquarium.
Rescher, H. (2003). Monument protection as an economic factor. Building culture as a tourist destination. *DStGB Aktuell—Stadt and community,* 10, 423–425.
Ministry of Environment and Energy, Thessaloniki Municipality. (2015). *Revision Study of the General Urban Plan of Thessaloniki Municipality* (in Greek).
Richards, G. (2001). *Cultural attractions and European tourism.* CABI Publishing.
Serraos, K., Gianniris, E., & Zifou, M. (2005). The Greek spatial and urban planning system in the European context. In G. Padovano & C. Blasi (Eds.), *Complessitá e Sostenibilitá, Prospettive per i territori europei: Strategie di pianificazione in dieci Paesi, Rivista bimestrale di pianificazione e progrettazione* (pp. 1–24). Poli Design Publishing.
Stein, N. C. (2013) *Using the visual to "see" absence: The case of Thessaloniki.* [Doctoral dissertation, Florida Atlantic University]. https://pqdtopen.proquest.com/doc/1426427268.html?FMT=AI.
Svoronos, N. (1996). *The commerce of Thessaloniki in the 18th century (in Greek).* Themelio.
UNWTO World Tourism Organization. 'Overtourism'?—Understanding and Managing Urban Tourism Growth beyond Perceptions, Executive Summary, Madrid: 2018. ISBN: 978-92-844-2006-3.
Vacalopoulos, A. (1972). *A history of Thessaloniki* (in Greek). Thessaloniki: Institute of Balkan Studies.
Vakalopoulos, A. (1985). History of Thessaloniki. In: *Thessalonikin Filippou Vasilissan.* (pp. 241–243) (in Greek). Thessaloniki: Studies on Ancient Thessaloniki.
Vickers, M. (1985) Hellenistic Thessaloniki. In: *Thessalonikin Filippou Vasilissan* (pp.487–490, 493–494, 496–497, 499) (in Greek). Thessaloniki: Studies on Ancient Thessaloniki.
Vlami, D. (2009) Entrepreneurship and Relational Capital in a Levantine Context: Bartolomew Edward Abbott, the "Father of the Levant Company" in Thessaloniki (Eighteenth-Nineteenth Centuries). *La Revue Historique, 6,* 129–164. https://doi.org/10.12681/hr.242.
Volkgenannt, U., & Rottger, E. (1995). *Thessaloniki bei Tag und Nacht. Wissenswertes iiber eine Vielseitige Stadt.* Clemon Verlag.
Vyzantiadou, M., & Selevista, M. (2019). Protection of cultural heritage in Thessaloniki: A review of designation actions. *Heritage, 2,* 717–731. https://doi.org/10.3390/heritage2010046
World Economic Forum (WEF). (2018). The Global Competitiveness Report 2018. Geneva.
Yaldiz, E., Aydin, D., & Siramkaya, S. B. (2014). Loss of city identities in the process of change: The City of Konya-Turkey. *Procedia - Social and Behavioral Sciences, 140,* 221–233. https://doi.org/10.1016/j.sbspro.2014.04.413
Yerolympos, A. (1992). Urbanisme et urbanisation en Grèce du Nord à l'époque des Tanzimat. In P. Dumont & F. Georgeon (Eds.), *Villes Ottomanes à la fin de l'empire* (pp. 47–74). Harmattan.
Yerolympos, A. (1996). *Urban transformation in the Balkans (1820–1920): Aspects of Balkan town planning and the remaking of Thessaloniki.* University Studio Press.

A Methodological Framework for the Conservation and Planning of Urban Spaces in Historical Centers Around Riverfronts. The Case of Arequipa, Peru

Carlos Zeballos-Velarde

Abstract

Many cities have been born around rivers, which have not only been a source of life and irrigation for crops but also have provided spaces for recreation and social gathering. However, growth and modernization processes have led to the social and environmental detriment of many of these areas, becoming marginal or neglected sectors of the cities. Nevertheless, in recent years there have been numerous examples of how waterfronts have been recovered for the enjoyment of their communities. The following paper explores a methodology in which a public space system in a historic center can be preserved and upgraded along with the development or improvement of its riverfront, acting as its structuring backbone. Taking the historic center of Arequipa, Peru, as a case study, the article proposes a territorial analysis model and a methodological approach that allows understanding the way in which the river landscape is conceived, used, and managed as a territory as well as its relationship with other public spaces in the historic center. The methodological approach is mixed, starting from a documentary review of how the open space system used to be integrated to the river in the past, a relationship that was lost during the modernization. Then, by reviewing other case studies of the continent and the analysis of interfaces around the river basin, a management system of public spaces is proposed, through 4 axes of action: the development of a monumental historical space and value system, economic productivity, environmental quality, and inter-institutional management.

C. Zeballos-Velarde (✉)
Universidad Católica San Pablo, Arequipa, Peru
e-mail: crzeballos@ucsp.edu.pe

Keywords

Preservation • Urban spaces • Riverfronts • Historical centers

1 Introduction

Public spaces in the central areas of Latin American cities have not been common or plentiful during colonial times. In general, many of these cities maintained the typical Hispanic colonial layout in the form of a grid, with a central square, which served both as a market and public space (Bielza de Ory, 2002). Other public spaces included the atria of churches, where the native communities were generally evangelized. Conversely, most gardens or orchards were private. It is in the wake of Hygienism, a movement that has its origin in Europe during the first half of the nineteenth century, when the health of cities inhabitants began to be taken into account more seriously. Under this influence, Latin American cities started to develop recreational spaces, such as parks and promenades. Some of these areas were located around riverfronts, which were transformed into promenades. However, due to the effects of modernization from the second half of the twentieth century, many rivers in Latin American cities have experienced abandonment and environmental deterioration, along with several other negative effects for their communities, including social, economic, cultural, and environmental issues.

These situations were caused by several reasons, such as the increase in urban pressure around rivers, an accelerated process of demographic growth of the cities, speculation of valuable areas, and consequently, the channeling and occupation of the river basins, sometimes even covering the rivers or drying them out. Also, a low-density urban expansion as well as the concentration of tertiary activities in the city centers and the displacement of housing toward the periphery. In the same way, the appearance of industrial uses

A. Versaci et al. (eds.), *Conservation of Architectural Heritage*, Advances in Science, Technology & Innovation,
https://doi.org/10.1007/978-3-030-74482-3_13

contiguous to the rivers (due to the availability of water) and other sources of pollution have led to environmental deterioration with direct repercussions on the water tributaries. Abandonment and pollution have had social effects as well, such as crime rise or lack of development. A decrease in the economic value is also common in these abandoned areas and their surrounding neighborhoods, also affecting the preservation of important historical heritage, accompanied by the lack of exploitation of the tourist potential of these areas. In short, in many cases, the physical, environmental, and social deterioration of the urban areas around rivers, and the discharge of waste resulting from production or consumption processes has been evident.

The aforementioned context leads to questioning the role that urban planners and decision-makers have played in Latin American cities, by allowing and even promoting the destruction of riparian spaces and their disarticulation with other public spaces of cities that have heritage value.

In any case, the displacement and lack of physical and social accessibility to rivers, the separation of cities from these natural habitats and the disappearance of important heritage around them have impacted their potential as democratic public spaces that could contribute to improving the public life of the community.

There are several experiences of cities worldwide that have incorporated rivers into the urban experience, some of them in historic centers. In several cities, this has been achieved by means of "waterfront development" strategies (Blanes-González, 2013) that account for the development of a city in the face of water resources (Dirk, 2014). In other cases, studies have been carried out to promote the development of public spaces (Belmessous & Berteli, 2014), the regeneration of city centers (Ertan & Egercioglu, 2016) as well as the synergy between urban design, the regeneration of ancient urban centers and the conservation of the built cultural heritage (Amit-Cohen, 2005).

2 Methodology

The methodology of the present research is mixed. It starts with a documentary exploration, which includes sources from historical literature in order to learn how the city once developed a system of public spaces around the river in its central area. Then, a field visit along with a review of scientific and technical literature of experiences from South American cities that have developed river recovery projects in urban basins, especially in areas of historical, monumental, or landscape value. Aside from these visits, on-site information gathering and interviews with the main stakeholders of the urban waterfront management processes, as well as the population in general.

Later, a comparative matrix of these cases has been elaborated, including the main social, environmental, physical spatial, and institutional management aspects concerning the riverfront.

Then, an integrated diagnosis finding the interfaces was used to understand the potentials and barriers that urban spaces face in order to be integrated into the Chili River. Finally, a conceptual model is proposed to create a system of urban spaces integrated into the riverscape.

3 The Importance of Learning from History: The Case of Arequipa, Peru

3.1 The Site

Arequipa is located at an altitude of 2500 m, in a valley formed by the Chili River and surrounded by three volcanoes: Misti (the only one with fumarolic activity), Chachani and Pichu Pichu. It is the second-largest Peruvian city with a population of 869,361 inhabitants (INEI, 2018), and both its historic center and riverscape are listed as a World Heritage site by UNESCO (2000).

3.2 Prehispanic Times

Skilled prehispanic settlers transformed the most arid desert in the world into the fertile valley of the Chili River, through a complex system of terraces and canals. Aside from that, these settlers worshipped the river as a deity, as several burial spaces were found facing the river (Galdos, 1990), developing a spectacular agricultural landscape and a complex irrigation system (Fig. 1).

3.3 The Colonial Era

This spectacular rural landscape was chosen by the Spanish conquistadores who, from 1540, built the city of Arequipa organized around a gridiron layout, as they used to do in other colonial cities. The main square was slightly off the center and surrounded by the main functions of the city: the cathedral, the jail, and the governor house. This gridiron layout housed the Spanish conquerors and their descendants, while the native population (called Indians by the Spanish) were kept in more organic separated settlements, organized around minor squares. During colonial times, the riverfront was considered a dangerous area and the city lost its connection to the riverscape (Gutiérrez, 1992).

Fig. 1 Panorama of the Chili River, an oasis amidst the driest desert in the world

4 The Rise of Hygienism and the Transformation of the Main Square

Before 1868, the Main square of Arequipa used to be a market place, and it had no pavement and no greenery, just a fountain in the middle of the space where the servants used to gather in order to collect water (Carpio Muñoz, 2006) (Fig. 2a). However, after the catastrophic earthquake of 1868 which devastated the city, the square was rebuilt following the concepts of Hygienism and landscape design brought to Arequipa by European immigrants after the Peruvian independence (Zeballos-Velarde, 2020) . The concept of modernity associated with closeness to nature transformed the Main Square from a place to "transit" and to "exchange goods" into a space to "stay" and to "observe" (Zeballos, 2007) (Fig. 2b).

Fig. 2 Main Square before **a** and after, **b** it's remodeling as a park

5 Understanding Urban Catalysis Development of Public Spaces Around the River

Urban catalysis is understood as the "positive impact an individual urban building or project can have on subsequent projects and, ultimately, the form of a city" (Attoe & Logan, 1989). An urban catalyst is an element that is modeled by the city and then, in turn, influences the behavior of its environment, which leads to a continual regeneration of an urban setting. "Catalysis involves the introduction of one element that modifies others. Catalysts are existing urban elements of value that are enhanced or transformed in a positive way. The new need not obliterate or devalue the old but can redeem it. The catalytic reaction is contained; it does not damage its context" (Attoe and Logan, op cit).

Thus, after the Main Square, and from 1868 to 1940 several other spaces were created as parks or areas of recreation, particularly around the river, which structured a system of public spaces, transforming the original perception of the city (Fig. 3). This system included:

(a) **Civic place (monumental scale)**:
The Main Square which traditionally agglomerated the main civic, religious, cultural, social, and political functions of the city.

(b) **Secular places (human scale)**:
Originally, San Francisco Square and San Lazaro Square were areas in front of the churches for evangelization purposes and later they became connected to the public space. They generally have an irregular geometry and were not enclosed. These spaces were frequently used as meeting points during religious ceremonies, and they had an intimate, human scale.

(c) **Playground place**:
Grau Park is located next to the Grau Bridge, but at a lower level, and also delimited by the San Lázaro stream and the Chili River. It is a space that, due to a difference in topography, provides a sense of intimacy and protection from the outside world, moving away from the noisy life of the city, but benefiting from the sound of the river.

(d) **Balcony place**:
The Bolognesi Boulevard and the Alameda Promenade are situated on top of a cliff, overlooking the river, from where it is possible to appreciate magnificent views of the Chili Valley. Given their picturesque conditions, these places were very popular for painting, photographing, or gazing at views of the city, the Chili Valley, the countryside, and snow-capped volcanoes in the background.

(e) **Integrating place**:
The San Lazaro Blvd used to be located along the edge of a seasonal stream, locally known as a *torrentera*, since during the rainy season it carries dangerous torrents of water. The space was used to unify two sectors that were historically separated, the gridiron layout of Spanish origin with the organic distribution of the Indian neighborhood, creating a promenade along the San Lázaro *torrentera*.

6 Deterioration: 1960–2020

In the last 6 dacades, due to industrialization and modernity, the riparian waters began to be polluted, until the Chili River has become one of the most contaminated rivers in Peru

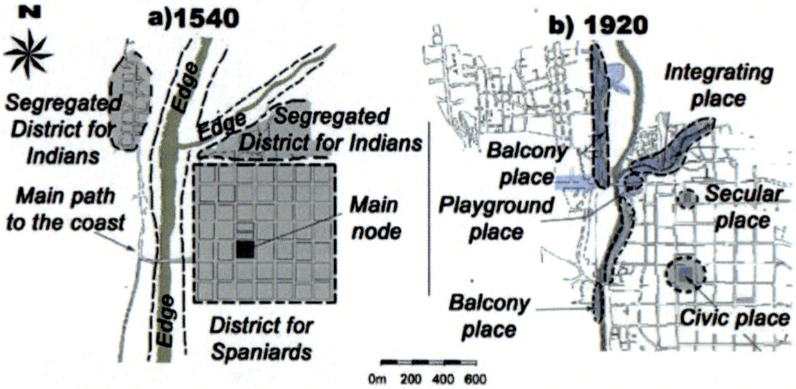

Fig. 3 Evolution of the public spaces in the center of Arequipa. **a** From 1540 to 1868 Arequipa used to be a very segregated city. **b** The creation of recreational public places from 1868 to 1940 promoted a new perception of the urban landscape and stressed the concept of a relationship between modernity and nearness to nature

(RPP, 2011). Industrial pollution by tanneries, contamination of agricultural land, discharge of drains and solid waste, air pollution, etc., was added to the loss of greenery and agricultural areas and to the exponential population increase, which caused a continuous process of environmental deterioration in the river and its surroundings (Zeballos, 2002). In the 1970s, a new rapid way was created along the river basin, effectively detaching the city from its contact with the Chili River. However, this area has an enormous environmental potential as some abandoned agricultural fields are located right in the city center, and they are an opportunity to develop public areas if purchased by the city government (Fig. 4).

From an urban planning point of view, the City Master Plan has considered the riverfront area as a landscape reserve, although a specific plan has not yet been developed to regulate it (IMPLA, 2016).

Also, important projects for the recovery of monumental heritage and social upgrade have been carried out in the so-called *tambos*, which are places built during colonial times where people used to live in poor conditions and they were renovated keeping in mind the needs and social networks of these communities.

7 Results from Study Cases in South America

From a broad view, it has been possible to account for initiatives in South America that seek to recover the environmental deterioration of the urban riverfronts, some of them, which have contact with monumental areas. Field visits to cities such as Lima (Peru), Asuncion (Paraguay), Buenos Aires and Rosario (Argentina), Porto Alegre (Brasil), Montevideo (Uruguay), and Santiago (Chile) have developed

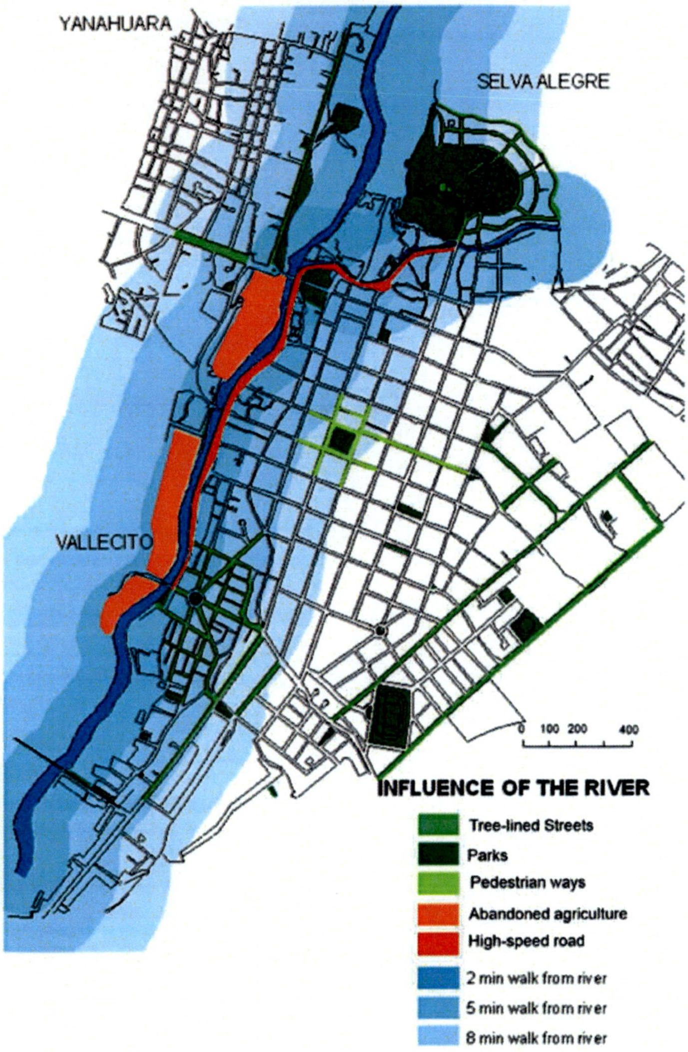

Fig. 4 Location and type of public spaces in Arequipa historic center and the barriers that prevent their connection to the river

several experiences in the case of urban environmental management in riverfronts, each of them with some particular strengths.

It is not the purpose of this paper to carry out a comparison between these different Latin American experiences, which are also quite heterogeneous in their approach and size of intervention, as well as their budgets and management. It is, rather, to know the answers or patterns of solution that those practices involved in order to solve the problems that concern the relevant aspects in the management of urban basins.

For this reason, Table 1 summarizes the main variables, the relevant problems found in the rivers and the patterns of solution assumed by the different experiences cited.

8 Discussion

8.1 The Diagnosis Using Interfaces

The method of interfaces identifies the most sensitive areas in an ecosystem, which are places of rich diversity that hold large amounts of information (Pesci, 1999). By means of the participation of the institutions and of the population itself, through workshops, it was possible to identify areas of homogeneous behavior, recognizing the most important problems that affect the development of the basin and spatially locate them.

Four subsystems of analysis were defined, which are closely interrelated, and are the same ones used to analyze the experiences in Latin America. These are:

Table 1 Summarizing main variables, relevant problems and their solutions

Variable	Problem	Solution	Example
Monumental and physical space value	Deterioration of built heritage	Restoration of monumental areas or buildings	– Parana Park, Rosario, Argentina – Puerto Madero, Buenos Aires, Argentina – Mapocho Park, Santiago – Rimac Park, Lima, Peru
	Lack of spatial relation between the city and the river	Creation, habilitation or recovery of metropolitan open spaces for social use	
	Existence of vacant spaces, abandoned buildings and incompatible uses		
	Location of slums around the river	Creation of low-income housing programs, with adequate services and public spaces for interaction with the river	– Coastal Strip, Paraguay – Regeneration of Lombas do Pinheiro, Brasil
Economic productivity	Deterioration of heritage. Social and economic marginality	Recovery of areas of landscape and cultural value	– Puerto Madero, Argentina – Coastal Strip, Paraguay
	Lack of means to execute projects of social and environmental interest	Enabling land for tourist use near the river, to finance social housing and recreation spaces	– Coastal Strip, Paraguay
Environmental quality	Contamination of water, by domestic and industrial or by solid waste	Provision of infrastructure and treatment plants to reduce the emissions	– Mapocho park, Chile
	Homes located in flood risk areas	Protection of the riverbanks and relocation	– Coastal Strip, Paraguay
Inter-institutional management	Lack of knowledge of the citizenship of the management initiatives	Communication strategies to involve public opinion	– Arroyo Miguelete, Uruguay

- **Natural Subsystem**: that studies how ecosystems were affected, such as the urban pressure on agricultural areas around the river, the presence of natural ecotones and their deterioration or contamination, and the degrees of environmental risk.
- **Economic Subsystem**: which, by means of a land use analysis, shows the quantity and quality of activities in the sector and exposes the potential tourist attractiveness of the sector.
- **Socio-Cultural Subsystem**: which indicates the scarce amount of public social activities around the river, as well as the important presence of historical and cultural monuments in the study area. In addition, it shows the presence of public and private institutions, committed to the development of the city, located in this area.
- **Physical-Spatial Subsystem**: that analyzes the characteristics of urban space, its quality of conservation, what areas are deteriorated, what the physical public accessibility to the river is and how natural and constructed edges in the riparian areas are formed.

These regulations give an account of rules that regulate spaces in general. A theoretical analysis of these rules in a spatial dimension (Burbano, 2009) admits that they can condition the use of space and the social practices that are sustained in it, therefore, influencing the appropriation of the riverfronts by the community, assuming that the water tributaries are producers of public space and that they intervene in the generation of territory. This approach admits analyzing that not all the rules are regulated, and these can result from the action and the use that people give to the spaces, so it is important to take into account the role of the community uses the riverfronts. From this orientation, the water system is considered fundamental for the study of territoriality (Puello-Bedoya, 2005).

The description of these subsystems, their conflicts, and potentialities as well as their aptitudes (patterns of solution) establishes a real framework of the situation in the study area and are the basis for proposing management models. The proposed management model takes the study area as a reference, based on an Integrated Territorial Diagnosis and is framed within the local rules and regulations.

The management model starts from identifying the interfaces and patterns of solution defined in the integrated diagnosis, in order to promote a set of actions, programs, and projects that aim to contribute to the environmental recovery of the Chili riverfront in monumental areas. On the other hand, the political-administrative actions that take place in the metropolitan basin by the various institutions must be channeled through management actions that strengthen them, give them continuity, encourage inter-institutional participation and allow them to be flexible before possible changes in the social or political context.

8.2 The Management of Rivers as a Public Space

The spatial approach to planning, spatial management and the definition of policies allow a perception of the riverscape in detail (Forman, 1997). This spatial approach considers four elements:

- In each landscape the presence of few but large groups of natural vegetation is required, in order to protect the richness of the species in this territory.
- In each landscape, long vegetation corridors, and mainly rivers, provide an exceptionally wide range of ecological benefits. These include the control of erosion and river flooding, the reduction of the dumping of harmful substances in the course of water, the favorable predisposition for the habitat of the ichthyofauna, and, finally, the protection of the movements of the fauna throughout the fluvial system.
- The connection between groups of vegetation, by grouping small spots of green area, is the key that allows the movement of the species through them.
- In each territory, these small portions or corridors of vegetation, provide many ecological benefits, such as the protection of diffuse species, reduce the volume of wind and reduce the phenomena of water and wind erosion.

However, the management of urban basins not only takes into account the landscape aspect but also concerted management, assuming a responsible vision that covers economic, environmental, and risk management, such as flood control, for example. From a vision that assumes riverfronts as areas that host public spaces, urban planning and management should deal with the construction and implementation of legal instruments and tools that promote, regulate, and control what happens in those spaces. From this approach, it is understood that the management materializes in institutional regulations.

From the integrated diagnosis, important driving forces or "central ideas" for the Chili river basin are also developed. A central idea is a future vision of a territory and to which the policies, fields of action, and programs or projects to be carried out ultimately lead. A central idea is also a communicational element, open and appropriable by the population so that it can provide identification and social and institutional support to a specific program that allows it to be rooted in the collective memory of the population.

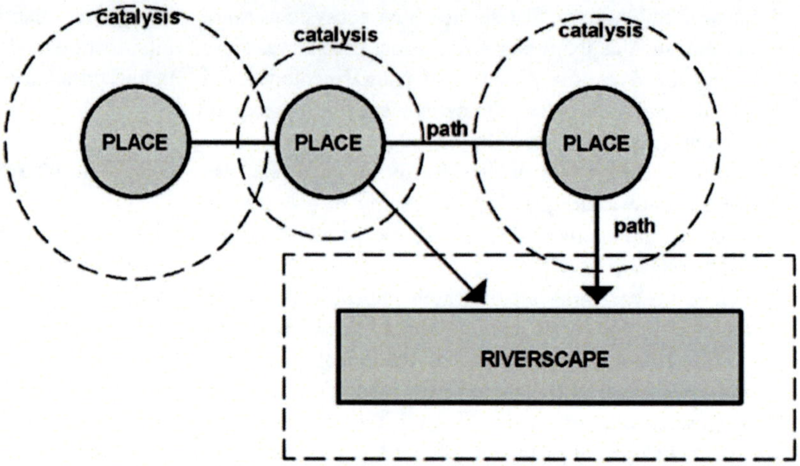

Fig. 5 Diagram of the proposed model of the process of urban catalysis in Arequipa

8.3 The Spatial Linkage

To achieve urban catalysis in Arequipa, spatial links have to be developed using subtle elements, since it was not necessary to destroy or renovate large areas of the city (Fig. 5).

The perception of public space was developed through a kinesthetic experience, that is, through the sensation transmitted by movements (Zucker, 1959). The urban components stimulated movement from one place to another, facilitating the walks, the orientation, the experience of reaching a goal and the linking between "here" and "there".

The most important urban connectors in the system are (Fig. 6):

(a) **Architectural landmarks**: composed mainly of church bell towers, which stand out in the predominantly horizontal urban landscape of downtown Arequipa. The most conspicuous examples located near the river are the towers of the Cathedral in the Main Square and the Churches of San Francisco and San Lázaro.
(b) **Frames**: they highlight a certain portion of the landscape, turning it into a significant place, inviting the pedestrian to walk toward it. The arcades of the Main Square, and in particular two large arches on each side of the Cathedral, which project toward the streets perpendicularly to them, are important references to the square.
(c) **Topography**: according to the morphology, the slope and orientation of the terrain.
(d) **Enclosure/exposure**: Referring to the emotions of mystery or surprise as well as the feeling of intimacy and enclosure that occurs after transiting through narrow streets and then encountering a wide, open space. In Arequipa, the narrow sloping streets descend into the impressive river landscape, which opens up as to wide and outstanding views.
(e) **Vegetation**: in a dry climate and a desert environment such as that of Arequipa, the inclusion of vegetation in the urban landscape was of great importance for the visual quality and comfort of the population, distancing itself from the dry urban landscape, characteristic of colonial streets in Hispanic America. The rhythmic sequence of trees encourages movement through the groves, and the clustering of vegetation emphasizes the importance of a destination.
(f) **Walkways**: the development new pedestrian or semi-pedestrian ways promote the flow of the population toward the riverscape and organize movement patterns.
(g) **New parks**: the inclusion of new large and medium-sized public spaces in the proximities of the river, encourages a spatial and social integration of the city and its riverscape.

9 Conclusions

- The proposed methodology combines three sources of information: the historical evidence, the compilation of study cases in Latin America, and the diagnosis by the method of interfaces, in order to produce a model that integrates the system of public spaces with the riverfront areas.
- There are initiatives in South America that had led to the environmental recovering of urban river basins near to the monumental areas, with different degrees of success. In order to achieve this objective, they have formed multisectoral entities that, along with citizen participation, propose, execute, and supervise environmental projects for the improvement of these areas. They have also developed programs of restoration and / or renovation of

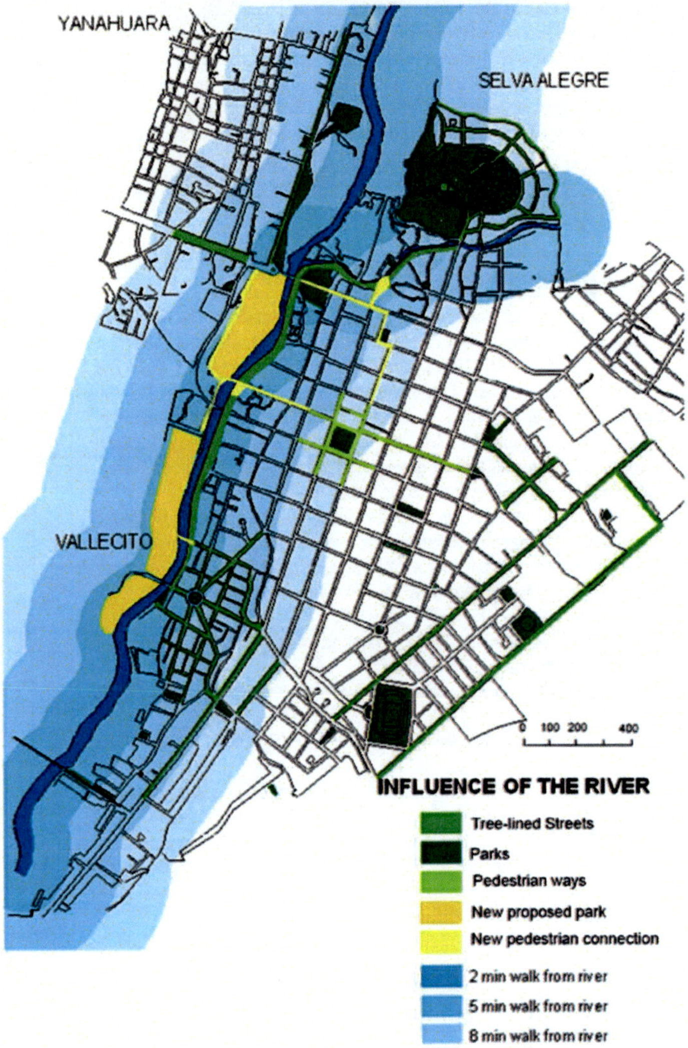

Fig. 6 Proposed urban structure of public spaces in the historic center of Arequipa

built heritage, and have provided spaces for social and public exchange, improving the local living conditions.

- In the mentioned cases, there has been an increase in the land value adjacent to recovered riverfronts, as well as the increase and development of economic activities related to tourism and recreation, showing the potential of these areas as economic drivers, revitalizing them, and renewing the identity of the inhabitants with their rivers.
- In the Chili River area that is in contact with the Historic Center, there exist various sources of contamination, both hydric and atmospheric, and solid waste. However, pollution levels are not irreversible, and providing infrastructure and control measures would achieve environmental clean-up.
- The study area houses the largest quantity and quality of historical monuments in Arequipa, as well as a strong link between the built historical heritage of Arequipa with the natural landscape legacy of the Chili River, as two indissoluble components of a landscape of universal value.
- The lack of spaces for social exchange in the Historical Center and the Chili River has produced a gradual loss of identity of the local inhabitant with his river. However, there are areas feasible to be converted into public parks near the river.
- The high density of institutions located in the study area that would support a Chili recovery program as well as the involvent of local population are important factors in the process of developing a socially legitimized riverfront development, if mechanisms are established to articulate efforts under the guideline of a coordinated and participatory management plan.

Acknowledgement The author would like to thank the Universidad Catolica San Pablo in Arequipa, for funding the project entitled "Análisis histórico de los espacios urbanos en centros monumentales usando herramientas de teledetección. El caso de Arequipa." (Historical analysis of urban spaces in monumental centers using remote sensing tools. The case of Arequipa) of which this chapter is the first outcome.

References

Amit-Cohen, I. (2005). Synergy between urban planning, conservation of the cultural built heritage and functional changes in the old urban center—The case of Tel Aviv. *Land Use Policy, 22*, 291–300.

Attoe, W., & Logan, D. (1989). *American urban architecture. Catalysis in the design of cities*. University of California Press.

Belmessous, F., & Berteli, T. T. (2014). Public spaces developments in Lyon (France) and Barcelona (Spain): a successful strategy against social exclusion and urban poverty? *Workshop 14 – Welfare Policy, Homelessness and Social Exclusion.*

Bielza de Ory, V. (2002). De la ciudad ortogonal aragonesa a la ciudad cuadicular hispanoamericanacomo proceso de inovación-difusión, condicionado por la Utopia. *Scripta Nova. Revista electronica de Geografia y ciencias sociale., 6*(106). Obtenido de http://www.ub.edu/geocrit/sn/sn-106.htm.

Blanes-González, M. L. (2013). Planning of the edges of the river Paraguay. *Bitácora, 23*(2), 31–41.

Burbano, A. (2009). La convivencia ciudadana: su análisis a partir de aprendizaje por reglas. *Revista Colombiana de Educación* (57), 28–45. Obtenido de http://www.redalyc.org/pdf/4136/413635251003.pdf.

Carpio Muñoz, J. G. (2006). Apuntes sobre la historia de Arequipa. (C. Zeballos, Entrevistador).

Dirk, S. (2014). Waterfront transformations and city/port interface areas in Hamburg. *Dimensión Empresarial, 13*(1), 9–20. Obtenido de http://www.scielo.org.co/scielo.php?script=sci_arttext&pid=S1692-85632015000100001.

Ertan, T., & Egercioglu, Y. (2016). historic city center urban regeneration: The case of Malaga and Kemeraltu, Izmir. *Social and Behavioral Sciences, 223*, 601–607. Reggio Calabria, Italy.

Forman, R. (1997). Approcio Spaziale. Revsita Urbanistica. *Ecologia Del Paesaggio e Pianificazione, 108*, 62–63.

Galdos, G. (1990). *Naciones Ancestrales y la Conquista Incaica, en Historia General de Arequipa.* (Vol. 1). Arequipa, Perú: Fundación M. J. Bustamante de la Fuente.

Gutiérrez, R. (1992). *Evolución Histórica de Arequipa.* Lima, Perú: Epigrafe Editores.

IMPLA. (2016). *Plan de Desarrollo Metropolitano de Arequipa 2016–2025.* Arequipa, Peru: Municipalidad Provincial de Arequipa.

INEI. (15 de Ago de 2018). *Arequipa alberga a 1 millón 316 mil habitantes*. Obtenido de Instituto Nacional de Estadística e Informática: https://www.inei.gob.pe/prensa/noticias/arequipa-alberga-a-1-millon-316-mil-habitantes-9903/.

Pesci, R. (1999). La Ciudad de la Urbanidad. La Plata Argentina: Editorial CEPA.

Puello-Bedoya, M. (2005). Agrópolis o el final de la ciudad-territorio. Aportes conceptuales para un planeamiento físico de base agropolitana. *Bitácora Urbano Territorial, 9*(1), 43–53. Recuperado el 1 de junio de 2019, de https://revistas.unal.edu.co/index.php/bitacora/article/view/18736.

RPP. (2011). *El río Chili de Arequipa es el más contaminado del Perú.* Obtenido de RPP Noticias: https://rpp.pe/peru/actualidad/el-rio-chili-de-arequipa-es-el-mas-contaminado-del-peru-noticia-379302.

UNESCO. (2000). *World heritage list*. Obtenido de Historical Centre of the City of Arequipa. http://whc.unesco.org/en/list/1016.

Zeballos, C. (2002). *Modelo de Plan de Gestión para la Recuperación Ambiental de Franjas Ribereñas en Zonas Urbano Monumentales: El Caso del Valle del Río Chili en Arequipa, Perú. Master thesis.* Arequipa, Peru: Universidad Nacional de San Agustin.

Zeballos, C. (2007). *Evaluation of the characteristics of urban landscape development in arequipa from 1868 to 1940. Tesis doctoral.* Kioto, Japón: Kyoto University.

Zeballos-Velarde, C. (2020). *Atlas Ambiental de Arequipa*. Arequipa. Universidad Catolica de Santa Maria.

Zucker, P. (1959). *Town and Square: From the Agora to the Village Green.* New York. Columbia University Press.

Conserving the Past, Ameliorating the Future—Situating Conservation of Architectural Heritage in the Urban Context—A Case of India

Prathyaksha Krishna Prasad

Abstract

A constantly morphing environment has resulted in the new encapsulating the old, thus forming layers of cultural heritage. It has to be noted that architectural heritage and its conservation cannot be assessed and executed in isolation. It is important to take into consideration the context. The recent thrust on urban conservation has paved way for management of heritage through various contextual lenses by multidisciplinary approaches. This research discusses the importance of urban conservation and the position and importance of architectural conservation in this comprehensive process. Using existing models of cultural heritage management, a detailed analysis of the urban heritage conservation process in the Indian context has been undertaken and demonstration of the conceived framework through different scales of sites articulating the prominent conditions prevalent in India is illustrated in this paper.

Keywords

Urban conservation • Urban heritage • Development • Management • Local communities • Participatory approach • Regeneration • Environment • Assessment • Significance

1 Introduction

Most historic cities have been a premise of 'traditional settlements superimposed by today's urban way of life'. (Krishna Prasad, 2016) Due to various constraints, many such historic cores around the country, namely walled city of Delhi (Shahjahanabad), Ahmedabad, Jaipur, Jodhpur, Colonial precincts of Madras, Calcutta and Bombay, are either developed to suit the growing aspirations, thus losing its historic character or left to decay. These historic cores have, over a period of time, become the cores of commerce to the corresponding new city that have grown and expanded around the old. It is clear that these historic cores have been contributing largely to the city's economy.

In India, the development front has observed a constant bias for conservation of urban heritage, 'given the need to catch up to the development quotient of today's world'. (Krishna Prasad, 2016) With an exponential real estate swell, burgeoning population, the need for affordable housing and infrastructural provisions, heritage is being viewed as an encumbrance for development. In such a scenario, urban conservation offers the scope of protecting heritage while allowing carefully structured development programmes that are heritage-inclusive, community-inclusive development plans. Traditional approaches to the conservation of cultural heritage in urban context, that does not address the social and economic dimensions of the problem, prove to be insufficient to ensure the survival and sustenance of historic settlements that are irreplaceable witnesses to the development of human civilisations.

In India, a country with alluring cultural heritage, the idea of heritage-driven regeneration is different for various core city conditions. This paper aims to present the assessment of values and potentials of cultural heritage in the urban context using historic area characterisation as a tool. Through this paper, an attempt has been made to bring out the importance of identifying the different levels of stakeholders and the different dimensions of planning in a vast country like India. Regeneration models driven by heritage facilitate sustainable development needs to start with grasping the underlying roots of the problem moving towards an approach that identifies appropriate objectives and subsequently measures performance against the objectives of the conservation plan.

P. Krishna Prasad (✉)
The Heritage Art Collective, Chennai, Tamil Nadu 600040, India

A. Versaci et al. (eds.), *Conservation of Architectural Heritage*, Advances in Science, Technology & Innovation,
https://doi.org/10.1007/978-3-030-74482-3_14

2 Literature Review

The term urban conservation is a fairly new concept in India, though there have been various initiatives nationwide by organisations like INTACH and Aga Khan Trust for Culture. However, the idea of conservation more or less pertains to buildings that are monumental (nationally/internationally significant buildings) or those that have a perceived economic value. In other words, 'some think heritage has nothing to do with economy and in fact is an impediment to economic development. It is viewed as a pastime for posh, clever people, nothing to do with ordinary people and social issues. It is also widely perceived that heritage has nothing to do with the environmental sustainability, except possibly sustaining its look'. (Loveday, 2012) This perception (Fig. 1) has rather led to environmental degradation than otherwise. Even the best perceived heritage plans have had the possibility of failure when they leave out the community and the stakeholders from the process.

The background research for this paper is greatly influenced by the concept of 'Heritage-led regeneration', a model that is multidisciplinary and inclusive, giving great importance to the deliverable of urban heritage to the overall development of the city. The project initiated by Michael Loveday of the Heritage Economic and Regeneration Trust, with Norwich as its key project, has now been practised in various historic cities globally. The project is oriented towards measuring the benefits of heritage and its key deliverables. The HEART model emerged from a percieved need to fill a gap in the over all management of heritage, aiming at a single strategic vision. The primary agenda was to recognise the failure points of heritage, measure performance against the objectives of the conservation context and evaluate benefites in terms of economic, social, cultural and environmental factors. The pilot project at Norwich initiated many such projects across the globe, namely the historic streets in China: the Southern Song Imperial Street in Hangzhou, the residential Pingjiang Street in Suzhou and the commercial Tunxi Old Street in Huangshan (Xie & Heath, 2019) pilot initiatives by the Prince's Regeneration Trust Canada (Prince's Trust, 2008).

However, in a diverse context such as India, there is a greater necessity to address the issues of heritage in a holistic manner, with communities playing an important role in the process. In the case studies below, the results of a holistic approach, with smaller architectural conservation projects taken up through partnership models within the larger urban conservation process and the positive impacts of community engagement in the process, are the points of significance.

In the present society, heritage, values, attitudes and belief systems are all manifested in material remains. However, heritage conservation has to be looked beyond just physical intervention, as a larger process that is supported by academic research, inquiry, policy-making and planning for sustainability. Conservation of architectural heritage becomes a successful process when the context/the setting, the local communities, stakeholders and the associated intangible aspects are all taken into consideration. The idea of a multidisciplinary and interdisciplinary approach has to be realised as a method to arrive at an inclusive development, where heritage coexists within the needs of the present-day society. This would establish a process that works sustainably to support the historic fabric and the communities associated with it, even after the conservation project is completed. Most projects fail at that level where the process is exclusive to a particular class only. The

Fig. 1 Diagrammatic representation of the perceptions of heritage and conservation in the development context. *Source* Krishna Prasad (2016)

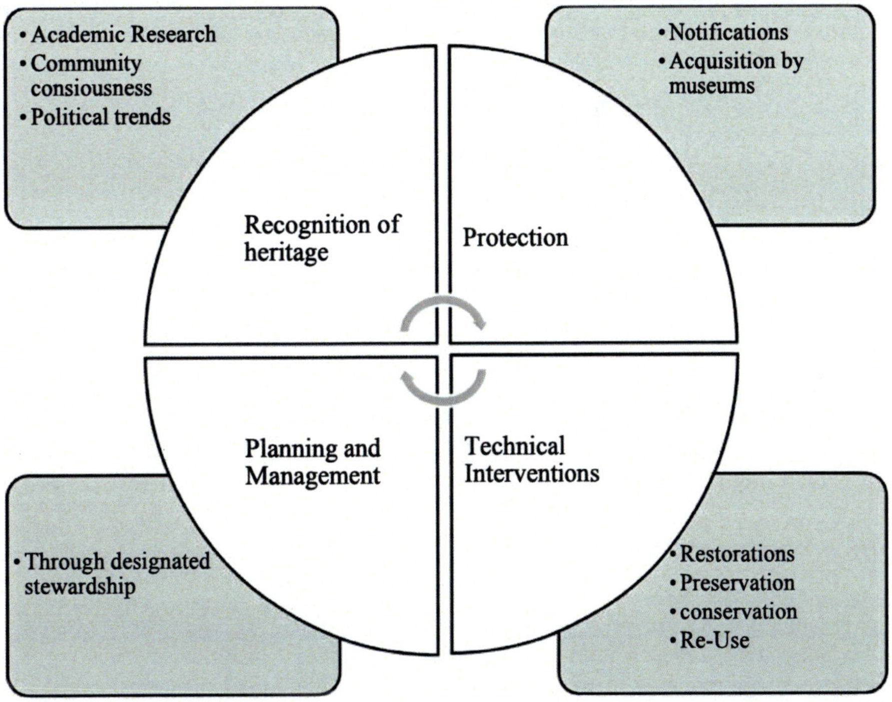

Fig. 2 Process of heritage conservation (Krishna Prasad, 2016, 52)

process of heritage conservation has to be looked at holistically with the social, cultural, environmental and economic aspects as quintessential points of study.

The main issue that is faced is designation of heritage, how cultural forms, tangible and intangible, qualify for designation and protection. The issues of designation and management for sustainability are closely related to the evolution of the society's self-image and the identity that the society wants to pass on to the generations to come (Fig. 2) (Krishna Prasad, 2016, 4). It is essential to identify and involve stakeholders, in as early as this step of the process.

Given the scenario of growing developmental pressures and issues associated with it, urban conservation looks at a holistic approach. The idea of aspirations has superseded the notions of environmental damage and economic depreciation, and sustainable development is rarely considered a practical approach. Urban conservation is a process that brings these aspects together to develop a sustainable end product, yet create and facilitate individual architectural conservation projects, the latter being sustained by the benefits of the former.

3 Research Methodology Framework

The primary aim of this paper is to address the process of urban conservation, the position of architectural conservation in the overall process, the importance of stakeholders in the overall process and evaluate the framework of the process using demonstrations (Fig. 3). Using three scales of project areas situated in the urban historic cores of India, an attempt has been made to demonstrate the process of value assessment, impact of local communitites in the process of conservation and deriving potential conservation projects that benefit the overall conservation plan. It is important to note the interrelationship between the scales of heritage in the urban context.

4 Paradigms of Heritage-Driven Regeneration in India

4.1 Revitalization of Urban Heritage Through Urban Renewal: The Case Pondicherry, India

Between 2002 and 2004, Pondicherry[1]achieved economic and environmental goals through heritage preservation initiatives under the Asia Urbs Programme. The process has helped address urban issues that have existed through revival and preservation of heritage. This has resulted in an ameliorated 'quality of life and has boosted community interest in heritage and its preservation (PEARL, 2015)'. A mutual understanding, lasting cooperation and cross-cultural

[1] Pondicherry, officially known as Puducherry (literally New Town in Tamil), is the capital and chief city of the Union Territory of Puducherry. The city is located in the Puducherry district and is surrounded by the state of Tamil Nadu with which it shares most of its culture and language. (OQ2).

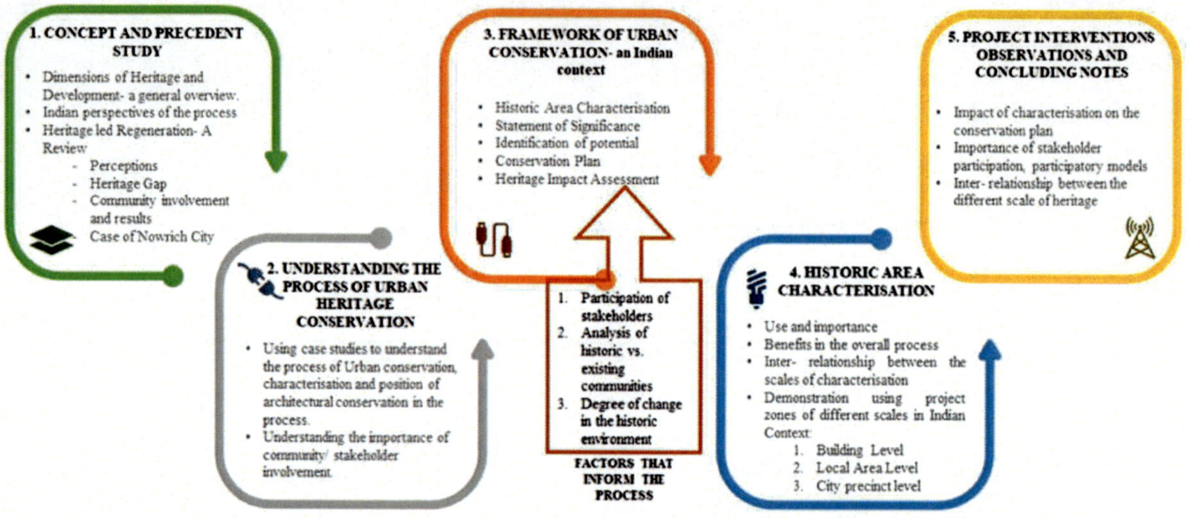

Fig. 3 Research method diagram. *Source* Author

association between Europe and Asia was achieved through this programme, bringing local governments, communities and organisations to work together to improve 'their' quality of life through regeneration and urban development. The Asia Urbs Programme was undertaken by the Pondicherry Municipality assisted by INTACH Pondicherry Chapter, partnering with the European Commission. The crucial objective of this project was to address environmental and urban issues of Pondicherry and enhance the resident's quality of life, also encouraging active stakeholder participation through 'decentralised planning' (otherwise known as localised planning/root-level planning) and 'management and strong exchange of information between the project partners (PEARL, 2015)'.

While considering a decentralised planning approach, it was imperative to keep in mind the heritage building stock—the domestic architecture that the city is known for. A great deal of importance was given to preserving the 'sense of place' and involvement of local communities in the heritage process. To achieve positive results for this strenuous task, a multidisciplinary approach was important. Rehabilitation of listed heritage buildings was an important objective with carefully laid out plans for developments in a sustainable manner that meant continuous use of the building aiding economy generation to facilitate maintenance and repair.

Building activities in the old town were closely monitored, including over weekends (as an action towards managing tourism and its effects on the built heritage), as an important task to ensure protection of the existing fabric. INTACH Pondicherry, the non-governmental organisation in-charge of the project did a meticulous study and analysis and came up with recommendations for preservation and building afresh (Fig. 4). This compilation (INTACH, 2010) was more like handbook with guidelines for authentic conservation and building new in heritage precincts so as to minimise gentrification.

One of the prominent projects as part of the Asia Urbs Programme in Pondicherry was the Urban Streetscape conservation of the Vysial Street (Fig. 5). The objective of this project was 'to showcase a typical Tamil streetscape and to persuade the house owners to appreciate their heritage homes (INTACH 2008)'. The project resulted in planning ideas that inspired other similar projects in the city. It was an important contributor for job creation involving craftsmen in the increased restoration activity. Jobs were also created through the tourism sector, which has intensified since the conclusion of the Asia Urbs programme. The overall project has been made accessible to all the stakeholders to strengthen and initiate public–private partnership programmes. The importance of shared heritage and an informed participatory process has proved to be an integral part for the success of this 'Asia Urbs Programme' project.

4.2 The Victorian Gothic and Art Deco Ensemble of Mumbai

A collection of a grand array of monumental Victorian Gothic Institutional buildings of the late 1800s (Fig. 6) bordering the Oval Maidan on the east and the Art Deco residential buildings of the 1920s on the west, along with other buildings of the same architectural style, constitute the Victorian Gothic and Art Deco Ensemble (Fig. 7), a UNESCO world heritage site inscribed on 30 June 2018, along with the formation of the Federation of Residents Trusts (FORT).

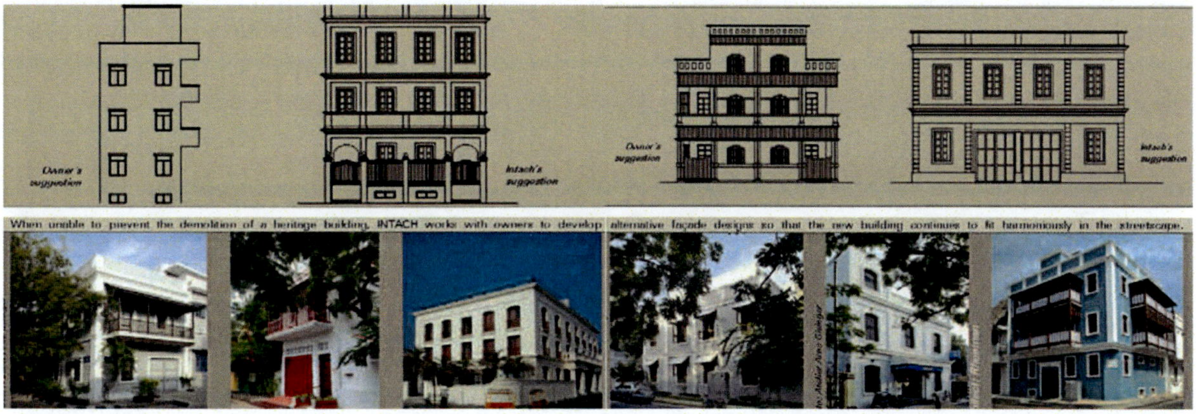

Fig. 4 Examples of building new in the designated historic precinct of Pondicherry–French town (INTACH, 2008)

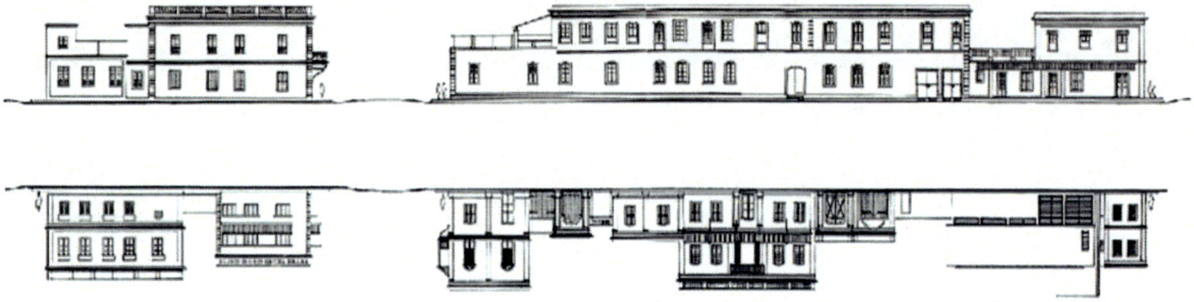

Fig. 5 Model street documentation—street elevation of Rue de le Marine, Pondicherry (INTACH, 2010, 14)

The process for the nomination and the work towards integration of heritage in the urban renewal framework, in Mumbai, started in the early 1990s. The year 2012 saw the nomination of The Victorian Gothic and Art Deco Ensemble to the UNESCO's tentative list. One of the earliest ventures was the formation of the Kala Ghoda Association (KGA), to improve a part of the public realm and to instil a feeling of shared memory, which is an integral part of intangible heritage. Situated at the nucleus of the Victorian Gothic cluster and an important junction of the Fort area of Mumbai, the precinct is called the Kala Ghoda Precinct. This was an initiative to use art of various media as a means to revive public memory of the historic precinct, and to this day the precinct is celebrated by the locals and people from different parts of the nation and world in the form of the infamous Kala Ghoda Art Festival.

Many citizen initiatives were developed, such as Foundation Nariman Point Church gate Citizens Association, Art Deco Mumbai Trust, Oval Cooperage Residents Association, to name a few. FORT is a trust that works towards sensitising locals and stakeholders through social media by publishing engaging information about the 94 structures within the world heritage precinct. It also uses digital platforms to elucidate the unique relationship between the citizens of Mumbai and their built heritage. A draft of 'planning policies and other measures were developed to ensure that the world heritage inscription is upheld in the future' (Bhosle, 2019) has also been created. The overall process has allowed many more stakeholders and locals to be a part of preserving their local heritage.

In June 2019, the announcement of proposal to demolish the Watson's Hotel[2] (Fig. 8a), the first and the oldest surviving cast iron structure in India and an integral part of the 'Victorian Gothic and Art Deco Ensemble' has brought about a lot of debate. The issue was clearly a lack of coherence between the government agencies. Issues of structural stability were cited by a report produced by Indian Institute of Technology Bombay, following which the Maharashtra Housing and Area Development Authority sought the court's permission to demolish it. The building being an integral part of the UNESCO site needed

[2] Watson's Hotel (actually Watson's Esplanade Hotel), now known as the Esplanade Mansion, is India's oldest surviving cast iron building. It is located in the Kala Ghoda area of Mumbai (Bombay). Named after its original owner, John Watson, the building was fabricated in England and constructed on-site between 1860 and 1863 (Wikipedia n.d.).

Fig. 6 Panoramic view of the Gothic Buildings along Oval Maidan, Mumbai. *Source* Author

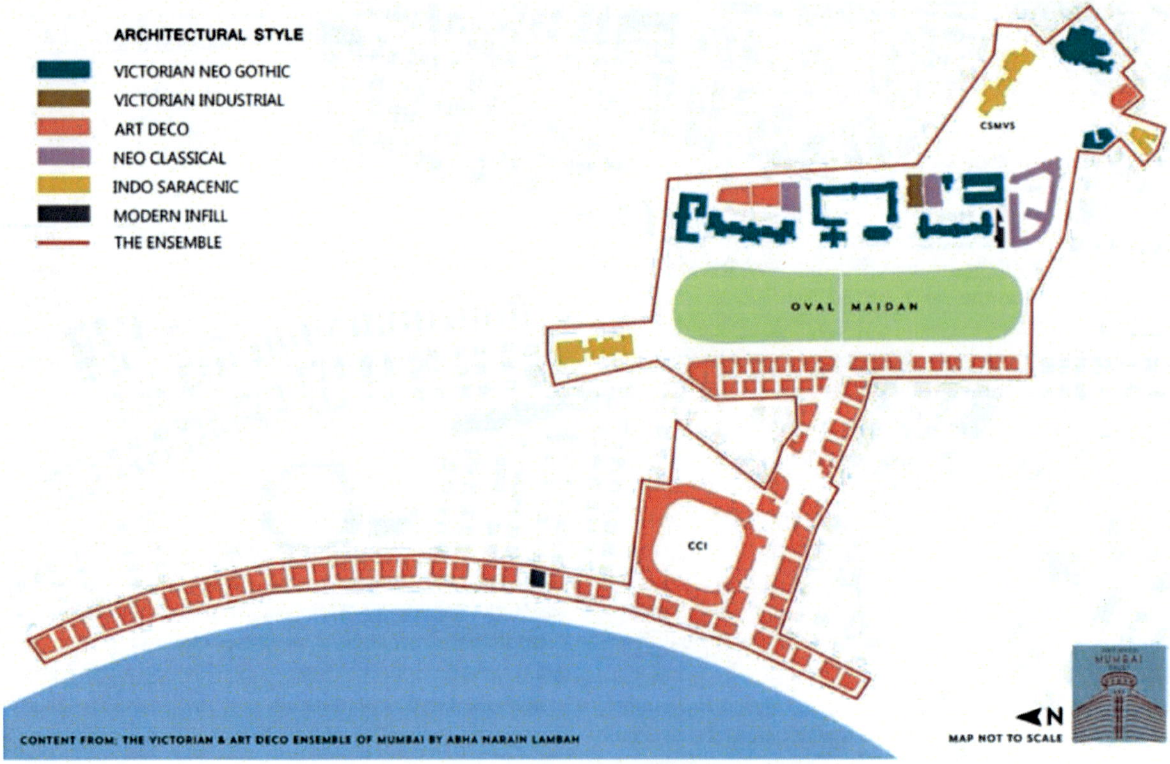

Fig. 7 Map showing precinct delineation and architectural styles of the buildings in the ensemble. *Source* Art Deco Mumbai (2018)

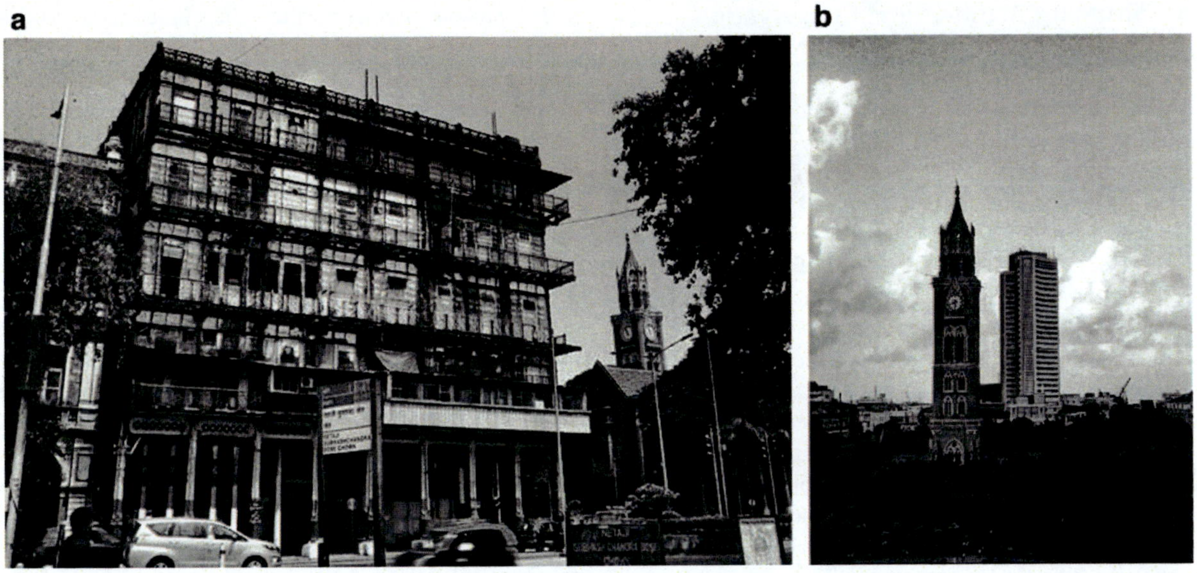

Fig. 8 **a, b** Watson's Hotel, Kala Ghoda precinct, Mumbai. *Source* World Monuments Fund (2006). Rajabai Towers, Mumbai. *Source* Author

assessment with great care, failing which could cost Mumbai its UNESCO world heritage status.

According to London,[3] cited by Fernandez, 2019, 'Esplanade's abused state is testament of its strength' (Fernandez, 2019). He also pointed out that the building being prefabricated in the UK and assembled on site, the replacement of parts would be easier than imagined. The building, its context and setting in an area of high heritage value, is crucial to act sensitively. With a great deal of public (residents) representations and constant support by conservation architects, the building is now under assessment and is soon to be conserved.

In addition to this, a recent partnership project between the Municipal Corporation of Greater Mumbai and the Kala Ghoda Association for the restoration works at Mumbai University Library building and the Rajabai towers (Fig. 8b), a public–private partnership model, which received UNESCO award for heritage conservation and is also a part of the 'Victorian Gothic and Art Deco Ensemble' UNESCO world heritage site (Sundarrajan, 2019). These are examples of some good models in conservation practice that have been achieved through a holistic multidisciplinary approach.

'Successful regeneration means bringing social, economic and environmental life back to an area. It transforms places, strengthens a community's self-image and recreates viable, attractive places which encourage sustained inward investment.' (McCallum, 2007).

The scenario of heritage conservation/protection of a historic urban area[4] is far more complicated than it was two decades ago. There is a pressing need to look at the overall context than just singular buildings. It has to be noted that colonial ideologies of conservation have to be localised in order to create a sustainable alternative by using heritage as a primary resource in a high-density urban development scenario, reducing the carbon footprints and thereby environmental damage. Urban conservation process helps instil this aspect in the regular conservation process and works multidisciplinarily as a unit to achieve the best possible regeneration. The case studies discussed above focus on the importance of stakeholder involvement, the efforts to educate and sensitise local communities that play a major role in the process of integrated heritage and development projects. We see that there are institutions and individuals that strive to preserve the rich heritage, and the impact though can be aptly named as pockets of excellence.

[3] Dr. Christopher W London is an American author, academician and architectural historian. In 1986–87, London researched the Watson's Hotel (Esplanade Mansion) building as part of his doctoral dissertation submission titled British Architecture in Victorian Bombay. This was for DPhil degree from Oxford University. He later included the research in his now famous book, Bombay Gothic that chronicled a larger study of the city (Fernandez 2019).

[4] Historic Urban Area: (From the ICOMOS Washington Charter) Historic urban areas, large and small, include cities, towns and historic centres or quarters, together with their natural and man-made environments. Beyond their role as historical documents, these areas embody the values of traditional urban cultures (Amoruso 2015).

5 Situating Architectural Conservation in the Process of Urban Conservation

Urban conservation process is a collection of a number of smaller processes. However, it has to be noted that the process holds good for a subject as big as a historic area (district) to a building or an artefact such as a statue of high heritage value found on-site, to something as small as an archaeological remain. So, the process of urban conservation (Fig. 9) can be successful only with the successful functioning of the various smaller processes within it.

A very important segment of this process is the historic area characterisation, also known as character appraisal in English heritage terms. This framework enables to assess the values and identify the potential of the various aspects that collectively constitute the historic environment. An effective aid in categorisation and value assessment and the historic area characterisation (Fig. 10) also enables ways to categorise the different levels of stakeholders that are involved and the different levels of development projects that affect the urban heritage (Fig. 9). In the Indian context, more singular heritage structures are situated within a historical environment, thus a conservation plan for the context would automatically imply to the singular building, even though the recommendations for the building's conservation would be specific to the building but sensitive to its environment.

The process of assesment of values based on the significance, (Table 1) informed by the stakeholders, is the key to formulate the 'statement of significance', which thereby aids in preparing the conservation plan. The value assessment coupled with the involvement of stakeholders also helps in identifying various potentials of integration of heritage in the development process. Carrying out a heritage impact assessment is a useful means for testing development proposals against the policies contained within the conservation plan, while the policies can directly inform the development of a maintenance plan for the day-to-day management of the site.

At this point, it is essential to understand that the above process is a representation made to explain the overall framework. The format can be adopted to suit the needs of the project and may include a section to identify where further information, research or analysis is requested. This becomes important when proposals are being tested against a conservation statement rather than the overall plan. The process holds good for the architectural conservation of a heritage building. Even though the factors of analysis are different, such as the style of architecture, age of the building, it is existing condition and use, the degree of change and so on, the identification of stakeholders, involvement of the existing tenants and the historical associations of the building inform the process of characterisation and value

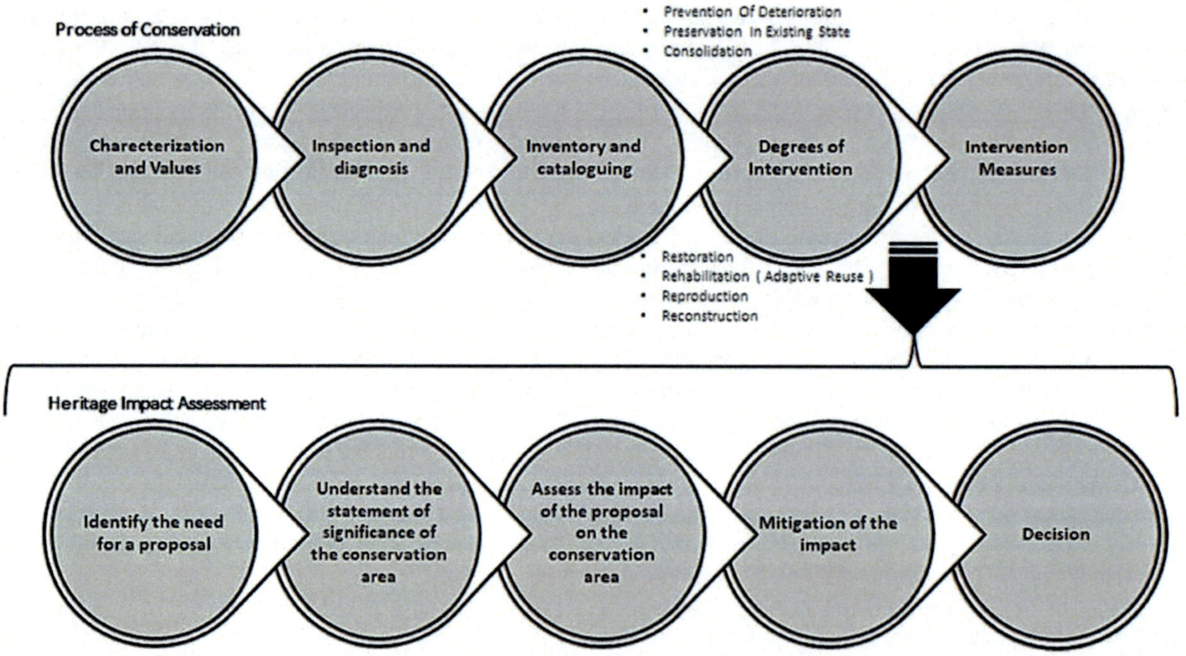

Fig. 9 Steps involved in a conservation process and heritage impact assessment (Krishna Heritage, 2012; Prasad, 2016)

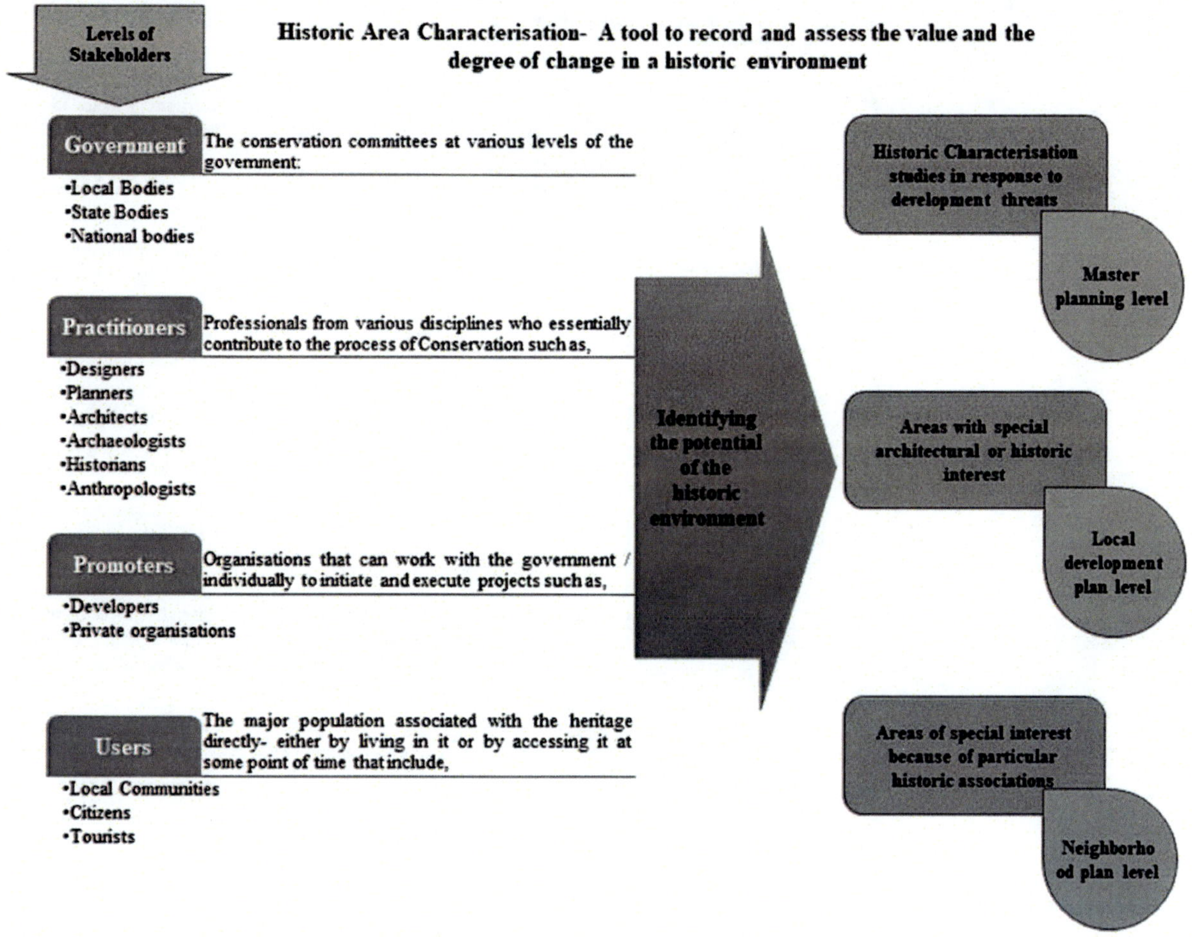

Fig. 10 Various aspects involved in historic area characterisation process (Krishna Prasad, 2016, 60)

assessment. Architecture is always rooted to its context. In the case of the historic/urban environment, the context is as important as the building itself. This maked architectural conservation an integral part of the urban conservation process.

6 Historic Area Characterisation: Using the Framework to Identify the Potential of Heritage Building/Precinct

A tool with great flexibility, historic area characterisation can be used to suit the needs of the heritage site under examination. It can aid in the value assessment of a single building to a precinct. This section attempts to show demonstrations of the use of historic area characterisation for specific project areas of different scales. The scale of the project dictates the format of the framework, the levels of stakeholders involved and scale of the conservation plan. This section demonstrates historic area characterisation through three different scales of historic fabric. The sites chosen for demonstration represent conditions pertaining to India; however, each site has its own unique characteristics.

7 Building Level Demonstration—Shrimat Bhaurangari Wada, Pune, Maharashtra

Situated in a densely populated old Pune precinct, this *wada*[5] is a very humble historic structure with high intangible value. The associations of the building to the beginning of the infamous '*Ganesh Visarjan*'[6] as a community festival/event and important associations to events that

[5] A wada (Marathi for house) is a traditional mansion typically in the western part of India.
[6] An Indian festival dedicated to Lord Ganesh celebrated in various states of India, a very important festival in Maharashtra.

Table 1 Criteria to be considered for listing of heritage assets—the framework in India (Chithra, 2010)

S. no.	Criteria for listing of heritage buildings	Assessment	Explanation
1	Historical significance	Date/period of construction	Age of the building as a determining factor to establish the historic significance
		Trends exhibited by the building	Significance realised sure to the reflection of a particular social, economic, political or cultural pattern, characteristic of the local settlements
		Events associated with the building	Significance built upon associations to an event of local, regional or national significance or a noteworthy historical event
		Persons associated with the building	Significance built upon associations to a notable person, group or institution of historical significance
2	Architectural significance	Design	Significance due to excellence in the artistic merits, uniqueness of design, compositions, craftsmanship or details such as decoration, colour, texture and massing of notable proportion
		Style	Significance due to its features that exhibit a particular architectural style and/ or period of construction
		Designer/builder	Significance due to the architect/engineer or other design professional relating to the period, era and reign of construction and is a work of local, regional or national importance
		Physical conditions	Present condition of the structure superior, minor structural repair or extensive repair or dilapidated
		Design integrity	Nature of reparability and worthiness to conserve
3	Cultural significance	Community context	Significance due to sentimental/symbolic associations with a community and is a significant part of the community identity

defined the course of the Indian Independence movement in Pune add intangible values to the building. The trust-owned building, however, does not appear in the heritage list prepared by the Pune Municipal Corporation, which makes the building more vulnerable to development and the effects of partial maintenance. It is the interest and efforts of the members of the trust and involvement of the public that has contributed to the fair maintenance of the building and its context. Figure 12a elucidates the importance of the context and its intangible value with respect to the building.

Figure 11 shows the preliminnary analysis to understand the values associated with the building and its context. The analysis of the tangible as well as the intangible (Fig. 12a) aspects of the building and its context results in the conservation area/historic area characterisation chart (Fig. 12b) that aids in assessment of values and in drafting the statement of significance. Since the building is not listed as heritage, this process also aids in the listing and grading of the building.

8 Demonstration at Precinct Level—Jogeshwari Caves, Mumbai

An outstanding example of ancient rock-cut architecture is currently situated in the urban area of Jogeshwari in Mumbai, and this historic site has both tangible and intangible values associated with it. The main objective is to revive the

Fig. 11 Structure plan for Shrimat Bhau Rangari wada—Picture of the 127-year-old paper mache idol that is still worshipped and the the original teak wood chariot that bears the idol during the Visarjan (Prathyaksha, 2019)

historical significance of the caves, by improving its visibility in the public realm and ensuring sustainable development, while taking into consideration the challenges of unplanned development. An important aspect under consideration was the community living on-site, the dwellings of the community, developed over time on the historic site. With such overlapping layers of heritage and development visible on-site, the conservation plan must establish an interdependency between the community and the historic site. This would enable the sustenance of the conservation plan, the historic fabric and the community associated with it.

A historic fabric of this scale requires the integration of both building-level and local area-level planning strategies. Figure 13 illustrates, in detail, the existing conditions and types of users and usage of spaces and degree of changes of the site and its context. The site along with its precinct is the subject of analysis to draw-up the historic area characterisation. Figures 13 and 14 show the interdependent process of study and analysis, where the site's history, architecture, existing conditions and stakeholders inform the development of the area characterisation chart that eventually results in formulation of the statement of significance and the conservation plan.

Fig. 12 a, b Diagram showing the intangible cultural significance of the public spaces adjoining the wada (Prathyaksha, 2019)

a

b

Sub divisions	Particulars	Significance		Value
		Tangible	Intangible	
Site	Shrimant Bhausaheb Rangari Bhavan	Architectural Aesthetics Historical	Social Cultural Religious Associative	High
Path	Main entrance from north	Aesthetic Architectural Historical	Associative Religious Cultural	Partial
Views	From the entrance	Aesthetic Architectural	Associative Cultural	High
	From left side	Aesthetic Architectural		High
	From backside	Architectural		Low
Access roads	From Chhatrapati Shivaji Maharaj road via Lal Mahal chowk			
	Maharishi Annasaheb Patwardhan chowk			
Planning	Maratha architecture	Architectural Aesthetics Historical	Social Cultural Religious Associative	High
Material	Brick masonry	Historical Architectural Archaeological		High
	Stone masonry	Historical Architectural Archaeological		High
	Timber	Historical Architectural Archaeological		High
Festivals	Ganpati		Social Cultural Religious Associative	High
	Dahihandi		Social Cultural Religious	Partial
Activities	Social gathering	Historical	Social Associative	Partial

9 Demonstration for Historic City Precinct—Old Madras Precinct, Chennai, Tamil Nadu

The historic precinct of Chennai, the Old Madras area, is an agglomeration of old and new, with various communities, historic market systems and institutions that are thriving even today. However, the physical condition of the cultural heritage is questionable due to high density and a very active public realm. Compared to the above examples, this site is larger in scale and consists of a wide range of tangible and intangible heritage. In the process of historic area characterisation, it is important to consider all the aspects that contributed to the vibrancy and realm of the precinct.

The framework for analysis of the precinct and its value assessment is very specific to the conditions of the precinct, in order to achieve a conservation plan that is specific and inclusive in all respect. The result of this process is used for designation of sub-precincts, grouping of buildings based on architectural styles and formulation of trails based on the narratives drawn through local communities and market systems. Figures 15 and 16 illustrate the framework used to assign values for various historic elements identified within the precinct, and Fig. 17 explains the types of values under consideration. The statement of significance, which is the result of this process, aids in identification of core project zones. One such project zone, illustrated in Fig. 18, highlights a prototype proposal that can be implemented in zones with similar characteristics within the precinct. In an urban-level regeneration plan such as this, there is a need for the government, heritage committee, NGOs and the citizens to participate.

A parallel can be drawn with the case study of Mumbai (Sect. 5.2), where building-level projects can be generated and executed through public–private partnerships within the larger conservation process, such that, even small interventions and strategies at a building level can enhance the overall precint. In some cases, it can even act as a precursor

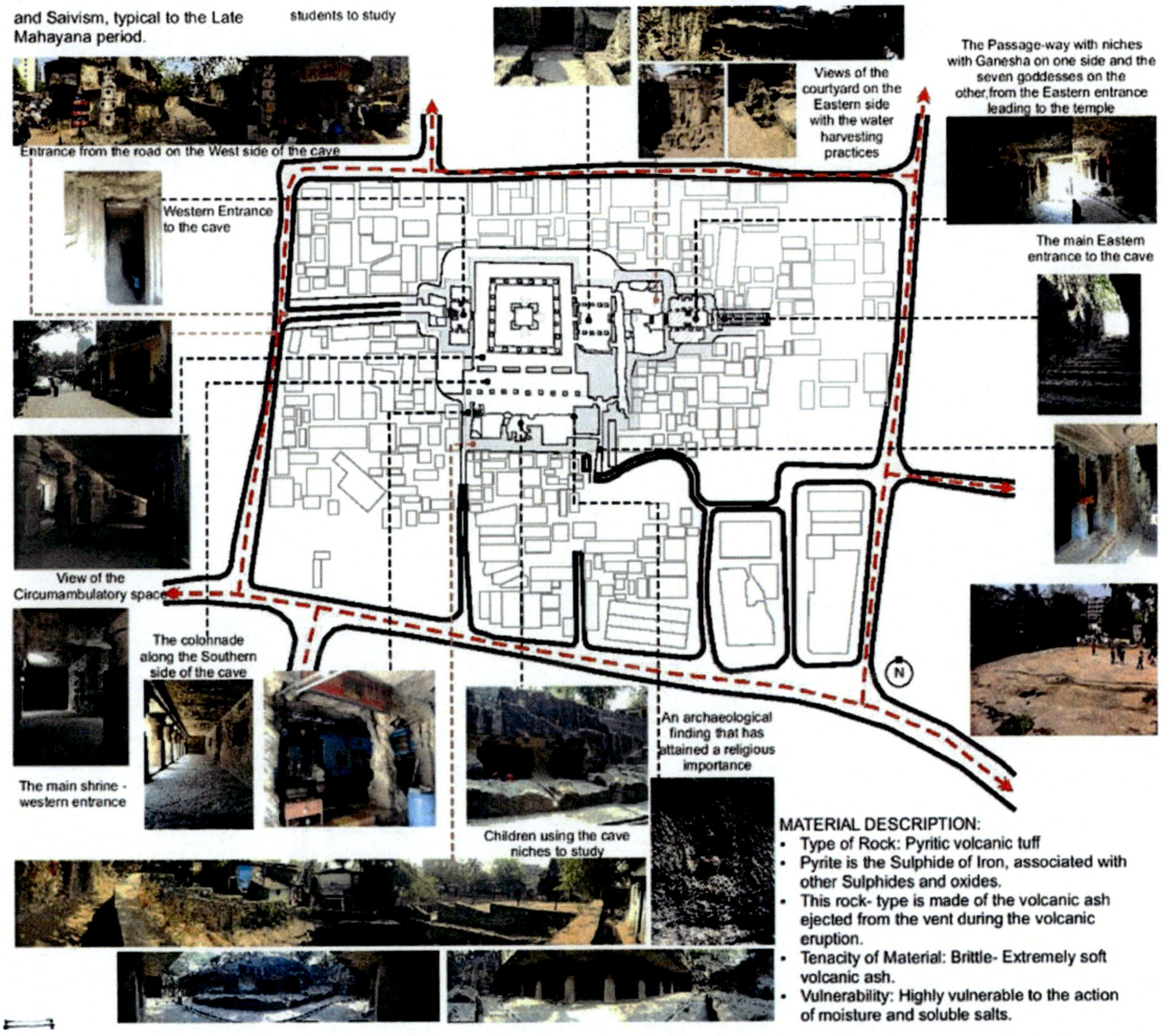

Fig. 13 Sturcture plan showing the site conditions of the Jogeshwari Caves precinct. *Source* Author

for many such similar projects. In overview, the building-level projects enhance the historic context and are eventually sustained by the startegies of the overall conservation plan where local communities and existing market systems play a major role as the key users.

The aforementioned demonstration areas display a variety of scales of historic fabric pertaining to Indian context. There are distinct yet overlapping layers of historic and urban environments that have been developing over time to suit the needs. These layers have to be comparatively studied in order to understand the degree of change and the authenticity of the heritage. This understanding is not only important for the overall urban conservation plan but also to address an individual architectural heritage in a particular context. The a

CONSERVATION AREA CHARACTERIZATION

	SUB DIVISIONS	PARTICULARS	SIGNIFICANCE: TANGIBLE	SIGNIFICANCE: INTANGIBLE	VALUE
JOGESHWARI CAVE PRECINCT	SITE	Jogeshwari Cave	Architectural Aesthetic Archaeological Historical	Social Associative Cultural Religious	High
	PATH	Eastern entrance to the Cave temple	Aesthetic Architectural	Religious Associative Cultural	High
		Western Entrance to the Cave Temple	Aesthetic Architectural	Religious Associative Cultural	High
	EDGES	Edge along the Eastern End of the Cave	Historical	Cultural Religious	High
		Edge along the Western End of the Cave	Historical	Cultural Religious	High
		Edges along the roof of the cave		Social Associative	Partial
	VIEWS	From the Duttatreya Mandri downward - the courtyard on the Eastern end	Architectural Aesthetic	Cultural Associative	Partial
		Jogeshwari Devi Mandir from the Eastern, Western and Southern Doorways	Architectural Aesthetic Historical	Religious	High
		From the top of the Southern side- the courtyard with the pillared hallway and the niches	Architectural Aesthetic	Cultural Associative Social	Partial
	ACCESS ROADS	From Western Express Highway			
		From JVLR			
	WATER BODIES AND SYSTEMS	The Nala on the Southern side of the site		Social	High
		The water- harvesting systems on site	Architectural Historical	Cultural	High
		The drainage pattern owning to the North- South gradient terrain	Natural		High
	PLANNING	Architectural accredition of the caves	Architectural Aesthetic Historical Archaeological	Social Associative Cultural	High
	MATERIAL	Pyritic Volcanic Tuff	Historical Archaeological Natural		High
	FESTIVALS	Shravan Somvar		Social Associative Cultural	High
		Maha Shivrathri		Social Associative Cultural	High
		Navrathri		Social Associative Cultural	High
	ACTIVITIES	Rituals	Historical	Social Associative Cultural	High
		Social gathering		Social Associative Cultural	Partial
		Students studying	Historical	Social Associative Cultural	High
		Playing		Associative Social	Partial

b

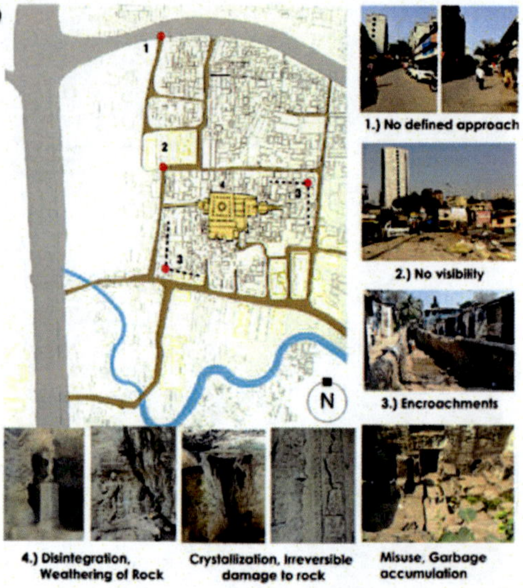

c

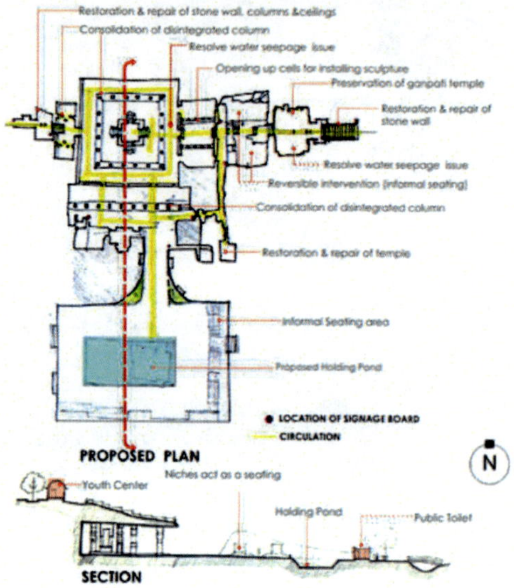

Fig. 14 **a**, **b**, **c** (Clockwise) Historic area characterisation table, identification of the problem areas and the proposal conservation plan for the site. *Source* Author

context (in this case, the urban context) and its related modern developments (namely housing and infracture) greatly impact the conservation process of an architectural heritage. The significance of the heritage building is influenced by its context and associations, and thus it is important to situate the architectural heritage conservation process in a larger urban conservation context, to make the conservation process, multidisciplinary, inclusive and holistic. It is important to note that the process illustrated in these examples is interrelated and can be altered to suit the needs of a particular historic precinct. Involvement of the local communities and stakeholders proves to be useful to carefully assess and arrive at the historic area characterisation chart and eventually the conservation plan.

a

NO.	CLASSIFICATION	NO.		NO	PARTICULARS	TANGIBLE					INTANGIBLE			
						ARC	AES	AR	HIS	EVI	AS	SOC	REL	COM
I	DISTRICTS	i	Fort District	1	Fort St. George	●	●	●	●	●	●	●	○	●
		ii	George Town (Blach Town) District	2	Muthialpet	○	⊙	⊕	●	●	⊕	●	○	⊕
				3	Peddanaickenpet	○	⊙	⊕	○	●	⊕	●	○	⊕
				4	Royapuram	○	⊙	⊕	○	●	⊕	●	○	⊕
				5	Chintadripet	○	⊙	⊙	○	●	⊕	●	○	⊕
				6	Sowcarpet	○	⊙	⊕	○	●	⊕	●	○	⊕
				7	Mannody	○	⊙	⊙	○	●	⊕	●	○	⊕
				8	Tondiarpet	○	⊙	⊙	○	●	⊕	●	○	⊕
						ARC	AES	AR	HIS	EVI	AS	SOC	REL	COM
II	STREETS	i	PATHS AND EDGES	1	First Line Beach Road(Rajaji Salai)	⊕	●	●	●	●	●	⊕	○	⊕
				2	China Bazaar Road (NSC Bose Road)	○	●	●	●	●	●	●	○	⊕
				3	Light House Road	○	●	⊕	●	●	●	●	○	⊕
				4	Old Jail Road	●	●	⊕	●	●	●	●	○	⊕
				5	Ebrahim Sahib Street	○	⊙	⊕	●	●	●	●	○	⊕
				6	Basin Bridge Road	○	⊙	⊕	●	●	●	●	○	⊕
				7	South Beach Road (Kamarajar Salai)	○	●	●	●	●	●	●	○	⊕
				8	Popham's Broadway	○	●	●	●	●	●	●	○	⊕
				9	Rattan Bazaar Road	○	⊙	⊕	●	●	●	●	○	⊕
				10	Esplanade Road	○	●	●	●	●	●	●	○	⊕
				11	Armenian Street	○	●	●	●	●	●	●	○	⊕
				12	Mannady Street	○	⊙	⊕	●	●	●	●	○	⊕
				13	Moore Street	○	⊙	⊕	●	●	●	●	○	⊕
				14	Angappa Naicken Street	○	⊙	⊕	●	●	●	●	○	⊕
				15	Burma Bazaar	○	●	●	●	●	●	●	○	⊕
				16	Linghi Chetty Street	○	⊙	⊕	●	●	●	●	○	⊕
				17	Sembudoss Street	○	⊙	⊕	●	●	●	●	○	⊕
				18	Anderson street	○	⊙	⊕	●	●	●	●	○	⊕
				19	Godown Street	○	⊙	⊕	●	●	●	●	○	⊕
				20	Govindappa Naicken Street	○	⊙	⊕	●	●	●	●	○	⊕

b

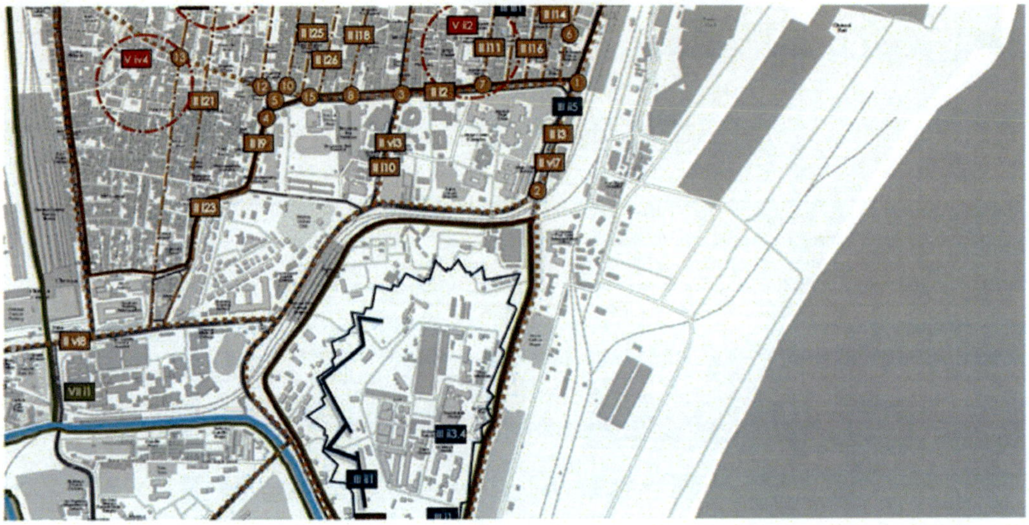

c

Fig. 15 **a**, **b**, **c** Sample historic area characterisation for the historic core of Chennai, India—tool to define precincts and assign values to historic elements within the historic core. **a** the characterisation table, **b** the corresponding map, **c** pictures as illustrations with respect to the table and the map (Krishna Prasad, 2016)

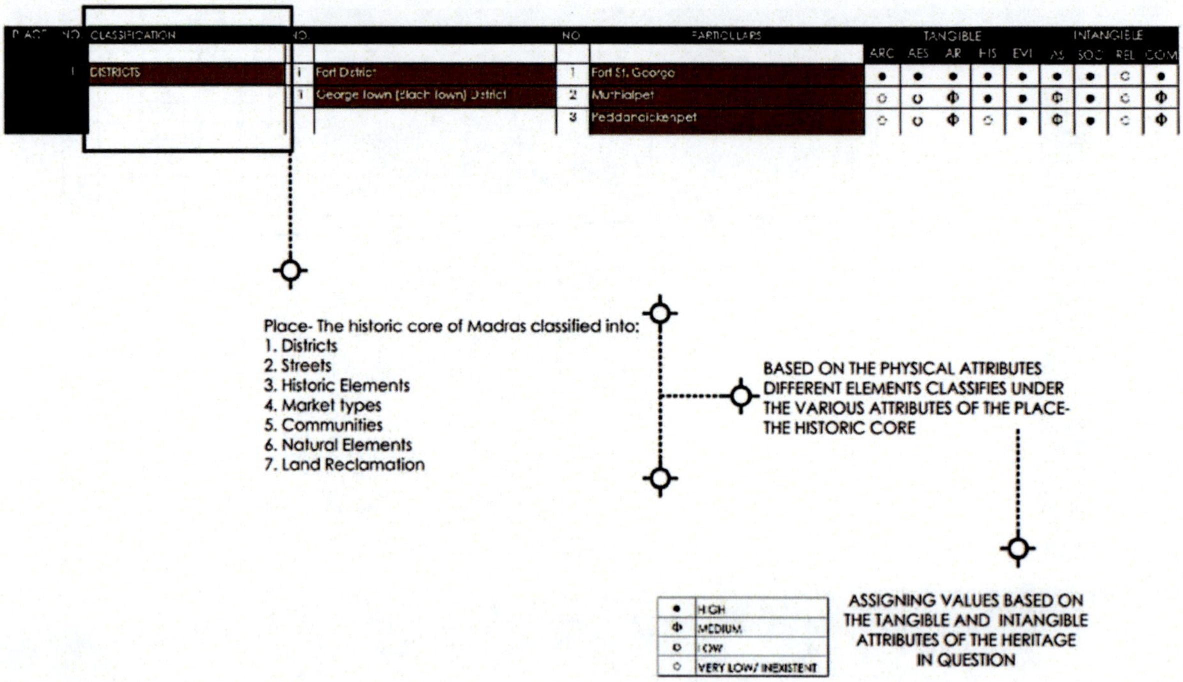

Fig. 16 Framework used for analysis and documentation of the values of the heritage stock (Krishna Prasad, 2016)

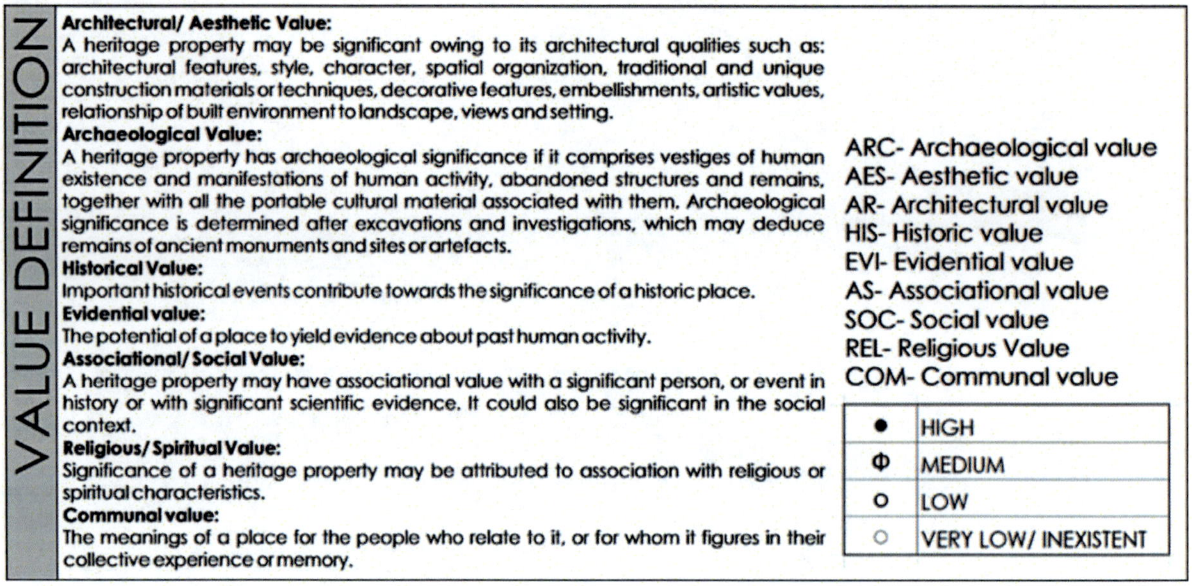

Fig. 17 Definition of values and abbreviations used in the characterisation chart (Krishna Prasad, 2016)

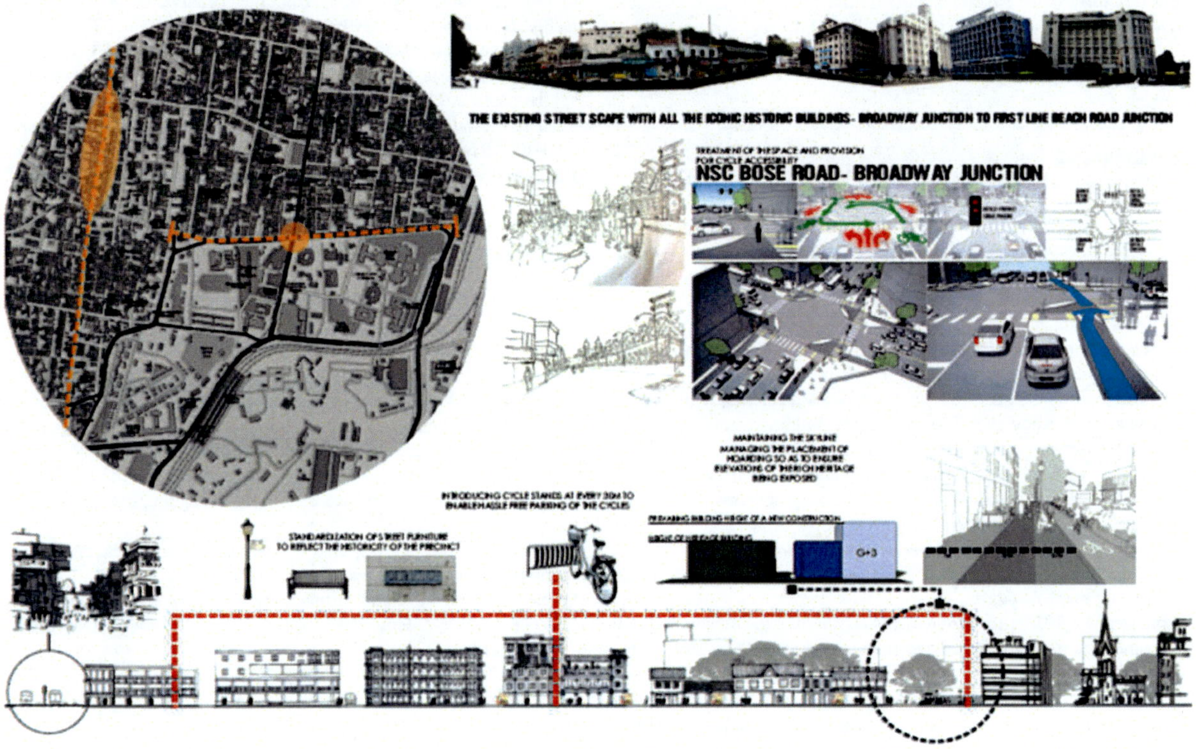

Fig. 18 Illustrated proposal for a major access road situated within the precinct (Krishna Prasad, 2016)

10 Conclusion

'Involving the wider community before the appraisal is too far advanced to exclude further influence on the outcome. It is usual to include a report on how community involvement and public consultation have been undertaken in the appraisal itself and explain how the input from the community was evaluated and how it has been taken into account in defining the special interest of, and recommendations for, the area.' (Ashfield, 2019).

Developing nations like India, where urban problems are a result of economic development and growing population, can see a standard pattern of development with new townships emerging on the fringes of the old town and agglomeration within the old town itself, resulting in a serious identity crisis. Urban conservation process, on the other hand, is a means of binding the local communities, its heritage and also allowing sensitive development which may result in sustainable development and reduced carbon footprint, thereby retaining the vibrancy of historic precinct. The process of urban conservation covers a strong documentation exercise involving listing, character appraisal, vulnerability analysis and designation of conservation zones, a result of this being a management plan that indicates problem areas and locates architectural heritage that needs immediate examination and intervention. Thus, architectural conservation is a fragment of the overall urban conservation/regeneration process that seeks to deliver social, economic, cultural and environmental benefits of heritage in the development process. 'This research also intends to further pursue the economic viability of the process so as to ensure the sustainable continuity of the local culture in the historic urban environment. Besides the snowballing effect on the investment towards heritage, the concept of urban conservation aims at promoting the property, land, capital, human resources, heritage resources and political commitment. Better condition of historic areas resulting from the conservation work carried out is believed to support the regeneration of the area as a whole. The profit from regeneration would provide capital for more conservation activities.' (Krishna Prasad, 2016, 138).

Architectural conservation and urban conservation go hand-in-hand. While urban conservation is an overall process, architectural conservation is its integral part with prevention (of deterioration), preservation (in existing state), consolidation, restoration, rehabilitation, reproduction or reconstruction as a degree of intervention, proposed based on all the research and documentation process involved in the study of a conservation area. It is essential to understand that heritage is an essential part of the built environment and using it as a resource for development is a sustainable approach.

References

Amoruso, G. (2015). Handbook of research on emerging digital tools for architectural surveying, modeling, and representation. *IGI Global.* https://www.igi-global.com/chapter/the-image-of-historic-urban-landscapes/133427.

Art Deco Mumbai, Trust. (2018). *Research—UNESCO.* https://www.artdecomumbai.com/research/list-of-buildings-architectural-styles-maps-of-the-unesco-precinct/.

Ashfield, District Council. (2019). https://www.ashfield.gov.uk. https://www.ashfield.gov.uk/residents/planning-building-control-and-land-charges/forward-planning/the-emerging-local-plan/.

Bhosle, P. (2019). https://www.mid-day.com. https://www.mid-day.com/articles/foundation-to-protect-victorian-gothic-and-art-deco-in-south-mumbai/20830157.

Chithra, Tmt. S. (2010). *Development regulation provisions for heritage conservation in CMA.* Chennai: Chennai Metropolitan Development Authority.

Fernandez, F. (2019). https://www.mid-day.com. https://www.mid-day.com/articles/esplanades-abused-state-is-testament-of-its-strength/21155192.

Heritage, English. (2012). *Understanding place: Conservation area designation, appraisal and management.* England: English Heritage.

INTACH, Pondicherry. (2008). Heritage conservation in Pondicherry.

INTACH, Pondicherry. (2010). *Architectural heritage of Pondicherry.* Published with financial assistance from the European Commission under the Asia Urbs Programme-Pondicherry in partnership with the cities of Urbino, Italy & Villeneuve-sur-Lot, France.

Krishna Prasad, P. (2016). *Conundrum of heritage and development—Will the two coexist to sustain?* (Vol. 4). Mumbai: KRVIA.

Krishna Prasad, P. (2019). *Expert speak.* https://www.sawdust.online/expert-speak/the-story-of-bhau-rangari-wada-pune/.

Loveday, M. (2012). Heritage led regeneration—An innovative model. In *Heritage and development—recent perspectives* (pp. 358–365). Indian National Trust for Art and Cultural Heritage. INTACH.

McCallum, D. (2007). "Regeneration and the historic environment". In M. Forsyth (Ed.), *Understanding historic building conservation* (pp. 35–45). Blackwell Publishing Ltd.

PEARL. (2015). *Urban heritage in Indian cities* (pp. 30–39).

Sundarrajan, V. (2019). https://www.thehindu.com. https://www.thehindu.com/news/cities/mumbai/mu-receives-award-for-heritage-conservation/article29216078.ece.

Prince's Trust, Canada. (2008). *Heritage regeneration initiative.* https://www.princestrust.ca/initiatives/heritage-regeneration-initiative/.

Wikipedia. (n.d.). https://www.en.wikipedia.org/wiki/Main_Page. https://www.en.wikipedia.org/wiki/Watson%27s_Hotel.

World Monuments Fund, Project. (2006). *Project-Watson's Hotel, 2006 Worls Monuments Watch.* https://www.wmf.org/project/watsons-hotel.

Xie, J., & Heath, T. (2019). *Heritage-led urban regeneration in China.* Routledge- Taylor and Francis Group.

Cultural Heritage and Contemporary Armed Conflicts in the Arab World

Mayas Nadim Ahmad Taha

Abstract

Wars and armed conflicts taking place in some Arab countries are one of the major reasons that have led to the deterioration and destruction of many world heritage buildings and sites in those countries. The wars have incurred a great deal of damage to large parts of world heritage structures. This is in addition to looting and theft-related damage, through illegal excavations and other forms of sabotage. In view of the importance of cultural heritage and the extent of damage to its components due to conflicts and wars, this paper extrapolates the mechanisms established by international conventions to protect cultural heritage. The paper covers issues such as the insufficient implementation of international preventive measures in some Arab countries where a large part of their heritage has been damaged. It highlights the realities of cultural heritage during armed conflicts in these countries and assesses how far they abide by the legal mechanisms. At the end of the research, there is a set of recommendations for the preservation of world heritage in areas of armed conflicts.

Keywords

Cultural heritage • Armed conflict • Conventions • Conservation • Arab world

1 Introduction

Built cultural heritage symbolizes the identity and history of people in general. It is an integral part of a nation's living memory and history. The built cultural heritage has been passed on from one generation to another. It is one of the basic elements that constitute the collective memory and cultural identity that characterize a society. So, any attack on any of its components can be an attack on the dignity and history of peoples.

Throughout history, mankind has suffered wars and other forms of armed conflicts that have inflicted a lot of damage to not only people, but also their personal property and those of the state, leaving lifetime scars on the symbols of their culture and civilization. One of the most recent examples is the case of most Arab countries that have witnessed, throughout the past nine years, large-scale armed conflicts. This has posed a serious threat to their cultural heritage and made it necessary for conducting a study on this matter, as there are no such studies in the Arab world. Since heritage is linked to people and their national and civilizational identity, this research is aimed at:

Highlighting the fact that one of the major reasons for the deterioration of cultural heritage in many Arab countries is their disregard for international conventions that stipulates preventive measures to protect cultural heritage in armed conflicts,

Exposing the destructive impact of wars on the heritage of some Arab countries and that such countries have not realized the objectives of the international conventions in terms of preserving the heritage in armed conflicts.

M. N. Ahmad Taha (✉)
College of Engineering & Architecture-Department of Architecture, Al-Yamamah University, Al-Riyadh, Saudi Arabia
e-mail: Mayas-84@hotmail.com; M_Taha@yu.edu.sa

A. Versaci et al. (eds.), *Conservation of Architectural Heritage*, Advances in Science, Technology & Innovation,
https://doi.org/10.1007/978-3-030-74482-3_15

2 Literature Review

Mechanisms of heritage protection in armed conflicts have been developed through many conventions and protocols. It is necessary to shed light on the development of such mechanisms throughout history. Most importantly, the extent to which countries with armed conflict have implemented those mechanisms and abided by them is a must and needs verification.

The concept of cultural heritage is broad, flexible and evolving as human life evolves. Many conventions have identified the concept of cultural heritage in line with the definition set out in the 1954 Hague Conventions (UNESCO, 1954).

Therefore, the definition can be found clear in *Article 1 of the 1954 Hague Convention.* The article reads: *For the purposes of the present Convention, the term 'cultural property' shall cover, irrespective of origin or ownership: (a) movable or immovable property of great importance to the cultural heritage of every people, such as monuments of architecture, art or history, whether religious or secular; archaeological sites; groups of buildings which, as a whole, are of historical or artistic interest; works of art; manuscripts, books and other objects of artistic, historical or archaeological interest; as well as scientific collections and important collections of books or archives or of reproductions of the property defined above; (b) buildings whose main and effective purpose is to preserve or exhibit the movable cultural property defined in sub-paragraph (a) such as museums, large libraries and depositories of archives, and refuges intended to shelter, in the event of armed conflict, the movable cultural property defined in sub-paragraph (a); (c) centers containing a large amount of cultural property as defined in sub-paragraphs (a) and (b), to be known as `centers containing monuments'* (UNESCO, 1954).

Armed conflicts are defined as follows: "armed fighting between the armed forces of two or more states for international or national interests which entail a set of rights and obligations" (AL-Mafraji, 2011).

2.1 Historical Brief About Legal Mechanisms of Cultural Heritage Preservation in Armed Conflicts

Before World War 1, some international conventions on cultural heritage preservation were endorsed. Yet, they had little impact. Subsequently, there were multiple attempts for more binding and effective conventions between WWI and WWII. After1945, a number of governments and NGOs to provide more conservation and protection exerted a lot of effort to endorse cultural heritage preservation in wars and armed conflicts. According to Mahmood (2000): "*Till this day, the international community hopes to reach common rules to control the military operations in armed conflict*" (Abdel-Ghani Abdel-Hamid, 2000).

The rules on this subject have evolved in the light of the establishment of international organizations, so the historical development of the legal preservation of built cultural heritage in armed conflict can be summarized as follows:

Article 17 of the The Brussels Declaration, 1874 stipulates: "In such cases all necessary steps must be taken to spare, as far as possible, buildings dedicated to art, scienceor charitable purposes." (Osgoode, 2017) "This represents decisive turning in the preservation of built cultural heritage in particular and cultural property in general" (AL-Mafraji, 2011).
In 1899, preservation of built cultural heritage in armed conflicts was marked by the holding of the First Hague Conference, which was confirmed by the Second Hague Conference of 1907 (AL-Mafraji, 2011).
In 1935, the Treaty for Protection of Artistic and Scientific Institutions and Historic Monuments, known as the Roerich Charter, was endorsed (Getty Conservation Institute treaty, 1935).
In 1949, the Fourth Geneva Convention was signed in to humanize the conduct of military operations. (UNESCO, 1949).
The greatest achievement was the conclusion of the "1954 Hague Convention" with its protocol for the protection of cultural property in the Event of Armed Conflicts (UNESCO, 1954).
Subsequently, the 1970 Convention for the Prohibition and Illicit Import, Export and Transfer of Cultural Properties was concluded (UNESCO, 1970).
It was followed by the 1972 Convention for the Protection of the World Cultural and Natural Heritage, as well as several recommendations adopted by Unesco for the protection of cultural heritage (UNESCO, 1972).
In 1977, additional protocols to the Geneva Conventions of 12 August 1949 was promulgated, including some articles relating to the protection of cultural property (ICRC, 2010).

Fifty years after the adoption of the Hague Convention, the Second Protocol was adopted on 26 May 1999. It was decisive in improving the situation of cultural property in general in armed conflicts (UNESCO, 1999).

2.2 Preventive Protection of Built Cultural Heritage in Situations of Peace

The protection of cultural heritage buildings in peace time is entrusted to the state on whose region it is located. The said state must take a range of measures to ensure that its cultural heritage is protected from damage that may be inflicted in the event of armed conflict. The discretion of those measures is left to the state itself.

In order to determine these measures, the Second Protocol of the Hague Convention of 1999 establishes preventive measures to be taken in situations of peace as follows (UNESCO, 1999):

Preparation of inventory lists for the cultural heritage sites.
Development of emergency for and disaster plans for transfer protection of cultural heritage buildings and sites.
Determining a competent authority charged with the maintenance and care of culture property.

Each state can intensify such preventive measures as it deems fit. Such measures are very important and not limited only to situations of armed conflicts, but also to situations of calamity and natural disasters.

2.3 Guidelines for the Protection of Cultural Heritage in Armed Conflict Areas

Pursuant to international conventions, there are three categories of heritage protection during armed conflict as follows:

A. **General protection**: or automatic preservation of complete cultural heritage and protection from partial usage of cultural heritage for any kind of military use. Figure 1 illustrates the shape of the logo of the general protection which, in turn, consists of several rules (UNESCO, 1999):

Rule 1: Heritage buildings should be labelled with distinctive signs, as set out in several articles:

Article 6 of the 1954 Hague Convention provides for a distinctive logo to identify and distinguish cultural heritage buildings by the parties of the conflict who may claim to be unaware of those buildings and their places.

Fig. 1 Logo of general protection (UNESCO, 1954)

Article 16 of the 1954 Hague Convention reads: "the nature of this logo, the choice, the status and visibility of the distinctive logo, shall be left to the authorities of the protection of the cultural heritage of each State." (UNESCO, 1954) It should be taken into account in the position of the signs to be seen easily during daytime whether by land or from the air. They should be visible and at regular and sufficient distances.
Article 17 of the 1954 Hague Convention prohibits the abuse of the logos.

Rule 2: Cultural heritage buildings must not be attacked during armed conflict. This has been reiterated by many articles, for example but not restricted to:
Articles 4&7 of the Hague Regulations, the Law of War of 1954 stipulates that: "*cultural heritage buildings should not be attacked in conflict situations*" (UNESCO, 1954).

Rule 3: Cultural heritage buildings should not be used for military purposes. The legal protection of cultural heritage buildings should be fully and effectively achieved. This has been reiterated by many articles:
Article 4 of the Hague Convention prohibits the use of heritage buildings for military purposes (UNESCO, 1954).
Article 8 of the 1999 Protocol II stipulates that "the conflict's parties should take as much caution as to provide protection to sites where cultural heritage buildings are located and to avoid the establishment of military objectives in their vicinity" (UNESCO, 1999).
Protocol I of the 1977 Geneva Convention, in article 1/53, prohibits "the usage of culture buildings in support of the war effort" (ICRC, 2010).
The 1935 Charter of Roerich states that "The use of cultural heritage buildings in military operations deprives them from the protection and respect prescribed in all international Charters" (Getty Conservation Institute Treaty, 1935).

B. **Special protection**: Buildings that listed in world heritage list will be under this protection, as set out in:
Article 8 of the 1954 Hague Convention stipulates that: "a limited number of bunkers for the protection of movable cultural property, memorial centers and other cultural property of great importance may be placed under special protection" (UNESCO, 1954).
However, full implementation of special protection can only be achieved by several conditions; pursuant to the 1954 Hague Convention, "*the most important of which being that cultural property & located at a sufficient distance from any major industrial center or military objective; that such property is not used for military purposes without the objection of any State to this: acceptance of making cultural property under international control*"(UNESCO, 1954). Figure 2 illustrates the shape of the logo of special protection.

Fig. 2 Special protection logo (UNESCO, 1954)

C. **Improved protection**: Adding special immunity on all cultural heritage, and on its usage. Article 10 of the 1999 Protocol II provides for the ethical conditions for granting enhanced protection in the following cases:
The building under consideration is considered a cultural heritage of the greatest importance to humanity.
The building had previously been the subject of protection by appropriate national legal and administrative measures and the state that owns it recognizes its exceptional historical value.
The building is not a legitimate target as it has not been used as a military site.

According to the above literature review, we can consider that international conventions have stipulated different mechanisms of protection for heritage buildings in armed conflicts. This stresses the necessity of applying those mechanisms by all states who are parties to the conventions.

3 Methodological Approach

This paper investigates the connection between the laws and international conventions related to the protection of cultural heritage in armed conflicts and the reality of the cultural heritage in some Arab countries that have suffered from conflicts in the past nine years. The paper's spatial scope is some of the Arab countries which have been affected by wars with more concentration on the status of some heritage sites in Syria. The time range is concentrated on the post-war period. In addtion, the methodology used in the research is an inductive descriptive and analytical method to examine the extent of Arab countries in compliance with the mechanisms stipulated by the laws of protecting cultural heritage.

The paper considers the urgent role of applying preventive measures recommended by international bodies and the necessity of promoting cultural awareness about the harm caused by neglecting these procedures in the Arab world. The lack of such awarness has been the major reason for damaging many heritage buildings and listing a lot of the remianing ones in the danger list. Figure 3 shows the ratio of world heritage sites in danger.

Therefore, in order to achieve the objective of the research, the paper will first explain and describe the realities of the cultural heritage in some of the countries which have, for many years, suffered from ongoing and destructive conflicts. The impact of such conflicts is evident in the

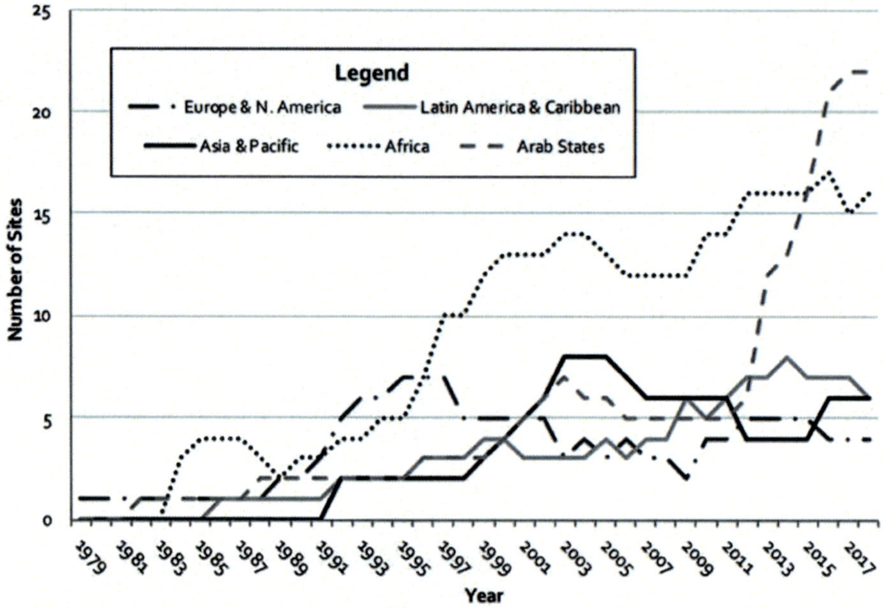

Fig. 3 Number of world heritage sites on the list in danger by regional group from 1979–2018 (Nicholas et al., 2019)

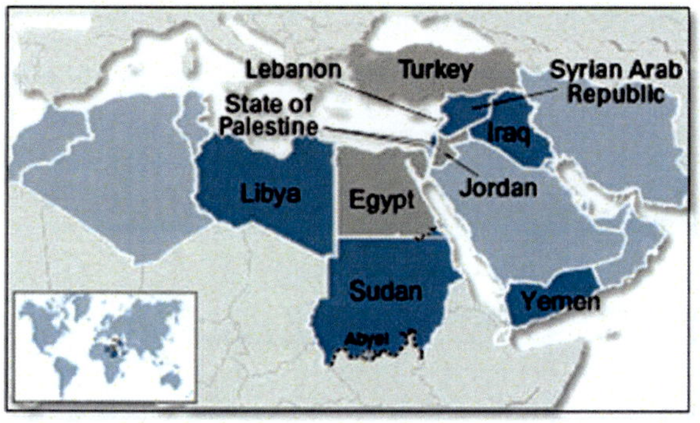

Fig. 4 Countries which suffer from armed conflicts in the Middle East (unicef.org, 2019)

destruction and vandalism of heritage monuments there. Figure 4 shows the countries which suffer from armed conflicts in the Middle East according to UNICIF reports.

The recent conflicts in Iraq, Syria, Libya and Yemen, along with the destruction of a lot of cultural buildings at many world heritage sites there, have highlighted the pressing need to study the current status of the cultural buildings at those countries. IT should also be noted that all of them ratified the 1954 Hague Convention on the protection of cultural property in situations of armed conflict. And all of those countries have been selected because they have suffered from armed conflicts for long periods. This is the main criteria of selection for consideration, and the study under this paper. (Table 1) sums up the realities of the cultural heritage in the above-mentioned Arab countries (unesco.org, 2019).

It can be concluded from the table that all the countries which have been selected for the study have ratified the 1954 Hague Convention on the protection of cultural property in armed conflicts.

According to Vecco, 2018, "*Since 2011, many of the Arab countries, saw popular revolutions in Syria, Libya, and Yemen inspired by the Arab Spring, the armed conflicts which presently afflict each of these countries and threaten their heritage have their immediate roots in the events of that year, the pressures on heritage.*"

The American School of Oriental Research (ASOR) established the Cultural Heritage Initiatives (CHI) in 2014. Since then, CHI has started to assess the state of cultural heritage in some Arab countries, particularly in Syria and Iraq. It has conducted the assessment on 6,662 heritage sites in Syria, Iraq and Libya kept the assessments updated and refined regularly. Utilizing the most recently available satellite imagery, each heritage site has been assessed for damage occurring since the start of the conflict and assigned a percentage of total visible damage. Figure 5 displays these assessments.

ASOR (CHI) assessment focuses on damaged sites. The majority of assessed heritage sites display no visible damage up to a total of 63%. The second-highest percentage of damage falls between 10 and 60%, at 26% of total assessed sites. (ASOR CHI, 2017).

This work considers the realities of the cultural heritage buildings in the above Arab countries through reports received from UNESCO and analysis of some satellite imagery to discover the reported damage. In Lebanon for example, the twenty-four-year civil war in addition to the devastating aftermath of repeated Israeli military operations, especially in 1982 and 2006, caused severe damage to many of the major heritage buildings in the central and the southern parts plus the capital of the country.

Lebanon, which has suffered from the civil war for many years, and which has five world heritage sites. It is noteworthy that the United Nations played a major role in preserving heritage sites after the Israeli invasion of Lebanon in 1982, in coordination with UNESCO, managed to return the documents and archives that had been looted after the occupation of the city of Beirut (Al-Rahaifa, 2012).

According to UNESCO, world heritage sites in Lebanon were not damaged by conflicts there. The major damage was hitting some old houses in Beirut and the old City of Baalbek. Figure 6 shows some effects of the war on the heritage buildings of Beirut.

Iraq is a country known as the living memory of the world with more than half a million archaeological sites and monuments of civilization and history. The losses in Iraq have perhaps been the most acute (Vecco, 2018). Cultural

Table 1 Analysis of the reality of the cultural heritage in the chosen Arab countries (unesco.org, 2019)

The country	The duration of ware	The date of ratifying on The Hague convention protocol 1954	World heritage sites	Special protection list	World heritage in danger
Lebanon	1975–1990	01 June 1960	5	0	0
Iraq	2003–2015	21 December 1967	7	0	3
Yemen	2015–till now	06 February 1970	4	0	3
Lybia	2011–till now	19 November 1957	5	0	5
Syria	2011–2019	06 March 1958	6	0	6

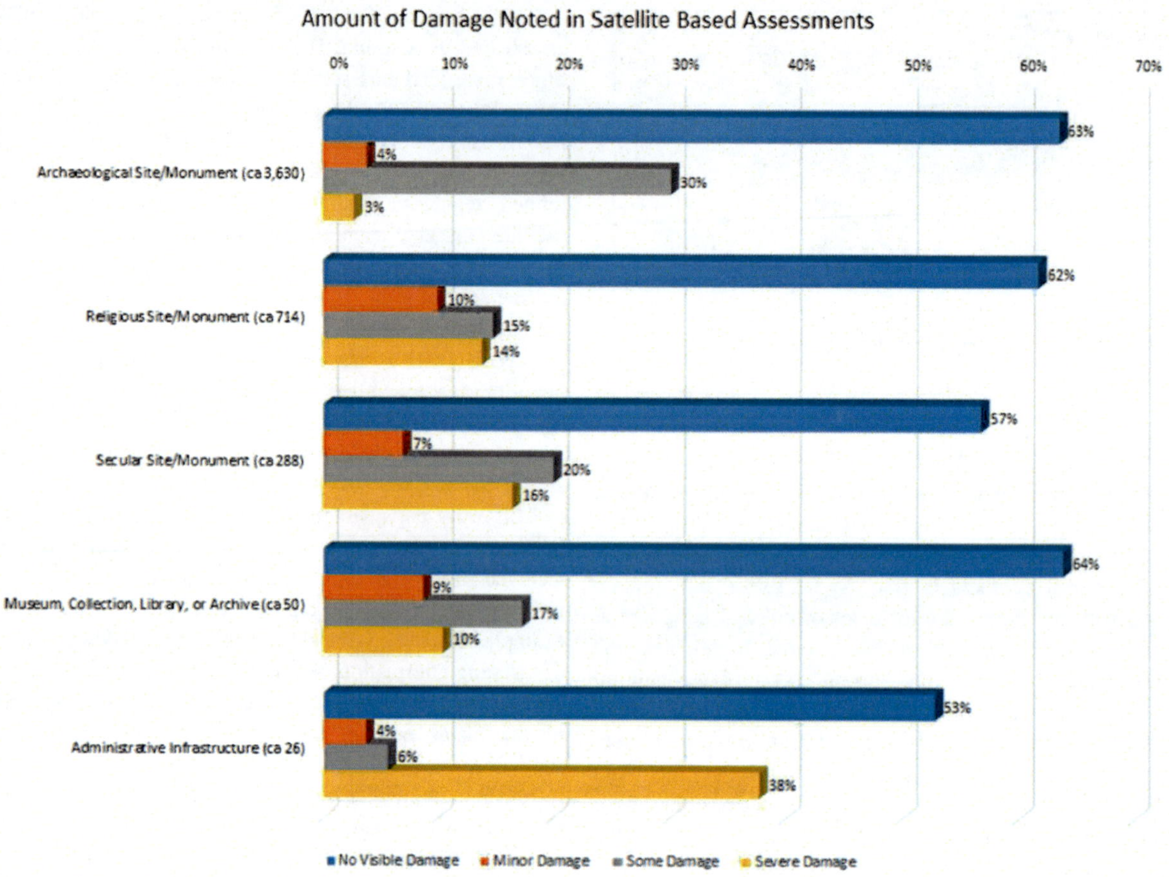

Fig. 5 Satellite-based assessments of heritage sites within the Cultural Heritage Initiatives (CHI) Inventory (mainly Syria and northern Iraq) according to site types and levels of assessed damage (Danti et al., 2017)

Fig. 6 Images showing the effects of the war on the heritage buildings of Beirut (Thekra & Ebrah, 2019)

facilities and places of worship have been systematically attacked to eradicate the past culture and civilizational history of this country. More than seven important archaeological sites have been used as military bases. Moreover, the majority of Iraq's antiquities have been looted.

The 2003 war in Iraq has exposed many cultural and heritage buildings to destruction and systematic looting, causing the loss of many historical pieces which have been embodiments of the Mesopotamian civilization (Nnamara et al., 2010). This is in addition to the systematic destruction of many heritage buildings and mosques in Mosul and the destruction of many artefacts in the Iraqi Museum in Baghdad.

Major heritage buildings have been used for military purposes. One of the most important buildings in Mosul is *Al-Nuri Mosque* whose minaret was destroyed in June 2017

Fig. 7 Images of al-Nuri Mosque and the Lighthouse of Mosul after the end of the ISIS war (AL-Nashef, 2004)

during the fights. The mosque dates back to the twelfth century (Al-Nashef, 2004). Figure 7 shows photos of the al-Nuri Mosque after the end of war.

According to Stone, (one of those who attempted to advise on the provision of site protection) quoted by Vecco (2018): "Remarkably, and essentially by luck, very little destruction of cultural property happened during the conflict part of the invasion… but as soon as the regime collapsed there was endemic looting of museums, libraries and archaeological sites across the country. We can't also forget the looting of Iraq Museum in Baghdad and the burning of the National Archive in that museum" (Vecco, 2018).

In Palestine, the negative effects and violations of the Israeli occupation are evident enough. Several historic buildings have been destroyed in Gaza City, which is an ancient city and has roots deep in history. *Qasr al-Deyafah*, which has an important historical value, has been completely destroyed by Zionsit shelling. In addition, the municipal court building has been partially demolished as a result of the Israeli shelling. (AL-Namreh et al., 2010), see Fig. 8.

So far, the Israeli authorities are still digging in the occupied territories and the city of Jerusalem breaking international laws and regulations.

In Yemen, which has three towns registered as world heritage sites, a large number of buildings in Sanaa have been destroyed in coalition airstrikes. The Marib Dam, another world heritage site, as well as other ancient sites have been damaged, according to UNESCO (UNESCO, 2015).

The civil wars the country witnessed have affected its ancient civilization landmarks and heritage. Furthermore, military operations have led to the destruction of much of its cultural heritage (The Guardian, 2019).

According to Harvey (2018), the director of the Yemeni General Organization of Antiques and Museums (GOAM) stated that: "*Fifty*-nine of the sites—just over 75%—had been damaged despite UNESCO providing a list of sites to avoid".

According to UNESCO: "*Three cultural World Heritage properties (Old Walled City of Shibam, the Old City of Sana'a, Historic Town of Zabid) are now inscribed on the List of World Heritage in Danger. The Old City of Sana'a and the historic center of Saa'da were hit by shelling and gravely damaged,*" (UNESCO, 2015); see Fig. 9.

According to UNESCO World Heritage Centre: "one of the most important historic cities in Yemen is Zabid, located about 75 km southeast of the coastal city of Hodeidah overlooking the Red Sea and controlled by the Houthis. Zabid, which includes 86 mosques, including the fifth oldest mosque, and it outdates Islam's creation in the seventh century, is considered an architectural jewel. It contains the highest concentration of mosques in the country and was Yemen's capital from the 13th to fifteenth century" (Bakir, 2013).

"Zabid was listed as a World Heritage Site by the United Nations Educational, Scientific and Cultural Organization UNESCO in 1993" (UNESCO, 2015).

Many of the most important heritage buildings have been targeted by military operations taking place in the city. This is in addition to the looting and theft of several rare artefacts.

The director of the Yemeni General Organization of Antiques and Museums (GOAM), Muhannad Al-Sayani, has reported that: "*Over 78 pieces of history had been damaged*

Fig. 8 Destruction of Qasr al-Deyafah palace building and the municipality building (AL-Namreh et al., 2010)

Fig. 9 Old historical city in Zabid and the destruction of the old city of Sana'a (Harvey, 2018)

Fig. 10 Ancient Roman amphitheatre of Sabratha, one of Libya's five UNESCO world heritage sites (https://www.unesco.org)

or destroyed; among them are ten archaeological sites, eight museums, ten mosques, two churches, 17 tombs, and six UNESCO world heritage sites. ISIS also damaged Qubbat al-Madhi, a historic mosque" (Harvey, 2018).

Libya has the most important Roman cities on the Mediterranean coast the amazing Roman ruins at Sabratha and Cyrene which are world heritage sites (https://www.unesco.org); see Fig. 10 showing the ancient Roman amphitheatre of Sabratha.

According to UNESCO World Heritage Centre: "*Libya is one of the richest countries in the Mediterranean basin in terms of archaeology and cultural heritage, with five enlisted UNESCO World Heritage Sites. The ancient Greco-Roman cities of Cyrene in the Eeast, Sabratha and Leptis Magna in the West have some of the best-preserved archaeological remains, presenting many amazing mosaics*" (https://www.unesco.org).

Libya still suffers from conflicts and wars that have started in January 2011 until now (Nature, 2011). Sufi mosques and tombs have been destroyed by Islamists. The Benghazi museum has been burgled, and notable increase in the trafficking of antiquities can be seen (Nature, 2011).

According to UNESCO: "*In June 2011 the UNESCO Director-General divulgated a call for the protection of two of the World Heritage Sites, the Old Town of Ghadamès and Leptis Magna*" (https://www.unesco.org).

To sum up, in view of the realities of the above Arab countries, it is clear that the conflicting parties there have not abided by the principles of international conventions to preserve heritage buildings during armed conflicts. The world heritage buildings in these countries have not been distinguished by the protection signs provided for in the international protocols, despite the fact that those countries have endorsed of all of respective protocols. Moreover, those countries have not complied with the rules of the Hague regulations on not to use their world heritage sites for military purposes. Many of such historical sites were battlefields, the thing that has resulted in great damage to many of them, particularly in Yemen and Iraq, in addition to Syria, as discussed below.

Unfortunately, most of the international rules have not been respected in many of the considered countries, especially in Iraq, Syria and Yemen. The result has been a great damage to and loss of antiquities. This is in addition to smuggling many of the art pieces during the conflicts. It can be noted that more than 12,000 artefacts were looted from Iraq as well as Lebanon and Syria. Major parts of the antiquities in the latter were looted and smuggled. The UNESCO estimates that out of 1,200 destructed archaeological sites analysed in Syria, at least 25% of have been looted.

Analysis demonstrates how the international heritage protection principles have not been respected in the above Arab countries during armed conflicts. This analysis stresses the necessity for Arab countries to abide by the rules of international humanitarian law, especially the Hague Convention for the protection of cultural property and its protocols as well as the Geneva Conventions and their protocols to prohibit attacks against cultural property and objects.

It can be recommended that in times of conflict, world heritage buildings in any country should be

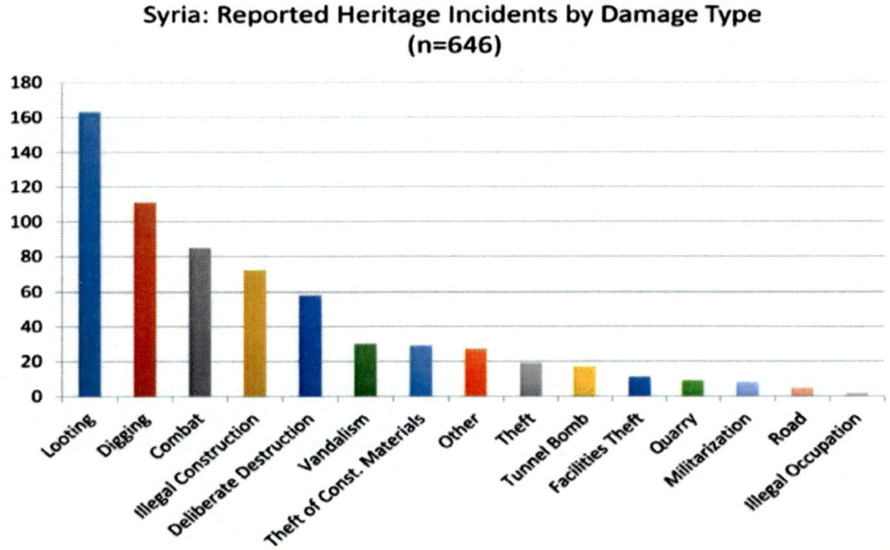

Fig. 11 Types of damage in Syria between 1 July 2014 and 30 April 2015, (ASOR CHI, 2017)

exclusivelyincluded in a list of "*enhanced protection*" adopted by the 1999 Second Protocol of the Hague Convention whileensuring that they wouldn't be attacked or used for military purposes.

4 Case Study of the Cultural Heritage in Syria During Armed Conflicts

The case study chosen for the research is Syria. The case study is aimed at monitoring the severity of damage to several archaeological sites registered as world heritage sites.

Since March 2011, the armed conflict in Syria has constantly escalated, leading to significant violence and degradation of humanitarian conditions. As a result, the destruction of Syria's exceptional archaeological, urban and architectural heritage has reached a deplorably high level and continued to seriously affect all of the six inscribed properties, the twelve sites inscribed in the tentative list and a wide number of highly significant cultural heritage sites all over Syria.

Considering the threats the built cultural heritage in Syria are facing, it can be concluded that the buildings and heritage sites in Syria have been exposed to severe threats during the war since 2011 till now due to the presence of many conflicting local and external parties. None of the conflicting parties have had any kind of awareness of the importance of the places which have been turned into mere battlefields. There are many sites with buildings and sites that have been registered as world or local heritage sites. And they have been exposed to deep and direct damage, among many other types of threats, which may be summed up as follows:

- Looting of cultural property, whether by local people (who used their sites and building in many wrong ways) or by conflicting parties (who used such sites and building as battlefields) and
- The destruction of monuments or cultural buildings and sites for ideological or propaganda purposes.

The result has been the ongoing loss of intangible and tangible heritage, traditions, customs, skills, etc. (Vecco, 2018). Figure 11 shows the types of damage many heritage sites in Syria have been exposed to.

Figure 11 shows damage types in Syria between 1 July 2014 and 30 April 2015, in terms of the number of incidents.

Damages due to combatting in addition to deliberate destruction of heritage sites by extremists are the most frequent acts with severe impact. Looting has also been of high impact on the heritage sites.

5 The Current Status of the Cultural Heritage in Syria

The current state of cultural heritage, especially the built cultural heritage, in Syria raises serious concerns. Many historical and archaeological sites have been affected by armed conflict, including those sites which had been listed on the world heritage list.

According to UNESCO: "*Damage to Syrian cultural heritage, whether transmitted or fixed, has been widely reported in the news, media and social media since the beginning of the armed conflicts in Syria in 2011. Several reports documented by UNESCO have shown the extent of damage to Syrian heritage sites*" (https://www.unesco.org).

As for the organizations involved in the protection of cultural heritage in Syria, there have been many attempts by the Syrian government along with NGOs in Syria to protect a number of sites. Great efforts have been exerted by the General Directorate of Antiquities and Museums (DGAM). The latter have transferred a large number of artefacts from conflict areas to safer museums and warehouses. They have also repaired many of the damage suffered by many artefacts and historical buildings (DGAM).

In the light of the challenges before preserving buildings and heritage sites in Syria, coordination at local and international levels is required along with the full support and supervision of UNESCO which is directly responsible for preserving such heritage.

5.1 Analysis and Discussion of the Case Study

Due to the difficulty in recognizing all the sites and buildings affected by the crisis in Syria, the search has been limited to sites located within the boundaries of the world heritage sites of the old city of Aleppo and Palmyra. Those two cities have been chosen for a number of reasons. The old city of Aleppo is one of the oldest continuously inhabited cities in the world. It may have been inhabited since the sixth millennium B.C. In addition to that, the damage suffered by the city of Aleppo has been severe, affecting about 70% of the old city. As for Palmyra, it is an example of an archaeological site with great monumental ruins. It has suffered severe damage up to 25%.

In the course of introducing and analysing the case study, two methods are used to explore the current situation of cultural heritage sites under consideration. First, reports by govermental and INGOs such as UNESCO along with those by the local authority responsible for heritage preservation in Syria, have been reviewd. Second, comparisons between old and new photos and maps of some buildings—of high heritage value—were torn down by the conflicting parties. The analysis assessed all the available satellite images for each chosen site, Aleppo and palmyra. In addition, some interviews with some used-to-be residents of the old city of Aleppo and whose buildings have been severely damaged have been made.

This residents' memory is always associated with the historical monuments of the Aleppo such as the old city, the Grand Umayyad Mosque, the old souks and their old houses along with many other buildings that have been utilized for commercial, educational and governmental activities.

5.2 The Old City of Aleppo

The ancient city of Aleppo is the oldest inhabited city in the world and was inscribed on the world heritage list in 1986 (https://www.unesco.org); Aleppo has been on the frontier of the current conflict since July 2012.

Severe damages hit the old city in the spring of 2013 (UNESCO, 2017).

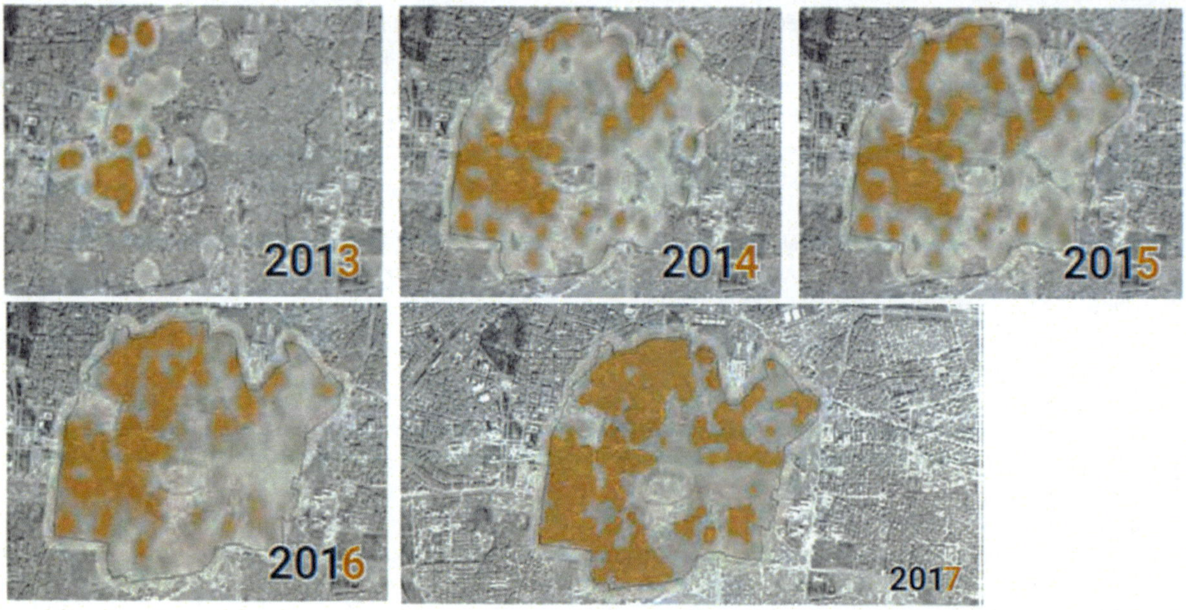

Fig. 12 Aerial photos showing the damage to historic buildings around the historic Citadel of Aleppo from 2013 to 2017 (UNESCO/UNITAR, 2018)

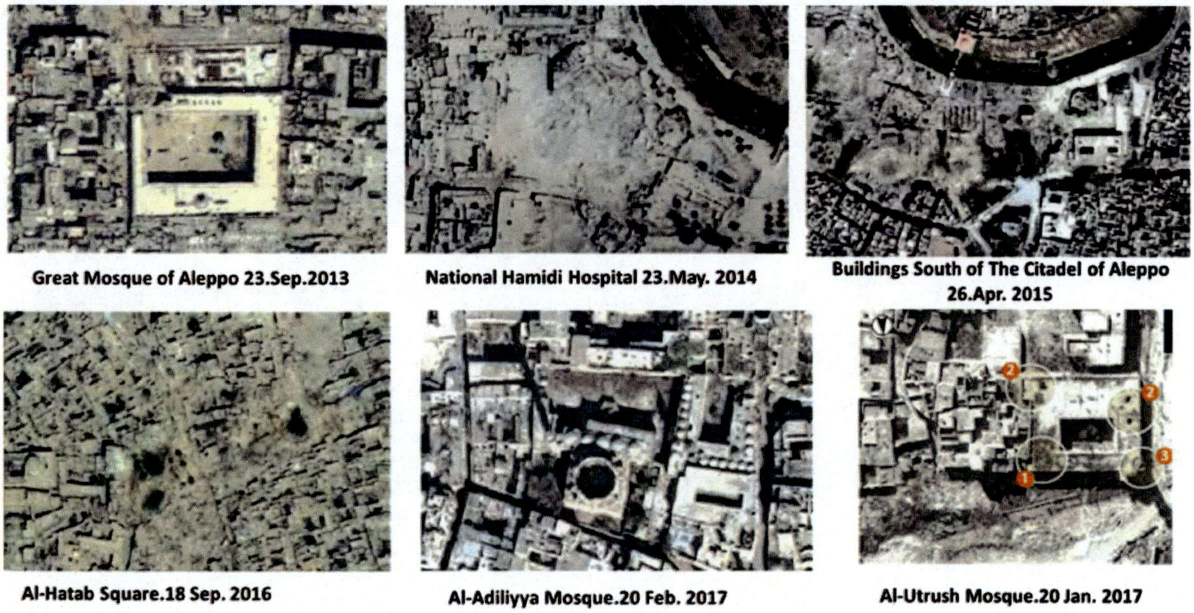

Fig. 13 Satellite photos showing the damage occurred to the world heritage site in Aleppo (UNESCO, 2018)

The minaret of the Grand Umayyad Mosque of Aleppo was destroyed. Major damages hit the area south of the Citadel of Aleppo, where many government buildings, such as the headquarters of the Ministry of Justice, the headquarters of the police, are located. This is in addition to the destruction of many historical buildings such as mosques, schools, khans among many others (UNESCO, 2017).

The UNESCO has also revealed severe damage in the area northn of the Aleppo Citadel through satellite images. The UNESCO team has reported: "*extensive damage at the Grand Umayyad Mosque, the Citadel mosques, churches, suqs, khans, madrassas, hammams, museums, and other significant historic buildings in Aleppo*".

Figures 12 and 13 show the state of historical buildings in Aleppo since 2013, till 2017.

The UNESCO has repeatedly assessed that "60–70% of the old city of Aleppo has been severely damaged, with 30% totally destroyed" (UNESCO, 2017) see Fig. 14.

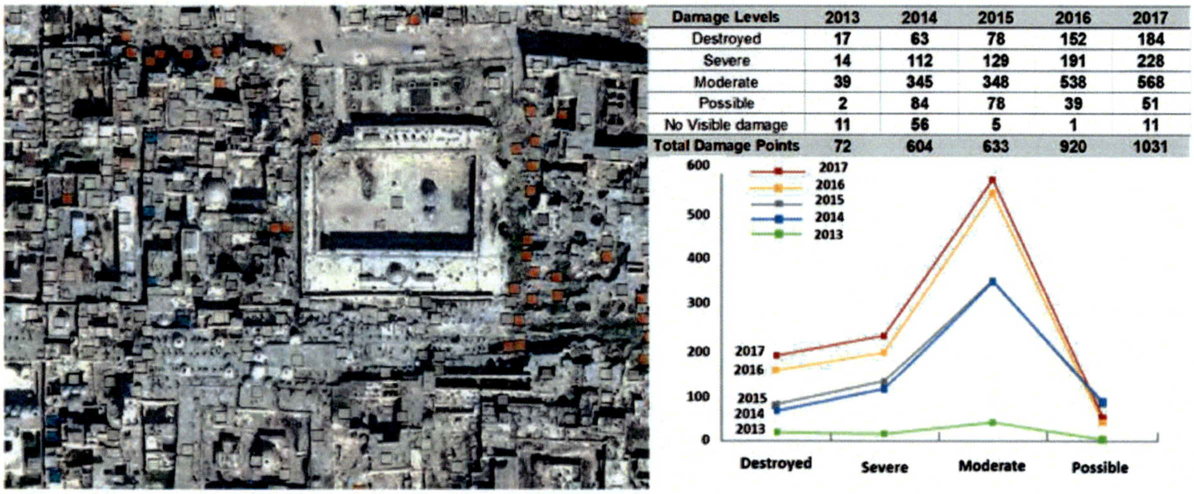

Damage Levels	2013	2014	2015	2016	2017
Destroyed	17	63	78	152	184
Severe	14	112	129	191	228
Moderate	39	345	348	538	568
Possible	2	84	78	39	51
No Visible damage	11	56	5	1	11
Total Damage Points	72	604	633	920	1031

Fig. 14 Number of damaged land plots in Aleppo between 2013 and 2017, (UNESCO, 2018)

Fig. 15 Extent of damage to the Grand Serail government building (left) and how it looks before the damage (right) (UNESCO, 2017)

The Grand Serail of Aleppo was the former seat of the governor of the Syrian city of Aleppo and located in the south of the Citadel of Aleppo. It was built between 1928and 1933 to serve as the main government building in the city. This building had been demolished between 2014 and 2015. The building is an example of local modern architecture. This was a heavy loss. Figure 15 shows the damage to the Grand Serail government building after the ware and how it was before that ware.

This is in addition to setting fires to the traditional heritage souks, leading to their destruction. DGAM has reported that "*the bombing of a tunnel near Aleppo Citadel resulted in the collapse of part of the defensive wall in the northeastern part of the castle. In addition, over 152 buildings inside Old Aleppo were destroyed and damaged during four years of the crisis.*" (DGAM) see Fig. 16.

The UNESCO has reported that the minaret of the Grand Umayyad Mosque of Aleppo was destroyed during fighting in April 2013, see Fig. 17, which shows the Great Mosque of Aleppo after damage comparing with the same mosque before damage. Earlier in September 2012, a huge fire in the covered souks damaged much of the eastern side of the mosque. This severe damage was visible by means of satellite imagery in 2014; see Fig. 18.

The destruction of the old city of Aleppo constitutes a great loss for humanity, not only in Syria, but all over the world.

Fig. 16 Khan al-Sabun: top left image (Before): 21 November 2010, Bottom Right image (After) on the destruction:16 January 2017 (UNESCO, 2018)

Fig. 17 Great Mosque of Aleppo; top right—the Minaret before destruction, December 2009, top left—the damage to the northern entrance, bottom left—damage to the great mosque and bottom right—the destroyed Minaret, January 2017 (UNESCO, 2018)

Fig. 18 Ancient city of Aleppo, total damage historical buildings, since 2013–till 2017 (UNESCO, 2017)

5.3 The Old Site of Palmyra

Palmyra is located in the middle of the Syrian semi-desert. According to UNESCO: "*Palmyra was listed on the World Heritage List in 1980*" (https://www.unesco.org). *The city* has faced a real threat of encroaching on all its archaeological buildings as well as the risk of destroying its monuments. This has been a disaster for Syrian culture and humanity in general. Palmyra was taken by armed groups on 21 May 2015 and remained under their control until 27 March 2016. During this period, invaluable losses to the property were inflicted.

The American Institute of Oriental Research of the University of Boston (ASOR) published a report dated 4-9-2015 showing the destruction of six old tower cemeteries in Palmyra. See Fig. 19.

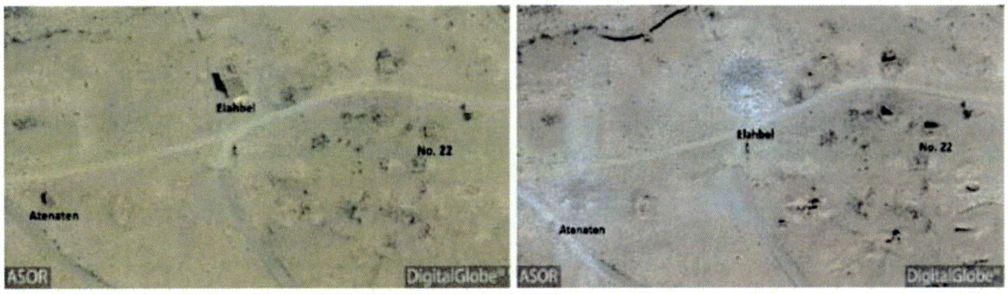

Fig. 19 Tomb of Elahbel right before blasting-left after (ASOR, 2017)

Fig. 20 Temple of Bell before and after the bombing (AAAS, 2014)

Fig. 21 Bombing of Baalshamin Temple (AAAS, 2014)

According to the report, the United Nations Research and Training Agency on 1 September 2015 (UNITAR) revealed aerial photographs showing the destruction of a large part of the Temple of Bell, see Fig. 20, which has long been regarded as a unique architectural icon.

The depicted damage also included the looting of sculptures from the unexcavated tombs at the Baalshamin Temple, which has been destroyed during the armed conflict in Palmyra. See Fig. 21.

"For conflicts bodies, tangible heritage -monuments, stones and tombs-area have been the main targets attacked by them. They have been in control over large areas there and thought they have to eradicate all the monuments considered *awful* from their point of view. *The biggest crime committed by them in Palmyra was when they blew up the temple of Baalshamin as part of their plans there*," said Director-General of Antiquities and Museums in Syria (UNESCO, 2016). Figure 22 shows the destroyed Triumphal Arch and the destroyed Cella of the Temple of Bel Baal.

UNESCO experts take preliminary stock of destruction in the world heritage site of Palmyra, through a mission on 24–26 April 2016. The mission was headed by Director of UNESCO's World Heritage Centre. It inspected both Palmyra's museum and archaeological sites, taking stock of considerable damage to the museum, where they found that most of the statues and sarcophagi that were too large to be

Fig. 22 Destroyed Triumphal Arch (left) and the destroyed Cella of the Tele of Bel (Baal), right (UNESCO, 2016)

Fig. 23 Damage inside the Museum of Palmyra (UNESCO, 2016)

Fig. 24 Aerial photo of the Baal Temple: April 2016, DGAM/ICONEM (UNESCO, 2016)

removed for safekeeping defaced, smashed, beheaded with their fragments left on the ground (https://www.unesco.org) (Fig. 23), showing the damage inside the Museum of Palmyra.

What happened in Palmyra as a result of the conflict is an awful crime against all humanity. Some of Palmyra's most important attributes have been destroyed. Moreover, the destruction of Palmyra as an iconic classical site is a severe blow to the Syrian people and the international community as well. Figure 24 shows the whole damage of Baal Temple according to UNESCO reports 2016.

There has been a lot of destruction in many places due to the conflict in Syria. The conflict has reached the Assyrian or classical heritage. Buildings and sites with religious significance have also suffered in northern Syria, including historic Islamic shrines (https://www.unesco.org).

6 General Conclusions and Recommendations

The threat wars posed on cultural heritage in any region is one of the most serious threats to the architectural heritage of all humanity. The international community has defined many instruments to address this issue. Yet, the aftermath of the recent conflicts in many Mideast countries has endangered their monuments and archaeological sites and affected their tangible and intangible heritage. The damage inflicted by conflicts upon all kinds of heritage is the worst that can affect human heritage. Doubtlessly, future generations in the conflict-hit countries would suffer a big loss due to the damage of their nations' tangible and intangible heritage.

Based on what has been considered in the research above, the following outcomes may be concluded:

- The damage and destruction have hit archaeological sites of high and exceptional value in conflict-stricken Arab countries.
- One of the most serious threats to human rights is the destruction of built cultural heritage as a result of wars and their destructive tools.
- The destruction of cultural property does not reflect a military necessity. It is rather associated with politics, as to destroy others' morale and humanitarian and spiritual symbols. Laws and legislations on the protection of cultural property and the built cultural heritage have been endorsed in many countries. Nevertheless, military necessities are still used as pretexts to justify violating such laws and regulations.
- Concerned organizations, whether local or international, and respective conventions cannot adequately protect the built cultural heritage in any country without the laws of

that country being rectified to meet the requirements of such protection; or without awareness of the value and importance of the built cultural heritage being entrenched in mindset of the people of the country.

- International conventions for the protection of the built cultural heritage do not include methods means or follow-up committees to assure their implementation in full and conformity with them at the built cultural heritage site or world heritage sites at least. Thus, major damage to world heritage sites cannot be avoided.
- There is a pressing need to raise the awareness of people on the importance of respecting the culture, heritage and history, and therefore, built cultural heritage.
- It has been noted that conflict-stricken countries lack the national means to protect their built cultural heritage, especially in cases of large-scale wars or the difficulty of accessing heritage sites at such conditions.
- The status quo of heritage sites and buildings indicates that conflict-stricken Arab countries lack and have—to a large extent—ignored the legal mechanisms stipulated in international covenants and protocols on the protection of such sites and building in armed conflicts.

Based on the above findings, the following recommendations are to be put forward:

- It is a must during peace time to label cultural sites with distinctive signs pursuant to Article 6 of the Hague Convention in order to ensure that they are neither attacked nor used militarily.
- There is a need to document tangible heritage, monitor any damage to any part of it, by means of tools of digital registration and recording such as satellite imagery, and pool them to facilitate assessment and remedial action whenever possible.
- It is a must to extend support, advice and training to local antiquities professionals and others in front lines and act to prevent illicit trafficking.
- All conflicting parties must take into consideration the high significance of the heritage buildings. They should respect the heritage of the region in which the conflict is taking place, regardless of any military necessity.
- Military attacks against heritage buildings should be avoided everywhere in armed conflicts.
- It is a must to keep military sites away from cultural sites because the presence of any military activity in the vicinity of cultural sites has been used as a justification to target them.
- The amount of preservation and protection that can be accorded to the built cultural heritage should be defined in the light of its importance and regardless of whether the country has ratified the conventions on the protection of cultural heritage in situations of armed conflicts or not. Built cultural heritage everywhere—according to all conventions on the protection of cultural heritage—is not only the property of the state in which it exists, but also the property of humanity.
- The contents of international conventions on the protection of cultural heritage buildings should be disseminated among citizens in order to raise their awareness of the need to protect and preserve cultural buildings. This can be done through visual, readable and audio-visual means. Educational institutions can play a fundamental and significant role in this regard.
- Those who work on plans for post-conflict reconstruction should be encouraged to include protection and rehabilitation plans and strategies for built cultural heritage in their plans on priority basis.
- It is highly important to make Arab countries that have not yet ratified the 1954 The Hague Convention and its Protocol I or II Protocol of 1999, ratify them.
- The implementation of the Hague Convention and its protocols in the Arab States should be monitored; and violations of the provisions contained therein should be prevented through some respective measures.

References

Abdel-Ghani Abdel-Hamid, M. (2000). *Protection of Victims of Armed Conflict in International Humanitarian Law and Islamic Law* First Edition, International Red Cross, Cairo, 2000, p. 57.

Al-Mafraji, A. M. S. (2011). *International protection of cultural property during armed conflict: Study in the light of the provisions of international agreements* (p. 50). Dar of legal books, Egypt.

Al-Nashef, Kh. (2004). *Destruction of Iraqi cultural heritage—Disaster classes*. Dar Al-Hamra for Printing, Publishing and Documentation.

Al-Rahaifa Salama, S. (2012). *Protection of cultural property, during armed conflict*. Dar Al-Hamed for Publishing. Amman.

American Association or the Advancement of Science AAAS. (2014). *Assessing the status of Syria's tentative world heritage sites using high resolution satellite imagery, part one*.pdf. https://www.aaas.org/sites/default/files/s3fs-public/content_files/AAAS-SyrianTWHS-122014.pdf.

ASOR : The American Schools of Oriental Research Cultural Heritage Initiatives. (2017). *The Recapture of Palmyra*. Allison Cuneo, Susan Penacho,Michael Danti, Marina Gabriel, Jamie O'Connell , With contributions from Bishera lIssa, Bijan Rouhani, and Kyra Kaercher. https://en.unesco.org/syrian-observatory/news/recapture-palmyra .

Bakir Abdul Rahman, Z. (2013). Identity in Yemeni architecture between modernity and heritage preservation in the age of urbanization and development. *Al-Andalus University Journal*, *5*(9).

Danti, M., Branting, S., Penacho, S. (2017). *Monitoring cultural heritage in Syria and Northern Iraq by geospatial imagery*. American Schools of Oriental Research Cultural, University of Central Florida, Heritage Initiatives. https://www.mdpi.com/2076-3263/7/4/95/htm. Accessed 21 Oct 2019.

Geneva Convention Relative to the protection of civilian persons in time of ware. (1949). https://www.un.org/en/genocideprevention/documents/atrocity-crimes/Doc.33_GC-IV-EN.pdf. Accessed 22 Oct 2019.

Harvey Fenton, J. (2018). *Yemen war erodes the Country's architectural heritage*. https://theglobepost.com/2018/03/06/yemen-war-heritag. Accessed 22 Oct 2010.

ICRC. International Committee of the Red Cross. (2010). *Geneva additional protocol I of 1977*, Switzerland. International Committee of the Red Cross. https://www.icrc.org/en/doc/assets/files/other/icrc_002_0321.pdf. Accessed 21 Oct 2019.

Nature. (2011). Libya's extraordinary archaeology under threat. *International Weekly Journal of Science*. https://www.nature.com/news/2011/110302/full/news.2011.132.html.

Nicholas, E., Brown, C., Lynn, M. (2019). The politics of peril: UNESCO's list of world, heritage in danger. *Journal of Field Archaeology*. https://doi.org/10.1080/00934690.2019.1600929. Accessed 19 July 2020.

Nnamara Jawad, N., Alqeeq, F. (2010). *The reality of the urban heritage of Gaza City in light of wars and natural disasters*. The second international conference for architectural conservation. http://www.cpas-egypt.com/pdf/NaderNamara/Researches/07.pdf.

Osgoode, H. (2017). The Brussels peace conference of 1874 and the modern laws of belligerent qualification. *Law Journal, 54*(3). http://digitalcommons.osgoode.yorku.ca/cgi/viewcontent.cgi?article=3159&context=ohlj. Accessed 21 Oct 2019.

Thekra & Ebrah. (2019). A documentary photo exhibition about the civil war in Lebanon that took place 13-4-2019. https://www.facebook.com/syndicate2/media. Accessed 21 Oct 2019.

The Getty Conservation Institute. (1935). *Protection of artistic and scientific institutions and historic monuments*, treaty series 899, signed at Washington April 15 ,1935. https://www.loc.gov/law/help/us-treaties/bevans/m-ust000003-0254.pdf. Accessed 22 Oct 2019.

The Guardian, independent journalism. (2019). *Manhattan of the desert: Civil war puts Yemen's ancient skyscrapers at risk*, Published article, independent journalism. https://www.theguardian.com/world/series/yemen-at-war. Accessed 22 Oct 2019.

The International Committee of the Red Cross. (2010). Protection of the cultural heritage of peoples in order to protect the human identity. *Periodical Journal*, (47), 9.

UNESCO. (2016). *Assesses damages to Syria's Palmyra world heritage site*. UN News. https://news.un.org/en/story/2016/04/527842-unesco-team-assesses-damages-syrias-palmyra-world-heritage-site. Accessed 21 Oct 2019.

UNESCO.ORG, United Nations Educational Scientific and Culture Organization (2016). UNESCO reports: *The UNESCO repid assessment mission to the World Heritage Site of Palmyra*. https://en.unesco.org/syrian-observatory/reports/palmyra. Accessed 7Oct 2020.

UNESCO.ORG, United Nations Educational Scientific and Culture Organization UNESCO. (2018). *Five years of conflicts, the state of cultural heritage in the ancient city of Aleppo*, UNESCO UNITAR. https://unesdoc.unesco.org/ark:/48223/pf0000265826/PDF/265826eng.pdf.multi. Accessed 7 May 2020.

UNESCO.ORG, United Nations Educational scientific and culture organization. (1954). *Convention for the protection of cultural property in the event of armed conflict with regulations for the execution of the convention*, 1954. http://portal.unesco.org/en/ev.php-URL_ID=13637&URL_DO=DO_TOPIC&URL_SECTION=201.html. Accessed 21 Oct 2019.

UNESCO.ORG, United Nations Educational scientific and culture organization. (1970). *Convention on the means of prohibiting and preventing the illicit import, export and transfer of ownership of cultural property*, Paris. http://portal.unesco.org/en/ev.php-URL_ID=13039&URL_DO=DO_TOPIC&URL_SECTION=201.html. Accessed 22 Oct 2019.

UNESCO.ORG, United Nations Educational Scientific and Culture Organization. (2018). *UNESCO and UNITAR-UNOSAT publication details damage to cultural heritage in Syria's Ancient City of Aleppo*, UNESCO reports. https://en.unesco.org/news/unesco-and-unitar-unosat-publication-details-damage-cultural-heritage-syrias-ancient-city. Accessed 21 Oct 2019.

UNESCO.ORG. United Nations Educational scientific and culture organization. (1972). *Convention concerning the protection of the world cultural and natural heritage*, in Paris. https://whc.unesco.org/en/conventiontext/. Accessed 21 Oct 2019.

UNESCO.ORG. (1985). *Conventions and Recommendations adopted by UNESCO on the protection of cultural heritage* (2nd edn.), Paris. https://whc.unesco.org/archive/convention-en.pdf. Accessed 21 Oct 2019.

UNESCO.ORG. United Nations Educational scientific and culture organization. (1999). *Second protocol (1999) to the 1954 Hague convention for the protection of cultural property in the event of armed conflict*. http://www.unesco.org/new/en/culture/themes/armed-conflict-and-heritage/convention-and-protocols/1999-second-protocol/. Accessed 21 Oct 2019.

UNESCO.ORG. United Nations Educational Scientific and Culture Organization. (2015). *Plan to safeguard Yemen's cultural heritage sites*. https://news.un.org/en/story/2015/07/504822-unesco-launches-plan-safeguard-yemens-cultural-heritage-sites. Accessed 22 Oct 2010.

UNESCO.ORG. United Nations Educational Scientific and Culture Organization. (2017). *Extensive damage in first emergency assessment mission to Aleppo*. UNESCO reports. https://whc.unesco.org/en/news/1619. Accessed 21 Oct 2019.

UNESCO.ORG. United Nations Educational Scientific and Culture Organization (2016). *World heritage sites in Libya*. https://whc.unesco.org/en/statesparties/ly. Accessed 21 Oct 2019.

Unicef reports. (2019). *The humanitarian crises in the Middle East and North Africa*. https://www.unicef.org/appeals/mena.html. Accessed 19 July 2020.

Vecco, M. (2018). Cultural heritage in times of armed conflicts in the Middle East: Much more than material damage. *Journal of Cultural Heritage, 33*.

Rehabilitation and Exploitation of Heritage Buildings. An Investment Approach

Analytical Comparative Case Study

Mohamed M. Shawky Abou Leila and Magdy M. ElBastawisy

Abstract

No question, heritage buildings are one of the cardinal stakes of sustainable economic and touristic resources in urban communities. Moreover, they revive the urban traces of their home city and by turn foreground the identity of their community. Mecca has been famed for its historic and monumental edifices. Unfortunately, these buildings are diminishing one after another, due to the lack of maintenance follow up processes, ending up losing their cultural value, social esteem, and economic worth. With the massive flow of pilgrims, the kingdom has opted to part with most of these heritage areas, to provide its increasing visitors with the utmost modern services. The promising 2030 plan and the exerted efforts to get it realized are exhibited in the cultural heritage renovation wave taking over the kingdom in general and Mecca in particular. The research paper aims at documenting the 2020 strategic initiative, where its second legislative code stipulates raising the awareness of these areas, in consideration of their preservation, rehabilitation, and development, to encourage the private sector to launch major investment projects right there. The research investigates the validity of the former international heritage renovation attempts to be either simulated or developed. The researcher selected a certain hypothesis, to serve the research goals, which is affirmative to the previous interrogation. Thusly, the research employs an inductive, descriptive, and analytical methodology. Then, the research exhorts the prevalent conception of heritage rehabilitation in international, pan Arab, and Saudi terms, respectively. The research is to proceed with elucidating the architectural and urban heritage typologies. Afterward, it explicates the aforementioned conception, as coded in unanimous international charters in terms of their enforced preservation policies. Furthermore, the research postulates the grounds of urban heritage renovation and architectural rehabilitation in proportion to their investment conception, mentioning the most poignant programs and their resultant economic fruition. The research provides an analytical overview and a comparative case study of rehabilitation and exploitation international model; Valle Longa in Naples; Italy in part and two Arab model; Khan Al Faring in Sidon; Lebanon, and Suleiman Palace in Al Tayssir, Mecca; saudia Arbia, to reach the research findings and recommendations.

Keywords

Investment • Rehabilitation • Exploitation • Heritage buildings • Valle Longa in Naples; Italy • Khan Al Faring in Sidon; Lebanon Suleiman Palace • Mecca

M. M. S. A. Leila (✉)
Department of Islamic Architecture, Department of Architecture, Faculty of Engineering, Um Al-Qura University, Mansoura University, Mecca, Mansoura, KSA, Egypt

M. M. ElBastawisy
Department of Islamic Architecture, Department of Architecture & Urban Planning, Um Al-Qura University, Port-Said University, Mecca, Port-Said, KSA, Egypt

1 Introduction

1.1 Heritage Definition

1.1.1 Heritage as an Arabic Lexicon

In the Arabic lexicography, the lexeme heritage "Turath" is a derivative of the morph "Irth" or "Mirath," the linguistic equivalence to "inheritance" in its English counterpart. This ultimately deciphers its Arabic definition, which is "the left legacy for younger generations" (Abou Leila & Albarqawy, 2019). In this context, heritage is the civilization and cultural norms of our ancestors. It shapes the identity of each community, no matter their span or ethnography, but acceptance, strength, and eternity within their communities. Heritage is those piled up priceless documents and edifices. It may be

A. Versaci et al. (eds.), *Conservation of Architectural Heritage*, Advances in Science, Technology & Innovation,
https://doi.org/10.1007/978-3-030-74482-3_16

also a sum of strategic techniques that could be developed to suit our modern age or urban associations that could be modified. Heritage may be our costumes and traditions that we long for and cling to even if they are behind the times.

The last period of the twentieth century has labeled any historic legacy as "Heritage," whether they are natural topography or manmade cultural premises that survived several historic epochs.

1.1.2 International Conception of Heritage (International Charter, 1999)

The International Charter of Cultural Tourism (Mexico, 1999) defined heritage as a broad general concept encompassing the natural and cultural topography in the account of the panoramic outlook of land, sites, and built environments allocations. It is also inclusive of the biodiversity of the former, yet thriving cultural practices that dictate the cognitive perception of its community. It extends to include the entire heritage traces that have challenged the span of time and its newfangled advancements shaping the core of the indigenous local, national, and regional identity as an integral component of modern life.

1.1.3 Pan Arab Conception of Heritage (Arab League, 2003)

In 2003, the Arab League defined urban heritage in its Arab Heritage Preservation Charter, as the collective sum of any archeological site or monumental edifice, which perpetuate their urban, architectural, economic, scientific, cultural, or functional values. This perspective abides by the international umbrella counterpart with respect to the special and temporal aspects of each country.

1.1.4 The Saudi Conception of Heritage (Abou Leila, 2012)

Heritage or the concept of urban monuments in Saudi Arabia addresses those buildings of religious, cultural, historical, or architectural worth. In other words, it pays tribute to mosques, ancient towers, palaces, and fortresses. This is a typical application to the aforementioned heritage-related conception to traditional villages and ancient neighborhoods. Urban heritage is the personification of cultural heritage as loaded items bearing the commemoration anecdote of a bygone nation and their survived events with respect to their social, cultural, economic, and environmental givens. It reflects the depth of the interaction of man with the surrounding environment, as denoted that urban heritage is not restricted on monuments, archeological sites, and cities, but also includes all the other elements set it up. These prolific civilizations have sparkled in the fields of science, literature, arts, and crafts. It represents the old neighborhoods within the genuine cities and villages, who keep up their tools and titles. At the level of architectural heritage, there was a confusion between heritage and monument, and it was defined as the building (monument) within or outside the boundaries of the urban form and reflects a special religious, cultural, and historical significance.

1.2 Heritage Typologies (Abou Leila, 2010)

It is acknowledged to be of several categories; urban architectural, monumental or cultural and natural heritage, and we deal with these typologies as follows.

1.2.1 Urban Heritage

This title is used to mark those interrelated space-blocks components, which have survived the successive and non-successive turnovers throughout a number of historical epochs, standing stern before the concurrent modernization maneuvers. Each entity that has earned hailing respect in a certain community for its authenticity as a visual reference portraying a named civilization is heritage. Heritage is anything that embodies the enlightenment and civilization of the former generations that are entitled to be a national wealth not only for a certain generation, but also for the future generations to hold on to as a public good.

1.2.2 Architectural Heritage

It is defined by the European Charter of Architectural Heritage (European charter, 1975) of 1975 that heritage is not confined to the iconic monuments solely. It also addresses the subordinate manmade constructions in ancient cities and villages in their natural environment. Architectural heritage is an integral factor of the cultural heritage of the whole world. It cannot be an exclusive reference to the landmark buildings and their context, but also it compromises all the areas, cities, and villages of historical worth. Thus, the Europeans authorities marked their architectural heritage to be proprietary. In general, all the European citizens owe their heritage sincere intimacy and intuitive custody against any threat. They have faith that their contemporary architecture is their future heritage that must be passed to their youngsters in the best of forms and quality.

1.2.3 Archeological Heritage

In 1990, the ICOMOS Charter (ICAHM, 1990) for the protection and management of the archeological Heritage defines it as whatever left traces from any humane existence, and any previously inhabited area that has once hosted humane behavior and activities. This involves the ousted or deserted buildings and the collective sum of underground and undersea residues. These are believed to be nonrenewable cultural resources.

1.2.4 Cultural and Natural Heritage

The cultural and natural heritage is a broad term including physical and abstract resources that narrate the historical development of a certain community in their society regarding their spatial and temporal context. This heritage has the upper hand in the modern era and the intellectual mindset of this society. Therefore, they must be disclosed to the world public physically, intellectually, and emotionally.

1.3 Heritage Preservation Processes

1.3.1 Heritage Preservation as in the International Charters

Abiding by the World Heritage Convention agreement (Nara Conference, 1994), preservation is the collective sum of the stipulated procedures in consideration of cultural heritage acquisition in terms of its history, semantics, and accentuating its components. These entail the commitment to periodical maintenance and amelioration processes if required. As for Budapest Declaration on World Heritage recommendations (Budapest Declaration, 1972), it is defined as reusing the historic buildings and ancient sites adding some probable usages, provided that neither the exterior façades; structure and traits; nor the interior spaces of these constructions get infringed.

The Appleton Charter of 1983 (Appleton Charter, 1983) established levels of intervention to preserve the architectural heritage, including the preservation of the form and materials that exist, the unit of the place, the restoration, the rehabilitation, the reconstruction and the redevelopment, and the activities necessary to maintain and stabilize the maintenance and renewal. The 1987 Brazil Panel defined the preservation of historical sites as "preserving and improving historical sites to express the past and strengthen the sense of citizenship." The New Zealand Document 1992 "ICOMOS New Zealand" (ICOMOS Newzeland, 1987) defined the preservation process as an intervention step-by-step process starting from non-intervention, maintenance, restore, reuse, and reconstruction.

In brief, heritage preservation is taking to countering all the ongoing atrocities, deformation, and deterioration practices toward heritage buildings. This mission must be realized in awareness of the needed economic, managerial, technical, and societal measures. In other words, heritage preservation compromises a couple of directly proportional processes. First, it aims at validating heritage buildings ensuring that none of the physical nor abstract propositions is distorted. Second, this process demands a master command to control the accordant changes adjoined to these progressive development processes of each respective building.

1.3.2 Grounds and Goals of Heritage Preservation

Calls for heritage preservation are grounded on:

1. **Social and historical grounds**: The building recalls the political, cultural, and social attributes of its historical epoch. Ancient buildings are not mere antiquities but eyewitnesses to incidental historical period with its norms, values, and worth.
2. **Economic grounds**: Urban heritage is an important tourism promotional function that augments the increase of economic revenues through the reuse and preservation of heritage buildings, namely cultural shrines or museums, and the artistic, technical and philosophical values of inscriptions and used materials for they cannot be passed unnoticed.
3. **Political grounds**: Heritage preservation recaps past events for present lessons. It recalls the former political dominance and the authoritative iron fist from time to time.
4. **Religious and ideological grounds**: This is clearly evident in places of worship, housing, and markets for urban components and others.
5. **Local grounds**: The building can acquire several important things, especially because it is associated with a political or artistic figure of social precedence, such as the house of Taha Hussein, dean of Arabic literature, which was converted into a museum in the Arab Republic of Egypt after its preservation.
6. **Subjective grounds**: The buildings have subjective importance in their architectural or structural attire, which necessitates their preservation.

On the previous grounds, heritage preservation is far from being a simplistic task, but of imperative adequacy and worth, by virtue of the history, norms, and content it bears. Undoubtedly, heritage preservation processes are employed to unravel the mystery of the entitled building characterization and its ancient civilizational compatibility with its surrounding special context. In addition, it is designated to raise the citizens' awareness of any futurist amendments regarding heritage buildings. Besides, heritage preservation processes are to convey humane esthetic and cultural values of their represented civilization to future generations. Finally, yet importantly, this process is expected to foster a sense of belonging and teamwork among its community members, as it all serves their public good.

1.4 Major Adapted Policies for Heritage Preservation (Abou Leila & Albarqawy, 2019)

- **Revival**: This implies bring the heritage building back to life, in terms of its target and functionality. This building may undergo some modifications, to serve the same function, but in pace with the modern age advancements.
- **Reuse and Adaptation**: This implies either restoring its former functional role or a feasibly alternative one, to cope with the modern age requirements. This policy varies its title by the enumeration of the undertaken adjustments and applied manipulations to the building. These adjustments may go further-reaching its architectural structure, to serve its apprenticed function. This may occur in using some agencies as small hotels, and this may be by adding some services in a harmless manner, or updating old heritage buildings without damaging their heritage architectural character and maybe by exploiting the building in an economic way if the new use of the building is appropriate. It can be rebuilt referring to the available drawings, photographs, and documents manifesting the disintegrated part.
- **Rehabilitation**: This implies taking certain procedures regarding the building, for its aptitude to carry out its functions. These procedures are to include slight changes, amendments, and additions, which would not affect the building components and traits that enclose its cultural worth or architectural value.
- **Renovation and Renewal**: Renovations of heritage buildings aim at creating a new form that corresponds to the pattern of the time of the process. Besides, these processes are to reach out to archeological buildings. The renovation process means equipping the old building to be used for contemporary functional use, and these renovations may provide the heritage building with modern technical facilities; air conditioning systems, electricity, telephone, fire extinguisher, alarm, Internet, as well as the renovation of water and drainage systems, healthcare services, and electricity networks or self-extinguishing technology. Undoubtedly, this should not affect the basic components of the building and its heritage characteristics, as well as the distribution of sites and gardens that may also include changing windows, doors, and painting walls for heritage buildings (Ibrahim, 1982).

1.5 Investment of Heritage Buildings (Abou Leila, 2015)

Amidst the chain of heritage preservation strategies, the idea of heritage building investment has flourished as a driving motif, thanks to being a prime promotion tool to cultural tourism and economic stake for its community. No wonder, tourism is the main pillar of the national economic prosperity in general and in those countries of heritage buildings in particular, especially in touristy centers. Most importantly, these buildings enable organizing cultural and heritage-oriented events and encourage their residents to shoulder various activities.

1.5.1 The Conception of Urban and Architectural Heritage Investment (Abou Leila, 2018)

Heritage buildings are the mainstream of national economic growth, where tourist development is the vein of contemporary sustainable development ideology. In spite of the supreme wealth of heritage buildings in the Arab countries, this treasure is destined to evanescence or negligence, due to either the impoverished societal awareness or bureaucratic reluctance or executive disinclination (Royal institution, 1993). Thus, it was urgent to preserve and sustain heritage buildings in general. It is necessary to formulate a preservation conception seeking a futurist aspiration of these buildings. Bilateral cooperation between the administrative and executive authorities to sort out this dilemma, to grant material, moral, and practical returns of heritage buildings. It is time to reside the internal prejudice and opt for clear administrative mechanisms and policies abiding by legal codes that take into account the different aspects of its value. Nowadays, ecosystem preservation came to surface for the novice idea of environmental investment or Green Investment in heritage buildings.

1.5.2 Economic Advantage of Urban and Architectural Heritage Investment

Confessedly, there are cardinal economic advantages out of urban and architectural heritage investment, as they are considered to be:

a. **Sustainable resource for national income**: The versatility and plurality of heritage buildings pave the way for sustainable and distinctive investment projects for uncurbed economic growth, where investors could spot, whatever heritage building to perform a certain function (Saudi Commission for Tourism & Antiquities, 2010).
b. **Job vacancies platforms**: Launching investment projects within a heritage site would create many job opportunities for its residents. Thus, this labor force will

stay in their residencies rather than treating it as dormitory towns, achieving stability.

c. **Private investment zones**: It is a globally accredited strategy, where the private sector is welcomed to take a role in heritage buildings investments, whether they are fully, partially, or not inhabited.
d. **Citizenly and societal participation mean**: These investments involve citizenly and societal engagement, whether they could take part in the investment flow or just take decisions concerning it.
e. **A revival of traditional professions and crafts motives**: These investments would retain the ancient ambiance of their respective sites. Hence, workshops and small factories for handicrafts such as inventory, carpets, Klim, baskets, folk heritage costumes, metal antiques, pottery, glass, ornaments, and decorative tools, as well as leather and woodwork. Ivory, burdy, and others will revive their deceased grandeur (Yusuf, 2014).

According to many international models, it is impossible to preserve the urban heritage outside the context of its native environment. Urban planning must cater to the protection of buildings, maintenance of archeological sites, monuments, and historical centers as development plan priorities (Budhqah, 2010).

1.5.3 Programs of Urban and Architectural Heritage Investment

There are several programs and adaptations as such:

1. Establish direct investment in those countries and villages of archeological areas: This could be obtained by developing pioneering models that have achieved great success in the realm of development across those villages in terms of their economic investment.
2. Establish investment partnerships to develop heritage villages.
3. Establish heritage development funds in villages funded by government institutions and sectors and private donations to contribute to the development, support, and establishment of services to help to invest in heritage villages.
4. Establish state-owned investment companies to manage some genres of heritage or archeological buildings such as castles, palaces, and houses, to invest in the field of accommodation and tourism according to binding codes, yet funded from various endowments…… etc.
5. Establish the heritage buildings reuse strategies.

   a. **Residential Units**: This is realized by either leaving it as permanent accommodations to its owners or leasing it for the private sector hotel investments, abiding by the latter's criteria. This requires a simultaneous periodical follow up maintenance programs to eternalize its perfect stature.
   b. **National Museums**: Heritage buildings are the most symmetric places for monumental displays when their inner spaces prove functional congruity.
   c. **Folk Food Restaurants**: These places could be functionally set up to serve folk dishes.
   d. **Painting and Plastic Arts Galleries**: These artistic works may be on the walls and along the corridors of these touristy buildings.
   e. **Folk Markets**: These products are of hailed value especially when they are allocated in their birthplace; for instance, Tsarist centers of some ancient cities, to which shoppers from many places come.
   f. **Heritage Artistry and Handcraft Products Shops**.
6. Exploiting Central Spaces and Heritage Urban Spaces.

   a. Launching investment projects and hosting popular events of organized mass attraction.
   b. Embracing recreational, cultural, and physical activities provided by their logistic services.
   c. Establishing commercial services for the visitors right there in accordance with the experts' and specialists' drafted technical criteria (Saudi Commission for Tourism & Antiquities, 2010).
   d. Maintaining the ancient shops encompassed in the heritage site to secure a constant financial profit, which is essential for the periodical maintenance of the heritage buildings within. Moreover, agoras may be built by the central axes' roads of the site (Ibrahim, 1982).

2 Analytical Comparative Case Study

2.1 Criteria for Building Selection in the Analytical Comparative Case Study

A selection standard has been drafted to compare international and Arab models for rehabilitation and exploitation. The case study buildings are residential multi-floored building of bearing walls system, with a large number of rooms and openings of the facades. These buildings are to be tackled in the light of the rehabilitation philosophy of the palace, creating new space functions represents a convergence between these two studies and the study of the stance in the city of Mecca.

2.2 International Rehabilitation Models and Attempts: Rehabilitation of Valle Longa in Naples; Italy

2.2.1 Introduction

The palace has been built in the early years of the eighteenth century as a double-floored villa, yet it has been preyed by several catastrophes one after another. In 1794, a mischievous earthquake took over it, but architect Camilo Sasso has transformed it into a palace, adding an extra floor to its original structure. In 1843, the palace has hardly survived a volcanic eruption, leaving it in mere distortion. Adding insult to injury, the palace was left in sheer maintenance negligence for the WW2 bombshells, to blow it out of the water. In 1980, another earthquake ended it all but for the staircase, heavenly expanse, and main façade. These ruins have been sold to Banca di Credito Popolare company to maintain and reuse it (Umran &George, 1997). Figures 1, 2, 3, 4, 5, 6, and 7 show Valle Longa before and after rehabilitation.

2.2.2 The Rehabilitation Philosophy of the Palace (Umran & George, 1997)

Most of the rehabilitation process has depended on restoring its former posture to commemorate its historic value and social grandeur. That is to say, most of the maintainer processes relied on reviving the functionality of the palace, leaving out some rooms for modern usages. Even the decorations have been confined to signalize the thrived architectural traces and highlight its historical characterization. The aforementioned ruined trinity has been maintained, as well as the warehouse and main foyer on the first floor, which has been rehabilitated to suit the new function.

2.2.3 The Maintenance Processes of the Palace

It was all about restoring and reconstructing the victimized parts of the building. At first, the collapsed bearing wall has retained its erection. Likewise, the razed roof has been elevated; together with the assemblage of the demolished underground basement. A prolonged conducted study in conjunction with the land topography beneath the central vacuum resulted in building a spacious room right there. This room serves as a locker to save huge funds. A couple of pillars that were built in the present century have been demolished to fortify the cylindrical vault holding the central wall of the first and second floors of the basement itself. In addition, they have been removed to unblock the original space of the room in the warehouse.

Fig. 1 Valle Longa Palace before and after rehabilitation and reuse as a bank. *Source* Umran and George (1997)

Fig. 2 Façade of valle longa after maintenance. *Source* Umran and George (1997)

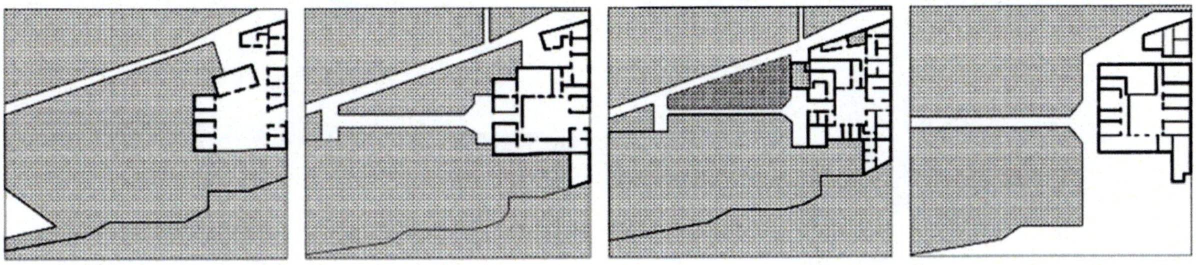

Fig. 3 Left to right illustration of the transformation of the building and its surrounding space. *Source* Umran and George (1997)

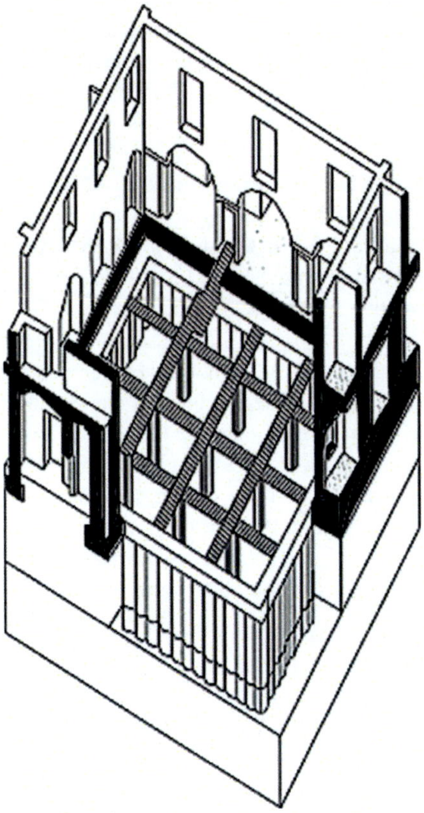

Fig. 4 Ground floor fortification structure to build the basement. *Source* Umran and George (1997)

2.2.4 Functionality Exploitation of the Palace

After the restoration of the cylindrical basement spacious room, a number of adjacent rooms have been attached to it. These rooms are to serve as a reception hall for the bank visitors and high-ranking deputies, to hold business meetings. A small lounge and a large meeting room have been established on the ground floor in the recovered four-room space. The restored space had been of disassembled floors and cellars among fallen down intersecting walls. Yet, a weight carrying horizontal structure has been applied to the two demolished walls. Furthermore, additional seats were added to this hall in the adjacent rooms with a closed TV network, and deposit boxes were placed in the basement (warehouse).

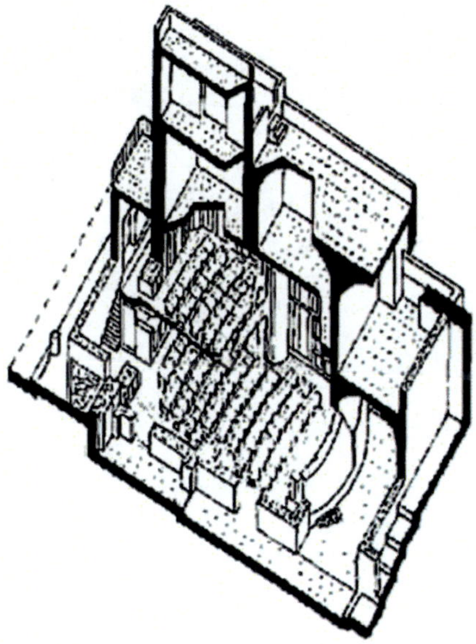

Fig. 5 Grand meeting hall in the ground floor (listening hall). *Source* Umran and George (1997)

2.2.5 Study Results

After demonstrating the rehabilitation profile of the palace, the researcher deduced the following:

1. The functionality renewal process has proceeded void of any harm neither to the historic character of the building nor its façades.
2. The installation process of modern tools and facilities has been manipulated to build the dome under the heavenly expanse without affecting the interior façades of the building or its foundations.
3. The functionality renewal process has been substantial for binding maintenance follow up coursework and promising validation.

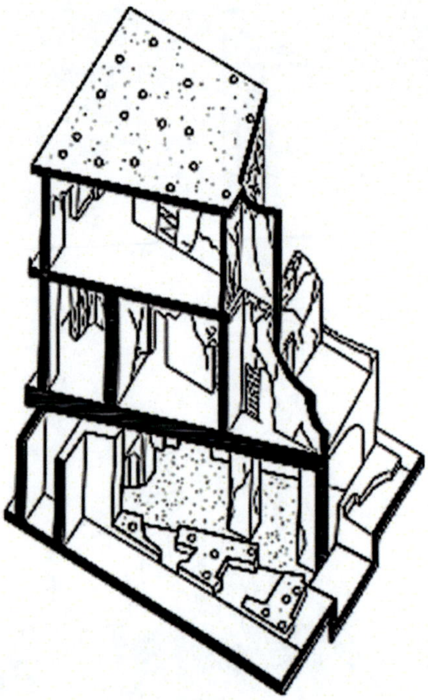

Fig. 6 Warehouse pre the maintenance process. *Source* Umran and George (1997)

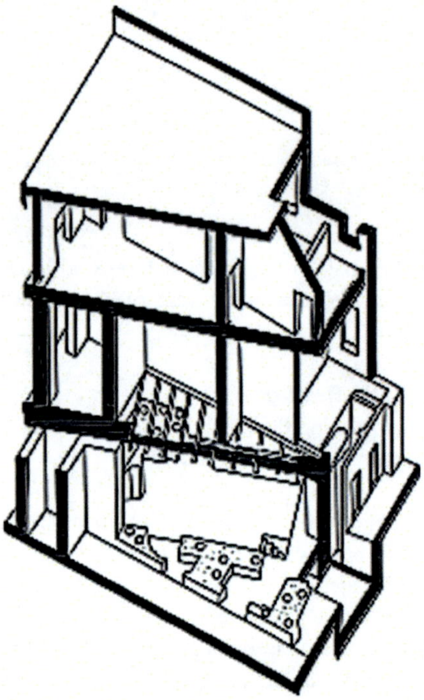

Fig. 7 Warehouse post the maintenance process. *Source* Umran and George (1997)

2.3 Arab Model (I): Rehabilitation and Exploitation of Khan Al Faring in Sidon, Lebanon (Al-Harithy, 2019)

The site dates back to Prince Fakhr Al-Deen in the early decade of the seventeenth century. It has been known as the French Khan (Lebanon Files web site, 2019). No wonder, it was a commercial center and a residential area for the French merchants; especially those crossing the sea from Marisela to Lebanon. In 1798, the Ottoman viceroy Ahmed Pasha Al Jazzar has transformed it into a military fortress in the reign of Napoleon Ponapart's invasion in Egypt. Later, it has been a monastery for nuns. Then, it has been used for scholastic purposes from an orphans' school to Ma'rouf Saad National School in the northern extension building attached to the site (Lebanon and middle east news elnashra, 2019).

2.3.1 Architectural Plan (Fouad, 2018)

The building occupies a flat approximately 58 × 58 m^2. The central courtyard flat is about 48 × 48 m^2 adjoined on its north by a stable with a flat 12.5 × 22.5 m^2. The building consists of two floors: ground and first floors.

- The ground floor: As shown in Fig. 8, it consists of 36 units and stables, where it reaches the sea and leads to a staircase next to each entrance to the first floor. Each unit leads to a door with an arch and a window with a square arch consisting of three columns of stolen welds. The portico opens onto the courtyard through pointed arches in two centers, and the stable consists of three corridors running parallel to the north wing.

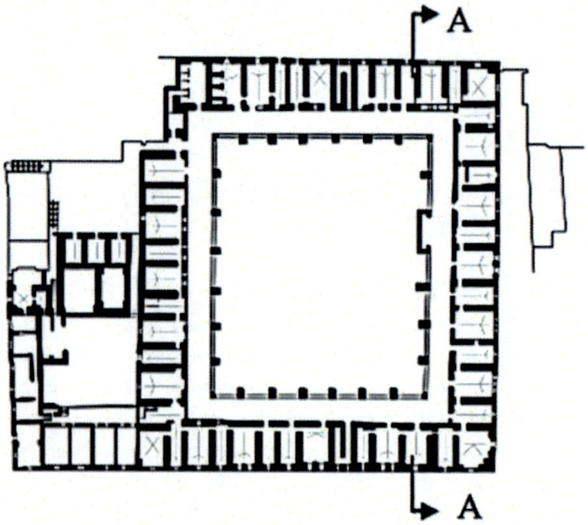

Fig. 8 Ground floor plan of Khan Al faring. *Source* Fouad (2018)

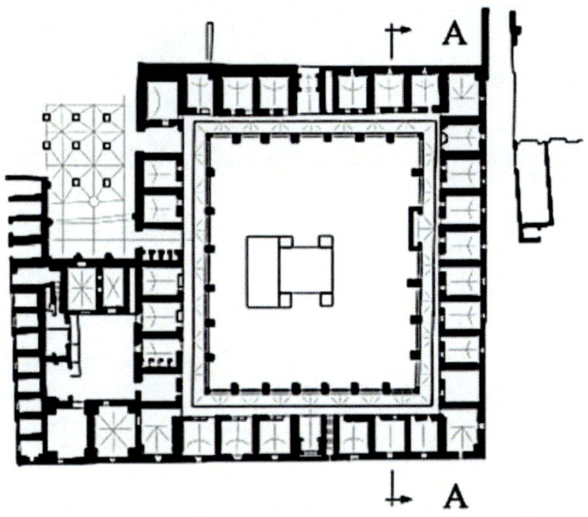

Fig. 9 First floor plan of Khan Al faring. *Source* Fouad (2018)

- The first floor: As shown in Fig. 9, it consists of 50 spaces, and the units open to a multi-door gallery circumscribing the courtyard, and most of the units have windows on the hallway with arches. The four galleries are separated from each other by pointed arches in the corners, and the chambers are opened toward the interior areas of some of them, but they are uneven in the spaces where a small chamber separates every two large rooms.

2.3.2 Structural Plan

The building adapted bearing walls system on continuous tape foundations. Limestone has been the used material in the construction and coverage of the ground floor and the first floor rooms with pointed cellars. Wood has been used in both of the bridges and roofs of the hallway, looking upon the courtyard on the first floor. The arches of the hallway overlooking the arcade. As shown in Figs. 10 and 11, a pillar was added to the rectangular space from the courtyard. These separately added supports are of the same level the first floor, as indicated by the study of the building structurally. The interior walls on the first floor are not right above the interior walls on the ground floor, leading to the placement of construction, which does not serve the load distribution. This has been proved evident amid the reconstruction process in the nineteenth century.

2.3.3 The Caused Infringements to the Building (National News Agency, 2019)

This building has suffered from successive catastrophes throughout its history. Amidst the years from 1798 to 1802, a massive fire devoured the building. No sooner, the building has been maintained in 1809, and it was doomed by the lamentable earthquake in Sidon in 1837 (Janobia web site, 2019). Away from the natural catastrophes, in 1840, the building has been sentenced to a life penalty by the allied British-Hungarian navy. The ferocious bombarding left Sidon to pay the dear price, particularly the harbor zone,

Fig. 10 Southern façade of Khan Al faring. *Source* Fouad (2018)

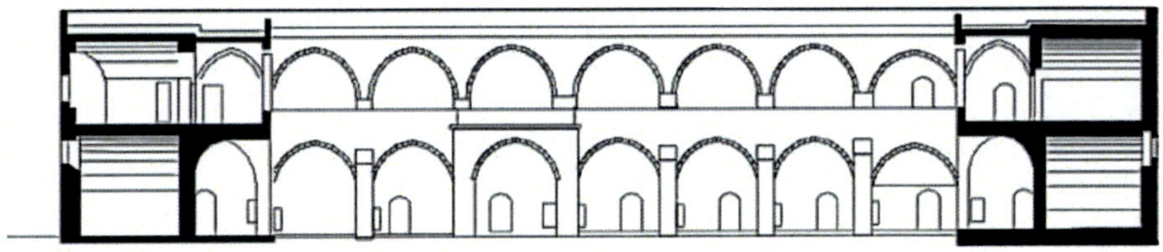

Fig. 11 Section A-A through the building. *Source* Fouad (2018)

where the ceiling of the first floor of the Khan had collapsed on the ground one. Worse part of the roof of the stable and many of the struts bearing the knots of the hallway overlooking the courtyard have collapsed. Thence, it was necessary to call for their restoration. The process coursework has lasted from 1881 to 1893. The wooden roof of the hall was erected, yet the building was damaged using it as a school, due to the leakage of drinking water and sanitation. In addition, the leakage of rainwater went from the roof to the walls and the wooden ceiling. The lack of abidance by adequate scientific restoration led to the appearance of new openings in the façades in deviation from their original function. Consequently, this created irregularities in the levels of openings and change their sizes resulting in increasing humidity rates within the walls of the ground floor and four meters higher.

2.3.4 Philosophy of Rehabilitation and Exploitation of Khan Al Faring in Sidon, Lebanon

Rehabilitation of the building for reuse corresponds to the nature of the monument and its parameters without making fundamental changes. Figures 12, 13, 14, 15, and 16 show Khan Al Faring after rehabilitation.

Fig. 12 A shoot of the main interior hall of the Khan, Khan Al faring: the grand Khan. *Source* Janobia web site (2019)

Fig. 13 Interior façade in the Khan. *Source* Wikimedia & Sidon (2009c)

Fig. 14 Interior view of the Arcade. *Source* elnashra news website (2019)

Fig. 15 Exterior supporting stakes to the arcade. *Source* elnashra news website (2019)

Fig. 16 Held festival in the khan after its reuse. *Source* elnashra news website (2019)

2.3.5 Rehabilitation and Exploitation Processes (Osman& Farahat, 2018)

1. The exterior and interior façades have been re-examined, and the original window shapes and sizes are defined to standardize the system of openings and other elements all over the façades (Center for the revival of the heritage of Islamic architecture in Cairo, 1996), reviewing the doors and interior windows.
2. The wooden ceiling in the first floor gallery was dismantled and rebuilt in the style of the roof of the ground floor gallery. Its origins are still allocated in the north-west and east wings, refrained from a unified roofing style, since it is applied to the ground floor in its original manifestation, but introduced to it right after the collapse.
3. The foundations of the added struts have been fortified and reconnected with the original struts, with the damaged and degraded building stones strengthened using appropriate tested materials.
4. A basin was built in the center of the courtyard according to old photos and drawings.
5. The courtyard floor was covered with gravel and limestone as commonly used until the beginning of this century in the streets of Sidon.
6. Water and sewage systems have been repaired.

2.3.6 The New Function of the Building

The building has been a fundamental Art and Culture Center (Al-Tawil, 2004) to spread the beams of enlightenment from Sidon to the entire neighboring outskirts.

2.3.7 Study Results

After the vivid dissection of the rehabilitation process of Khan Al Faring in Sidon, Lebanon, it is crystal clear that the process has been assigned to experts and specialists of the named discipline. The process reveals the exquisiteness of the analytical architectural and structural research on the building to get the full image of the familiar posture of the building, giving the chop to all the simultaneous unauthentic adjunctions. That is to say, it eliminated the appendages on the building, the roof of the wooden hallway. The restoration of the building to its original status, as modern techniques were shown in adding strengthening stones and mortar, building materials tests, insulation work, etc. The rehabilitation of the building for exploitation purposes corresponds to its nature without making fundamental changes to it (Al watan newspaper, 2019).

2.4 Arab Model (II): Suleiman Palace in Al Tayssir, Mecca

2.4.1 Historical Background

Ibn Suleiman palace; As shown in Fig. 17; dates back to the first authorized Minister of Finance in the reign of King Abdul Aziz bin Abdul Rahman Al Faisal Al Saud; the Founder of the Modern Saudi State. Albeit, this palace was built 80 years ago, where Ibn Suleiman has bought it from

Fig. 17 Exterior shoots of the Al Suleiman palace displaying its precious architectural details. *Source* Al madina newspaper (2019)

its former tenants Al Yuquri tribe; one of the most prestigious and wealthiest of all in Mecca till the time being. The deal was not after diplomatic ends, but he bought it to be a residence for his family, friends, and fellows.

2.4.2 Components and Description of the Palace

The palace is 1400m^2, and the palace consists of four floors and includes numerous rear extensions that were a shelter for servants and entourage, as well as a freshwater well, including generators to light the house, when the electric flow goes off all over Mecca and the Land of Hejaz, which was quite revolutionary in that period.

One of the most important decorations and inscriptions at the west side top of the palace is the image of a car model 36 or 46. As for the south side of the top as shown in Fig. 18, there is a picture of a table and next to it the law of justice codes right beside the Holy Quran and sided by the flag of monotheism in Islam. The palace consists of stunning decorations carved from luxurious wood (Al madina newspaper, 2019).

2.4.3 The Construction of the Palace

This palace was implemented in 1351 Ah, where it was built in a fancy Islamic architectural style, gathering the Islamic architecture styles merits, such as arches, columns, cranes, crowns, and ornate columns. The construction was carried out by the famous contractor at the time, Master Al-Maki Hassan Wazir. Ironically, the man was illiterate, for he has accomplished his mission using a stick. As for the tutors and laborers of the palace, most of them were inhabitants of the Jaroul suburb (Abkar, 2009).

2.4.4 The Occupancy of the Palace and Its Historically Different Functions

After Sheikh Suleiman moved from Mecca, he was hired by the Ministry of Interior Affairs. It then became the headquarters of the courts of Mecca until it became the headquarters of a public library belonging to the Holy Mosque, and the palace was put up for sale for 100 million riyals two years ago (Al-Riyadh Newspaper, 2019).

Fig. 18 Details of the knots and inscriptions covering the façades of the palace. *Source* Al-Riyadh Newspaper (2019)

Fig. 19 Interior entrance of the Palace. *Source* Al-Riyadh Newspaper (2019)

2.4.5 The Status of the Palace

Lamentably, the palace is in short of the least historic authentications, architectural documentation, or typographical archives, though its nearness to the Harram, the worth of its residents, and flow of visitors. In conjunction with its pathetic stance, it was to receive a fatal blow in the duration of the Harram expansion plans in sheer negligence from the Ministry of tourism, but for serendipitous interferences has rescued it from certain eradication, as shown in Fig. 19.

3 Research Findings

The research paper has expounded the definition of heritage in terms of its Arabic lexicography, international, Pan Arab conceptions, and Saudi Arabian vision in particular. Then, it has encapsulated heritage typology on the urban, architectural, cultural, and monumental or archeological basis. The research has proceeded to emphasize the signification and significance of heritage preservation vindicated by various international charters and their stipulated policies in this realm. Besides, it hypothesized the eligibility of heritage investment projects; architecturally and urbanely and their inevitable economic advantage, pinpointing their most commanding programs. On these grounds, the research recalled the international Italian and regional Arab models of heritage rehabilitation and exploitation; Valle Longa in Naples and Khan Al Faring in Sidon, Lebanon and finally; Abdullah Suleiman Palace in Al Tayssir, Mecca, ending up in the research findings and recommendations.

The research has sorted out all the discrepancies regarding the actual existence of former heritage rehabilitation and exploitation attempts and their validity as investment projects to be simulated or adjusted in Mecca. Fortunately, the research has affirmed this hypothesis and called for its simulation and modification to rescue the forsaken heritage buildings in Mecca, rather than being on the verge of extermination. Both of the Italian and Lebanese models have been cited in the research according to analytical studies and other diagnosed criteria that have been unfolded in the research. Both models have proved their applicability to the case study in Mecca. The research has given prominence to the historical background, architectural, structural plans, and its usages and status throughout the research. The search has brought to the fore the most discernable predicaments as such:

- Architectural and urban heritage preservation policies reflect the actual economic, social, and cultural set up of each country, revealing its priorities.
- Lack of architectural and urban heritage authentication and documentation for future preservation projects ends up in ultimate citizenly unawareness of their civilization and history, for these historic buildings to be vulnerable to cleansing.
- The excelling success of cultural and architectural heritage preservation projects in some countries has been in their economic advantage, motivating their civil society and local communities, to get engaged in the preservation and prosperity moves.

4 Research Recommendations

The research deduces some recommendations as follows:

- Prioritization of the urban and architectural heritage preservation projects among the Saudi authority's considerations for its economic and touristic benefits. This is currently adapted in accordance with Vision 2030.
- Drafting general strategies to realize heritage zones and historic buildings preservation and exploitation, enforcing heritage building reuse policies, which will serve the entire community and benefit it.
- Authentication and documentation of heritage buildings are necessary, for these archives must enclose the classification, registration, and documentation of the urban and architectural heritage buildings and sites. This is but to safeguard their historical glory and civilizational worth aided by integrated documentation mechanisms, to record and monitor the targeted parameters.
- Bilateral cooperation between public and private sectors is of urging necessity. This step would assist in saving heritage buildings and sites from the ongoing modernization strife. Additionally, it will emphasize the importance of surface and deep public participation in the preservation and revival of heritage buildings.
- Authorization of local national bodies to contribute to the institutional preservation of architectural heritage.
- Manipulation of the general awareness by the action of media to raise awareness of the importance of heritage buildings.
- Documentation and accentuation of the Meccan heritage buildings, physically and spatially, to eternalize its eroding originality.
- Promotion for tourism, revolutionizing new worth-visiting sites. This is attained by developing the palace, reviving and exploiting, to ensure its placement in the Meccan touristy zones map.
- Preservation of monuments from obliteration and elimination by the General Authority for tourism and antiquities.
- Conversion and reuse of these buildings, to assure the authentic Meccan architecture preservation by formulating architectural reconciliation solutions to revive these spaces

References

Abkar, A. M. (2009). *Visualizations of Mecca Heritage in the 14th AH century* (2nd edn), Quran Science Association, Damascus, Manar Publishing and Distribution Center, ISBN 9786030014200

Abou Leila, M. S. (2010). Sustainable documentation of heritage sites: case study, mansoura city in the Arab Republic of Egypt. *A Published Research in Sustainability and Technicality Conference*, King Saud University, KSA.

Abou Leila, M. S. (2012). *Heritage preservation in Egypt and the Kingdom of Saudi Arabia: A comparative study of Urban and architectural heritage preservation legislations in Egypt and Saudi Arabia*. Published research in the 3rd International Conference on Urban Heritage Preservation-Dubai.

Abou Leila, M. S. (2015). *Green investment in heritage environments' case study Al-Hosseniya' Mansoura City in Delta of Egypt*, Eco city World Summit 2015 held from 11 ، 13 October 2015, the Abu Dhabi National Exhibition Centre, Abu Dhabi, United Arab Emirates.

Abou Leila, M. S. & Albarqawy, W. (2019) *Urban and architectural heritage preservation methodologies in the Arab World*. A book under publication; Umm Al-Qura University Ibrahim, Abdelbaqy. (1982). Civilization values consolidation in the contemporary Islamic City Foundation, Institute of Architecture and Planning.

Abou Liela, M. S. (2018). *Environmental management a doorway to green investment of heritage environments*, published paper in Green Heritage International Conference Chance—Change—Challenge held in BUE Cairo.

Al Tawil, H. (2004). *Rehabilitation of traditional centers in arab cities: case study the lebanese model*. The Traditional Arab City Conference, Syria.

Al-Harithy, H. et al., (2019). Saida Urban sustainable development strategy, local expert team, cultural and natural heritage, http://www.medcities.org/documents/22116/135803/3.+Saida_+Diagnosis_Cultural+and+Natural+Heritage.pdf/a63dafac-c6cc-471e-bed3-c5f09e7e12af. Accessed on 25 Oct 2019.

Al Watan newspaper online. (2012, September 9). An article "demands attention to the historic palace" Sulaiman. https://www.alwatan.-com.sa/article/151026/مطالب-بالاهتمام-بقصر-السليمان-التاريخي Accessed on 10 Aug 2019.

Al-Madina Newspaper. (2013, January 25). Ibn Sulaiman Palace. A unique archaeological landmark built by my mother from Mecca. https://www.al-madina.com/article/205126. Accessed on 5 Aug 2019.

Al-Riyadh Newspaper. (07 Nove 1435–11 Nov 11, 2013). Issue 16576 Article "87 years old and mediates the new central in Mecca Palace of the First Minister of Finance in the Kingdom. Is turning into a tourist complex. http://www.alriyadh.com/883012. Accessed on 5 Aug 2019.

Appleton Charter for the Protection and Enhancement of the Built Environment. (August 1983) http://www.icomos.org/charters/appleton.pdf. Accessed on 5 June 2019.

Budapest Declaration on World Heritage. (1972). http://whc.unesco.org/en/decisions/1217/. Accessed on 5 July 2019.

Budhqah, F. (July, 2010) *Urban Heritage in Ancient Arab Cities: Technics, Planning, and Identity* (3rd edn). Palestinian Geographical Conference, Al-Najah National University, Nablus, Palestine.

Charter for the protection and management of the archaeological heritage. (1990). Prepared by the international committee for the management of archaeological heritage (ICAHM) approved by the ninth General Assembly in Lausanne in 1990, https://www.icomos.org/images/DOCUMENTS/Charters/arch_e.pdf. Accessed on 2 Sept. 2019

Ghoussayni, F., Ali, A. & Bayyati, A. (2018). Rehabilitation of historical cities in the mediterranean using modular patterns—Case study of Sidon, *Lebanese Science Journal*, *19*(3). https://doi.org/10.22453/LSJ-019.3.410431 (National Council for Scientific Research) Lebanon 2018© lsj.cnrs.edu.lb/vol-19-no-3–2018.

Ibrahim, A. (1982). *Civilization values consolidation in contemporary islamic cities foundation*, Institute of Architectural and Urban Studies.

ICOMOS New Zealand. (1987). http://icomos.org.nz. Accessed on 1 July 2019.

International cultural Tourism charter. (1999). Principle 1.

Janoubia. (n.d.). https://janoubia.com/2018/12/12/خان-الافرنج-لمحة-تاريخية/. Accessed on 5 Sep 2019.

Lebanon and middle east news elnashra. (n.d.). https://www.elnashra.com/news/show/887984 /خان-الإفرنج-صيدا-معلم-سياحي-اثري-ومركز-ثقافي-رائد Accessed on 5 Sep 2019.

Lebanon files. (n.d.). http://www.lebanonfiles.com/news/346423. Accessed on 5 Sep 2019.

Nara Conference, on Authenticity in Relation into the World Heritage Convention held at Nara Japan. (November 1994).

Osman, K. A. & Farahat, B. I.(2018). The conservation of the waterfront of Saida: A model for tourism and culture-led revitalization in valuable areas. *Housing, and Building National Research Center HBRC Journal*, *14*(3), 321–333, https://doi.org/10.1016/J.HBRCJ.2017.02.003.

Royal Institution of Chartered Surveyors. English Heritage. (1993). *Investment property databank the investment performance of listed buildings: A joint research project.*

Sidon. (2009). https://commons.wikimedia.org/w/index.php?curid=6857553.

The European Charter of the Architectural Heritage. (1975).

The national news agency. (n.d.). http://nna-leb.gov.lb/ar/show-report/357/nna-leb.gov.lb/en. Accessed on 5 Sep 2019.

The restoration was supervised by the consultant office at the Center for the revival of the heritage of Islamic architecture in Cairo, the period of completion of the work: May 1993 - 1996.

Urban Heritage Preservation Charter. (2003). The Arab League.

Umran, H., & George, D. (1997). *Archeological buildings: maintenance and preservation*. Ministry of Culture Periodicals, Damascus

What is behind caring for urban heritage? (2010). Saudi Commission for Tourism & Antiquities (SCTA) Urban Heritage Initiatives, Saudi Commission for Tourism & Antiquities (SCTA), Riyadh. P. 17.

Yusuf, Muhammad Mahmoud Abdullah. (April, 2014). Urban Heritage investment programs and their funding tools based on arab and international models. *Scientific Bulletin, Journal of Urban Research. Faculty of Regional and Urban Planning (FRUP)*, (2) (Cairo University).

Towards a Sustainable Use of Architectural Heritage: How to Assess Accessibility and the User's Role

Maria Luisa Germanà and Elvira Nicolini

Abstract

Scientific communities and institutional bodies are supporting many shared efforts towards improving accessibility to architectural heritage, in the awareness that it provides the basis for inclusion and enhancement. Accessibility is related to the wider tendency toward a conscious and sustainable use of the architectural heritage, considered as a multifaceted tool for its conservation. This has also highlighted the user's role, which is paramount in order to reach and increase sustainable processes in the field of architectural heritage, as well as in all the technological processes. The user should be active in his/her relationships with the built heritage, in a multi-sensorial way and in a secure and comfortable manner; digital innovation offers many important tools for this aim. The first section of the paper summarises the evolution of the controversial relationships between use and conservation of the architectural heritage. The conflict between the requirements of using and the constraints of the conservation still poses many theoretical and design challenges, but a more responsible role of the 'visitors' (better considered as 'users') can help the built heritage to leave its *ivory tower*. The second section, after outlining the *performance-based approach* as a basic analysis and early-design methodology centred on the user's exigences, firstly describes the meanings and consequences of the kinds of user and the dimensions of use in architectural heritage; secondly it describes the main economic, social, and environmental aspects of sustainability that regard the use of built heritage. The remaining sections of the paper are devoted to the results of multidisciplinary and multiscale studies that address the accessibility of architectural heritage from many points of view. In particular, an assessment method is tested to support decision-making processes oriented towards a quality-oriented design, focused on the accessibility to the architectural heritage. Ancient architectural heritage is here considered a sort of *borderline case*, in which the general critical conditions of the built heritage are emphasized. The inclusion of users, especially by reinforcing their awareness and active role, is proposed as a tool in the sustainable exploitation of architectural heritage, to be applied in many aspects of management, such as community engagement and planned maintenance.

Keywords

Architectural heritage • User's role • Performance-based approach • Sustainability • Accessibility

1 Use and Conservation of Architectural Heritage: Beyond a Controversial Relationship

Use and conservation have been the two poles of a controversial relationship, ever since the very idea of architectural heritage arose. After evidence of the consequences of the *First Industrial Revolution* on traditional settlements had become clear, the dramatic differences between an ever more distant and non-replicable past and an increasingly aggressive present, over the last two centuries, elicited a sort of defensive approach to the inherited built environment, which is still recognisable today. In fact, terms such as *defence*, *protection*, and *safeguard* recur in almost all reasoning regarding the cultural heritage, whether in the bureaucratic apparatus or in scientific literature, whether in the technical and professional field or with the general public and media.

M. L. Germanà (✉) · E. Nicolini
Dipartimento di Architettura, Università di Palermo, Palermo, Italy
e-mail: marialuisa.germana@unipa.it

E. Nicolini
e-mail: elvira.nicolini@unipa.it

A. Versaci et al. (eds.), *Conservation of Architectural Heritage*, Advances in Science, Technology & Innovation,
https://doi.org/10.1007/978-3-030-74482-3_17

The controversial relationship between conservation and use evolved following developments in the theories of restoration (see ICOMOS, 2004). In the first stage of this evolution, the aim of preserving prevailed over the issue of exploiting the architectural heritage; as Alois Riegl argued in 1903, a conflict often arises between the historical significances of the built heritage and the various kinds of its contemporary values, which include the *value of use*. In the second phase, a distinction was made between the *dead monuments* (such as ruined archaeological buildings, further from contemporaneity and devoid of any use) and *living monuments*, in which uses that were not very different from the original ones (in order to avoid excessive alteration) were considered acceptable (CSABA, 1932). In the third phase, the conservation of the physical essence of architectural heritage maintained the role of the main aim of any intervention, whether or not this was any longer considered an end in itself. This derived from the awareness that the lack of any use, after a certain amount of time following a conservative intervention, leads to decay, similar to disuse in cases of abandonment. So, the opportunity for historic buildings and areas to have new functions, consistent with the demands of contemporary life, became a prerequisite for their conservation (CoE, 1975). In the fourth phase, after establishing that the use of the architectural heritage is not just a necessary evil, but rather a precondition for survival, the emergence of a process-centred approach to interventions on the built heritage has proved the need for considering the theme of use in all phases of analysis, conservation and enhancement (Della Torre, 2012; Germanà, 2014).

The paradigm of *adaptive reuse* synthetises the current idea of using the built heritage as an apparently obvious choice, but the conflict between the requirements of using and the constraints of conserving still poses many theoretical and design challenges. The lack of use has been considered a form of *immaterial incompleteness*, to be tackled by researching appropriate design and management solutions, balanced between conservation and utilization (Germanà, 2015). Nowadays, the focus is on the role of the users (visitors and general public, individuals and communities) in both the scientific and operational fields, as a key factor of conservation and enhancement of the architectural heritage, with technological innovation offering new possibilities in improving this role in various ways and aspects (Germanà 2019a, b). For all these reasons, a less controversial relationship between use and conservation is realistically foreseeable in the near future. This relationship can help the built heritage to leave the *ivory tower* where it has been placed for a long time: a niche field, devoted only to cultured people and where only specialist experts have a say. The only *legitimized relationship* between experts, heritage site and *visitors* has been too often only a "top-down relationship, in which the expert 'translates' [...] the site and its meanings to the visitor. The very use of the term 'visitor' also facilitates the construction of passivity and disconnection" (Smith, 2006: 34).

2 More Than Appropriate, a Sustainable Use of Architectural Heritage

Use can be considered as the main distinguishing feature of architectural heritage, in comparison to any cultural heritage that has nothing to do with the construction; in fact, since the Vitruvian triad, utility (*Utilitas*) has been one of the three pillars whose coexistence generates the very idea of Architecture. With reference to architectural heritage, this theme could be taken into consideration from two points of view, related to the analytical and the design level respectively; on the one hand, full knowledge of the artefacts should include their original utilisation and their successive transformations; on the other, the conservation interventions should take account of the contemporary and potential future uses of the built heritage (Germanà, 2015).

The quality-oriented methodology outlined below takes into account both these points of view, because it can be applied both as an ex-post assessment tool and as preliminary guide in the early design phase. The theoretical basis of this methodology is the *performance-based approach*, developed and codified in the second half of the last century as a fundamental of the ACE (Architecture, Construction and Engineering) sector, focusing on the role of the user as of primary importance and "thinking and working in terms of ends rather than means" (Gibson, 1982). Even though the origins of the performance-based approach (dating back to a period featuring huge quantitative and qualitative transformations) were firmly focused on the new buildings, a similar approach is also applicable to the architectural heritage, keeping in mind the specificity of this field.

The term usability (along with other requisites such as safety, manageability and maintainability) in architectural heritage acquires specific meanings, without losing the same relevance that it has within generic built environment. For this reason, the *performance-based approach* seems to be the appropriate basis for a holistic vision and unitary strategies, to which European official documents have lately been alluding: "*Baukultur*, as an aspect of cultural identity and diversity, holistically embraces every human activity that changes the built environment, including every built and designed asset that is embedded in and relates to the natural environment. *Baukultur* calls for contemporary creation and the existing buildings, infrastructure and public space, including, but not limited to, monuments of cultural heritage, to be understood as a single entity (...)" (EU Ministers of Culture, 2018). This theoretical turning point leads to strengthening the weak link between experts and users also

Table 1 Meanings and consequences of kinds of user and dimensions of use in architectural heritage (reworked from Germanà, 2015, 2019b)

Kinds of user	Meanings and consequences
Direct The direct user is a person (resident or visitor) that is in physical contact with the built heritage, as individual or as part of a community	The physical contact could be intentional or unintentional, sporadic or continuous, individual or collective. It always generates consequences on the material essence of the built heritage, both positive (using is a premise for caring), and negative (using could produce overusing and, consequently, extensive damage)
Indirect The indirect user is anyone knowing certain specific architectural heritage and, thanks to this knowledge, appreciating its value and meaning	A certain level of awareness is the prerequisite for indirect use, which can also happen through artistic works, such as literary narratives, cinematographic representations, etc. The main consequences are potential and limited to an incremental added meaning of architectural heritage, thanks to the artistic interpretation or narration by some other, direct or indirect, user
Dimensions of use	*Meanings and consequences*
Tangible/quantitative The quantitative aspects of the use depend mostly on the physical circumstances of the built heritage (typology, conditions), and its surroundings	The tangible dimension of use can improve the inclusion of users in the management processes. Quantitative aspects of use and direct users are related issues in the main topic of accessibility, a set of requisites aimed at optimising the using conditions for each person (in opposition to the limitations possibly deriving from age, gender, health and from physical or intellectual capacities). An accessible heritage encourages tourism, taking into account the increasing percentage of elderly people
Intangible/qualitative The level of understanding and awareness of the users influences the qualitative aspects of the use	The very idea of heritage would not exist without this dimension of use, which motivates its particular interests. The meaning of the intangible dimension of use depends on the cultural context and changes continuously in intergenerational passages. It plays a determining role in forming the identity of the individual and communities

in the field of architectural heritage, both at collective and at individual level, helping to highlight many important ethical issues in management of cultural sites that are open to the public (ICOMOS, 2017; Germanà 2021). In fact, the issue of using the built heritage evidence the crucial relationships with local communities, in terms of quality of life and social justice: "ethical issues arose in their contemporary form when cultural heritage managers found that they had to forge links with groups and communities, as they were required to do so by regional planning and heritage legislation" (Egloff, 2019: 101).

Dramatic changes, which the *Fourth Industrial Revolution* has accelerated over the last few decades, have been recognisable in every feature and dimension of exploitation of the architectural heritage since the actual idea arose (Germanà 2019a). In comparison to the physiological reuse of existing buildings in the pre-industrial era, of which many examples recur in architectural history, contemporary reuse poses both theoretical and design challenges, due to the conflicting goals of utilization requirements and conservation demands. Many researchers have faced this problem, highlighting the need for strategic and verified solutions for reuse (De Medici & Pinto, 2012).

The theme of exploiting the architectural heritage, keeping in mind its specific *conspicuous historical, archaeological, artistic, scientific, social or technical interest* (CoE, 1985), has been the focus of previous studies, distinguishing the various ways and meanings of use in this field, these often being reciprocally interlinked (Table 1).

Furthermore, use has been proposed as a term of reference for designing and assessing interventions on architectural heritage; paying particular attention to the current or expected uses has been outlined as a primary prerequisite for their reliability. Firstly, in the analysis of the status quo, the existing conditions of utilisation of the architectural heritage depend on the contextual conditions (the environmental surroundings and the anthropic context, the relationships between the urban fabric and the impacts of all human activities, the socio-economical background). In addition, analysis of the existing conditions of use embraces the typology, shape and dimensions of the interior and exterior spaces and their mutual connections, paying attention, lastly, to the building materials. Secondly, in the management phase, the issue of use is equally important: on the one hand, the demanding levels of use require more intensive and frequent maintenance; on the other, inclusive utilisation provides the opportunity of involving users in the conservation processes, improving their active role and awareness and guaranteeing responsible use. Thirdly, in enhancement strategies, utilisation plays a pivotal role, being the main objective; in fact, any intervention should aim to improve use, given that the meaning, value and physical consistency of the built heritage cannot actually be increased. In this regard, moreover, the relationships between direct and indirect uses are dense and depend on more complex and a wider range of factors that embrace the tangible and intangible dimensions of use; for this reason, the issue of sustainable use becomes strategic in designing and maintaining balanced solutions (Germanà, 2015).

Within this framework, the methodological contribution described below finds its theoretical basis in the concept of sustainable use of architectural heritage, overstepping the limitations of the appropriateness of current use in historical values and meanings; this only focuses on the balance between the conservation and transformation of the tangible aspects. In fact, the well-known dimensions of sustainability, to be achieved in an integrated manner (as stated by the *2030 Agenda for Sustainable Development* in 2015), are all relevant also considering the use of the built heritage.

Before defining sustainable use, the multiple reasons that can render the use of architectural heritage unsustainable are summarised as follows, emphasizing that these may regard both indirect and direct use, and their effects range from quantitative to qualitative consequences. With reference to direct use, all the quantitative and qualitative consequences can be observed in the physical consistency of the built heritage (Table 2). With reference to indirect use, which happens without any physical contact, the issue of appropriateness of sustainability regards the potentially misleading identification of the built heritage, based on the mediated memories currently increased by digitalization. The future evolution of this aspect of using the architectural heritage is not easily foreseeable and depends on new kinds of memory, which are emerging as an effect of the global non-historical hypertext, a gigantic almost undifferentiated cluster of information in which, Past, Present and Future fluctuate within our digitalised world (Castells 2010, p. 403).

The economic, social and environmental dimensions of sustainability regard the use of architectural heritage in different and interlinked ways.

- The programming phase of the interventions can guarantee sustainable use, considering it within the framework of feasibility. On the one hand the sustainable use of the architectural heritage can contribute to economic added value, leveraging on virtuous processes initiated as local economic spin-offs. On the other hand, sustainable use can contribute to covering the costs of management and maintenance. The economic dimension of sustainability leads to encouraging mixed public and private use, integrating investment both in interventions and in management.
- Sustainable use is not limited to accessibility, which is focused on avoiding the barriers to direct use or to full understanding. Over the last decades the tendency has been growing to prefer inclusion as a keyword, which shifts the focus to the potentialities of an active role for the users and on the mutual relationships between people and heritage. The social dimension of sustainability leads to integrating the sphere of the experts with the sphere of the users, involving local communities thanks to appropriate communication and engagement.
- The exploitation of the architectural heritage is environmentally sustainable if: it spares the natural resources and does not produce damaging emissions or waste and if it follows circular processes that reduce or eliminate any impact, preferring durable functional solutions. In

Table 2 Consequences of unsustainable direct use

Aspects	Unsustainable use	Meanings	Consequences
Qualitative	Abuse	Illegal utilisation; unauthorized occupation (squatting); uncontrolled privatization	High possibility of damages limitations in public use
	Misuse	Inappropriate use, misleading with regard to the identarian significances	Average possibility of damage limitations in understanding
Quantitative	Uselessness	Conservation seen only as an end in itself Abandonment	Unreliable (Useless) conservation vandalism
	Overuse	Excessive use, in terms of duration and intensity (e.g. mass tourism, or high-impact uses)	High possibility of damage limitations in understanding

addition, the interventions on the built heritage, and especially those aiming to improve accessibility, should utilize recycled and renewable building materials (or those inspired by local tradition). Using the built heritage can imply a wide range of needs for comfort, in accordance with the duration of the visit to the site and the interaction with the natural environment (orientation to the solar path and exposure to weather conditions). Passive heating and cooling solutions, also inspired by traditional buildings, integrated with active systems supplied by renewable energy sources, certainly contribute to a sustainable use of the architectural heritage.

3 Design for Accessibilty to Architetural Heritage

Reconnecting with the theme of sustainable use of the architectural heritage and the conflict between contemporaneous use and conservation, the issue of accessibility will be specifically addressed in this section, with reference to the possibilities of reaching a given building, to entering with ease and exploiting the spaces and appliances in conditions of comfort, safety and autonomy. There are also immaterial factors affecting accessibility, linked to information and communication, which condition the multi-sensorial experience of the visitor. In fact, accessibility has been defined as "the quantity of information that it is possible to receive from the cultural or environmental heritage site with which one comes in contact" (Quagliuolo, 2003, p.17; transl. by the Author).

UNESCO has attributed great importance to accessibility requisites, stating that the management plans of the sites included in the WHL (World Heritage List) should guarantee it, especially when considering the impact of tourism. "To understand the socio-economic impact of tourism on World Heritage Sites it needs to identify the factors that might have a stronger impact on them. The primary factor is *accessibility*: the more accessible a site is, the more it can be visited; according to the type of site, the number of visits can vary a lot, with either positive or negative consequences" (UNESCO UNITWIN, 2018, p. 85). The *Outstanding Universal Value,* which prompts recognition by UNESCO, attracts mass tourism-commercial policies, causing invasive usage that increases the vulnerability of the heritage. "Sustainable conservation and safeguarding intangible cultural heritage in a local tourism context can be achieved only by fostering awareness, in-depth knowledge and understanding among local communities of the significance of their heritage and diverse influences that have come together to create—and continue to create—a unique culture. Building intergenerational capacity among local community members, especially in young people, to support them in engaging with and interpreting their heritage and in communicating successfully with visitors, has the dual benefit of enhancing visitor experiences and strengthening their own sense of self-worth and identity" (ICOMOS, 2014, art. 1.2. a–b).

Given that every individual has a right to the cultural heritage (CoE, 2005), for a sustainable use of architectural assets (as presented in the previous Sect. 2) the active participation of the individual is necessary, based on a *sense of belonging* that, not only consents acknowledgment of its cultural value, but also provides the drive in wanting to protect it. The growth in his/her feeling for a shared asset on the part of the individual is proportional to the extent of access and utilization of the actual asset, with the condition that fruition is a holistic and interactive experience.

Sustainable use of the architectural heritage, of which accessibility is a key factor, derives from a project that bears in mind the following:

- There is a need to surmount the concept of accessibility seen only in terms of overcoming fixed obstacles, creating for all visitors conditions of well-being and encouraging interaction with the place itself.
- One must apply "a multi-layered approach to planning and interpretation. Physical, intellectual, emotional and economic access need to be reconciled within bespoke strategies for interpretative planning and quality assurance mechanisms" (ICOMOS, 2014, art. 1.3.b).
- There is a need to consider the possible impact of the intervention on the environmental, economic and social development of the local context in which it finds itself, by involving public and private players.

In the case of ancient architectural heritage, the accessibility project demands great attention in finding a balance between satisfying the visitor's needs and respecting the strictures of conservation. "Dealing with accessibility to cultural heritage means combining two values: the right that every human being has to fulfil his experience through the encounter with the past and the consciousness that the conservation of cultural heritage represents on our present having passed through our past" (Arenghi et al., 2016, p. 15). The need for this juxtaposition leads to a high degree of uncertainty in the intervention processes on the item in question and the surrounding area. From this there emerges the need to launch strategies for the evaluation of impact and the possible future benefits in cultural, social and environmental terms. Considering that the ancient built heritage is extremely varied (from museums to archaeological excavations, monuments, historical centres, etc.) it is not possible to single out a homogeneous *modus operandi*, but it is necessary to evaluate cases individually as to whether and how to intervene. However, it is opportune to individuate a method employing technicians and a site manager backing the success of the intervention, taking account of multi-disciplinary factors and positioning the actual user at the centre. The acknowledgement of compatibility should result from a complete assessment of the benefits that might be obtained through the intervention, in terms of safeguard and enhancement of assets and, at the same time, accessibility.

4 Needs and Involvement of Users of Architectural Heritage

The needs of the user, as seen in preceding Sect. 2, are at the centre of the *performance-based approach*, which today can be seen in the most wide-ranging *human-centered vision* characterizing the most recent shared positions with regard to technological processes and, in particular, the cultural heritage. With reference to this vision, the *performance-based approach* is coherent with *Design for all* or *Universal Design* (Mace, 1985), a design approach for spaces, environments and objects that can be used by the individual, regardless of their age and psycho-physical capacity. The logic of *Universal Design* has been clarified in seven basic principles that place the well-being of any user at the centre of the question. "These seven principles may be applied to evaluate existing designs, guide the design process, and educate both designers and users about the characteristics of more usable products and environments. The principles are: equitable use, flexibility in use, simple and intuitive use, perceptible information, tolerance for error, low physical effort, size and space for approach and use" (Story et al., 1998). In 2008, the Italian Ministry for Cultural Assets published the guide-lines for accessibility to sites of cultural interest, confirming the need for this approach in the field of architectural heritage. This governmental document, recently updated with the guide-lines for the PEBA (*Piani per l'Eliminazione delle Barriere Architettoniche*—lit. *Plans for the Elimination of Architectural Barriers*), is coherent with the principle that it is no longer a question of eliminating or surmounting, but of re-discussing in dialectic fashion the actual basics of the design activity, which should be capable of stimulating the potential of the project and not restricting it, whilst taking into consideration the needs of real people as a point of departure (MIBAC 2008, p. 14).

Thus, with the aim of inclusive use of the architectural heritage, the methodology, above all, organized the needs of the user with regard to accessibility, applying a broader meaning to the term "use", imagining a space that is not only accessible and exploitable in comfort and safety, but also encouraging cultural cohesion and inclusion, as well as social interaction.

The user is any individual, with or without disabilities and regardless of psycho-physical conditions, exploiting a determined space, the characteristics of which might hamper, totally or partially, the correct use and interaction of the senses. Starting from the bio-psycho-social model, according to which environmental factors affect the ability of the individual in a given moment and place (WHO 2001), the user for whom accessibility should be guaranteed might be any individual, with varying physical and economic capacity, with various previous educational experiences, with boundless expectations, but with a sole intention: to enjoy a site of cultural interest. Beginning from the needs of the user, in concrete terms, the goal of accessibility demands the following actions:

- To involve local administrations and the tourist industry, making it easier to reach the site with both public and private transport and guaranteeing parking in the immediate vicinity of the entrance.
- To individuate, in the site, entrances and itineraries accessible to all, avoiding uneven surfaces and obstacles and adding aids to accessibility (signs, handrails, etc.).
- To encourage the exploratinal aspects as well as the cognitive, involving the visitor actively during his/her visit, studying more valid technology and itineraries in order to enhance learning and orientation.
- To activate strategies for user well-being and safety, providing for services, rest-areas or seating, manageable and functional appliances, monitoring the environment from the thermal, acoustic, olfactive and atmospheric points of view.

At the collective level, one might think, for example, of the concept of *capacity building* for the involvement of the local community in understanding the historical urban landscape in UNESCO, 2011), but apart from this, there is an individual level of user involvement, focusing on a personal interpretation of the asset, development of awareness of associated values and consequent participation in conservation. This brings to mind the concept, quite common in museums, of *interpretation centres* (Tilden, 1977; Tugas et al. 2005), which lead to the creation of significant processes for the visit-experience, focusing on the actual visitor. The interpretation centres have the aim of enhancing the heritage worth preserving, in order to hand it down to future generations, providing the visitor with the tools and technology needed to interpret the intangible aspects that need to be resolved.

The architectural heritage, with its interpretation centres, maintains a constant relationship with the user throughout the visit, utilizing multi-sensorial mediation devices, such as: scenographic presentations that enable the visitor to immerse himself emotionally in the experience; digital technology, which enable one to individualize the experience, by offering the possibility to select from a rage of available contents and adapt them to the visitor's profile; tactile devices; didactic and creative areas geared towards effective learning and sensitizing the visitor to specific themes in parallel to the itinerary. The involvement of users may also be based on monitoring the degree of satisfaction during the visit, which will help the site managers to improve accessibility, consenting a long-term assessment of the effectiveness of the intervention carried out, with their responding to requirements, and then initiating an immediate intervention in the case of a shortcoming. Feedback from the user is vital, also for future growth; it is a virtuous process that stimulates creativity in the community and triggers circular dynamics of co-operation, bringing the community closer to the public asset that has been activated in order to be enjoyed. Monitoring can profit from digital technology, which includes automatic tracking systems and feedback that record the visitor's behaviour in specific parts of the itinerary. This fosters a participative approach, since the visitor can communicate to the administrators of the site any suggestion along the way. Other forms of more traditional data collection, such as interviews or compilation of questionnaires, can generate a greater willingness to adapt itineraries in accordance with the data collected.

A monitoring form was prepared to individuate the various requisites for accessibility (Table 3); this can be supplied either digitally or in paper form and depending on the various feedback items or following a dramatic increase in demand, it may be used to suggest what requisites the site should possess. By submitting the monitoring forms for perusal by various groups of stakeholders, this could lead to the prompting of numerous points of interest. The greater the number and the variety of the groups interviewed, the more objective the final data deriving from a statistical comparison. Potential results might thus prepare the ground for bringing to light the multiple aspects of accessibility requirements, in such a way that the shared, suggested improvements might then be applied.

5 Constraints on Accessibilty to Architectural Heritage

With reference to the principles of *Universal Design* as expressed previously, bearing in mind the heterogeneity of users, in order to be accessible, a site must fulfil various

Table 3 Requisites for accessible use

Fields			Requisites
2.1 Material accessibility	2.1.1 Spatial distribution	2.1.1.a Horizontal paths	Reachability of the site
			Suitable dimensioning (current regulations)
			Practicability of surfaces (disconnections, slipperiness, wide grid in pavement gratings, unsecured doormats)
			Absence of barriers
			Presence of rest areas
		2.1.1.b Vertical paths	Suitable dimensioning
			Connection between uneven surfaces with suitable gradients
			Absence of spiral staircases or corner steps
			Presence of handrails
			Suitable dimensioning
			Presence of auto-levelling and damping systems
			Facility of access to commands (active systems)
	2.1.2 Services		Presence of point of welcome, bookshop
			Presence of refreshment area and assistance point
			Presence of phone and wifi signal
			Presence of adequate toilets
	2.1.3 Security		Conformity with fire regulations
			Absence of risk of falling
			Warning signs

(continued)

Table 3 (continued)

Fields			Requisites
2.2 Immaterial accessibility	2.2.1 Way-finding		Presence of signage
			Presence of multilingual captions
			Diffusion of useful-to-visit material
			Presence of tactile and/or sound support (LOGES system)
	2.2.2 Management		Accessibility during various weather conditions
			Presence of tracking and feedback systems
			Adequate maintenance conditions
			Wide range of visiting times
	2.2.3 Economic efficiency		Ticket-price reductions
			Useful-to-visit material
			Service-price reductions
	2.2.4 Comfort	2.2.4.a Thermo-hygrometric	Protection from sunlight
			User-flow controls
			Ventilation controls
			Temperature and humidity controls
		2.2.4.b Visual	User-flow controls
			Presence of braille aids
			Absence of visual barriers
			Suitable lighting
		2.2.4.c Acoustic	User-flow controls
			Presence of sound-absorbing systems
		2.2.4.d Respiratory	Ventilation controls
			Air quality controls
	2.2.5 Interaction	2.2.5.a Digital	Presence of multimedia devices
		2.2.5.b Virtual	Presence of immersive multimedia devices
		2.2.5.c Tactile	Presence of touch-sensitive systems
	2.2.6 Personalization of experience		Presence of creative, didactical and laboratorial spaces
			Presence of community engagement activity
			Presence of post-visit evaluation tools

requirements, which also take into account the heterogeneity of the built heritage. These are usually places that are intrinsically foreign to the requisites of accessibility, in which structures added over time, in successive periods, coexist and with diverse purposes, the fruit of varied technological cultures; they are not always accomplished in terms of compatibility (Sposito & Germanà, 2003, p. 81; Arenghi & Pane, 2016, p. 58). This goes some way to explaining the notable operational difficulties that hamper interventions aiming to improve environmental and technological performance. In 1992, Italian legislation extended the principle of accessibility to architectural heritage, but introduced the criterion for which solutions could be reversible, thus validating elements that were precarious and rather unsightly, not to say rather impractical and harmful for the pre-existing.

Many researchers have criticised this criterion of reversibility (Picone, 2004), highlighting the need for solutions to accessibility that would be lasting, but without impairing the identity of the historical pre-existence. This would justify the need to support the project for accessibility to architectural heritage by utilizing instruments of evaluation, so as to be able to deal with the requisites needed to satisfy the demands of the user and the existing performance, in an overall picture of emphasizing the levels of quality. Wherever there are no performance features, several alternative interventions need to be assessed whilst, at the same time, reflecting on the impact that fresh solutions might cause for the pre-existence.

The multi-criterial approach represents a method of assessment that describes the interventions with regard to the countless constraints geared towards safeguarding the identity of the heritage by maintaining its specific features. Any presence of unfavourable elements interfering with the acquiring of knowledge about the asset can be considered in terms of discontinuity in morphological, constructive and symbolic relationships. The system of strictures is arranged into morphological-dimensional constraints, perceptive-cultural constraints, material-constructive constraints taking into account methodologies that had previously been tried out for a historical urban space (Nicolini, 2016).

An evaluation form was created for this (Table 4) in order to verify adherence to these strictures, by analysing intervention choices, either ongoing or planned, with reference to accessibility. Impact criteria were individuated for the three categories of constraint under examination, with the aim of ascertaining whether the choices had produced or might produce alterations, and the degree of tolerance for them to be acceptable. This systemic model is an auxiliary instrument in identifying critical points and unfavourable perceptions that adjustment interventions induce in the observer. "Conservation requires the retention of an appropriate setting. This includes retention of the visual and sensory setting, as well as the retention of spiritual and other cultural relationships that contribute to the cultural significance of the place. New construction, demolition, intrusions or other changes which would adversely affect the setting or relationships are not appropriate" (ICOMOS, 2013, art. 8).

The morphological-dimensional constraints safeguard the relationships of the elements that characterise the edifice, conserving the morphological structures of particular importance; legibility of forms, proportions, typical elements of a historical edifice, testimonies to the formal culture and the material characterizing a determined historical–geographical context; respect for the character of a geometrical-spatial configuration, and therefore maintenance of the hierarchical relationships between the spaces and conservation of the dimension and the morphology.

The perceptive constraints refer to the experience obtained by the users during the visit, which can be favoured or hampered by the conditions of accessibility. The perceptive constraints are linked to the cultural ones; in fact, the historical character of the site triggers in the visitor an awareness of intangible values that identify it as a specific place. The need to maintain in the community an interest that consists in conservation of the aesthetic value, of traditional and historical instances, is expressed as a cultural constraint.

The cultural constraint safeguards the symbolic value that communities usually associate with the specific site, in terms, for example, of a theatre of historical and legendary events, or

Table 4 Constraints deriving from the objective of conservation and impact criteria

Constraints	Impact criteria	ID
Morphological-dimensional	Alteration in height	MD1
	Alteration in volume	MD3
	Alteration in the continuity of relations with natural elements	MD4
	Alteration in the continuity of relations with architectural elements	MD6
	Alteration in the roofing profiles	MD7
	Alteration in the urban section profiles	MD8
	Alteration in the building envelopment ("full/empty" relationship)	MD9
Perceptive-cultural	Discontinuity of significant relationships	PC1
	Reduction in recognisability of identifiable elements	PC2
	Reduction in viewpoints and itineraries	PC4
	Reduction in perception of togetherness	PC6
	Obstruction from extraneous elements in spaces and/or on elements	PC7
Material-constructive	Alteration of technical-constructive elements	MC1
	Alteration in material elements	MC2
	Alteration in decorative elements	MC9

Table 5 Impact degree for assessment of the comparison between constraints and impact criteria

Impact degree	Descriptors	Value
I. Very low impact	The impact is hardly noticeable since the intervention safeguards the indivisibility and participates coherently in the order of the historical-cultural elements characterising the architectural system	1
II. Low impact	The impact can be clearly individuated	2
III. Medium impact	The impact is perceptible and its presence could have an unfavourable effect on conservation of legibility and recognisability of the quality of the architectural system	3
IV. High impact	The impact is recognizable and its presence upsets the continuity of the typological, morphological and linguistic-formal features	4
V. Very high impact	The impact is clearly observable and its presence bestows considerable alterations on the elements and relationships that make up the organization of the architectural system and context	5

as object for literary, pictorial or traditional celebrations, or even sites of great attraction for tourists. The perceptive constraint takes into account the image that the visitor has of the site and the expectations he/she has with regard to it. The material-constructive constraints are geared towards respect for construction techniques and materials, with the aim of conserving authenticity. It is not only the construction materials that are safeguarded, but also the traditional techniques as distinctive expression of that specific geographical and historical context. The constraint is geared towards respect for mechanical-structural behaviour, deriving from solutions that are the fruit of the knowledge of that period. With reference to the material, one should finally mention the decorative aspects, which range from the pictorial to sculpture, of esteemed aesthetic quality, as well as testimony to the work of local craftsmen. These elements identify architectural heritage and render it recognizable to the local community.

The degree of impact of interventions for sustainable use of the architectural heritage is attributed to each of the above-mentioned constraint categories (Table 4), according to indications describing their level of presence and participation in safeguarding. In this way, the descriptors (Table 5) consent the classification of degree of impact in accordancce with five levels, increasing in proportion to the degree in which the new intervention interferes unfavourably with conservation of site-quality. The overall assessment is the result of the predominant quantitative value emerging from the qualitative evaluation of the interventions carried out and compared with the constraints imposed.

6 Multi-criterial Analysis

"Accessibility can be measured like any other quality (…). Accessibility is a function of the quality of experience that can be implemented on the occasion of a visit and comprises physical, economic and cultural accessibility" (Quagliuolo, 2003, p. 15; transl. by the Author). Measurability enables us to verify and improve this important aspect of the sustainable use of the architectural heritage.

By following a line of thought in accordance with a needs-performance approach it is possible to define an instrument that is auxiliary to the decision-making process, in which the needs of the user are the precondition for defining the requisites of the project; a need is whatever is demanded by the normal execution of an activity by the user; a requisite is the meeting of a need by means of a series of features that characterise the built environment in determined conditions of use, context and solicitation. With reference to a site with high-level historical characteristics the requisites must also respect the constraints for safeguarding the asset and the project orientation tools take on ever greater importance. In many cases, the visitor's needs coincide with the asset's conservation needs; for example, disproportionate flows of visitors might be potentially harmful to heritage because of greater wear and, allied to variations in temperature and air-quality, could bring about problems of overheating, sight and breathing in the visitor.

Whilst acknowledging that solutions to accessibility in historical contexts cannot be standardized, a multi-criterial grid with objective parameters was prepared so that it could operate in many contexts. The grid aims to evaluate the degree to which the visitor's needs are met and the classes of impact of the adjustment interventions, both planned and executed. The objective is to provide a quantitative response to two qualitative aspects: accessibility and vulnerability of architectural heritage. The method proposes a two-step assessment: Regarding the user—to assess whether the site is "usable" in sustainable fashion. Q.v. Section 4 and Table 3. With regard to the architectural heritage—for each area that meets the needs of the visitor, to assess whether, and to what

Table 6 The summary fact-sheet

Fields	Presence	Criterion	ID	Level	Description/notes
Spatial distribution	Yes/no	Morphological-dimensional			
		Perceptive-cultural			
		Material-constructive			
Services	Yes/no	Morphological-dimensional			
		Perceptive-cultural			
		Material-constructive			
Safety	Yes/no	Morphological-dimensional			
		Perceptive-cultural			
		Material-constructive			
Way-finding	Yes/no	Morphological-dimensional			
		Perceptive-cultural			
		Material-constructive			
Management	Yes/No	Morphological-dimensional			
		Perceptive-cultural			
		Material-constructive			
Economic efficiency	Yes/no	Morphological-dimensional			
		Perceptive-cultural			
		Material-constructive			
Comfort	Yes/no	Morphological-dimensional			
		Perceptive-cultural			
		Material-constructive			
Interaction	Yes/no	Morphological-dimensional			
		Perceptive-cultural			
		Material-constructive			

extent, the solutions respect or alter conservation of the site. Q.v. Section 5 and Tables 4 and 5.

The summary fact-sheet (Table 6) combines the two steps, providing an overall picture regarding the presence of aspects, elements and solutions geared towards accomplishing site accessibility and the degree of their integration. By entering in the summary fact-sheet a value among the five pre-defined levels of impact (Table 5), an overall value will be obtained with a weighted average that takes into account the significance of each criterion addressed in reaching the objective, i.e. sustainable use of the historical architectural heritage. This weight represents the element that will vary from case to case in function of the features emerging in the specific context.

7 Conclusions

The research carried out contributed to clarifying the fact that the sustainable use of architectural heritage, a precondition for lasting and effective conservation, can never be the fruit of pure chance; on the contrary, there is a need for a project oriented towards quality, opportunely backed up by the inclusion of the numerous aspects that characterise sustainable use in its economic, social and environmental dimensions. Among these aspects, the described methodology focuses on accessibility, a transversal theme for sustainable use, which concerns material and immaterial components of the heritage and which has been the object of numerous previous studies. The principal critical area highlighted thus far in accessibility to architectural heritage is the need to find the right balance between the demands of the visitor and the priority need for conservation.

These requirements link up with the research described here, which consists in applying a multi-criterial approach in the assessment of solutions for accessibility (observable in the actual state of affairs or in planning provisions) as a basis for a method that aims to comprise both analysis and decision-making.

It is foreseen that the research will continue into the near future, linking up with wider-ranging studies regarding technology in living spaces. The intention is to apply the methodology to several Sicilian case-studies, with the involvement of various categories of stakeholder and then broadening its scope to other aspects of sustainable use in architectural heritage.

Acknowledgements This paper has been edited under the coordination of M. L. Germanà. She has written sections nos. 1–2. E. Nicolini has written sections nos. 3–6. Elvira Nicolini is a triennial researcher funded by the Italian Ministry for the University (PON R&I 2014/20 AIM D.M. 407/2018).

References

Arenghi, A., & Pane, R. (2016). The addition in conservation project for the accessibility to cultural heritage, *Techne. Journal of Technology for Architecture and Environment, 12*, 57–64.

Arenghi, A., Garofalo, I., Sormoen, O. (2016). *Accessibility as a key enabling knowledge for enhancement of cuktural heritage*. FrancoAngeli Editore, Milano.

Castells, M. (1996 1st, 2010 2nd). *The rise of the network society*. Wiley-Blackwell, Hoboken.

CoE, Council of Europe. (1975). *The declaration of Amsterdam*. https://www.icomos.org/en/and/169-the-declaration-of-amsterdam.

CoE, Council of Europe. (1985). *Convention for the protection of the architectural heritage of Europe* (Granada Convention). https://www.coe.int/en/web/conventions/full-list/-/conventions/treaty/121.

CoE, Council of Europe. (2005). *Framework convention on the value of cultural heritage for society* (Faro Convention). https://www.coe.int/it/web/conventions/full-list/-/conventions/rms/0900001680083746CoE.

CSABA (Consiglio Superiore per le Antichità e Belle Arti). (1932). *Carta Italiana del restauro* [Italian Chart of architectural restoration]. http://ilportaledelrestauro.blogspot.com/2009/02/le-carte-del-restauro-carta-italiana.html.

De Medici, S., & Pinto, M. R. (2012). Public cultural heritage properties and reuse strategies. *Techne. Journal of Technology for Architecture and Environment, 3*, 140–147.

Della Torre, S. (2012). Renovation and post-intervention management. *Annales, Series Historia et Sociologia, 22*, 533–538. https://zdjp.si/wp-content/uploads/2015/10/ASHS_22-2012-2_Della-Torre.pdf.

Della Torre, S., Pracchi, V., & Treccani, G. P. (2007). Orizzonti di riferimento nel progetto di restauro. In A. Arenghi (Ed.), *Design for all. Progettare senza barriere architettoniche* (pp. 191–193). UTET, Torino.

Egloff, B. J. (2019). *Archaeological heritage conservation and management*, Archaeopress Publishing Ltd, Oxford. http://www.archaeopress.com.

EU Ministers of Culture. (2018). *Towards a European vision of high-quality Baukultur* (Davos Declaration). https://davos-declaration2018.ch/media/Context-document-en.pdf.

Germanà, M. L. (2014). Conoscenza, conservazione, valorizzazione: criticità, processi e approccio unitario. In S. Della Torre (Ed.), *Proceedings of Preventive and Planned Conservation International Conference* (pp. 21–31). Nardini, Milano. http://www.polo-mantova.polimi.it/fileadmin/files/Download/Volume1/Germana_M.L.__Conoscenza__conservazione__valorizzazione.pdf.

Germanà, M. L. (2015). The 'use' in the reliable interventions on the Mediterranean architectural heritage. In I. L. M. Palmero (Ed.), *Proceedings of ReUso 2015. II Congreso Internacional sobre Documentación, Conservación, y Reutilización del Patrimonio Arquitectónico* (pp. 150–157). Editorial Universitat Politècnica de València. https://riunet.upv.es/handle/10251/56060.

Germanà, M. L. (2019a) Technology and architectural heritage: Dynamic connections. In D. Hawkes et al., (eds.), *Conservation of architectural heritage. A Culmination of Selected Research Papers from the Second International Conference on Conservation of Architectural Heritage (CAH-2)* (pp. 77–92), Springer Nature Switzerland Advances in Science, Technology & Innovation (IEREK Interdisciplinary Series for Sustainable Development). Springer, Cham Print.

Germanà, M. L. (2019b). The user's perspective in architectural heritage. In M. Bisson M. (Ed.), *3rd International Conference on Environmental Design* (pp. 507–512). Palermo University Press, Palermo. https://iris.unipa.it/retrieve/handle/10447/379245/773126/MDA%202019%20Estratto%20ML%20Germana%20Users'%20prespective.pdf.

Germanà, M. L. (2021). Ethics, use, and inclusion in managing archaeological built heritage: The relationship between experts and visitors/users. https://doi.org/10.1007/978-3-319-51726-1_3496-1.

Gibson, E. J. (1982). *Working with the performance approach in building*. Report of Working Commission W060, CIB Publication 64. Rotterdam. www.irbnet.de/daten/iconda/CIB_DC23969.pdf.

ICOMOS. (2004). *International charters for conservation and restoration*. http://openarchive.icomos.org/431/1/Monuments_and_Sites_1_Charters.pdf.

ICOMOS. (2013). *The Australia ICOMOS charter for places of cultural significance* (Burra Charter). https://australia.icomos.org/publications/charters.

ICOMOS. (2014). *The florence declaration on heritage and landscape as human values*. https://www.icomos.org/images/DOCUMENTS/Secretariat/2015/GA_2014_results/GA2014_Symposium_FlorenceDeclaration_EN_final_20150318.pdf.

ICOMOS. (2017). *Salalah guidelines for the management of public archaeological sites* (adopted by the 19th ICOMOS General Assembly, New Delhi, India, 15 December 2017). http://icahm.icomos.org/wp-content/uploads/2018/07/GA2017_6-3-3_SalalahGuidelines_EN_adopted-15122017.pdf.

Mace, R. (1985). *Universal design, barrier free environments for everyone*. Designers West, Los Angeles.

MIBAC (Italian Ministery for Cultural Heritage). (2008). *Linee guida per il superamento delle barriere architettoniche nei luoghi di interesse culturale*. Gangemi Editore, Roma. https://www.beniculturali.it/mibac/multimedia/MiBAC/documents/1311244354128_plugin-LINEE_GUIDA_PER_IL_SUPERAMENTO_DELLE_BARRIERE_ARCHITETTONICHE.pdf.

MIBAC (Italian Ministery for Cultural Heritage). (2018). *Linee guida per la redazione del Piano di eliminazione delle barriere architettoniche (P.E.B.A)*. http://musei.beniculturali.it/notizie/notifiche/linee-guida-per-la-redazione-del-piano-di-eliminazione-delle-barriere-architettoniche-p-e-b-a.

Nicolini, E. (2016). *Metodi di Valutazione per la Conservazione Integrata, Recupero, Manutenzione e Gestione del Patrimonio Architettonico, Urbano ed Ambientale*, Ph.D. thesis, Università Federico II Napoli.

Picone, R. (2004). *Conservazione e accessibilità. Il superamento delle barriere architettoniche negli edifici e nei siti storici*. Arte tipografica, Napoli. Summary. http://www.federica.unina.it/architettura/laboratorio-di-restauro/superamento-barriere-architettoniche-edifici-siti-storici/.

Quagliuolo, M. (2003). Accessibility as acquisition of information from the cultural and environmental heritage: Physical, economic and cultural aspects. In M. Quagliolo (Ed.), *La gestione del patrimonio culturale. Accessibilità ai beni culturali e ambientali. Atti del VII Colloquio Internazionale* DRI, Roma. https://www.lettere.uniroma1.it/sites/default/files/2244/7_Accessibilita%20beni%20culturali.pdf.

Smith, L. (2006). *Uses of heritage*. London: Routledge.

Sposito, A., & Germanà, M. L. (2003). L'accessibilità nei siti archeologici. In M. Quagliuolo (Ed.), *La gestione del patrimonio culturale. Accessibilità ai beni culturali e ambientali. Atti del VII Colloquio Internazionale* (pp. 80–89). Roma, DRI. https://www.researchgate.net/publication/285589223_Laccessibilita_dei_siti_archeologici.

Story, M. F., Mueller, J. L., & Mace, R. (1998). *The universal design file: Designing for people of all ages and abilities*. Raleigh, North Carolina State. https://www.researchgate.net/publication/230853086_Universal_design_in_education_Principles_and_applications.

Tilden, F. (1977). *Interpreting our heritage*. Chapel Hill: The University of North Carolina Press. https://is.muni.cz/el/1421/podzim2017/MUI_338/Interpreting_Our_Heritage__Chapel_Hill_Books_pdf.

Tugas, et al., (Eds.), (2005). *Centri di Interpretazione del Patrimonio culturale. Il manuale Hicira.* Barcelona; Diputació de Barcelona. https://www.diba.cat/c/document_library/get_file?uuid=b52899e4-8b2f-466d-9838-9407b8fb6338&groupId=99058.

UNESCO. (2011). *Recommendation on the historic Urban landscape.* https://whc.unesco.org/uploads/activities/documents/activity-638-98.pdf.

UNESCO UNITWIN. (2018). *Tourism management at UNESCO world heritage sites*. Università della Svizzera italiana, Lugano. https://www.fun-mooc.fr/asset-v1:Paris1+16012+session01+-type@asset+block/MOOC_TMatUWHS_manual.pdf.

WHO (World Health Organization). (2001). *International classification of functioning, disability and health (ICF)*. Geneva. https://www.who.int/classifications/icf/en/.

Heritage Conservation and Community Development Through a Sustainable Management Approach: The Case of Abu Mena

Hossam Ragab Mahmoud Ahmed

Abstract

In the past, cultural heritage sites tended to be individual monuments and buildings that were considered "standalones." Today, the setting and surrounding landscape are also recognized as a part of the cultural heritage. Heritage conservation is the process devoted to preserving the cultural heritage for the future. Abu Mena is a cultural heritage site that used to be on the UNESCO World Heritage List. However, after severe deterioration in terms of its structural integrity, it was placed on the UNESCO Red List. The primary cause for its structural damage was the significant rise in the groundwater level caused by the flooding irrigation technique utilized by the farmers who transformed the region around Abu Mena into agricultural lands. The purpose of this study is to put forward some propositional action plans to be taken into account by heritage managers to satisfy both heritage state of conservation and the people's well-being and welfare. It also presents a look into the issue of conflict between the local people's interests and cultural heritage conservation, where a balance needs to be struck. The research is undertaken using literature review analysis and field visits and interviews with relevant parties in the Abu Mena area. The study concludes that there is a great potential in implementing Farmer Field Schools in the surroundings of Abu Mena Heritage site to secure the sustainable use of heritage while developing sustainable agricultural practices. Finally, the major conclusion that has emerged from the present study is that placing a certain heritage site on the UNESCO Red List could act as a catalyst to put more effort to protect the site and re-consider its management.

Keywords

Heritage conservation • Sustainability • Balance • Community • Development • Agriculture • Endangered heritage • Living heritage

1 Introduction

The World Heritage Resource Manual-Managing Cultural World Heritage stated that the range of what is regarded as heritage has been widened considerably over the last half-century. At the beginning of this manual, the new broad definition of heritage has been set and, thus, acted as a base for what comes after in terms of cultural heritage management. To clarify, heritage properties tended in the past to be individual monuments and buildings and were considered as a "standalone," with no particular relationship to their surrounding landscape. Today, the setting in which those monuments and buildings are located is also recognized as a part of the heritage as a whole, as the setting is affected by its interaction with humanity. Further, the World Heritage Convention recognizes that heritage can be defined as "monuments, group of buildings, and sites" as opposed to the singular definition previously subscribed to. In practice, a broad set of typologies has developed that includes: urban centers, archeological sites, industrial heritage, cultural landscapes, and heritage routes (UNESCO/ICCROM/ICOMOS/IUCN, 2013) As a result, the tasks required from a cultural heritage manager have not become limited to managing the individual heritage assets in terms of preserving their fabric and maintaining their values. Additional competencies and qualifications are also required; e.g., conserving the values imputed to the heritage by the people and understanding and working with the local development projects undertaken nearby. Site managers must now be able to assess the potential impacts on the

H. R. M. Ahmed (✉)
Helwan University (Egypt) and Brandenburg Technological University (Germany), Cairo, Egypt

A. Versaci et al. (eds.), *Conservation of Architectural Heritage*, Advances in Science, Technology & Innovation,
https://doi.org/10.1007/978-3-030-74482-3_18

heritage places and make sure that management plan they set does not hurt the well-being of community.

In 1979, Abu Mena was inscribed as a cultural world heritage site on the UNESCO World Heritage List. A few years later, in 1988, a large-scale irrigation project was put into operation as a part of a national land reclamation project in the surrounding region of Abu Mena. Consequently, a dramatic rise in the level of the groundwater table occurred. This rise led to severe deterioration of the heritage site. As a result of the flooding irrigation technique utilized by the farmers around the site, the World Heritage Committee, at its 25th Session (Helsinki, Finland), decided to place Abu Mena on the UNESCO Red List as an endangered property (UNESCO, 2013).

The agriculture lands around the site of Abu Mena were part of a large national land reclamation project initiated by former President Mohamed Hosni Mubarak in the 1980s. Moreover, agriculture and land reclamation have been key items on the agenda of Egypt's policy makers in recent history. However, this initiative negatively affects many Egyptian heritage sites. This conflict between heritage preservation and community development has been a growing issue in Egypt. There have been many attempts at rectifying this divide; however, these have only led to unsustainability and failure.

2 Methodology

The research process involved many procedures. To collect the data of this study and formulate the proposed approach, a desk-based research was carried out so as to review the existing body of literature. The importance of the research being of a multidisciplinary nature meant that a number of semi-structured interviews were conducted with the stakeholders, e.g., archeologists, water and sanitation engineers, agricultural engineers, and the local individuals. Self-guided field visits to the site were also made by the author to acquire the sense of the place as well as to acquire that indispensable understanding of the natural setting. All the gathered information was then funneled through a qualitative analysis to interpret the data and present solid findings. The results of the research are applied to a single-case study herein which was chosen for its special circumstances and features.

3 Literature Review

The World Heritage Convention recognizes that heritage can be defined as "monuments, group of buildings, and sites." In practice, a broad set of typologies has developed that include: urban centers, archeological sites, industrial heritage, cultural landscapes, and heritage routes (UNESCO/ICCROM/ICOMOS/IUCN, 2013). As a result, the tasks required from a cultural heritage manager have not become limited to managing the individual heritage assets in terms of preserving their fabric and maintaining their values. Additional competencies and qualifications are also required, e.g., conserving the values imputed to the heritage by the people and having a say in the local development projects undertaken nearby by assessing their potential impacts on the heritage places and making sure that they will be compatible with the well-being of the tangible heritage products.

Heritage conservation in general is the discipline devoted to the preservation of cultural heritage for the future (Poulios, 2014). As such, a conservation plan must seek the correct path for each individual case. Different approaches are used to determine which path is most suitable to the site such as material-based approach, value-based approach, or living heritage approach. In order to serve the purpose of this paper and to do the site justice, the living heritage approach is the most preferable. The continuity is the key aspect of this approach: Continuity of the function of the heritage site, continuity of the physical existence of the heritage site's community, and continuity of maintaining both the social and physical space of the site. This kind of heritage conservation approach is seen in the Great Meteoron Monastery, Greece (Fig. 1). This is an example of preventive conservation that educates people on how to behave at the heritage sites and sustains the continuity of both social and physical space of the site.

The World Heritage Convention emphasized the importance of linking cultural heritage with the life of the community by obligating all state parties to the convention to allow the cultural and natural heritage a sufficient role in the prosperity of people (UNESCO, 1972). This ethnological–anthropological relationship has been further acknowledged by UNESCO by adding the 5th "C" for the word

Fig. 1 Dual-language sign at the entrance of the Great Meteoron Monastery, Greece. *Source* Ubiquity Press Ltd. Gordon House 29 Gordon Square London WC1H 0PP, London

"Community" to the already-existing 4 Cs of the Budapest Declaration: Conservation, Credibility, Communication, and Capacity Building (UNESCO, 2002).

As a result, maintaining an adequate balance between heritage preservation, sustainability, and community development has been deemed as one of the strategic objectives of the Budapest Declaration, so that World Heritage properties can be protected congruent with the quality of life of the surrounding communities (UNESCO, 2002). It has been observed that the main motivation for listing cultural heritage properties on the UNESCO World Heritage List has changed a great deal over time to incorporate this sustainability of capacity building. Table 1 is a conclusion of analyzing the reasons behind which certain heritage properties have been inscribed on the World Heritage List. As seen, the concept of authenticity was understood in the 1980s in terms of design and material, while the later nominated heritage sites are considering the socioeconomic impact on their communities and using it as a main motivation for their inscription rather than the celebration of their Outstanding Universal Value (Rebanks, 2009).

UNESCO recommended that a general policy of water economy combined with the need to reclaim the desert in order to create a prosperous agricultural should be of supreme interest of Egyptian authorities. The preservation of archeological sites is also a part of this policy in a larger context that associates water management with all the initiatives aiming at economic development and environmental protection (Cleere, 2005).

There are approaches which consider agriculture not only as the main pillar of economic prosperity but also as an integral part of the cultural landscape management. These approaches look at the farmers as the main guardians of the landscape. Also, they (the farmers) are interpreters of the past, the biodiversity infrastructure, and recreational and tourism resources (Raap, 2017).

For example, Germany is a good representation of this beautiful interdependence between natural countryside and tangible cultural heritage (Fig. 2). Generally, 80% of the German territory is dominated by a rural character. Also, it is quite impossible to keep and sustain cultural heritage without taking into account the measures that sustain and keep agriculture. Additionally, it would be equally impossible to maintain agricultural practices without conserving and managing heritage. The local identity of Germany is well determined when such a balance is maintained (Raap, 2017). Similarly, and at the Egyptian level, agriculture is considered the most identifying feature of Egyptian society. One cannot help but be amazed by the stunning serenity that combines cultural heritage with agriculture in many Egyptian archeological landscapes (Fig. 3).

Additionally, what makes agriculture is vitally important in Egypt is that questions of food security have been raised due to the rapid population growth by an annual average rate

Table 1 Timeline depicting how the main motivation for listing cultural heritage property on the UNESCO list has changed over time

Best practice WHS case study	Date of inscription	Focus shift
Head-smashed-in buffalo jump	1981	These sites were not initially focused on WHS-driven socioeconomic impacts. They have, instead, reinvented themselves since the mid- 1990s, learning the lessons of other destinations
Canadian rocky mountains	1984	
Roman frontiers/Hadrian's Wall	1987	
Town of bamberg	1993	Sites like these were the innovators in starting to give WHS status a socioeconomic focus and experimenting with how this could add value to existing initiatives
Collegiate church, castle and old town of quedlinburg	1994	
Volklingen ironworks	1994	
The old and new towns of Edinburgh	1995	
Laponia area	1996	Sites like these represent the emerging minority of WHSs that have from the start of the process had a clear socioeconomic motive and a new perspective on using natural or cultural heritage as a key economic driver
Portovenere, Cinque Terre and the Islands (Palmaria, Tino, and Tinetto)	1997	
Blaenavon Industrial Landscape	2000	
Dorset and East Devon Coast 'Jurassic Coast'	2001	
Derwent Valley Mills	2001	
Vegaoyan—the Vega Archipelago	2004	
Bordeaux, port of the moon	2007	

Source World Heritage Status Is there opportunity for economic gain? Research and Analysis of the Socio Economic Impact Potential of UNESCO World Heritage Site Status, n.d, P. 19

Fig. 2 Green castle of Storkow surrounded by greenery, Brandenburg, Germany

Fig. 3 Temple of Kom Ombo surrounded by agriculture Kom Ombo, Aswan, Egypt

of 2%. In addition to the population growth, Egypt consists of more than 95% desert, which makes the arable land less than 450 m^2 per person. This share increases significantly in developed countries which reaches 4000 m^2 per person (Seada, 2016). Because of the River Nile, Egypt has great potential in agriculture and land reclamation. The country's dependence on food imports has been one of the biggest in the world (some 70% of the wheat is imported) which is why reclamation has been an issue at the forefront of Egyptian policy making (RUF, 1993). In 2009, the Egyptian Ministry of Agriculture announced a land reclamation goal to reclaim an additional 3 million feddans by 2030. However, that plan never came to fruition due to the political and economic unrest in the years that followed.

Therefore, sustainable agriculture has been recognized as the main factor of the Green Egyptian economy. Egypt is at a crossroad while the main challenges it faces according to the global competitiveness report in 2015 are "Green Economy," "Human Development," and "Innovation" (Seada, 2016).

4 The Case Study: Abu Mena, Egypt

Abu Mena is an ancient Christian city that dates back to the 4th Century and is referred to as the city of St. *Menas* in *Maryut* (Frankfurter, 1998). The city of Abu Mena (Fig. 18.4) lies between *Wadi El-Natron* and the city of Alexandria itself (Cleere, 2005). The huge city composes the church, the baptistery, basilicas, public buildings, streets, monasteries, houses, and many workshops. This city was built over the tomb of the martyr *Menas*, who was an officer in the army of Diocletian. He did not follow the instructions to kill the Christians and he converted to Christianity and was subsequently tortured and beheaded. The tomb itself was lost until miraculous healings were reported in the area resulting in its discovery. Then, the site was developed around his final resting place becoming a significant pilgrimage destination for Christians worldwide (Frankfurter, 1998).

Fig. 4 Map illustrates the location of Abu Mena. *Source* ICOMOS (2009) the joint ICOMOS- World Heritage Centre Reactive Monitoring Mission to ABU MENA (Egypt), P. 10

Abu Mena was inscribed on the WHL in 1979 as a cultural heritage property under criteria IV. The justification went on to state that *"No other place in Egypt has such a vast and fine collection of marble sculpture in general of the highest quality manufactured in Alexandrian workshops. In the same time the early Christian architecture in Abu Mena, even in its ruined state, gives an excellent idea of the architecture of that time ~ which unfortunately is* [now] *covered completely by the modern town* [of Alexandria]." (UNESCO, 2013).

"It is the unique artistic character, of Abu Mena as the place where truly Egyptian architecture of the Christian period meets with the architecture of Europe and Asia Minor that makes it outstanding" (UNESCO, 2013).

A few years after the inscription, particularly in 1988, a national land reclamation project in the surrounding region of Abu Mena was put into operation. The main crop cultivated by the local people was sugar beet which required a massive amount of water. As a result, the irrigation techniques utilized there was flooding technique. Consequently, a dramatic rise in the level of the groundwater table was observed. This rise led to severe deterioration to the world heritage site. As a result of this negative effect of the flooding irrigation technique, the World Heritage Committee, at its 25th Session (Helsinki, Finland), decided to place Abu Mena on the UNESCO Red List as an endangered property (UNESCO, 2013).

In the case of Abu Mena, the key stakeholders include the monastic community and the Egyptian Youth. The former owns a considerable land share, and they are working on their cultivation to produce a food supply through their sale points all over Egypt (Ghoneim, 2017). In the case of the latter, these lands were given to the Egyptian Youth during the 1980s as compensation for not being able to provide them with a job. The lands were given to the youth for a monthly fee of EGP 400 (Ghoneim, 2017). Some people were keen to settle and to work on cultivating them, while others preferred to sell their share to others who take over the responsibility of cultivating these lands.

Based on the site visit and the interviews with the relevant stakeholders, the community around Abu Mena is a rural community in the first place with agriculture as the fundamental occupation of the people. There is an intimate relationship between people as they all have common experiences, purposes, customs, and traditions form the basis of unity in the village. It is observed that the community is characterized by simplicity. There is a mini store providing basic groceries. There is also a clinic that is powered by an imported workforce of doctors from Cairo. There are some vehicles that link the villages around the heritage site of Abu Mena with each other. On the road leading to the heritage site of Abu Mena, there is a big sign, upon which "the sugar beet association" is written, an indicator to the agricultural lands without any indication to the heritage place (Fig. 5).

The SWOT Analysis for Abu Mena Heritage Site

Strengths: The land surrounding Abu Mena is fertile, and it affords the locals a great opportunity to make a living from agriculture. This feature strengthens the economic value of the heritage site of Abu Mena.

Weaknesses: The Lack of Site Management: It is quite clear that there is no management plan set for the heritage site of Abu Mena. There is no adequate management plan set for any Egyptian cultural heritage site, as the heritage management is quite newly introduced in Egypt (ElShazly, 2017). In the case of Abu Mena, neither archeological survey nor conservation survey exists. Also, on the level of conservation, there is no formalized conservation recording plan at the property (UNESCO, 2013).

Opportunities: The projects of cultural heritage conservation would enhance the inhabitants' rights, social cohesion, mutual understanding, as well as civic participation. Such projects would, too, raise public awareness on issues like discrimination, inter-ethnical violence (Facchinetti, 2014). Throughout Egyptian history, taking care of the religious buildings has been always a guaranteeing policy applied by many rulers to get close to the Egyptians and to gain their trust. Hence, the conservation of Abu Mena will potentially act as a catalyst for the social cohesion and will restore the social marginalization of the Coptic society in Egypt, especially at the present time, where it is observed that tension raised in the Egyptian society due to the terrorist attacks that targeted the monasteries and churches. For instance, the restoration of the Rab'a Mosque by the Egyptian army in 2014 has remarkably strengthened the sense of appreciation by the locals to the military leaders.

Threats: The heritage site of Abu Mena is of great significance especially to the Coptic community in Egypt. Therefore, the deterioration (or the loss) of this cultural heritage will result in the loss of the cultural identity and will harm the sense of national pride of the Coptic minority. From this point, heritage conservators are playing a key role in restoring, not only the cultural heritage fabric but also the cultural identity. They, thus, help the politicians reach their ultimate goal of enhancing the social cohesion and avoiding violence.

The Ground Water Problem in Abu Mena

Some people argue that the agricultural activities in the region of Abu Mena are not sustainable, as these activities depend on flooding irrigation. In light of the serious water shortage in Egypt, irrigation techniques, and therefore, the agricultural practices are not sustainable (Sonbol, 2017). Consequently, and as emphasized by the UNESCO-ICOMOS monitoring mission to Abu Mena 2005, the supreme interest of the Egyptian authority should be to adopt a policy that ensures water economy, agricultural practices, and heritage preservation. This suggested policy is considered the most critical action needed (Cleere, 2005).

The issue of the ground water has been tackled by the Egyptian authorities through some unsustainable dewatering projects. This has been the most critical solution for Abu Mena as (Soliman, 2017) stated. The dewatering projects using the pumps have proved to be unsuccessful before because of that the lack of maintenance to these pumps after they were installed (Mikhail, 2017). Also, (Sonbol, 2017) mentioned that the dewatering method is not sustainable and might do more harm than good.

Earlier, a dewatering project has been developed by the ministry of culture with active cooperation with the Ministry of Agriculture and the Ministry of Water Resources and Irrigation. The prime goal of this project was draining the groundwater by means of ditches to be drawn to centralized tanks. Then, the water in these centralized tanks will be raised by pumps to be delivered again to the main canal that provides the agricultural lands with intensive flooding for irrigation (Cleere, 2005).

It is worth mentioning that the project was well planned, as it was conducted in the light of some key information

Fig. 5 Signs erected next to one another on the way around Abu Mena. *Source* Author (2017)

such as the local topography and the elevation. The total cost was measured to be 2.3 M€ and was to be fully funded by the Egyptian government, all in all taking 3 years to fully take form (Cleere, 2005). However, the project was not implemented. Instead, an alternative ill-conceived project was conducted. This project was about drilling pumping wells inside and around the archeological site of Abu Mena. This action has depended on vertical drainage by erecting vertical pumping wells. These pumping wells are equipped with subsurface pipes that take the water along an open-air canal which conveys the water to a weir, then to the Mediterranean Sea (ICOMOS, 2009, p. 200). This project was implemented after accurate observations from different experts that the vertical drainage will bring significant results rather than the horizontal drainage. In the end, this project did more harm than good in the following ways:

- First, there was no hydrological survey before implementing the project, which resulted in the explosion of the water reservoir while installing one of the pumps which caused more damage to the site.
- Second, the lack of maintaining and conserving the pumping wells after they have been erected has resulted in the damage of them all after a short period of time (Mikhail, 2017).

Despite the fact that this dewatering method has proved unsustainable, a similar project of dewatering using the pumping wells is already underway. A group of heritage experts is involved. What distinguishes this project is that it is based on a reliable scientific base. Many institutions with different backgrounds are involved such as the National Authority for Remote Sensing and Space Sciences (NARSS) and the Faculty of Science, Helwan University. This project is implemented under the supervision of the Ministry of Antiquities, and it is expected to bring rewarding results (Soliman, 2017).

5 The Proposed Approach for a Sustainable Balanced Management

Generally speaking, the purpose of placing a certain property on the Danger List is to draw attention to protecting the site and implementing a conservation-based management approach rather than blame or punishment (ICOMOS, 2009). There are some sites that have been on the Danger List for far less time and could be considered "successes" for the Committee actions. For example, Tipasa (4 years), Hampi and Dubrovnik (7 years), Timbuktu (11 years), and Bahla Fort (16 years). Kathmandu Valley was on the In-Danger List for only 4 years (2003–2007) before (ICOMOS, 2009). Furthermore, Cologne Cathedral has been listed as an endangered heritage site in 2004 due to the very same reason Abu Mena was listed; local development which should have had a negative impact on the heritage site. However, Cologne Cathedral succeeded to set an example of a successful heritage management, so that it did not remain on the Red list for more than 2 years after managing the potential community development projects (in that case: sky scrapers) in a way that does not harm the visual integrity of the heritage site (Zacharias, 2009).

The management proposed here is primarily focusing on the state maintaining the cultural heritage of Abu Mena while not neglecting or stopping the surrounding community's agricultural practices.

The Vision Statement for the proposed heritage management approach is that Abu Mena is a site of great cultural and religious importance. This importance is appreciated by the Egyptian authority, so that the site is well equipped to attract a considerable number of tourists on both local and international levels. The management model of Abu Mena is a rewarding one and is not only limited to managing the tangible archeological remains but also is exceeded to cover the management of the local activities undertaken nearby the site (particularly, the Agriculture) within a comprehensive modern heritage management scheme. The management plan is based on a participatory basis. Abu Mena is, thus, setting a management model for many Egyptian and regional heritage sites and landscapes where a conflict raises between heritage conservation and community development.

Action Plans

Before presenting the proposed action plans, the following part displays the Behaviour Change Wheel that illustrates how people's behavior could be changed. This is very important at this phase of the research since the action plans that will follow are tackling the complicated behavior of the farmers in the region around Abu Mena and how those behavior patterns could be changed in a way that secures sustainable heritage use and agricultural practices alike. The Behaviour Change Wheel is a new method created by Michie et al. (2011) to understand the process needed to change a certain pattern of people's behavior.

According to the Behaviour Change Wheel (Fig. 6), it is suggested that every behavior is driven by a number of factors as the red area shows. Importantly, and to change this behavior, there is a set of intervention functions and policy categories that need to be considered. Putting this into the context of this particular research, the target behavior is "the unsustainable agricultural practices and using flooding irrigation." Following is a couple of action plans that focus on the intervention functions as a tool to change the farmer's behavior. The intervention function is understood here as raising awareness, delivering tailored training and education

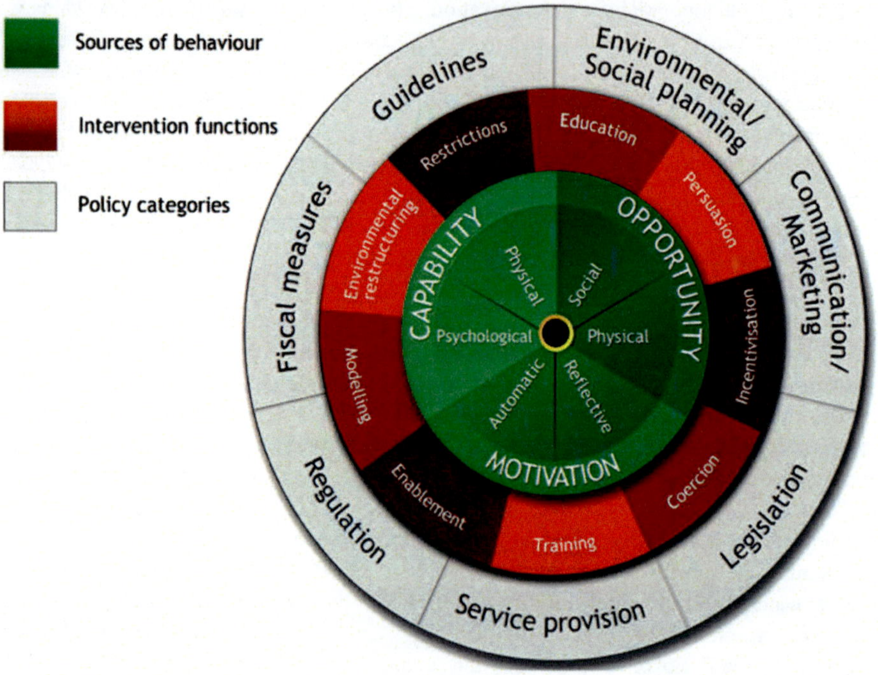

Fig. 6 Behaviour Change Wheel: a new method for characterizing and designing behavior change interventions implementation science. *Source* http://www.behaviourchangewheel.com/

to the local communities persuading them to give up on the flooding techniques that are unsustainable.

Since the dewatering projects in Abu Mena have proven their failure because of the lack of their maintenance as shown above. The action plans proposed below represent more sustainable solutions. They are addressing the agricultural practices and irrigation adopted by the local community around Abu Mena.

5.1 Farmer Field Schools

The sugar beet is currently the most common crop cultivated around Abu Mena (Cleere, 2005). Other crops are also found there such as tomatoes (Ghoneim, 2017). Both types of crops are of a magnificent economic value for the locals, but they require so much water to be properly irrigated that the groundwater is threatening the heritage site. So, trying to solve the problem by adopting dry crops is an indispensable solution here. This can be done by the so called Farmer Field School (Fig. 7).

Generally speaking, the Farmer Field School (FFS) is defined by FAO (Food and Agriculture Organization of the United Nations) as a group of 20–25 farmers meet once a week in a local field setting and under the guidance of a trained facilitator. In groups of five, they observe and compare two plots over the course of an entire cropping season. One plot follows local conventional methods, while the other is used to experiment with what could be considered "best practices." In this field-based setting, farmers are able to investigate a wide range of topics, such as management of soil fertility and water resources (FAO, 2020).

The only negative point about implementing the dry crops is that they might not have the significant economic value as the sugar beet (Soliman, 2017). Therefore, it is suggested that the dry crops are to be cultivated at least within the boundary of the heritage site and in the buffer zone (after they are correctly set) (Cleere, 2005). However, deciding on the type of dry crops is a big issue that has not been tackled in this paper. So further research is needed to determine which dry crops will be the most suitable and will satisfy the needs of the local people. Furthermore, strong communication and convincing skills are needed to put forward a valid economic argument to the farmers (Gouda, 2017). It is suggested that the responsibility of this action plan will be given to the Ministry of Agriculture and Land Reclamation with collaborative input from international institutions and agencies already located in Egypt and might be keen to put this action plan into operation. Among such institutions is the *Deutsche Geselschaft für Internationale Zusammenarbeit* (German Development Agency), which is referred to as GIZ.

Fortunately, there are 22 FFSs implemented successfully in Egypt as of winter 2015/16. The Egyptian Ministry of Agriculture and Land Reclamation and GIZ are currently undertaking a project that aims at institutionalizing this

Fig. 7 Farmer Field Schools in Egypt. *Source* GIZ (2017)

group-based learning process as a standard approach of the extension services. The name of this project is The Egyptian–German Water Management Reform Programme (WMRP) Also, The International Fund of Agricultural Development (IFAD) adopted the same concept in Upper Egypt with 36 FFSs which has increased the chances and the likelihood of a broad-based impact (GIZ, 2017).

Last, the following statement has been said by Mabrouka Mohamed, an Egyptian female farmer, attended a FFS in Egypt: "I have been going to the Farmer Field School every Tuesday during the past two years. The benefit was great. I learned about farming, how to cost-efficiently run my household. The FFS introduced us to improved irrigation, and taught us about saving water. Now we save a lot of water that was wasted before. The water is now contained in the pipes and irrigation has become much easier." (The Egyptian German Water Management Reform Program, 2019).

5.2 Changing the Irrigation Techniques

While talking about more sustainable solutions, changing the irrigation techniques can not be neglected. It is seen that the irrigation methodology used by the farmers around Abu Mena is the flooding. The UNESCO-ICOMOS monitoring mission to Abu Mena in 2005 stated that the groundwater is the most pressing problem affecting the site (Cleere, 2005). Some people argue that the agricultural activities in the region of Abu Mena are not sustainable, as these activities depend on the flooding irrigation. In light of the serious water shortage in Egypt, irrigation techniques, and therefore, the agricultural practices are not going to last anyway (Sonbol, 2017).

Generally speaking, KfW (the German State-owned Development Bank) sees that the best way to manage water shortage threat more efficiently and sustainably is to improve the irrigation system. (KfW, 2015). There are also other challenges, e.g., climate change, rising temperature, and rates of evaporation, increased salinity of the groundwater and surface waters brought up by rising sea levels. All those challenges should alert us to fix the problem of the scare freshwater resources in Egypt before it is too late. (KfW, 2015) Consequently, and as emphasized by the 2005 UNESCO-ICOMOS monitoring mission to Abu Mena, the supreme interest of the Egyptian authority should be to adopt a new policy that ensures water economy, agricultural practices, and heritage preservation (Cleere, 2005).

The responsibility of applying this action plan will be given to the Ministry of Irrigation and Water Resources, while the funding issues are yet to be decided. However, the research argues the Egyptian Government already drafted the first National Water Resources Plan in 2005 which required implementation of Integrated Water Resource Management (IWRM). It identifies risks and areas where change is needed and defines the foundation for coordinated, more efficient, and sustainable water management. This project was commissioned by the Federal Ministry of Economic Cooperation and Development (BMZ).

If this action is adequately implemented, the results will be satisfying to the main two stakeholders: the heritage experts and the farmers. Economically, this action will save a considerable amount of water that might be used by future generations. Socially, using this type of irrigation will set a sustainable management example for the heritage managers and farmers all over Egypt. It will also save the right of the water for future generations. Environmentally, this action will encourage other land reclamation projects to be implemented in other areas of Egypt.

6 Conclusion

Agriculture is a tool for economic growth in Egypt, and in fact it is considered the most identifying feature of Egyptian society. It also could have an important role in heritage conservation, as there are approaches which consider agriculture not only the main pillar of economic prosperity but also as an integral part of cultural landscape management. Hence, Egypt has great potentials in agriculture and land reclamation due to the River Nile and the desert land which represents 95% of Egypt's total area. However, Abu Mena is a cultural heritage site that used to be on the UNESCO World Heritage List, surrounded by a large-scale irrigation project. Thus, it was placed on the red list, due to its deterioration caused by the negative effect of the flooding irrigation techniques utilized by the farmers around the Abu Mena Site. The research revealed that there is no adequate management plan set for Abu Mena cultural heritage site, as the heritage management is quite newly introduced in Egypt. In particular, there is a need to start putting a management plan into operation for Abu Mena and all Egyptian heritage sites. Furthermore, the management plan should place the activities undertaken by the local people at the heart of the heritage management system. Unfortunately, the issue of the groundwater in Abu Mena cultural heritage site has been tackled by the Egyptian authorities through some unsustainable dewatering projects, due to the lack of maintenance of dewatering pumps after installing. Accordingly, more sustainable solutions need to be tackled.

There is an opportunity in the conservation of Abu Mena Cultural heritage site, as it will potentially act as a catalyst for social cohesion and local development. Moreover, there is a massive potential in implementing Farmer Field Schools in the surroundings of Abu Mena Heritage Site, to secure the sustainable use of heritage on one hand, and sustainable agricultural practices on the other hand. Especially after having so many successful examples conducted all over Egypt. We must emphasize that cultural heritage management in Abu Mena should consider the well-being and prosperity of local communities as much as the well-being of the heritage site. This could be realized through striking a balance that is to be determined by a multidisciplinary team where every single stakeholder will have an input to reach the common good. Finally, this study illustrated that placing a certain heritage site on the UNESCO Red List does not necessarily lead to a negative outcome, as it could act as a catalyst to put more effort to protect the heritage site while delivering local development needs through re-considering its management.

References

Cleere, M. B. (2005). *The UNESCO-ICOMOS monitoring mission to ABU MENA (Egypt)*. UNESCO.

ElShazly, Y. (09 de January de 2017). *Former head of the international relationships*. Ministry of Antiquities . (H. Ragab, Entrevistador).

FAO. (02 de May de 2020). *Integrated production and pest management programme in Africa*. Fonte: fao.org: http://www.fao.org/agriculture/ippm/programme/ffs-approach/en/.

Facchinetti, S. (2014). *Cultural heritage management in Myanamr: A gateway to sustainable development*. Myanamr, European Institute for Asian Studies.

Frankfurter, D. (1998). *Pilgrimage and holy space in late antique Egypt*. Brill Academic Publishers.

Ghoneim, R. (03 de April de 2017). Officer, ministry of agriculture (H. Ragab, Entrevistador).

GIZ. (2017). *Farmer field schools*. Cairo: Deutsche Gesellschaft für Internationale Zusammenarbeit (GIZ) GmbH.

Gouda, N. (18 de November de 2017). Water and sanitation engineer (H. Ragab, Entrevistador).

ICOMOS. (2009). *World Heritage in Danger Compendium II*. ICOMOS.

KfW. (2015). *Integrated water resource management (IWRM)*. Frankfurt: Federal Ministry for Economic Cooperation and Development.

Khalil, M. (Diretor). (2019). *The Egyptian German water management reform program* [Filme Cinematográfico]. Egypt.

Mikhail, N. (12 de April de 2017). Central administration for the archaeological deocumentation, Ministry of Antiquities. (H. Ragab, Entrevistador).

Poulios, I. (2014). *The past in the present a living heritage approach—Meteora, Greece*. London: Ubiquity Press.

Raap, E. (2017). *Farming the historic landscape*. Brussels: Government of Flanders Department of Culture, Youth and Media Cultural Heritage Division.

Rebanks, J. (2009). *World heritage status: Is there opportunity for economic gain? Research and analysis of the socio-economic impact potential of UNESCO World Heritage Site status*. Lake Ditrict World Heritage Project.

RUF, T. (1993). The history of agricultural development. In G. M. Craig (Ed.), *The agriculture of Egypt* (p. 207). Oxford: Center of Agricultural Strategy, University of Reading.

Seada, T. (2016). *The Future of agriculture in Egypt*. Heliopolis University for Sustainable Development.

Soliman, M. (22 de April de 2017). Sector of Islamic and coptic antiquities. Ministry of Antiquities. (H. Ragab, Entrevistador).

Sonbol, G. (29 de June de 2017). Head of the administration of antiques restoration. Ministry of Antiques . (H. Ragab, Entrevistador).

UNESCO. (1972). *Convention concerning the protection of the world cultural and natural heritage*. UNESCO.

UNESCO. (2002). *Budapest declaration on world heritage*. UNESCO.

UNESCO. (2013). Report on the joint world heritage center/ICOMOS reactive monitoring mission to the world heritage site of Abu Mena, Egypt. Paris: UNESCO.

UNESCO/ICCROM/ICOMOS/IUCN. (2013). *Managing cultural world heritage*. Paris: The Federal Office of Culture FOC, Swiss Confederation.

UNESCO. (n.d.). *Rice Terraces of the Philippine Cordilleras*. Acesso em 20 de June de 2020, disponível em https://whc.unesco.org/en/list/722/.

Zacharias, D. (2009). *Cologne cathedral versus skyscrapers—World cultural heritage protection as archetype of a multilevel system*. Cologne: Max Planck Yearbook of United Nations Law Online 10.1.

From Conservation to Revitalization. Ambiguous and Critical Conditions in the Heritage Conservation Experiences of Erice

Giuseppe Abbate, Maria Luisa Germanà, Raffaele Savarese, and Ferdinando Trapani

Abstract

The plans and the evolutionary trends of policies and practices implemented in Erice *old town*, in the south of Italy, in the western part of Sicily, can be considered a case study of *total conservation* as highly representative of Italian urban planning of the last 60 years. This paper aims to deal with the actual problems and the perspective of sustainable development of the historical centre of Erice, by adopting the approach of urban and territorial regeneration. This paper also highlights the historical and cultural context in which the Gubbio Charter (issued in 1960) is affirmed; i.e. the most appropriate approaches to historical urban centres; what we can learn from Erice's integral conservation experience thanks to and/or without the direct influence of plans and other updated kinds of policy; some reflections regarding the rationale of differentiation in the urban/territorial functions and the actual asset and town management issues. In the final part of the paper, the authors propose a scenario in which the future of historical centres will no longer be based only on the quality of the safeguarding methods of restoration and recovery, nor on the ability of the architects/experts involved. The evidence of Erice's recovery process, intended as an example of the protection and restoration models applied in Italy and in the foreign school of architectural restoration, shows that the social, economic, environmental and cultural aspects of the regeneration of the historical centres are intertwined and depend on the developmental conditions of the entire city. The experience of the complete recovery of the historical centre of Erice, where the restoration and accessibility of the buildings and the historical urban fabric was perfectly successful, did not, however, resolve the trend towards abandonment. The latest experiments to enhance the historical centre in a completely fresh and creative way, with the improvement of the accessibility system, could form the basis of a new urban and territorial role for the ancient part of the city. Without an integrated approach, it will be very difficult to regenerate historical centres; for this reason, it is indispensable to leverage on the interconnections between cultural built heritage and the entire urban context. This article is the result of common work among all authors. In particular, Ferdinando Trapani edited paragraphs n. 1, 2, 3, 6, 7, 8 and 10; Raffaele Savarese edited paragraph n. 4; Giuseppe Abbate edited paragraph n. 5, and Maria Luisa Germanà edited paragraph n. 9.

Keywords

Cultural heritage safeguarding • Historical centres • Urban planning • Cultural heritage enhancement • Urban regeneration

1 The Approaches to Historical Centre Heritage

Cultural heritage is linked to people and reflects their national identity and cultural civilization. The safeguarding and ways of enhancement/sustainable use and reuse are provided for by international laws, through international conventions and protocols drafted in this regard. This paper is focused on the integral conservation of the cultural heritage (ICCH) approach with regard to historical centres.

G. Abbate · M. L. Germanà · F. Trapani (✉)
Dipartimento Di Architettura, Università Di Palermo, Palermo, Italy
e-mail: ferdinando.trapani@unipa.it

G. Abbate
e-mail: giuseppe.abbate@unipa.it

M. L. Germanà
e-mail: marialuisa.germana@unipa.it

R. Savarese
Circolo Culturale Istrice, Palermo, Italy
e-mail: architetto.raffaele@libero.it

A. Versaci et al. (eds.), *Conservation of Architectural Heritage*, Advances in Science, Technology & Innovation,
https://doi.org/10.1007/978-3-030-74482-3_19

There are many legal mechanisms dealing with the protection of cultural heritage buildings and sites to protect this built heritage (Monti & Cerroni, 2019).

According to Rodwell (2003, 2006) and Niglio (2012), with the exception of the WHL, UNESCO has focused on the concept of heritage, from protection to enhancement (Mucci, 1997) and processes of selection for historic centres; here, we list some of the most important documents regarding cultural heritage protection.

Having as our objective a clear interpretation of the relationship between old and modern urban elements and their functions, we need to highlight the principal sources of knowledge and practices to be applied.

It is important to mention the contribution of Patrick Geddes, as a cultural starting point for the actual problems of protecting, conserving and presenting not only the historical monuments and h architecture, but of the entire historical fabric and, above all, the historical cities and the local residents as a whole. According to Geddes, with his conservative surgery (Tyrwhitt, 1947), historical centres are the best places to observe directly the rules of civic science, rather than reading economic-social surveys or going to libraries (Geddes, 1915).

Regarding the Italian approach and its international references to the interpretation of the role of the historic centre against a background of contemporary lifestyle and modern architecture, there is a long history of contributions characterized by uncertainty and extensive ambiguity, but open to future developments (Cutolo & Pace, 2016).

Other more specific theoretical approaches regarding the safeguarding of the heritage come from Camillo Sitte (1889) and others, but we are obliged to consider, in the practical sense, the extensive theoretical-operative production of Gustavo Giovannoni (Neri, 2001), who explored the figure of the 'total architect' as the best professional figure to guide and carry out the protection and reactivation of the historical parts of cities. Giovannoni provided the first theoretical approach to cultural heritage management, based on an equilibrated interaction between the old architectural styles and modern functions, with the 'diradamento' (thinning out) as opposed to the gutting of the historical urban fabric (Giovannoni, 1931). Giovannoni was the first to conceive of the relationship between the town and country and the new role of transportation in linking up lives in the community in historical centres, as part of the entire urban context, and taking into account the external environment.

He was the leader of the national commission for the protection of the natural environment in Italy (1939). Unfortunately, the theories of Giovannoni and his aesthetic approach to the old architectonic styles were largely falsified, with the justifying evidence of extensive destruction of a huge part of the historical heritage.

Examples of the poor application of the theories of Giovannoni are the master plans for Caltanissetta (1943), Trapani (1944, 1950) and Palermo (1955), in which Edoardo Caracciolo was an important player. The outcome was probably affected by his deep feeling for the great suffering of the populations after the Second World War and the ensuing economic and social deprivation (Trapani, 2014).

In December 1960, the journal Urbanistica published the Proceedings of the Conference on the Safeguarding and Rehabilitation of historical-artistic centres, held in Gubbio on 17-8-19 September of that same year. The conference, promoted by several architects, urban planners, jurists, restoration scholars and several municipalities (Ascoli Piceno, Bergamo, Erice, Ferrara, Genoa, Gubbio, Perugia and Venice), produced a final declaration, called 'Carta di Gubbio', a declaration of principles on the preservation and rehabilitation of historic centres:

- nderlining the importance of specific local conditions;
- declaring priority to proceed with a survey and classification of historical centres and areas to be protected and restored;
- deeming it necessary for protected historical centre areas to be included in the master plans, since their safeguard must be considered a precondition for development of the modern city.
- Edoardo Caracciolo was one of the producers of the Gubbio Charter and this fact influenced his choice in designing the master plan for Erice.
- the following approaches to conservation action must be prevented/prohibited:
- rebuilding of old architecture and historical buildings to be disrupted (lit. 'come era-dove era'—'where it was, as it was');
- adding other styles to existing ones;
- camouflage;
- demolition of modest buildings;
- thinning out of the urban fabric by isolation of historical buildings from the historical urban fabric.

According to Giovanni Astengo, the term safeguard regarding the whole historical town must be recognized not only as passive defence but, above all, from the operational point of view of rehabilitation and restoration.

The following approaches to conservation action must be implemented:

- consolidation of the essential structures of buildings;
- elimination of the recent utilitarian superstructures harmful to the environment and hygiene;
- rebuilding of real estate units to provide functional and hygienic dwellings, with adequate sanitary facilities, or

other destinations for economic or public activities, or small-scale equipment compatible with the environment, whilst preserving rooms and internal elements to which the historical–critical investigation has assigned value.
- redevelopment, wherever possible, of open spaces for gardening and orchards;
- the establishment of intangible and non-building obligations.

The Gubbio Conference Committee became, in the 'Associazione nazionale centri storico-artistici (ANCSA)' National Association for Historical-Artistic Centres.

The ANCSA aims:

- To promote multidisciplinary studies and research for the safeguard and restoration of historical cities;
- To collect and coordinate the results of studies and research carried out in different locations (public bodies, universities, etc.);
- To promote meeting and debating initiatives, as well as experimental interventions;
- To provide advice for the public and/or private agencies involved in terms of critical reflections and technical support;
- To experiment and promote appropriate legislative activities and regulatory measures at the general level;
- To promote and carry out any other activities considered relevant from the social perspective.

The Gubbio Conference offered the opportunity to take stock of the situation in the historic centres, the regulatory insufficiency of town plans, urban planning technique, and simultaneously compare the concrete experiences of certain plans and the related approach to the theme of intervention in historical centres.

Probably one of the first examples of the gutting-plan interventions for the historical centres is the Projecte de Reforma y Ensanche de Barcelona by the engineer Ildefonso Cerdà y Suñer, with plans for several new ways of gutting the entire urban fabric of *old towns* (Ingrosso, 2006: 28).

According to Lowenthal (1996), heritage is a very problematic issue, considering the ways of interpreting its role in actual human life; since heritage can be understood as a myth to be feared or something to destroy, we have to remember 'that heritage is not fixed but changes in response to our own needs and is integral to our creative involvement with history' (Lowenthal, 1996: 250, reprint 2009).

The main thing attracting visitors or local people is the sense of space; this feeling burrows deeply into one's mind and creates roots of imagination throughout one's lifetime in a sort of topophilia (Bachelard, 1957) and characterizes historical centres with a specific kind of atmosphere for one's personal imagination.

Regarding the case studies from the south of Italy, where a huge part of the WHL UNESCO national heritage is concentrated, it is very important to adopt specific guidelines for analysis and strategies to enhance the cultural heritage and historical centres in particular (Abbate, 2002, 2004, 2012, 2014; Cannarozzo, 2010, 2016).

It is very important to promote strategies to stop the phenomenon of gentrification and heterotopia, which are frequently the outcome of policies focused merely on the protection of the stones/materiality, without any attention to the fluidity of the living memory in the concrete urban facts (Houdek, 2014: 58). Michel De Certeau's notion of 'pedestrian rhetorics' (1980) permits us to understand the local methods applied to town construction and reconstruction from the ground up; this is completely different from the traditional urban planning approach of analysing and recovering the existing and ancient city (De Certeau, 1980).

'I can conclude by saying that (...)' it is important 'to recognize that the critical spirit to which our culture refers, guiding a new feeling that requires the preservation of the building tradition of historical cities; it is enhanced and justified as an active participation in human progress by the ability of historical centres to represent the fundamental cornerstones of living phenomena inspiring new urban forms for our civilization'. At the same Gubbio Convention, 1960, Edoardo Caracciolo, inspired by Giovanni Astengo, emerged as one of the main players with his presentation of the Erice master plan.

Today, thanks to the theories and best practice or failures, it seems that we have the real possibility of knowing precisely what it is right to do and what not. But it is not so simple in all cases. It is very difficult to explain why in Italy, where we have had theoretical and practical declarations regarding protection and conservation since the nineteenth century, extensive historical heritage is now in danger and for several reasons (ANCSA-CRESME, 2017). One reason is the complexity of the decisions to be taken at all levels: from the single material or immaterial piece of art to the entire *old town* and its cultural landscape, all of which has to be protected and enhanced in a specific way. The theoretical approach of the Boito Charter focused on a gradual process of intervention from a very low degree of transformation to the maximum degree of substitution, only when there was the danger of memory-loss of the cultural heritage. In spite of the theories, rules and technicalities, it is not easy to resolve questions that are not only limited to the sphere of action of architectural restoration, but to the town planning field as a whole.

Protection and the enhancement of the cultural heritage have to be based on economic, environmental and social analyses and forecasts. This is very important for correct construction and a sharing of the action plan of interventions. In the case of the historic heritage in Italy, it is possible to propose certain actions to address and resolve problems of abandon, as often happens in the historical centres in the poorer regions of southern Italy, or super concentration in the cultural tourism capitals of Venice or Florence (ANCSA-CRESME, 2017: 9):

- reinforcing the safeguard of heritage within the framework of Regional town-planning laws;
- initiating policies to encourage residency;
- regulating tourist utilization of local assets;
- managing and planning tourist flows on occasions of excessive pressure;
- facilitating control of commercial activities;
- promoting the development of compatible creative and productive activities;
- promoting the use of advanced management technology;
- drawing up risk-prevention and crisis-response plans;
- providing systematic tools for understanding the dynamics of change taking place in historical centres.

Particular technology, such as Information and Communication Technologies (ICT), is necessary in order to develop, in coordination with a new data-analytics management-unit, a digital infrastructure for the promotion and marketing of what is on offer; there needs to be updated data about supply/demand, for the main destinations and products of local tourism, in order to help coordinate promotional activity at the Regional/local level and contribute to a sustainable new system for the booking of tourist services and products (such as museums, concerts), through connection to other booking systems.

Actually it seems that the problem might lie in the interpretation of the transitional phase from abandon to actually attracting the public and/or from exploitation to managing concentrations of people; the success or failure of the protection and enhancement plan depends on the levels of development and culture at the management/organizational level, and entrepreneurial innovation of the regions in which the historical centres are located.

Following this line of thinking, we might consider the contribution of the European Community, as focused on an integrated vision for future policies, using several different strategies, in the economic, social, environmental and cultural sense, in which the declarations regarding cultural diversity are located.[1]

[1] EU 2007/C 306/01, Art. 2, par. 3: https://eur-lex.europa.eu/LexUriServ/LexUriServ.do?uri=OJ:C:2007:306:FULL:EN:PDF.

2 Research Questions

In this paper, bearing in mind the introduction regarding the international debate and the final target of the findings from the Erice experience of integral conservation of the heritage (ICH), we propose:

- In the case of the historical centre conservation, can integral conservation of the ICH be developed correctly, respecting acknowledged and scientifically recognized models of intervention, or is there a risk of losing cultural resources?
- What happens to a historical centre when it loses most of its inhabitants?
- Do the ICH produce positive results in terms of social (equity-inequality) impact?
- Do the ICH produce positive results in terms of economic and financial impact?
- Do the ICH produce positive results in terms of environmental impact?
- Do the ICH produce positive results in terms of cultural impact?

The methodology described in this paper is based on consideration of the historical data regarding the study case, the players involved (the technical office of the Municipality of Erice) and the progress of the actions emerging in the social, economic and cultural contexts (Figs. 1 and 2).

3 Ancient Erice

We have no data regarding the origins of Erice (Blasetti Fantauzzi & De Vincenzo, 2012), but on the basis of texts by Thucydides, Erice was linked to the civilization of the Elymians in the fifth-century B.C. At that time, the Temple of Aphrodite was very famous, and, according to Polybius, was the richest in ancient Sicily.

After a long period of decline, which lasted from late antiquity to the high Middle Ages, when much of the remains of the sanctuary were lost, a small church dedicated to Santa Maria della Neve was built in the area, perhaps in conjunction with the construction of the castle by the Normans (XI–XII sec.) (Fig. 3). The ancient myth of Erice, based on fertility, was related to the Goddess of Love/Fertility, but the Normans reconstructed the site mainly as a military fortress for territorial control over the entire western part of Sicily (Maurici, 1992).

Fig. 1 Map of Sicily with the location of Erice

Fig. 2 Erice seen from above

In more modern times, the area around the castle was further tampered with, starting from the construction of the current access ramp (in the sixteenth century), which replaced the ancient drawbridge, filling up the moat that divided the fortified lower part (known as the castle of 'Balio') from the main body of the fortress.

Further restoration and interventions were carried out in the nineteenth century; lastly, archaeological excavations

Fig. 3 Castle of Venus on a pre-existing Punic site

Table 1 Resident population in Erice on 1881–2011

Town (part)	Site	1881	1961	1971	1981	1991	2001	2011
Monte San Giuliano	Top	1,770	1,575	1,138	942	651	512	–
Ballata	Hill	291	1,500	1,121	975	867	533	–
Casa Santa	Flat	–	3,682	18,482	22,157	26,425	25,149	–
Erice (total)		2,860	18,021	21,979	25,275	29,420	29,338	27,970

Source Master plan (Municipality of Erice)

were carried out by the Soprintendenza di Trapani (institutional public authority for the protection of cultural heritage and environmental resources).

From the original population levels of the historical centre (1,770), the population has now fallen to 1,000 inhabitants (Table 1).

The only centre of excellence in this part of the region is represented by the Majorana Scientific Centre. The Ettore Majorana Foundation and International Centre for Scientific Culture was founded 50 years ago by the outstanding Italian physicist, professor Antonino Zichichi (Fig. 4). Every year the Centre of Erice hosts several workshops covering different scientific disciplines (e.g., physics, medicine, chemistry, life sciences etc.) for hundreds of participating scientists. Since its foundation, this centre has represented an important meeting place for scientists from all over the world. Embracing 127 schools, covering all branches of science, the centre is situated in the old, pre-mediaeval city of Erice where three restored monasteries provide an appropriate setting for high-level intellectual undertakings.

As regards the culture of planning in Sicily, it is important to recall Edoardo Detti, the famous Italian academic and urban planner, who founded the International School of Environmental Sciences at the Ettore Majorana Centre for Scientific Culture in Erice from 1972 to 1978. He established and directed the school, fostering studies on urban design, urban and regional planning, landscape architecture, but above all the protection and enhancement of the cultural and environmental heritage, with 'conservation' as the main theme. The interdisciplinary groups of undergraduates and graduates endeavoured to tackle and examine in depth the problems related to the urban situation in the area of Erice and Sicily, producing texts, memoirs, talks at conferences and articles in newspapers.

4 Notes on the Historical Evolution of the Territory of Erice

Urban and territorial evolution is determined by the context of natural transformations that are independent of and dependent on human action. The common opinion or widespread judgment about the perception of a landscape is limited to recent cultural facts. Instead, we need to work on an in-depth territorial analysis, and for this purpose, it is necessary to take into account the local and international factors in which the case study is inserted.

Fig. 4 Ettore Majorana centre for scientific culture

Two premises: the kingdom of Sicily, emerging from the dismemberment of the large Norman/Swabian kingdom (Abulafia, 1988; Collenuccio, 1929; Kantorowicz, 1927), was based around the administrative division between states and feudal cities.[2] Moreover, the state-owned cities had a reference territory, of greater or lesser size, also containing feuds or states (i.e. groupings of feuds) under their jurisdiction and administered by the citizens (Emanuele Gaetani, 1754).

The countryside was an ambivalent, productive and cultural entity suitable not only for producing food that might be scarce in the cities, given that the presence of urban gardens is documented at least until the sixteenth century (Bresc, 1972), but also for food production (and not) for regional and international trade.

Until the fifteenth century, extensive international trade and the value of the feuds seem to have been connected not only to arable land, sheep farming, the production of cheese and wool, but above all to the production activities related to forestry: timber, game, rabbit hair, herds of cows with the production of *caciocavallo* cheese, breeding of war horses, pigs and salted meat.

Wheat represented a kind of trade of great political-strategic interest and was therefore subject to strict administrative control (despite smuggling) and is the best known by historians[3] (Trasselli, 1982a; Peri, 1952–53, 1978). The cultivation of sugar cane along all the coastal areas of Sicily and the destruction of forests for the use of wood as fuel had a great importance, especially from the 14th to the eighteenth centuries, i.e. with considerable geographical and time differences. The period of decline had no clear explanation beyond what is indicated by Trasselli (1982a). Tax documentation is generally poor, both as regards evasion and more or less official exemptions. In this period, sugar represents a kind of great luxury reserved for the wealthy classes.

The power of Erice (city-state) until the sixteenth century was such as to be able to impose on the feudal lord of the State of Baida (Ventimiglia-del Bosco), resident in Trapani, the obligation of six-monthly residence in the fief. This eventually gave rise to the construction of the present castle, through exercise of the right to 'ranteria'[4] (Internicola, 2003; Savarese, 2018). A short time before, a landing place had been erected at Guidaloca (Aymard, 1993), west of Castellammare, as a base geared towards the cutting down of the existing woods in the neighbouring mountain range (M.te Sparacio), all to the advantage of the sugar industry (Trasselli, 1982b; Giuffrida, 2012), but with the resulting destruction of the forest economy. Following the abolition of feudal law (1812) and the seizure of the fiefdoms, including those of the peasants (communia) by the ruling classes, Bourbon administrative reform intervened, and, taking note of the decline of Erice (Mount San Giuliano), subjected it to the jurisdiction of Trapani (Carvini, 1670/1680; Arancio, 1844).

In other words, Erice lost its surrounding possessions, a vast territory, which at first benefitted Trapani and Castellammare and, subsequently, gave rise to several autonomous municipalities, with villages created for the benefit of the navy, industry and agriculture. Bourbon reformism had accelerated the agrarian evolution of the territory with the

[2] The kingdom of Sicily included, at its height, through different legal norms, central and southern Italy including Abruzzi and Molise, part of North Africa (Maghreb), the Kingdom of Jerusalem, etc.

[3] But it should be remembered that, beyond the myth, once Carthaginian power had been destroyed, Tunisia became the granary of the Roman Empire, along with the Maghreb areas further north of southern Sicily.

[4] The right to 'ranteria' allows the feudal lord or the dominant city, to take the animals deemed 'missing' and impose on the owner, if necessary, the payment of a fine for the charges for maintenance and surveillance. With the collapse of the state/ administrative entity it is easy to see the transformation into rustling.

planting of specialized crops suitable for the regional and international market and encouraging rural settlement. The transportation of produce substantially took place by sea, favouring the small landing places along the coast and, above all, Castellammare. It is no coincidence that the first drivable road through farmland was opened in 1800 and only in 1850 was it continued, in the opposite direction, towards Trapani. Then, the Bourbons, via reform taking into account the on-going epochal changes, transferred territorial control to Trapani (probably also under the influence of the English protectorate), administratively establishing the progressive decline of Erice, which was deprived of a part of its territory (Emanuele Gaetani, 1754). These considerations are supported by the progressive increase in the population distributed throughout the small rural and/or coastal settlements, such as Bonagia, which, at the end of the 1700s, was exporting tuna to Venice (Savarese, 2018) whilst there was a decrease in the population of the town perched 800 m above.

The relentless decline of Erice, without any surrounding agricultural hinterland, and threatened by the economic and industrial development of Trapani, led, after the war, not only to the establishment of autonomous municipalities,[5] but to the crystallization of its historical centre.

After the First World War, with the arrival of fascism (1922), the administrative system of the Provinces was defined through the urban and architectural establishment of the centres of power. Country people were forced to live in rural towns, a little known phenomenon in Sicily, where, except for particular instances, serfdom had never existed. In the areas around Trapani and Marsala, especially with the intensive development of vineyards (1800s), scattered houses and small villages sprung up, which also gave rise to new settlements, all to the detriment of Erice.

So, in this case too, there was no urbanization and Erice continued its long decline a decline that would safeguard its historic centre.

5 Caracciolo's Approach to Historical Heritage: A Change of Direction with the Plan for Erice

Starting from a selection of master plans and reconstruction plans drawn up by Edoardo Caracciolo between 1943 and 1955 and focusing on plans for the relevant historical centres, we might argue that the proposed design solutions do not appear to be very respectful of the historic urban environment as a whole, for hygienic-sanitary reasons, to which Caracciolo always paid special attention (Bonafede, 1997; Trombino, 2000). It is no coincidence that in his essay published in 1954 (Caracciolo, 1954b), entitled 'Il problema dell'abitabilità e del risanamento' (The problem of habitability and sanitation) Caracciolo still focused on the concepts of habitability, ventilation, sunlight, lighting and sizing, stating that greater hygienic well-being and a decrease in the mortality-rate constituted fundamental characteristics of today's social environment and urbanism. This approach impels Caracciolo to propose or endorse demolition and gutting, with the opening of new roads in the plans drawn up for Caltanissetta, Marsala, Trapani and Palermo. Fortunately, not all these plans were carried out.

In the case of the PRG of Caltanissetta (1943), Caracciolo believed that in the historical centre there were no sanitary shortcomings that might demand radical urban intervention. In the master plan report, Caracciolo wrote: 'It is a well-known fact that the so-called demolitions are only grievously necessary when there are specially infected housing units, which in Caltanissetta there are not; therefore the hygienic justification for such operations collapses' (Trombino, 2003).

Despite these statements, the master plan for the historic centre also included some thinning-out and replacement of buildings in order to improve the road network. In particular, the master plan for the Angeli medieval district stipulated: the enlargement of via S. Domenico with the consequent alignment of buildings on only one of the two fronts; the demolition of entire blocks to create, in the resulting space, a scenic green staircase traversing the whole neighbourhood. On the other hand, further, extensive gutting was planned in the eighteenth-century S. Rocco district, in order to create a large new urban axis between Corso Umberto I° and Viale Testasecca, which constitutes the extension of Via Cavour (Fig. 5). However, these plans were not realized.

In the case of the reconstruction plan for Marsala (1946),[6] Caracciolo proposed, for the historic centre and the adjacent port area, where the largest number of destroyed buildings were concentrated, a series of thinning-out operations and the widening of some roads. The proposed planning solutions for the ancient city were only partially correlated with the frame of reference of war damage and dictated rather by reasons of health and hygiene. In general, the planned thinning out did not envisage redevelopment with new structures; in this way, Caracciolo rejected a model of intervention that had been widely applied in other cases, in which the reconstruction plans were transformed into extraordinary tools of land enhancement. The reconstruction plan, although formally in force for many years, was completely disregarded, with the exception of a number of road

[5] The municipalities of Valderice, Custonaci, Buseto Palizzolo and San Vito Lo Capo became autonomous in the postwar period.

[6] The Marsala reconstruction plan was commissioned for Caracciolo in 1945, at the same time as the task of drawing up the master plan.

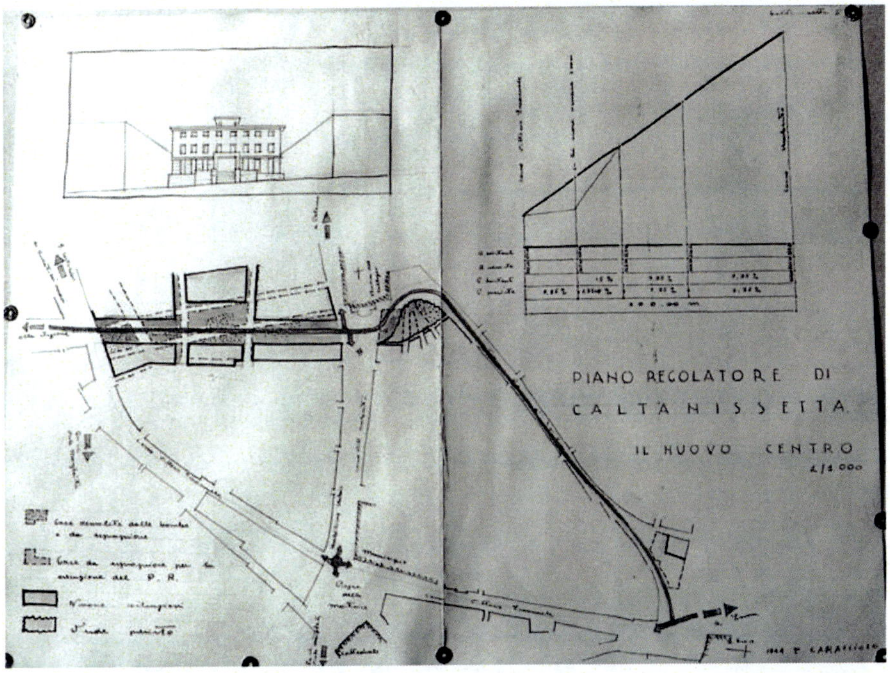

Fig. 5 Master plan for Caltanissetta. Drawing by Caracciolo

rectifications and the laying-out of squares financed by the Ministry of Public Works.

In the case of the Reconstruction Plan for Trapani (1950),[7] wide-scale urban renovation for hygienic and sanitary purposes was envisaged for the restoration of the old neighbourhood of San Pietro, of medieval origin, adjacent to the port area and seriously damaged during the war; a straight road of nineteenth-century inspiration was superimposed on this historical medieval fabric.

The master plan projections were fulfilled with the opening of a new urban axis (Corso Italia) that cuts across the entire district from one side to another (Fig. 6). Impressive multi-storey buildings[8] were built along this new axis. However, the construction of Corso Italia irreparably disfigured the historical centre, leaving unresolved (if not exacerbating) the problems of the insalubrious areas situated behind the new buildings (Fig. 7).

The enhancement of the waterfront, another item in the master plan, entailed the substitution of historical buildings with a sequence of modern multi-storey buildings endowed with porticos, completely out of character with the surrounding buildings.

In the case of the Palermo city development plan (1956–1962), the plans for the historical centre were contained in the so-called Piano Particolareggiato di Risanamento dei Quattro Mandamenti (detailed recovery plan of the four main districts). Regarding the complex history of the drafting of the Palermo city master plan, Caracciolo was appointed member of the Drafting Committee with six university professors, experts in urban planning. In this case, therefore, Caracciolo was not the only one making decisions, but, with regard to the planning choices for the historical centre, he never actually distanced himself. The plan for reorganization envisaged the historical centre of Palermo being crossed by new roads in north–south and east–west directions, with the consequent transformation of the urban structure and the substitution of existing buildings. Some buildings of historical and architectural value were to be safeguarded, but it was mainly several historical facades and some architectural and decorative elements that the plan intended to keep, even embedding them in the new buildings. In this case too, the disemboweling was not carried out.

In the case of the Erice Master Plan (1957), as compared to previous plans, there is no doubt that Caracciolo demonstrated a changed approach—certainly more mature—with regard to the issue of intervention in historical centres; he put into practice what had already been described in 1954 in the essay 'The problem of habitability and rehabilitation' in *Tre lezioni di Urbanistica* where he wrote 'The thinning out obtained through the demolition of historic buildings—with the construction of roads with high buildings on the margins—are to be considered, today, technical and administrative errors' (Caracciolo, 1954a).

[7] Also in the case of Trapani, as in that of Palermo, the approval of the law on reconstruction plans interrupted the formation process of the master plan.

[8] In the indications the new Corso Italia would have been the extension of Via Virgilio. However, the section connecting Via Virgilio was never built.

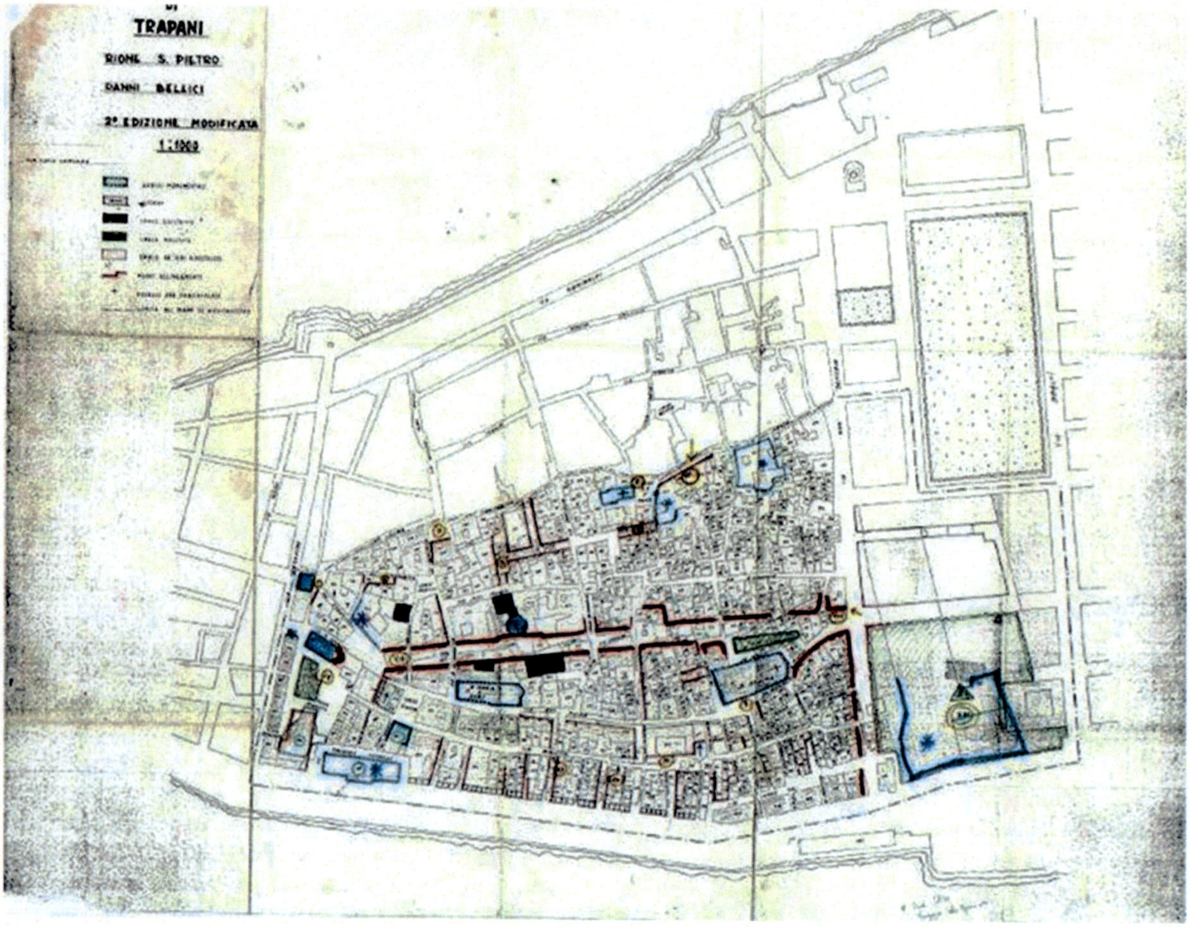

Fig. 6 Reconstruction plan for the S. Pietro neighbourhood (Trapani)

Caracciolo's interest for Erice was already evident in 1934, on the occasion of his second degree, in architecture, with a thesis relating to a project for a hotel included in the 'Urban planning of Mount San Giuliano'. Luciana Natoli Di Cristina defined this interest as 'a lasting love of a lifetime' (Natoli Di Cristina, 1964), and it continued over the years, as demonstrated by his participation in the competition for the drafting of the Erice Master Plan in 1938, for which Caracciolo obtained the first prize; there were also numerous writings on Erice relating to its very long urban history and a close analysis of the built environment (Fig. 8), with particular attention to the typological and morphological characteristics of the historical building heritage (Caracciolo, 1938, 1939, 1954a, 1956) (Figs. 9 and 10).

The master plan for Erice, from the methodological point of view, derived directly from the application of the principles (based on integral conservation) of the Gubbio Charter and was considered merely a sound point of departure for an integrated action plan based on economic, social, environmental and, not only, cultural interventions.

In particular, in the case of Erice, contrary to what the same urban planner envisaged in plans for other Sicilian cities, the integral conservation consisted of the following planning choices:

1. Maintenance of overall urban morphology;
2. Maintaining the relationship between public spaces, semi-private (internal courtyards) and private dwellings;
3. Survey of stratifications of architectural styles and typological evolution with consequent protection for the legibility of the historical changes that have occurred;
4. Allowing small interventions inside the housing units to adapt them to contemporary lifestyles without altering the external configuration that would have an impact on the urban landscape (Fig. 11).

Fig. 7 Corso Italia (Trapani)

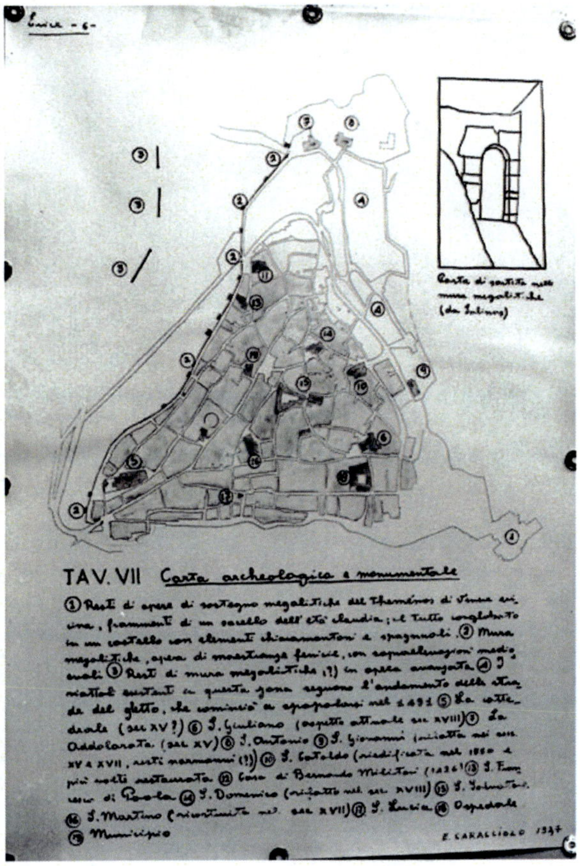

Fig. 8 Erice: archaeological and monumental interpretation drawing by Caracciolo

Furthermore, the Erice master plan seems to find an important reference in Giovanni Astengo's plan for Assisi (Astengo, 1958), which already in those same those years constituted an exemplary case of how to deal with the issue of conservation of the existing city.

In the same way as in the plan for Assisi, the construction of a new expansion district is envisaged for Erice, at a certain distance from the existing city, in order to maintain, as far as possible, the perfection of the surrounding landscape. This residential area was to be built following the contours of the land around the existing sports field. (Figs. 12 and 13).

Lastly, the plan also provides for: the elimination of vehicular traffic from the historical centre, identifying areas outside the city walls to be used for parking; the recovery of the building heritage of the entire historical centre; the conservation of priceless scenery, also via the reforestation of Mount Erice (Caracciolo, 1960).

The city awarded Caracciolo honorary citizenship and, in 1962, the Associzione Nazionale Centri Storici Artistici (ANCSA), after his death, awarded Caracciolo the gold medal in memory of his exceptional merits achieved in the field of conservative restoration in historical centres. In the same year, the ANCSA instituted the Caracciolo Prize, to be awarded both to a study of significant methodological value and to those public administrations that distinguished themselves in having promoted and guided studies and urban recovery interventions.

The master plan for Erice drawn up by Edoardo Caracciolo and Francesco Puletto was never approved by the public authorities involved, but has constituted a fundamental experience in the methodological evolution of recovery interventions in historical centres.

6 Erice Today

The somewhat drastic fall in the population of Erice mirrors what has happened in other towns throughout southern Italy.

Since the end of the fifteenth century, a large part of the population, initially amounting to about 6000 inhabitants, has abandoned the city of Erice and descended into the valley; only 1000 inhabitants remain in Erice.

Since the second half of the nineteenth century, with the construction of two new local connecting main roads and, subsequently, also the cable car, Erice has emerged from its ancestral isolation.

Fig. 9 Erice: examples of typical courtyards

From the post-war period onwards, Erice became an international tourist destination, the pressure on private individuals to build second homes increased dramatically, with the risk of widespread illegal construction.

Considering only its historical centre, Erice has lost a lot of its inhabitants: from 1,138 in 1971 to 512 in 2001.

The new Erice master plan presents a forecast of population increase.

The projected demographic increase regarding 2011 was of more than 32,000 inhabitants, whereas the national ISTAT Survey suggests more than 28,000. The hypothetical new population will prefer to live in the modern and coastal parts. However, the historical centre of Erice could remain deserted in the future because the master plan is not sufficient to attract new people and/or convince them to stay (Tables 2 and 3).

It might be very important to know where local people may actually want to live.

7 New Plans

A survey of the plans drawn up after Caracciolo's revealed that three new plans had been drawn up, as follows.

The detailed historical centre plan, developed by Matteo Tusa and his group, was adopted by the municipality in 1991 and rejected in 1997. This plan contained all the detailed representation of the real estate units and their historical, typological assets in accordance with the integral approach used by Edoardo Caracciolo. All the principal buildings were devoted to a public function, to empowering and fostering tourism and cultural sectors in general.

The detailed historical centre plan (1988–20002) developed by Bruno Gabrielli (1932–2015), from the University of Genova, incorporated the analysis of the above-mentioned plan and formulated several proposals for reinforcing the public services in the *old town*.

Fig. 10 Erice: images of the urban landscape

Fig. 11 Erice: an image of the urban landscape

Fig. 12 Abandoned sports field around which the new residential complex (holiday homes) was planned

The municipality's new master plan is the new tool for the future growth of this important small town: 4,810 new inhabitants in the next ten years and more than 8,000 in the next twenty. It is very difficult to justify these forecasts objectively bearing in mind the evidence from the recent negative trends of population growth.

8 New Approach to the Conservation Through the Revitalization

In recent years, the municipality and several cultural and tourism operators in Erice have tried to curb the demographic and economic decline, above all by impelling private investment towards a form of cultural tourism that respects the whole more than individual instances. The municipal administration has strengthened and qualified environmental services with a green planning design strategy and artistic creativity in order to render the livability of the *old town* more user friendly and less marginal.

The most important initiative by the private sector is the MEMS, Erice Montagna del Signore Museum (lit. Museum of the mountain of the Lord), which has 'a museum system that proposes the wide-scale enhancement of ecclesiastical cultural assets, both mobile and immovable, scattered throughout the urban area of Erice. It develops along an itinerary that involves the whole medieval town and currently, the proposed tour circuit includes eleven sites, with an integrated route that presents contextual knowledge of the peculiar urban structure of Erice and visits to the main ecclesiastical monuments'. The aims of the project are the conservation and exploitation of ecclesiastical cultural heritage. The MEMS project has religious tourism objectives as constituting living testimony to the Gospel, in order to nourish the spirituality of the Christian faithful; it is divided into two areas of intervention: a. the protection, conservation, restoration and enhancement of the architectural and artistic heritage of the churches of Erice; b. safekeeping and longer opening hours for the churches themselves, so as to enable the faithful and general visitors to deepen their knowledge. The promoting bodies of the MEMS project are: (a) Ecclesiastical entities (Diocese of Trapani, Parish of Maria SS. Assunta, Parrocchia San Cataldo Vescovo, Ente Chiesa SS. Salvatore); (b) associations (La Montagna del Signore, EriceLab) and co-operatives (Santa Maria della Grazia, Fe.Ar.T. Erice, Meeting Point).

The most important proposal by the Erice Municipality for enhancing its ancient town is a new wellbeing park within the framework of green planning regarding the areas surrounding the historic walls, with a blueprint for the tourist-cultural and environmental itineraries of the Elim Walls and the restoration of the ex-arrival station of the Cable Way with additional cultural facilities. In the project presented to the local people and the stakeholders for the participation of Erice in BIAS 2020 (Biennale Internazionale Arte Contemporanea Sacra e delle Religioni dell'Umanità), there were several potential initiatives regarding art, theatre, dance, cinema and performances by several music groups (Compagnia Moto Armonico Danza, Collettivo Mandala, Ettore Majorana Foundation and Centre for Scientific Culture, Associazione Italiana Agricoltura Biologica, Istituto alberghiero Ignazio e Vincenzo Florio, Club Alpino Italiano, Il Club del Libro).

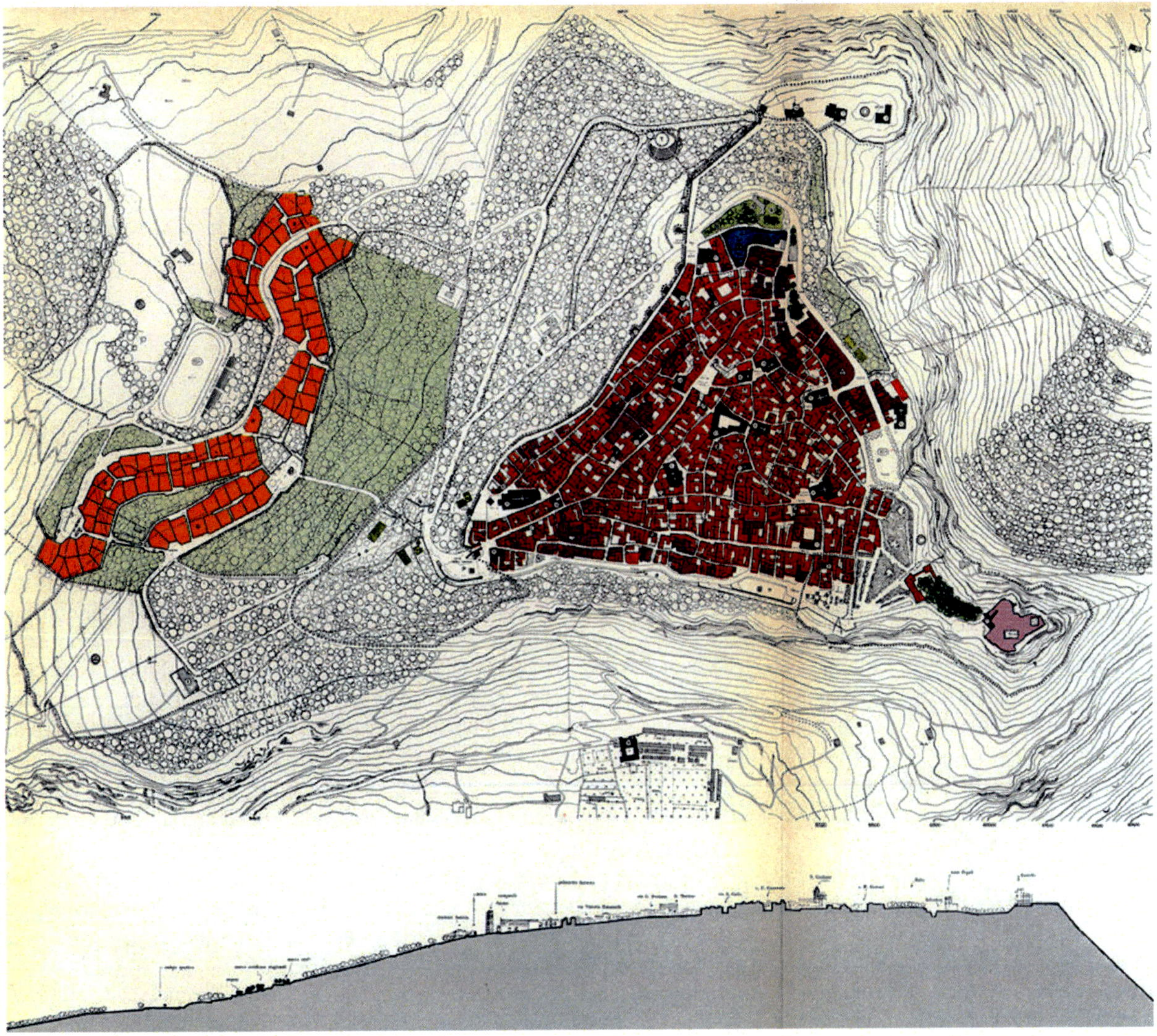

Fig. 13 Master plan for Erice

Table 2 Resident population in Erice in 1881–2011

	1971	1981	1991	2001
'Erice Vetta' population	1,138	942	651	512
variation				
1981–1971	–	17%		
1991–1981			-30.89%	
2001–1991				-21.35%

Source Masterplan (Municipality of Erice)

The project itemized several creative activities that integrated green planning with strategies for health and well-being. Walking and dancing in the open air or in forests can bring several benefits to one's organism, as these contexts present ideal conditions to prevent and cure non-communicable chronic diseases (Barp & Bolla, 2005; Edwards & Tsouros, 2006):

- Breathing the forest air brings benefits in combatting pulmonary and vascular disease;

Table 3 Demographic projection for Erice in 1961–2031

Master plan population increase	10 years	4,810
	20 years	8,488
Population by ISTAT survey	1961	18,021
	2001	29,338
Projected demographic increase by master plan	2011	32,167
	2018	34,148
	2021	34,997
	2025	36,128
	2031	37,826

Source Official website of the Erice Municipality

- Moving through the woods helps prevent and combat diabetes problems;
- Trees have a therapeutic power on the psyche;
- Living the woods as anti-stress therapy: i.e. being elsewhere.

The green planning project for the Erice Forest represents the first public initiative in reviving the *old town,* starting from February 2020.

The research will monitor the on-going process of sharing and people's feelings over a certain period of time.

9 Accessibility in the Conservation and Revitalization Processes: Erice as a Generalizable Case

The identity of the historical centre of Erice is characterized by the specific site, already inhabited for millennia. This *old town* dominates the surrounding area from a hilltop at about 800 m above sea level. This peculiar position, which can be considered the main reason for actually first settling here and a factor in the medieval urban fabric, also influences the current conditions of this case-study, making it appropriate to focus on accessibility, an issue that can be generalized to many other old Italian hill-towns (e.g., Taormina on the eastern coast of Sicily).

Accessibility has been balanced with other factors in the built environment, the importance of which does not depend on origins, historical features or the cultural value, because quality, both on the building- and the urban-scale, should be an objective concerning human settlements, regardless of the specific features. For this reason, the need for a holistic approach has been outlined, in order to embrace the entire contemporary built environment as a complex phenomenon, with a broader framework of management, oriented towards feasible and long-term-oriented strategies (Germanà, 2013). The importance of accessibility to the urban settlement has been related to 'access', one of the five 'dimensions of performance' of the *good city* form proposed by Kevin Lynch, referring to 'the ability to reach other persons, activities, resources, services, information or places, including the quantity and diversity of the elements which can be reached' (Lynch, 1984:118).

The need for accessibility arose in the second half of the twentieth century, following the sensitization for social inclusion of disabled people, and becoming a pillar of the *Universal Design* or *Design for All* (EIDD, 2004). This explains why the ancient built environment is rather inaccessible in the vast majority of cases. Elements such as stairs, differences in height, steep sloping paths, narrow and winding streets and uneven road surfacing make it difficult for elderly or disabled persons to enjoy the ancient built environment (Fig. 14). Regarding public spaces in historical centres, the public and collective dimension of the fruition increases the importance of accessibility, as it concerns two categories of users: local inhabitants (for whom this requisite assumes—above all—the moral value of inclusion) and general visitors (for whom it also implies the positive economic effects of cultural tourism or added value for recreational activities).

Conservation obviously demands accessibility in an ancient town with the characteristics of Erice; in fact, 'many constraints arise as a consequence of the intrinsic inaccessibility of the ancient built heritage and the need for conservation poses cultural challenges, as well as those of design and technique' (Germanà et al., 2020). However, a partial solution could be sought, at least for the main itineraries, by creating circumscribed paving easier for walking or accessible by wheelchair, as demonstrated in other historical cities (Conti & Petriccione, 2018). So far, the only measure to facilitate the fruition of Erice by disabled people has been the purchase of an electric vehicle by the municipality (Vultaggio, 2019).

Apart from these general difficulties as a historical centre, Erice—as a hill-town—has the additional problem of accessibility, due to the winding and steep access road, from the plain below and especially from the larger town of Trapani. In 2005, a new cableway replaced the previous one, which had been inaugurated at the end of the 1950s and was

Fig. 14 Porta Trapani, the gateway into the old town of Erice, is quite near the arrival station of the cabway

Fig. 15 View of Trapani from the cableway

no longer functional. The cable car allows you to reach the historical centre of Erice in 10 min from the outskirts of Trapani, whilst enjoying magnificent scenery (Fig. 15). For this reason, the Erice cableway has become a tourist attraction and not just a device for vertical mobility.

Travel experience sharing sites (such as www.tripadvisor.it) have posted complaints about the costs and discontinuity of the service, as well as the fact that it is not really as easily accessible to everyone, as it should be. The fact is that the cableway, like any other vertical mobility device (such as elevators, escalators, platform lifts and mechanical ramps) has multiple features that go beyond mere physical mechanisms and interact with the anthropic context, strongly influencing the conditions for the fruition of the built environment (Germanà et al., 2020). Considering the role of the cableway in enhancing the historical centre of Erice, its efficiency should be a priority, to be pursued with every management tool possible.

10 Considerations Regarding the Results

Bearing in mind the above-mentioned research issues, we may arrive at certain considerations.

In the case of historical centre preservation, the integral conservation of the heritage (ICH) should be developed correctly whilst respecting acknowledged and scientifically recognized models of intervention, and not exposing the cultural heritage to risk of damage, since we do possess a sufficient number of theoretical studies and practical experience of success/failure; however, there are several problems to be tackled in managing and interpreting the transition phases of place-based policies and/or practices at local and regional level.

When a historical centre loses most of its inhabitants, it is possible to attract people from outside by adopting certain strategies; however, it is necessary to prepare an integrated plan of economic, financial, environmental and social (and not only cultural) interventions and not only based on building restoration and rehabilitation.

In the case of Erice, as highlighted by the decline of its population, the ICH has produced no results in terms of social (equity-inequality) impact. In other cases, the ICH or partial interventions on the heritage may also bring about gentrification (Aygen, 2013).

The ICH produces positive results in terms of economic and financial impact for the tourist agencies and operators alone, but in the case of Erice, tourist numbers are also in decline.

The ICH produces positive results in terms of environmental impact, especially in the case of Erice, located at the top of a hill, because the municipality was encouraged to continue protecting the surrounding resources in the same way as the historical centre.

The ICH produces positive results in terms of cultural impact, because it is easier to enhance the protected heritage, including it in various kinds of cultural activity more smoothly than other parts of the town and the Region.

It is very difficult to defend the functional separation between the historical centres and/in the city; every part of the city plays a specific role and has a specific identity, but in the transitional phases of their evolution it is impossible to adopt abstract models or theoretical approaches to policies for their protection and regeneration. For every city and its many parts, every action plan has to be different.

The scale of intervention for the protection and enhancement of historical centres has to be multiple: from the scale of the single building to the entire scenario, also with regard to small villages.

The comparison between ancient and modern architecture or the urban fabric is not a conflict in the theoretical sense, but is an occasion in which architectural design, with opportune technology, can actually include conservation and not the contrary (De Vita, 2015). This is true not only at the architectonical level, but also the urban. In the case of ICH, the insertion of a modern idiom in ancient urban settings demands more complex and extremely specific strategies. The quality of architectural solutions has to be related to the complex problems to be tackled and resolved. In this case too, there is the issue of participation as regards the choices of the local people and their preferences. Therefore, every decision regarding the ancient (philological reconstruction) or modern (new design of intervention) requires public debates.

Bearing in mind the tough competition in the world of sustainable global tourism, the cultural resources of Erice are today simply not appealing enough. We might consider Erice in terms of its historical buildings and archaeological resources not only as a single unit, but also in terms of integration of the historical quarter and the active geographical setting (a place from which it is possible to look out at Sicily in all directions) and passive (Erice can be perceived as a notable landmark in the western part of Sicily). This may be considered as both a strength or/and a weakness: it can be considered a point of strength to present the historical centre to visitors as a kind of 'time machine', whereas, on the other hand, weakness is evident in the lack of famous, evocative and attractive single cultural elements. There is a need to find new communicative solutions to attract more visitors, not only in summer but all the year round. In both scenarios, a profound transformation of traditional tourism is needed. Storytelling or a new 'place-telling' might have a role in reanimating a place that is ever more similar to a ghost town. Now, an agency or an individual, either in Erice or from the outside, needs to reinterpret and reassemble the local cultural resources in a new way, with an eye to quality of life, not only in terms of the sense of mystery (the ancient cult of Venus), but a quiet and elegant medieval atmosphere, an incredible sense of power and spectacular views, etc.

Edoardo Caracciolo, a principal figure in the founding of modern urban planning in Sicily, had imposed his theory of urban planning based on the centrality of history and its role in the present, in the transformation of city spaces and the surrounding territory; this was the point of arrival for studies that originated in the 1930s and ended a few years after his professionally untimely death (Leone, 2014). There is no contemporary architecture project without profound historical knowledge of the context and its evolution over a period of time.

Edoardo Caracciolo had been impelled by strong social demands for transformation after the disruptions of the Second World War; he had to opt for extensive demolition and only in the case of Erice *old town* did he decide to follow the academic and cultural road to total conservation.

In the end, it is possible to make a few remarks:

- Caring for historical centres means managing the urban/regional centrality and not the outlying districts;
- Every age has its own kind of territorial centrality (economic, environmental, cultural and social) in relation to the real power held by the players/agency in transforming the city space over time;
- Caracciolo was one of the first planners in Europe who had the occasion to experiment with the total conservation of urban architectonic heritage;
- Today, we know perfectly well how to conserve, safeguard and restore the physical urban heritage;
- We do not know how to recreate the old centrality without profound changes in the historical morphology.

Historical centres often constitute an occasion for social and economic development at urban and regional level. This occasion provides the essential conditions for initial recovery, with regeneration at a later stage, within the framework of the overall development of the whole city and the region. But the condition of regional development is represented by the adoption of a place-based approach and not a place-neutral approach (Barca et al., 2012). In many situations, the success of the regional development policy depends on the effectiveness of the place-based policies and practice, initiated by the local stakeholders with the support of the regional authorities and institutions. This may be the case of Erice ICCH, but there have been no initiatives at the regional level, and the municipality has been left quite alone to manage its very significant cultural heritage with little hope of success.

Now we need to think ever more seriously about the total/integral conservation of the historical/cultural heritage in alignment with other regeneration and revitalization strategies, based on private intervention and the supply of innovative technology. Public intervention is not sufficient, and therefore, we have to find other sources, with the help of all players and agencies, to revive and sustain the new city and to attract new resources from outside the local area.

References

Abbate, G. (2002). *Il ruolo dell'analisi tipologica nel recupero di centri storici. Il caso Palermo*. Publisicula.

Abbate, G. (2004). Trapani: Un territorio straordinario dal futuro incerto. *L'universo, 1*, 4–25.

Abbate, G. (2012). Centri storici meridionali e riuso dell'esistente. *Planum-the Journal of Urbanism, 25*, 1–7.

Abbate, G. (2014). L'intervento nei centri storici: l'ANCSA e il caso Erice. In N. G. Leone (Ed.), *Edoardo Caracciolo. Urbanistica, architettura, storia* (pp. 131–138). Milano: Franco Angeli.

Abulafia, D. (1988). *Frederick II. A medieval emperor*. Allen Lane for Penguin.

Amico, V. (1855). *Dizionario topografico della Sicilia*. Palermo: Pietro Morvillo.

ANCSA-CRESME (2017). *Centri storici e futuro del Paese. Indagine nazionale sulla situazione dei Centri Storici*. Roma.

Arancio, F. (1844). *Guida statistica su la Sicilia e sue isole adiacenti. Con carta coroidrografica doganale statistica. Anche di Malta e Gozo*. Palermo: Virzì.

Astengo, G. (1958). Assisi: Piano generale e piani particolareggiati di primo intervento. *Urbanistica, 24–25*, 2–124.

Aygen, Z. (2013). *International heritage and historic building conservation: Saving the world's past*. Routledge.

Aymard, M. (1993). Uno sguardo sulla Sicilia: Le coste e i territori. In M. Sgarlata (Ed.), *L'opera di Camillo Camilliani* (pp. 99–117). Roma.

Bachelard, G. (1957). *La Poétique de l'Espace* (English edn). Paris: Presses Universitaires de France.Bachelard, G. (1964). *The poetics of the space*. New York: Orion Press.

Barca, F., McCann, P., & Rodriguez-Pose, A. (2012). The case for regional development intervention: Place-based versus place-neutral approaches. *Journal of Regional Sciences, 52*(1), 134–152.

Barp, A. & Bolla, D. (2005). *Spazi per camminare. Camminare fa bene*. AUSSL 9 Veneto-IUAV. Venezia: Marsilio.

Blasetti Fantauzzi, C. & De Vincenzo, S. (2012). Nuove indagini alla cinta muraria di Erice. Le campagne di scavo 2010 e 2011. *Journal of Fasti Online*. http://www.fastionline.org/docs/FOLDER-it-2012-272.pdf . Accessed 20 May 2020.

Bonafede, G. (1997). *La pianificazione in Sicilia. 1944–1990*. Palermo: La Zisa.

Bresc, H. (1972). Les jardins de Palerme (1290–1460). *Mélanges de l'Ecole française de Rome. Moyen-Age, Temps modernes, 1*(84), 55–127.

Cannarozzo, T. (2010). Centri Storici e città contemporanea: dinamiche e politiche. In G. Abbate, T. Cannarozzo, G. Trombino, *Centri Storici e territorio. Il caso Scicli—Historical towns and their hinterland. The Scicli case study* (pp. 9–22). Firenze: Alinea.

Cannarozzo, T. (2016). *Palermo e l'urbanistica sociale di Edoardo Caracciolo. Scritti e Piani, 1930–1960*. Palermo: Caracol.

Caracciolo, E. (1938). Edilizia minore ericina. In P. Ajroldi, E. Caracciolo, V. Lanza, *Galleria mediterranea. Rilievi di edilizia minore siciliana. Catalogo della mostra (Palermo, Palazzo De Seta, 1–15 febbraio 1938)* (pp. 7–10). Palermo: Scuola tipografica Boccone del Povero.

Caracciolo, E. (1939). *Edilizia ericina*. Palermo: Scuola tipografica Boccone del Povero.

Caracciolo, E. (1950). Ambienti edilizi sul Monte Erice. Proposta di scheda per la storia dell'urbanistica. *Archivio Storico Siciliano*, III (IV), 183–257.

Caracciolo, E. (1954a). La città sul Monte Erice. *Casabella, 201*, 42–48.

Caracciolo, E. (1954b). *Tre lezioni di urbanistica*. Palermo: De Magistris.

Caracciolo, E. (1956). La casa ericina. In *Atti del V Convegno Nazionale di Storia dell'architettura. Perugia 1948*. Firenze: Noccioli.

Caracciolo, E. (1960). Erice: Conservazione e valorizzazione di un patrimonio eccezionale. [Atti del Convegno di Gubbio]. *Urbanistica, 32*, 85–88.

Carvini, V. (1670–1680). *Erice antica e moderna sacra e profana*. Biblioteca Comunale di Erice.

Collenuccio, P. (1929). *Compendio de le istorie del Regno di Napoli*. Bari.

De Certeau, M. (1980). *L'invention du quotidien*. Vol. 1, *Arts de faire*. Paris: Union générale d'éditions.; De Certeau, M. (1984). (English edn) *The Practice of Everyday Life*. Berkeley, CA: University of California Press.

Cutolo, D. & Pace S. (Eds.). (2016). *La scoperta della città antica. Esperienza e conoscenza del centro storico nell'Europa del Novecento*. Roma: Quodlibet.

Conti, C. & Petriccione, L. (2018). Le superfici orizzontali e l'accessibilità dei beni culturali. Materiali e tecniche per la realizzazione di percorsi inclusivi in contesti storici. In G. Biscontin, G. Driussi, (Eds.), *Intervenire sulle superfici dell'architettura tra bilanci e prospettive* (pp. 248–258), Venezia: Arcadia Ricerche.

De Vita, M. (2015). *Architetture nel tempo. Dialoghi della materia, nel restauro*. Firenze: DidaLabs-Firenze University Press.

Edwards, P. & Tsouros, A. (2006). Promoting physical activity and active living in urban enviroments. The role of Local Governments. *WHO*. http://www.euro.who.int/__data/assets/pdf_file/0009/98424/E89498.pdf . Accessed 23 April 2020.

EIDD (2004). Stockholm declaration. European Institute for Design and Disability. http://www.designforalleurope.org . Accessed 23 Apr 2020.

Emanuele Gaetani F. M., marchese di Villabianca (1754). *Della Sicilia nobile*. Palermo: Stamperia dei Santi Apostoli per Pietro Bencivenga.

Geddes, P. (1915). *Cities in evolution: An introduction to the town planning movement and to the study of civics*. Williams & Norgate.

Germanà, M. L. (2013). L'accessibilità della città storica: aspetti gestionali tra specificità e strategie unitarie. In F. Castagneto, V. Fiore (Eds.), *Recupero, Valorizzazione Manutenzione nei Centri Storici. Un tavolo di confronto interdisciplinare* (pp. 22–25). Siracusa: Lettera Ventidue.

Germanà, M. L., Pons Izquierdo, J.J., Lecardane, R., Trapani, F. (2020). Vertical Mobility: a multi-faceted Tool for enhancing Architectural Heritage. In D. Trabucco et al. (Eds.), *Vertical mobility for accessibility* (pp. 32–39). Conegliano: Anteferma Edizioni.

Giovannoni, G. (1931). *Vecchie città ed edilizia nuova*. Torino: UTET.

Giuffrida, A. (Ed.). (2012). *Imprese industriali in Sicilia (secc. XV–XVI)*. Palermo: Associazione Mediterranea.

Houdek, M. (2014). When heritage preservation meets living memory: Constructing the Medina of Fez as a world heritage heterotopia. *Theses, 52*. https://surface.syr.edu/thesis/52 . Accessed 7 Apr 2020.

Ingrosso, C. (2006). *Barcellona. Storia dell'architettura e della città*. Aracne.

Internicola, G. V. (2003). *Baronia di Baida e territorio di Castellammare*. Alcamo.

Kantorowicz, E. (1927). *Kaiser Friedrich der Zweite*, Berlin: Ergänzungsband; Kantorowicz E. (1988). (Italian edn) *Federico II imperatore*, Milano: Garzanti.

Leone, N. G. (Ed.). (2014). *Edoardo Caracciolo. Urbanistica, architettura, storia*. Milano: Franco Angeli.

Lowenthal, D. (1996). *The heritage crusade and the spoils of history*. New York: Free Press. Reprint: Lowenthal, D. (2009). *The heritage crusade and the spoils of history*. Cambridge: Cambridge University Press.

Lynch, K. (1984). *A theory of good city form*. The MIT Press.

Maurici, F. (1992). *Castelli medievali in Sicilia*. Palermo: Sellerio.

Monti, L., & Cerroni, R. (2019). How the Widespread Presence of Historical Private Real Estate Can Contribute to Local Development. *Review of European Studies, 1*(11), 183–193.

Mucci, F. (1997). La valorizzazione del patrimonio mondiale culturale e naturale: Significato e strumento di una tutela sostenibile. In M. C. Ciciriello (Ed.), *La protezione del patrimonio mondiale culturale e naturale a venticinque anni dalla Convenzione dell'UNESCO* (pp. 269–290). Editoriale Scientifica.

Natoli Di Cristina, L. (1964). Edoardo Caracciolo, primo urbanista siciliano. In G. Pirrone (Ed.), *Edoardo Caracciolo. La ricostruzione del Val di Noto*, Quaderno n. 6 (pp. 7–21). Palermo: Facoltà di Architettura Università di Palermo.

Neri, M. L. (2001). Giovannoni Gustavo, biographical entry. In C. Olmo (Ed.), *Dizionario dell'architettura del XX secolo, III* (pp. 82–84). Umberto Allemandi.

Niglio, O. (2012). *Le Carte del Restauro. Documenti e Norme per la conservazione dei Beni Architettonici ed Ambientali*. Roma: Aracne.

Peri, I. (1952–1953). Città e campagna in Sicilia. Dominazione normanna. In *Atti della Accademia di Scienze Lettere e Arti di Palermo, IV*(XIII), II.

Peri, I. (1978). *Uomini città e campagne in Sicilia dall'XI al XIII secolo*. Roma-Bari: Laterza.

Rodwell, D. (2003). Sustainability and the holistic approach to the conservation of historic cities. *Journal of Architectural Conservation, 1*(9), 58–73.

Rodwell, D. (2006). Managing historic cities. The management plans for the bath and Edinburgh World Heritage Sites. *Journal of Architectural Conservation, 2*(12), 41–62.

Savarese, R. (2018). *Storia di palazzo Bonagia a Palermo*. [mimeo].

Sitte, C. (1889). *Der Städtebau nach seinen künstlerischen Grundsätzen*, Wien: Verlag von Carl Graeser; Sitte, C. (1945). (English edn) *The art of building cities: city building according to its artistic fundamentals*. New York: Reinhold Publishing Corporation.

Trapani, F. (2014). Il disegno sociale del piano. In N. G. Leone (Ed.), *Edoardo Caracciolo. Urbanistica, architettura, storia* (pp. 164–170). Milano: Franco Angeli.

Trasselli, C. (1982a). *Da Ferdinando il cattolico a Carlo V. L'esperienza siciliana 1475–1525*. Soveria Mannelli: Rubbettino.

Trasselli, C. (1982b). *Storia dello zucchero siciliano*. Racalmuto: Sciascia.

Trombino, G. (2000). *L'urbanistica in Sicilia negli anni della ricostruzione*. Roma: Officina.

Trombino, G. (2003). Caltanissetta. *L'Universo, 5*, 580–594.

Tyrwhitt, J. (Ed.). (1947). *Patrick Geddes in India*. Lund Humpries.

Vultaggio, P. (2019, October 5). Una carrozzina elettrica per aiutare i disabili che vogliono visitare Erice. *Quotidiano di Sicilia*. https://qds.it/una-carrozzina-elettrica-per-aiutare-i-disabili-che-vogliono-visitare-erice . Accessed 21 May 2020.

Importance of Architectural Heritage Conservation

In Part III of this book, the authors briefly highlight the importance of raising awareness with regards to the importance of heritage value, establishing a link between architectural heritage preservation and sustainable tourism development in traditional urban neighbourhoods and heritage sites.

In this part, the challenges of loss of culture, heritage and inherited buildings are discussed and the need for their Conservation and Management for future generations is stressed upon. As a result, there is a current and increasing research interest in the preservation and restoration of historical monuments and buildings, urban renewal policies and identity preservation. New and creative strategies and methodologies for the conservation of sensitive zones, heritage villages, lost heritage buildings after wars as well as oral heritage are being investigated and developed by developers, architects and professional education programs.

Accordingly, the first chapter in this part, with the title of "A Review of Existing Literature in Saudi Domestic Architecture" explores reasons that have led to the disappearance of Architectural Heritage in the Kingdom of Saudi Arabia as well as current attempts and approaches to conserving its architectural heritage and local built environment. In their work, the authors provide a review of existing literature to examine Saudi Arabia's Architectural Heritage design, used materials and strategies. Similarly, the author of "Inspiring Traditional African Art & Architecture in the Modern African Hotel Design Field" takes us through a specific site to detail how the rich heritage of African culture can be integrated into modern architectural design. Following in their footsteps, the chapter on "Syrian Cultural Heritage Conservation and Management for Future Generations: Issues and Challenges" examines local perspectives towards heritage planning, management and planning in Syria in addition to governmental laws and policies, thus identifying a gap in knowledge and a lack of standards when it comes to conservation and reuse of heritage sites.

Moreover, other approaches, with potential of further development, that make way for a better understanding of heritage zones/ regions/ cores and their heritage assets and can be used as starting points for a clear roadmap for policy-making in heritage sensitive areas are presented as well in this part. In the chapter entitled "Urban Conservation of Heritage Sensitive Zones in India: A Methodological Approach", concepts of urban planning and architectural conservation of heritage assets are integrated to facilitate a sustainable and sensitive development of cities. On that same note, the author of "Protection and Regeneration of Urban Historical and Cultural Area in New Period—A Case Study of Pingshijie Area in Nanjing" tackles the problems presented by a lack of legal protection for historical and cultural areas in China. The author focuses on a traditional area, south of Nanjing, while presenting ways in which the area can be protected, its style and features maintained, resources utilized and infrastructure improved, as to support facilities.

Presenting a number of case studies from around the world including Scotland, Albania, Bangladesh, India, KSA, Syria, this part also showcases Egyptian case studies where City Branding is used as a new strategy to achieve a sustainable urban image and preserve the non-physical elements such as culture, economic and social aspects and activities. In "Historic Preservation as a Tool for City Branding Case study Khedivial Cairo", the connection between historic and valuable images of urban spaces and city branding is defined. The authors also propose a framework for evaluating the level of success of historic urban spaces from urban aspects to improve their identity. The chapter entitled "Vjosa River Valley; Strategies for Sustainable Tourism" focuses on the development of tourism as one of the leading potentials and contributors to the sustainable development of the Vjosa river valley in Albania. Looking to the latest government strategies, the chapter goes through a deeper understanding of the built and natural environment in the area and provides the reader with a background on the development of the Albanian rural economy through the development of sustainable tourism strategies in the Vjosa river valley area.

The part not only focuses on cultural and historical heritage and architectural buildings but also goes as far as looking into the conservation of oral heritage in a built form. In the chapter entitled "Conserving Oral Heritage Through Architecture", light is shed on the distortions and manipulations transmitted through generations which emphasizes, once again, the importance of increasing awareness to conserve a city's heritage, record history in writing, and turn oral heritage into structural monuments, memorial and cultural spaces that live on as physical evidences and authentic expressions. The authors use their very own monument design as an example of the ways in which Architecture can play a major role in reviving Oral Heritage. This part is meant to inform the reader that the obligation to conserve the architectural heritage of our communities, its values and identity is now more imminent than ever. The world is left with a need to formulate or further develop guidelines for design and planning and adjust planning policies in order to achieve sustainable development, meet the needs of the present, protect future rights all while preserving our past.

A Review of Existing Literature in Saudi Domestic Architecture

Mohammed ALghafis, Magda Sibley, and Eshrar Latif

Abstract

The Kingdom of Saudi Arabia (KSA) has lost much of its traditional appearance and identity since the mid-twentieth century, mainly due to economic development and urbanization, with the introduction of new technologies, regulations, and building materials. The features and elements of heritage buildings in Saudi cities have suffered from the impacts of modernity, with heritage domestic buildings fading into history and extinction. This paper examines Saudi architectural heritage design in terms of building materials and strategies related to various geographic areas in the five regions of the KSA (Central, Western, Eastern, Northern, and Southern). It then discusses the reasons that led to the disappearance of the architectural heritage of the KSA, followed by explanation of current attempts to conserve the architectural heritage in the local built environment.

Keywords

Architectural heritage • Domestic buildings • Built environment • Design strategies • Saudi Arabia

M. ALghafis (✉)
Research Student in Welsh School of Architecture, Cardiff University, Al-Mulida, 52571, Saudi Arabia
e-mail: M.alghafis@qu.edu.sa

M. ALghafis
Teacher Assistant in Architecture Department, College of Architecture and Planning, Qassim University, Al-Mulida, 52571, Saudi Arabia

M. Sibley
Reader in Architectural History and Theory in Welsh School of Architecture, Cardiff University, Cardiff, UK
e-mail: Sibleym@cardiff.ac.uk

E. Latif
Welsh School of Architecture, Cardiff University, Cardiff, UK
e-mail: Latife@cardiff.ac.uk

1 Introduction

For the past fifty years, the built environment in the Kingdom of Saudi Arabia (KSA) has lost its vernacular appearance and identity because of the economic boom and enormous changes, for example, urbanization, modernization, the introduction of new technologies and new building materials. The features of heritage buildings in the KSA cities have suffered from modern impact. As a result of this, many of the heritage buildings have either gone into extinction or totally restructured to have a modern look. Cities in the KSA have continued to face the loss of its originality in the built environment. In addition, some changes have continued to be made to these buildings to meet individual needs; thus, many vernacular buildings continue to lose their identity.

The continuous growth of buildings within the city of Saudi Arabia without marking out building conservation areas has contributed significantly to the loss of historical buildings and the demise of the Saudi Arabia Architectural Identity. Sighting an example with the UK, building conservation areas are marked within the city to protect historical buildings. This has helped to conserve many of British historical buildings as well as retaining its architectural identity. The building conservation strategy in the UK has helped to protect its historical buildings from extinction, even with the rapid development rate.

The traditional primary materials in Saudi Arabia are stone, mud, thatch and timber. These materials have been used wisely by the early builders to tackle various environmental challenges. Thick mud walls between 600 and 1000 mm were a response to the extremely hot desert sun causing a high temperature between 35 and 40 °C to keep the indoor spaces cool. Conventional building materials are appropriate and have a less negative impact on the environment than modern materials. They are considered as a breathing structure 'eco-friendly,' but most of the new

A. Versaci et al. (eds.), *Conservation of Architectural Heritage*, Advances in Science, Technology & Innovation,
https://doi.org/10.1007/978-3-030-74482-3_20

buildings do not create humidity because they are constructed from impervious materials (Oxley, 2015).

The paper will review the historical background of Saudi built environment. Then, it will attempt to study the Saudi architectural heritage design which differentiates itself by using several materials and strategies based on the various geographic areas in five regions (Central, Western, Eastern, Northern and Southern). It will discuss the reasons that led to the disappearing of the architectural heritage in the KSA, and current attempts and ways to conserve the architectural heritage in the local built environment.

2 Historical Background of Saudi Built Environment

Since late 1800 until now, with the Ottoman Empire over some centuries, urban planning in Saudi Arabia has experienced various phases of development. There has been in recent Arabian history a number of 'watershed' moments. Firstly, after 1918, over the post-Ottoman age of Allied administration, 1932 saw the establishment of the KSA. Farming was the most significant method for transport and trade before the new motorcar and oil industry; thus, there were agricultural business activities due to settlements developed rapidly (Rahmaan, 2011).

Secondly, the oil industry contributed to a great driver of the Saudi economy. Urban planning regulation from the West was utilized to the growing KSA environment in the biggest places, such as Al-Riyadh, Jeddah and Al-Dammam cities. For example, Al-Malez project (Fig. 1) completed in the late 1955s, a-500 acre development in Al-Riyadh considered one of the first post-1950 (Al-Hathloul, 2017).

In 1968, Constantine Doxiadis started changing and planing Al-Riyadh from a small regional capital town on the green Wadi Haneefa (linear oasis) (Fig. 2) into the biggest capitals in the Middle East, based on Western modern planning standards. Evolving from this, in 1992, the Saudi government with ARAMCO (the Arabian American Oil Company) imported the modern design, planning and techniques due to the high demand for buildings (Al-Hathloul, 2017). In the next part, it focuses on the architecture of regional architecture in the KSA.

3 A Brief Review of Regional Architecture in the KSA

The KSA is the largest country in the Middle East, located in Asia. It is surrounded by Iraq and Jordan to the north, Kuwait to the northeast, Bahrain, UAE, and Qatar to the east, Yemen to the south and Oman to the southeast. In the west, it is bounded by the Red Sea. In the south, the Arabian Gulf to the east and the Indian Ocean. The capital of the KSA is Riyadh, 900 km from the Red Sea and 400 km from the Gulf.

The KSA has four distinct regions: The Northern and Central (Najd), the Western region (Hejaz), part of Eastern and the Southern Arabia (Asir), with mountains area to the south-west part.

These regions vary with each other in many aspects, mostly in climate. For example, in the Red Sea, the coastal place (the western region) has a humid climate with an average temperature of 28 °C. In contrast, places in central Arabia (Najd region) have hot arid weather in summer with an average of 38 °C (weather spark 2019). These regions are summarized in Fig. 3.

Central Region

The climate of the central region is hot-dray (Fig. 4), located in the central of KSA and surrounded by about two deserts. Soil and mud were the most available building material in this region collected from the wades (dry valley) after the season of rain (Facey, 1997).

Water, straw and soil were used to make a mud brick. Stones, tree trunks and palm trees were used as construction materials of the buildings, and these trunks covered with soil to form the roof.

Typically, there were one or two central courtyards in a house with triangular windows. These small openings are not only providing ventilation but also control solar radiation. In most of the day, a house keeps on shade due to the height of the walls, and the external windows were small, maintaining the privacy and providing ventilation.

Houses in the central region were located close to each other, and narrow streets were there that provide shade. Consequently, the exposed surface area of the building was limited; that provide insulation against heat (Al-bakri, 1997). The planning and design pattern were no based on concept, but it was based on results in need of the people.

Western Region

The climate of the western region is hot humid (Fig. 5), located beside the Red Sea coast. It has affected by pilgrims who have come to visit this place since it has two holy mosques. About two to three million pilgrims move from all around the word to toward this region of KSA, resulting in exchange of the knowledge, experience, ideas and the techniques of the building (Al-Oraier, 2005).

New experiences were imported in the western region due to the existence of pilgrims who came to this region to settle or do business, contributing to new skills. For instance, in the Ottoman Empire, the architecture was influenced by Turkish soldiers (Rahmnan, 2011).

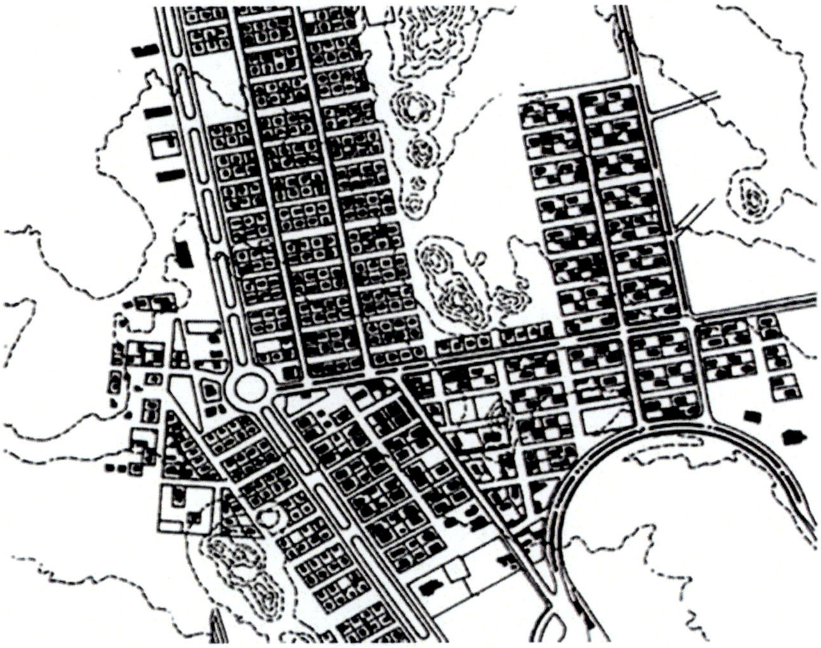

Fig. 1 Project of Al-Malez completed late 1950s. *Source* Al-Hathloul (2017)

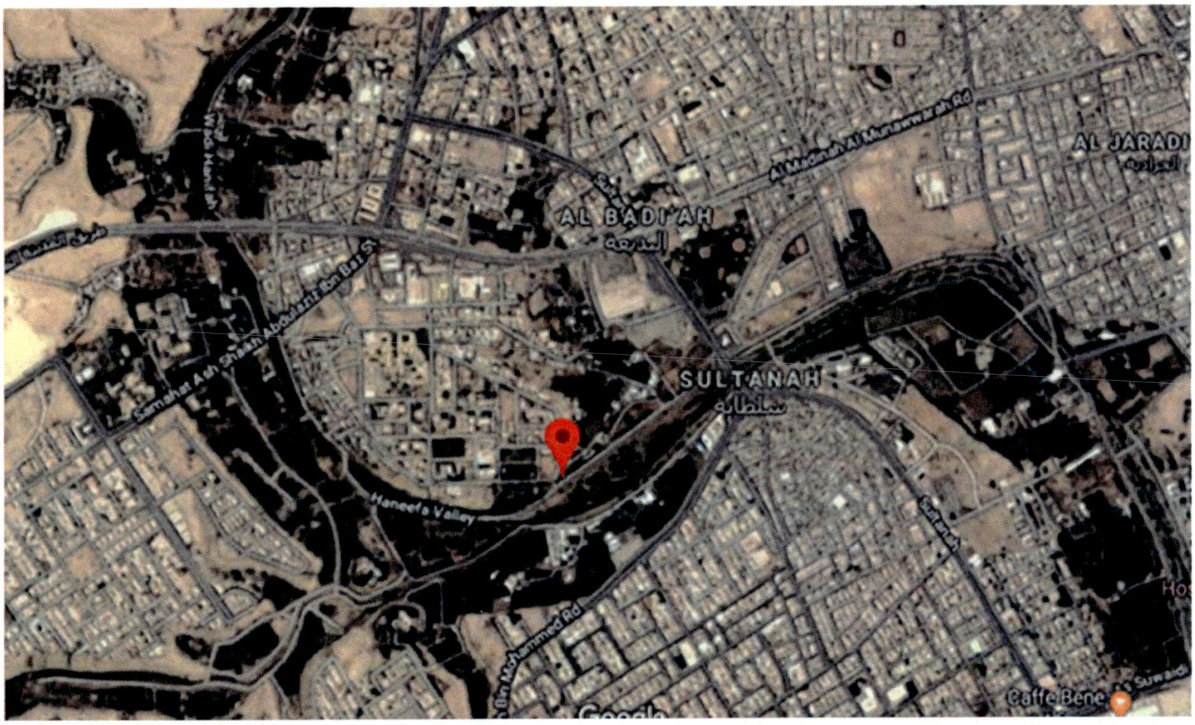

Fig. 2 Wadi Haneefa in Riyadh city. *Source* Google map

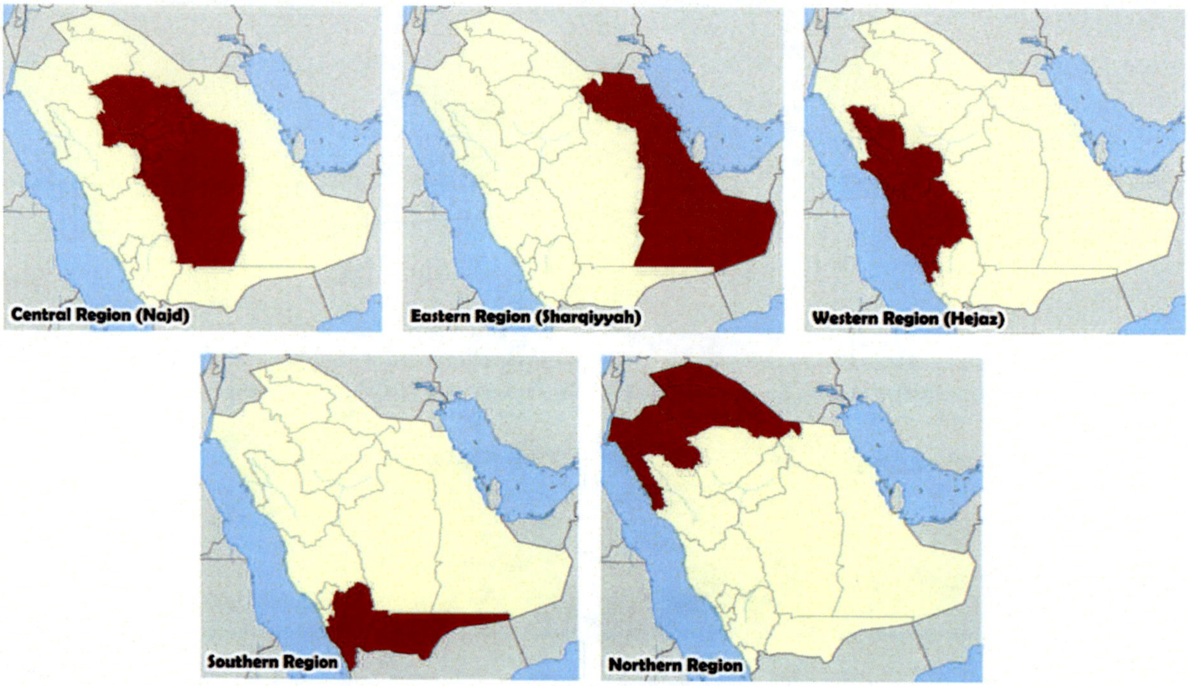

Fig. 3 Five different regions in the KSA. *Source* Google maps

Fig. 4 Average temperature in Central Saudi. *Source* Weather spark (2018)

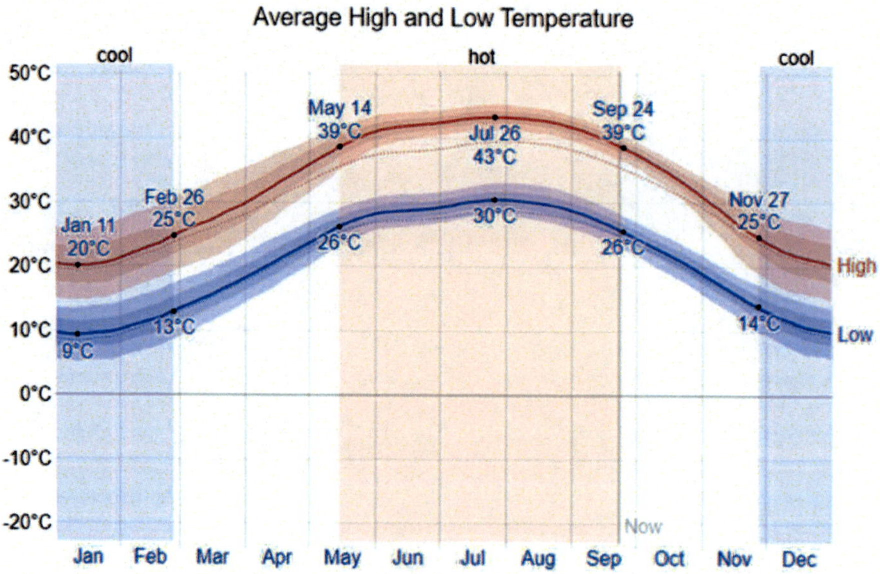

As a result, mashrabiya was used in a multi-story building, as Talib (1983) defines that as a decorative wood over some time. Jeddah, for example, is one of the cities which has hot, humid climate and the vernacular houses in Jeddah used Mashrabiyas were not tight, compared to central region to allow air movement between the buildings and reduce of high humidity. A smaller structure consisting of strips of wood in the building have made handcrafted, in merging of different pieces of wood.

Mashrabiya has many remarkable functions. It achieves privacy, and it is the best environmental strategies which allow cool breezes in the summer's heat. Mashrabiyas, therefore, were also applied to break up the glare of the sun and provide shade.

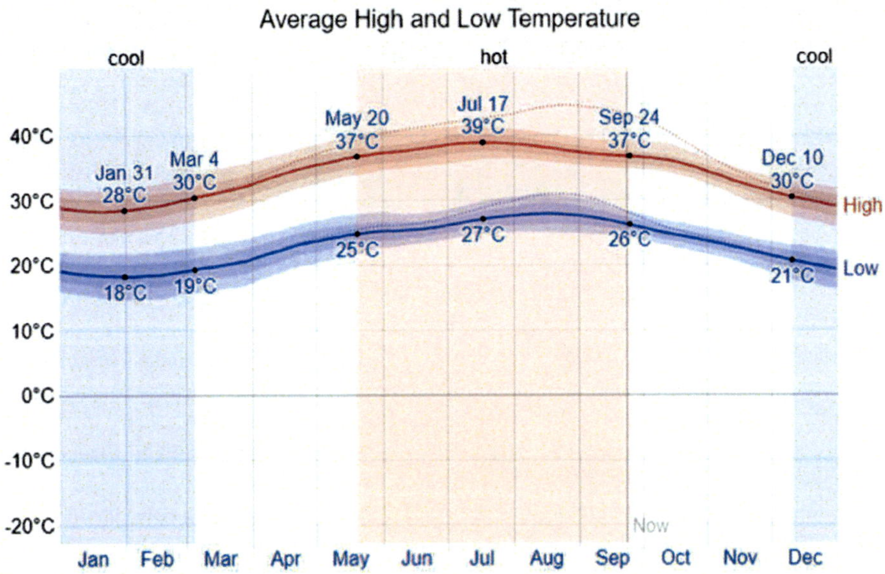

Fig. 5 Average temperature in the Western Saudi. *Source* Weather spark (2018)

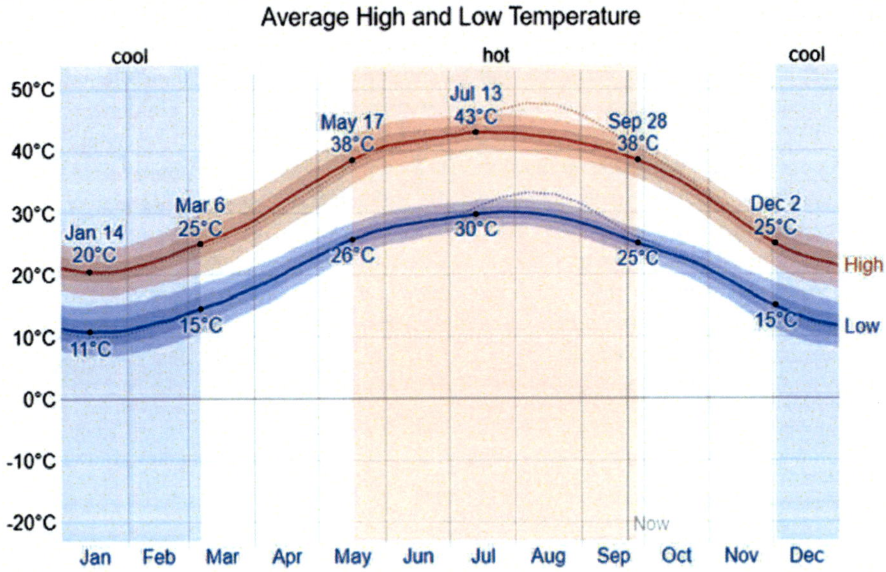

Fig. 6 Average temperature in the Eastern Saudi. *Source* Weather spark (2018)

Eastern Region

The climate of the Eastern region is quite like this found in Western place (Fig. 6). It is located along the Arabian Gulf, and the old building was built using two types of materials; mud bricks and rubble stones that are available in large quantities in the Eastern region.

Residents who live in the area dealt with the highly humid hot climate by using wind towers (Malaqf). It is the main traditional feature in the region that leads to minimizing the internal air temperature. According to Fathy (1973) who describes the Malagf, (Natural energy and vernacular), it is a kind of 'wind catch' that allow wind to flow from many directions, might be four or two, preventing wind from flowing in one side and out in the other hand. These wind towers found in Hafuf are which located in Eastern cost, coming from Iran and these features used as air conditioning in the past. This system contributes to cool and refresh the air inside the buildings (Fathy, 1973).

The wind tower was used in the houses of Iran and coast places of KSA. It was constructed higher than the roof. The concept of it is to allow wind flow and catch the sea breezes to cool interior rooms, creating ventilation and refresh the air and making a comfortable environment in the summer months. However, in the winter months, the wind tower should be closed to prevent cold air and retain the warm air into the building.

Northern Region

Houses of Northern region are quite like those found in the Central area (Fig. 7). They were built of mud bricks and stone. Mud walls of courtyards and narrow streets are the features that applied in the Northern region. Furthermore, room hanging was used to provide shadow in the street level and pedestrian (Al-Oraier, 2005).

Southern Region

Due to the different topographies, Southern area has about two climates (Fig. 8): hot humid along the coast of the Red Sea, but cold in the mountains. One of the most common old buildings in this area is the reed house because of the humid climate. However, the buildings in the mountains were built of mud and stone.

The architecture in Southern area was influenced by, of course, climate, available materials and the procedures of local materials' construction of Yemen and maybe part of Africa (Talib, 1983).

As mentioned, vernacular architecture in Saudi Arabia has various styles and Fig. 9 summarized the architectural heritage diversity in five regions: Northern, Southern, Central Western and Eastern region. Courtyards were used in the past in Central, Northern and Eastern provinces. In the next part, it focuses on the reasons that led to disappearing these vernacular features and elements.

4 Reasons Led to the Disappearing of the Architectural Heritage in Saudi Arabia

There are some reasons which might contribute to disappear of architectural heritage, such as economic development, the adoption of Western building regulation and dramatic population expansion in the KSA.

Economic Development

As mentioned before, the rapid economic growth in the last six decades has brought in a massive expansion of population in the KSA urban cities. The oil industry and modernization have contributed to dramatic changes in some sides, for instance, physical, social and economic. Between the years 1950 and 1992, people changed from rural places to cities, and that movements have led to a dramatic increase in the most significant cities from 10 to 75% (Babsail & Al-Qawasmi, 2015).

According to Al-Hathloul (2017), the rapid change in the economic condition of the people of Saudi contributed to the migration of people from traditional rural areas to urbanized regions close to the city. With the continuous growth of modern buildings, building users or occupants cannot be blamed for making this kind of choices. Modern architects are supposed to combine the new style with the vernacular style but not jettison it.

Building Regulations and Policy

Building rules and codes were copied from the new western design. For example, Doxiadis project in Riyadh started in

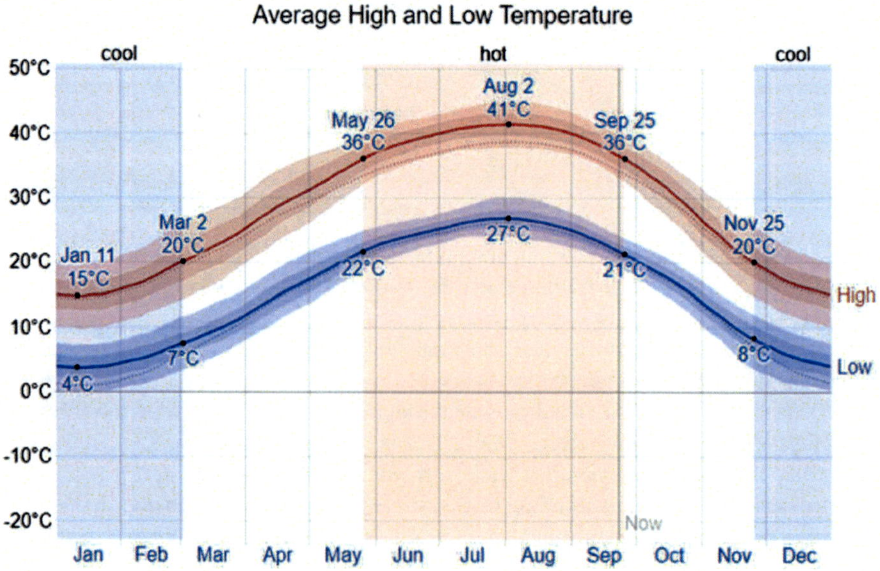

Fig. 7 Average temperature in the Northern Saudi. *Source* Weather spark (2018)

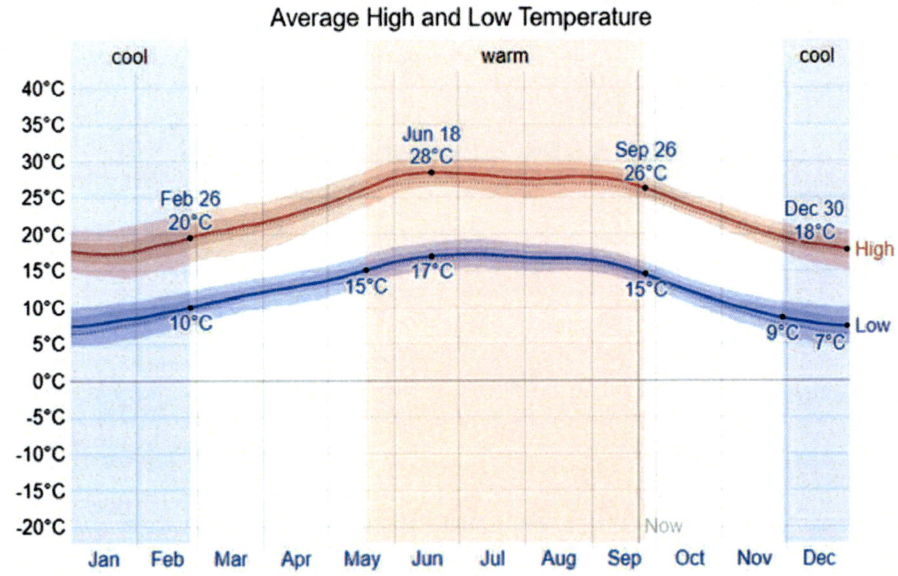

Fig. 8 Average temperature in the Southern Saudi. *Source* Weather spark (2018)

Diversity of Architectural Heritage	 **Central Region**	 **Western Region**	 **Eastern Region**	 **Northern Region**	 **Southern Region**
Basic Information	**Riyadh** is the capital of the Cntral region. It has a continental climate; the summer season is long, arid. The winter season is dry, cool and mostly clear.	**Jeddah** is the capital of the Western region. It has a a maritime climate; the summer is long, muggy, windy and partly cloudy. On the other hand, the winter is very short, windy and warm.	**Damam** is the capital of the Eastern region. It has a a maritime climate; the summer is long, muggy, windy and partly cloudy. On the other hand, the winter is very short, windy and warm.	**Hail** is the capital of the Northern region. It has a continental climate; the summer season is long, arid. The winter season is dry, cool and mostly clear.	**Asir** is the capital of the Southern region. It has a humid cold climate; the summer season is moderate cold and long, humid. The winter season is cold and mostly rain.
Evaluation	Cultural Openness Building Strength Building Height	Cultural Openness Building Strength Building Height	Cultural Openness Building Strength Building Height	Cultural Openness Building Strength Building Height	Cultural Openness Building Strength Building Height
Building Materials	**Roof:** Palm leaves and wood (Athel) **Wall:** Mud, Adobe mud layer and gypsum **Floor:** Athel tree trunks Palm leaves, clay layer.	**Roof:** Trees trunks **Wall:** Stone and tree trunks **Floor:** Tamarisk trees trunks, straw mats, palm leaves.	**Roof:** Palm trunks **Wall:** Coral stone and mud layer. **Floor:** Palm fronds, mud and trees trunks.	**Roof:** Palm leaves and wood (Athel) **Wall:** Mud - Adobe **Floor:** Palm leaves and wood and clay.	**Roof:** Trees trunks **Wall:** Mud / Stone **Floor:** Trees trunks and clay layer.
Discussion	The house has courtyard with thick mud walls and triangular openings. Central space was applied to achieve the social and family privacy and it is the best sustainable approach which allows natural ventilation as well as daylight and avoids direct sunlight.	Multi story stone houses with Mashrabiyas. They were applied to achieve the social and family privacy and it is the best sustainable approach which allows natural ventilation as well as daylight and avoids direct sunlight.	Courtyard houses with wind tower and thick walls and roof. Tower forms were applied to take advantage of environmental aspects such as natural stack-effect ventilation, the vertical air movement providing air movement from outside to inside by pulling cool air in at lower levels	Courtyard house and normal windows opening with thick mud walls and roof. Central space was applied to achieve the privacy and allow natural ventilation, daylight and avoids direct sunlight.	Multi-storey stone/mud house with rain protection reed. The building is high to allow natural stack-effect ventilation, the vertical air movement providing air movement from outside to inside by pulling cool air in at lower levels.

Fig. 9 Architectural heritage diversity in the KSA. *Source* Mortada (2003) adopted by author

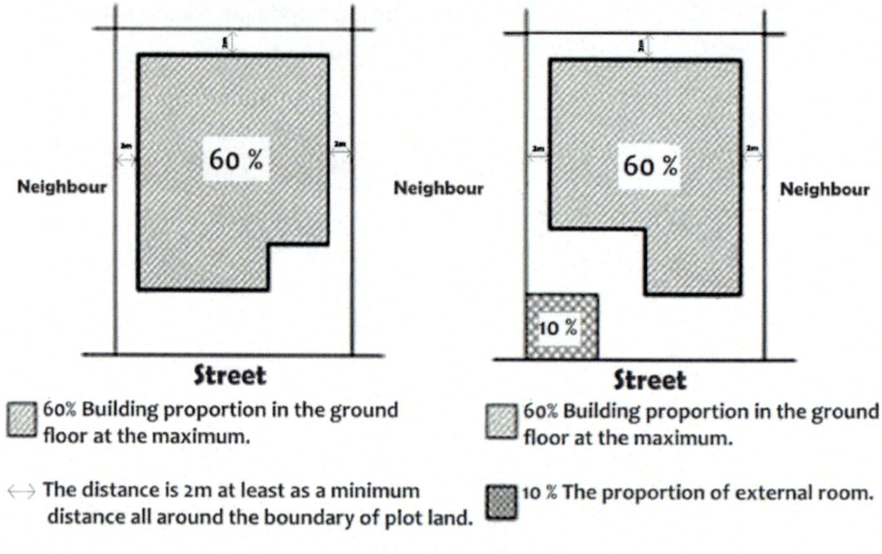

Fig. 10 Examples of the old rules; buildings must be separated from others. *Source* Ministry of Municipal and Rural Affairs (2018)

1975 is resulting in a new system and design to the capital of Saudi Arabia and then, followed by the other regions (Al-Hathloul, 2017). In addition, Doxiadis program imposes the height of houses and the streets' width. Thus, these new regulation and changes have driven KSA cities to lose their original pattern and follow the western layout.

The previous rules of house design in the KSA determine that houses must be separated from other neighbors to create windows opening toward outward (Fig. 10). The minimum distance between two houses is at least two meters as a minimum distance, according to the Saudi Building Code (2018). These building regulations might help to disappear of architectural heritage in the KSA. The next section concentrates on the ways to activate and conserve the architectural heritage in the KSA.

5 Ways to Conserve the Features and Elements of Architectural Heritage

The final part discusses the potential of conserving heritage features and elements in modern buildings by analysing the new regulations, policy, and the benefits of incorporating architectural heritage attributes into contemporary design.

New Regulation

In September 2018, the Ministry of Municipal and Rural Affairs in cooperation with the Saudi Building Code established a new regulation in designing houses. The regulation known as the Saudi Building Code 201 contains 35 sections of regulations, and one of them is about buildings area and height as well as area limitation (SBC, 2018).

The new regulation will allow people to build a house without creating distance between next to each other in the ground floors (SBC, 2018) (Fig. 11). Furthermore, the proportion of building in land reaches up to 80%, compared to 60% in 2018 and before. Owners, designers and developers could rethink and create a courtyard house that suits their requirement. Perhaps, it is a golden opportunity to activate traditional features in modern design.

On the other hand, poor design quality and thermal performance have contributed to depend on air conditioning in buildings. Most of the contemporary houses have designed with large windows look outward, resulting in the external walls exposed to the sun caused heat gain; consequently, a vast amount of energy is consumed. Due to the new Saudi Building Code 201, the question arises if people are going to use new regulations fully or will they activate some features and elements used in past architecture. Saudi householders should consider these questions.

Higher Electricity Bills

Saudi Electricity Company (SEC) has increased bills since 2018 due to a significant increase in the use of air conditioners. There has been rising in electricity bills, and 70% of electricity is consumed by using air conditioning equipment (Demirbas et al., 2017). The use of electricity leads to a large amount the CO_2 emissions, such as a fossil energy sources station burning power like many Saudi electric stations. For example, heating, cooling, lighting and cooking all contribute to the fuels burning except using renewable electricity.

In addition, not using passive design strategies contribute to high-energy consumption. The current wrongful design

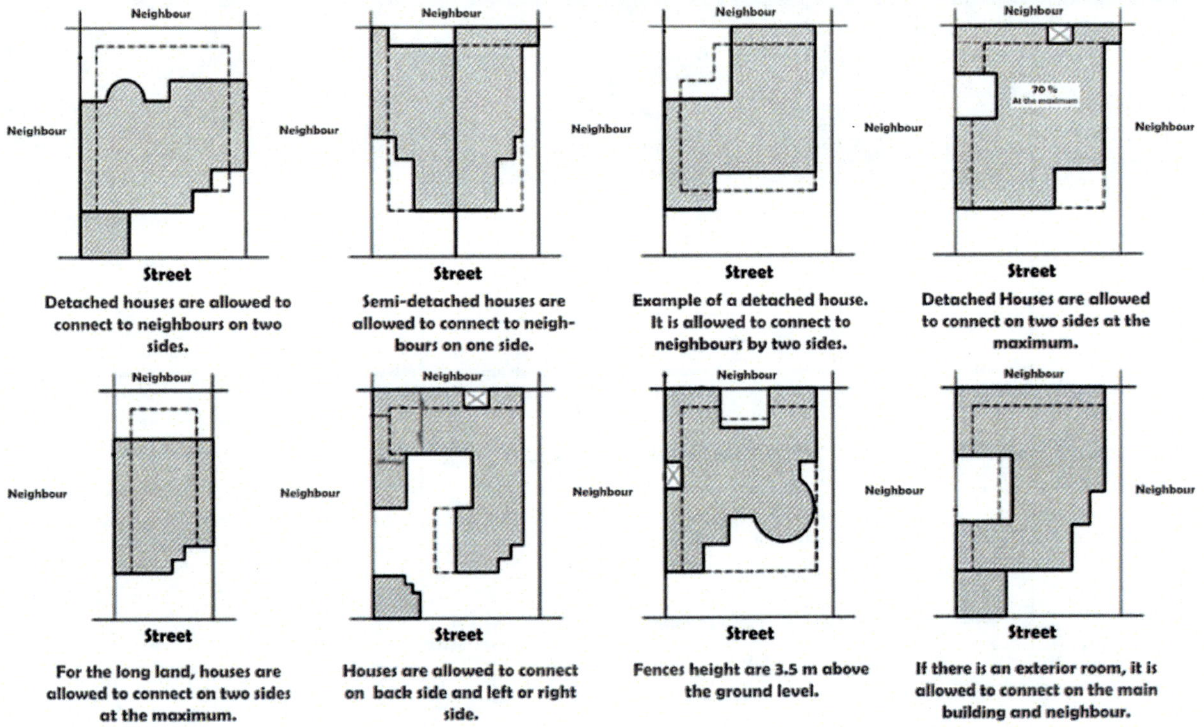

Fig. 11 Examples of new rules. They are more flexible, allowing people to build without creating distance between neighbors. *Source* Ministry of Municipal and Rural Affairs (2018)

strategies exacerbated the high degree of energy consumptions and subsequent carbon emissions buildings. Renewable energy ought to be used in the present time to avoid these issues, for instance, at the power station, using gas rather than coal, using alternative energy sources, using eco-friendly design approach like a courtyard, shading systems, etc. Designers ought to look for these environmental design strategies which might help to energy consumption reduction in buildings.

Incorporating Architectural Heritage Attributes into Modern Design

As mentioned, conventional buildings in Saudi Arabia have many styles; Figure 12 summarizes the main KSA architectural styles and attributes in the five places: Northern, Southern, Central, Western and Eastern region.

Courtyards were applied in the past in Central, Northern and Eastern regions to obtain family, social privacy and allow natural light and ventilation as well (Facey, 1997; Ragette, 2003).

In Southern and Eastern regions, tower form was used to achieve passive design features, for instance, passive stack ventilation (PSV), the movement of vertical air providing air movement from outside to inside by pulling fresh air in at lower levels. Tower forms provided shade at street level and allowed the family to live vertically above shop spaces at ground level. If the residents wealth improved, the building could expand upward.

Additionally, Mashrabiya was also used in Western to fulfill the privacy, and it is the best environmental aspects which allow daylight and avoids direct sunlight. Wind tower (Malaqf) was used in the eastern region to allow wind flow and catch the sea breezes to cool interior rooms, creating a comfortable environment in the hottest months (Al-Naim, 1998).

There are some clear examples of architectural heritage attributes have been interpreted in modern buildings. For instance, Francois Zevaco's courtyard houses in Morocco, Residence Andalous in Tunisia by Serge Sautelli and Tuwaig Palace in Saudi Arabia by Baseam Shihabi and Nabil Fanous. These examples (Fig. 13) have been conspicuously successful in achieving people's social and climatic requirements; these buildings won the Aga Khan Award for Architecture (Edwards et al., 2006).

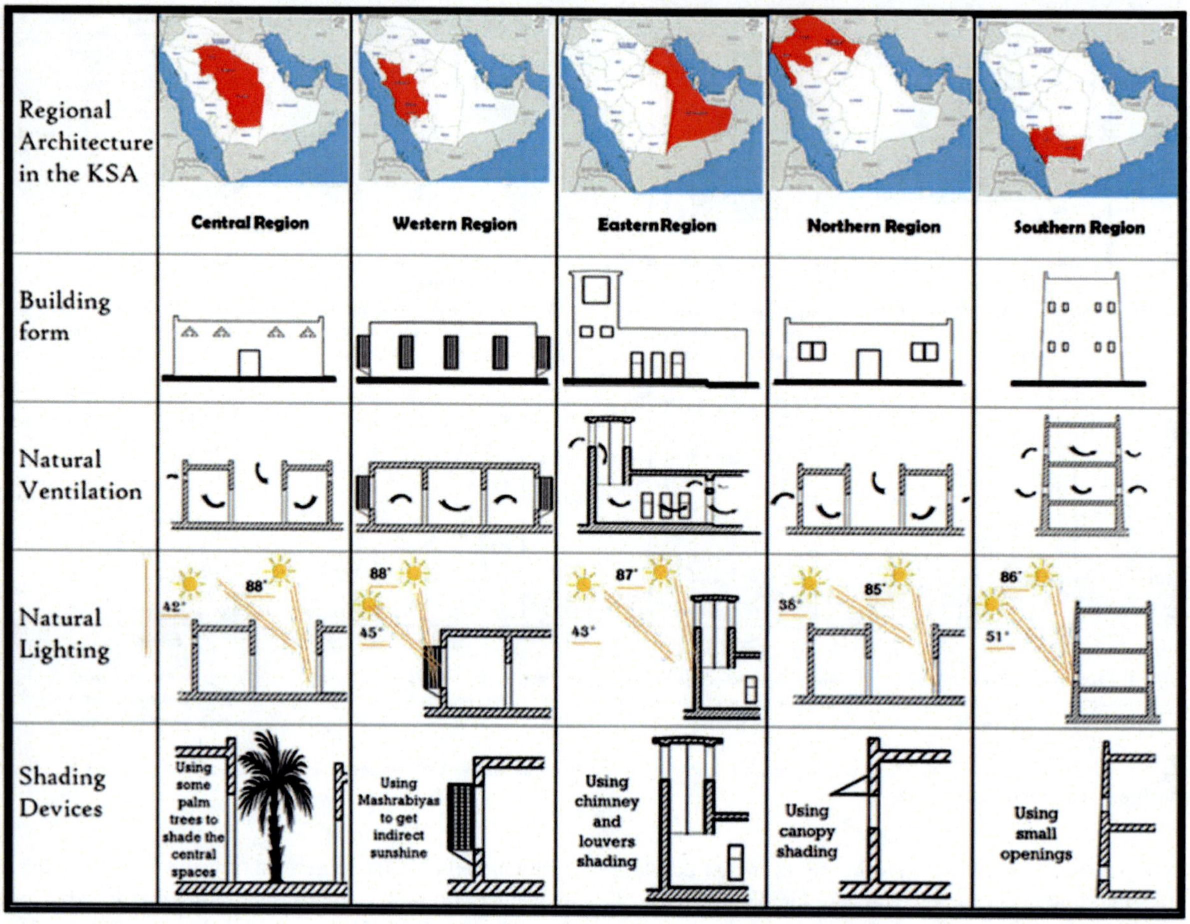

Fig. 12 Architectural Heritage Attributes and passive design strategies in the past. *Source* Author

Fig. 13 Courtyard houses in Morocco (left), Residence Andalous in Tunisia (middle) and Tuwaig Palace in Saudi Arabia (right). *Source* Adapted from Ragette (2003)

6 Closing Remarks/Conclusion

Each community over the years has produced an architecture which suits its requirements Edwards et al. (2006) stated that the design of vernacular houses achieved both social and climatic requirements. Some architectural heritage attributes could be incorporated into the modern design in order to connect the present to the past uninterrupted continuity.

The paper has presented the historical background of Saudi built environment. Then, it has attempted to study the Saudi architectural heritage design which differentiates itself by using several materials and strategies based on the various geographic areas in five regions (Central, Western,

Eastern, Northern and Southern). It has also discussed the reasons that led to the disappearing of the architectural heritage attributes in the KSA, and current attempts and ways to conserve these heritage attributes in the local built environment.

It was found from the literature review that materials and passive design strategies based on the various geographic areas are the key features of vernacular Saudi building, and some of these features could be implemented. It was observed that using a new regulation in designing houses that allows people to build a house without creating distance between next to each other; it is possible to create a comfortable modern building, while retaining the essence of its traditional frame. Owners, designers and developers could activate some conventional features and build, for example, a central courtyard house that suits their social and climatic requirements to create a comfortable modern building.

Lastly and most importantly, learning from past architecture is important because it improves the modern built environments. However, variations in climate, geographic, and context should be considered, while transferring concepts, mainly when foreign architects and designers are practizing in Saudi Arabia. In the same way, Saudi architects are supposed to combine the new style with the vernacular style but not jettison it.

References

Al-Bakri, U. A., (1997). *Natural ventilation in traditional courtyard houses in the central region of Saudi Arabia* , Ph.D. thesis, Cardiff: Welsh School of Architecture, Cardiff University.

Al-Hathloul, S., (2017). Riyadh development plans in the past fifty years (1967–2016). *Current Urban Studies*, *5*(01), 97–120. https://doi.org/10.4236/cus.2017.51007.

Al-Naim, M. A., (1998). *Continuity and change of identity in the home environment: development of the private house in Hofuf, Saudi Arabia*, Ph.D. thesis, Cardiff: Newcastle University.

Al-Oraier, F., (2005). *Thermal analysis of traditional adobe dwellings in Riyadh City, Saudi Arabia*, , PhD. thesis, Cardiff: Welsh School of Architecture, Cardiff University.

Babsail, M. O., & Al-Qawasmi, J. (2015) Vernacular architecture in Saudi Arabia: Revival of displaced traditions. In C. Mileto, F. Vegas, L. Garcia Soriano, & V. Cristini (eds.) *Proceedings of the international conference on Vernacular Heritage, Sustainable and Earth Architecture, Architecture: Towards a Sustainable Future*, 'Vernacular Architecture in Saudi Arabia: Revival of Displaced Traditions' (pp. 99–104). Valencia, Spain. London: Taylor & Francis Group.

Demirbas, A., Hashem, A. A. & Bakhsh, A. A. (2017). The cost analysis of electric power generation in Saudi Arabia. *Energy Sources, Part B: Economics, Planning, and Policy 12*(6), 591–596. https://doi.org/10.1080/15567249.2016.1248874.

Edwards, B., Sibley, M., Land, P. & Hakmi, M. (eds.), (2006). *Courtyard housing: past, present and future*. Taylor & Francis.

Facey, W. (1997). *Back to Earth: Adobe Buildings in Saudi Arabia*, Riyadh: Al-Turath.

Fathy, H. (1973). *Architecture for the Poor: An Experiment in Rural Egypt*. The University of Chicago Press.

Ministry of Municipal and Rural Affairs (MOMRA) (2018). Developed System for building. https://www.momra.gov.sa. Accessed 21 October 2019.

Mortada, H. (2003). *Traditional Islamic principles of built environment*. Routledge.

Oxley, R. (2015). *Survey and repair of traditional buildings: A conservation and sustainable approach*. Routledge.

Ragette, F. (2003). *Traditional domestic architecture of the Arab region*. Edition Axel Menges.

Rahmaan, A. U. (2011). *The Imperatives of Urban and regional planning: concepts and case studies from the developing world*, America: Xlibris Corporation.

Talib, A. K. (1983). Review of climatic design concepts and details in traditional architecture in various climatic zones-Saudi Arabia. In *Proceedings of the Second International PLEA Conference*, 'Review of climatic design concepts and details in traditional architecture in various climatic zones-Saudi Arabia' (pp. 233–243). Greece: Elsevier.

The Saudi Building Code National Committee (2018). Saudi building code-general. https://www.sbc.gov.sa/en/Pages/default.aspx. Accessed 16 June 2019.

Weather Spark. (2019). Average weather in Saudi Arabia. https://weatherspark.com/y/101171/Average-Weather-in-Jeddah-Saudi-Arabia-Year-Round. Accessed 8 Aug 2019.

Syrian Cultural Heritage Conservation and Management for Future Generations: Issues and Challenges

Mahmoud Abdulkader Alghafri and Mohannad Ali Mohammad

Abstract

Syria has many architecture and heritage sites but because of the crisis, the heritage came up with a disaster. The main reason behind this study was to analyze the process which should be applied in heritage conservation plan through and after the crisis. Along with that, it also examines the government sector policies and contributes in making plans for the preservation of the mentioned resources. The methodology adopted for this study involved a mixed method to measure Syrian local community image and perception to heritage resources, (the quantitative methodology). The qualitative approach was triggered to examine the local perspective toward heritage exactly the management and planning. Syria lacks in law of conservation process in times of crisis and to the lack of standards conservation and reuse of heritage sites. The study showed that the methods of preserving of heritage are still lacking the level of scientific and technical required. The results indicated the conservation process which used by the government is not clear to the vision of architectural restoration of heritage sites. Contrary to expectations, the results indicated that the successful conservation processes will be more useful in improving cultural heritage protection, and the level of living for the local community, which raises the local and regional economy level in general. The recommendations according to the results implemented cultural heritage conservation process prominently for locals, which can be used to architectural patrimony sites. Additional attention needs to be given to cultural heritage conservation in the time of struggles.

M. A. Alghafri (✉)
Department of Tourism Management, Faculty of Tourism, Damascus University, Kursk State University, Kursk, Russia

M. A. Mohammad
Department of Tourism Management, Faculty of Tourism, Far Eastern Damascus University, Federal University, Vladivostok, Russia

Keywords

Cultural heritage assets • Preservation • Involvement • Effective management • Unused value • Syria

1 Introduction

The conservation and revitalization of architectural heritage and historical areas is an important requirement for countries that aspire to participate actively in humanitarian giving on the cultural and historical level, reflecting their specificity, identity, and civilization. This process requires adequate knowledge of local beliefs and customs and the surrounding social, economic, and environmental conditions (Alghafri, 2019a, b, c, d), in order to develop appropriate strategies for their protection. Architectural and Cultural heritage—it belongs for all people in the world (UNESCO, 1972). It is not only the international organizations that have the responsibility of protecting, conserving, and managing heritage sites (HS) (Versaci, 2016), but also the local governments and community (Alghafri, 2019a, b, c, d). Because of the Syrian crisis since 2011 until now, the heritage in all its forms has been destroyed in a high level (Alghafri, 2019). While the Syrian government has been attempting to restore, protect, and conserve the heritage, but faces great difficulties in human and financial resources (Alghafri, 2019a, b, c, d).

1.1 Problem of the Research

The problem of the research was the Syria that lacks to the law of conservation process in times of crisis and to the lack of standards conservation and reuse of heritage sites. In addition to that, it has not included the social and economic dimensions in conservation process.

A. Versaci et al. (eds.), *Conservation of Architectural Heritage*, Advances in Science, Technology & Innovation,
https://doi.org/10.1007/978-3-030-74482-3_21

1.2 The Motive of the Research

The main reason behind this research is to examine methods which are applicable in heritage conservation plan (HCP) through and after the crisis. Moreover to find a mechanism how to protect and preserve HS, it also examines public sector policies and develops plans to conserve these resources without prejudice to their value.

1.3 Research Data

It is dependent on two levels. Initial information deals with questionnaires and interviews, whereas auxiliary information deals with reviews and others. Data were collected in two famous Syrian cities that are Damascus and Aleppo. Two batches of the questionnaires were distributed among the local community and managers.

1.4 Research Method

Data were analyzed using the SPSS Version 25. The random sampling technique for the questionnaire survey (Q1-local respondents) and the target sample (Q2-heritage managers) were the sampling methods in this study. It was examined for the reliability of the frequency in the experiment in order to the Cronbach's Alpha (α). In example, $\alpha \geq 0.9$ means excellent, $0.9 > \alpha \geq 0.8$ indicates very good, $0.8 > \alpha \geq 0.7$ means good, $0.7 > \alpha \geq 0.6$ means moderate, and $0.6 > \alpha \geq 0.5$ indicates weak (Hair, et al., 1998). In experiment, Cronbach's Alpha coefficients were used for each constraints used in order to achieve the accuracy in the constraints. As per system, accuracy constraints of 0.70 could be considered enough, although majority of researchers considered 0.60 as sufficient as per Table 1 (Yates, et al., 2008; Zulfiqar & Bhaskar, 2016).

All measurements that were inputted in every part in the questionnaires were examined by five-point Likert's scale and resulted to be enough, and reliability assumption reached its target and was perfect for additional study. The Pearson's correlation was used to examine to identify vital connection between two inputs in this study. The correlation coefficient highlights value between the negatives and positives ($-1< r >1$). A perfect connection of (1 or −1) triggers variables can be decided clearly by recognizing the numbers of the other variables. Finally, the correlation of (0) triggers was no relationship variables. The researchers took decision to obey (Cohen, 1988) methodology that can examine the connection as follows: In examples, 0.1 to 0.29 or −0.1 to −0.29 means weak, 0.1 to 0.29 or −0.1 to −0.29 means moderate, 0.1 to 0.29 or −0.1 to −0.29.

1.5 Research Methodology

In this study, both qualitative and quantitative methods were used with different participants. The survey period lasted 22 days in February 2019. 300 samples of the survey were distributed in the cities of Damascus and Aleppo and 291 responses were received from the local community. 25 samples of the survey were distributed to stakeholders (tourism officer, academician, conservation officer, and NGO activist) and 19 responses were recorded.

2 Research Results

2.1 Descriptive Statistics (Mean/Percentage)

Factors resulted to enrich high levels of local community (LC) respondent's point of view for cultural and architectural heritage conservation and management. In example, 98.0% of the LC stated that conservation management (CM) inspired variety of cultural activities by LC and flexibility of Syrian law encourages attraction of foreign capital and investment. In addition, the involvement of LC in cultural and architectural heritage conservation projects is to manage it efficiently (Table 2).

Also, 97.6% create experiences in conservation process, because there are many global heritage sites in Syria, while 97.0% of the conservation process created better job opportunities, i.e., 95.6% of the locals encourage inbound and outbound tourism through CM of cultural and architectural heritage sites (AHS). It was agreed by 95.9% of the locals with the good CM of cultural and AHS triggers to create investment opportunities which injected massive summation of cash.

Table 1 Inputs for this study

Variables	No. of items	Cronbach's Alpha Society	Resistance
Cultural and architectural heritage conservation and management	13	0.570	Weak
Importance of counting this unused values of architectural and cultural heritage	6	0.670	Moderate

Table 2 Cultural and Architectural Heritage Conservation and Management (Average: 3. 96)

Inputs	Local community		
	Average	SD	(%)
Good administration of cultural and architectural heritage sites (AHS) provides good jobs	4.16	0.86	94.7
Investing in the cultural and AHS is a waste of money	2.53	1.00	45.8
The condition of the society reached to next level in comparison with LC by cultural and architectural heritage conservation (AHC)	4.00	1.00	87.2
Create experiences in conservation process; because they are many global heritage sites in Syria	4.31	0.68	97.6
Good conservation management of cultural and AHS helps create investment opportunities which provide a huge of money	4.23	0.82	95.9
The conservation management actions triggered massive improvement in usage	3.87	0.97	88.1
The absence of professionalism in the field of conservation distorted general perspective of HS	3.91	1.03	86.9
The conservation process provided a goal for local cultural heritage restoration agenda	3.63	1.29	74.1
The conservation process created more job opportunities	3.89	0.76	97.0
The CM encouraged more choices for culture movements by locals	4.28	0.75	98.0
The flexibility of Syrian law encourages attracting foreign capital and investment	3.89	0.76	98.0
Encourage inbound and outbound tourism through CM of cultural and AHS	4.03	0.78	95.6
The involvement of LC in cultural and AHC projects those saves time and effort	4.34	0.66	98.0
Number of participants	*291*		

Legend: High = 4.00–5.00; moderate = 3.00–3.99; and low =1.00–2.99

2.2 The Views on the Unused Values of Heritage

Significantly there is no substances in conserving unused values from point of view of locals. Nevertheless, respondents wanted the importance of protecting the unused value which was in good progress (above 98%). Table 3 shows probability of the survey field on views toward importance of preserving the unused values on the CH. For locals, importance of gain when saving the unused values could make changes to other people in experimenting will be adequate values and justification on the CH in the near future.

2.3 The Views of the Managers

The study in Table 4 resulted 94.7% of the stakeholders agreed with significance of combating the CH from undecided development without concerning of the cost. This result has been consistent with quantitative survey of (Q1) where they indicated good reaction toward the CHC. Results that triggered stakeholders were conscious in creating a shield in their heritage. Interviews made sure that ministry differently internal issues to counter locals for immovable heritage. Moreover, research found that 84.3% of the

Table 3 Points of conserving unused values of the architectural and cultural heritage in Syria (total average 4.24)

Variables	Local community		
	Average	SD	(%)
Significance of conservation of the Syrian cultural heritage places (SCHP) so that the LC can continuously make gain in future	4.27	0.70	98.1
Significance of conservation of the SCHP so that outsiders can judge and appreciate its importance of CH in future	4.26	0.73	98.0
Significance of conservation of the SCHP to keep the uniqueness of CH	4.17	0.70	97.9
Significance of conservation of the SCHP to boost the identity of this historic country	4.17	0.70	97.1
Significance of conservation of the SCHP to contribute cultural and past result of the site	4.26	0.68	98.0
Significance of conservation of the SCHP to give us the opportunity to preserve the verification for future	4.31	0.66	98.1
Number of participants	*291*		

Legend: High = 4.00–5.00; moderate = 3.00–3.99; and low =1. 00–2.99

Table 4 Information of the managing bodies

Inputs	No	(%)
The responsibility to combat CH		
Yes	18	94.7
No	1	5.3
Total	19	100.0
Sufficient funding support		
Yes	3	15.7
No	16	84.3
Total	19	100.0
Generating income		
Yes	4	21.1
No	15	78.9
Total	19	100.0

stakeholders are happy about the fund that was not enough for its HS preservation and management.

For government involvement in profit enriching scheme for the protection, 78.9% of critics mentioned lack of aims, concept, or projects injected through management for boosting revenue for preserving SCH. Only 21.1% of them highlighted that managing bodies injected an uprising revenue program or for the SCH. According to the results, it indicates managing bodies did not produce any positive output and concept for improving revenues for the locals.

As per research, 81.8% of the managers found it valuable for striking concreteness for conservation, biodiversity, and accessibility. Syrian local community and potential economic use of world heritage sites (WHS), (Table 5). 54.5% of the managers agreed to keep in record the fund for rebelling WHS with its construction, also any smooth zone, from unrequired improvement to protect WHS from climate fluctuation but confirming negative growth is not at all the expenditure of integrity. Simultaneously, 72.7% of the stakeholder decided for using the capital in rebelling WHS for being accurate through appreciable management.

3 Discussion of Research Result

Cultural and AHC management is a difficult process but good conservation management of cultural and AHS helps to create investment opportunities which provides massive amount of money (Ahmad, 2014) that triggered a more positive field for the job market. In addition, the involvement of LC in cultural and AHS projects saves time and effort (Alghafri & Mohammad, 2018; Alnafeesi, 2013). The Syrian government faces major challenges to preserve the cultural and AHS. Nevertheless, Syrian war prevented the goals of the conservation process. Furthermore, non-involvement of LC in conservation management plan, transmigration and demographic direction, inefficient funding and human resources (Fitri, et al., 2015), the deficiency of experts in field of HCM, inappropriate CM process and lack of public awareness and support; creating a negative curve in cultural and AHS.

As interpreted in Table 6, outcomes of the analysis are between the interest groups' insight about AHC management.

Table 5 Authenticity and conservation from the Syrian cultural and architectural heritage

Variable	Managers (11)	
	No	(%)
Preserve the WHS, including any area around this heritage site, without any suitable improvement	6	54.5
Evaluation between the needs of conservation, biodiversity, access, the interest of LC and the potential economic use of the WHS	9	81.8
Preserving the WHS through positive management (where appropriate)	8	72.7
Preserving the WHS from weather fluctuations which ensures that mitigation doesn't triggers genuine expenditure or honesty	6	54.5
Others, more specific	0	0

Table 6 Statistically significant relationship between the LC's point of view toward cultural and architectural HC and management

Variable		Syrian cultural heritage protection and management for future generations	Highest Syrian cultural heritage protection and management for future generations
		Local community	Local community
Cultural and architectural heritage conservation and management	Pearson's correlation	0.171**	0.051
	Sig. (two-tailed)	0.161	0.956
	No	291	291

**Correlation is significant at the 0.01 level (two-tailed)

As a result, insignificant hypothesis of: "Is there any statistically significant relationship between the local community's perceptions about the cultural and architectural heritage conservation and management," which answered and was not accepted for the numerical value of p that resulted to be higher than $\alpha = .01$ and 0.05. So, Syrian government must focus on heritage conservation to save it for future generation.

Table 7 comes up with the connection between the LC views on the significance of preserving unused values of cultural and AHS in Syria. Thus, the insignificant hypothesis of: "Is there any statistically significant relationship between the LC views on the importance of preserving unused values of cultural and AHS in Syria," was accepted due to p that was smaller than $\alpha = .05$.

As explained in Table 8, results of multiple straight retraction examinations for any relevance of local answers are between the attitudes toward cultural and AHC management. The significant of preserving the unused values of the cultural and AHS in Syria is toward the highest CHC and management for future generations. From the analysis and the variables, the Score A1 and Score A2 did not affect the Syrian CHC and management for future generations due to $P > \alpha = 0.05$.

Syrian government who had the primary duty to protect and conserve cultural and AHS had the site under their control, by coordination between international organizations and involvement of LC in conservation plan. As a result, needs and interests of the LC have the importance for establishment. The consultation with LC in AHC plan will be more efficient. Hence, it is worthwhile to focus on the preservation and management of heritage buildings in a manner that preserves their original architectural character and their durability by using them appropriately (Versaci, 2016). In addition, addressing the LC and sensitizing them to

Table 7 Statistically significant relationship between the local community's point of view toward significance of conserving of not using cultural and architectural heritage in Syria

Variable		Syrian cultural heritage protection and management for future generations	Highest Syrian cultural heritage protection and management for future generations
		Local community	Local community
The significance of preserving the unused values of the cultural and architectural heritage in Syria	Pearson's correlation	0.129*	0.402
	Sig. (two-tailed)	0.043	0.953
	No	291	291

*Correlation is significant at the 0.05 level (two-tailed)

Table 8 Relationship between Score A1, Score A2 the highest Syrian cultural heritage conservation and management for future generations (local respondents) excluded variables[a]

Model	Beta In	t	Sig	Partial correlation	Collinearity statistics
					Tolerance
(Score A1) cultural and architectural heritage conservation and management	0.012^b	0.212	0.824	0.011	0.809
(Score A2) the significant of conserving the value that is not used for cultural and architectural heritage in Syria	-0.019^b	−0.294	0.771	−0.015	0.925

[a]Dependent variable: Highest Syrian cultural heritage conservation and management for future generations

Table 9 Issues arise in cultural and architectural heritage management faced by international organizations, local government, and local communities

No	Stakeholders	Issues
1	International organizations	• Non-strong support from international cultural organizations such as UNESCO • The influence of international media negatively affects the Syrian crisis and, consequently, the image of Syria in the minds of tourists • Lack of resources and financial support • For a long time, the role of international organizations was limited • Non-compliance with the law of Hague In 1945 and 1999 that published by UNESCO; on the safeguarding CH in crisis
2	Local government	• Lack of clear policies and plans for CM of cultural heritage • Lack of specialized personnel in HCM • Destruction of a large number of heritage sites due to war • Collapse of heritage sites due to lack of restoration • Lack of a special conservation office and staff for CHS • Non Involvement of LC in conservation management plan
3	Local community	• Lack of awareness of LC in the importance of heritage • Urban crawl on heritage sites • Small contribution of the private sector to the protection and conservation of WHS • Lack of financial resources, technical knowledge and skills of the local community engaged in job creation • Not high interest from partners to support the LC in finding work • Potential partners are not ready to support and fund conservation projects • Illicit trafficking in cultural objects • Continued migration of capital and the social elite, such as university professors in tourism and heritage due to the crisis

the importance of preserving the Syrian AHS and involving it in making decisions concern their heritage (ICOMOS, 1990). Table 9 shows positive growth in the deficiencies in CHM faced by international organizations, local government, and LC.

Certainly, investment in architectural heritage sites leads to many local benefits resulting from the reestablishment and employment projects for a number of heritage villages, historic city centers, popular markets, citadels, and historical palaces, which are spread in the areas of Syria. Most of which are still in good condition (Gholitabar, et al., 2018). They can be rehabilitated and invested. Thus, providing employment opportunities for LC in those areas, and contributing to the reduction of the phenomenon of population migration, as a result of their association with the jobs afforded to them by the investment projects of AHS (UNESCO, 2013).

For the investor and the LC, increase the employment of people in their sites; to participate in the practice of activities, events, and investment services in the development of AHS (Guzman et al., 2017). Employment opportunities resulting from investment in rehabilitation and conservation projects of AHS are very large; any project in this area provides, for example, job opportunities for engineering consultancy offices that prepare studies and project plans, and oversee their implementation, as well as job opportunities for institutions (Lussetyowati, 2015). Preservation companies work in the field of conservation heritage and encourage the LC to invest in the restoration of conservation heritage buildings, and other investment areas in support of these projects (ICCROM, 2016), such as transport services, crafts and handicrafts, sale of local products, accommodation, parking, and the establishment of some commercial centers or cultural, add value due to their stay next to the architectural heritage buildings to set up more than anywhere else.

The social benefits generated by investment in AHS are as important as the economic benefits. The social benefits are derived from economic benefits, which are to provide employment opportunities for LC, which increases their permanent association with it, thus increasing their contributions to development (Mohammad, 2019). The project had the importance of investing in development for undeveloped regions in order to increase the availability of job opportunities that contribute to the reduction of unemployment rates, as well as its contribution in reducing migration to the big cities (Gamini, et al., 2013).

Fundamental constraints in projects of HCM in Syria are constructed for funding. HCM, which was backed by government, private institution, or international organizations. It was insufficient for restoration, rehabilitation, and management of AHS. Under the Ministry of Culture is integral, national heritage was managed in Syria. In terms of CH assets, as a CH area, preservation is linked with other sectors in Syria, majority of them were misused (Poudel, 2014). For conserving CH, process to take decision must be authentic between government and community, supervision by UNESCO or other organization to bring a positive growth in the valuation of CH and confirms its compatibility with the principles and guidelines as mentioned. Due to the ongoing war, it was challenging to apply HCP in many regions in Syria.

4 Conclusion

As per research, the methods of preserving of AHS have deficiency in the level of scientific and technical requirements for the non-presence of coordination in terms of controlling the heritage resources between the locals and other interest groups in continuing Syrian crisis since 9 years. In addition, Syria has not sufficient financial support, experts, and human resources. Creating institutional frameworks and promoting a given for conservation of architectural heritage public authorities should be aware of the fact that the strength of developed local economic through successful heritage management and conservation. Additionally, CHC from local point of view ensures a new approach for HS planning and management in Syria. Therefore, the careful planning of CHM and conservation leads to sustainable regional development. Contemporary, the successful conservation processes will be more useful in improving CH protection and the living standards for LC affecting local and regional economy in a positive way. Conserving WHS for uprising and knowledge is very important. Furthermore, the interference for local looking at CH is important for keeping their heritage. In conclusion, the construction of the scientific interference in the conservation of Syrian heritage in times of crisis is required. When taken into account, it will navigate Syrian model HC improvement leading to a growth in the overall CH and tourism in territory.

Acknowledgements We would like to thank the experts and governmental organizations which involved in the validation documents for this research project and they have given me much help: Directorate General of Antiquities and museums (DGAM), Damascus, Syria. Faculty of Tourism in Damascus, Ministry of Tourism, Syria, Kursk State University library in Kursk, Russia. Without their highly coordination, participation and input, the validation survey could not have been successfully conducted without their help.

References

Ahmad, A. (2014).Cultural heritage tourism in Malaysia: Issues and challenges. In: *4th International Conference on Tourism Research (4ICTR)* (Vol. 12, pp. 2–8).

Alghafri, M., & Mohammad, M. A. (2018). The future investment of cultural heritage in Syria. In: *International Scientific and Practical Conference "Strategy for the Development of Border Territories: Traditions and Innovations". Kursk State University* (Vol. 5, pp. 308–314).

Alghafri, M. (2019a). Cultural heritage and sustainable tourism development in Syria. In: *Materials VII International Scientific and Practical Conference "Economic Science and Practice", Krasnodar* (Vol. 7, pp. 42–45).

Alghafri, M. (2019b). Cultural heritage management as a factor in the socio-economic development of territories. In: *Materials International Scientific and Practical Conference "Actual Issues of Economics and Management", April, St. Petersburg* (pp. 3–8).

Alghafri, M. (2019c). The effect of cultural heritage in local economic development (for example, Syria). *Economics and Society., 2*(57), 3–9.

Alghafri, M. (2019d). The role of investment projects in the renaissance of the Syrian cultural heritage. *Theory and Practice of Modern Science., 2*(44), 423–430.

Alnafeesi A. M. (2013). The role and impact of preservation architectural heritage on domestic tourism. In: *Proceedings of International Conference on Tourism Development* (pp. 93–100).

Cohen, J. (1988). *Statistical power analysis for the behavioural sciences* (pp. 11–12). Lawrence Erlbaum Associates.

Fitri, I., Ahmad, Y., & Ahmad, F. (2015). Conservation of tangible cultural heritage in Indonesia: A review current national criteria for assessing heritage value. *Procedia-Social and Behavioral Sciences, 184*(1), 71–78.

Gamini, W., Thompson, J., & Young, C. (2013). *Managing cultural world heritage* (p. 152). UNESCO.

Guzmán, P. C., Roders, A. P., & Colenbrander, B. J. F. (2017). Measuring links between cultural heritage management and sustainable urban development: An overview of global monitoring tools. *Cities, 60*, 192–201.

Gholitabar, S., Alipour, H., & Costa, C. M. M. D. (2018). An empirical investigation of architectural heritage management implications for tourism: The case of Portugal. *Sustainability, 10*(1), 93.

Hair, J., Anderson, R. E., Tatham, R. L., & Black, W. C. (1998). *Multivariate data analysis* (p. 23). Prentice Hall.

ICOMOS. (1990). Charter for the protection and management of the archaeological heritage. ICOMOS. Lausanne.

ICCROM. (2016). Conservation of cultural heritage in the arab region. ATHAR programme. ICCROM, UAE (p. 143).

Lussetyowati, T. (2015). Preservation and conservation through cultural heritage tourism. Case study: Musi Riverside Palembang. *Procedia-Social and Behavioral Sciences, 184*, 401–406.

Mohammad, M. A. (2019). Factors affecting community participation in heritage resources management in Syria. *Journal of International Economic Affairs, 9*, 705–712.

Poudel, S., Nyaupane, G. P., & Budruk, M. (2014). Stakeholders' perspectives of sustainable tourism development: A new approach to measuring outcomes. *Journal of Travel Research., 55*(4), 465–480.

UNESCO. (1972). *Convention concerning the protection of the world cultural and natural heritage* (p. 19). UNESCO.

UNESCO. (2013). Creative economy report 2013: Widening local development pathways (p. 19). New York and Paris: UNDP/UNESCO.

Versaci, A. (2016). The evolution of urban heritage concept in France, between conservation and rehabilitation programs. *Procedia-Social and Behavioral Sciences, 225*, 3–14.

Yates, D. S., Moore, D. S., & Starnes, D. S. (2008). *The practice of statistics*, 3rd edn. (p. 35). Freeman.

Zulfiqar, A., & Bhaskar, S. B. (2016). Basic statistical tools in research and data analysis. *Indian Journal of Anaesthesia., 60*(9), 662–669.

Urban Conservation of Heritage-Sensitive Zones in India: A Methodological Approach

Shahim Abdurahiman and A. K. Kasthurba

Abstract

Most developed nations utilise heritage infrastructure and artefacts as potential resources for their urban growth to showcase their valuable cultural, historical and socio-economic significance. In India, urban heritage resources are often under-utilised and often remain abandoned or poorly maintained. The historical cities in India along with its network of heritage resources, need to be commensurate with its rapid urban development. In India, current regional Master Plans incorporate a judicious attempt to protect and conserve the architectural character of the region, by delineating a boundary around this area and tag it as 'heritage zones', 'special area zones' and 'protected zones'. These heritage-sensitive zones were focal areas of human activities within the cities and are ever-evolving with enormous scope for development. Development of these heritage-sensitive urban areas offers challenges for city planners and heritage conservationists. Heritage-sensitive urban development needs to be inclusive despite the compulsive developments creating a negative impact on the heritage areas. The objective of this paper is to integrate concepts of urban planning with architectural conservation involving heritage assets with concepts of authenticity and integrity, for sensitive urban development. This paper attempts to adopt a methodological approach for urban conservation of historic zones to facilitate sustainable development of cities, incorporating heritage resources. The objective of the study was achieved through literature study and review of case studies with expert opinions concerning urban development in historic urban areas. The approach is broadly divided into two stages which facilitate a step-by-step phasing. The first stage consists of documentation of heritage assets to establish the heritage value by understanding the urban context, identifying the urban heritage assets, establishing the significance by measuring its authenticity and integrity. The second stage mainly deals on how to incorporate it within the urban planning framework. The approach can be further adapted in context-specific heritage areas. It aims at reviewing and understanding the potential for a change in the overall approach in looking at heritage zones/regions/cores and their heritage assets. This study will aid in the betterment of the preparation of development masterplans. It would create a template for policy-making in heritage-sensitive areas and listing of urban heritage resources within a city.

Keywords

Heritage city • Urban conservation • Urban development • Heritage zones • Urban heritage

1 Introduction

Urban heritage, a repository of history and invaluable knowledge systems, constitutes an irreplaceable resource for any city (UNESCO, 2011). Going beyond conventional perceptions of 'monuments', urban heritage encompasses living habitats, historic environments and cultural assets, both tangible and intangible, which undergo continuous transformation (Labadi & Logan, 2016; Udeaja et al., 2020). Globally in developed nations, urban development policies are formulated to highlight the architectural identity of the city and to generate economy utilising these heritage resources (Licciardi & Amirtahmasebi, 2011; Serageldin, 1999). In India, development schemes and masterplans often view urban heritage as a burden, rather than inspecting them as an asset that needs to be protected and nurtured for future generations (Steinberg, 1996; Udeaja et al., 2020). To ensure an equitable and culturally sustainable development

S. Abdurahiman (✉) · A. K. Kasthurba
Department of Architecture and Planning, National Institute of Technology Calicut, Calicut, Kerala, NIT Campus P.O. 673601, India
e-mail: kasthurba@nitc.ac.in

A. Versaci et al. (eds.), *Conservation of Architectural Heritage*, Advances in Science, Technology & Innovation,
https://doi.org/10.1007/978-3-030-74482-3_22

trajectory, it is imperative to conserve and manage these heritage assets to enhance the quality of space and life in any historic settlements (Zancheti & Hidaka, 2011).

Urban experience varies for different cities based on its character and context of these elements ranging from large public spaces to landmarks (Zancheti et al., 2006). Heritage experience in an urban area facilitates dynamic understanding and learning involving historical elements that constitute the city. The urban heritage elements embodied in cities have an impact on the urban experience to create personal meanings to urban spaces. Hence, the heritage resource needs to be seen as an asset which helps to enhance economic value and grandeur in urban development plans (Park, 2016). Although the urban public values the heritage asset, there is a lack of awareness in utilising them to its full potential for enhancing the urban quality (Greffe, 2004). Conservation of urban heritage involves protection and preservation of the historical assets for the present-day and forthcoming generations. It requires inputs from a wide range of professionals with a multidisciplinary approach.

Urban conservation is a muti-speciality team discipline involving various specialists from various disciplines with additional skills and values. A thorough understanding of the historical context and the values embedded are imperative for a better understanding of the historical setting and its sustainability. Over the years, society has become aware of the need to protect and conserve the cultural heritage within its historical context to bring about the essential vitality to the urban setting. Many professionals within varied planning frameworks have worked on 'solutions' or ways of zoning these heritage-rich areas to facilitate urban development by preserving their memories, without harm to these historical shreds of evidence. The paper aims to adopt a systematic methodological framework for urban conservation of historic zones to facilitate sustainable development of cities, incorporating heritage resources. The research paper intends to integrate concepts of urban planning with heritage conservation of heritage assets with concepts of authenticity and integrity, for sensitive urban development.

2 Literature Background

2.1 The Concept of Sustainability in Urban Conservation

Understanding the transition of value from historical to present helps in shaping the future. The historical urban landscape encompasses various components such as monuments, social complexes and living heritage which is embedded in various layers of culture and aesthetics. Rapid urbanisation often poses a threat to the sense of place and identity of communities. Resource value and obligation towards society are two primary stimuli for conservation (Lichfield, 1988). The resource value is vital for a sustainable method in conserving the existing, as it is a sustainable approach by saving resources than the construction of new ones, especially when the quality and value of the existing is superior to the new ones (Elnokaly & Elseragy, 2013). Conservation of heritage sites in urban areas catalyses sensitive transformation by improving the urban setting and environment to its community context. Contemporary architecture practices are often based on subjective perceptions which disregard a continuum (Yung et al., 2016). Adaptive reuse of existing old buildings is a vital tool in sustainable urban conservation (Kong & Yeoh, 1994), to indirectly control urban sprawl. Preventing the destruction of obsolete property and infrastructure can help to provide a stand-in space for future needs.

The notion of sustainable development is one of the fundamental agendas that has been practised globally over the past few decades (Steinberg, 1996). The concept of urban conservation concept that has gained its relevance since the mid-twentieth century. The integration of 'urban heritage conservation' and 'sustainability' can be brought together in a common philosophy and practice by setting out the relationship between heritage conservation and sustainability, which will maintain and enhance cultural identity. Conservation and preservation are to be handled keeping in consideration the traditions and identities in the right composition, constant and meaningful. Heritage assets are urban resources for sensitive development, simultaneously building the heritage in a much more socially sensitive way respecting the needs at the same time enhancing the quality of life for a sustainable future that aims for social equity, cultural diversity and the right to memory, within an interdisciplinary and interinstitutional approach (Figueiredo, 2014).

2.2 Urban Heritage and Heritage Zones

Urban conservation is an integral part of contemporary heritage policies. Historic cities stand out from modern cities as places of significance. They have become the nation's icons that facilitate global cultural tourism and places for the enjoyment of a different and distinctive lifestyle. Urban Conservation is a practice of understanding and managing heritage areas for the future in a much more sustainable way, while urban morphology reflects the area through the physical changes and manifestations evolved through time (Whitehand & Gu, 2007). Urban conservation deals with the conservation of tangible and intangible heritage assets in the built environment or urban cultural landscape that hold heritage value and significance. The tangible includes single entity properties like buildings, a cluster or group of structures, public open spaces, streets, water bodies like tanks, gardens or landscapes and other historic infrastructure. The Archaeological Survey of India (ASI) and their divisional state counterparts protect the heritage of India. It includes monuments, archaeological sites, remains and ruins, widely constituting the tangible fabric, which provides a distinctive national identity and significant character to towns or cities. Development projects for urban areas seldom take into account the protected and unprotected components of urban heritage. Among the urban planners, urban heritage is often limited to 'monuments' that mainly constitute religious structures, palaces, fort walls and public intuitional buildings. The historical setting- historic settlements and city centres are blindly excluded at times. They equally contribute to the 'urban heritage'. Furthermore, apart from the tangible aspects, there is a lack of initiative of incorporating and understanding of the non-tangible assets, which is the soul for the built environment.

3 Case Study: Status of Heritage Sites in India

Most of the heritage structures in Indian urban cities are in a dilapidated condition and demands a sensitive approach to preserve and conserve these structures by the public and private sector. According to United Nations estimates in 2018, about one-third of Indians, i.e. 461 million, live in Indian cities, which is statistically expected to get doubled by 2050 (Sankhe et al., 2010; UNESCO, 2015). The task of providing a better quality of life in historic urban areas is a challenge faced by city administrators. Protecting, preserving and conserving heritage structures in cities, in such a scenario would be less prioritised. As of 2020, the United Nations Educational, Scientific and Cultural Organisation (UNESCO) has recognised 38 World Heritage Sites in India. In 2016, Nalanda Mahavihara or old Nalanda University of Bihar was enlisted as a UNESCO World Heritage Site. In 2019, Orchha located in Madhya Pradesh was enlisted in the UNESCO World Heritage sites tentative list. In 2015, the Government of India launched the National Heritage City Development and Augmentation Yojana (HRIDAY) to improve infrastructure and amenities around heritage sites.

The ASI protects 3650 ancient monuments and sites with an allocated fund of 975 crore rupees in 2018–2019 budget. Twelve cities were selected and 70 projects worth 422 crore rupees were approved.[1] Very few urban cities in India have attained the World Heritage tag, Ahmedabad being the first in 2017. In July 2019, Jaipur was enlisted as a World Heritage City. Figure 1 shows the Hawa mahal and Sireh Deodhi Bazaar, located at Jaipur, which is well maintained. With the enlisting of cities under the UNESCO world heritage, it ensures effective conservation, maintenance and management of heritage sites and properties, a concern that has been debated by local development authorities the possibility of progressive infrastructural development with the enlistment. In 2015, the nomination of Delhi's Shahjahanabad and Lutyen's Bungalow zone was withdrawn by the central government. Instead, now proposals have been made for the redevelopment of the central vista.

Apart from the protected and listed monuments and sites, there are heritage sites and precincts that hold more significant value to its local community or even to the whole. Recent events have shown a lack of protection and false approaches to them. In June 2019, Patna's Gole market was demolished by the local authorities for the proposed Smart City project. The market is more than a century old and is Patna's first planned municipal market and is embodied with local heritage values. The proposed plan of the project incorporates a multistorey commercial building, a contemporary municipal market along with retail shop areas. Ironically, Gole Market was also recognised as a heritage property in a Bihar government publication—Patna: A Monumental History. Regardless of the historical and heritage value of the structure, the demolition drew meagre public protest despite their strong sentimentality. Figure 2 shows the dismantled historic Gole market to make way for a seven-storied commercial complex under 'Smart City Project'. Similarly in 2018, the first public hall of Patna—the Anjuman Islamia Hall which is 133-year-old, was demolished to construct a modern commercial complex.[2] The lack of concern for heritage structure has been discussed and debated. However, local historians and heritage specialists have raised the issue as an attempt to obliterate the layer of colonial history.

In the city of Thiruvananthapuram in Kerala, the state government was unaware of Kowdiar's status as a heritage zone while allotting land in 'conservation area', for developing and constructing the 'Kalam Knowledge Centre and Space Museum'. The matter eventually raised a conflict between the art and heritage commission and government institutions over clearance for various projects. The recommendation of art and heritage commission to modify the plan at Kowdiar is the latest instance of lack of coordination between government departments. Over the past decade, the land around Kowdiar has been sub-divided and sold as plots defying all norms mandated by Kerala municipality building

[1] Ministry of Culture, Ministry of Housing and Urban Affairs, PIB.

[2] https://www.thehindu.com/news/national/other-states/patnas-more-than-a-century-old-heritage-market-demolished-for-smart-city-project/article27956017.ece.

Fig. 1 Hawa Mahal and Sireh Deodhi Bazaar, Jaipur (*Source:* © www.indiahighlightvoyages.com)

Figure 2. **a**, **b** The dismantled historic Gole market in Patna (*Source:* ndtv.com)

rules as well as directives of heritage commission.[3] Figure 3 shows the Kowdiar region which was the main domain of the Travancore Rajas. The region encompasses the heritage structures including the Kowdiar Palace shown in Fig. 4.

Heritage Conservation and development ought to be seen inclusive within the Smart City Programme, to safeguard and preserve our heritage assets. Several attempts have been made on Indian cities, to address the issues of heritage precincts or areas in the master plan. One such is in the Delhi 2021 master plan, under the chapter, 'Conservation of Built Heritage'. Jaipur has taken a step forward in preparing a Heritage Management Plan and integrating it along with the master plan. The lack of proper planning procedures and strategical system in the protection of urban heritage has resulted in the poor management of urban heritage resources.

As mentioned earlier, the ASI under the 1958 'Ancient Monuments and Archaeological (Sites and Remains) Act, protects, maintains and regulates the heritage sites and properties. The states have their respective state regulations to protect and regulate the assets within their legislative boundary, but they are only limited to the monuments, sites and remains. Buffer zones are demarcated to prohibit encroachments and regulate development activities. The protection and regulation of the urban heritage are seldom supported efficiently by the local government bodies and local policy-making. However, the identification and demarcation heritage zones are weakly discussed and presented in the city development plans with no strong guidelines that can ensure sensitive development. Attempts for an improved system of

[3] https://timesofindia.indiatimes.com/city/thiruvananthapuram/govt-unsure-about-heritage-zones/articleshowprint/67731031.cm.

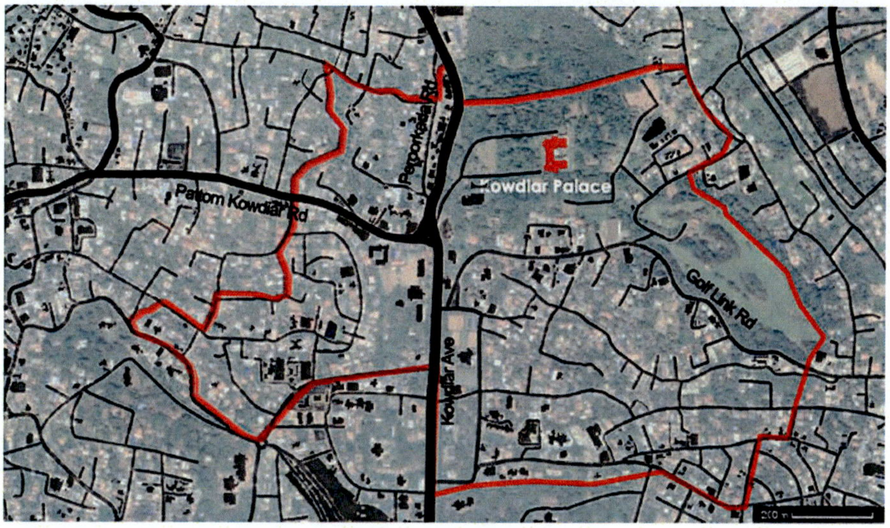

Fig. 3 Kowdiar Region, Thiruvananthapuram (*Source* Adapted from Google Maps)

Figure 4. a, b The Kowdiar Palace (*Source:* hellotravel.com)

sustainable planning and governance with the focus on the development of Heritage precincts have been carried out through the 'Jawaharlal Nehru National Urban Renewal Mission' (JNNURM) scheme.[4]

4 Material and Methods

The research methodology adopted for the study is by carrying out extensive literature survey and expert opinions from case studies concerning the status of heritage-sensitive areas in India and their management. The scope of the research is to develop a methodological approach in carrying out sensitive urban conservation in historic precincts. The step-by-approach has been established and broadly divided into two stages, as shown in Fig. 5. The first stage documents the heritage assets to establish its value by understanding the urban context, identifying the urban heritage assets, establishing the significance by assessing its authenticity and integrity. The second stage deals with steps on how to incorporate heritage assets within the urban planning framework. Further sub-sections briefly explains the steps to device the methodological framework.

4.1 Understanding the Urban Context

Urban heritage can be broadly classified as—tangible heritage and intangible heritage. The tangible heritage includes

[4] Supplement Toolkit on City Development Plan with Focus on Heritage. Jawaharlal Nehru National Urban Renewal Mission (JNNURM), Ministry of Urban Development & Ministry of Housing and Urban Poverty Alleviation, Govt. of India, 2006.

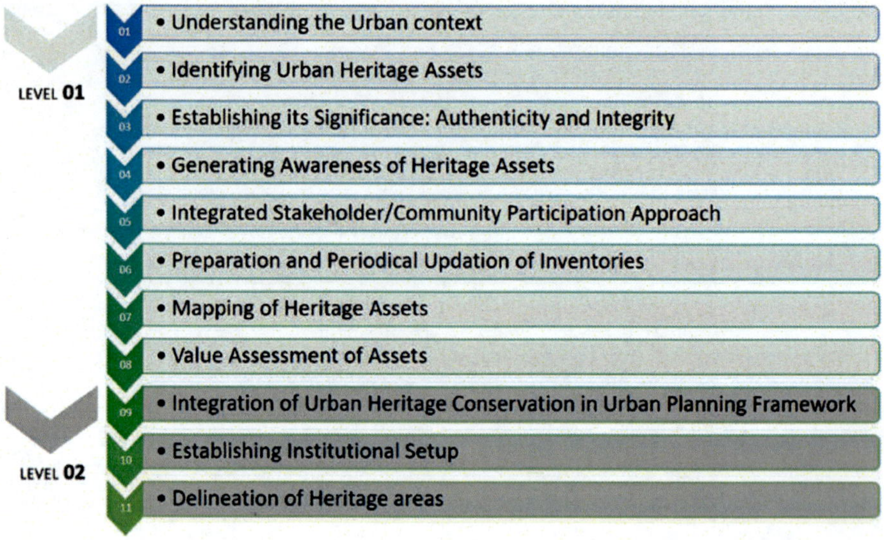

Fig. 5 Methodological Framework (Author)

the physical components of heritage value and significance. The intangible heritage comprises the immaterial components that are associated to the tangible heritage such as visual and performing art, culture, music, literature, culinary tradition language, dialects, folklore, legends, myths, traditional knowledge, spirituality, movable artefacts, handicrafts, traditional medicine, rites, rituals, festive events, etc. Understanding the tangible and intangible, and their interrelation, within the Urban Setting is essential. Historical and documentary research is to be carried out for an understanding of urban context and value of the individual assets embodied within.

4.2 Identifying and Listing Urban Heritage Assets

Preliminary identification of a heritage asset in a city involves the recognition of the historical value associated with it. This can be developed or gathered through oral history or even by physical evidence. Surveys need to be conducted along with the collection of secondary data. Three stages of the survey, i.e. (1) exploratory, (2) systematic and (3) confirmatory, can be conducted.

4.3 Establishing Its Significance: Authenticity and Integrity

A quantitatively and qualitatively assessment of the heritage asset's authenticity and integrity will be carried out to determine and establish the significance of the heritage asset.

According to the Nara Document of Authenticity, the authenticity of any cultural heritage fundamentally depends on the its type and cultural context, provided they plausibly express values in certain attributes such as '*form and design; materials and substance; use and function; traditions, techniques and management systems; location and setting; language and other forms of intangible heritage; spirit and feeling; and other internal and external factor*'(ICOMOS, 1994).

The integrity of the urban heritage assets is '*a measure of the wholeness and intactness of the heritage assets and its attributes*'. To examine the integrity of a cultural heritage, an assessment of the asset regarding the intactness of the elements is essential. All the elements necessary to substantiate the Outstanding Universal Value (OUV) needs to be expressed. Secondly, the cultural heritage should display a representative and acceptable value of significance (ICOMOS, 1994).

4.4 Generating Awareness of Heritage Assets

Public awareness is essential and crucial to preserve, protect and maintain our heritage. The communities, in time, often get accustomed to their surroundings and built environment; and tend to become less conscious about it. Creating awareness will help strengthen their sense of belongingness with their assets and develop a more harmonious relationship with it.

4.5 Integrated Stakeholder/Community Participation Approach

Engaging the community as the primary stakeholders in the development and planning process is essential for sustainable development. The stakeholders will include mainly the local community, people associated with assets and even the migrants who act as engines and keeps the place alive. Extensive dialogue and participation with the stakeholders will result in a better understanding of the community needs, existing urban resources and their potentials. Engaging the community in Programmes and activities will add up to the experience and in understanding the local knowledge system.

4.6 Preparation and Periodical Updation of Inventories

Listing/Inventories help construct a databank of heritage assets with all necessary information such as its present status and condition; and typology of properties. Mapping of these properties and sites in the Development Plans is essential in order to identify and delineate a Heritage area or special Zone for planning purposes. With these, site-specific guidelines and regulations can be prepared. It will eventually help determine the potential and prioritise the reuse of the urban heritage assets in a much more sensitive manner. Periodical monitoring and updating of the heritage assets are also to be looked upon.

4.7 Mapping of Heritage Assets

Mapping of heritage assets and resources helps identify the significant spread and concentration of identified heritage assets within the city/town wards. These wards should be recognised as heritage zones in the existing development plans. All the identified heritage assets will be enlisted and marked on the heritage zone maps separately. All data collected and worked out must be digitised and made public, to make it much more resourceful for all.

4.8 Value Assessment of Assets

A strategy will be formulated and developed to evaluate the value of the urban heritage assets, especially those that are in a vulnerable condition. Heritage properties must be assessed categorical based on the inventory prepared. Periodical supervision actions are recommended along with its updation in its respective inventory.

4.9 Integration of Urban Heritage Conservation in Urban Planning Framework

To integrate Urban Heritage in the Urban Planning Framework, the Development Plans should provide details of all the listed heritage assets—both tangible and intangible along with the criteria inscribed for this purpose, which would also include the Acts under which it is being designated. The development plans will address details including provisions, regulations and limitations of the heritage Acts for protection, preservation and conservation of the heritage assets, which would be based on the heritage grading assigned, property ownership rights and responsibilities. The development plans will address the rules and regulations to carry out activities such as demolition and reconstruction, additions and alterations, façade modifications, height alterations, building setbacks, adaptive reuse, displaying of external features and signage, etc. The CDP will provide details of the competence of heritage Acts for meeting the contemporary issues and challenges faced by heritage property owners.

4.10 Establishing Institutional Setup

The primary task is to enlist all the institutions that are involved and accountable directly or indirectly in heritage protection and management. This enlisting will be carried out for all urban areas separately. A preliminary review report consisting of detailed analysis with all necessary details on their obligation and responsibilities relating to the upkeep, protection and maintenance of heritage, maintenance of archival and public records, will be drafted. The city development plan (CDP) of respective cities can incorporate all the drafted details. A functional matrix between different organisations and institutions will be analysed for the improvement in coordination in the conservation and management of the heritage. The CDP needs to focus and analyse the role of the ULBs directly or indirectly involved in heritage protection and conservation. Establishing a Heritage Cell in the Urban Local Body to take responsibility for heritage-related issues can help aid in better functioning. The role of non-governmental organisations and other associations in heritage protection, maintenance and development needs to be closely examined too. The ownership status of heritage properties is to be documented and recorded. The drawbacks of the functioning of the current institutional system are to be identified. Improvements will be carried out on the existing systems for the efficient functioning of the whole institutional setup.

4.11 Delineation of Heritage Areas for Effective Urban Conservation

The city development plan, worthwhile respecting all the listed heritage assets, must utilise the heritage assets in urban development taking into consideration the values embedded in them and delineate areas based on the measure of sustainability. Special norms and regulations are to be imposed on 'asset-specific' rather than being 'area-specific'. This creates a new approach for conserving heritage assets within the context instead of isolating them amongst the urban-scape.

5 Discussion

Although most of the heritage buildings are protected by the ASI and State Archaeology departments, there are still a large number of unprotected urban heritage assets yet to be identified and mapped. While preparing city development plans (CDP), it is highly essential to incorporate Conservation strategies. These conservation plans should prioritise on the areas identified or documented as heritage precincts of the city. These zones should be linked administratively with the defined existing Wards of the area, thereby enabling context-specific policy-making and investment decisions in all areas of the city. Socio-economic and cultural indicators need to be identified to prepare a CDP for urban heritage protection, urban conservation and urban development.

The methodological approach discussed in the first stage constituting of steps 1 to 8 has enabled to establish an initial framework to aid in determining the value and significance of the heritage asset through identification and value assessment of the urban heritage asset. It has helped identify heritage assets in a much more contextual manner respecting their respective community stakeholders, thereby making the delineation of areas or zone much more context and site-sensitive leading to sensitive urban development. Utilising the assets for an unhindered urban growth that equally values the assets is achieved through the second stage that constitutes step 9 to 11. The paper opens up to the opportunity of changing the perspective towards urban heritage. It thus initiates a discussion on exploring different approaches of going about with urban development, thus creating a platform to challenge the existing system of heritage zoning in various cities. However, specific prerequisites for the proper and functional implementation of the above methodology are to be noted beforehand. Individual assets must be beforehand studied with prior knowledge to undergo a timesaving approach and helps justification regarding its relevance. As mentioned in the methodology, each urban asset can be effectively read and studied in the preliminary stages through a primary oral history narration, reading the urban landscape or by an effective urban survey with prime focus on the assets and its setting. Consistency regarding collected data is to be ensured by applying several alternative procedures. The detailed study and value assessment will help decision-makers to prioritise assets while policy-making and development of plans, in addition to its interdisciplinary management and monitoring.

6 Conclusion

Heritage-sensitive urban development needs to be inclusive despite the compulsive developments creating a negative impact on the heritage areas. The notion of urban heritage being a liability and burden for urban development must shift to a perspective where it is appreciated as a community asset embodied with values. The paper underscores the marginalisation of history from the urban planning domain and urges for a more integrated form of planning where a heritage precinct is not regarded as an area to be isolated from the rest of the city, but like the rest of the city, it lives in the here and the now by organically evolving. Revival and preservation of urban heritage assets and urban development must symbiotically coexist; and if planned sensitively, it can lead to sustainability and economic prosperity. The methodological approach described in this paper can be further adapted in context-specific heritage areas. It aims at reviewing and understanding the potential for a change in the overall approach in looking at heritage zones/regions/cores and their heritage assets. The study will aid in the improvement of the drafting of city masterplan, and it would also create a template for policy-making in heritage-sensitive areas and listing of urban heritage resources within a city. The paper sets up an objective and practicable methodological approach for effective and heritage-sensitive urban conservation. The paper provides scope for further in research and iteration in the adopted methodology framework by considering the economic dimension to discuss the economic feasibility and financial arrangements in the domain of urban conservation.

Acknowledgements We have great pleasure in acknowledging our gratitude to all the experts in the professional and academic field for their expertise and valuable insights. We would also like to express our gratitude to the IEREK and Springer Publications. We are immensely obliged to all the reviewers for their sincere and valuable comments, and any errors or inaccuracies in data or information is solely under the author's responsibility.

Conflicts of Interest The authors declare no conflict of interest.

References

Elnokaly, A., & Elseragy, A. (2013). Sustainable heritage development: learning from urban conservation of heritage projects in non western contexts. *European Journal of Sustainable Development*, *2* (1), 31–54. https://doi.org/10.14207/ejsd.2013.v2n1p31

Figueiredo, V. G. B. (2014). Patrimônio cultural, cidade, sustentabilidade: Qual o papel da legislação urbanística na preservação e no desenvolvimento? *Ambiente & Sociedade, 17*(2), 91–110. https://doi.org/10.1590/s1414-753x2014000200007

Greffe, X. (2004). Is heritage an asset or a liability? *Journal of Cultural Heritage, 5*(3), 301–309. https://doi.org/10.1016/j.culher.2004.05.001

ICOMOS. (1994). *The Nara Document on Authenticity. Knowledge Creation Diffusion Utilization.*

Kong, L., & Yeoh, B. S. A. (1994). Urban conservation in Singapore: A survey of state policies and popular attitudes. *Urban Studies, 31* (2), 247–265. https://doi.org/10.1080/00420989420080231

Labadi, S., & Logan, W. (2016). Approaches to urban heritage, development and sustainability. *Urban Heritage, Development and Sustainability: International Frameworks, National and Local Governance* (January 2015), 1–20. https://doi.org/10.4324/9781315728018

Licciardi, G., & Amirtahmasebi, R. (2011). Monitoring and evaluation in cultural heritage projects at the world bank, 1–8.

Lichfield, N. (1988). Economics in urban conservation. *Economics in Urban Conservation*. https://doi.org/10.2307/635463

Park, S. hoon. (2016). HabitatIII and the New Urban Agenda. *The Korean Association of Space and Environment Research, 58*, 5–8. https://doi.org/10.19097/kaser.2016.26.4.5

Sankhe, S., Vittal, I., Dobbs, R., Mohan, A., Gulati, A., Ablett, J., et al. (2010). India ' s urban awakening: Building inclusive cities, sustaining economic growth. *McKinsey Quarterly*.

Serageldin, I. (1999). Very special places: The architecture and economics of intervening in historic cities (ed. Culture in Sustainable Development). Retrieved from http://www-wds.worldbank.org/external/default/WDSContentServer/WDSP/IB/1999/09/14/000094946_9905260827175/Rendered/PDF/multi_page.pdf

Steinberg, F. (1996). Conservation and rehabilitation of urban heritage in developing countries. *Habitat International, 20*(3), 463–475. https://doi.org/10.1016/0197-3975(96)00012-4

Udeaja, C., Trillo, C., Awuah, K. G. B., Makore, B. C. N., Patel, D. A., Mansuri, L. E., & Jha, K. N. (2020). Urban heritage conservation and rapid urbanization: Insights from Surat, India.

UNESCO. (2011). *Recommendation on the Historic Urban Landscape. Records of the General Conference - 31st Session.*

UNESCO. (2015). UNESCO global report on culture and sustainable urban development. Concept note. In: *International Conference on "Culture for Sustainable Cities"*.

Whitehand, J. W. R., & Gu, K. (2007). Urban conservation in China: Historical development, current practice and morphological approach. *Town Planning Review, 78*(5), 643–670. https://doi.org/10.3828/tpr.78.5.6

Yung, E. H. K., Lai, L. W. C., & Yu, P. L. H. (2016). Public decision making for heritage conservation: A Hong Kong empirical study. *Habitat International, 53*, 312–319. https://doi.org/10.1016/j.habitatint.2015.12.004

Zancheti, S. M., & Hidaka, L. T. F. (2011). Measuring urban heritage conservation: Theory and structure (part 1). *Journal of Cultural Heritage Management and Sustainable Development, 1*(2), 96–108. https://doi.org/10.1108/20441261111171666

Zancheti, S. M. S. M., Lira, F. B., Silva, A. D. F., Braga, A. C., Gameiro, F. G., Costa, L. S., et al. (2006) The patina of the city. *City & Time, 2*(2), 11–22.

Historic Preservation as a Tool for City Branding Case Study: Khedivial Cairo

Mona Y. Shedid and Noha H. Hefnawy

Abstract

Historic urban spaces are considered one of the important components of the visual image that have significant effect on the urban environment. Those spaces are identified as planned open spaces which are usually located in the city center, and it can be a gathering place for community members and tourists, a suitable place for open markets, festivals, or political events. Today, many cities aim to strengthen the valuable urban character of the historical urban spaces to achieve a sustainable urban image by integrating several elements like economic, cultural, social aspects and activities via some new strategies including urban branding strategy. City branding is a novel aspect of urban communication that improves marketing of the city image in several ways by transforming the visual image of the city into a brand image. Historic preservation can help cities keep their unique character and diversity of urban parts in downtown areas and consequently lay a foundation upon which they can form their unique urban space brands that attract talent. Simultaneously, historic properties can be used to marketplace branding. This study aims to define the connection between historic and valuable images of the urban spaces and city branding, in order to achieve a proposed framework for evaluating the level of success of historic urban spaces (Opera Square) from the urban aspects to improve the identity of the square.

Keywords

City branding • Culture • Identity • Visual image • Opera square

M. Y. Shedid · N. H. Hefnawy (✉)
Faculty of Engineering At Benha, Benha University, Benha, Egypt

1 Introduction

The globalization created a competition between cities for a share in the world's tourism, businesses, investment, and attention. Due to the fact that some cities form the economic and cultural backbone of nations, they are increasingly becoming the focus of this international competition for funds and fame through the transformation of their urban environment into a branded visual image.

A main character for urbanization for any city is its visual image, with successful urban branded spaces as one of its main elements, thus drawing tourists and residents. Urban spaces are commonly located in the center of the city, which can be a place for tourists and residents to meet and celebrate, for open markets or political events.

The identity of a place is probably synonymous with its uniqueness, character, historic context, and roots. It is crucial for city branding. Downtown living has become an important marketing tool in the city's branding. It is of utmost importance for a downtown that it retains is character. Historic buildings and neighborhoods can contribute strongly to the character of a downtown.

In this context, historic preservation can help cities retain their unique character and diversity in downtown areas, thus laying a foundation to develop their unique place brands. Simultaneously, historic areas can be used as a marketing resource for place branding.

2 Research Problem

Successful urban branded spaces are rare to find in many Egyptian cities, despite their profound importance in the formation of the urban visual image and attraction. Thus, a set of research questions were developed to help to explore the relation between the historic and valuable images and identity of the urban spaces and city branding in order to achieve the proposed framework.

A. Versaci et al. (eds.), *Conservation of Architectural Heritage*, Advances in Science, Technology & Innovation,
https://doi.org/10.1007/978-3-030-74482-3_23

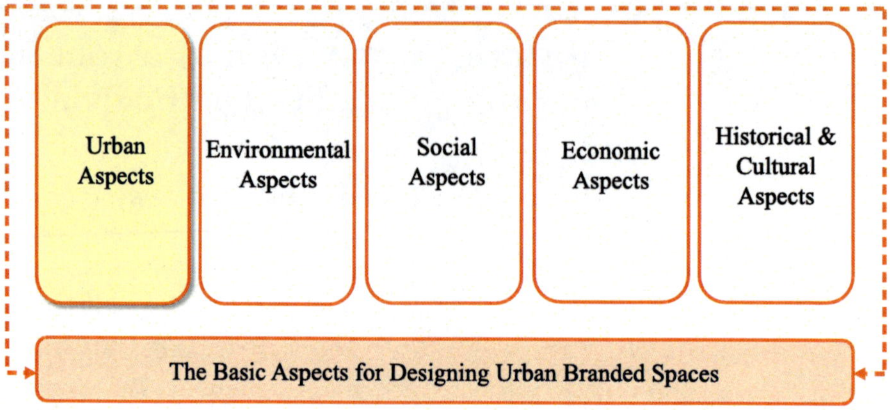

Fig. 1 Basic aspects for designing urban branded spaces

These questions can be summarized as follows:

- What are the main basic aspects for designing urban branded spaces?
- What are the main criteria to construct urban branded spaces, especially historical urban spaces?
- What are the attributes and concepts that show significant positive impact to improve the identity of the square and to transform the Opera Square to a branded urban space.
- Can the strength points in the studied area (Opera Square) convert the square into an urban branded space?

3 Objectives and Hypotheses

The paper aimed to achieve a proposed framework for evaluating the level of success of Opera Square from the urban aspects to improve the identity of the square and thus enhance the branding strength of the square; as these urban spaces will be considered as a nucleus for the development of the surrounded urban area, and so improve the city branding.

4 Methodology

The research methodology is qualitative and is based on two main parts:

The first part:

Reviewing literature of the identity and unique image of urban spaces and city branding in books, scientific journals, and research projects. A theoretical framework was adopted; various qualities of branded urban spaces as suggested by the most leading urban theorists and planners are categorized into 5 main aspects that should be taken into consideration while planning to convert an urban space into an urban branded space; see Fig. 1 (Abdelaal & Hussein, 2012). The research study will focus only on the urban aspect, which includes a set of dimensions, which have a group of indicators, and each indicator will be defined through several key design concepts. These concepts were used to conduct the proposed framework.

The second part:

This part is based on the empirical study and includes the following steps:

- Data collection and site documentation.
- Data analysis of the strength and weakness points occurred in the study area.
- Evaluation of the indicators of the urban aspects with the different urban elements in the study area.
- Formulation of the proposed framework and the objective of the research study; see Fig. 2.

5 City Branding: Overview and Definitions

City branding arose as a field of study in the 2000s, and today, destination branding is on the radar of most developed nations, as well as increasingly among developing countries seeking an enhanced image to attract direct investment, tourism, and trade and improve international relations. However, in recent years, branding discussions have moved toward city brands, reflecting the growing importance of cities in national and global economies and as attractors of investment and people.

Branding of a place is critical due to several reasons, most commonly to promote economic growth. This is due to the fact that a strong brand can achieve a shift in the perception of a place that has a poor image among external and internal constituents, form a vision for potential and future of the

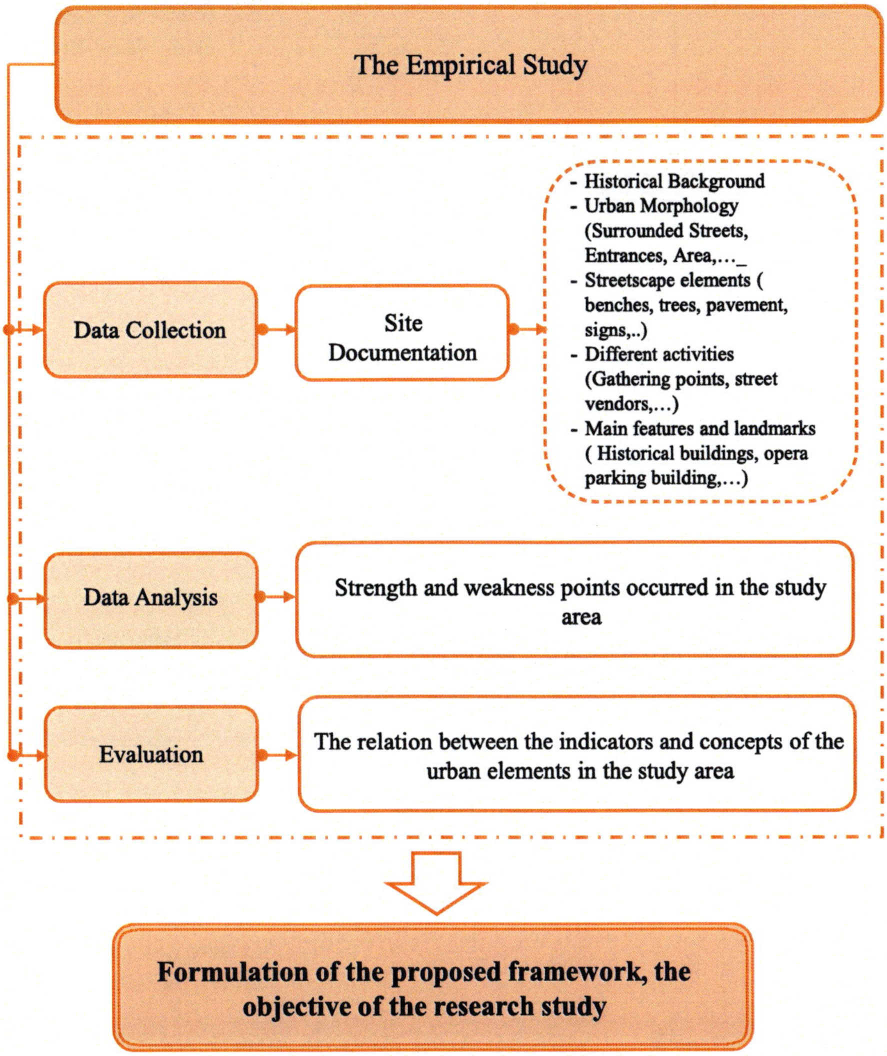

Fig. 2 Second part of the research methodology

community, and enhance its local, regional, and/or international awareness and position. It can even transform unfavorable stereotypes related to a place more appealing.

Brand image is the way customers currently view a brand and signify what it currently stands for. This overall impression is a product from many sources (Vanolo, 2008).

Kavaratzis and Ashworth (2005) stated that places are brandable entities if they can be differentiatable from each other by their characteristics. Due to the intensely growing competition for investors, skilled workers, and tourists, substantial efforts to differentiate places have frequently been noticed in recent years. Hence, place branding today refers to the application of proper marketing strategies to differentiate cities and countries from the competition, regarding economic, social, political, and cultural aspects.

So from this point of view, city branding concept can be defined as "the practice of developing a strategy that underlines the value the place offers to its target publics, namely residents, workers, employers, investors, tourists, and so on" (Biçakçi, 2012).

5.1 Place Branding: Overview and Link to Identity

"Identity" is defined as the properties of an individual or group which differentiate them from others. Conversely, Urban identity is defined with natural and artificial urban elements and sociocultural characteristics of urban environments. The urban identity has 2 major components: "social" (socioeconomic, sociocultural, and psychological) and "environmental" (natural and artificial) (Örer, 1993).

Thus, place branding depends upon its identity, experience, and image. The common conception is that if the brand

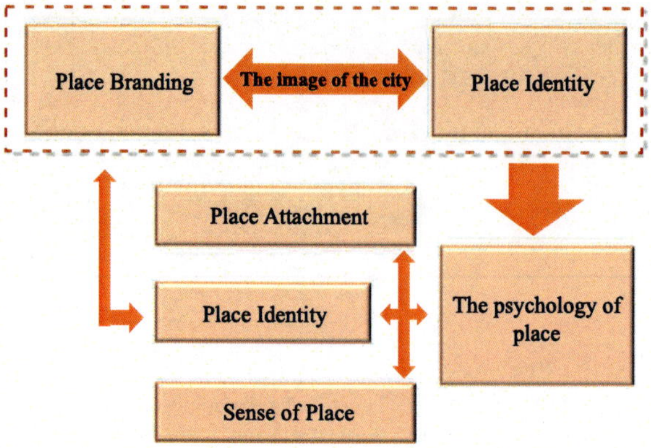

Fig. 3 Relation between place branding and place identity

is not based on identity then the branding efforts can only lead to a brand "estranged" to the place, especially to its residents.

Place identity was defined as a subdivision of self-identity by environmental psychologists, consisting of cognitions about the physical world. Three main constructs that can be used to account for the psychology of place were identified by environmental psychologists: "place attachment," "place identity," and "sense of place" (Fig. 3).

They also formulated a three-part definition of place: location—the geographical space providing the setting for social interactions; locale—the setting for informal and institutional social relations; and sense of place—the local set of feeling that subjectively and emotionally attaching people to places. See Fig. 4 (Knez, 2005).

6 Historic Preservation, Place Identity, and Place Branding

The focus of this part is on the identity of urban structure in the city since it aims to show the contribution of historic areas to place brand. A city should positively discriminate itself from its competitors to identify its own characteristics to be able to ensure that it becomes a competitive and effective place brand in the global marketplace. This is where a city can benefit from the preservation of its cultural heritage which in return would promote an authentic sense of place that would be a great aid to attracting the people to visit and utilize this place.

The downtown area contains the historic buildings and old long-standing inner-city neighborhoods. These provide a unique streetscape in addition to defining and shaping the new development. A healthy downtown retains its historic fabric, while simultaneously allowing new development and the retrofitting historic buildings with modern facilities. A distinctive downtown exhibits a harmonic mix between historic elements, modern facilities, and sensible rehabilitations joining preservation and modern life.

There is always a competition between preservationists and developers over downtown areas. Comprehensive historic preservation can be only guaranteed if a city government considers the integration of historic preservation plans into community master plans.

Paradoxically, historic preservation is a very modern activity, as well-preserved historic buildings can reflect a city's modern culture in addition to its ancient past. Therefore, a lot of cities nowadays try to market their cultural facilities—such as art museums and concert halls—as proof of cultural

Fig. 4 A holistic approach between place branding and place identity

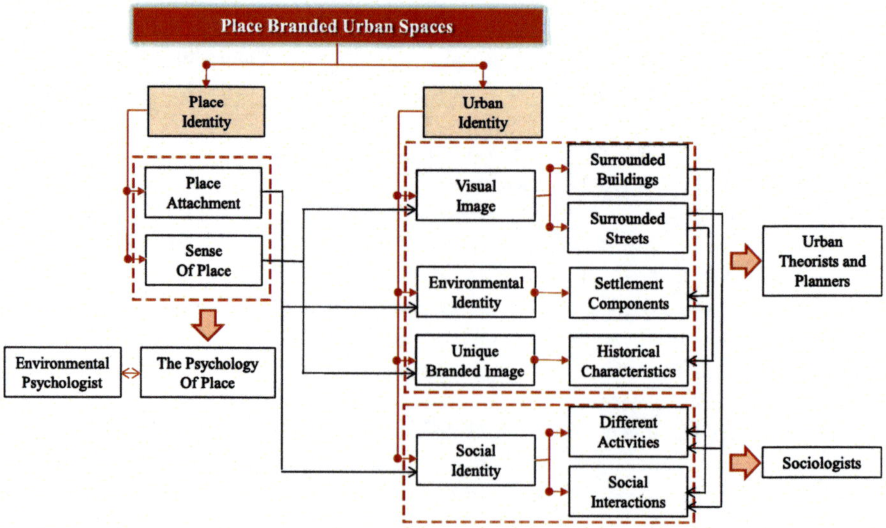

richness to lure both tourists and new inhabitants. In summary, historic city centers play a pivotal role in differentiating a city from its competitors and to highlight livability and attractiveness, thus drawing talent to the city (Brien, 2012).

7 The Basic Urban Aspects, Dimensions, and Indicators to Design a Historical Branded Urban Space

Branding of places links the urban image to the elements of the urban form, and historical and cultural identity. It aims to develop new and creative way of communicating the image of the city to the rest of the region or the world. From this point of view, the aim of this part is to find the relation between various qualities and aspects of branded urban spaces. The research will subdivide the urban aspect into several dimensions, and each dimension has a set of indicators. These indicators will be defined through a group of concepts to reach the proposed framework, as shown in Table 1 (Abdelaal & Hussein, 2012 & Hefnawy, 2018).

8 Case Study: Opera Square

8.1 Reasons for Selecting "Opera Square"

There are many reasons for selecting Opera Square, and the most important of them are:

Table 23.1 Relation between various qualities and aspects of branded urban spaces

Aspects	Dimensions	Indicators	Concepts
Urban aspect	Visual image	Surrounded buildings	Continuities and homogeneities of facade materials, texture, details
			Homogeneities between buildings and streetscape elements
			Skyline
			Urban spaces surrounded by physical form
		Surrounded streets	Has a clear start point and end point
			Has a strong entrances and leaves
			They lead to unique and dominant landmarks
			Locate the urban space on main routes in order to make the movement economy more efficient
			They oriented toward interesting views
	Environmental identity	Streetscape elements	Linking streetscape elements with the identity and character of the urban space
			Singularity and continuity of streetscape elements
			Supporting social activities
			The condition of streetscape elements
	Social identity	Different activities	Concentration and variation of activities to enrich the space with life and movement
			Enough and comfortable gathering spaces
			The urban space users feel comfort and safe
			Designed spaces for street vendors
			Design plazas that allow for diverse functions and consider the needs of users
	Unique brand image	Unique building	Presence of historical buildings
			The condition of buildings
			Homogeneities of the character and style of the historical buildings
		Significant arch	Create urban spaces surrounded by landmarks (especially the historical one)
			Create landmarks that are distinguished by their dominance and singularity
			Remarkable for their prominent position as seen from far and near distance
		Events	Hold events and cultural activities
			A designed space for the cultural events

- The location of the square in historic districts in Cairo's center. It is one of the very important historic central squares in Cairo because of its historical value:
- It directly overlooked the oldest opera house in Egypt, which had a historic, architectural, and functional value (until it was burned to the ground).
- The square also includes the statue of Ibrahim Pasha and the Azbakiya Gardens.

The square suffers from the traffic bottlenecks, which have degraded the visual image and increased environmental pollution. Figure 5 shows the main elements and the buildings surrounded Opera Square.

8.2 Historical Background and Urban Morphology

Opera Square and Ibrahim Pasha Square are all names of the square that witnessed one of the most important cultural and artistic renaissance features of Egypt. Built during Khedive Ismail Era, its name was changed repeatedly as it was first known as "Teatro Square" followed by "Opera Square" and later on "Ibrahim Pasha Square" after moving Ibrahim Pasha Statue from EL-Attaba Square, and finally after 23 July revolution in 1952, it returned to be "Opera Square." Figure 6 shows the basic information, accessibility, and location of the square.

Fig. 5 Main elements of the Opera Square

Fig. 6 Basic information and map showing location and accessibility of the Opera Square

Opera Square	
Access Points	Adly Pasha Street/ Abd El-Khaliq Tharwat Street/ EL-Gomhureya Street/ and Azhar Bridge
Area	About 3500 m2
Activities & Uses	Mixed Uses (Residential/ Commercial/ Entertainment/...)
Main Features	Azbakiya Gardens/ Ibrahim Pasha Statue/ Opera Parking Building/ and historical buildings

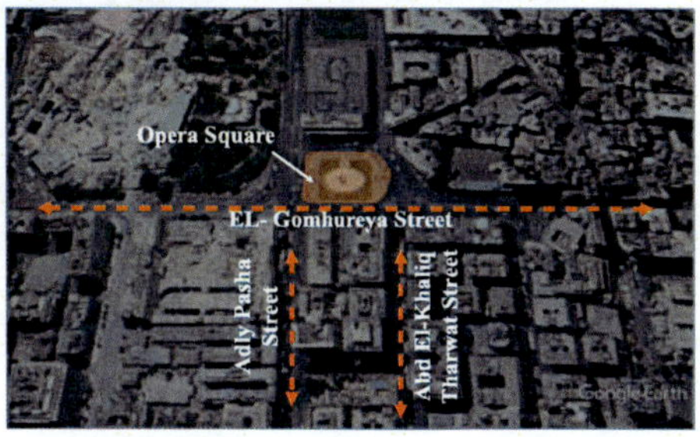

8.3 Site Documentation

The documentation of Opera Square was categorized into five main categories: surrounded buildings (style, color and texture, and skyline), main streets, streetscape elements, different activities, and main features of the square, as shown in Fig. 7.

8.4 Analysis and Evaluation

The response of Opera Square visitors was assessed through a field survey, observations (Fig. 8), and questionnaires.

A questionnaire was distributed among 30 persons of the street visitors, to evaluate the Opera Square from the urban aspect. It consisted of two main parts, where:

The first part includes:

- Questions about the main indicators and dimensions of the urban aspects.
- A three-point rating scale was used to answer the questions. The scale ranged from 1 to 3 (1—poor, 2—average, and 3—good). The mean values of the ratings of the questions were calculated for each indicator of the urban aspects.
- The scale was divided into three-part score, where the first part is from 0 to 4 (poor), the second part from 5 to 7 (average), and the third part from 8 to 10.

The second part includes:

- Three direct questions about the problems that faces the people in the square and their recommendations to improve the advantages of the square from the urban aspects.

Table 2 expresses the main points of strength and weakness in Opera Square based on the observation, while Fig. 9 evaluates the relation between the urban aspects and the different dimensions, indicators, and concepts of an historical branded urban space based on the questionnaires.

9 Conclusion

The research study reached the following conclusions:

Historic preservation aims to retain a city's real cultural assets, and with a good plan to preserve and use them appropriately, the cultural resources can significantly contribute to the city's livability, thus helping a city to develop an effective place brand. Therefore, historic preservation can be an effective mean for branding.

The paper concluded a proposed framework which is useful in developing and enhancing a historical branded urban space. It revealed the key design aspects, dimensions, indicators, and concepts of successful historical urban branded spaces, as shown in Fig. 10.

The visual image elements of any urban space could have one or more of these characteristics: singularity (though it is unique in shape, color, proportion, surface and in contrast to its surroundings recognizable), dominance (through its size, height, and other physical characteristics improves image quality), clarity and simplicity (simple forms are easily recognizable), exposure, and unity or equivalence (elements should be in relation to each other).

Linking the visual image to the cultural, economic, social activities is of profound importance, and these activities should originate from the place and encourage its translation by different advertising elements.

Emphasize the squares identity, through using a group of buildings and landmarks, giving them a unique identity, and making the appropriate treatment for building facades overlooking these squares to fit their architectural style.

Fig. 7 Site documentation of the Opera Square

Indicators		Site Documentation
Surrounded Buildings	Pictures	
	Comments	<u>- Style:</u> Building A: constructed in 1899 with classic style. Building B: built in the beginning of 20th century. Building C: constructed in the middle of 20th century, style of Neo-Baroque. <u>- Color and Texture:</u> All the buildings around the square are homogeneous in texture and color. <u>- Skyline:</u> The average height of the buildings is ground + 6 floors.
Main Streets	Pictures	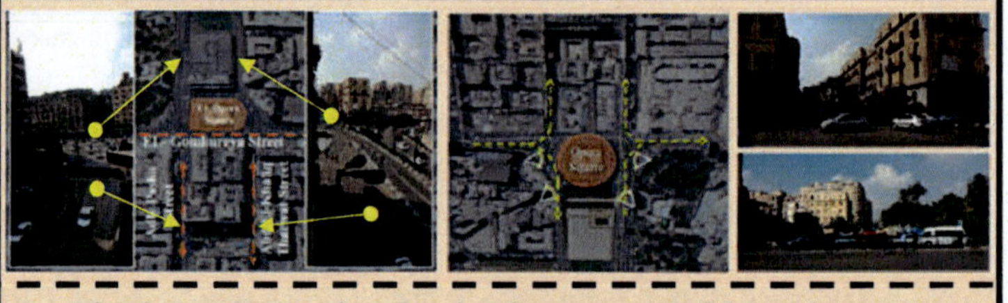
	Comments	- The Opera Square is a focal point surrounded by main streets, - The surrounded streets are oriented to the square. - The square doesn't have a strong start and end points (there is no identified entrances and leaves.
Streetscape Elements	Pictures	
	Comments	- Streetscape elements are in a bad condition. - They are not in harmony with the unique character and identity of the square, except some of the lighting fixtures.

Fig. 7 (continued)

Indicators		Site Documentation
Different Activities	Pictures	
	Comments	- There are not a designed place for street vendors. - There are a variation of mixed used activities such as shops and restaurants. - Lack of designed gathering spaces. - The seating areas are not comfortable: not shaded, design of the benches (back to back: not suitable for social interaction)
Main Features	Pictures	
	Comments	- Ibrahim Pasha Statue is distinguished by its dominance and singularity. - Azbakeya Garden - Three main historic buildings, they have an unique and significant character. - The Opera Parking building, but The building is not in harmony with the surrounding historical buildings.

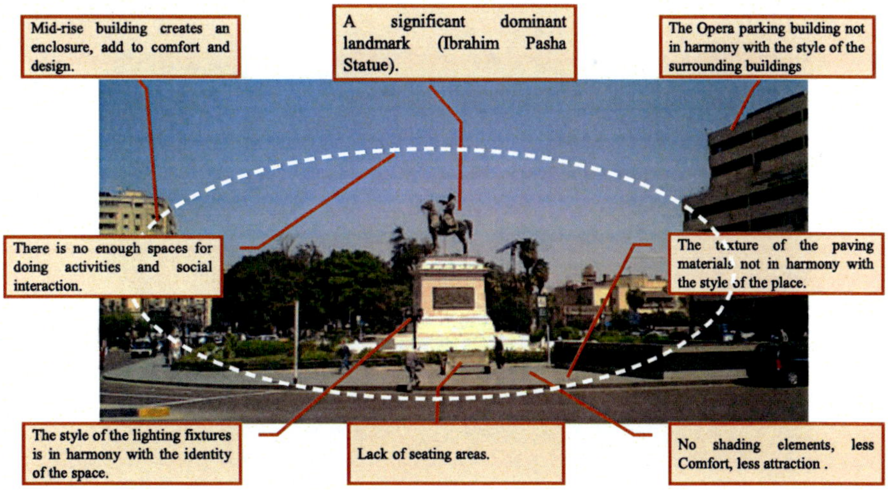

Fig. 8 Main observations at Opera Square

Table 23.2 Analysis of the Opera Square

	Items	Picture	Description
Points of strength	Ibrahim Pasha Statue		Landmark that is distinguished by its dominance and singularity of shape, size, height, location, and visibility
	Azbakiya Gardens		Constructed in 1872, and comprises an artificial knoll planted with rare trees
	Historical buildings		Building A: constructed in 1899 with classic style Building B: built in the beginning of twentieth century Building C: constructed in the middle of twentieth century, style of Neo-Baroque
	Opera Square	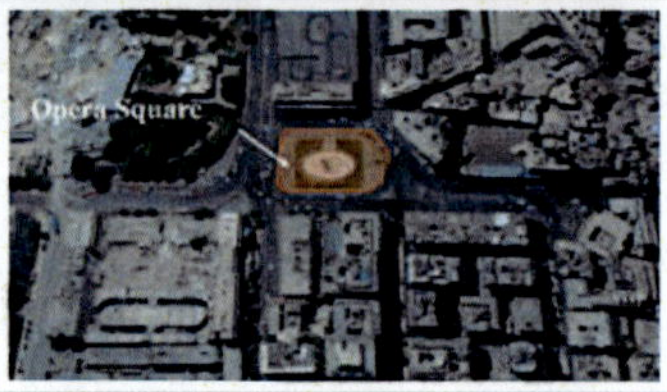	Strong node supported by strong landmark (Ibrahim Pasha Statue) and surrounded with historical buildings The presence of metro stations makes the place more accessible

(continued)

Table 23.2 (continued)

	Items	Picture	Description
Points of weakness	Opera Parking Building		The building is not in harmony with the surrounding historical buildings
	Square flooring		Floorings are not in good condition, as some tiles are broken
	Seating places		Seating areas are not enough for doing different activities and social interaction
	Lighting fixtures		Lighting fixtures are not in good condition
	Shading areas		There are not any shading areas, which makes people less comfort
	Green areas		Green areas are in bad condition
	Street vendors		There is no place designed for street vendors

Aspects	Dimensions	Indicators	Points: Strength	Points: Weakness	Concepts	Evaluation: Good	Evaluation: Average	Evaluation: Poor	Pictures
Urban Aspect	Visual Image	Surrounded Buildings	●		Continuities and homogeneities of facades materials, texture, details	●			
					Homogeneities between buildings and streetscape elements		●		
					Skyline	●			
					Urban spaces surrounded by physical form	●			
		Surrounded Streets	●		Has a clear start point and end point.	●			
					Has a strong entrances and leaves			●	
					They lead to unique and dominant landmarks	●			
					Locate the urban space on main routes in order to make the movement economy more efficient	●			
					They oriented towards interesting views	●			
	Environmental Identity	Streetscape Elements		●	Linking streetscape elements with the identity and character of the urban space		●		
					Singularity and continuity of streetscape elements.			●	
					Supporting social activities			●	
					The condition of streetscape elements		●		
	Social Identity	Different Activities		●	Concentration and variation of activities to enrich the space with life and movement	●			
					Enough and comfortable gathering spaces			●	
					The urban space users feel comfort and safe			●	
					Designed spaces for street vendors			●	
					Design plazas that allow for diverse functions and consider the needs of users			●	
	Unique Brand Image	Unique Building	●		Presence of historical buildings	●			
					The condition of buildings		●		
					Homogeneities of the character and style of the historical buildings	●			
		Significant Arch.	●		Create urban spaces surrounded by landmarks (especially the historical one)	●			
					Create Landmarks that are distinguished by their dominance and singularity	●			
					Remarkable for their prominent position as seen from far and near distance	●			
		Events		●	Hold events and cultural activities			●	
					A designed space for the cultural events		●		

Fig. 9 Evaluation of the relation between the urban aspects and the different dimensions, indicators, and concepts of an historical branded urban space

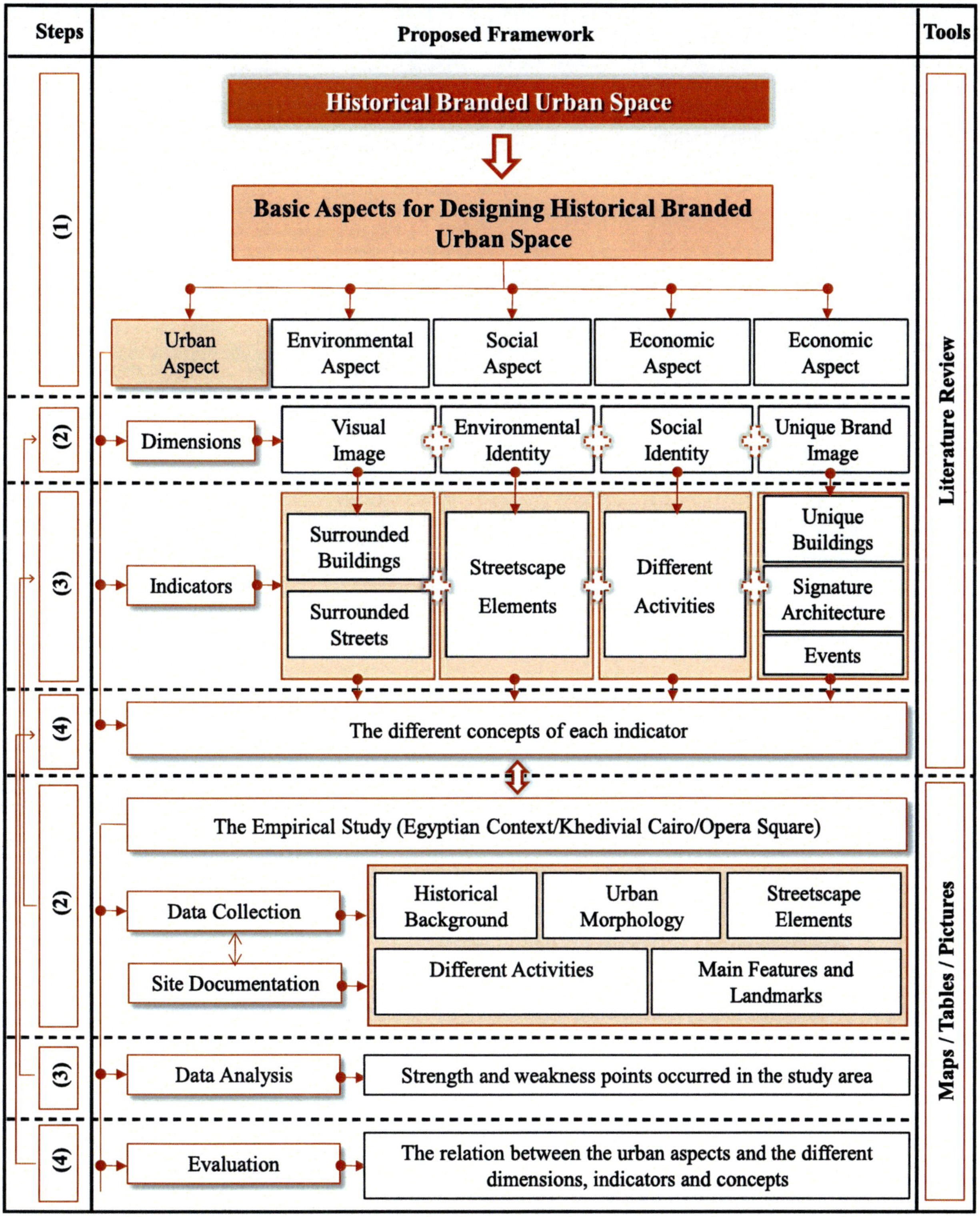

Fig. 10 Proposed framework of historical branded urban space

The need to link squares visually to the surrounded urban environment, by enforcing the paths and entrances leading to it.

The importance of conducting activities that are related to the history and the visual image of a place (as celebrations) and promoting its translation through various advertising methods to suit the desired branded image.

References

Abdelaal, M., & Hussein, R. (2012). Branding urban spaces as an approach for city branding -case study: Cairo City, Egypt, world academy of science, engineering and technology. *International Journal of Economics and Management Engineering, 6*(11).

Biçakçi, A. (2012). Branding the city through culture: Istanbul, European capital of culture 2010. *International Journal of Human Sciences, 9*(1).

Brien, O. (2012). Livable historic city cores and enabling environment: A successful recipe to attract investment to cities. In *The Economics of Uniqueness: Investing in Historic City Cores and Cultural Heritage Assets for Sustainable Development, edited by Guido Licciardi and RanaAmirtahmasebi.* Washington, D.C.: Urban Development Series. ISBN: 9780821397060.

Hefnawy, N. (2018). The potential effect of urban image on city branding, Phd. Faculty of Engineering, Cairo University.

Kavaratzis, M., & Ashworth, G. J. (2005). City branding: An effective assertion of identity or a transitory marketing trick? *Tijdschrift Voor Economische En Sociale Geografie, 96*(5), 506–514.

Knez, I. (2005). Attachment and identity as related to a place and its perceived climate. *Journal of Environmental Psychology, 25*(2), 207–218.

Örer, G. (1993). İstanbul' un Kentsel Kimliği ve Değişimi (The urban image and its change in Istanbul), İ.T.Ü.-Fen Bilimleri Ens. Y.L Tezi, İstanbul.

Vanolo, A. (2008). The image of the creative city: Some reflections on urban branding in Turin Cities. *Cities, 25*(6), 370–382.

Conserving Oral Heritage Through Architecture

Ar. Renu Khanna and Er. Nakul Khanna

Abstract

"Emotive & experiential architecture can make you travel through time and space."

It is the need of the hour to increase awareness to conserve our heritage and save what is left of history. This may be through architectural buildings, written recorded history or an oral heritage which is passed on from one generation to another. In the past, a lot of attention has been given on conserving architectural heritage; however, no thought was given to conserving of oral heritage in a built form. Oral heritage is liable to distortions and manipulations as it gets transmitted through generations. Architects and conservationists can play a major role in conserving such oral heritage by creating monuments, memorials and cultural spaces. The approach used by the architects and developers of the monument should be rational and systematic in nature. The objective of the architect should be to put soul in the monuments through narrative architecture. For conversion of oral heritage into structural monument, the architect has to hold in-depth discussions with local communities, examine available literature, verify the contents and look into any physical or structural available evidences for more authentic expression. One such example is a monument designed by the authors, where the sacrifices and valour of a great saint soldier Baba Banda Singh Bahadur have been glorified. This emotive, experiential and narrative architecture is an attempt to concretize oral heritage. Prior to this monument, Baba Banda Singh Bahadur's contributions were almost forgotten. After its construction, it has now caught the attention of the younger generation. Thus, architecture can play a major role in reviving the oral heritage

Keywords

Oral heritage • Narrative architecture • Experiential architecture • Baba Banda Singh Bahadur • Chappar Chiri

1 Introduction

Oral and intangible heritage is human communication where knowledge is transmitted orally through speech, songs, drama and other art forms from one generation to subsequent generations. These facts are vulnerable to distortions and manipulations; hence, they need to be preserved in architectural built forms. Architecture can play an important role by providing structural evidence based on in-depth study of oral narrations available and by depicting its true version. Ramleela is an oral heritage in India where the story of lord Rama, his wife Sita, brother Laxman, his saviour Hanuman and the demon Ravana has been enacted every year in a festival called Dussehra since thousands of years.

Lord Rama with the help of Hanuman rescued his wife Sita from the devil Ravana who had kidnapped her (Fig. 1). Ramleela celebrates the winning of goodness over the demon within us. As no tangible architectural structural evidence is available, hence a big controversy regarding Rama's birth city has been going on for years (Tripathi, 2019). On the contrary, it was Pharaoh Hatshepsut's inscriptions on obelisks (Fig. 2a), the architectural built forms which were discovered after noticing her erased images (Fig. 2b) and works which brought to light her excellent rule and contributions (Highfield, 2007). Hence, architecture is an important tool in conserving oral heritage.

Ar.Renu Khanna (✉)
Fellow of Indian Institute of Architects, Mumbai, India

Ar.Renu Khanna
Member Council of Architecture, New Delhi, India

Er.Nakul Khanna
Masters in Structural Engineering, Chartered Engineer, The Institute of Engineers, Kolkata, India

A. Versaci et al. (eds.), *Conservation of Architectural Heritage*, Advances in Science, Technology & Innovation,
https://doi.org/10.1007/978-3-030-74482-3_24

Fig. 1 Oral heritage—Ramleela drama

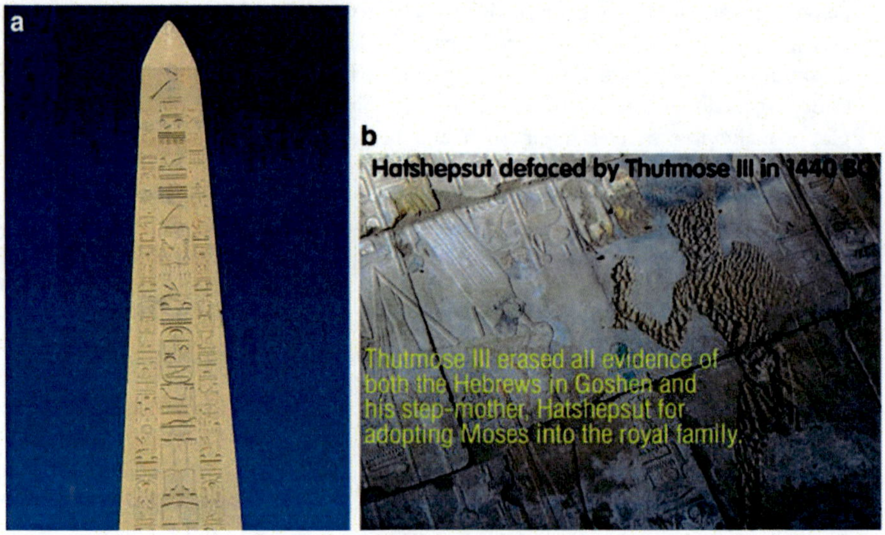

Fig. 2 **a** Pharaoh Hatshepsut Obelisk Luxor and **b** Pharaoh Hatshepsut erased image

1.1 Aims and Objectives

The aim of this present communication is

(a) To emphasize the need to preserve oral heritage for the coming generations.
(b) To highlight the role of architecture in conserving and promoting oral heritage.
(c) To construct memorials, monuments, museums and cultural spaces in order to preserve the oral heritage.

1.2 Oral Heritage

Heritage is always initially oral. For greater viability, it has to be converted into structural monuments, memorials and cultural spaces. It is important to conserve oral heritage for coming generations without distortions.

1.3 Journey of Oral Heritage

Although oral heritage existed in each and every village since times immemorial, there was no globally recognized platform. In 1998, UNESCO launched a new international entity "Proclamation of Masterpieces of the Oral and Intangible Heritage of Humanity". This proclamation talks about the sociocultural spaces and traditional folk forms that adds value to the cultural expression of that society. The key objectives of the convention were as follows:

a. The convention intended to mobilize opinion towards recognition of the value of oral and intangible heritage. It also highlighted the need to protect, revitalize and sensitize the value of oral and intangible heritage.
b. To assess and list the oral and intangible heritage sites across the world.
c. To persuade countries to set up national inventories of oral and intangible heritage. Also, to prepare legal and administrative measures to protect them.
d. It intended to encourage the mass participation of local artists, traditional art for the revival of the intangible heritage.

The fundamental purpose of this proclamation was to safeguard, advocate and also boost the cultural heritage. It further intends to set up a global platform for cooperation among various institutions, government bodies, NGOs.

1.4 Challenges in Preserving Oral Heritage

It is a great challenge to maintain the viability and vitality of world's oral and intangible heritage. Though important, it is difficult and much problematic to maintain this oral heritage. The main hindrances are:-

(1) Possibility of distortion of facts.
(2) Lack of government support.
(3) Ethnic discrimination.
(4) Switching interest of young generation to contemporary types of entertainment (Fig. 3).

2 Methods of Conserving Oral Heritage

For boosting the oral heritage, national communities are promoting local language and creating buildings like memorials (Fig. 4), monuments, museums, community centres and storytelling through audio–video, sound and light shows and organization of regular festivals and workshops.

Oral heritage is ephemeral and vulnerable to distortion and, therefore, needs to be preserved.

2.1 Conservation is Vital

Conservation of heritage is vital. All countries take pride in their heritage, its appreciation by younger generation and its international recognition. Best method to preserve oral heritage is by its architectural conversion. "The architectural heritage will survive only if it is appreciated by the public and in particular by the younger generation. If not preserved, then it is liable to be lost forever. Educational programmes for all ages should, therefore, give increased attention to this subject"(Embaby, 2014).

The UNESCO (1972) World **Heritage** highlights cultural heritage monuments, buildings and sites that are of outstanding universal value from historical prospective, art and science. These monuments become living heritage, a thing of pride and glory. One such example of officially classified monument historique in 1936 is the Obelisk of Luxor at the Place de la Concorde in Paris (Fig. 5).

For conversion of oral heritage into structural monument, the architect has to hold in-depth discussions with local communities, examine available literature, verify the contents and look into any physical or structural available evidences for more authentic expression. A sense of identity and continuity is achieved by conservation of heritage through monuments in a fast changing world for the younger generations. They become the architectural heritage of that era. The objective of the architect should be to put soul in the monuments through narrative architecture which can attract global tourist and also empower the locals. These folklore and monuments help the locals to generate revenue through tourism and create employment besides reviving the old heritage.

A number of earlier publications have highlighted the importance of conserving intangible heritage as a revenue-generating tourism resource (Ruhanen & Whitford, 2019; Kim, et al., 2019).

2.2 Approach

The approach used by the architects, researchers and archaeologists should be systematic, scientific and objective in nature. It involves identifying such oral heritage worldwide. In-depth research to find the origin, existence and value of the oral heritage must be done.

We propose architectural built monuments and memorials for preserving the oral heritage. The architecture of such monuments or memorials should be narrative and experiential.

Fig. 3 Switching interest of young generation to contemporary types of entertainment, i.e. malls

2.3 Methodology

The architect must acquire in-depth knowledge of that oral heritage which needs to be preserved.

1. This knowledge can be acquired through historians and available literature.
2. Discussions with locals of that region associated with that folklore or the oral heritage to find more authentic version which may not have been documented.
3. Try to recreate those circumstances which existed then. The design can be shown to historians and locals for content approval.
4. The monument is then constructed which converts the oral heritage into a living architectural heritage.

3 Case Study—Oral Heritage Made Living Heritage by Constructing a Monument Cum Memorial

3.1 Fateh Burj—Memorial for Baba Banda Singh Bahadur, Chappar Chiri, India, Designed by Authors

The memorial for Baba Banda Singh Bahadur at Chappar Chiri, Mohali, India, is one such example of promoting oral heritage through narrative, emotive and experiential architecture. The architect evolved the architectural concept of this monument by reading many books on him Data (2012), Sagoo, (2001), Singh, (1999, 2000), Dhillon (2004). Baba

Fig. 4 Memorial for remembering War Martyrs

Fig. 5 Monument—the Obelisk of Luxor at Place de La Concorde in Paris, France

Banda Singh Bahadur, a titanic commander of the Sikh community, gained victory in a decisive battle against Wazir Khan, the commander of the Mughal army, on 12 May 1710 AD. The battle of Chappar Chiri is the only battle in the long span of seven centuries that marked the victory of indigenous Indians over a foreign invader. Fascinated by the intriguing history of the place and the demand of the locals, it was decided to imbibe this glorious legend of antiquity in an architectural design. The aim was to recreate the nuances of the time of the war and also hint upon the strategies used by the valorous Sikh Commander Baba Banda Singh Bahadur (Fig. 6) that led to this momentous victory (Singh, 2000). This sensitive design makes a visit to this monument an unforgettable one. Baba Banda Singh Bahadur was a Rajput by birth, an ascetic (baba) by choice and playful by temperament, but whose personality and vision underwent a complete metamorphosis at a mere glimpse of the Tenth Master Guru Gobind Singh ji whom he considered as his master, and from whom he received initiation in the Khalsa-fold (Singh) and provided political leadership (Bahadur) to the Khalsa during the most turbulent period of its history.

3.2 Oral Heritage to Be Conserved

After the battle of chapper chiri (12 May 1710 AD) Baba Banda Singh Bahadur became prominent in Sikh history for almost 8 years i.e, 1708AD-1716AD. Baba Banda Singh Bahadur is among the most amazing characters in Indian history. He was transformed into Guru Gobind Singh's (the initiator of Sikh religion) most faithful disciple despite being an ascetic. Guru Gobind Singh went all the way to a different state Maharashtra (Nanded) and appointed Baba Banda Singh Bahadur as his deputy, when the guru got seriously injured, sacrificed his four sons in the battle and could not lead his Sikh army to fight the Mughal forces. He gave Banda his sword, a mighty bow, arrows from his own quiver, his battle standard and his war drum as an evidence of this appointment. Banda started his journey from Nanded (where Guru Gobind Singh passed away) as the leader of a small group of Sikhs, which, before he reached Punjab, his small group of Sikhs had grown into a formidable army. Over the next few years, his pitched battles and skirmishes became great legends. His personal heroic courage and skill in warfare brought his name among the first of many legendary Sikh generals. This battle of Chappar Chiri was a landmark Sikh victory establishing Sikh rule from Lahore to sanctum sanctorium of Delhi.

It is said that when General Wazir Khan came to know that Baba Banda Singh Bahadur had arrived close to the Mughal camps, Sirhind, he chose a site for war which was

Fig. 6 Baba Banda Singh Bahadur

plain on one side and had rough terrain on the other. This monument is built on the actual site where battle had been fought 300 years ago. Keeping the comfortable plain site for himself, General Wazir Khan left the rough terrain for Baba Banda Singh Bahadur's forces thinking that the Sikhs would lose the battle. When Baba Banda Singh Bahadur reached the site, he climbed the highest mound and examined the enemy's line-up. He decided that his war strategy would be to allow the enemy to come forward and face this rough terrain, while his own forces engaged in guerrilla tactics. The strategy was highly successful. The Mughal soldiers on horses rode towards the Chappar (ditch), but could not cross it and fell in the muddy water. The few that managed to make it across got stuck in the jhiris (Jharis or bushes). The Sikh soldiers had positioned themselves behind the Tibbas (dunes) and won the war through guerrilla warfare. After discussions with present-day locals, i.e. the third generation of the warriors, the architect was informed that even after hundred years of battle some unblasted cannon balls were found at the site. Baba Banda Singh Bahadur had also foreseen that the cannon balls would not blast as the mounds (Tibbas) or sand dunes would resist the cannon balls from exploding and rather absorb them. The approach used by Ar Renu Khanna (authors) in reviving the oral heritage of highlighting the sacrifices and war strategies of Baba Banda Singh Bahadur has been described below.

3.3 Making of a Monument

The architect evolved the architectural concept of this monument from the pages of history. The architect read many books on Baba Banda Singh Bahadur, met historians and had a feel of the site of war, and when came to know that the families of martyrs and those who fought the war were allotted land there, the architect then also met them and got the first-hand information from the third generation of the warriors, besides reading many books on him.

The architect then designed and recreated the whole landscape of Chappar (pond), Chiri (bushes and shrubs) and Tibbas (dunes) in the monument cum memorial similar to that which existed 300 years back (Fig. 7). This was done to give the same experience to the visitor as it was at the time of battle. The majestic victory tower and recreated natural landscape of that era kept intact give a feeling of history filled with grandeur. The statues on top of Tibbas have façade lighting which in the evening gets reflected in the pond and gives a mesmerizing view to the visitor.

The six Tibbas (dunes) are made of RCC dome structures covered with earth landscaping having statues of Sikh Commander Baba Banda Singh Bahadur and his five generals which make the visitor experience the guerrilla warfare (Fig. 8). These Tibbas (RCC domes) have cafeteria and an auditorium below.

Fig. 7 Recreated landscape of pond (Chappar) and mound (Tibba) at memorial cum monument

Fig. 8 Experiencing guerrilla warfare behind mounds (Tibbas) and the 328-ft.-high victory tower (Fateh Burj)

The statues on top of Tibbas have façade lighting which in the evening gets reflected in the pond and gives a mesmerizing view to the visitor (Fig. 9).

Apart from recreating the war scene, the architect also wanted to push the boundaries and create art which would leave an imprint on the minds and hearts of people for ages to come. Baba Banda Singh Bahadur belonged to a hilly area Jammu and Kashmir and had the habit of observing situations from the highest point to get a wider perspective. Therefore, we proposed a victory tower befitting this trait of Baba Banda Singh Bahadur. While one walks past the entrance of the memorial and enters a sprawling lawn, one is struck by the magnificent and towering Fateh Burj. It stands 328 feet high with its three levels of victory walks, making it the tallest victory tower in India and the eighth tallest in the world. A Sikh dome is placed on top with religious symbols (Khanda and Kalash). There are triangular openings which symbolically refer to the openings in the walls that were used for guns and canons positioning. The imposing monument is an architectural marvel glorifying the valour and the victory of the great Sikh commander (Bahga, 2020).

Fig. 9 Beautiful evening view of lighted statues on Tibbas with their reflection in the pond

3.4 Architecture and Multimedia Reviving History

An open air theatre faces the victory tower. This has been provided for creating interesting sound and light shows, wall mappings using the stark and bold walls of the victory tower to narrate the story of Baba Banda Singh Bahadur through multimedia (Fig. 10). This monument is non-invasive with regard to both the historical heritage and the natural environment. As the tourist approaches the monument, he can perceive an abstract and also a projected world, as well as a natural space that recovers the lost native vegetation to recreate the memory of the 300 years old landscape. The victory tower and also the layered abstract system (sound, lighting, vegetation, paths and soils) translate the oral heritage into a unique network of expressions for its conservation. Victory tower (Fateh Burj) has colour wash lights which make it a beacon (Fig. 11).

In 2011, the year of its inauguration, the victory tower and its grand architecture was seen an example of perfect integration in its natural environment, in architecture of India. Standing at victory walk, one can get aerial view of the landscape of Chappar Chiri and Tibbas from the triangular opening of the victory tower (Fig. 12). Now, it has become a reference of heritage architecture that has given shape to the local folklore and vocal tradition (Bahga, 2020). Globally, when there is worry about lack of architectural sensitivity, the region of Punjab in the state of India has shown an admirable effort to protect our cultural heritage. The flowering of such collective conscience leads to the creation of architectural monuments that belong to time and space. Such architectural monuments are the affirmation of our innate artistic elements. They are components that shape our culture.

The creation of useable buildings under the mounds was a challenging task in itself. Reinforced cement concrete domes of 10-m radius were constructed for housing cafeteria and auditorium for immersive experience. They had statues of generals on top (Fig. 13).

To add icing on the cake, the pond was positioned to receive complete reflection of the 328-feet-height victory tower. This has become a great picture point where visitors can capture their own image along with the victory tower having full reflection in water, pond and mound having Baba Banda Singh Bahadur's statue all in one frame (Fig. 14).

4 Conclusion

Oral heritage needs to be conserved by creating monuments, memorials and cultural spaces which provide a definite structural identity not vulnerable to distortion of facts. The architect

Fig. 10 Multimedia wall mapping on Fateh Burj

Fig. 11 Colour wash lights on Fateh Burj

creating the monument needs to have a thorough understanding of narrated events and proof of their authenticity. The architect and the developers should be able to visualize the uniqueness of past events that gave birth to the oral heritage and translate them in the form of architectural monuments.

By creating a befitting monument, the architect (Renu Khanna) has very well converted the valour, vision and the war strategies of Banda Bahadur into a permanent structural form. The 328-ft-high victory tower along with the recreated landscapes which Baba Banda Singh Bahadur exploited to win the battle serves as a major tourist attraction. Youngsters stop by to gaze upon the victory monument and acquaint themselves with its glorious heritage. Stopping by at the site, one can often spot families and youth alike taking selfies with the victory tower in the background and treasure this leaf from the past (Fig. 15).

Fig. 12 Ariel view of landscape of Chappar Chiri and Tibbas from the triangular opening of the victory tower

Fig. 13 Plan and section of Tibbas how building planned under Tibbas (mound)

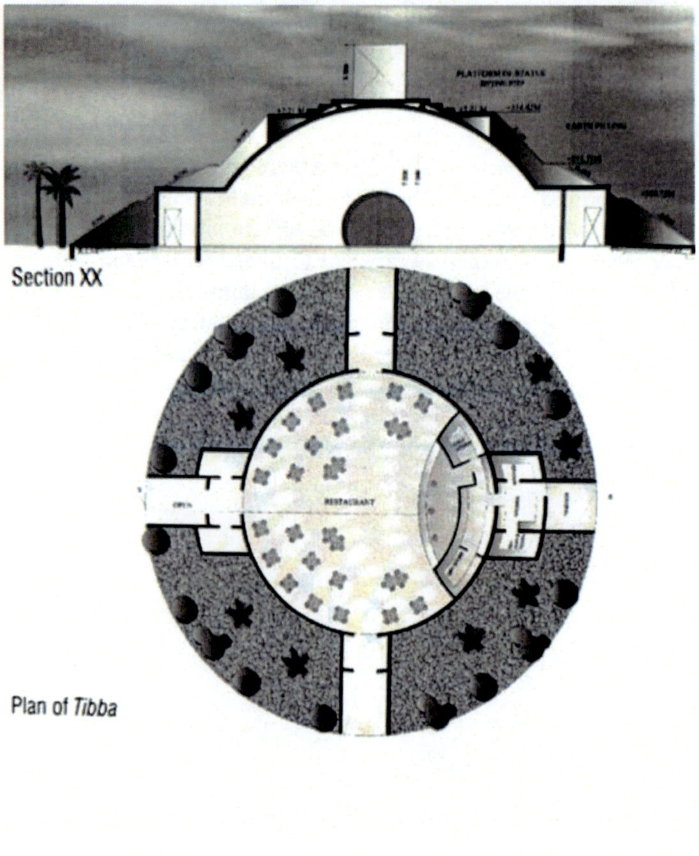

Fig. 14 Reflection of victory tower in pond

Fig. 15 Youngsters visiting Fateh Burj

References

Bahga, S. (2020). Baba Banda Singh Bahadur war memorial, Mohali: Recreating Battle field of Chappar Chiri (Architecture News). Retrieved from https://worldarchitecture.org/India.

Data, P .S. (2012). Mahabali Banda Singh Bahadur Publisher Hay House. Amazon.com: Books. Originally published in 1988.

Dhillon, H. (2004). First Raj of the Sikhs the life and times of Banda Singh Bahadur.

Embaby, M. E. (2014). Heritage conservation and architectural education: An educational methodology for design studios. *HBRC Journal, 10*(3), 339–350

Highfield, R. (2007). How I found queen Hatshepsut. Retrieved from https://www.telegraph.co.uk/news/science/science-news/3298587/How-I-found-Queen-Hatshepsut.html.

Kim, S., Whitford, M., & Arcodia, C. (2019). Development of intangible cultural heritage as a sustainable tourism resource: The intangible cultural heritage practitioners' perspectives. *Journal of Heritage Tourism, 14*(5–6), 422–435.

Ruhane, L., & Whitford, M. (2019). Cultural heritage and Indigenous tourism. *Journal of Heritage Tourism, 14*(3), 179–191.

Sagoo, H. K. (2001). *Banda singhbahadur and Sikh Sovereignty.* Google books.

Singh, G. (1999). *Life of Banda singh Bahadur based on contemporary and original records.* Originally published in 1935 Google books.

Singh, S. (2000). *Life and exploits of Banda Singh Bahadur.* Publication Bureau Punjabi University Patiala.

Tripathi, A. (2019). Faith, evidence prove Masjid was on Ram's birthplace. deccanherald.com/national/national-politics/.

UNESCO (1972). World Heritage Convention. Retrieved from https://www.igi-global.com/dictionary/the-importance-of-being-honest/48665.

UNESCO (1998). Proclamation of masterpieces of the oral and intangible heritage of humanity. UNESCO. Retrieved from. https://www.lacult.unesco.org/docc/124628english.pdf

An Inquiry and Documentation of Architectural Styles of Residential Building Facades in Old Rajshahi, Bangladesh

Md. Asaduzzaman, Z. H. M. Monjur Murshed, and Md. Nazmul Hoda

Abstract

Rajshahi is one of the historic cities in Bangladesh, which was a portion of the Pundra located in ancient Bengal. Rajshahi has experienced rapid urbanization from the last few decades. As a result, heritage buildings and traditional identity are quickly disappearing with time; in the process, the city is facing an identity crisis. Preservation of the old city is essential for conserving traditional identity of historic cities like Rajshahi. Architectural heritage buildings are associated with identity of a city. There is no significant research about the identification and analysis of architectural styles in old Rajshahi. The purpose of this research is to conserve the traditional identity of heritage buildings through a documentation of comparative analysis of architectural styles. The methods of this research are, firstly, identifications and documentations of residential building's facade with architectural features and categorize the buildings based on different time periods of development. Finally, a comparative analysis will prepare for concluding with identifying architectural characteristics and influences of selected residential heritage building's facade in old Rajshahi. This research will help to conserve the architectural styles of historic building facades through documentations and give an analytical comparison of different periods, which will help to identify the local characteristics of this old part of Rajshahi.

Keywords

Residential building facade • Architectural styles • Documentation • Old Rajshahi

Md.Asaduzzaman (✉) · Z. H. M. Monjur Murshed · Md.Nazmul Hoda
Department of Architecture, Rajshahi University of Engineering & Technology, Rajshahi- 6204, Bangladesh

1 Introduction

Rajshahi is the 4th largest metropolitan city, located in the northwest of Bangladesh. It was the part of the Gaur Kingdom under the Pundra region of ancient Bengal. Rajshahi was dominated by various Maharajas, Rajas and Zamindars ("800 Years of Rajshahi City," 2013). The root of the title of "Rajshahi" might have come from Hindu Rulers and Zamindars as *Raj* and the Persianized *Shahi*, both of which mean "Royal" or "Kingdom." Rajshahi as a regulatory area was set up in 1772. As a Rajshahi town stands by the river Padma, the Dutch, the French, the British and various other foreign powers began to use Rajshahi as a commercial hub. The Portuguese came here to fulfill their various kinds of interest and built *Boro Kuthi*, the first colonial building, just by the bank of the mighty River Padma. The town was mainly signified in 1825, when the East India Company shifted the administrative headquarters here from Natore (Dr. Salim Sabrin, 2013). The original Rajshahi town was on the Mouza of Srirampur, Dorga Para, Sagar Para and Ghoramara; a large part of the city is now under the River Padma (Doza, 2012).

The first permanent massive two-storied brick building *"Boro Kuthi"* was constructed by Dutch silk traders in the early eighteenth century, which was built in the Mughal period. The study area is located in the older part of Rajshahi. Land use pattern of the old part of Rajshahi is predominantly residential with commercial mix. There are total twenty-one buildings selected for analysis located in Sagar Para, Bose Para, Sheikh Para around the Saheb Bazar area (the central commercial hub of this region) (Fig. 1). Residential buildings broadly grouped into four different time periods as colonial, British Raj, post-partition period (pre-independence) and post-independence of Bangladesh. This paper is focused mainly on analysis of architectural styles of residential building facade in the old part of Rajshahi in these time periods.

A. Versaci et al. (eds.), *Conservation of Architectural Heritage*, Advances in Science, Technology & Innovation,
https://doi.org/10.1007/978-3-030-74482-3_25

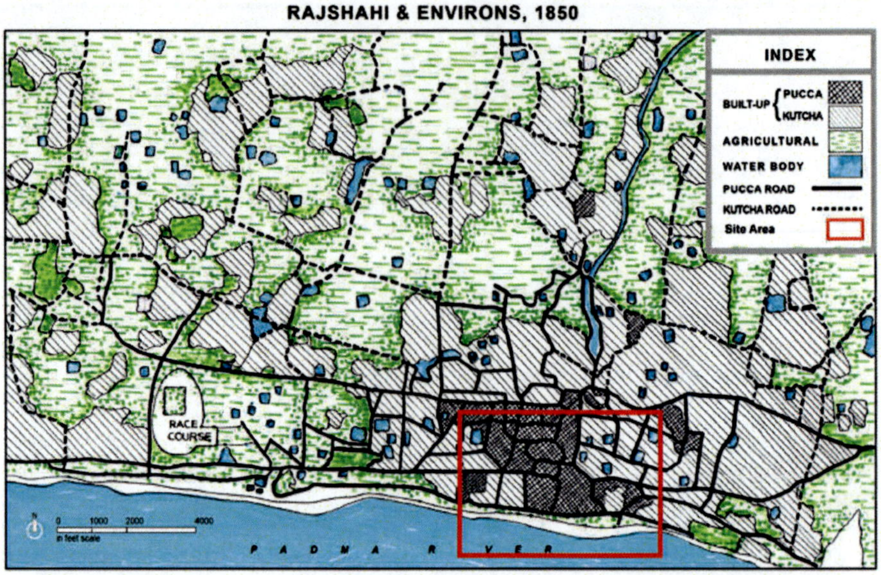

Fig. 1 Map of the old city of Rajshahi (*Source* Rajshahi and Environs 1850, redrawn by Authors)

2 Objectives

The objective of the study is to analyze documentation and identification of the architectural styles and characteristics of residential building facades in the old part of Rajshahi in relation to different time periods. Initially after finding out the architectural elements and their influences on residential buildings in chronological order, this paper concludes with a comparative analysis of major distinctive characteristics of facade within these selected buildings in the old town of Rajshahi.

3 Methodology

A total of 21 residential buildings' facade in the old Rajshahi area was surveyed and inspected in this study (Fig. 2). The determination of the building's façade was based upon key criteria which are as taken after:

1. Different time periods (Mughal, colonial, British Raj, post-partition period (pre-independence), post-independence period.
2. Architectural styles (different influences).

A few building overviews and assessments covering location examination study were carried out to set up the distinctive architectural styles of residential buildings in old Rajshahi. The study too incorporates comparative analyses between the periods covered in this study.

The architectural styles of the building's facade are also studied and classified in order to have a better understanding of the changes in the buildings from colonial to after independence. Since there is no central body for recording characteristics of private buildings in this city, the recognizable proof of the private ancient buildings overviewed within the think about was based on the taking after methods (Fig. 3):

- Collecting any conceivable record at different offices, local institutions and preservation bodies such as chronicles, libraries and colleges.
- Conducting an academic survey in old part of Rajshahi under studio work conducted by authors.
- Carrying out spoken interviews with user group and people expected to have some information about those residential buildings.

4 Architectural Styles of Residential Building Façade of Old Rajshahi

The Mughal Period (1576–1757 AD):

The first permanent massive two-storied brick building *Boro Kuthi* at old Rajshahi was constructed by Dutch silk traders in the early eighteenth century (Fig. 4), which was built in the Mughal period. Later, it was used by the British headquarter. And now, it is used as a clubhouse by the Rajshahi University. These were probably used as a prison as well as for storage of silk goods.

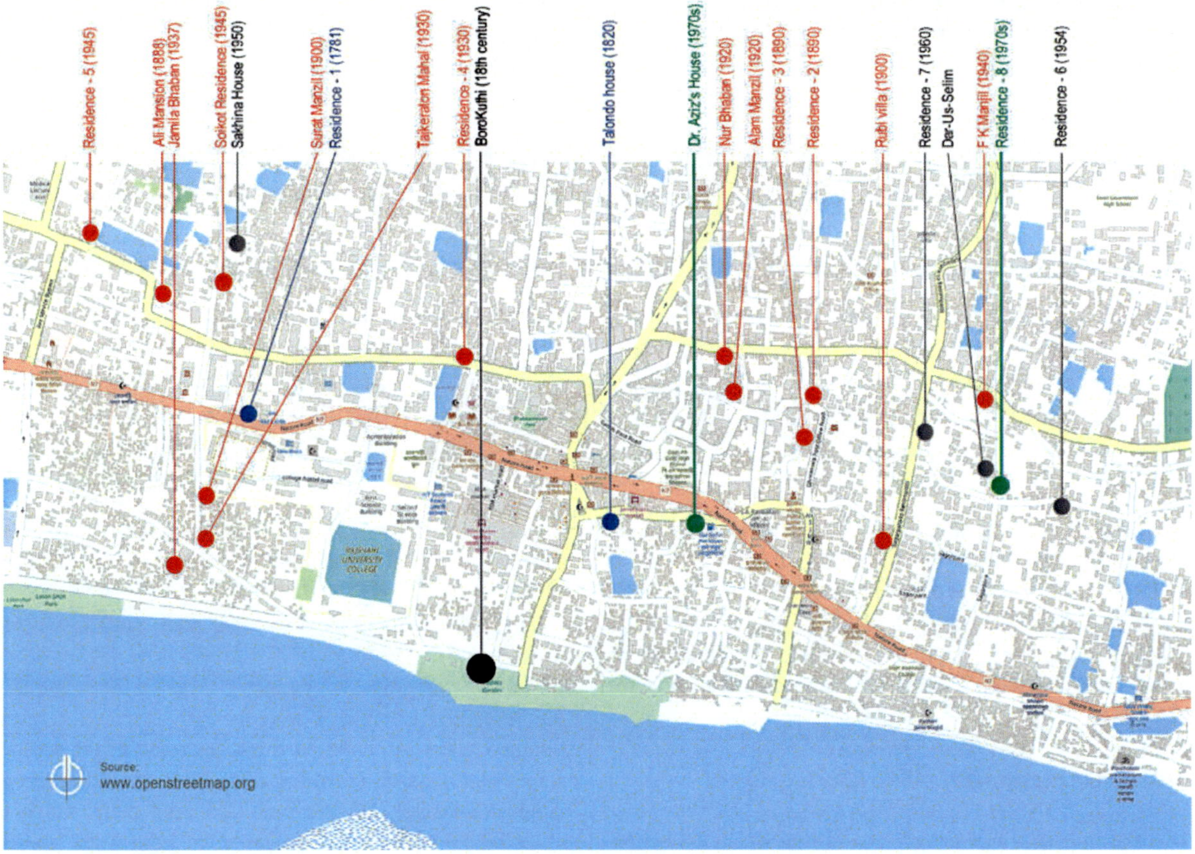

Fig. 2 Location map of selected residential buildings in the old Rajshahi (*Source* Open street map, drawn by Authors).

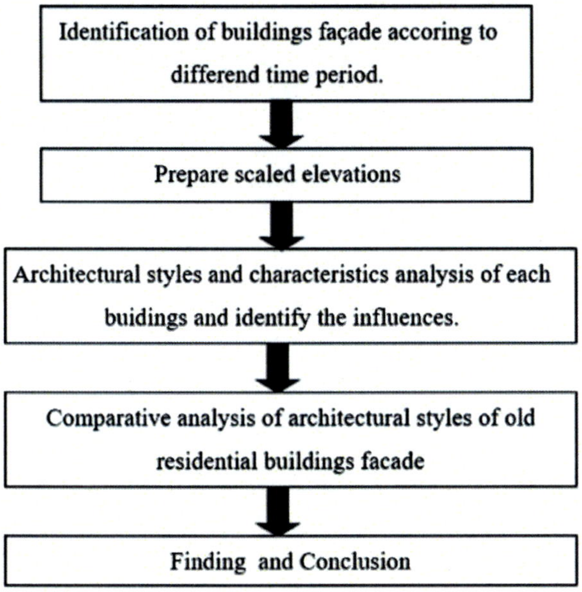

Fig. 3 Illustration of methodology (*Source* By Authors)

The bastion-like octagonal towers on either side liberally provided with embrasures for musketry, boldly the otherwise monotonous appearance of the building (Husain, 2007).

The façade of this building is symmetrical and white color brick-plastered building. This building is simple in design with flat roof. The octagonal dome tower like minarets and semicircular arch opening with venetian blind represents the influence from Byzantine and Mughal architecture.

British Colonial Period (1757–1858 AD):

By definition, colonial architecture is the architectural style borrowed from a country of origin and then integrated into the structures located in far-off regions. This particular architectural style evolved when colonists created a fusion by blending the architectural vocabulary of their country of origin with the design principles of the region they colonized (Afzal, 2018).

There are two selected buildings' façades in this period. The architectural style is closer to Mughal and Byzantine architecture. The *Residence-1* was constructed in 1781 AD with the simple geometry and symmetrical façade. There are

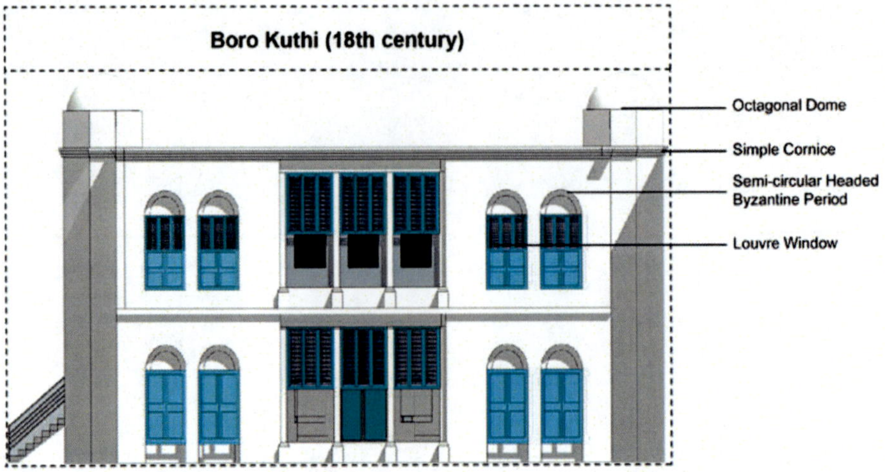

Fig. 4 Facade of the two-storied *Boro Kuthi*—eighteenth-century works (*Source* Drawn by Authors)

semicircular windows with a pilaster and a simple cornice. The façade of *Residence-1* is illustrated in Fig. 5.

The *Talondo House* was built in 1820 AD as a Zamindar house in this area. It is used as *Kuthi Bari*. Semicircular opening is also found with wooden door. The geometry of this building is simple, and wall is plastered with color shown below. (Fig. 6).

British Raj Period (1858–1947 AD):

The British period was not only one during which the British alone were responsible for constructions, but also one during which magnificent palaces were built by the local ruling class, in its style unique to Bengal (Ahmad, 1986). There are thirteen buildings selected under this period. The architectural style of residential building façade of that time was various architectural influences. The architectural style of first three selected building (1888 AD–1890 AD) façade is simpler geometric and semicircular opening of *Ali Mansion* reflecting Byzantine architecture. Rest two buildings *Residence-2* and *Residence-3* are representing Greek colonnade architecture, which is shown in Fig. 7. The windows were simple rectangular in geometry with detailed cornice.

The architectural styles of residential buildings were constructed from 1900 to 1937 AD and were highly decorative and symmetrical in geometry (Fig. 8). *Surat Manzil*, *Rubi Villa* and *Alam Manzil* were the early twentieth-century examples where red exposed brick was used in the facade. Parapet with kalasha crowning also shows the influence of Indian architecture. Pilaster was also used there which is influenced by Roman Corinthian pillar with floral capital (Fig. 9). This time period also showed the influences of neoclassical architecture.

Surat Manzil and *Jamila Bhaban* are the finest examples of British Raj period. These examples are representation of mix characteristics of Indian, Roman as well as Byzantine influence. The cast-iron railing details of *Jamila Bhaban* also reflect the influence of modern period (Fig. 10). Uses of balustrade parapet with crowning were visible in this period.

We have identified three buildings' façade from later phase of British Raj (1940 AD–1945 AD). The titles of these

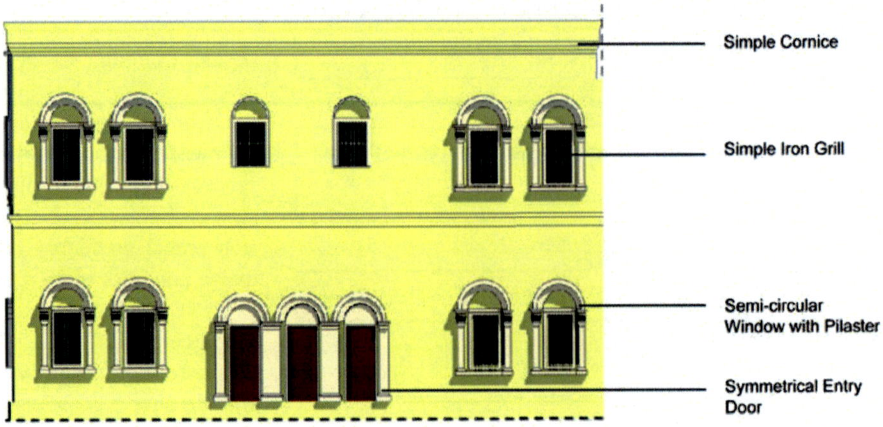

Fig. 5 British Colonial *Residence-1*(1781 AD) facade details (*Source* Drawn by Authors)

Fig. 6 British Colonial *Talondo House* (1820 AD) (*Source* Drawn by Authors)

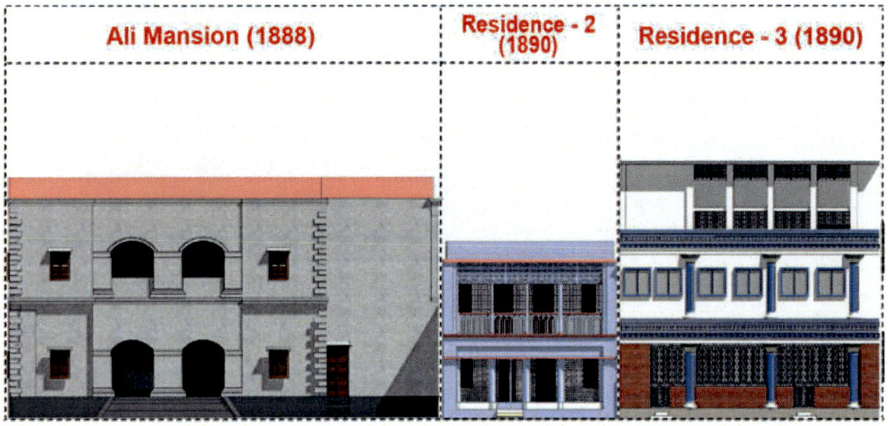

Fig. 7 British Raj (1888 AD–1890 AD facade details (*Source* Drawn by Authors)

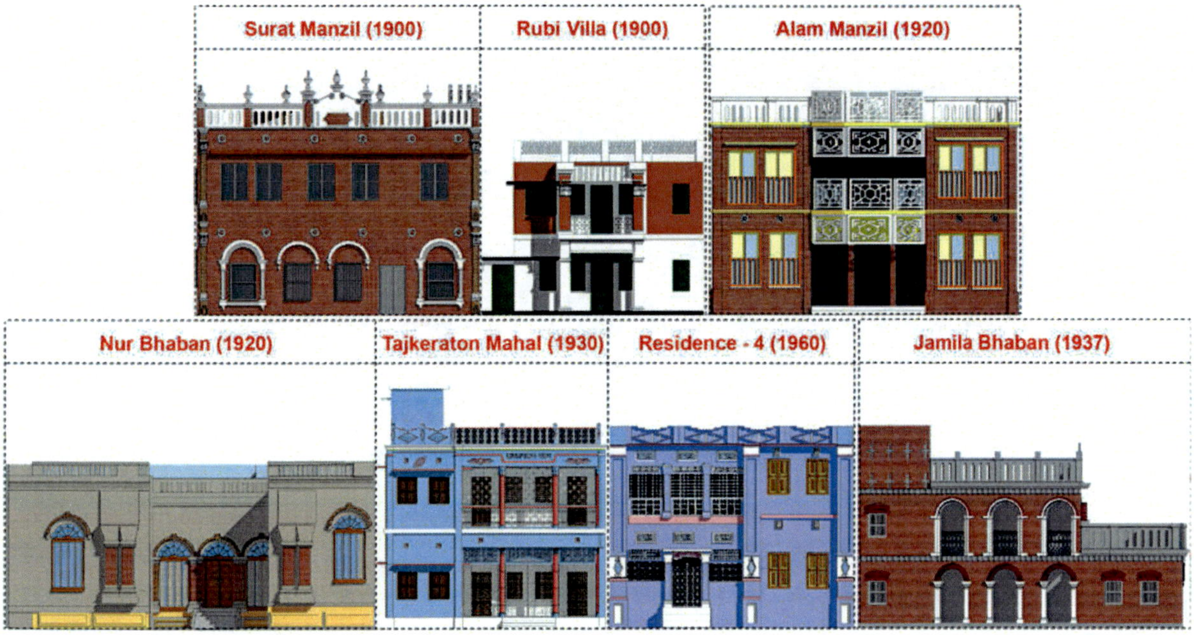

Fig. 8 British Raj (1900 AD–1937 AD) facade details (*Source* Drawn by Authors)

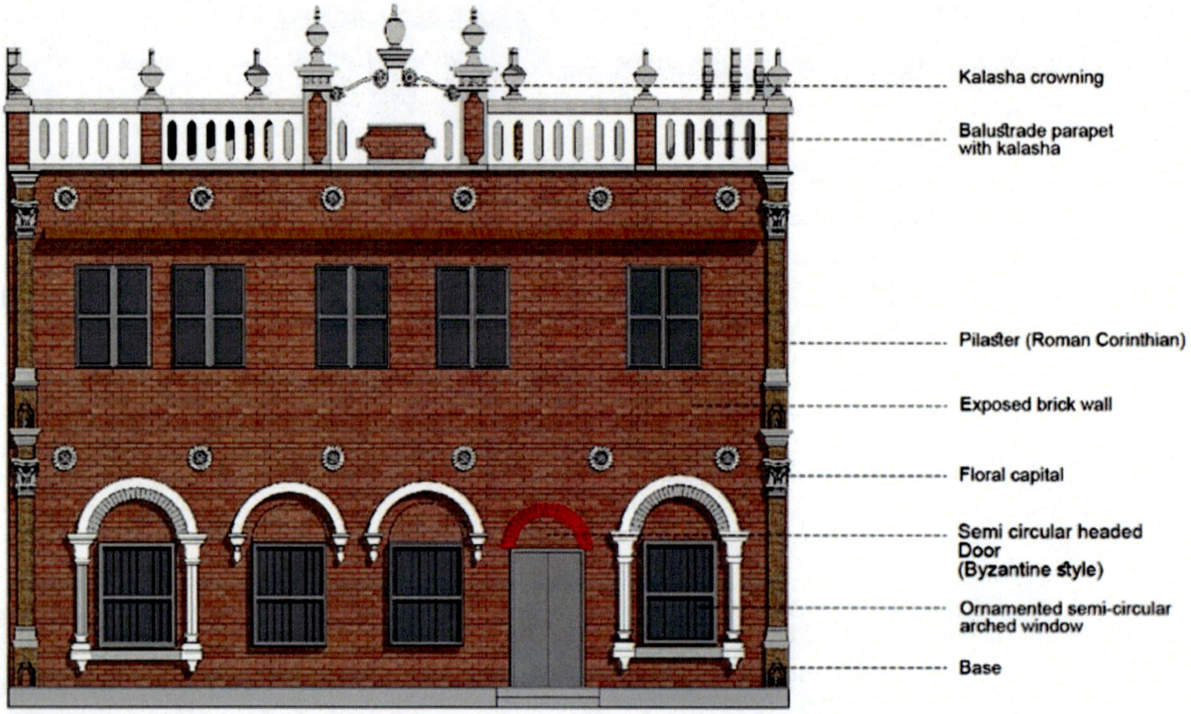

Fig. 9 British Raj period—*Surat Manzil* (1900 AD) facade details (*Source* Drawn by Authors)

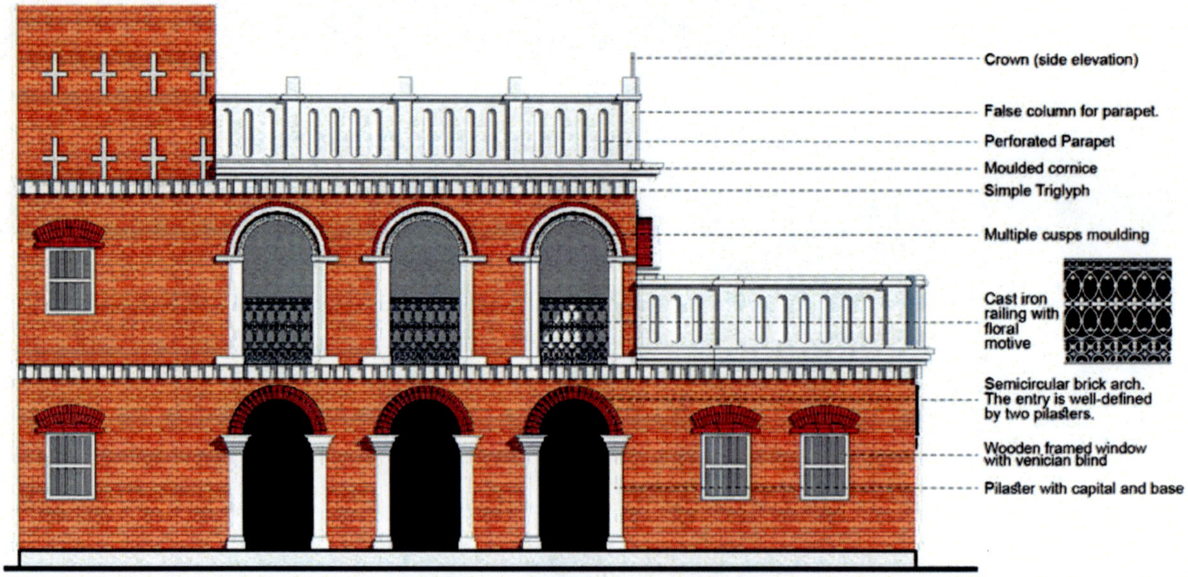

Fig. 10 British Raj period—*Jamila Bhaban* (1937 AD) facade details (*Source* Drawn by Authors)

buildings are *F K Manzil*, *Soikot Residence* and *Residence-5* (Fig. 11). The architectural style of the façade of this period of time was a presentation of straight lines. Windows are arranged in groups, and shading devices were made by concrete. The door and windows were made with metal frame infilled with glass. Those architectural elements were influenced by Art Deco style.

There are various column details from selected building's façade of British Raj period (Fig. 12). The diverse structural elements are found in residential buildings at old Rajshahi.

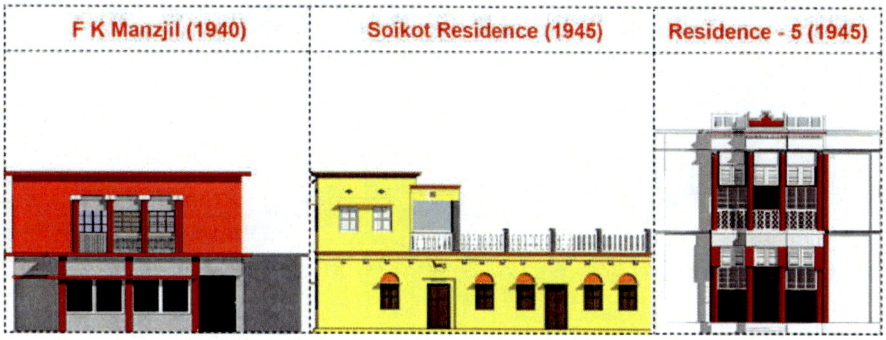

Fig. 11 British Raj (1940 AD–1945 AD) facade details (*Source* Drawn by Authors)

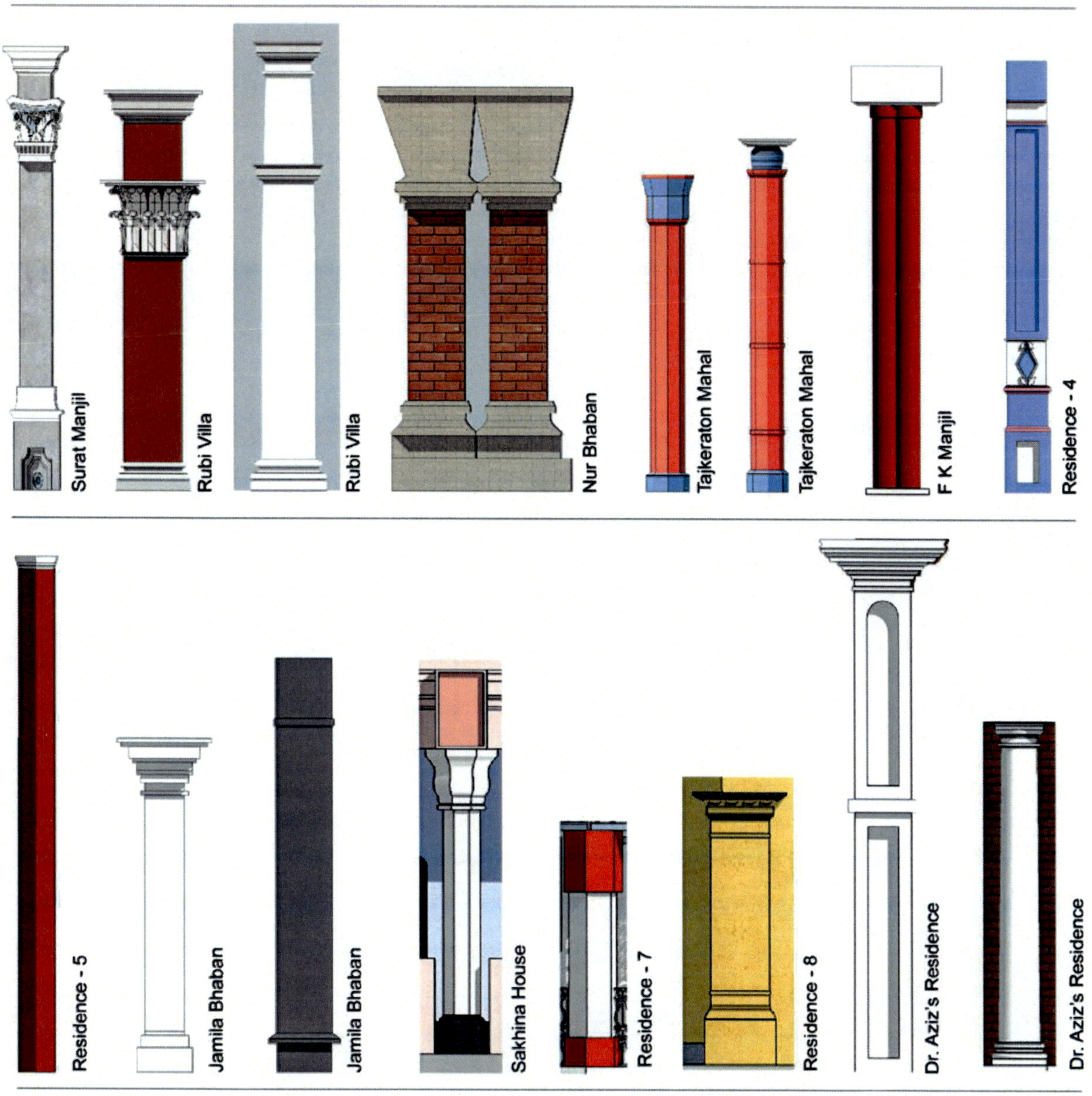

Fig. 12 British Raj period—various types of orders used in decorative column (1888 AD—1945 AD) (*Source* By Authors)

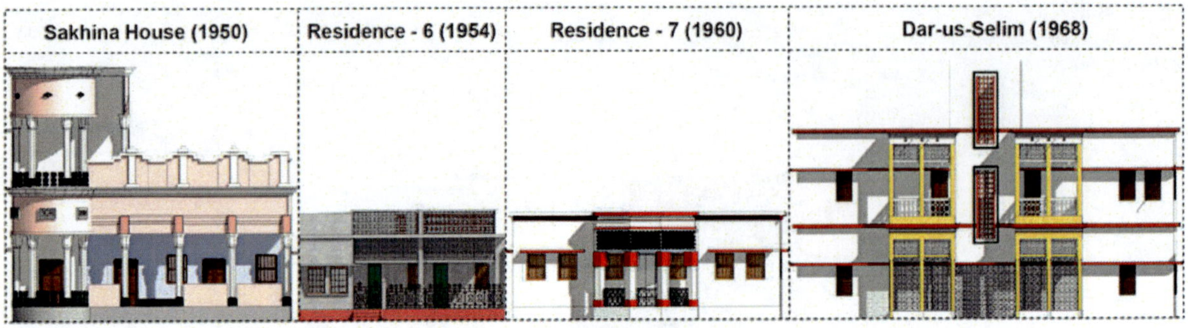

Fig. 13 Post-partition period (pre-independence) (1947 AD–1971 AD) facade details (*Source* Drawn by Authors)

Post-partition Period (Pre-independence) (1947 AD–1971 AD):

The time period between 1947 and 1971 AD is also defined as early modern time in the world perspective. There are four buildings' façade selected from this time span. They are *Sakhina House*, *Residence-6*, *Residence-7* and *Dar-us-Selim Residence* (Fig. 13). Those building facades are influenced by simple geometric patterns and clean design. The construction material of those buildings is of reinforced concrete. The windows and doors are commonly used of steel frame infilled with wood. There is no ornamentation found. The decoration was simple geometric. Use of air ventilation slots is also a prominent feature on the façade.

Sakhina House was constructed in 1950 AD with reinforced concrete. The geometric shape is simple in nature, but intricate geometric decoration was found. The lotus kalasha capital of the column is influenced by Indian architecture. Cast-iron railing reflects the early modern time's use of metal for decoration. The use of floral motive ventilator, simple geometric drop wall with curve cornice and use of parapet represent influence of modern time. The façade itself is asymmetrical (Fig. 14).

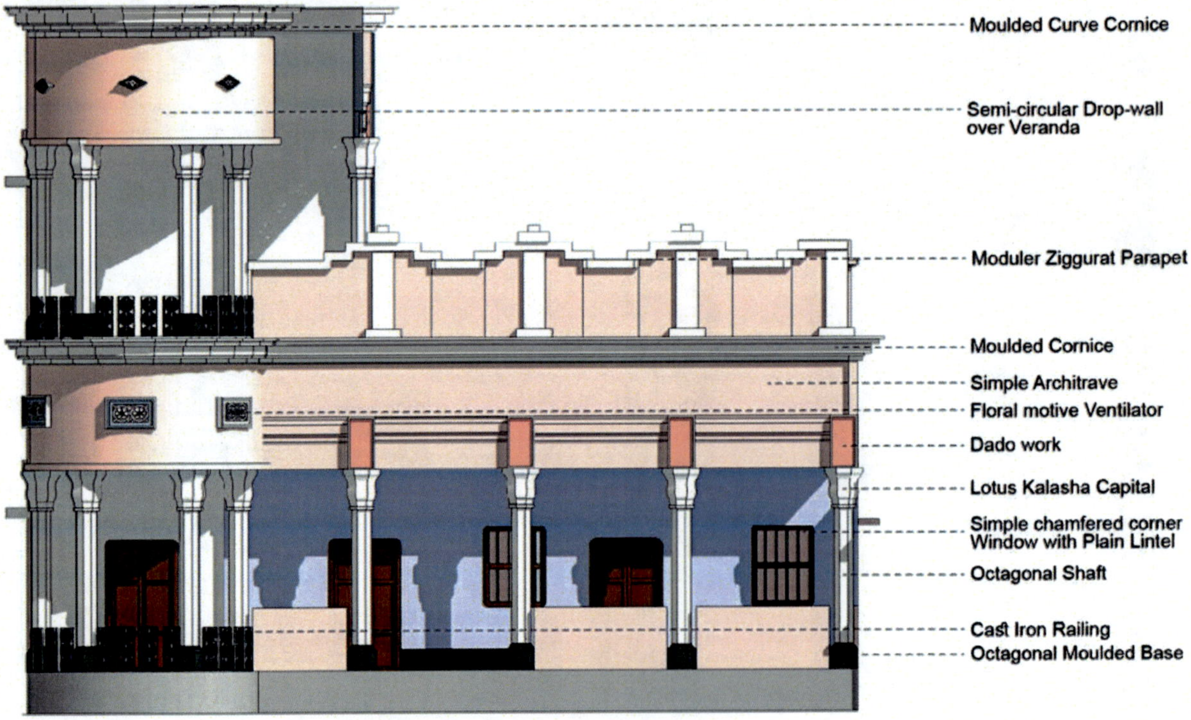

Fig. 14 Sakhina House (1950 AD) facade details (*Source* Drawn by Authors)

Fig. 15 *Residence-6* (1954 AD) facade details (*Source* Drawn by Authors)

Uses of glass with metal frame, metal railing and concrete modular *Jaali* parapet are found in *Residence-6*, a 1954 AD building's façade (Fig. 15). The column is a simple circular one, without any ornamentation. The façade is more simplified than the previous example.

The *Residence-7* was built in 1960 AD (Fig. 16). The building elevation is symmetrical with clean fenestration. The *Jaali* drop wall and sunshade are made from concrete. A pentagonal simple geometric entry approach is also created. The built form is also equipped with simple windows which are adorned with plain lintel and wooden frame. Molded cornice was also used in the roof.

The *Dar-Us-Selim House* was influenced by early modern architecture of old Rajshahi City (Fig. 17) which is constructed on 1968 AD. Use of precast screen wall, drop wall and railing is influenced by the industrial products of early modern time. The uses of metal grill were common in the residential building's facade in this time period. Extended/continuous lintel over window is also created as shading devices.

Post-independence Period (1971 AD–1975 AD):

After independence of Bangladesh, amid the construction time period of 1971–1975 AD, two residential buildings' fenestration is selected for this paper. They are *Residence-8* and *Dr. Aziz's House* (Fig. 18). Although the time period refers to a modern time in the global perspective, that building façade was inspired by diversified historical features and architectural characteristics of colonial, Mughal and Byzantine details while featuring the modern style.

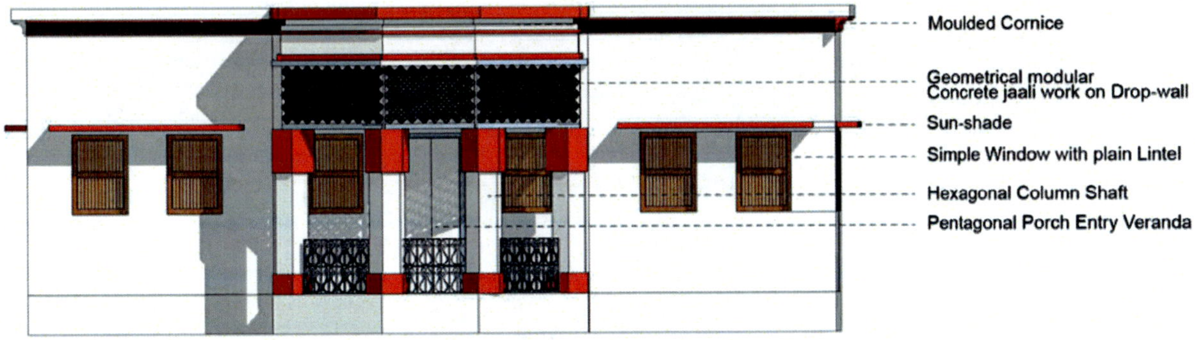

Fig. 16 *Residence-7* (1960 AD) facade details (*Source* Drawn by Authors)

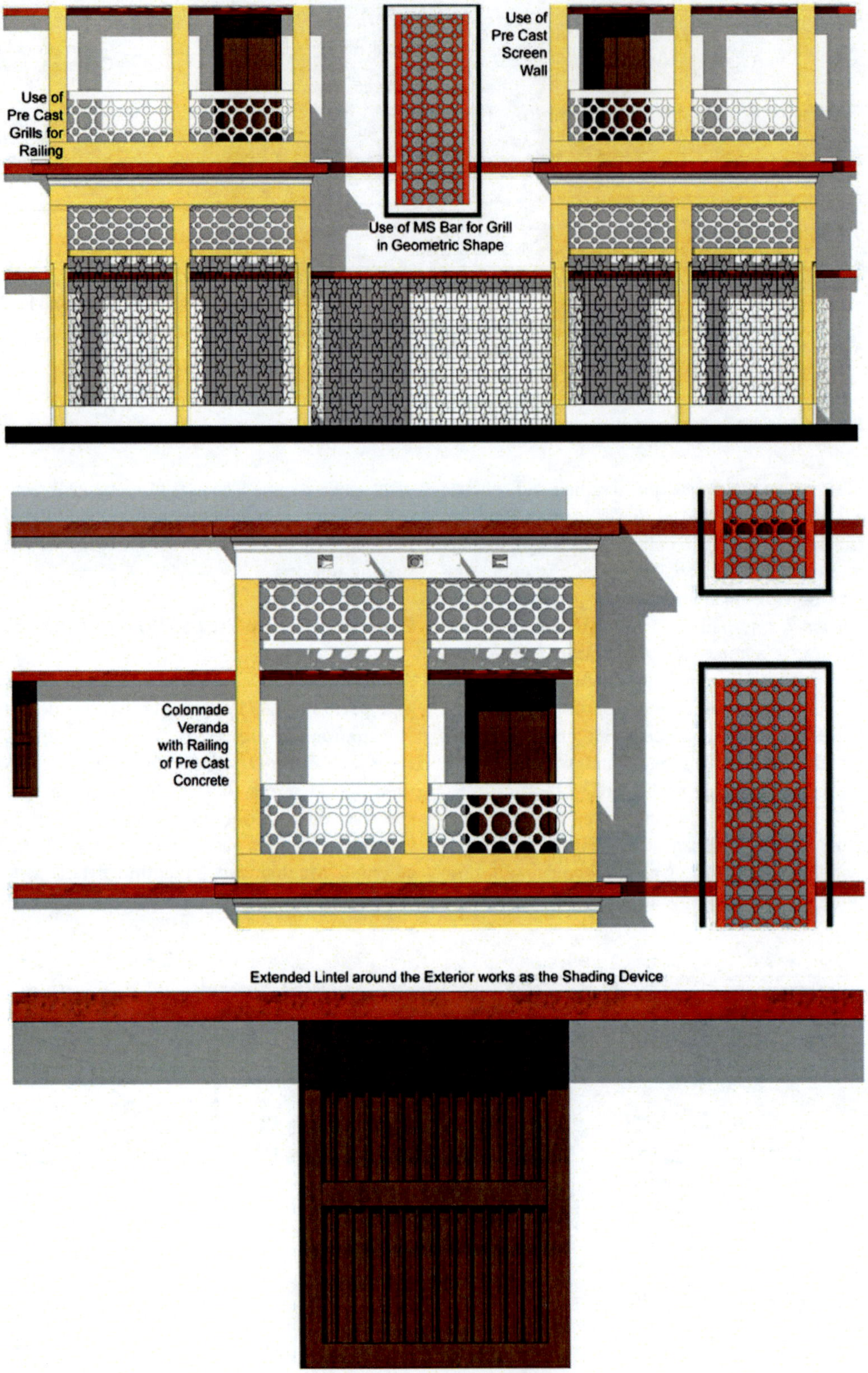

Fig. 17 *Dar-us-Selim Residence* (1968 AD) facade details (*Source* Drawn by Authors)

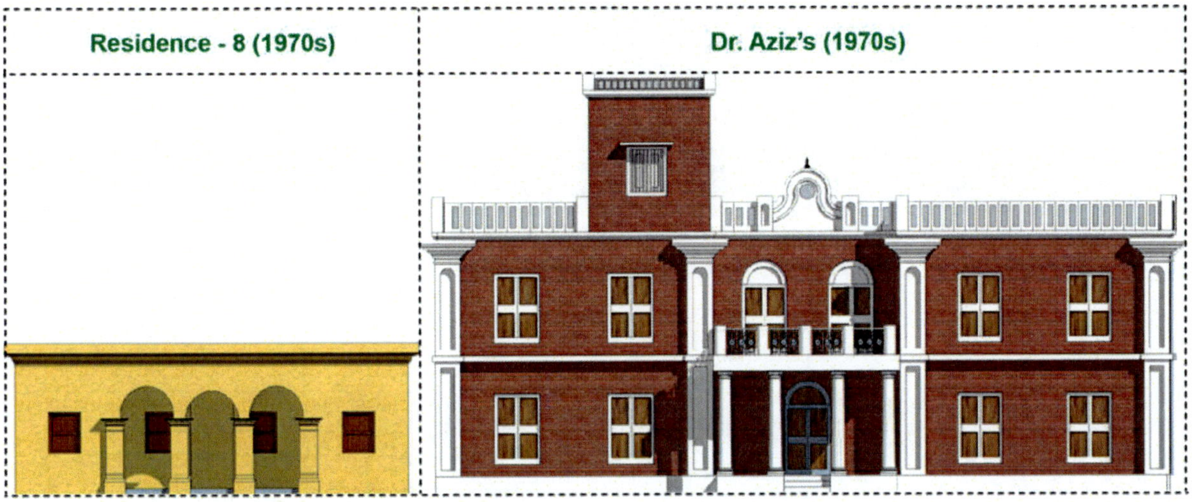

Fig. 18 Post-independence period (1971–1975 AD) facade details (*Source* Drawn by Authors)

Residence-8 has simple façade with semicircular symmetrical entry. The columns are simple rectangular in shape (Fig. 19). The use of ornamented stepped column capital is visible. The extension of roof slab with low height parapet with cornice details is also featured in this building.

Dr. Aziz's House was built in 1974 AD with British colonial influences in modern style (Fig. 20). Column and other details were brought from colonial architecture with the use of metal frame infilled with glass windows. This building used smooth shaft with plain capital for the column. Then, the uses of precast railing are influenced by modern style, while the pilaster and the use of crowning in parapet signify the British Raj influence for decoration and ornamentation.

5 Comparative Analysis of the Residential Building's Façade in Old Rajshahi (Table 1)

This paper illustrates a comparative analysis of the residential houses constructed during different times in the Rajshahi history and studies the changes in the residential building's facade from colonial, British Raj, post-partition (pre-independence period) and post-independence period of Bangladesh. The common characteristics of all these residential houses' facade have been discussed in the paper to provide a better understanding of the architectural styles of these buildings and the factors which influenced these styles. Usually, there are seven key features which rule the architectural styles of the residential house's façade in old Rajshahi. They are opening, entry, column, grill, railing, cornice and symmetricity.

6 Conclusion

Residential building's façade of those periods ought to be considered as territorial legacy mostly due to their chronicled and architectural values. All historic structures that are historically and architecturally significant should be included in a gazetteer to prevent them from being destroyed. Building preservation ought to be seen as a way of protecting specific aspects of history and development of Rajshahi. These buildings have brought about within the arrangement of a special character and picture to the urban texture which may offer assistance to secure the ancient building's façade of old Rajshahi. It is imperative to recognize that the ancient residential buildings are too verifiably and architecturally significant, especially in complementing the chronicled advancement of old Rajshahi.

Due to the rapid and mass development of new buildings and uncontrolled conservation and renovation works, the architectural styles of residential building facade are slowly disappearing from the urban cityscape. Conserving and preserving these residential heritages' building façades at old part of Rajshahi City is very important to prolong their architectural styles for future generation to see and get inspired.

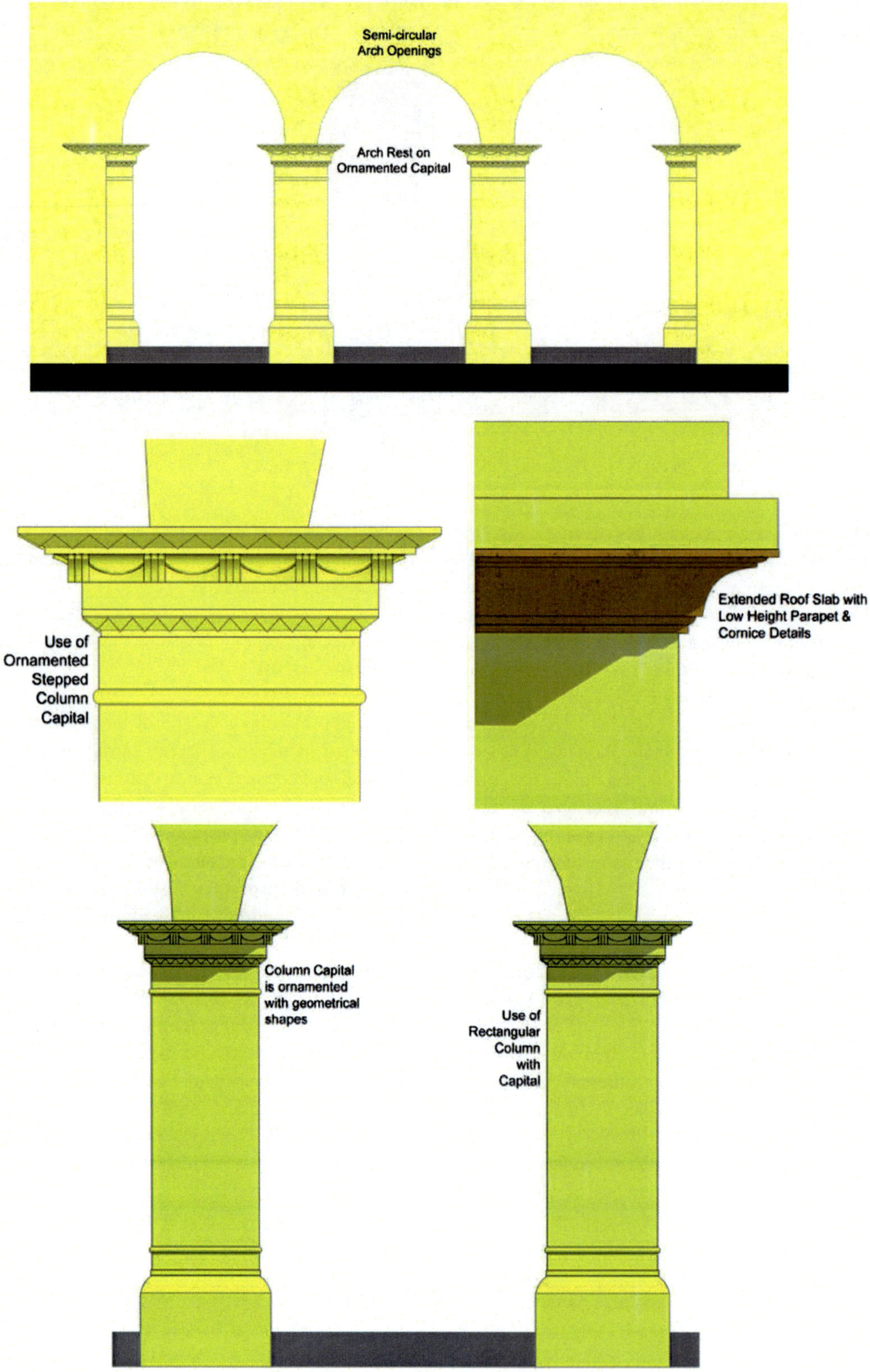

Fig. 19 Post-independence period (1972 AD) facade details (*Source* Drawn by Authors)

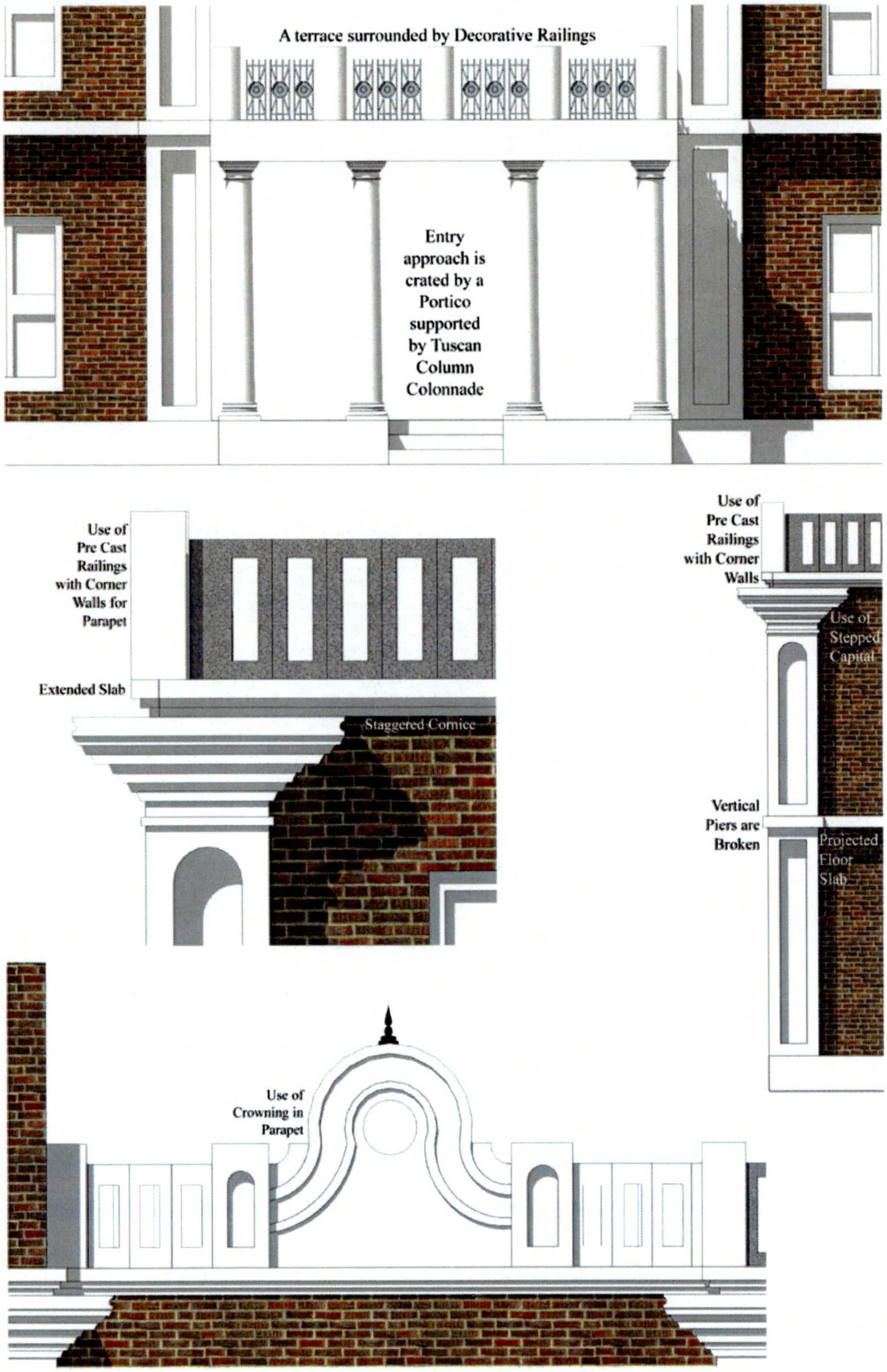

Fig. 20 Post-independence period (1974 AD) facade details n(*Source* Drawn by Authors)

Table 1 Comparative analysis of building façades on diversified aspects (*Source By Authors*)

The Mughal period 1576–1757 AD	British colonial period 1757–1857 AD	British Raj period 1858–1945 AD	Post-partition (pre-independence period) 1947–1971 AD	Post-independence period 1972–1975 AD
Name: *Boro Kuthi*	Name: *Residence-1*	Name: *Surat Manzil*	Name: *Residence-7*	Name: *Residence-8*
Early eighteenth century	1781 AD	1900 AD	1960 AD	1972 AD
Façade: Symmetrical	*Façade*: Symmetrical	*Façade*: Asymmetrical	*Façade*: Symmetrical	*Façade*: Symmetrical
Major influence: Mughal and Byzantine	*Major influence:* Byzantine and Roman	*Major influence:* Indian, Roman, Byzantine and neoclassical	*Major influence:* Early modern with Roman	*Major influence:* Modern and Byzantine
Opening: Semicircular headed with venetian blind	*Opening*: Semicircular headed with pilaster	*Opening*: Semicircular headed with rectangular pilaster	*Opening*: Rectangular and simple geometry	*Opening*: Rectangular window and semicircular door
Wall: Brick work with plaster	*Wall*: Brick work with plaster	*Wall*: Exposed red brick	*Wall*: Brick work with plaster	*Wall*: Brick work with plaster
Ornamentation: Simple with windows and cornice details	*Ornamentation*: Windows and cornice details	*Ornamentation*: Parapet, window, ventilator and cornice details	*Ornamentation*: Drop wall and railing details with concrete *Jaali*	*Ornamentation*: Simple rectangular window with simple column
Special feature: Octagonal tower in either side like minaret	*Special feature:* Symmetric three door entries	*Special feature:* Balustrade parapet with kalasha crowning and pilaster with floral capital	*Special feature:* Continuous lintel as shading devices and octagonal geometry for entry approach	*Special feature:* Symmetric three colonnade entries
Facade color: White	*Facade color:* Yellow	*Facade color:* Red brick	*Facade color:* White with red	*Facade color:* Yellow
Plaster: Plain	*Plaster:* Plain	*Plaster:* Molded	*Plaster:* Plain	*Plaster:* Plain
Column: No exposed column	*Column:* No exposed column	*Column:* No exposed column	*Column:* Hexagonal shaft	*Column:* Geometrical with smooth shaft

References

Ahmad, N. (1986). In Sanday, J. (Ed.) *Buildings of the British Raj in Bangladesh*, 1st edn. The University Press Limited (UPL).

Afzal, F. (2018). *British Colonial Architecture in Bengal | The Daily Star*. Reteieved February 19, 2018, from https://www.thedailystar.net/in-focus/british-colonial-architecture-bengal-1536610.

Doza, S. B. (2012). Longing of a small town: The tale of a traditional town in the northern part of Bangladesh. In: *International Conference on Tradition, Heritage, Art & Architecture of Pakistan*.

Husain, A. B. M. (2007). *Architecture : A history through the ages*. Asiatic Society of Bangladesh.

Salim S. (2013). Socio-political background and recreational development of modern Rajshahi. *RDA Magazine/ The Guardian*. http://www.theguardianbd.com/socio-political-background-and-recreational-development-of-modern-rajshahi/

years of Rajshahi city. (2013, June 6). *The Indipendent*. http://www.theindependentbd.com/paper-edition/others/art-a-culture/90653-800-years-of-rajshahi-city.html

Vjosa River Valley; Strategies for Sustainable Tourism

Irina Branko

Abstract

Along the Adriatic and Ionian seas, lays Albania in a relatively small territory, but a diverse environment. Its population has been most of the time of a rural character spread all over its surface through small villages, which date from the XIV to XIX century. Studies show that these urbanized forms differ from place to place, as per the main characteristics of the territory and inhabitant's lifestyle. By evaluating the natural environment, using local materials, through their know-how, the rural population of Albania has always been able to establish a dialog with the environment and collecting its fruity. During the socialist regime, the rural inhabitants have gone through poverty and a difficult period. It was mostly due to the change of the property rules and creation of the cooperative system. With the fall of the regime after five decades, the migration of the rural population phenomena begun. Many rural areas are abandoned, and together with them, the history, the culture, and the natural resources. Based on the latest government strategies on the rural development, especially on the *Academia 100+ Villages Program*, this paper aims to go through a deeper understanding of the build and natural environment of a particular area in the country—the *Vjosa* river valley. The development of this study together with the creation of a database on history, people, and know-how, combined with European rural development models, hopefully will serve as a background for the development of Albanian rural economy through the development of sustainable tourism strategies in the area.

Keywords

Sustainability • Tourism • Local • Rural • Environment

I. Branko (✉)
Departament of Urbanism, Faculty of Architecture and Urbanism, Polytechnic University of Tirana, Tirana, Albania

1 Introduction

The preservation of cultural and natural heritage is of great interest nowadays. With modernization, more and more people do tend to leave the rural areas and flee toward the cities. This tendency brings the lifestyle generalization of people in different provinces, different regions, and sometimes of different countries. This phenomenon has two primary risks: (1) Increase of cities population and city challenges in developing sustainable development strategies and (2) loss of local identity, through the chain process of migration, abandonment of the territory, loss of know-how, history, ethnography, and loss of interest on the natural environment.

It is now 30 years since Albania has overthrown the totalitarian regime and is working on building sustainable development and economy. There have been many signs of progress as well as failures, of course. From the experts in the field, the development of tourism remains one of the most important economic factors in the economic growth of Albania, due to the: (1) Size of the territory and population and (2) rich natural and cultural heritage it has.

Within this context, the aim of this research is: (1) To create a complete database of this particular region in the country, population, natural resources, cultural tangible and intangible heritage, and their transformations during the years; (2) to generate a methodological approach for other possible case studies; and (3) the primary purpose is to conclude with strategies for a sustainable development, based on tourism, for a better local economy and a better connection with the national natural and cultural heritage.

2 Background

Geographically speaking, Albania is one of the countries surrounding the Mediterranean Sea and lays along the Adriatic and the Ionian waters. Although with a small area,

A. Versaci et al. (eds.), *Conservation of Architectural Heritage*, Advances in Science, Technology & Innovation,
https://doi.org/10.1007/978-3-030-74482-3_26

only 28,748 km^2 is characterized by a very diverse territory, change of topography from west to east side which varies from sea level until it reaches an altitude of *2751 m (Top of mount Korab).* Change of topography, combined with the water resources in the territory, creates a large variety of climates and microclimates, therefore very diverse biodiversity. Nowadays, in Albanian territories, there can be count a large number of habitats and unique species.

Today, Albania has a population of *2,862,427 inhabitants* (INSTAT, 2019). We should underline that large-scale socioeconomic changes have been taken place these last 30 years, due to the radical changes in the political system and the transition from a totalitarian regime to democracy. It was characterized mostly by the population movement as an internal migration, directed toward the capital Tirana. There was emigration toward neighboring countries such as Greece and Italy but also toward other European countries and North America. The most affected areas from all these movements remain the rural areas.

Until the beginning of the XX century, the rural population constituted the vast majority of the Albanian population (Thomo, 2016a, 2016b). The locals relied their living on their know-how, having a strong relationship with the environment. Due to historical conditions, the life of inhabitants in rural areas has been developed within a closed circle and based on their home, garden, property, and village. Sometimes, it was developed in the province or a larger group. In different regions, different lifestyles were developed based on tradition, territory, and environment. The rural settlements, as developments, are the product of this lifestyle resulting in spread developments on the whole regions. The cities of the time had a low impact on rural life, basically limited only to economic exchanges (Thomo, 2016a, 2016b).

There are two significant moments in the history of rural development to be underlined. The first is related to the change of the above conditions during the socialist regime, where the private property returned to the state. Afterward, the statal cooperatives were founded. From this moment, the agricultural economy and livestock began to be managed by the state. This period was associated with extreme poverty in the rural areas and the isolation of the population. The second significant moment is related to the end of the totalitarian regime and the development of the free economy. This second period found an unprepared rural population and resulted in a massive migration.

Although at a lower rate, the migration phenomenon is continuing. The results of the high rate migration are the abandonment of settlements, degradation of agricultural infrastructure, therefore a decline of the economy. If this would be still the trend, Albania will be facing an even considerable risk, that of cultural heritage loss through the loss of settlements, important social and cultural objects, loss of know-how, and the culinary tradition, furthermore loss of spiritual values.

The fall of the socialist regime was accompanied by other negative phenomena that have started since 1990 and continues—the development of informal settlements. Informality is still a non-solved problem in the largest cities such as *Tirana* or *Durrës* but also in other areas. As the cooperative system disrupted, agricultural land was often misused and even changed the land use from agricultural to urban. Agricultural and natural lands were the most damaged from all this sudden development. With all of its effects, migration did not mark just negative ones but also has contributed to our economy through the investments made, usually at a family level.

Due to a rich background of natural and cultural heritage, a lot of government strategies have been proposed for developing rural areas. At national level, the *National General Plan* offers a referring strategic framework toward the territorial sustainable development for the next years with a main purpose the assurance of a balanced social and economic development, its responsible resources management, the environmental protection, and rational use of territory (National Territory Planning Agency, 2016). As per the vision of the *National General Plan*, the economy in the area should be based on these main pillars: (1) Development of Agricultural and Rural Sector and (2) Development of Tourism Sector by (3) Environmental Revaluation and Conservation and their integration, thus bringing the social aspect improvement.

There have been many national and international programs implemented to support these orientations. One of them was the *Academia 100+ Villages Program*, a governmental initiative with the main aim that of developing rural Albania. The program took place during August–December 2018 period, coordinated and funded by the government and managed by *National Territory Planning Agency* (NTPA) with the participation of universities, state agencies involved in the rural and tourism development, and experts from different fields. The program aimed development of economic models for 100 villages. The villages were chosen based on the cultural and natural possessions criteria. This way, most of the proposals used these primary resources on building economic models, which would furthermore generate energy for the surrounding territory. This program would be with the final focus—that of the rural development based on cultural and natural heritage.

3 Study Area and Data

3.1 Position and Natural Resources

Our study area is located in southeast of Albania in a very particular area next to the *Albanian-Greek* border (Fig. 1). Its terrain morphology is characterized by two parallel, long ridges, *Trebeshinë-Dhëmbel-Nemërçkë* and *Shëndëlli--Lunxhëri-Bureto*, and flowing waters of *Vjosa* and *Zagori* rivers in between. In the area also, one of the biggest national parks (Protected area; II Category) such as *Bredhi i Hotovës-Dangëlli* and other natural parks (Protected area; IV Category) such *Zheji* (Qiriazi, 2017) and later *Zagoria* was added. This particular landscape of a mountainous terrain rises on both sides of the *Vjosa* river, one of the longest rivers in the country, with a total length of 275 km. *Vjosa* river enters the Albanian border from Greece, where it is known as *Aoos* river. It has great importance since it is one of the few wild rivers left in Europe and has remarkable national and international significance as per the biodiversity and a large number of habitats for many species (Qiriazi, 2017). On both sides of *Vjosa*, there are a large number of small rivers, springs, and waters which contribute to biodiversity and people life. The river has always been important to locals and has served for agricultural irrigation and personal use. Although in two different countries, the river and the terrain define just one spatial continuity.

The list of natural resources in the area continues with the presence of canyons such as *Lengaricë* canyon, thermal springs of *Bënjë*, rivers other than *Vjosa* such as *Lengaricë, Lumicë, Zagori, Drino,* etc., many streams, and springs of fresh water (Fig. 2).

3.2 Timeline and Material Heritage

The studies showed that the area has been inhabited since the VI-III century BC period. Ruins of settlements, such as those found in *Antigone*, belonging to the beginning of III century BC (Baçe, 2016), castles such as the one in *Këlcyrë* with traces dating the same period (MC & ICM, 2014), rebuild during the XI century AC (Karaiskaj, 2016b), or the one in *Limar* (ICM, n.d.) on the northeast side of *Mount Lunxhëri*, add a great value to the national material and cultural heritage of the area. During the XI-XV centuries, two types of cult objects were developed: (1) The one with one nave and (2) the other type with cross-shaped plan layout with dome. Both types are found in the region (Meksi, 2016). Belonging to this period, we can mention *The Prophet Ilia Church*, dating the XI-XII century, in *Bual* (Figs. 3 and 4), and *Saint Mary Church* dating the XII century, in *Kosinë,* both villages of Përmet (Fig. 5).

Another important period in the field of construction was the XV-XX century. During this period, there were build new castells, such as *Ali Pasha's castell* in Tepelena (XIX Century) or rebuild previews ones such as the *Këlcyra castell* (Karaiskaj, Architecture on XV-XX Centuries; The Architecture of Fortifications, 2016a). Starting from second half of the XVI century until the beginning of XIX century, a large number of Christian cult objects were build, a considerable number of which is conserved until nowadays in Central and South Albania. They were of three types: (1) With one nave; (2) cross-shaped plan and a dome; and (3) the Basilica (Thomo, 2016a, 2016b). We can mention a few of these types, such as *Saint Sotiri,* in Tremisht (XVI Century); *Saint Mary*, in Seranjperat, Përmet (XVII Century); *Saint Mary*, in Leusë (XIX Century) (Figs. 6 and 7); *Saint e Premte*, in Përmet (XVIII Century) (Fig. 8); *Saint Kolli*, in Lipë (XVIII Century) (Figs. 9 and 10) (Thomo, 2016a, 2016b).

During this period, there were also many engineering constructions completed. One of the most important engineering works of the time is *Egnatia* road which connected east and west through the *Shkumbin* valley, but also Via *de Zenta*, starting from *Shkodër* toward the north-eastern Balkans, roads which passed thrugh the *Vjosa* and *Drino* valleys, and many others (Karaiskaj & Meksi, The engineering Constructions in the Medieval Period, 2016). There were also build aqueducts and *stone bridges* over the rivers. Today, we count quite a few stone bridges with bows, ruins of them, but also some of them still in use, over rivers Zagori, Lengaricë, Sarandopor, and Vjosa, furthermore—(1) Ruins of *Hoshtevë Bridge*, *bridges of Limar and Nivan* over river Zagori; (2) *Ruins of Muço Hysa, Petran, Badëlonjë* and *Përmeti bridges* over river Vjosa; (3) *Katiu bridge* over river Lengaricë; (4) Ruins of *Berati bridge* over Sarandopor river; and (5) *Bridge* over *Bënja* stream (Fig. 11) (Albanian Institute of Islamic Thought & Civilization, 2019). It was this precise period that on both sides of the *Vjosa* river, many rural settlements flourished.

3.3 Non Material Heritage (Intangible Heritage)

The area is very known for its rich folk, such as folk-poems, folk-dance, but is the most known for the existence of *Iso-polyphony.* Iso-polyphony music, with its two genres (*Iso-polyphony toske* and *Iso-polyphony labe*), is proclaimed by UNESCO as "A masterpiece of oral and intangible heritage of humanity," since 2005 (UNESCO, 2008).

The different provinces of the region have developed their traditional clothing as well with influences from the climate and social life. This type of clothing was in use by locals until the beginning of the XX century.

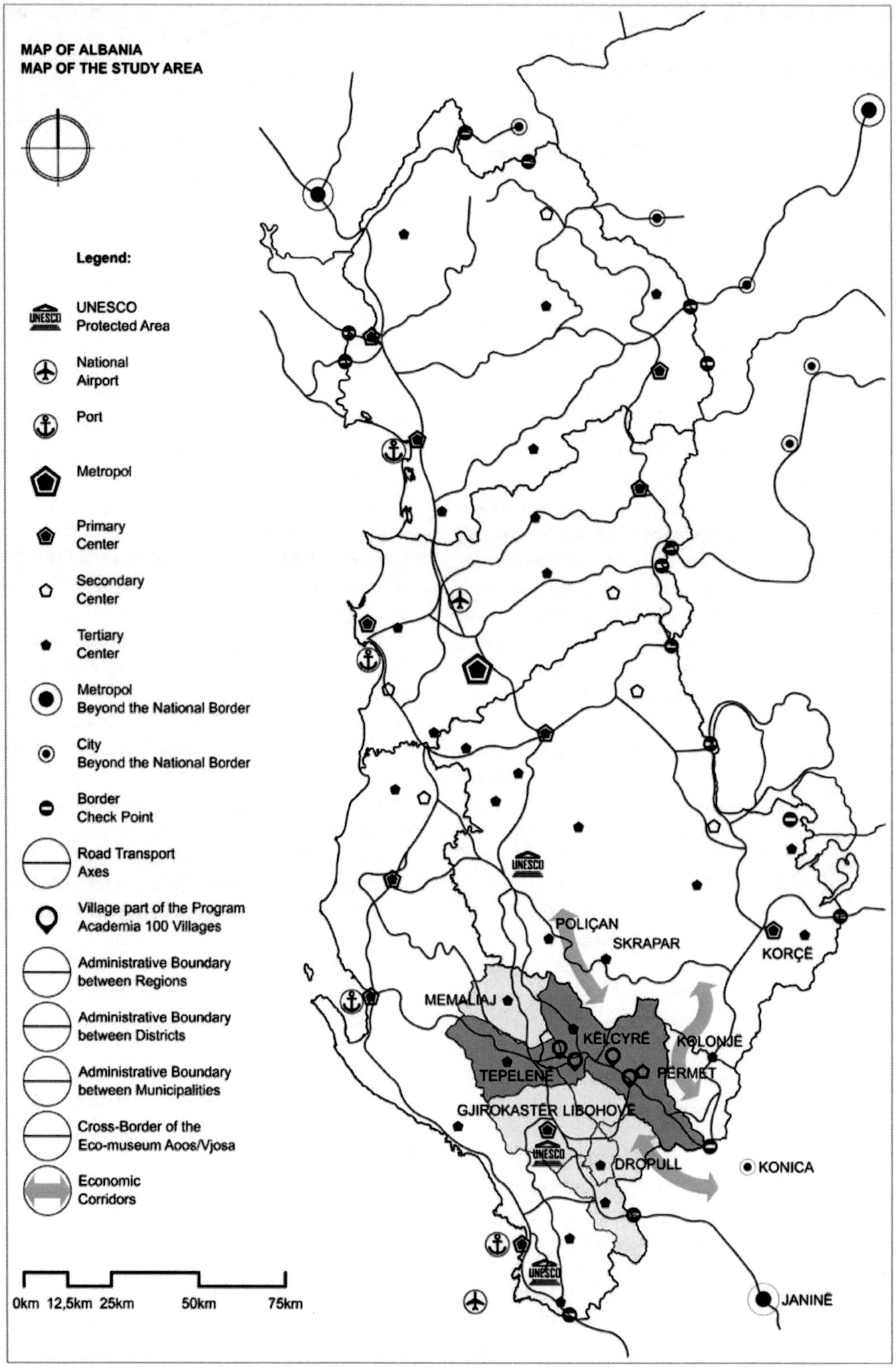

Fig. 1 Map of Albania; Identification of the study area; *Source* Initial data from NTPA; Elaborated: Author

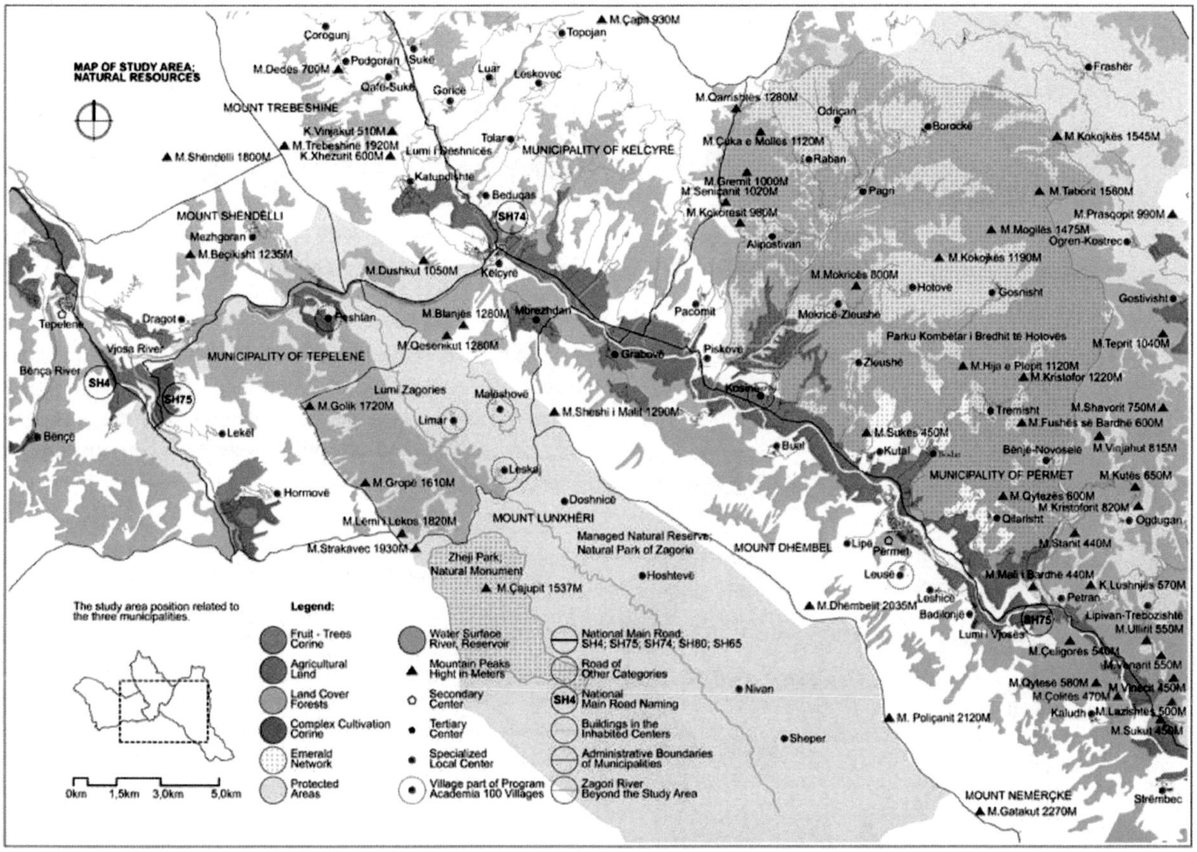

Fig. 2 Study area; Map of natural resources; *Source* Initial data from NTPA; Elaborated: Author

Fig. 3 Plan of *The Prophet Ilia Church*, Bual; *Source* "Architectural History in Albania; From its beginning until 1912"; Elaborated: Author

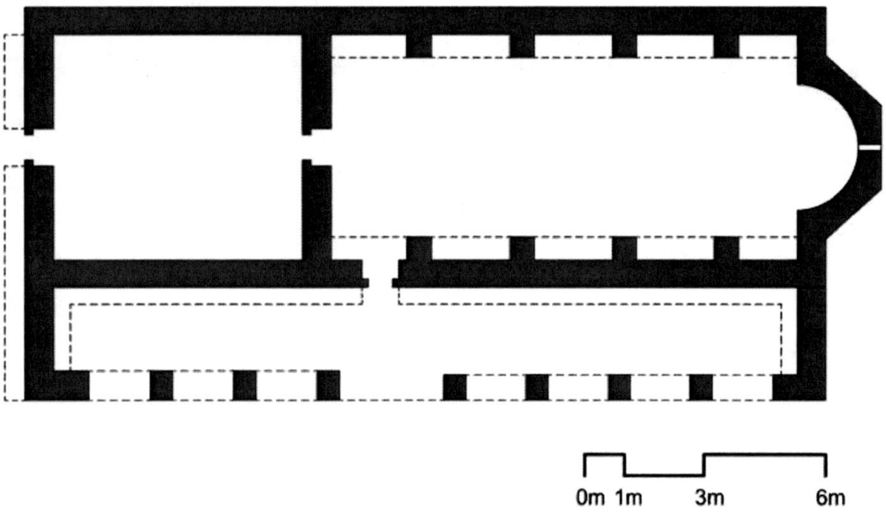

The region is very well known for being the birthplace, and where many famous people in the Albanian social life of the XXI century lived their early life. They not only have left a written inheritance of extraordinary value but have contributed to the survival and conservation of Albanian language during the rule of the Ottoman Empire. Notable characters such as *Naim Frashëri, Sami Frashëri, Abdyl Frashëri, Andon Z. Çajupi, Nonda Bulka*, and many others

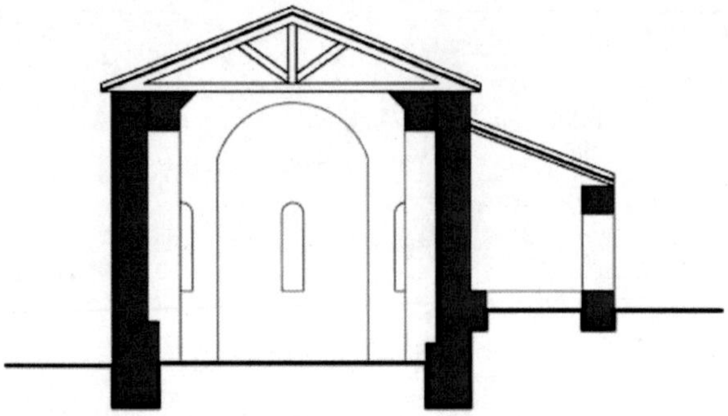

Fig. 4 Section of *The Prophet Ilia Church*, Bual; *Source* "Architectural History in Albania; From its beginning until 1912"; Elaborated: Author

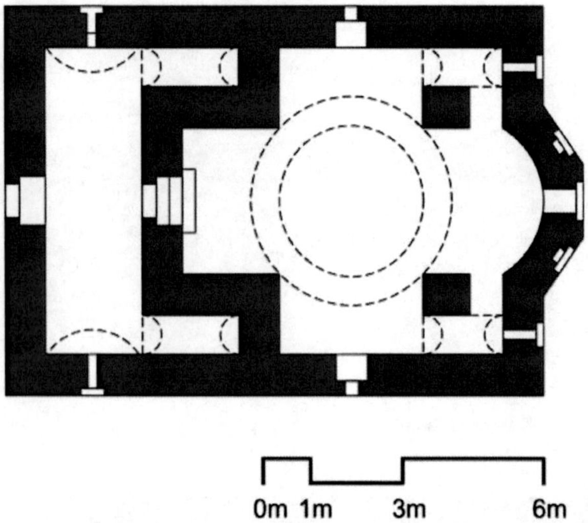

Fig. 5 Plan of *Saint Mary Church*, Kosinë; *Source* "Architectural History in Albania; From its beginning until 1912"; Elaborated: Author

have described in detail the beauties of the region, the people's lifestyle, the traditions, the economic situations of the period they lived contributing thus, in the creation of a better picture of national historical developments.

3.4 Tradition and Culinary

The area is very well known for its delicious dishes, but also in the production of local products such as those from *dairy,* but also *gliko, raki,* and *wine* production. The area is also known for the cultivation of particular *grape cultivars*, known as autochthonous, being an additional asset, as nowadays their conservation is of great interest.

4 Method

The site visit was essential for the study. It is needed to be understood the absence of a database on our territory during the last 30 years, which includes settlements, economy, environment, but especially demographic movements and actual population. We knew that in the study area, the migration had been at high levels, that is why the visit was planned and realized in the summer period. It marked a period in which it was hoped the migrants would have returned to their localities.

The visits completed by the working team were following the NTPA and the local authorities' agenda. Due to the technological advancements, all the moves in the area were tracked by GPS and afterward were saved in applications such as *View Ranger,* etc. for being downloaded after.

4.1 Interviewing the Local Interest Groups

The interest groups were the local authorities at three municipalities *(Municipalities of Tepelenë, Këlcyrë, and Përmet)*, at the administrative unit level and village administration level. Different important actors operating in the area such as CESVI were also part of the interviews realized, together with all categories of people met on site, such as families, children, businessman, and tourists. Meeting with the local people again was part of the NTPA agenda. This process served for creating a database. In the data, the following was included: (1) Quantitative data such as population; the number of houses; the number of children; the number of services; the number of businesses; the quantity of production; the quantity of agricultural land; and data on the livestock; (2) qualitative, such as the history of the area; migration of the population; quality of services; business typology; type of vegetation and/or production; product handling; data on natural resources; and data on cultural heritage.

4.2 Development of Maps

The maps provided by NTPA served for an initial familiarity with the region. During the site visits, with the information collected, updates were completed were possible. It was a necessary step to interpret the environment in terms of surface coverage, enrichment of the map with toponyms, highlight the significant places furthermore, and build a more solid connection between the map and the image of the territory. Sections of the territory developed afterward,

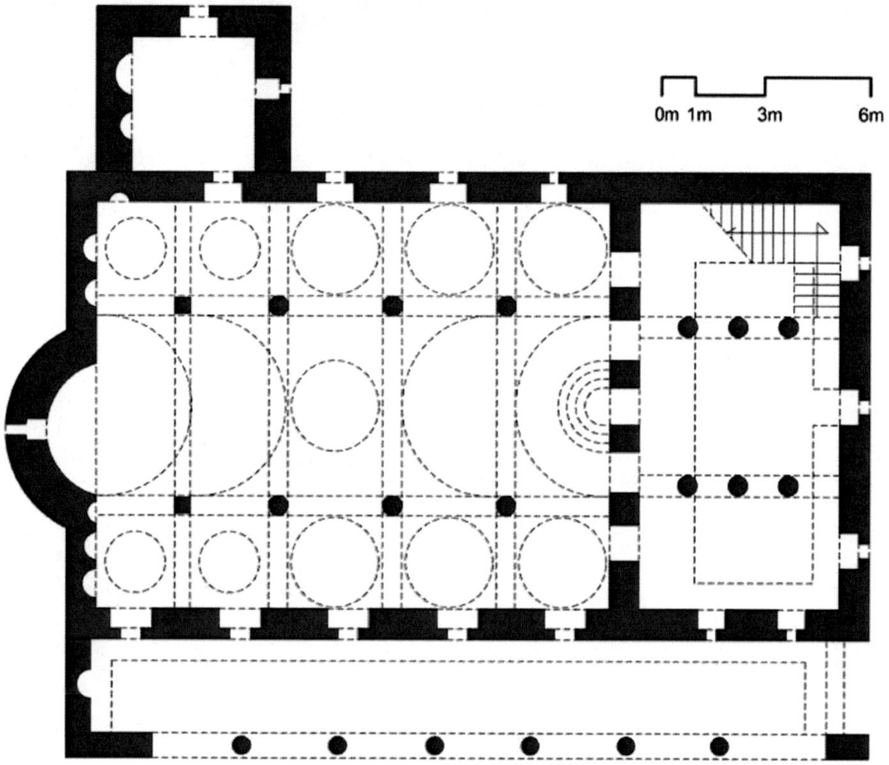

Fig. 6 Plan layout of *"Saint Mary Church,"* Leusë; *Source* "Architectural History in Albania; From its beginning until 1912"; Elaboration: Author

Fig. 7 Section of *"Saint Mary Church,"* Leusë; *Source* "Architectural History in Albania; From its beginning till 1912"; Elaboration: Author

served for better understanding of it, getting to know more data such as altitude or orientation. It would give a closer understanding of each region characteristic in terms of climate, microclimate, relation with the habitat, or with water sources. The working team completed the visits and interviews in four villages, all part of the program.

The villages' positions were over an enormous territory. All the data collected, including the one given by the experts, the use of literature, and the information served by locals, was later used for updating the existing maps and database of the area.

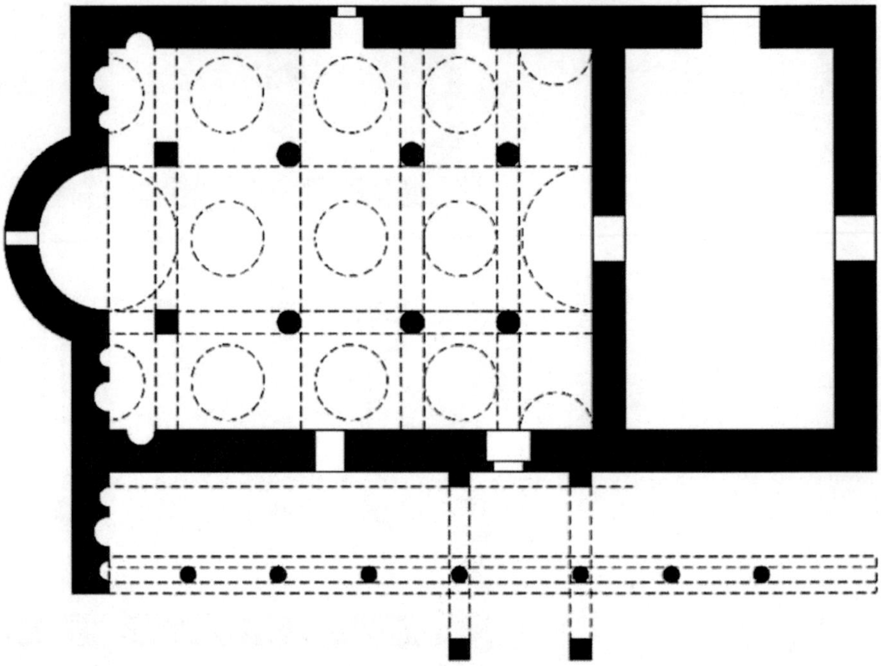

Fig. 8 Plan layout of *Saint e Premte Church*, Përmet; *Source* "Architectural History in Albania; From its beginning till 1912"; Elaborated: Author

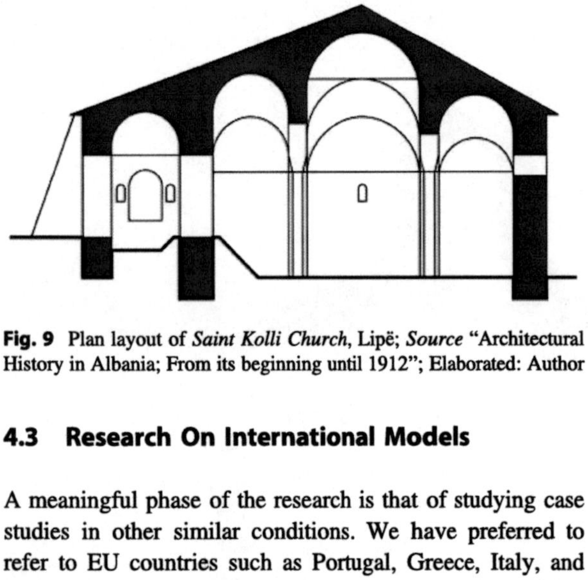

Fig. 9 Plan layout of *Saint Kolli Church*, Lipë; *Source* "Architectural History in Albania; From its beginning until 1912"; Elaborated: Author

4.3 Research On International Models

A meaningful phase of the research is that of studying case studies in other similar conditions. We have preferred to refer to EU countries such as Portugal, Greece, Italy, and France. The positivity about this selection was: (1) Being part of the EU countries, the models represent the most advanced strategies and applications on the field, and (2) being areas that belong to the same climate conditions, the case studies have similarities with our context in terms of natural resources, material heritage, sometimes similarities in mentality and lifestyle, and sometimes in the economic aspect.

4.4 Consulting the National and EU Legislation

Getting to know the national and international legislation was a very important stage of the research, which created a general framework of national and international development trends, especially in tourism and rural development. The research also included the study of the regulatory plans in several levels, national and regional (General National Spatial Plan and General Plans for the municipalities involved in the study area). As mentioned before, the study area is at the *Albanian-Greek* border and therefore bears even meaningful importance the understanding of the overall *cross-border* development trend, so that the strategies of this particular research could fit properly.

5 Results

5.1 Build Environment

The results of the built environment were impressive. There are small urban and rural settlements, integrated with the natural topography, with proportion, and colors in dialog with those of the natural landscape. It was a result of the use (found in the vertical structure and roof) of the local construction materials such as "black stone," also known in the foreign markets as *"alba-nera" stone*. These settlements

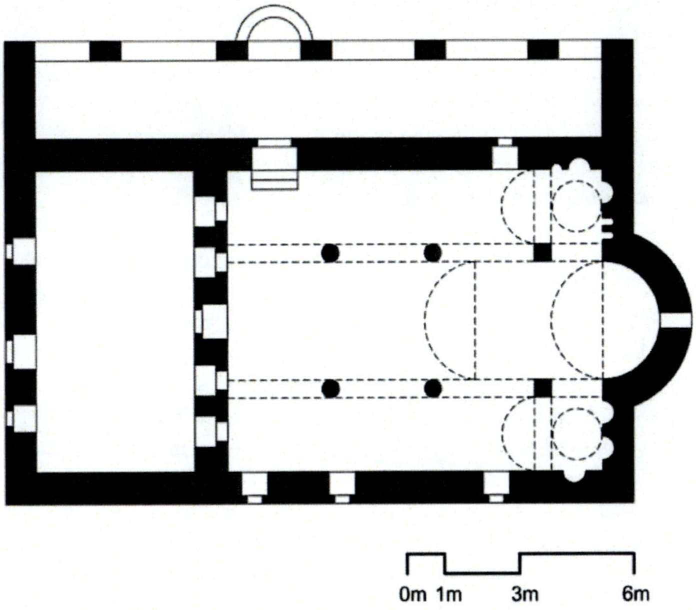

Fig. 10 Plan layout of *Saint Kolli Church*, Lipë; *Source* "Architectural History in Albania; From its beginning until 1912"; Elaborated: Author

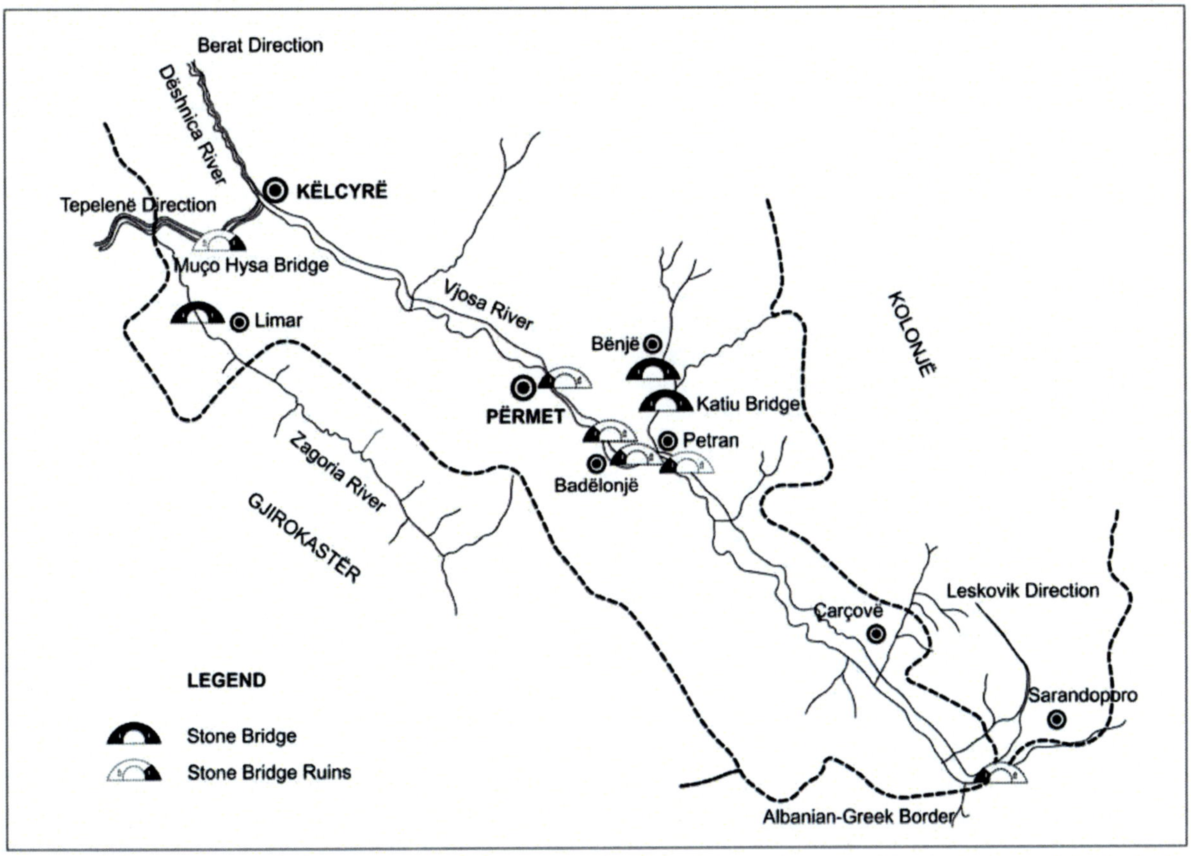

Fig. 11 Map showing all *Stone Bridges* or their *Ruins* on the rivers of the Përmet area; *Source* http://www.aiitc.net; Elaborated: Author

flourished around the XVIII and XIX centuries, and each of them had a history to tell. A considerable number of these settlements had been a battlefield during the Greek occupation in 1914, but also in 1941 during the *Italo-Greek War*, and again in 1944 during the World War II. Many damages were done, some of the settlements were burned totally and rebuilt on later periods, and others still preserve the same structures. The rural settlements nowadays are at about one-third of the size they were during their flourishment back in the XIX century.

The buildings were organized in one to two floors, with a garden. The two floors buildings always contained the main living areas on the first floor; other areas, more of a service use, were positioned on the ground floor. In the past, the livestock used to be kept on this level. The houses had their garden always rich in fresh vegetables, fruits, and olive trees. This particular element showed a meaningful connection still existing between the locals, territory, and its fruits. The houses were surrounded by high walls, which served as a fortress to the family's property, and had the main entrance mostly wooden arched port. In a few cases, wooden carved ceilings were still conserved, a fact which proved the economic well-being in the area around 1930, as it was told, due to the massive emigration of the population toward the USA, France, and Turkey.

The study area resulted rich in important *cult objects* for our national heritage, as they are preserved objects by law. A few of these objects we could mention are: (1) The *Saint Mary Church* in *Leusë* (Figs. 6 and 7), valuable for its internal structure, its colorful frescoes, its iconostasis made of wood carved with floral motifs; (2) The *Saint Mary Church* in *Kosinë* (Fig. 5) is valuable for its age, shape of its plan layout, and the techniques used in the construction of the walls; and (3) The *Saint Mary Church* in *Bënjë* village also marks an important typology of the churches found in Albania.

There was no lack of engineering works as well. From the *stone bridges* still used and in stable shape, we can count (1) *Limar* and *Nivan* bridge on the *Zagori* river and (2) Katiu bridge on *Lengaricë* river. All this together with the water sources and cobblestone alleys were another fact proofing the local's know-how and their mastery in crafts, especially wood and stone processing.

While in other parts of Albania, the informality is a very spread out phenomenon, in this particular region, there were almost no such interventions. The urbanized areas resulted to have been untouched in years. One of the most important centers of the area, *Përmet* city, had remained also at the same shape and size as it has been in the past. The developments consisted mostly of renovations of a few (1) central buildings such as hotels, banks, community objects, and (2) public spaces, such as *Përmet* and *Peshtan* main squares. On the one side, low economic development had happened during the last 30 years; the positive side effect of this was that the region was as yet dominated by small urban or rural development inserted in the natural environment. In particular, the rural areas were the most abandoned ones.

5.2 Non-build Environment

The non-build environment creates beautiful, diverse landscapes. The big difference in levels, starting from 150 m and arriving at 2200 m, creates many microclimatic areas with their particular ecosystems and habitats. One could reach from the Mediterranean to subalpine and even alpine zones. There were around 120 types of medicinal and aromatic herbs in the area. Territories such as National Park of *Bredhi i Hotovës—Dangëlli* or *Mountain Golik* are both very well known for their rich biodiversity. The locals have always been in a relationship with the land's natural resources. They had collected wood, herbs, and animal feed for their living.

A large number of water resources found in the area have often been of interest to hydropower developments. *Vjosa* river, its ecosystems and habitats have been threatened these last years by these developments. While local environmental groups could not save *Lengaricë* river from the construction of a hydropower plant (a branch of Vjosa river, and part of National Park of Hotovë and Dangëlli), there is still hope *Vjosa* river will continue to remain one the last wild rivers in Europe.

The area is distinguished by a richness in ecosystems, also aerial, terrestrial, and water habitats. It has a large number of cultural and natural monuments protected by the national laws, although much more is needed to get done for their protection and conservation.

5.3 Demography and Social Life

It was evident the shrinking of the population in the area. Most of it had moved toward bigger cities such as *Tirana, Durrës, and Gjirokastër* but also toward other countries such as Greece and the USA. The old-age population prevailed in these areas. The youth usually tended to leave with the purpose of their education, have a social life, but also to create possibilities to find a better job market.

In some of the most distant provinces, the particular social phenomenon had started to develop these last years that of new agencies who match made new couples. Usually, the local men were more connected to the land, and the young women were more likely to leave the area. In the conditions of a larger male population, marriages between young women from north Albania and local young men had become to take place.

There was an overall lack of interest in continuing the traditions of being farmers or shepherds, mostly due to low infrastructural, services and economic opportunities these areas could provide.

5.4 Economy; Development of Bottom—Up Economies

In this area, the economy has always based on products of the territory that includes personal gardens, agricultural lands, and livestock. It was more than evident that the topography and natural conditions defined the local population orientation toward agriculture or livestock. As a result, for the villages with better accessibility to the road and water resources, agriculture and processing of agricultural products was the largest economic resource. The deeper areas, usually characterized by a higher altitude, were more livestock oriented. From the observations completed during this research, it is quite important to emphasize the high quality of agricultural land and rich biodiversity pasture. Both facts define a great asset to the area. For helping their economy, the local population was also engaged in bee-keeping and collection of medicinal and aromatic herbs. The diverse ecosystem of the region favors these two branches of the economy very much.

One very distinguished evidence found in the area during the site visit was the existence of a few local actors. (1) CESVI—an Italian non-profit organization with its center in *Përmet*—has been operating in the area since 1997. It is an organization oriented on world solidarity. It has implemented a broad number of projects in the area with its leading goal the sustainable development by supporting the local resources. Some of CESVI's previews projects relied on training residents in tangible and intangible heritage conservation, and promotion of local products and culinary, build environment, and the beautiful nature of the area. (2) *Pro-Përmet,* a consortium of local businesses in the areas of *Përmet* and *Këlcyrë*, created and supported mainly with the initiative of CESVI. This consortium operates today in the local market by branding and marketing the local products, by promoting this way the preservation of their territory and environment. (3) It was identified the existence of a few initiatives from a low number of local people in the area who had understood the values of the territory and had started their small businesses such as offering accommodation, local culinary products, and handicrafts, or tours through the beauties of this territory. Usually, the tours offered included hiking, bicycling, horse riding, and rafting, but rides like off-road safari were quite often organized.

5.5 Eco-museum

In 2014, a *cross-border memorandum* had been signed, the Eco-Museum Vjosa—Aoos (Hoti & Ago, 2012). The Eco-Museum covers an area of approximately 3540 km^2 in both counties. This interregional program developed in the period 2007–2013 was financed by both EU and national funds and was compiled by institutes and actors of both countries; nowadays, it is still not approved by the Albanian government. The program had the primary purpose of the creation of a *cross-border entity* based on natural and cultural similarities and values on both sides of the border. As a main concept, Eco-Museum was developed in France in 1971 by *George Henri Riviere* and *Hughes de Varine*. Its purpose is to create a museum without walls, managed by the local communities for showing and preserving leading elements of natural and cultural heritage (Qiriazi, 2017). As per the final result, the visit in the Eco-Museum Aoos—Vjosa was conceived into four main thematic roots: (1) Water and culture; (2) movement and roots; (3) fruits of nature; (4) earth and water, each of which shows a healthy connection of territory, water, people, production, culture, heritage, and history on the area (IPA- Cross-border Program, 2013).

6 Discussions and Conclusions

Vjosa River valley is one of the most attractive areas in Albania, being a perfect combination between the natural environment, history, cultural heritage, culinary, local traditions, and being a *cross-border* spatial and cultural continuity. There are too many potentials the area offers, which could contribute to a broader tourist destination. As mentioned above, there are a few sporadic initiatives for bringing such development, mainly encouraged by CESVI. However, they did not cover the whole territory and the important actors in the area, and that is why they do not mark a strong enough element to be able to change the local economy.

During the site visits and interviews completed came clear the connection the local population had with the territory. Due to degradation or lack of infrastructure, difficult accessibility, lack of services such as schools, health services, lack of investments, and substantial interventions, most of the population has abandoned their land resulting in an elevated rate of migration. The people still living in the area had many difficulties in building sustainable economies.

As per the agricultural land, after the fall of the regime according to the law, No. 7501 was given back to the local inhabitants. According to this law, the agricultural land was defragmented into small plots, and now in the lack of a better organization or private cooperatives, there were difficulties in using it effectively. On the other hand, livestock products such as milk, meat, or wool often could not be traded due to the low-quality infrastructure. It is necessary to emphasize that the quality of these products in the area is of high quality. It is due to the presence of natural ecosystems in the area.

It is clear to us that tourism development is one of the leading potentials of the area, but we cannot see it as the only pillar on which the local economy will have to rely. A more sustainable economy is that of reuse the resources the territory offers, sustainably use them without harming the natural environment. Therefore, a combination between the tourism, agriculture, livestock, processing of products economy could be a perfect combination of area development.

There is an encouragement for the bottom-up economies lately, which is a good initiative but, according to the trends of interventions on the territory by the locals, this bears a few risks which might be at high levels if they are not managed in time and in an appropriate way, risks such as (1) the development of similar economies in the whole territory; (2) the damage of natural and cultural heritage; and (3) the inappropriate use of the environment. As the main conclusion, our main proposals would consist of the following strategies:

- The building of a general legal framework as per the EU requirements, including the creation of Local Action Groups (LAG-s);
- Create a strategy on the agricultural and livestock economy were complementary economies should be provided; the economic movements and developments will start to increase the population interest;
- The standardization of local products for enlarging the market with the leading goal—international markets;
- Create a holistic database of tangible and intangible resources;
- Raising the awareness on the environment and cultural heritage of the area; raising the connection of the youth to nature;
- Create a general strategy for each region with high potential in the tourism, therefore Reinforcement of a Destination Management Organizations;
- The cross-border marketing of the area;
- Ensure the necessary funds through the state budget, donators, EU, etc.
- Necessary investments such as cultural heritage preservation, accessibility infrastructure, and agricultural infrastructure;
- As per the built environment, the development of strategies on their conservation, and the design of building standards for the area.

All of the above would contribute as an initial background on the sustainable development of the *Vjosa* river valley.

Acknowledgements This research was made possible due to the Albanian Government initiative known as *Academia 100+ Villages Program*. The program was coordinated at the national level by Mrs. Elva Margariti, in cooperation with National Territory Planning Agency, directed by Mrs. Adelina Greca, other state agencies, and also universities. The universities participating in this program included professors, students, and experts from many different fields. This particular area of study treated in this research was one of the 22 lots, being visited and studied during the August–December 2018 period, and it is known as Lot 17.The author would like to express her gratitude particularly to the NTPA for making UPT, FAU part of the program, and also for their excellent organizing work during the process, which concluded as a satisfying, successful, and quite efficient experience. Another special thank the author expresses to the students of FAU for participating in the *Academia 100+ Villages Program*.

References

Albanian Institute of Islamic Thought and Civilization. (2019). *Old bridges in Përmet district*. Retrieved from www.aiitc.net: http://www.aiitc.net/index.php/en/arkitekture/item/828-ura-te-vjetra-ne-rrethin-e-permetit-i.html.

Baçe, A. (2016). Urban planning and architecture of the residential builidngs. In M. Aleksander, B. Apollon, R. Emin, K. Gjerak, & P. Thomo, *Architectural history in Albania; from its beginning until 1912* (p. 107). Tirana: Kristalina-KH.

Hoti, E., & Ago, R. (2012). *Inter-local strategic plan of the Vjosa Valley Përmet District*. Tirana: Albdesign.

INSTAT. (2019). *Population of Albania, January 1st, 2019*. Retrieved from www.instat.gov.al.

IPA-Cross-border Program. (2013). *Vjosa-Aoos Ecomuseum*. Retrieved from www.ecomuseum.eu: http://www.ecomuseum.eu/.

Karaiskaj, G., & Meksi, A. (2016). The engineering constructions in the medieval period. In A. Meksi, A. Bace, E. Riza, G. Karaiskaj & P. Thomo, *Architectural history in Albania, from its beginning until 1912* (p. 647). Tirana: Kristalina-KH.

Karaiskaj, G. (2016a). Architecture on XV-XX centuries; The architecture of fortifications. In A. Meksi, A. Bace, E. Riza, G. Karaiskaj & P. Thomo, *Architectural history in Albania; from its beginning until 1912* (pp. 514–518). Tirana: Kristalina-KH.

Karaiskaj, G. (2016b). The Medieval fortifications. In A. Meksi, A. Bace, E. Riza, G. Karaiskaj & P. Thomo, *Architectural history in Albania, from its beginning until 1912* (pp. 396; 397). Tirana: Kristalina-KH.

MC and ICM. (2014). *Declaratation of Protected Area*. Tirana.

Meksi, A. (2016). Christian Cult Constructions; From VII-XV century. In A. Meksi, A. Bace, E. Riza, G. Karaiskaj & P. Thomo, *Architectural history in Albania; from its beginning until 1912* (pp. 442–443; 457–459). Tirana: Kristalina-KH.

National Territory Planning Agency. (2016). *Albania 2030, The National General Plan, The first national document on territorial Planning 2015–2030*. Tirana.

Qiriazi, P. (2017). *The natural heritage of Albania; values, risks and management*. Tirana: Kristalina.

Thomo, P. (2016a). The Architecture of the Christian Cult. In A. Meksi, A. Bace, E. Riza, G. Karaiskaj & T. Pirro, *Architectural history in Albania, from its beginning until 1912* (pp. 557–558; 578; 580–581; 589; 593). Tirana: Kristalina-KH.

Thomo, P. (2016b). The rural settlements and residences. In *Architectural history in Albania; from its beginning till 1912* (p. 751). Tirana: Kristalina-KH.

UNESCO. (2008). *Albanian folk iso-polyphony*. Retrieved from www.unesco.org: https://ich.unesco.org/en/RL/albanian-folk-iso-polyphony-00155.

Investment of Heritage Villages in Saudi Arabia—Case Study of Al-Khubara Village in Qassim

Mohamed Ibrahim Elbelkasy and Mohsen Mohamed Ibrahim Mustafa

Abstract

The kingdom of Saudi Arabia is rich with many heritage sites that reflect the desert architectural and urban character that has been characterized throughout ages. This character represents one of the most important models to be a sample for sustainable development of the desert urban communities. The Saudi Commission of Tourism and National Heritage had exerted great efforts in restoring many villages in order to preserve this heritage. In return, heritage festivals held in these sites reflected welcoming interaction between inherent character and the local community. Because modernization crept to these communities in the last quarter of the twentieth century. It was the main reason for neglecting the original urban and architectural features that were suitable to the environment of those heritage villages. The research main goal is to study the investment and development of one of the heritage villages in Al-Qusaim, which reflects distinguished features of the desert architectural and urbanism of this region. The Saudi Commission of Tourism and National Heritage had finished the project of restoration and conservation of Al-Khubara village. This research represents a proposal for reuse and investment scenarios through evaluating the rehabilitation and economic feasibility of this village. The research methodology is based on a theoretical approach that involves the definitions of the heritage sites investment and reuse projects and their fundamentals, while the analytical and practical approach finds out the best functions that suit the nature of the village, in addition that choosing the best evaluation technique regarding economic feasibility. Finally, the applied methodology comes to help in the economic evaluation process of the project.

M. I. Elbelkasy (✉) · M. M. I. Mustafa
Umm Al-Qura University, Mecca, Saudi Arabia
e-mail: mmmostafa@uqu.edu.sa

M. I. Elbelkasy
Kafr El Shiekh University, Kafr Elsheikh, Egypt

M. M. I. Mustafa
Menoufia University, Al Minufya, Egypt

Keyword

Urban heritage • Urban heritage investment • Reused

Introduction

The kingdom of Saudi Arabia aims to be an economical model leader through its 2030 vision, and it depends on the variety of its none petroleum economic resources as a basic income resources, so the crown prince launched project NEUOM as one of the initiatives of the future of the investment which will make the kingdom one of the most important universal investment and tourism destination; therefore, preservation of the heritage and historical sites should be integrated with the futuristic vision and development projects as the heritage and historic backbone of these projects.

Heritage investment is one of the heritage conservation main sustainable factors which deals with economic, environment and community; also, heritage investment should be a mean of income diversity which the kingdom wants to achieve through its future vision. Heritage investment increases national income from tourism; investment process incomes can increase the community awareness which will reflect on the heritage conservation process.

Al-Khubara village is one of the heritage villages which presents distinguished heritage characteristics of Al-Quasim; it reflects urban heritage and architectural features and a model to village deal with developing desert village. The village was an old commercial center of this area; it contains several stores and houses that can be a commercial base if an investment project was launched; the conservation project of the village implemented by the Saudi Commission of Tourism and National Heritage is one of the most important investment factors (The Saudi Commission of Tourism and National Heritage, others, 2010).

A. Versaci et al. (eds.), *Conservation of Architectural Heritage*, Advances in Science, Technology & Innovation,
https://doi.org/10.1007/978-3-030-74482-3_27

Research methodology depends on the theoretical method which presents a literature review about heritage sites investment to set a definition of the heritage investment, economic heritage values, investment role in the conservation process, and methods of economic evaluation for the investment of heritage sites to select an appropriate application method to the case study. The case study depends on analytical and applied methods which study reasons for choosing the case study, apply the investment methodology to analyze factors of investment engaged, and evaluate the proposed investment project to the village, finally result and recommendations (Fig. 1).

1 Heritage Areas Investment

The heritage area is one of the most important economic resources as human, financial, and nature resources; hence, it is very sensitive investment field, investment in heritage different from other investment fields in value of the commercial resources; it has social and cultural values which set a culture and civilized standards beside the economical standard but we cannot determine a constant value for these standards. When dealing with the environment and therefore heritage areas, we cannot measure the economic resources with market standards directly because the heritage values and benefits are intangible and moral values similar to the values of nature environment (Seraj-Aldine, 2002). Hence, we used an environmental economic field as an approach to deal with economic values of heritage conservation and heritage investment in heritage areas.

1.1 Concept of Heritage Investment

Heritage investment concept can be set through the concept of the environmental investment, because the similarity between the heritage and environmental values, so the concept of the heritage investment is the investment that achieves the highest returns considering sustainability. Sustainability is one of the most important issues which should be considered when dealing with an investment of the heritage resources because they have social and cultural dimensions exceed the economical dimension.

1.2 Considerations of Heritage Areas Investment

When dealing with an investment of heritage areas, we should consider the following points:

- Achieve the highest returns from the investment project.
- Best use of the heritage resources and considering sustainability, using the heritage resources carefully because exhausting these resources to gain money only will affect the environmental system.
- The heritage investment process should aim to develop all the environment's elements, one of them is the local community who considers the heritage resources one of his capital.

2 Investment as a Supporter of Heritage Conservation

Investment is one of the most important supporter of heritage conservation and reuse, but investment returnes should be calculated accurately to compare investment returns and the social and urban changes in the heritage environment due to investment, which can be different from the traditions and culture of the local community. The organizations and commissions that are interested in heritage conservation

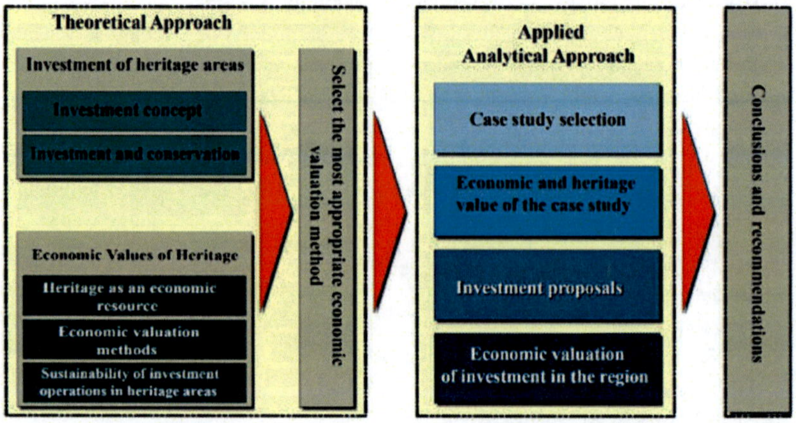

Fig. 1 Research methodology

should be considered different factors that affect the relationship between investment and heritage (Ready & Navrud, 2002).

The investment of cultural heritage can be a supporter to economic of the local community, not only for using cultural values but also for increasing job opportunities and capital cycle (Ibenholt & Bowitz, 2009), which can be a concept when dealing with heritage transfer that the conservation of cultural heritage is kind of the investment or additional investment to the recent one.

Evaluation of heritage conservation should be considered different factors affect the investment which can be important to affect conservation decisions (Tuan & Navrud, 2009). Sometimes, concepts of heritage conservation procedures can be changes according to investment vision; also, evaluation of heritage conservation projects can determine the intervention priorities in heritage sites, and it can also determine the governing framework of conservation projects (Del Saz Salazar & Montagud Marques, 2005).

3 Economical Values of Heritage Areas

Heritage areas contain several cultural and social values like historical, spiritual, aesthetics, social values; these values cannot be evaluated with traditional economic valuation methods. Economical valuation of the heritage areas aims to determine the priority of conservation and investment process and the most effective method to deal with. Therefore, the conservation policies should be real and applicable, also determining investment input and output to determine the economic cost and benefits, which make the economists interested in evaluating the economic values of the heritage sites with the concept of the macroeconomic value (Samir, 2008), which can be divided into three methods as following:

1- Value of direct use: In this method, the value of the building can be determined through the building function; this method follows market changes, sell and buy movement which will affect the economic value of the building. Thus, this value can be increased by studying market movement and determine the best use of the building according to the economic importance of building new use. In the case of reuse building in a new use different from its use which was built for we should be considered that there is no conflict between the new use and the heritage, culture and historical value of the building.

2- Value of indirect use: The indirect use value can be defined as the value derived from the site's cultural and aesthetic properties; the indirect use value comes from the cultural role of the assets and heritage resources, for example, heritage building role in informing us about the construction methods and styles. Aesthetics and historical values are the main factors in evaluating and measuring indirect use value so that it is difficult to determine its economic value.

3- Values unrelated to use (Elbelkasy, 2017): Values unrelated to use are the values appears as a result of the heritage conservation; it can be also defined as the value which the people get when they know that heritage areas exist even if they will not plan to visit it; this method is one of the most difficult methods in evaluating economical value of the heritage sites, although it is one of the main factors in evaluating culture sites.

4 Economical Valuation of Heritage Areas Investment

Economic valuation of the heritage areas investment depends on two main factors, one of them is heritage conservation economics, and the second is benefits from the investment that depend on a comparison between cost and benefits of the investment; moreover, social and cultural values should be considered.

Several approaches and techniques used to measure the cost to benefits of the heritage investment for studying the feasibility of conservation and investment projects. The research does not only depend on economical methods but also depends on the relationship between economic, heritage resources, and heritage environment in the economic evaluation of the heritage investment.

The research divided the economical evaluation of heritage areas into three factors, evaluate assets and heritage resources, evaluate economic activities according to conservation and investment, and measure total economic values for the investment process.

4.1 Valuation Assets and Heritage Resources

Evaluation of the assets and heritage resources approach depends on determined the economic value of the heritage resources using one of the following methods:

- Cost-effectiveness: The value of this method of the heritage resources is determined according to study benefits of the reuse or from the fees of the site visit to determine how visitors appreciate the visit.
- Preventative expenditure: This method assumed that the expenditure of preventing goods can determine the cost

of this good (Turner et al., 1993); this approach is sufficient when used in heritage resources assessment because it is priceless.

- Market alternative: This approach depends on the measure of goods linked to heritage environment by market economics and then the heritage component can be separated from the market price (can be know from the individual preference of goods or market service) which will be alternative to heritage component value (Al-Laham, 2007).
- Benefits transfer: Using a project assessment as a guide to evaluating another similar project, for example, the benefits of visiting one of a heritage site can be used as a benefit to another one.

4.2 Valuation of Accompanying Economic Activities

This valuation depends on accompanying activities determining the economical values through economical changes in heritage areas activities; this assessment can be divided into two methods as following:

- Productivity change: In this method, economical benefits are assessed according to studying changes in heritage areas economical activities productivity.
- Reality assessment study: In this method, studying the effect of heritage conservation and investment project on area economic, for example, changes in rent value or services in this area.

4.3 Valuation of Total Economic Value

Total economic value assessment approach depends on to determine the total economic value of the conservation and investment heritage area; through comparing cost and benefits, this approach can be divided as follow:

Contingent valuation method: This method depends on measuring values from individuals themselves, by asking them if they will pay to get this service or good, or they have the ability to lose this service or good; therefore, we can determine the real value of compensation due to losing this service or goods. It is the only method that can determine the non-use value, and it helps to determine indirect use value.

Cost–benefit method: This method depends on comparing cost and benefits to measure feasibility of investment heritage sites, in spite of the ability of this method to determine the feasibility of investment heritage project, but it cannot be applied in case of intangible values as, social and cultural values, so it is recommended to apply with contingent method.

5 Sustainability of Heritage Areas Investment

The sustainability of heritage investment is one of the most important factors to success in investment because it increases project financial returns besides the social and cultural values of the conservation which is the base of the investment process.

Heritage resources are a very sensitive resource; therefore, sustainability factors should be considered, as developing social, environmental, and cultural values beside economical development. The investment of heritage sites should consider the following factors (Fig. 2):

Sustainability of heritage areas can be done by developing the local community which is embedded in the investment process and project suggestion to be appropriate to the local community's culture and traditions.

Studying the impact of the proposed project on the community and the predict changes especially the touristic development projects, due to the conservative traditions of the heritage communities.

Environmental changes due to investment which can affect the heritage areas should put in consideration because it could increase the pollution from projects or visits increase.

The government should encourage the investment of heritage areas project and motivate the private sector to participate in it through a package of procedures that encourage investment in this sector.

Increase the investors and local community awareness of the importance of the investment projects and social and

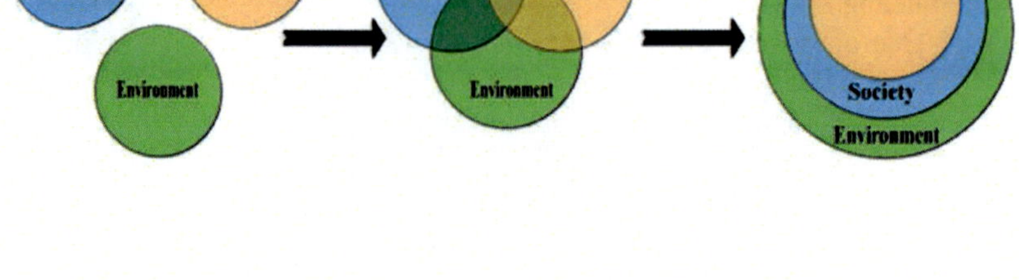

Fig. 2 Interaction of heritage investment sustainability factors

cultural sides in heritage areas which will increase the intangible benefits of the investment.

6 Selecting the Appropriate Method for Economic Evaluation

Selecting the appropriate method for economic valuation in the research's point of view the following factors:

- Studying factors of sustainable economic (feasibility study), social (select suitable projects according to public participation), and environmental (intangible values).
- Project feasibility study (cost/benefits).
- Adequate economic evaluation methods for the research area.

According to the previous standard, the researchers propose using cost–benefit analysis methods to determine the feasibility of the proposed project after studying points of strength, weakness, opportunities, threats (SWOT analysis) and suggest a plan for the public participation in the investment project.

7 Case Study

Case study methodology is related to the following points (Fig. 3):

- Select stage: In this stage, standard for selecting the case study will set which should be integrated to research goals and kingdom futuristic vision 2030.
- Analysis and applied stage: In this stage, research studying and analyzing the heritage site properties, the investment possibility of this site, applied one of the investment proposals and study the expected changes which occurred as a result of the investment process.
- Evaluation stage: In this stage, the investment project will be valuated according to the proposed economic evaluation method.

7.1 Select Stage

Select stage divided into two sectors as the following:

7.1.1 Standard of Case Study Selection

Kingdom of Saudi Arabia has an important economic transfer to create none petroleum economic; according to the kingdom vision 2030, a very important project proposed, one of them is NEOUM project which will make an economic, social and cultural booming; therefore, the case study should present the historical and cultural aspect, so the standard of selecting case study can be summarized in following points:

- The case study should reflect the distinguished architecture character and related to respected history from the local community.
- It should be near to the tourist developed areas according to kingdom vision 2030.
- The case study should have heritage events and festivals which will help in choosing investment field and evaluation methods.
- The case study should reflect the desert development model to be a guide model in dealing with future sustainable development in kingdom heritage sites.

7.1.2 The Case Study Selection

According to the previous standard ,the north of the middle area of the kingdom is the suitable area; it is near to NEOUM project area (Fig. 4), which represents civilized and historical dimension in kingdom history; the historical village in this area reflects the environmental and technical approach which is used to developed desert areas, and

Fig. 3 Case study methodology

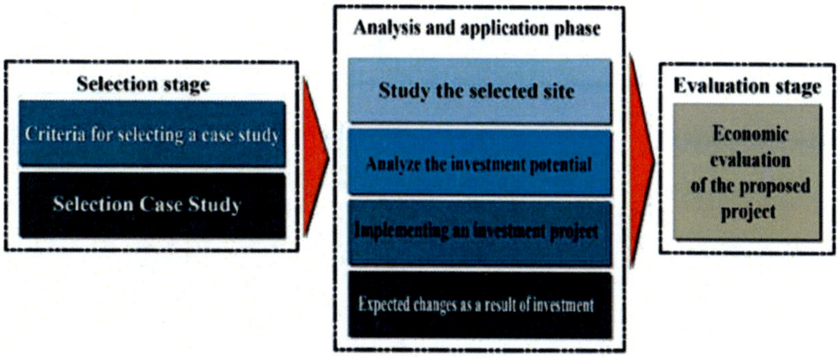

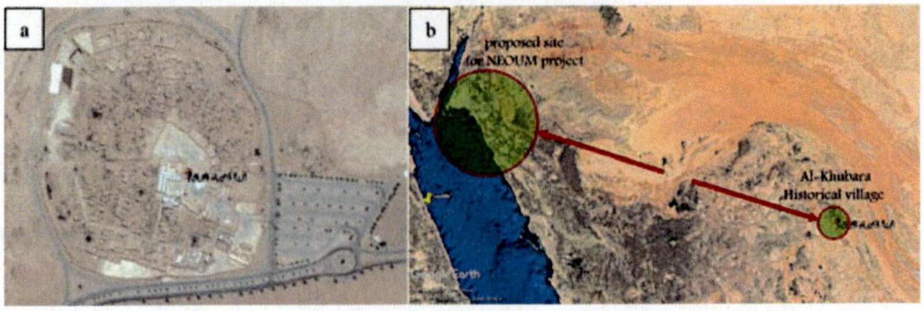

Fig. 4 **a** Case study layout **b** relation between case study and NEOUM project area ology

Al-Khubara village is one of the most important villages in this area which distinguished with economical base related to its location in the middle of the Al-Qusim Al-Madina road. The Saudi Commission of Tourism and National Heritage implemented a restoration and rehabilitation project to the village. and it is completed and opened in 1430 H; according to previous standards, Al-Khbara historical village is good example to be a case study.

7.2 Al-Khubara Village Historical Study

Al-Khubara village is located in the south of Al-Quasim city on the of Al-Quasim—Al-Madina road in a high area on the bank of Al-Rama valley, which was full of water in the season of rain, so village community was farmers beside there commercial and grazing activities. The village is surrounding by a wall of four gates, the village is divided by four main streets, and the village community leaves the village to a new city which was a reason for village damage along consecutive three decades (Saleh, 2010).

Al-Khubara village is distinguished with its urban and architectural characters which prevail in Nagdian architecture (Fig. 5); the village was built according to the Islamic Abbasid style as circular planning surrounded by a wall; central area of the village was the service center and it had (Saleh, 2010)four main commercial radian axes. The village contains 400 houses build in clay, and the village area is 120.000 m^2 established in 1115H, 1703.

The Saudi Commission of Tourism and National Heritage in cooperation with Al-Quasim area and ministry of municipal and rural affairs made a project for restoration and rehabilitation of the village with public participation, and the project was an opportunity to make heritage festivals which can raise the community heritage awareness and invite the investor to invest in heritage conservation projects.

7.3 The Village Investment Opportunities

Village restoration and revival projects made investment opportunities but these opportunities were seasonal activities that affect the heritage conservation sustainability; the research used quadrilateral analysis (SWOT analysis) to present the appropriate investment projects to village properties and kingdom vision 2030, which facilitate sustaining the village conservation project.

7.3.1 Strength

- Good condition of the village buildings after restoration, the village reflects the middle area of the kingdom architectural and urban heritage values.
- The village is near to important regional roads (Al-Quasim—Al-Madina road), and the village is prorated near to NEUOM project.
- Investment in heritage activities which was launched due to the heritage festivals and the community interaction with these festivals (Fig. 6).
- Reuse of the village buildings in new activities like (heritage museum—traditional handcraft affairs) which can help touristic marketing to the village (Fig. 7).

Fig. 5 Architectural character and restoration of the village

Fig. 6 Limited reuse of the village in heritage festivals and seasonal activities only

Fig. 7 Building good condition after restoration and reuse of the village building

7.3.2 Weakness

- The conservation and reuse of the village are not sustainable due to seasonal use.
- There are no financial resources to cover the periodical maintenance.
- There is no vision to reuse the village, and it is limited seasonaly.
- There is no local, regional, and global propaganda to introduce local heritage, urban, and architectural values.

7.3.3 Opportunities

- Heritage festivals in Al-Quasim region like dates and kleja festivals which will increase investment opportunities in the village (Fig. 8).
- Local community awareness and his participation in conservation and rehabilitation of the village increase local community investment and appropriate proposed activities to the local community.
- Variety of transportation (Al-Quasim airport—regional roads) which serve the village will help marketing the heritage investment projects at regional and global levels.

7.3.4 Threats

Social changes can be occurred according to investment projects that can have a negative reflection on a conservative community like region local community.

Neglection of reuse and investment architectural heritage can be a factor of damage and loss of this heritage.

The analysis shows that touristic development due to local heritage festivals beside urban and architectural heritage will be the main investment hub in the village, and proposed development projects according to kingdom vision 2030 can help investment heritage projects in the village in a global touristic map that can be a success reason and increase financial income of these projects.

Studying social and environmental effects for the investment projects should be used in evaluating touristic projects that aims to attract global tourism, the local

Fig. 8 Heritage festivals in Al-Qusaim region

community should participate in choosing investment projects that suit the community, and increasing social awareness can help decrease negative effects of the proposed investment projects if it occurs.

7.4 Investment Projects

According to previous analysis, research suggested touristic projects to be the economic and investment base, the village has 56 stores used in festivals, we can use them as heritage products stores along the year, village building can be reused as environmental heritage hotel started with 100 house and will increase gradually to cover all village's houses when the kingdom vision 2030 complete. Service uses can be added like restaurants, cafes, specialized museums and handcraft workshops which will be touristic attraction factors.

7.5 Changes According to Investment Projects

Changes due to investment projects can be divided into positive and negative one as the following:

1- **Positive changes**
   - Setting economic base to cover the periodical maintenance.
   - Increase local community income that embedded in the investment process due to investment income increasing.
   - Conserving and marketing local urban and architectural heritage at local, regional, and global levels.
   - Increasing local community awareness of heritage conservation and heritage investment, developing local community by participating in the investment process, heritage handicraft training for the local community and job opportunities that heritage investment will provide.
   - The proposed heritage environmental hotel will depend on renewable energy that will decrease operation costs and disseminate the culture of using renewable energy in the local community.

2- **Negative changes**
   - Local community changes due to touristic investment that aims to attract foreign tourist who has different manners and traditions from local community traditions.
   - Environmental effects due to the increase in visitors and pollution from transportations and use

7.6 Village Heritage Investment Evaluation

The investment process affects several issues related to heritage conservation and reuse of the heritage buildings. Evaluation can be divided into heritage conservation, reuse evaluation, evaluation on local community level and economical and environmental evaluation.

- Heritage conservation and reuse evaluation: Heritage investment of the village helps local community to knew the real value of the heritage, and how they can use it in increasing income, in spite of the investment activities, was seasonal activities, which encourage complete the restoration of all building in the village and complete village infrastructure which sustain the village conservation.
- Evaluation on the local community level: Interaction between local community and heritage festivals that occurred is one of the success indicators of investment and reuse of the village that ensure sustainable conservation of the village. Inform the local community of positive and negative effects and embedded them in choosing an investment project also sustain the conservation of the village.
- Economical evaluation: Due to could not get any information about investment in hotel in the region of the study, so the economical evaluation and economical study of the proposed investment project depends on data from the Saudi Commission of Tourism and National Heritage study for hotel investment opportunities in Al-Baha region and statistics of the Saudi commission of statistics, information center of touristic researches 2019, using cost/benefits analysis

8 Economical Study for the Proposed Project (Reuse and Operation)

The research suggested a partnership system between the governorate and privet sector or invertors in the touristic field, the contract time proposed to be ten years. The feasibility study methodology summarized in the following points:

Table 1 Total cost of the proposed project

Statement	Cost/riyal	%
Restoration project cost	50,000,000	68%
Additional buildings cost	7,500,000	10.2%
Infrastructure cost	2,500,000	3.4%
Furnituring cost (250 room × 24,000)	6,000,000	8.2%
Pre operational cost	7,500,000	10.2%
Total cost	73,500,000	100%

- Studying the investment costs.
- Project time schedule.
- Financial prediction.
- Financial indicators.

8.1 Estimated Investment Cost of the Project (Table 1)

Estimated cost calculations depend on previous data and hypotheses and cost village restoration and revival project that costs 50 million Saudi riyals (Ghaly, 2013).

8.2 Project Estimated Time Schedule

Estimated time to project execution is one year, the first six months are for initial studies, additional buildings architectural design, interior design, and furniture hotel spaces. The second six months are to implement all development stages; the real operation will start by the second year.

8.3 Financial Prediction Study

This study contains the study hypotheses of the financial prediction and financial benefits of the project.

1- The financial study depends on rent stores, rent hotel rooms according to the occupational rate of the information and touristic research center statistics 2019, assuming that the project administrate will be able to manage all project components in the best way.
2- Assuming incomes: The assuming incomes contain annual income which called annual cash flow; this cash flow is according to:
   - Investment benefits from the rent of stores and hotel rooms, the service income (restaurants- café -..etc.)
   - Returns from annual events for ten years.

Retunes from the rent of stores and hotel rooms:

- The income from store rent (20 stores) is assumed according to study market requirements, annually.
- The calculation of hotel rooms is based on the occupational rate of 68.5% for rooms (150 rooms) and 65.5% for units (suites) (100 suites) according to the information center of touristic research 2019.
- Mean of the rent rate 367.8 riyals for room and 292.4 riyals for suite according to the information center of touristic research 2019.

Retunes from annual events:

- The income from booths rent (booths for selling food and drinks—selling souvenirs—selling heritage products) assumed according to study market requirements.
- The area for commercial activities calculates as 30% of the main plaza (5000 m^2), available area 1500 m^2.
- Boothe's area will be rent during events only the price of renting boothes is 50 riyals/m^2 according to the similar market study.
- The events continue for 120 day/year.

8.4 Annual Operation Cost (Outcomes)

See Table 2.

8.5 Investment Project Returns

Estimated returns will be 51,099,113 riyals and net profit is 33674350 riyals with a percentage of 66% of project returns. Tables 3 and 4 show total project net profit /year.

Returns from the hotel (rooms—suites)

Tables 3 and 4 show that there are two respected methods to calculate the returns of the proposed project; one of them is

Table 2 Predicted the cost of annual operations

Statement	Cost riyal/year	%
Cost of the hotel operation	8,668,782	59%
Hotel administration expenses	2,600,635	18%
Real estate developer Administrative expenses	1,000,000	6%
Depreciation 3% from cost	2,425,000	17%
Total	14,694,417	100%

Table 3 Returns from hotel (rooms—suites) according to the commission of tourism and heritage, 2009, Al-Baha region

Level of occupation		No. of days	Perc. of occupancy	Net days	Room price/day	Perc. of discount	No. of rooms	Room income/year
High	Spring holiday	15	90%	13.5	850	0%	150	1,721,250
					2000	0%	100	2,700,000
	End of the year holiday	60	100%	60	850	0%	150	7,650,000
					2000	0%	100	12,000,000
	Fest holiday	15	100%	15	850	0%	150	1,912,500
					2000	0%	100	3,000,000
	Week end	70	75%	52.5	850	30%	150	1,561,875
					2000	30%	100	3,150,000
Low	Reminder days of the year				850	50%	150	1,115,625
					2000	50%	100	5,250,000
	Ramadan				850	50%	150	286,875
					2000	50%	100	450,000
Total returns/riyal								40,798,125

Table 4 Returns from the hotel (rooms—suites) according to the center of touristic research and information 2019

	No. of days	No. of units	Occupancy percentage	Mean price (Al-Qusim region)	Total/year
Rooms	365	150	58.5%	317.7	10,175,533.88
Suites	365	100	60.2%	310.3	6,818,221.9
Total returns/riyal					16,993,755.78

based on the properties of the environmental hotels prices in another region (Heritage, 2009), and the second study based on the prices of hotels in the same region but this study did not consider the properties of the heritage hotel (The Saudi commission of statistics, 2019); therefore, the research takes the mean of two methods to calculate the returns of the project.

Mean of the returnes = (40,798,125 + 16,993,755)/ 2 = 28,895,940 riyals/year

Returns from restaurants, café and stores (Table 5)

Total net profit of the project (Table 6)

8.6 Financial Indicators

Financial indicators are an important tool in project analysis that shows the profits, operation efficiency, and assets management.

Return on sales:

The projectability to gain profits from operation incomes, in the proposed project the return on sales run to 66% which is a very good indicator.

Return on invested capital

The indicator of the invested capital return is 45% from the start of the second year; it will grow up in the following eight years.

Table 5 Returns from services

Services	Income calculation	Total income	Service income
Restaurants and cafe	40% of total income	28,895,940	11,558,376
Other services	5% of total income	28,895,940	1,444,797
Stores	20 × 10,000	200,000	200,000
Booths (for 120 day)	120 × 50 × 1500	9,000,000	9,000,000
Total returns			22,203,173

Table 6 Total net profits of the projects

Incomes list	Total income/riyals	%
Incomes		
Income from hotels (rooms—suites)	28,895,940	56.5%
Income from services (food—drinks -stores—booths)	22,203,173	43.5%
Total income	51,099,113	100%
Expences		
Total expenses	14,694,417	
Profit before (Zakat—VAT)	36,404,702	
VAT 5%	2,554,956	
Profit after (VAT)	33,849,747	
Zakat 2.5%	846,244	
Total net profit/year	33,003,503	

Payback period

The payback period of the capital is 2.2 after the first year of operation and 3.2 years after the start of the investment.

Cost/benefit analysis

The cost/benefit analysis shows the feasibility of the investment; it calculates according to the real operation period nine years from project start to the assumed operation period ten years.

Cost/benefit = total project returns for nine years/total project cost = 4 (good investment indicator).

9 Conclusion

The research conclusion can be summarized in the following points:

Results

- Preservation and investment heritage sites are very important factors of conservation sustainability besides the social and environmental factors.
- The local community should be a part of the investment process and should participate in projects suggestion and choosing the most appropriate projects to them, which will be a success factor of the investment.
- The financial benefits are not the only goal of the investment of the heritage sites but the preservation of cultural, social, and aesthetic values of the heritage areas is one of the investment goals in heritage.
- The project achieves the financial benefits needs to make the investment successful, but the cost/benefit analysis calculates only the financial benefits; therefore, another method should be applied (conditional assessment) to determine the benefits of values of indirect use and values unrelated to use. (out of the research scope).

Future vision

- Set an investment plan for heritage conservation projects in the kingdom integrated into the kingdom vision 2030.
- Prepare an investment study for environmental hotels in Al-Quasim region to make the result of the financial evaluation more accurate, the research study based on a study in the Al-Baha region.
- Set study for environmental and social impact assessment of the investment projects in heritage sites before implementation which will decrease the project's negative effects.

- Activate the local community NGOs and embedded them in investment projects.
- Set laws and rules for investment in heritage sites contains financials motivations (loans—taxes exemption) which will increase the investment benefits and motivate the private sector to participate in these projects.

References

Al-Laham, N. (2007). *Touristic planning for heritage sites using environmental impact assessment.* cairo - Egypt: Nile for Printing, Press and Distribution.

Elbelkasy, M. (2017). *Investment heritage areas*. Noor Publisher Press.

Del Saz Salazar, S., & Montagud Marques, J. (2005). Valuing cultural heritage: the social benefits of restoring of old arab tower. *Journal of Cultural Heritage, 6*(1), 69–77.

Ghaly, A. (2013, 8 14). Al-Khubara heritage festival received 3000 visitors daily. *Alsharq Newspaper,* issue no. 917994.

Heritage, T. S. (2009). *Investment opportunities study for hotels in Al-Baha region*. Commission of Tourism and National Heritage Press.

Ibenholt, K., & Bowitz, E. (2009). Economic Impact of cultural heritage - research and perspectives. *Journal of Cultural Heritage, 9* (3), 326–337.

Ready, S., & Navrud, & R. C. (2002). *Valuing cultural heritage: Applying environmental valuation techniques to historical buildings, monuments and artifacts.* USA: Edward Elgar Publishing.

Saleh, M. (2010, 6 27). Old and new Al-Khubara. *Al-Jazirea Journal, edition no. 13786.*

Samir, H. (2008). *Conservation economics of heritage zones to a methodology for economical assessment for conservation and reuse programs*. Ph.D. thesis, Cairo, Egypt: Faculty of Engineering, Cairo University.

The Saudi Commission of Tourism and National Heritage, others. (2010). *hayat Al-Khubara. Conservation project for heritage village of Al-Khubara.* KSA: Commission of Tourism and National Heritage Press.

The Saudi commission of statistics, t. r. (2019, 4 10). www.mas.gov.sa/dashboard. Retrieved from www.mas.gov.sa.

Seraj-Aldine, I. (2002). *Revitalizing historical cities.* Alexandria. Egypt: Bibliotheca Alexandrina Press.

Tuan, T. H., & Navrud, S. (2009). Capturing the benefits of preserving cultural heritage. *Journal of Cultural Heritage*, *9*(3), 326–337.

Turner, R., Pearce, D., & Bateman, I. (1993). *Environmental economics: An elementary introduction.* The Johns Hopkins University Press.

Protection and Regeneration of Urban Historical and Cultural Area in New Period—A Case Study of Pingshijie Area in Nanjing

Yifan Cheng

Abstract

The urban historical and cultural area has witnessed the development and change of the city. It is the window of urban history and culture, and to a certain extent, serving as a card reflecting the characteristics of the city. In China, however, historical and cultural areas are not legally protected. Due to the development of urbanization and industrialization, urban land is increasingly tense. Driven by market capital, the protection of historical areas is particularly difficult. In addition, how to change the function of the protected historical area in the new period is also a problem worthy of discussion. As the last traditional area in the south of Nanjing, Pingshijie area is of great significance to the city, and its protection and renewal research has lasted for many years. According to the latest planning progress in this area, this paper analyzes and summarizes the four aspects of the existed research results: how to protect the historical area, how to continue the style and features, how to activate and utilize resources, and how to improve the infrastructure supporting facilities, aiming to provide some useful references for the protection and regeneration of the contemporary historical and cultural area.

Keywords

Historical and cultural area • Protection • Regeneration • Planning and practice

Y. Cheng (✉)
School of Architecture, Southeast University, Nanjing, 210000, China
e-mail: 341162516@qq.com

1 Introduction

1.1 Interpretation of Historical and Cultural Area

The historical and cultural area is an indispensable part of the city. The traditional culture, street texture, architectural art, and environment of the area are of certain value. Its protection and regeneration are also important contents of the sustainable development of the city.

At present, many cities in China have issued protection laws and regulations related to blocks with historical value (Yiran & Heping, 2018). Among them, the definitions of historical and cultural area and historic areas are not the same. Since the research object of this paper is in Nanjing, the concept of the research object is defined according to the relevant documents of Nanjing city. According to the *Protection plan of historical and cultural city of Nanjing (2010–2020)*, it includes three protection levels: historical and cultural area, historic area, and general historicarea (Nanjing Municipal People's Government, 2019). The above-mentioned three protection level areas all retain a certain historical street pattern and historical buildings with features. These areas can not be copied, and they are important witnesses of urban development. In today's promotion of cultural competitiveness, they have become a platform for cities to create and publicize their own culture.

It should be noticed that the historical and cultural area is different from the historic area. Firstly, the protection and renewal of the former are not restricted by the clear laws and regulations in most cases (Yunyan et al., 2018), thus so many areas are demolished in the process of urbanization. Secondly, the street texture and architectural facades of historical and cultural area have been destroyed for many years, providing some difficulties to the investigators. Thirdly, the latter's architectural functions are relatively simple, while the former's are complex, including but not limited to residential, commercial, office, industrial. Finally,

A. Versaci et al. (eds.), *Conservation of Architectural Heritage*, Advances in Science, Technology & Innovation,
https://doi.org/10.1007/978-3-030-74482-3_28

the protection and planning of historic area are more rigorous, with restrictions on the form, height, and style of buildings, while the protection and planning of historical areas are relatively loose, with some simple aims such as reflecting the historical life scenes, and therefore there is a large space for transformation.

1.2 Research Significance

Nanjing is the first batch of "famous historical and cultural cities" announced by the State Council of China. It began to exist 2500 years ago, with rich history and culture. There are many distinctive cultural areas in the city. As the former capital of the Republic of China, Nanjing is the birthplace of many modern Chinese history. The pattern, architecture, and culture of the city are all related to that period. As people say, "If you want to learn the culture of Sui and Tang Dynasty, you can come to Xi'an; If you want to learn the culture of Ming and Qing Dynasty, you can come to Beijing; If you want to learn the culture of the Republic of China, you can come to Nanjing (Yue et al., 2013)."

Culture is the epitome of history and the name card of a city. It is of great significance to enhance the competitiveness of the city and even the national's. In new period, the protection and regeneration of urban historical and cultural areas are put on the agenda. Some scholars believe that "the historical and cultural areas (its historical remains) have been changed a lot" and "are not suitable for the protection in a strict way" (Jinghui, 2006). Therefore, in China, many cities including Nanjing have promulgated and implemented relevant laws and regulations on the protection of historical and cultural areas, and the protection of them has attracted more and more attention.

The protection and regeneration of Pingshijie area in Nanjing have lasted for many years. This paper summarizes the practical results and abstracts four aspects of the strategies for the protection and regeneration of architectural heritage, aiming to provide useful guidance and reference for the protection and design of architectural heritage in China and even in the world.

2 Overview of Pingshijie Area

2.1 History

The Pingshijie area is located in the south of Nanjing. The plot starts from Zhongshan south road in the east, Dingxin road in the west, Shengzhou road in the south, and Nima lane in the north. It consists of two parts, Pingshijie area and Nanbuting area, with a total land area of 167,305 square meters. Among them, the main part of Nanbuting area is Ganxi house, a national key cultural relics protection unit. The total land area of Pingshijie area is 136,074 square meters. This time, the research scope is 122,916 square meters in the west of Dabanxiang. In the block, there are many architectural remains built in the middle and late Qing Dynasty, the Republic of China, the middle and late of last century, which reflect the typical characteristics of architectures in various periods of Nanjing, witnessing the development of modern Nanjing (Fig. 1).

2.2 Protection Work of Pingshijie Area

According to *Nanjing City's Protection Plan of Historical and Cultural City (2010–2020)*, 22 historical and cultural areas have been identified. Pingshijie area is one of the 22. In accordance with the reply to the *Nanjing City's Protection Plan of Historical and Cultural City (2010–2020)*, issued by the provincial government on November 15, 2011, and the

Fig. 1 Current situation and analysis of cultural relics and buildings

relevant provisions of the *Nanjing City's Regulations on the Protection of Historical and Cultural City*, this regulatory protection planning and design is formulated to improve the Nanjing's protection system of historical and cultural city, inherit the traditional regional culture of Nanjing, and improve the urban functions.

In 2012, the Architectural Design and Research Institute of Southeast University and the School of Architecture, Southeast University put forward a planning and design plan of "72 lane of Jinling" based on the results of preliminary planning and design. On the basis of the current roads, the scheme appropriately widens and opens up the main roads and other branches in the area, clarifies the road structure, migrates the historical relics together, and displays them intensively, while the pedestrian system in the core protection area is designed maintaining the original pattern and shape of the streets and lanes in the area. In terms of area's developing orientation, it is proposed to build a Cultural Park composed of 72 professional pavilions, including Xiyuan (Kunqu Opera), The First Theater (film), Gongda Calligraphy, and Painting Academy (calligraphy and painting).

On December 19, 2014, the office of Nanjing Urban and Rural Planning Commission organized an expert consultation meeting on the "Detailed Planning Plan for the Construction of Pingshijie Area in Nanjing." The experts and relevant leaders at the meeting discussed the planning and design results seriously and thought that the whole scheme work was timely and necessary with rich contents. The ideas of "small-scale, unit by unit, gradual regeneration" and other ideas proposed for the protection of the area were basically feasible and put forward suggestions for further modification to guide the further promotion of the planning and design work.

The 2016–2020 protection planning scheme, on the basis of summing up the previous design research and upper planning, systematically carded the cultural relics protection, spatial pattern, historical and cultural resources in the area, and made a detailed planning for the specific protection scope and protection mode.

3 Protection and Regeneration Strategies

3.1 How to Protect the Historical Area

3.1.1 Protecting the Area Texture

The map of Nanjing in the early period of the Republic of China had fully reflected the overall pattern of Pingshijie Area. The main streets and lanes had experienced a hundred years of development and are still able to continue completely (Fig. 2). The map of Nanjing in the early period of

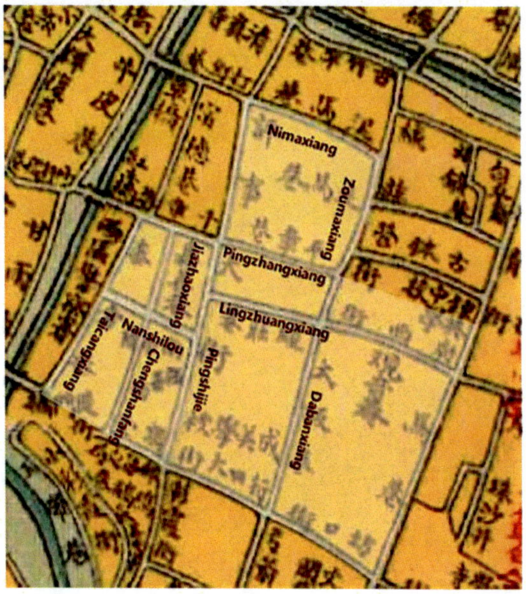

Fig. 2 The map of Nanjing in the early period of the Republic of China

liberation showed that Pingshijie area is the most densely built core area in Nanjing (Fig. 3).

Historical place names are of great significance. In Pingshijie area, in addition to 15 historical streets and lanes, there are still some traditional place names around the area, such as Pingshijie (Fig. 4), Daqiao, Dingxinqiao, Nanshilou, Hongtuqiao. These historical information, together with the existing internal historical elements of the area, constitute the environment background of the whole historical area. There are more than ten historical streets and lanes that can be verified in the area, of which Pingshijie is the main street. After a long history of development, part of the streets and lanes are occupied by residential buildings, and the texture of the streets and lanes is not obvious (Fig. 5). According to the trend of traditional streets and lanes, the plan rearranges the block division, respects the traditional space scale, and forms a number of small-scale block spaces. The new street space follows the development strategy of "small-scale, unit by unit, gradual regeneration," which is convenient for the access of people flow between plots and provides technical support for the restoration and regeneration of historical area (Fig. 6).

The original streets and lanes and more than 60 historical relics inside the area are the main controlling factors of the whole regional planning. The overall planning of the area pays attention to the distribution of historical relics and especially emphasizes the control of their style on the style of the whole area. According to the above principles, the whole area is divided into more than 30 small-scale units for

Fig. 3 The map of Nanjing in the early period of liberation

Fig. 5 Previous texture of the streets and lanes

Fig. 4 Old photos of Pingshijie

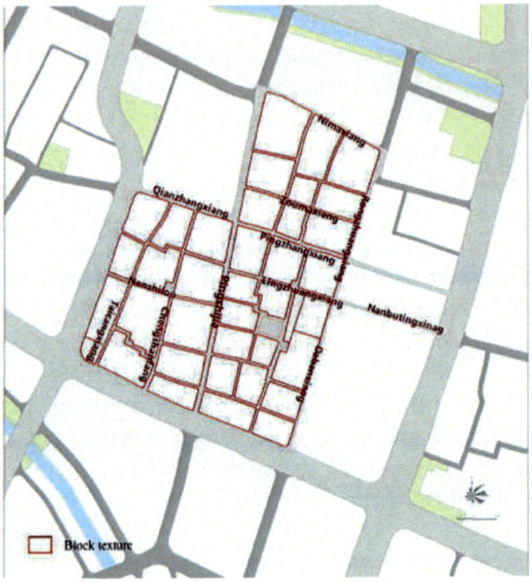

Fig. 6 Texture after

gradual renewal. At the same time, the streets and lanes should be dredged as much as possible to form a grid network system (Fig. 7).

According to the road level, the interface types around the area can be divided into four categories:

- the urban interface around the area;
- the main street interface including Pingshjie and Dabanxiang;
- the traditional historical street interface;
- the new branch lane interface;

From the analysis of the original block texture, it can be seen that the building layout presents the characteristics of "perpendicular to the street." As for the buildings located in the internal lane, they mostly face south.

Therefore, the texture of the planned buildings is strictly based on the layout characteristics of "multi entrance courtyard space" in the area. For different street and lane interfaces, while controlling the width of the street and lane,

Fig. 7 Architecture layout

the building layout facing the street and lane should follow the following architectural principles:

- the main street facing facade of the new building should face the city interface;
- the main street facing facade of the new building should face the historical street;
- the new building whose facade facing the side of the historical street, its layout should follow the historical texture with the north–south orientation;
- the new building whose facade facing the new branch lane, its layout depends on the actual needs;

3.1.2 Protecting the Historical Relics

According to the results of several rounds of investigation, the Nanjing Cultural Heritage Administration announced that there were 64 existing historical relics in the area. At present, many of them are being repaired, among which 5 buildings' repairs have been accomplished (Fig. 8).

3.1.3 Protecting the Traditional Style

The residential buildings in the area are usually closed to the outside, with few windows. Therefore, the entrance of the building becomes the key point of the facade treatment. The traditional brick lintels, wooden doors, and windows, as well as the European style doorways in the period of the Republic of China, are preserved. The bulding method of stone windows is similar to that of lintels, which are decorated with brick, and some of which are carved with wood. The two-story attics and the facades along the street are mostly equipped with wood carved hollow balustrades or iron railings. The facades or gables of the building and the entrances are often decorated with traditional decorative details. Some of them are traditional ones, and some of them have traces of European style. The exterior walls are mostly cavity walls. The indoor floor is mainly paved with masonry, cement water mill, and wood. Most of the first floor buildings are paved with masonry, and a few are paved with wood floor (Fig. 9).

3.1.4 Protecting the Traditional Culture

The significance of Pingshijie area is not only reflected in the layout of ancient streets and lanes, ancient buildings, ancient wells, ancient trees, and other material heritages, but also in some aspects of intangible cultural heritage, such as traditional handicrafts, folklore, and so on, which have been passed down for generations. The former residents of the area engaged in different commercial activities, and there are many kinds of traditional skills in the area. Many skills are closely related to the place names and the history and culture of the south of Nanjing. The inheritance and development of these traditional handicrafts reflect the cultural of the city. Different cultural forms are integrated with daily life, showing multiple colors (Yunao & Long, 2012).

3.2 How to Continue the Style and Features

After a comprehensive study of the area, we put forward the following updated orientation. First of all, it is one of the most important historical areas in Nanjing. Secondly, it

Fig. 8 Rendering drawings

Fig. 9 Details of architectures

opens a window for people to see traditional culture. Finally, it will become a tourism destination catering to the future's diverse needs.

3.2.1 Function and Culture

The project is based on the history and culture of the south of the old Nanjing, focusing on the local unique lane culture, street culture, architectural culture, Hui nationality culture, opera culture, folk culture, etc., which are different from the Confucius Temple and the Laomendong not far away from the area. It has a high degree of cultural identity, facing the residents of the main city and tourists. It is a historical and cultural leisure area, integrating cultural and creative retail, designer retail, innovation industry, cultural exhibition hall, featured catering, cultural tourism, leisure, and entertainment, featured accommodation, mixed residence, and other functions.

3.2.2 Style of Architecture

More than 60 historical buildings have been restored in accordance with traditional features. Architectures around the relics and the main historical streets will be designed according to the historical style and the traditional architectural elements. Some modern multi-storey buildings should be desinged in a new way.

(1) The original repair and protection buildings are mainly the historical relics in the area, as well as some new buildings adjacent to the them. Most of them are brick and wood structures, which are repaired according to the principle of repairing the old as the old, forming a complete traditional architectural complex. The amount of this part accounts for about 20% of the whole area (Fig. 10).

(2) The appearance of the newly built traditional architectures is unified with the traditional style and the interior space is suitable for various functional requirements. In this part of buildings, the space can be slightly larger than the

Fig. 10 The original repair and protection buildings

Fig. 11 The newly built traditional buildings

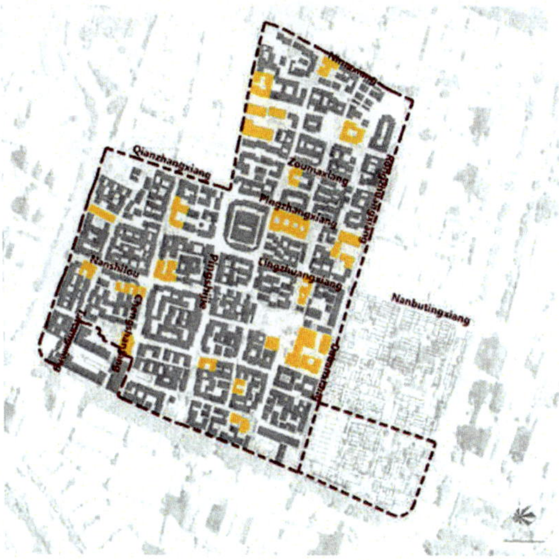

Fig. 12 The new buildings

original building, but the increase proportion should not be greater than 20% of the original building volume. The amount of this part is controlled at 50–55% of the whole area (Fig. 11).

(3) In order to reflect the flexibility and diversity, some new buildings are considered to be set up in the area. Material selection and construction methods reflect the characteristics of the present age. For this part of buildings, the space should not be greater than 25% of the original one. The amount of this part is controlled within 25–30% of the whole area (Fig. 12).

According to the regulations, the heights of building cornice in the area are controlled within 7 m. However, due to the large scale, the highly unified cornice height may result in a single urban interface and the fifth facade may seem monotonous. At the same time, according to the *Regulations of the building height in the old area of Nanjing*, the height of the cornice of some newly built buildings in the historical and cultural area can be properly controlled between 9 and 12 m, but the floor area of the building base shall not exceed 15% of the construction land area (Nanjing Municipal People's Government, 2017). There are many modern multi-storey buildings in the area, which also provide the possibility for the break of height.

3.3 How to Activate and Utilize Resources

3.3.1 Traffic Strategy

The network around including one main road (south Zhongshan road), three secondary roads (Shengzhou road, Jianye road, and Dingxin road), which will be the main channel for traffic distribution in the future. The internal roads are mainly secondary roads and streets, which are arranged in grid shape, and should be systematically combed in combination with historical texture, current conditions, planning requirements, etc. At present, there are two entrances and exits in each of the four directions of the area. In the future, the mode of single entry and single exit can be planned to improve the travel efficiency.

The main strategy of traffic organization: pedestrian and vehicles are separated. According to the analysis of the current traffic flow, it is suggested to take south Zhongshan road, Shengzhou road, Dingxin road, and Jianye road as the outer ring. In the area, it is suggested to take Baiyi'an, Nimaxiang, Qijiawan as the north ring, take Pingzhangxiang as the south ring, take rongzhuangxiang, and Dabanxiang as the east ring, and Pingshijie as the west ring. Pingshijie is the main street of the area, only pedestrians have access to it. In case of emergency, fire trucks can be allowed to enter in. Dabanxiang is planned to be a mixture of pedestrians and vehicles, which can meet the requirements of traffic and commercie. During night and holidays, Pingzhangxiang and Nanbutingxiang will be adjusted to pedestrian street, then the whole area will become a pedestrian block.

It is necessary to increase the development and utilization of the underground space and fully organize the underground transportation system.

3.3.2 Underground Space Design

The utilization of underground space is the key point of the construction. The principle of it is to reduce the existing ground building density as much as possible, create a good neighborhood atmosphere, and integrate auxiliary functions into the underground part for systematic utilization. In line with the basic principle of protecting historical relics, the

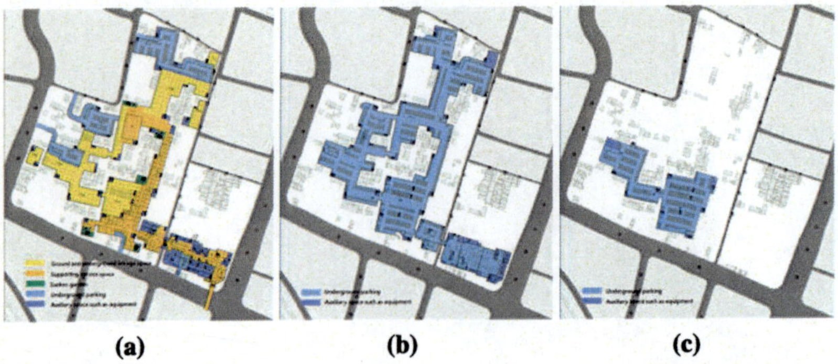

Fig. 13 a B1 plan b B2 plan c B3 plan

excavation of the basement avoids these buildings as much as possible. In the process of project implementation, all these buildings need to go through architectural mapping and structure safety appraisal.

There are two motor vehicle entrances and exits on the south, west, and north sides of the area. The underground second floor is designed as a parking garage, which provides additional guarantee for relieving the ground traffic pressure and increases the accessibility of the area. According to the existing planning, 1250 vehicles can be parked in the underground parking area (Fig. 13).

Considering the time sequence and efficiency of underground space construction, it is suggested to adopt the mode of gradual development by stages. The construction above ground is separated from the basement in structure. According to the existing planning, the construction batch is based on the sequence from east to west, from south to north. Each time, a number of small plots are selected to form a group, and the underground part is constructed synchronously.

3.4 How to Improve the Infrastructure Supporting Facilities

Backward infrastructure is a problem that exists in almost every historical and cultural area. The backwardness of water supply and power supply facilities and the inadequacy of sewage discharge facilities are not consistent with the modern life. Therefore, the infrastructure needs to be improved. After a series of investigation and research, we give the following solutions.

According to the analysis of the current pipeline: d1000 sewage pipe will be laid on the north side of Shengzhou road, the drainage direction is from east to west, and the starting point is at the intersection of Pingshijie; d500-d600 rainwater pipe will be laid on the south side of the road, and the drainage direction is from east to west. D500 sewage pipe will be laid on the west side of the central line of Dingxin road, with the drainage direction from south to north; d600 ~ d1000 rainwater pipe will be laid on the east side of the central line, with the drainage direction from south to north. In Dabanxiang, d600-d1000 rainwater pipes are laid, which are discharged into Qinhuai river from south to north; and d300-d1000 sewage pipes are laid, which are discharged into intercepting ditch from south to north. In Pingshijie, d400-d1000 rainwater pipes are laid, which are discharged into Qinhuai River from south to north, d300-d400 sewage pipes are laid, which are discharged into the d1000 sewage main pipe of Shengzhou road from north to south (Fig. 14).

Specific design of sewage scheme: making full use of the current sewage pipelines in Dabanxiang and Pingshijie, and lay municipal pipelines reasonably in combination with the development of underground space. Due to the control of the buried depth of the underground space, the lifting pump is laid in a reasonable position, the excavation depth is moderate, the construction is convenient, the buried depth of the pipeline meets the requirements and other pipelines can be avoided (Fig. 15).

Fig. 14 Analysis of municipal pipeline in surrounding road network

Fig. 15 Sewage scheme design

Fig. 16 Aerial view

4 Experience Gained in the Protection and Regeneration Work

It can be seen from the practice of Pingshijie area that the protection and regeneration of architectural heritage is a continuous and dynamic process. Instead of blindly pursuing the speed, it should be developed coordinating with the modern cities. In order to make heritage play a role in today's world, we must make a reasonable plan for its future development according to the actual situation: what to keep, what to innovate and how to maintain vitality are all unavoidable issues. How to protect the historical area, how to continue the style and features, how to activate and utilize resources, and how to improve the infrastructure supporting facilities, are the four aspects considered in this protection work, covering all aspects of heritage from protection to reuse. Although there are still some deficiencies in the work, from the perspective of several demonstration areas that have been put into use, the final effect is satisfactory (Fig. 16).

5 Conclusions

The protection and regeneration of historical and cultural area is an important part of urban renewal. The purpose of the protection and regeneration is to carry the historical context and restore the vitality of streets. As an entity bearing the relationship between human and environment, modern and history, the historical and cultural area truly reflects the unique culture of the city. At the same time, it is a special kind of architectural heritage. The general practice of protecting architectural heritage is not necessarily suitable for it. In recent years, the development of cultural tourism and leisure industry is rapid, and the historical areas in cities across the country have set off a trend of commercial development based on historical and cultural protection. The protection and regeneration of Pingshijie area is a new attempt based on the previous experience, which generally follows *THE VENICE CHARTER 1964* that "all their information must be transmitted without any distortion" (IInd International Congress of Architects and Technicians of Historic Monuments, Venice, 1964). The local unique architectural style and the cultural features of the specific historical period are preserved for future generations to appreciate, which can be used as a reference case for the protection and regeneration of the historical area in the future.

References

IInd International Congress of Architects and Technicians of Historic Monuments, Venice (1964). Adopted by ICOMOS in 1965.

Jinghui, W. (2006). Urban planning and cultural heritage protection. *City Planning Review, 1*, 57–59.

Nanjing Municipal People's Government (2017). Regulations of the building height in the old area of Nanjing. http://www.nanjing.gov.cn/zdgk/201810/t20181022_573505.html. Accessed March 22, 2017.

Nanjing Municipal People's Government (2019). Protection plan of historical and cultural city of Nanjing (2010–2020). http://ghj.nanjing.gov.cn/ghbz/ztgh/201705/t20170509_874088.html. Accessed December 27, 2019.

Yiran, Y., & Heping, L. (2018). Value assessment and planning strategy of traditional scene area: Datianwan Chongqing. *Planners, 2*, 73–80.

Yue, H., Xiaohua, Z., & Jie, S. (2013). Nanjing Huiyuanli historical district preservation planning. *Planners, 9*, 77–82.

Yunao, H., & Long, C. (2012). *Cognition, protection and revival – a study on the historical and cultural heritage of Pingshijie*. Nanjing Normal University Press, Nanjing.

Yunyan, L. et al. (2018). Multiple balances of comprehensive evaluation of building values of historical and cultural feature areas under new norms——Take Chongqing Cuntan as An Example 1, 92–99.

The Impact of Symbolic Values in Regaining Lost Heritage Buildings After War: A Case Study from the Modernist Architectural Heritage in Mosul

Aseel Fatahullah Abdullah Al-Ta'ai and Saba Sami Mehdi Al-Ali

Abstract

Architectural heritage endures damage and loss due to the violence that accompanies wars and conflicts inside cities. The damage could be due to the exposure of heritage buildings to negative or positive events and experiences, which remain in the memory of the city's population, and give heritage buildings symbolic meanings. This phenomenon raises the question about the impact of such periods of violence on the decision of how to regain the lost building. This paper aims at investigating this relationship, and clarifying the peculiarity of the symbolic values in architectural heritage as a reflection of the collective memory of the city. After clarifying the values attributed to architectural heritage, the research, relying on previous studies, moves to explore the theoretical basis for the relationship between architecture, violence, collective memory and the mechanisms of regaining heritage buildings upon loss. Thus, a case study from Mosul is examined, represented in the building of "the National Insurance Company", a modernist building, by the late Iraqi architect Rif'at Chadirji. The building suffered during the period of violence (2014–2016) during the control of ISIS of Mosul and the liberation war that followed. The case is investigated through a questionnaire/sample survey from the Mosul community on the values they cherished in the building and their preference for the mechanism of its regain. The research found that negative symbolic values deposit in such architectural heritage due to the community's collective memory influences public opinion towards restoring lost heritage with renovation projects despite its significant original architectural values.

Keywords

Heritage symbolic value • Violence of war • Collective memory • Mosul National Insurance Company building • Rifat Chadirji

1 Introduction

Architecture is known to establish a relationship with the community that uses it and deals with it directly or indirectly. A building, especially if public, is a record of events and experiences stored in the collective memory shared by its inhabitants and would extend to outer community. This collective increases in those buildings of cultural or historical importance as they convey the memory of successive generations, and thus form a part of the identity of the society and the city as well. Consequently, a damage to such buildings affects people's memories. Damage can be tangible, such as a total or partial destruction of buildings due to violent operations, or intangible represented by the loss and distortion of their values, as well as their memory. This paper is concerned with the phenomenon of change in the symbolic values attributes in heritage buildings as a result of abuse in a period of violence and war, and the extent of impact of such phenomenon on the decision making to restore or reconstruct the lost building (see Fig. 1)[1].

The research then sheds light on the values of architectural heritage and the dynamics between architectural heritage and violence. It explores the concept of afflicting the collective memory during war, and then explores mechanisms for recovering lost architectural heritage. A case study of a building in Mosul affected by the war is presented, by using a

A. F. A. Al-Ta'ai (✉) · S. S. M. Al-Ali
Department of Architecture, Al Nahrain University, Baghdad, Iraq
e-mail: dr.sabasami@eng.nahrainuniv.edu.iq

[1] Rif'at Chadirji is an Iraqi architect (1926–2020). Studied architecture in UK before the fifties of the last century, have several built projects in Iraq, most well-known of them are Tahrir square's freedom monument, the unknown soldier monument.

A. Versaci et al. (eds.), *Conservation of Architectural Heritage*, Advances in Science, Technology & Innovation,
https://doi.org/10.1007/978-3-030-74482-3_29

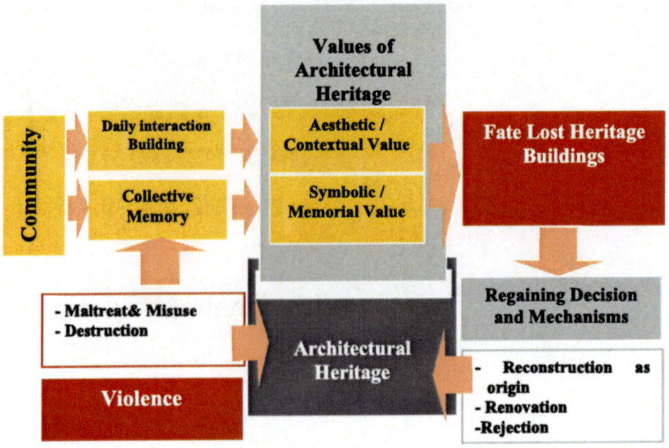

Fig. 1 Research concept (by authors)

questionnaire for a sample survey of the population, regarding the values of the building, and what mechanism is preferred to regain it. The paper aims at providing a clearer vision to the peculiarity of symbolic values and people's attitude to it, and highlights the necessity to consider such values, whilst taking decision regarding architectural heritage.

2 Methodology

A theoretical base to the relevant concepts is established using an analytical descriptive approach to previous literature. Then, to support and validate the theoretical proposition, the researchers present a case study; conduct a questionnaire using the qualitative and sensory scales, then analyse the results down to the conclusions and recommendation.

3 Theoretical Discussion

3.1 Values of Architectural Heritage

The importance of architectural heritage stems from all the significant values it holds being tangible or intangible. To identify these types, a set of charters and studies were reviewed such as the Burra Charter (1999), the Australia ICOMOS Charter of (1999) Mason (2002), Drury et al. (2008) and Saydem (2013) and a classification of these values is suggested, as follows:

- **Historical Value**: Either explanatory, which helps explaining the past through a shared experience of a place, or associative, which is attributed to places that are closely related to the work of people who made important discoveries.
- **Aesthetic/Architectural Value**: Includes aspects of perception including consideration of the shape, scale, colour, texture and materials of the fabric, as well as the smells and sounds associated with the place and its use.
- **Contextual Value**: heritage either displays harmony with nature, where the architectural blocs merge with the natural elements on the site, or may exhibit complementarity with the urban scene, or salience as a landmark in the city.
- **Scientific/Academic Value**: this depends on the importance of relevant data in terms of its rarity, quality or representational quality, and the degree to which the place may contribute with substantial information.
- **Functional Value**: presented by high-functional performance through the suitability of the form for the function. It is also evidenced by the building's ability to adapt to functional change.
- **Economic Value**: the value of the use of a material heritage refers to the goods and services that flow from it.
- **Social Value**: the reflection of thought, belief, or social values of interest to society (perpetuating a sense of identity), as well as enabling and facilitating social ties, and other relationships.
- **Sentimental Value**: (Religious, National): they stem from the positive influences of religious heritage buildings such as simplicity and purity of the soul and it may also include feelings of wonder and unease raised by visiting heritage places.
- **Symbolic/Memorial Value**: reflects the memory of the place in relation to the people associated with it, provokes events of the past in the memory and collective identity, alternatively, it may be understood through interpretation.

All these values are subject to change, due to human and natural risks, and opportunities as well. Some might weaken or even disappear, whilst others might develop stronger and new ones may emerge. The change of these values can affect the will a community might display to the conservation of an affected heritage building and to regaining it when it is lost.

The focus of this paper is mainly around values that are mostly related to people; namely, architectural and contextual values which people appreciate as long as the building is present in their daily life, and symbolic values which people attribute to heritage buildings as much they witness events that are stored in community's collective memory.

3.2 Architectural Heritage and Violence

Wars and conflicts are sources of destruction to buildings and cities due to the violence that accompanies them. However, studies stress that "what is destroyed is not only buildings and structures but also architecture and places that carry meaning and significance for its inhabitants and people. The impact of war on architecture transcends its physical presence and impacts its symbolic meaning as a representation of cultural identity" (Mahgoub, 2008). It is also acknowledged that there is a relationship between the destruction of a significant building in a community and the eradication of the community itself. In Bevan's words: "The continuing fragility of civilized society and decency is echoed in the fragility of its monuments" (Bevan, 2007).

On top of the list of places that are most targeted in purpose during wars comes "Traditional buildings that store and preserve a huge collective memory transmitted from one generation to another". (Hoteit, 2015) Abusing architectural heritage causes damage to its tangible and intangible values, and consequently affecting its relation to people and its significance for them.

3.3 Building Collective Memory and War

Humans collect and store personal memories throughout their lives. However, collective memory of a community is something different from personal memories. It is known to be that which includes all memories that are commonly shared by the individuals of that community. "It involves the integration of various different personal pasts into a single-common past that all members of a community come to remember collectively" (Zerubavel, 1999). It can be noted that memories could be good memories, meaning that remembering them provides joy, pride, self-esteem or any other positive feelings. They can also be bad memories which may, when remembered, trigger negative feelings like anger, shame or pain. Either way, sharing feelings collectively helps community members to bind and be in solidarity. With time, this contributes in propping national identity.

Christian mentions Halbwachs' view upon a community's identity in that it is generated when the collected traces of integrated (but separate) memory create feelings of individual identity that when shared in numbers creates feelings of group identity. He argued that this is the essence of the transmission of generational identity stating that: "We create future identity of those who succeed us with the creation of present memory today. Each generation reconstructs past memory to adapt it to present identity in an endless process of creating "collective effervescence" (Christian, 2012).

Paez and Liu explain that collective events are more effective in generating collective memory when they: (1) influence social change in the long run and are socially relevant in the present; (2) are emotionally loaded; (3) elicit abundant social sharing; (4) are socially rehearsed by mass media; and (5) are associated with collective behaviour and commemoratives rituals that can be narrated coherently by institutions and individuals (Paez & Liu, 2015). From this view point, it can be said that wars and conflicts can generate strong and enduring memories to a community. It is noticed that this capacity became a tool used intentionally by conflicting parties as part of a psychological war. Violent events provoke deep positive or negative emotions that are collectively recalled long time after the conflict is finished.

3.4 Abuse of Architectural Heritage and Its Impact on Collective Memory in War

"The impact of crises affects the way people produce, understand and inhabit spaces and places. Places can change their social function: a mosque becomes a guest house or a dispensary, a school becomes a barracks or a prison. [...] Even with those changes, populations are expected to re-inscribe all those social places into known social meanings through coping strategies or use and reinvention of traditions" (Piquard & Swenarton, 2011). Yet, as mentioned previously, not only the owning community generates meanings in its heritage; but also, dismantling of places in conflicts can be a tool in the hand of the enemy, imposing new meanings to be delivered to the community. Bevan stresses that in conflicts there is an emphasis on identifying the enemy group, which leads to its devaluation and then its dehumanization. The latter is "an essential step towards making it acceptable to dismantle an enemy's heritage" (Bevan, 2007).

Heritage buildings dismantling comes in variety of manifestations; it can be direct and systematic destruction, aiming at complete erasure of the community's memories and presence. Or it can be maltreat and misuse of significant buildings aiming at achieving psychological oppression. As stated by Piquard and Swenarton: "in every case of symbolic violence, the dehumanization of people is accompanied by an anthropomorphic attitude towards buildings that sees them as something to be "wounded" or "humiliated": for example, a sacred or symbolic building used as a stable, latrines or place of torture" (Piquard & Swenarton, 2011). Such aggressive actions of destruction or maltreat, being performed on public will be registered in the community's collective memory. Relying on Halbwachs, Hoteit describes that places where collective events occur "become a vital part of the memory, and accordingly whenever seen by the individual or the group will stimulate the recalling of the memory or the event that occurred therein". (Hoteit, 2015) With this process, the significance of the heritage that is presented in its recognized symbolic values take change in view of its owning community.

3.5 Regaining Lost Architectural Heritage After War

Where the war leads to the loss of the architectural heritage, the primary purpose after war will be to regain this heritage. Here, the regaining process is defined as: the retrieve of a lost or damaged heritage building of various degrees in its various tangible and intangible values and with or without its original function. It is conducted through several mechanisms that may partially change the building. The purpose is to maintain the efficacy of the building in the city, whilst preserving its values or acquiring other added values.

By reviewing charters such as the ICOMOS New Zealand Charter of 2010, which outlined the mechanisms of intervention to recover the lost architectural heritage, they appear to include: Reconstruction, renovation, reproduction and restoration.[2] The first two are of interest to the present study, as they both are consistent with the research and case study purpose. They could be defined as follows:

- **Reconstruction**: Means building again as much as possible to a previous documented model using available original materials as much as possible, in addition to new materials, to replace the lost materials. It includes total reconstruction and partial reconstruction. It should be executed on the exact original location of the building. An example is the total reconstruction of the Cathedral of Christ the Savior,[3] which is located on the northern bank of the Moskva River in Russia (Fig. 2). Some controversial decisions were made in the process, the most significant was changing the material used in recreating the marble reliefs on the interior walls. The original pieces were made of light marble, but were then replicated using bronze. The dark, bronze reliefs contrasted with the light, original marble ones; thus the modern element was criticized for not aligning with the style of Russian church architecture. The use of bronze is symbolic of other modern aspects of the cathedral (Buckler & Hazzard, 2016) Howard (2020) (Fig. 3).
- **Renovation**: It is the reviving of the structure from its poor condition and the production of new parts in the framework of an old building. The renovation processes can be hidden, whilst improving the building's existing atmosphere (Fig. 4). Or it can be radical or severe, as it is very similar to partially building remodelling. An example of this was the KaDeWe store (Kaufhaus des Westens),[4] which was regained with a partial renovation mechanism as the store became a symbol of a new beginning for the destroyed city of Berlin. The process of renovation included; expanding the store to seven floors, with the opening of a special international floor in it (taking the character of early modern buildings), in addition to other renovations that took place later (Zeunert, n.d., p2; Watkins, 2019).

4 Case Study: The National Insurance Company Building of Al-Chadirji

4.1 Importance of the Building

The building was located near the Iraqi city of Mosul (the right side of Mosul) in the Dawassa district was designed and built by Iraqi architect Rif'et al.-Chadirji in 1966. It occupied an important corner of the (Governorate

[2] Whilst reconstruction indicates rebuilding in the same location, reproduction signifies rebuilding the lost heritage in another location and context. As for restoration, it is used for partial damage.

[3] The cathedral is considered one of the most important churches that revived Russian architecture. Established in 1860 AD, it was built to commemorate the victory of the Russian forces over Napoleon's invasion, as a national monument to express the nation's gratitude to Christ Jesus for saving Russia. The cathedral was destroyed by powerful explosions, which were approved by the Soviet state in 1931 and turned into rubble (Fig. 1). In 1994, the government agreed with the Russian Orthodox Church to rebuild the cathedral in 2000 (Buckler and Hazzard 2016)

[4] KaDeWe have been a well-known commercial space in Berlin for over a hundred years. This shopping center is the largest of its kind in continental Europe. Established in 1907 AD, the store was distinguished by modernist attempts at the structural side, through stone material and the achievement of small window openings. It was also inspired in its design by the surrounding residential buildings, as it was built in a residential area. Previous evidence suggests that the store in 1930 had become an important point in the Berlin city plan, with minor changes taking place inside it during the Nazi rule. In 1943 a US plane bombed the store during a raid on Berlin, which led to burning it on the ground (Fig. 4), and it can be said that the Berlin store is in a state of continuous renewal and change, as it is reshaped almost every ten years, starting from 1956 until 2018, when it laid plans to fully renovation and modernize the store (Zeunert p1-p4).

Fig. 2 Demolition of the cathedral under Stalin in 1931. *Source* Alamy Stock Photo

Fig. 3 Reconstructed Cathedral of Christ the Savior. *Source* Courtesy of Termsit Siriphanich

Fig. 4 Demolition of theKaDeWe Storein 1943. *Source* Alamy Stock Photo

intersection) near the governorate building (Fig. 5). It used to present harmonic integration in its urban scene with the modernist statue of (Fatat Um ilRabi'ein—the Girl of the two springs city) (Fig. 6)[5] which made it a sign of the emerging importance of Mosul as a modern city, as it was one of the earliest multi-storey buildings in the city, and carried a modern architectural style with a local flavour. Although the building is not listed locally as heritage building, it falls within the attention of the Docomomo Iraq chapter,[6] being one of the very few buildings realized in Iraq by Chadirji.

4.2 Life Cycle of the Insurance Company Building

The building used to be a part of Mosul's community daily life since its construction as it changed its function several times as below:

- 1966, was built to be headquarter of the national insurance company in Mosul which provided insurance services on life and properties of citizens.
- 1975, changed to headquarter of the Iraqi department-stores company,[7] and named locally "The Orosdi-Back building".[8]
- 1986, the building was transformed to the "Directorate of Retirement Affairs".[9]

Since its control of the city, ISIS was aware of the strategic importance of the building as it was in a central location of the city as well as a landmark with high altitude. Hence, the building was not only used a point of control and monitoring, but also as a platform to execute convicted victims by throwing them from various points of the building. These were mostly young people sentenced for a variety of accusations.

The building has suffered from vandalism by ISIS members, and it was also badly affected by months of fighting and fierce liberation battles to expel ISIS from Mosul, which ended in the summer of 2017, (Fig. 7). The building was finally targeted by two messiles that put down most of its structure and severley damaged its foundations. Hence, it became a threat to public safety, so a decision to demolish it was made, and about a month after the start of the demolition work of the building, only three damaged floors remained (Fig. 8), until in 2019, it was completely removed from its location.

[5] One of the earliest modernist statues of the city, designed by the Mosulian sculptor Fawzi Ismaiel in 1975. It was removed from its place by ISIS as well as all other figurative statues of the city.

[6] Founded in Baghdad in 2017.

[7] According to conversation with Dr. Momtaz Hazim, Department of Architecture, Mosul University in 25 December, 2019.

[8] A metaphor of an older Jewish department-stores company "The Orosdi-Bach" which used to have branch in Baghdad until the 1970s.

[9] Ibid.

Fig. 5 The National Insurance Company building 1970s, with its urban context. *Source* Mosul eye (2019)

Fig. 6 Statue of "The Girl of the two springs city" opposite the building. *Source* Mosul Demolishes Iconic Building Used by IS for 'Gay' (2019)

Fig. 7 The Insurance building before and after the destruction. *Source* National Insurance Company (2019)

Fig. 8 The remaining three floors of The Insurance Building after demolition, demolished later. *Source* Website N(7)

4.3 Symbolic Values Associated with community's Collective Memory

The 2014 ISIS war played a crucial role in the history of the Insurance Building and in the size of the memories which the Mosul community constructed to it. ISIS used the building as theatre for their tactic of mistreating and killing people to strip them of their humanity as a punishment example to whoever thinks of violating ISIS ideology. Those events were conducted in public gatherings, and media constantly reported them, which affected people's emotional conscience. The building acquired several names such as: the "execution platform". The symbolic significance emerged that after being a place dedicated to securing lives of the people of Mosul, and this building became a place for ending youth's lives.

4.4 Questionnaire Procedure

To evaluate the effect of collective memory on the building's symbolic value and to investigate its relation to the decision concerning the reconstruction of this landmark, a questionnaire was distributed to a sample of two categories of people living in Mosul:

First: 20 questionnaires to a selected sample of specialists in architecture, art and heritage.

Second: 40 questionnaires to a random sample of ordinary people.

Both categories were asked about: their awareness of the values of this building. The symbolic values were investigated through asking people about their stored memories associated with the events that ISIS held in the building (qualitative and sensory questionnaire), as well as their opinions about how to regain the building. (Qualitative &

sensory questionnaire). (*See a concise of the form of questionnaire in the Appendix 1*).

4.5 Results and Discussion

- Results describing both categories in terms of age, education and residency are shown in (Appendix 2) it's worth mentioning that most respondents are residents in Mosul (90%).
- Results of values appreciated in the insurance building are illustrated in (Fig. 9).
- Results of the detailed questionnaire **for symbolic / Memorial values** are illustrated in (Fig. 10). It shows that 71% of public had negative memory about the building associated with ISIS violent actions in the building. Whilst only 29% did not store such memory. For the specialists who were also residents in the city 95% of them agreed to have negative memory about the building.

As for positive symbolic values attributed to the building which are associated with attitudes of (steadfastness and heroism), results showed that 65% of the public had good memories about the building, whilst 35% did not have any. The experts on the other hand, showed approval of having positive memories in a percentage of 85%, whilst only 15% did not support such value.

- The results of the questionnaire regarding **regaining mechanisms** are illustrated in (Fig. 11) it shows that:17.5% of the public supported the reconstruction of the building as it was, whilst 62.5% supported the integration of reconstruction and renovation, and 17.5% supported renovation of the building, with only 2.5% did not support any of the two mechanisms.

As for the specialists, only 5% was for reconstructing the building as it was originally with 85% supporting the integration of reconstruction and renovation (70% interior and exterior renovation with additional functions, and 5% interior renovation with additional functions, and 10% supported the addition of functions only), with 10% supported renovation of the building (5% interior and exterior renovation with additional functions, and 5% interior renovation with additional functions).

Hence, we note the following:

- The National Insurance building scored high in estimation of its values for specialists and the public alike. Architectural and contextual values are amongst the highest scores in the scale, which insures the significance the building has in the conscience of Mosul's community. As a result, most of them responded positively to a decision of regaining this lost heritage building. However, for public, symbolic values made the least score compared to other values. Such a result could not be interpreted as a kind of indifference to violent events but more as a lack of awareness of the correlations between the violent events and the building itself.
- Negative memories for specialists were higher than those scored for public, such high score by the experts might be interpreted in terms of their professional pre-interest in the building itself as an architectural heritage which makes them more attentive to events associated with it.
- The negative symbolic values the building has in people's memory and specialists' memory is superior to the positive memory. Such result could be seen as a justification for both groups to support integration of reconstruction with partial renovation in the building. Renovation could be seen as symbolic action against the abuse the building witnessed, as well as a sign for a better future life, a matter that may improve public collective memory about the building.

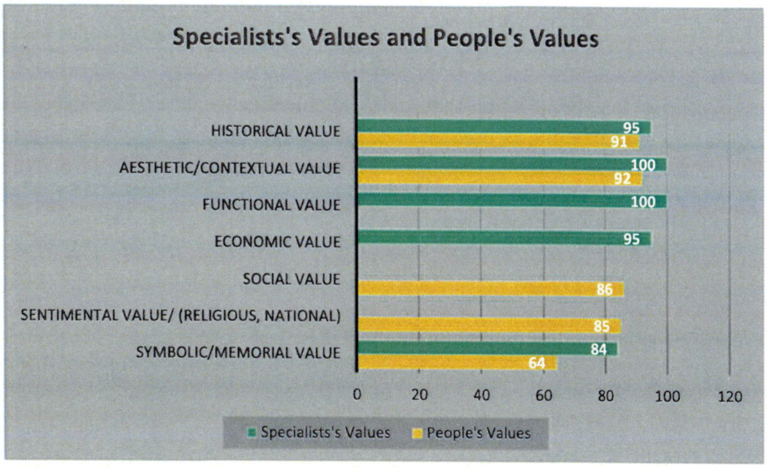

Fig. 9 Values of Insurance building attributed by specialists and people (by authors)

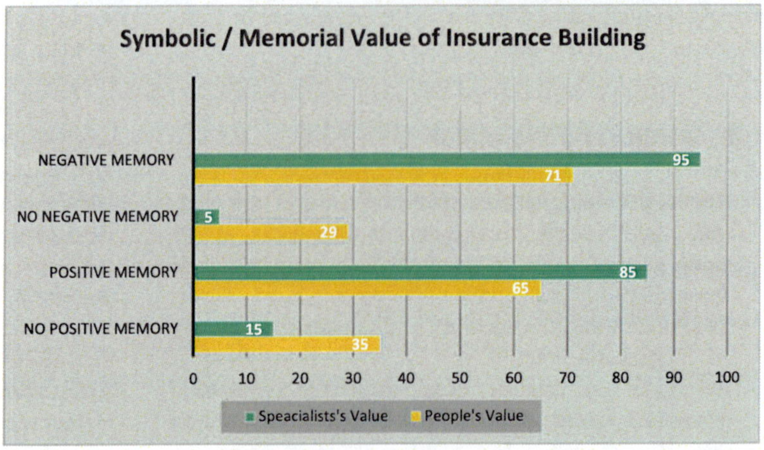

Fig. 10 Negative and positive events in the memory of Insurance building (by authors)

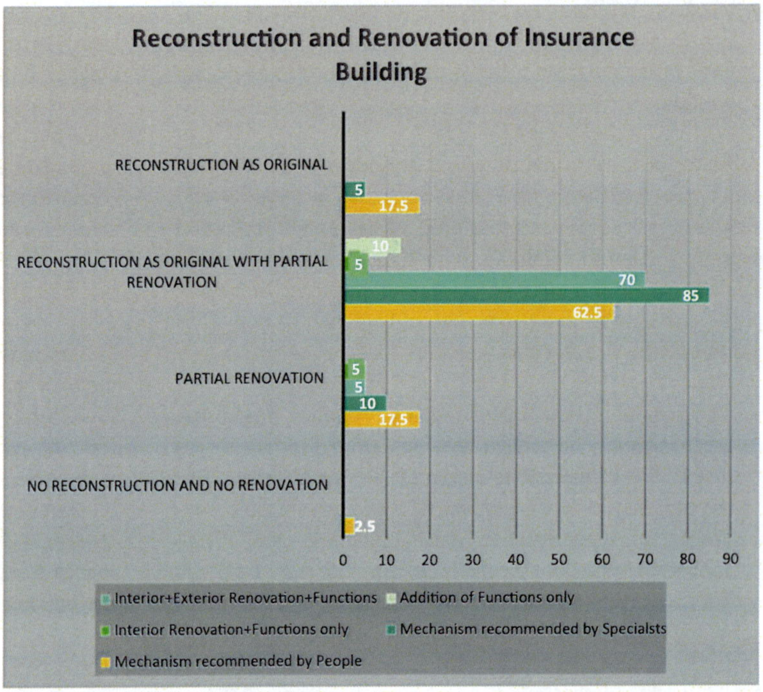

Fig. 11 Reconstruction and renovation mechanisms to regain the Insurance building (by Authors)

- It is worth noting that most specialists responded positively to the idea of renovation of the interior, whilst keeping the exterior as original. Such a choice could be related to the awareness to its high architectural and contextual values scored by them, conserving the original appearance is a priority to preserve the building's significance as identity of the city and its community.
- It is also noted that reconstruction as original was relatively more requested amongst public than specialists, and this could be interpreted in terms of specialists'

superior ability to visualize a wider scope of renovation possibilities.

5 Conclusions

Heritage buildings of high significance can be subject to abuse and maltreat during war times. Such events affect public collective memory and may change symbolic values attributed to the building into negative once. With this collective memory people can direct decisions to regain the lost heritage towards choices away from faithful reconstruction. Negative collective memories are effective, as much as positive memories, in choosing a regain for lost heritage building associated with them. This regain might be seen as response of self-esteem recuperation against psychological abuse. In this regard, the use of the renovation mechanism might appear preferable to the community when the symbolic/memorial value attributed to the building in question is strong, no matter how highly appreciated the original architectural and aesthetic characters it held, especially if the memory is negative and highly emotional, as was the case of the National Insurance Company building, presented in this study.

New design could symbolize new life, and thus expected to improve public collective memory about the building. However, when such renovation is chosen by responsible parties, the task implies a delicate balance between recuperating original values which once won the community's awareness and sympathy, and inserting new ones which could satisfy peoples' aspirations of a better future.

6 Recommendations

- Local residents should be allowed to participate in the decision-making process concerning their lost heritage. However, such participation needs prior involvement in workshops and informative activities to ensure their awareness of the values and qualities of this heritage. The survival of the building and the continuation of interaction with it are the most important elements and conditions for the existence of identity. The role of collective memory appears as a flexible and dynamic process that develops and adapts to its main role in preserving cohesion and enhancing national belonging.
- The regaining of the building must give priority to the social fabric that is a main condition for its existence, and dealing with the collective memory must be flexible and rational in order to conform to the emotional and spiritual needs of people.
- The important city landmarks must be preserved to the maximum possible, because they are important reference points to the memory, and they have a large symbolic value, despite their renovation or change. This means that regaining must be based on regulatory controls and planning studies.
- The high significance the insurance building by Chadirji scored by the public as much as by the specialists draws a recommendation by the authors in favour of its reconstruction, and to have it thereafter officially listed as a heritage building of national significance.
- Priority needs to be given to merging reconstruction and renovation in regaining the insurance building of Chadirji to what the building's image reminds of the violent event, and with a new image that inspires hope amongst the residents of the city of Mosul.

Appendix 1. A Concise of the Form of the Questionnaire

Note: for the purpose of curtailment, some details that were included in the form are not listed here. They are: an introduction to the subject, a definition to the concept of regaining lost architectural heritage and definitions of values of architectural heritage. Also, some values included several detailed indicators.

Questionnaire:

1.a General Information for Specialists only:

- What is your specialty?

Architect Urban Planner Archaeologist Art Historian

- How many years of experience?

Between 10-20 year Between 20-30 years Between 30-40 years 40 years and over

1.b General Information for Public only:

- What is your educational background?

Primary Secondary College Student Graduate Postgraduate Other

- What is your age group?

Between 20-30 year Between 30-40 years Between 40-50 years 50 years and over

- How much time did you spend in mosul?

Permanent Residency Resident for a Short Time City Visitor

2. Values of the Insurance Building by Chadirji:

*values are measured by one of four degrees:
Strong moderate weak none

1- Do you find a historical value in the building?
2- Do you find an Aesthetic, Architectural/Contextual value in the building?
3- Do you find a social value in the building?
4- Do you find a functional value in the building?
5- Do you find an economic value in the building?
6- Do you find a sentimental value in the building?
7- Do you find a symbolic/memorial value in the building?

- Do you find a positive memory (steadfastness attitude and heroism in war) in it?
- Do you find a negative memory (painful and violent attitude in war) in it?-

3. Mechanisms to Regaining the Insurance Building by Chadirji

*Each indicator of mechanisms indicators is determined in four degrees:
strongmoderate weak non-compatible

1- Do you prefer adopting the mechanism of reconstruction as original detailed image?
2- Do you prefer adopting the mechanism of partial renovation in a way that brings novelty in some parts of the building?

*The indicators below were only asked to specialists.

- Do you prefer partial renovation in the interior design?
- Do you prefer partial renovation at some of the external details?
- Do you prefer that the renovation goes for adding including functions only in the building?

Appendix 2. Results Describing Sample in Terms of Age, Education and Residency:

- Education: 50% postgraduate studies (20 persons), 42.5% graduates (17 persons), and 7.5% university students (3 persons).
- Age group: 37.5% (15 persons) between 20–30 years, 20% (8 persons) between 30–40 years, 17.5% (7 persons) between 40–50 years, and 25% % (10 people) 50 years and over.

- Residency in Mosul: the majority of the respondents were permanent residents of 90% (36 persons), and resident in the city for a short period by 7.5% (3 persons), and the city visitor by 2.5% (one person).

References

Bevan, R. (2007). *The destruction of memory: Architecture at war*. Reaktion books.

Buckler, J., & Hazzard, S. (2016). *Reconstruction of the Cathedral of Christ the Savior: A Return to Russian Roots*. http://hum54-15.omeka.fas.harvard.edu/exhibits/show/tsereteli/reconstruction-of-the-cathedra.

Christian, P. J. (2012). *History, memory & conflict: The collective memory of Maurice Halbwachs*. https://www.researchgate.net/publication/265379898_History_Memory_Conflict_the_collective_memory_of_Maurice_Halbwachs.

Drury, P., McPherson, A., & Heritage, E. (2008). *Conservation principles: Policies and guidance for the sustainable management of the historic environment*. https://pdfs.semanticscholar.org/6bc5/a138ffefa4bc39ba357a2abc380a693e5504.pdf?_ga=2.243911685.1045342837.1598528726-1159332813.1598528726.

Hazim, M. (2019). https://www.facebook.com/mumtazhazim .

Hoteit, A. (2015). War against architecture, identity and collective memory. *International Journal of Development Research, 5*(2), 3415–3420.

Howard, R. (2020). *Germany, Berlin. Ruins of the KaDeWe department store. 1943*. https://www.pinterest.ph/pin/697424692265045918/.

ICOMOS New Zealand Charter, for the Conservation of Places of Cultural Heritage Value. (2010). https://www.icomos.org/images/DOCUMENTS/Charters/ICOMOS_NZ_Charter_2010_FINAL_11_Oct_2010.pdf.

Mahgoub, Y. (2008). The impact of war on the meaning of architecture in Kuwait. *Archnet-IJAR. International Journal of Architectural Research*, 2(1), 233–246. https://archnet.org/publications/5109.

Mason, R. (2002). Assessing values in conservation planning: Methodological issues and choices. *Assessing the Values of Cultural Heritage* (1), 5–30. https://www.getty.edu/conservation/publications_resources/pdf_publications/pdf/assessing.pdf.

Mosul Demolishes Iconic Building used by IS for 'Gay' Killings. (2019). https://www.channelnewsasia.com/news/world/mosul-demolishes-iconic-building-used-by-is-for-gay-killings-11122258.

Mosul Eye (@MosulEye). (2019). *National Insurance Company, Mosul 1966–2017*. https://mobile.twitter.com/MosulEye/status/891616162748018688.

Paez, D., & Liu, J. H. (2015). Collective memory of conflicts. *A Chapter in "The Collective Remembering of Conflict and its Role in Fueling an Ethos of Conflict in Society"*. https://www.researchgate.net/publication/300777905_The_Collective_Remembering_of_Conflict_and_Its_Role_in_Fueling_an_Ethos_of_Conflict_in_Society.

Piquard, B., & Swenarton, M. (2011). Learning from architecture and conflict. *The Journal of Architecture, 16*(1), 1–13. https://www.tandfonline.com/doi/pdf/https://doi.org/10.1080/13602365.2011.557897.

Saydem, M. W. (2013). *Reviving heritage architectural values in contemporary local architecture, case study - Gaza City*. Master thesis, College of Engineering. Department of Architecture. Islamic University. https://iugspace.iugaza.edu.ps/handle/20.500.12358/19756?locale-attribute=en.

A short history of National Insurance Company Administrative Offices (Mosul). (2019). https://mobile.twitter.com/omardemosul/status/964921059391754240/photo/3.

The Burra Charter, the Australia ICOMOS Charter for places of cultural significance. (1999). http://icomosubih.ba/pdf/medjunarodni_dokumenti/1999%20Povelja%20iz%20Burre%20o%20mjestima%20od%20kulturnog%20znacenja.pdf.

Watkins, K. (2019). OMA to Renovate Berlin's Historic KaDeWe Department Store. https://www.archdaily.com/780602/oma-to-renovate-berlins-historic-kadewe-department-store.

Zerubavel, E. (1999). *Social Mindscapes*. Harvard University Press.

Zeunert, E. (n.d.) Kaufhaus Des Westens: A German State for over 100 Years. *Undergraduate Journal for Art and Culture, 1*. 1–4. https://carleton.ca/arthistory/wp-content/uploads/Zeunert_KaDeWe.pdf.

Inspiring Traditional African Art & Architecture in the Modern African Hotel Design Field

Alshaimaa Mortada Abozaid Ali

Abstract

This is a study of the impact of traditional African culture and art on the interior Design of African tourist establishments, and the environmental requirements and requirements of providing thermal comfort can be achieved through the return to nature as old African did by using materials from the surrounding nature. These materials have become an important part of the African architectural and art, especially in the sub-Saharan part.

Architecture, art and, interior design are the result of a number of influences in the nature of people, and there are many factors affecting architecture and interior design in Africa. the arts, culture, and traditions of the African peoples and highlight the differences stemming from the difference of climate and the surrounding nature and influences from different cultures and the impact on modern interior design, which is inspired by the African architecture a lot of elements of colors and furniture pieces, etc., of natural materials. Traditional African art has reflected on the personality of the interior design of modern African hotels, while the experience of the tourist destined for Africa, looking for an experience full of nature and culture, the interior design of the hotel on the African character feel the guest isn't separated from African culture without abandoning the degree of luxury that he needs.

Hotel interior design includes a wide range of coverage and Africa still uses a small portion of it. By developing a design that enhances the capabilities of hotels. It helps hotels attract more. The development of African hotel design is important for the African economy in conjunction with the global rise in tourism destined for Africa. As for the development of the tourism industry in Africa, it's important to be more concerned about the design of the hotel with an African character.

Keywords

Africa • Art • Tradition • sub-Sahara • Hotel • Tourism • Mud architecture • Vernacular • Yoruba • Ashanti • Furniture • Game reserve • Safari • Heritage • Pottery • Environment • Ecotourism • Lodges • Eco

1 Introduction

The continent of Africa, that continent whose people suffer from poverty and disease, is a continent rich in folk heritage and arts for its peoples, and despite the suffering of its people from poverty, but they feel beauty and express it within them through their clothes with delightful colors and seats that are not without a sense of beauty, and even the spoons that they use are not without wonderful forms with an esthetic touch, the African practiced all kinds of art since ancient times. And the oldest arts in Africa is rock art. There is nothing in African arts designed to satisfy tastes or provide abundant pleasure, for the purpose of art in this part of the world is to express the significance of human life. Artistic creativity within the African continent is not limited to a specific category of people or a special occasion but is practiced through practicing the daily life of the African (Osama, 2005).

Also, African art is one of the arts of human heritage rich in many colors, symbols, and various formative relationships, due to the special geographical and climatic nature, historical origins, and cultural heritage that contributed to the formation of this art. In addition to the religious beliefs, customs, traditions, and economic conditions that influenced him and had a role in the diversity of African arts (Osama, 2005).

Whereas Africa is a land of great topological diversity (Thelma, 1974), and therefore the African environment is a stage for the African movement and its activity, and a major source for the development of its conscience and feelings. We find a clear reflection of the characteristics of the African

A. M. A. Ali (✉)
Alexandria University, Alexandria, Egypt

A. Versaci et al. (eds.), *Conservation of Architectural Heritage*, Advances in Science, Technology & Innovation,
https://doi.org/10.1007/978-3-030-74482-3_30

environment on the African architecture, so every region of Africa tends to use materials from nature around it in its own architecture. The Egyptian countryside, for example. Where the vegetation is largely confined to the thin pastoral cover, the people are often nomadic, using tents of animal hides and woven hair for shelter. In cultivated areas and less forests, weeds are also used as building materials. Hardwoods are used in forested areas for construction, such as bamboo palm and raffia. Land and mud are also the main building resources. The African environment was also strongly and explicitly reflected on African art, so we see the African artwork showing man face to face with nature.

2 Study Area

Because interior design and architecture mainly originate from the place where a person lives and this is evident in the interior design of buildings in Africa—specifically the research deals with sub-Saharan Africa or what is known as black Africa -. It is geographically separated from the southern edge of the Sahara, which is consistent with the landscape and uses materials from the surrounding nature, some of which may threaten With the extinction due to the prevailing trend toward the use of certain materials and reinforcing steel due to the cheapness and prevalence of it.

Hence, the study of interior design on the African character is a study of black Africa and the arts, cultures, and influences it carries, as well as the study of traditional African architecture and its connection to the surrounding nature and its permanent and continuous connection with it.

3 Traditional African Arts

The African artist used the symbolic and engineering elements in many of his motifs. They are characterized in West Africa by their accurate colors and the esthetic values represented in their shapes and elements. And the African symbols not only depended on formative formulations, but also a way to express beliefs and social life, and the most important characteristic of it is simplicity, spontaneity, and honesty of expression. The patterns of decorative designs and symbols in African arts may not affect our minds, just as African sculpture affected Western artists.

And because of the many common historical and environmental influences, there are many similarities in the arts across the continent, but there is also a great diversity. Artistic styles differ from group to group. European art had influences on African artists, especially during colonial times, and those effects were not necessarily negative or disastrous. As African art was lively and powerful, it did not die. And still full of richness and diversity (William, 1973).

If we looked closely at the mask's inscriptions, we would have seen mastery of creative artistic styles on many levels. Perhaps the most striking of this creative skill is the ability to simplify the amazing ideas of vibrant and originating in the original forms and natural images. It appears to us in most of the mask's inscriptions that the artist has intentionally bypassed the outward appearance of the natural shapes to penetrate into their essence. African art is a way of living in the world. In all its forms, whether it is masks, sculptures, homes, fabrics, pottery, poetry, music or dance, it makes the invisible.. visible (Abiodun, 2014).

The meaning of the confrontation between life and death is revealed. Paul Klee was influenced by African art, who said, "The task of art is to make things invisible visible" and that the African artist is not considered an artist at all. It could be the farmer who plows, or the talented blacksmith (Thelma, 1974).

As the African forest is full of trees that produce excellent wood, the artist used wood and tree bark to make statues and masks that represent the mainstay of African plastic art. Besides wood and copper, it also used ripening clay, which researchers call "terracotta."

The African artist has achieved beauty in the things that he uses in ordinary life or daily life, such as swords and tools used in fighting and hunting, as well as seats see Fig. 1 and spoons see Fig. 2 and headrests shape see Fig. 3 that they took from Egypt, which are very similar The head restraints found in the tomb of Tut's[1] tomb.

African arts have combined a number of interrelated spiritual and material values, in addition to the diversity of its materials and its close connection to its natural and social environment and the use of these arts to provide different means of living, such as pots, plates, spoons, drinking methods and benches, beds, necklaces, decorating and beauty elements for humans and animals, a weapon for defense, hunting tools, etc. What a person needs in a rural environment in which life is based on agriculture, animal products, and hunting, as well as other doctrinal and legendary purposes related to the tribe or clan or to festivals, seasons, customs, and traditions. As a result of the high artistic and esthetic values that African arts carry, they spread and increased demand for them, as well as successive special exhibitions in most of the world.

3.1 The Impact of the Environment on African Art

African art is one of the arts of human heritage rich in many colors, materials, and elements that have a variety of formal

[1] https://en.wikipedia.org/wiki/Tutankhamun.

Fig. 1 Bench with wood carving from Congo

relationships, due to the geographical and climatic nature of the African continent. The African environment was still full of many materials that the African artist dealt with in the production of his art and the tools he needed to help him direct this art.

The heterogeneity between people and their contrasting geography led to the choice of many types of materials. Most of the tools are initially plant-based, mostly wood-based, and use fibrous bark, raffia, cotton, reeds. They are commonly used in rattan, bamboo, and resins. The animal materials that are used are horn, leather, wool, feathers, shells, ivory, and teeth. While the minerals used are bronze, copper, iron, aluminum, stone, clay, glass, and dyes, and less frequent due to the export needs of the gold and silver trade.

4 The Features of African Architecture and Its Effect on Interior Spaces

African architecture is a multi-rhythm architecture, just like their music. The rhythm of architecture varies with different cultures. The culture of Africa is a mixture of indigenous people with strong connections to neighboring and exotic cultures. And due to the multiculturalism within the continent, we find astonishingly unique architectural diversity while maintaining its unity and interdependence across the continent. Therefore, the architectural history of this continent is rich and varied.

It can be deduced that the most important architectural features of African countries. In Africa, homes are built with thick walls, in order to preserve the sun during the day and to resist cold air at night. Often, thatched roofs are used widely in African architecture. Architecture in Africa also depends on natural materials such as wood, clay, and rocks, so we find that in North Africa: it uses stones and rammed earth. While the tropical zone “Africa’s forests”: dry land and mortar are used, West Africa depends mainly on clay. Central Africa uses perishable wood and plant materials. While South Africa relies on stone, straw, and wood.

We find that African architecture has most references including ancient Egyptian, Nubian, Swahili, and Moroccan architecture, and many other buildings that are taught independently sometimes without studying the relationship between them and the native African architecture, but the original African architecture without any other cultural influences found in the African vernacular architecture African architecture Vernacular extends to the beginning of the human race. Some African colloquial architecture seems to have evolved among the ethno-linguistic groups—the Afro-Asylum groups, the Nilo-Saharan, Niger, the Congo, and Khoisan. And among the Afroasians. African giant architecture also uses a variety of materials: grass, leaves, limestone, sandstone, coral, twigs, bamboo, raffia palms, solid wood, mangroves, papyrus, semi-candy soils, highly clay soil or lateral soils Sewn goatskin, woven camel hair, clay. Clay can be mixed with shea butter, livestock blood, or livestock manure.

And by studying the African exterior architecture, we find a clear impact on the internal architecture, as Africa is a fairly large continent, extending from countries such as Morocco and Egypt in the north to Namibia, Mozambique, and South Africa in the south. Certainly, each country has its own unique type of interior design; however, interior design patterns from this region usually include more neutral neutral color tones which are colors derived from African architecture colors and their main materials such as clay, straw, and others mentioned above. And the decor of Africa is mainly inspired by nature and it is the same basic element that the African interior architecture depends on its composition, in addition to the colorful fabrics, wild animals, and various types of traditional wood carvings.

In the following points, a number of African architectural elements and their uses in designing the interior spaces of a number of African hotels will be covered.

Fig. 2 Spoonful from Somalia/wood

4.1 Shapes and Materials for Roofing in African Architecture

It is noticeable in Africa that most houses built on the ground have a roof, either of straw or mud. Reeds, herbs, bananas, and bamboo leaves were used as elements to form straw. The mainframe of these houses was a circle of long straight columns whose bases were firmly integrated into the ground. The columns were bent inward so that their tops could be joined together. Then, the frame was reinforced either by fastening through the collar makers or by weaving the split columns inside and outside the legs. One of the poles of the centers, or sometimes in large houses, is a ring of columns, and sometimes it strengthens the main tree branches (Susan, 1978).

The shape of the ceiling was usually strictly related to the shape of the walls. The ceilings above the circular walls are conical see Fig. 4. The ceilings above the rectangular walls are sloping or hierarchical. Bamileke Homes make a particularly interesting exception because they have conical roofs that rest on square walls.

Thatch roofs are a feature still enjoyed by African architectural design, and the resulting ceilings are a basic feature of the interior design with an African character, and are widely and clearly spread in the roofs of African hotels and resorts, Reflected as a distinctive interior ceiling for the interior spaces and the wooden beams in the ceilings gain a special and warm African style distinctive and associated with nature consistent with the general features of the African character in the interior design and find African roof shapes on a large scale in the architecture of African hotels and resorts where wooden beams appear from bamboo and straw in their colors where they represent an authentic and influential part of the African character in the interior design, and it is an impressive cover It is used to ensure warmth in the winter and cold in the summer and The straw roofs last from 10 to 14 years, but it does not

Fig. 3 Wooden headrest of the Daejeon people, Mali

Fig. 4 The conical roofs are made of straw for the African Nadebel architecture

provide good protection from the risks of infection from birds and great risks from fire.

For example, the use of thatched roofs in hotels, we find the design of hotel "Red Pepper House" in Luma-Kenya is designed by Urko Sanchez Architects and it is located on an area of 1500 square meters. The design of the hotel represents a form of organic architecture related to nature and combines traditional crafts with modern requirements. Where the process of building the hotel in accordance with local traditions and doing so using local building systems as shown in Fig. 5a and b.

The design did not overlook the interest in nature, and this was clearly represented in the design of the roof from the raw materials of the surrounding environment. The vertical projections form see Fig. 6. In the Swahili architecture of the Kenya coast (Vizy & Kerry, 2003), "the makuti roof" is used as a roof-top structure or as a separate structure as a temporary construction. Here it has been expanded to include the separation planning of rooms under one space that protects from the sun and rain, and in these outdoor places, one can share a close relationship with nature. The use of thatched roofs also contributes to controlling the temperature, as the weather in this area is very hot during the day and at night. Cross ventilation allows air to flow through space. This naturally cools the room. The winds that pass over the sea will bring a cool breeze to the hotel. The traditional "Makuti"[2]roof provides a sun visor and is also a good heat insulator (Angelika & Deidi, 2008).

4.2 The Floors in African Interior Architecture

The floors in most African homes traditionally have terracotta floors, and we may find the idea that the terracotta floors are very soft and unhealthy. But this is far from the truth as terracotta floors are largely prepared correctly. Where the clay floor is solid as cement, and a good solid floor is obtained by hitting the clay with a wooden beater during preparation. The clay is then mixed with charcoal, as is the case in the Zulu and other small houses throughout southern Africa. Sometimes it is mixed with cow dung and then ashed. In addition, many tourist resorts on the African character tend to use traditional clay lands to bring about

[2] https://en.wikipedia.org/wiki/Makuti_thatching.

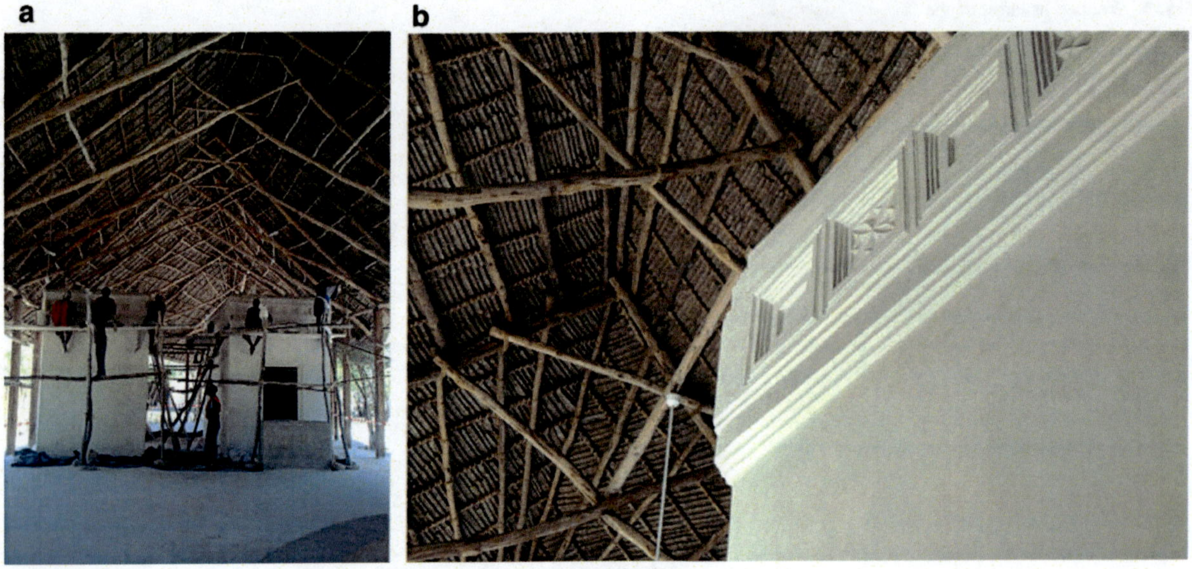

Fig. 5 **a** The stages of the work on installing the roof from straw according to traditional methods **b** The traditional African ceiling of "Red Pepper House"

Fig. 6 Red Pepper House Lodge vertical section

integration in the interior design on their African character see Fig. 7 (Susan, 1978).

We also find that traditional African wood floors, especially South African ones, are widespread and widely used and are considered to be the most modern now. The traditional South African floor is made of wood shavings (first layer) and sawdust (top layer)—compact and attached to either natural resin or Another material.

And most modern interior designs for spaces of an African nature tend to use wood floors due to their ease and availability from traditional clay floors, and those wooden floors must be unpolished or glossy and left in their closest natural condition to emphasize the African character. With the use of some pieces of carpet with traditional African inscriptions and rough texture, which reflects a close feeling of the feeling of walking on the African grass. The floor

Fig. 7 Use the clay floor in Leobo Private Reserve

becomes a strong and influential component of the African design of the interior.

We find the use of wood floors see Fig. 8 in a variety of ways while maintaining their connection to nature and their compatibility with the interior design of Nairobi's Giraffe Manor, located about 20 km from Nairobi—Kenya, and we find that the relationship between tall ceilings with exposed wooden beams and the floor forms the design unit and its uniqueness, as well as the use of leather and brown wood furniture Carved, as wood floorboards earn a lot of charm and richness to this hotel.

4.3 African Architectural Openings

The African windows are characterized by their small size see Fig. 9 so that they are merely small scales that allow the entry of sunlight in a small amount to maintain the moisture of the inner space and allow the entry of air and light.

windows and doors in Africa are simpler and more used for stucco, engraving and moldings see Fig. 10a and b. Despite the simplicity of doors and windows, the African people do not give up their printing of the African artistic character at times by adding drawings and engravings to the frame of the windows in line with the patterns that characterize the house as a whole or through works Wood carving, as indoors, also to drive away evil spirits and prevent thefts.

It is noticeable in the architectural and interior design of modern hotels, the trend to rely on relatively large openings, due to modern solutions for ventilation and temperature control through air conditioners and other modern methods, where large windows in hotels allow to link the surrounding nature to the interior design see Fig. 11a and b. And enjoy a clearer view of nature and African wildlife.

4.4 African Interior Wall Materials and Colors

The walls are the main component that constitutes the African interior space and are the dividing line and the connection between the interior and the nature surrounding the space.

The relationship between nature and organic life clearly shows us how to use building materials. Respecting nature and its natural materials requires that the architect, interior designer not attack the nature. For example, it does not

Fig. 8 The wooden floor in the Nairobi Giraffe Manor hotel

damage the grains or natural ripples in the walls and does not obscure their features Nor are her natural qualities, which are created by the Creator, in industrial dyes and paints that are man-made, but he must leave them to their nature to show their beauty, type, method of emptying and ripples.

The natural texture also gives many suggestions that may be different from the nature of the material itself, so we see that the African nature is full of picturesque touches that result in shade and light, as it abounds in many levels of infinite beauty and dazzling.

Often, the African interior walls, especially in the sleeping rooms, are filled with a mixture of mud and cow dung, as it was found that this is especially good for repelling chiggers and other pests. Some Nubians still excavate graphite in the walls, assuming it is a water-repellent device (Susan, 1978). We find that along the East African coast, the Swahili civilization sees Fig. 12 that flourished between the twelfth and nineteenth centuries produced many stone buildings but are of a fundamentally different nature. The building style that developed there was based on walls and columns built from coral debris, which were studded together and affixed to locally produced cement and lime wash (Ndugu, 2013; Susan, 1978).

Colors of the interior walls in African interior design, we find that the African design style stems clearly from Africa, where the bright colors are combined with earthy colors, and it is no secret to anyone that the primary colors of this pattern are usually warmer than what gives the space on the African character warm and distinctive to it.

There are practically no cool colors in this style at all. The use of these warm colors makes any area appear to be full of sunlight. The African style in the interior is characterized by the warm shades inherent in Africa's landscape. Most of them find different types of brown from light to dark, as we find colors derived from green grass, swamps, olives, sand, and red bricks. Similarly, in African colors, we find suitable contrasting combinations, for example, black, red, and gold. Such a range causes bonding to the desert, at night, and setting fires, which incite tired travelers to warm up in the dark.

The base color in African design is endless deserts, bustling forests, sunburned plants, and star nights. The shades of yellow (sand, saffron, honey, amber) see Fig. 13, green (marshes, grass), and brown paint (terracotta) are reflected. In addition to orange, red, orange, and black colors, it is important to note: the use of black color should be reduced. It is permissible to use a blue and a turquoise color, simulating the clear sky of Africa, in small quantities only.

4.5 African Interior Staircase

The stairs in African architecture is part of the building, connected to it, and most often it is from the same building material. Because it is part of African architecture in general, it takes its attributes in a tendency to simplicity and the use of straight geometric lines. It reflects on the interior architecture, where the staircase is one of the most important and simplest design elements at the same time. In many cases, natural wood is used in the design of stairs in the African style and recently in the interiors, African Textiles are used with their distinct patterns and types, especially in the form of carpets placed on stairs to give it the African character in its simplest form.

Fig. 9 Small window openings spaces in Mali, Bandiagara Village

It also adds to the features of staircase in the African interior spaces that are of relatively small width as shown in Fig. 14. This is due to the use of clay and primitive natural materials, whose simple methods of construction do not allow the work of stairs with a large width, as is the case in modern architectural methods.

An example of the stairs using the simple and distinctive African character "Leobo Lodge in Waterberg" is a new marvel designed by Silvio Rech and Lesley Carstens that redefines the esthetic of bush-filled bushes, which are distinctive and exceptional interiors. Where working with teams of local craftsmen and building with available materials,

Fig. 10 a Traditionally carved Dogon window shutters, Sanga, Dogon Country, Mali **b** African door wood carving

Fig. 11 a Large windows at "Gondwana Private Game Reserve" **b** Large windows at "Leobo Lodge in the Waterberg"

Fig. 12 The interior walls in the Swahili civilization of the East African coast

Fig. 13. The harmony of Singita Pamushana Lodge's interior walls with nature

often unconventional. The interior and exterior stairs Fig. 15 are part and parcel of the design and its local materials are mud color that tends to read in harmony with the colors of nature connected to it and unique of its kind.

We also find another example of staircase in African hotels in Shompole Lodge in Kenya, where each room was built using natural white quartz stone, light crusts, and other materials. The stairs in the hotel vary between wooden and

Fig. 14 Interior stairs of houses in Kargue, Mali

white stone, which enrich the interior design of the hotel Fig. 16.

5 Thermal Comfort in African Hotels

Thermal comfort was considered the primary concern of travelers during their stay in hotel rooms. It is of importance as it was for the design of hotels within the African continent due to the relatively high temperatures experienced in the African continent. As well as different moisture rates, these factors must be addressed internally to achieve the maximum comfort for inmates. Whether there are air conditioners or heaters of all kinds to achieve thermal comfort and use of building materials and structural and architectural design. The wind factor also affects the interior designs of hotels in terms of selecting services sites, as well as the locations of water to benefit from them, and the transmission of odors to the hotel's interior spaces has faded. As well as the exploitation of green spaces and high trees to repel winds loaded with dust, receive the desired winds, spread flower beds to benefit from them in softening the hotel's interior climate, and spreading the pleasant aroma of the tender plants. Due to the hot climate of the African continent, the amount of sun radiation that falls on it is high, so the interior design must take advantage of the appropriate amount of it

Fig. 15 A staircase of red clay for the ladder in "Leobo Lodge in Waterberg"

to provide natural lighting and block the harmful and excess part of it to provide the appropriate comfort for the guests.

The African has resorted since ancient times to its environment and relied on it to achieve the thermal comfort of his residence, which is recently resorted to sometimes as architectural solutions in the design of African hotels, and where we find that the climatic effects on local African architecture are large and can be very complicated. We find that most homes in North Africa have a patio with a fountain or pond. As the air is cooled by water spray and evaporation is withdrawn through the building via the natural ventilation created by the building model. Likewise, the word Southern

Fig. 16 Simple wood staircase design in Shompole Lodge in Kenya

Africa often contains very high thermal mass and small windows to keep the inner space cold, and to attract air through the interior spaces.

6 African Art Reflection on the Interior Design of Modern African Hotels

The interior spaces of hotels on the African style are experiencing their richness character, mixed with African arts, materials, and crafts. And in Africa in particular, where many old hotels are redesigning to meet the needs of new and developed travelers in previous years, and by balancing their esthetics and design while providing guests with the services they need.

With regard to the interior design of hotels of an African character, the local character, art and culture must be used in most cases using the elements of African heritage furnishing, as one of the most important features of traditional African furniture Fig. 17 is that the furniture itself is made of wood, and it is characterized by simplicity, roughness, and heaviness, and has engineering shapes. And the furniture unit is complemented by drawing or sculpture. And with the diversity of African cultures, its wood sculptures for furniture pieces varied and developed, thus winning each piece of personal furniture independently. We find the traditional African beds that have its unique personality with its wood carvings that distinguish it and if it is not used in its traditional form now it is used to add a sculptural element to any

Fig. 17 Bidjogo Royal Stool, Orango Grande Island, Bissagos Islands, Guinea Bissau

interior design on the African style, for example; it can be used as a coffee table or a seat anywhere Fig. 18. Various elements of African art are also used in interior design, so we find the use of units of pottery or statues and African masks and murals that contain African wood carvings Fig. 19 in addition to the use of textiles with traditional African patterns (Abdullah, 1987).

Fox (2002) and Worcester (1999) both emphasized the positive impact that the local traditions could have on hotel décor. According to "Prad Elias", head of the Elias Design

Fig. 18 African bed as a coffee table at Lelapa Lodge

Fig. 19 The traditional African wood carving at Lopesan Baobab Hotel

Group in New York, embracing the city's culture will increase the design success.

7 Example of African Hotel Design "Sandibe Okavango Safari Lodge"

It is located in the heart of the Okavango Delta, which is a UNESCO World Heritage site. It was designed by a South African architectural firm, "Nicholas Plewman Architects" and "Michaelis Boyd" studio in the United Kingdom. As it was built entirely from biodegradable materials, it has no physical impact on plants and animals. The animals continued to live in complete disregard for the emerging buildings see Fig. 20. The designers have chosen the "Pangolin—Armadillo Africa" and the endangered animal as an inspiring animal because of its shy, elusive, and completely harmless nature see Fig. 21, as the building appears as if it was an organic growth from its location on the banks of the river.

The 24-bed hotel is almost entirely made of wood the curved shape of the main building—which was created by arched beams of giant pine wood and the wood that makes up the building is water-resistant by acrylic and rice plate layer. The curved structure is translated into the hotel's interior design to appear like a rib cage made of wood see Fig. 22a and b. The team also drew inspiration from the carpentry of the "**Bayei**"[3]indigenous people in Botswana.

The hotel contains a main building and a sub-building. The first floor of the main building contains the main area which includes the reception and dining hall. The dining hall is a rectangular multi-level open structure and is covered with a large curved roof that blends well with the environment.

The main has a modern organic design on a large open space with another hall and comfortable seats in addition to the distinctive long wooden buffet bar. A group of stairs also take you up, as there are comfortable seats surrounding the bar with a fireplace for the winter evenings. This upper part extends and flows with an open surface and extends to the high outdoor viewing area. We find that all twelve Bedrooms in the boutique hotel "Sandibe Okavango"—are elevated from the ground to allow better views of the "Okavango" Delta with large opening windows see Fig. 23.

The furniture in the hotel is largely modern organic furniture while sticking to crafts and arts in the Okavango Delta in Botswana. This is evident through the reliance on wicker furniture in Rattan Furniture at the hotel. The large dining area is dominated by a distinctive large table carved from the root of a solid tree see Fig. 24. The furniture was supplied from South Africa. The modern hotel furniture is not without distinction from the use of traditional African furniture see Fig. 25.

Several types of wood have been incorporated into the structure and interior design. We find each of the South African pine wood for the walls, and the rice, in addition to

[3] https://yourbotswana.com/2017/02/26/batswana-tribes/.

Fig. 20 The animals ignore the buildings of the hotel, "Sandeepi Okavango"

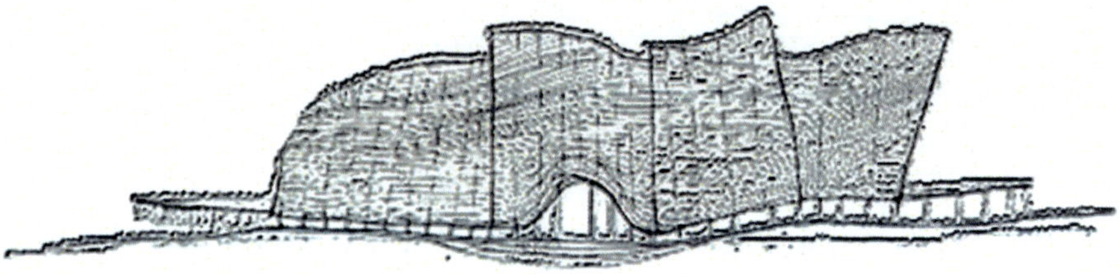

Fig. 21 The concept of "Sandibe Okavango Safari Lodge" Inspired by the animal "Pangolin—Armadillo Africa"

the Australian eucalyptus for the floors, where the arched and curved wooden structure of the building is translated to the roof of the interior spaces, which results in a mixture of colors, texture, and elegance and creates a unique atmosphere rich in positive pulses. The interior is completely built using local light natural materials in the area and is completely concrete free.

Due to the hotel location. And environmental standards that comply with it. All wastewater and soil are collected and pumped through an approved biological treatment plant that makes liquid waste safe and reliable for discharging into a highly sensitive environment. This is in addition to the completion of wastewater treatment and removal of harmful waste to the environment. And try to reduce them. By using biodegradable organic materials that do not have a detrimental effect on the environment, such as the distance from the use of plastic materials and bottles, and recourse to recyclable materials, even though there are difficulties that are the distance from the recycling facilities from the hotel.

The building's architectural design made it open on every side, allowing air to flow into all spaces through these wide openings. It also allows a clearer vision of the wildlife in the natural environment surrounding the hotel.

8 Conclusions

Despite the mysterious future of the African continent in the loud voice of conflicts and wars, and despite all the circumstances that its inhabitants suffer from, the African continent abounds with many tourism potentials. It is a continent rich in beaches that are unparalleled, African art and heritage, as it is rich in monuments and folklore arts. The tourism sector in Africa has witnessed an unprecedented growth in the numbers of foreign tourists arriving on the continent, as the number doubled from 24 million during the period 1995–1998 to 48 million during the period 2005–2008 and reached 56 million during the period 2011–2014,

a b

Fig. 22 **a** The hotel's interior design to appear like a rib cage made of wood **b** The hotel's interior design to appear like a rib cage made of wood

With the number expected by the year 2030 to reach more than 200 million tourists (New York University, The Africa Travel Association & World Bank 2011).

Also, it is not possible to overlook the importance of designing hotels on the African style in relation to the African tourism industry in general, so achieving comfort for tourists is an important component of the tourism process that in turn influences the achievement of African economic growth (The World Tourism Organization, 2015). The modern interior design of hotels. Because an important part of the success of the design of the hotel is related to the surrounding environment and culture, and that the design of the hotel is not separated from the local culture and heritage of the city on which it is located, but it must be an authentic part of its design that is the local arts and culture in order to complete the tourist experience. In order for this to happen, the interior design, which is inspired by African architecture and art, must receive greater attention by studying the symbols of African arts and their significance.

While paying attention to the development of hotels, the mutual influence of both tourism on the African environment and vice versa should not be neglected. Do not overlook hotel designers. Preserving the natural African environment,

Fig. 23 The large sliding wooden doors inside the sleeping pavilions

Fig. 24 A huge table carved from the root of a solid tree in the spacious dining area of the "Sandeepi Okavango"

Fig. 25 Using traditional African furniture as decorative furniture

which abounds in wildlife, including rare animals and plants, because the African nature is closer to the concept of modern Sustainable architecture, and it must be studied in this way and not just a group of masks on the wall. And, of course, an African person loves to see beauty in everything that falls on his eyes, and therefore behind a tremendous artistic heritage, diversified richness. It must be preserved and exploited in his favor (Morad et al., 2015).

References

Abdullah, M. A. (1987). *Tourist architecture*. Kuwait Foundation for the Advancement of Sciences.

Abiodun, R. (2014). *Yoruba art and language: Seeking the African in African art*. Cambridge University Press.

Angelika, T., & Deidi, V. S. (2008). *African interiors* (25th ed.). Taschen America Llc.

Clare, A. G. (1994). *Tourism planning: Basics, concepts, cases* (3rd ed.). Taylor & Francis.

Ilse, P. (2011). "Interior designers: Unacknowledged role players in South African retail design", University of Johannesburg, Faculty of Art, Design and Architecture & Visual Identities in the Art and Design. Janis, J., & Bernadette, B. (Eds.), *Sixth International Design Education Conference* (pp. 3–7). https://www.defsa.org.za/papers/interior-designers.

Morad, A., Ahmed, A., & Alaa, M. E. (2015). *Tracing history of the Green architecture and sustainability movements*, [Project paper submitted in partial fulfillment of the requirement for the Bachelor of Art in Design Degree, Ain Shams University]. https://www.researchgate.net/publication/312040906_Tracing_History_of_the_Green_Architecture_and_Sustainability_Movements.

Ndugu, N.(2013). *Contemporary Swahili inspired interior design for guest houses, case study: ST. Augustine Nyeri Guest House, Nairobi.* [Project paper submitted in partial fulfillment of the requirement for the Bachelor of Art in Design Degree submitted to the School of the Arts and Design, University of Nairobi]. https://cae.uonbi.ac.ke/sites/default/files/cae/cae/CONTEMPORARY%20SWAHILI%20INSPIRED%20INTERIOR%20DESIGN%20FOR%20GUEST%20HOUSES.pdf.

New York University, The Africa Travel Association & World Bank (2011). *The State of Tourism in Africa*. https://yaw-atnyarko.squarespace.com/assets/publications/01.29_TheStateofTourisminAfrica.pdf.

Osama, E. (2005). *African art, cairo, the Egyptian general book authority*.

Susan, D. (1978). *African traditional architecture*. Holmes & Meier Publishers INC.

The World Tourism Organization. (2015). *Tourism in Africa: A tool for development*, affiliate members regional report. https://webunwto.s3.eu-west-1.amazonaws.com/2019-09/AM-Regional-Report-Volume-four-Tourism-in-Africa-A-Tool-for-Development.pdf.

Thelma, R. N. (1974). *Contemporary African arts and crafts: On-site working with art forms and processes*. Allen and Unwin.

Vizy, E. K., & Kerry. H. C. (2003). *Connections between the summer east African and Indian rainfall regimes*. American Geophysical Union. https://www.researchgate.net/publication/251274808_Connections_between_the_summer_East_African_and_Indian_rainfall_regimes.

William, B. (1973). *African art in cultural perspective*. W. W. Norton & Company .

Advancing Technologies and Innovation to Conserve the Architectural Heritage

In the last years, the new challenges in the conservation and promotion of architectural heritage have certainly induced the research to develop new strategies based on advanced technologies and innovative approaches belonging to the digital world. The advantages reached through the application of technologies in cultural heritage conservation are manifold, both for technical purposes and for promoting a multiplicity of activities, e.g. related to tourism, teaching and more. By the technical side, advanced technologies have been mainly developed in the framework of modeling, to have increasingly reliable tools to be used in the design processes of new buildings and of restoration works for the reuse of existing structures.

In this context, in the Chapter "Architectural Innovative Technologies for Improving the Built Environment: A Technical Framework" the Authors present a comprehensive examination of the advantages provided by the use of innovative technologies in architecture, taking into account the energy buildings environmental impact, the thermal comfort requirements and the improvement of the built environment. Significant innovations can be found also in the field of architectural surveys, from which detailed digital models are obtained. In the Chapter "Importance of Digital Techniques of Documentation for the Conservation of Cultural Heritage" the Authors describe the fundamentals of photogrammetry, also through the new UAV devices, and laser scanning based techniques, calling attention to the importance of the documentation making in the process of conservation of cultural heritage. An interesting application of such technologies is proposed in the Chapter "Towards Developing a Heritage-Based Architectural Methodology to Design New Buildings in Historic Areas, The Case of Beith in Scotland", where the Authors illustrate how a large scale survey, carried out on urban areas through laser scanning and photogrammetric techniques, allow to obtain specific architectural information about the urban setting. The proposed methodology aims at capturing historic architectural features and properly using data acquisition, in order to achieve new building designs integrated with sensitive characteristics of their surrounding settings. The approach is applied to the conservation area of Beith, Scotland, UK. The development of data sciences and computer vision, strengthened during the last years, has determined the growth of new frontiers in the promotion of cultural heritage, through the use of interactive digital systems, which are becoming of common practice for all the people. A fascinating application is described in the Chapter "Strategic Architectural Interior Design Methodology for Recruitment the Hieroglyphs Line in Developing Ancient Areas by Hi-Tech Technology To Conserve the Egyptian Heritage", in which the Author proposes advanced technology solutions to translate the ancient Egyptian language, expressed by the hieroglyphs lines and its marks, into the other common languages through the interactive design of windows display. This innovative solution is proposed in the context of the construction of the "Grand Egyptian MuseumGrand Egyptian Museum", allowing the preservation of the heritage environment in the pyramid area, employing a new concept design of commercial units.

Similar strategies can be successful also for archaeological sites, providing virtual environments useful for the interpretation of ruins, with their reconstruction through virtual reality, for promoting the sites by means of digital documentation and many other aspects, which allow a proper safeguard of the archaeological sites. In the Chapter "Building Virtual Environment for Safeguarding Archeological Ruins Case Study—El-Sheikh Ibada (Antinoopolis)" the Authors present the more appropriate recent technologies to be used in this framework, offering specific guidelines that can help the interested reader to find suitable solutions. In particular, the methodology is applied to the case study El-Sheikh Ibada (Antinoopolis) archeological site. In the Chapter "Heritage and Sustainability: Motives behind the Use of Colours Derived from Natural Materials in Al-Qatt" the conservation of traditional art of interior wall decoration, Al-Qatt, native to the Asir region of Saudi Arabia, is treated. The Authors present in detail the peculiarity of this art, and in particular the natural materials from which the necessary colors must be derived. The work highlights the importance of sustainable heritage preservation concerning several factors, among them the community identity, aesthetic purposes, economic issues and environmental and health concerns.

Towards Developing a Heritage-Based Architectural Methodology to Design New Buildings in Historic Areas, the Case of Beith in Scotland

Maryam Al-Irhayim and Adrian Friend

Abstract

New building designs often do not serve sensitive sites like conservation areas (Heritage England and CABE, 2001) According to the publication, only new building designs that integrate with the sensitive characteristics of its surrounding settings are successful; this is a result of keen designers making effort to conduct site analysis and careful contextual character appraisal (2001). It is believed that "this lesson has universal applications" regardless of location (2001). This paper presents the current stage of an ongoing research that aims to develop a methodology for objectively capturing historic architectural characteristics, using data acquisition and evaluation based on quantitative data analysis. The research employs the case study of the conservation area in Beith, a town in the North Ayrshire council district within Scotland, UK. Preparing the needed information and employing the right tools are two of the main challenges of this research/practice field. Although desktop studies can gather appropriate information on historic characteristics, yet they proved limited. Similarly, big data sources, such as Lidar satellite information, digital mapping data sets, and city council planning documents, proved to be unsuitable, due to the data sets usually gathered not displaying a high level of accuracy to comprehend the historical features. Hence, this research conducted a desktop study to identify the quality of the data to be initially used. Later a Faro 3D laser scanner was used to gather higher-quality data. The photogrammetric model produced by the scanner proved able to capture accurate information relating to the scale, position, and proportion of the architectural characteristics of the area. These produced precise measurements help leading to further assimilations, such as correct identification of urban patterns, as well as the different architectural elements that compose them, and defining the combination of rule sets connecting them. This methodology will enable a more thorough comprehension of the overall historic architectural characteristics of the area, and it aims to aid the design of more appropriate new buildings that consolidate their historical settings towards a sustainable future development.

M. Al-Irhayim (✉) · A. Friend
University of Central Lancashire, Preston, England
e-mail: maryam_al-irhayim@hotmail.com

M. Al-Irhayim
Manchester School of Architecture, Manchester, England

Keywords

Architecture • Heritage • New buildings in historic areas • 2D scanning • Pattern • Façade • Design process

1 Introduction

With architecture as the physical manifestation of time in cities and towns, new buildings will need to develop a sustainable method to integrate themselves within the existing historic fabric of their surroundings. The subjective understanding of what is deemed "historical" and the regulations imposed to how to design in and around historical areas has created a lack of a formal methodology to approach gap sites. Historic organization bodies and planning authorities encourage "context-based designs" to be followed by architects. Through the implementation of formal prescriptive planning guidelines that control the height, proportion, and characteristics of new designs, this has created confusion on what is deemed "appropriate" design. Therefore, to better understand this site-specific typology, this research recognizes that better evaluation techniques are needed to assess the historic character within their surroundings. This research paper introduces a developing methodology of data capture, using 3D scanning to create a digital library of specific historic patterns, relying on generative design tools to evaluate and subsequently integrate

A. Versaci et al. (eds.), *Conservation of Architectural Heritage*, Advances in Science, Technology & Innovation,
https://doi.org/10.1007/978-3-030-74482-3_31

discovered data into knowledge-based design outputs. Therefore, this research intends to better understand the true parameters of site-specific design, as well as integrating data-centric design decisions, into the architectural design process to be used in designing new buildings in historic areas.

The first section will introduce the overall architectural discussion surrounding new buildings in historic areas. Outlining the issues that impact the architectural design and evaluating morphological planning regulations imposed within the UK. The second section evaluates existing research methods and developing technologies being utilized within architecture and other professions. Outlining, the proposed research methodology will be undertaken, leading to the third section, which introduces the case study's site, current planning implications, and morphological prescriptive on future designs, as well as, showcasing a series of preliminary experiments prior to the site visit. The fourth section will evaluate the observations and findings of the digital data gathered, from the site surveying tools that are used to capture the historic characteristics, including a limited desktop study and a 3D infrared laser scanner. Section 5 will evaluate the findings observed about the overall historical character of the area concluding with the contradictions found alongside planning guidelines. Lastly, an overall discussion of the current implications of the research on the wider context explores the next steps to continue developing the research.

2 Architectural Literature: New Buildings in Historic Settings

The UK designate significantly more buildings for protection than most countries Larkham (as cited in While, 2007). The protection of buildings is a topic that is discussed within Rem Koolhaas's 2010 exhibition at the Architectural Biennale in Venice titled "Cronocaos". The exhibition considered provocative in which it considers issues about preservation, such as claiming that 12 per cent of the earth's surface has been landmarked by groups like UNESCO, moreover bringing light to the recent historic designation of buildings that are just 30 years old and the equivocal nature this has on ancient buildings to stir debate (as cited in Ouroussoff, 2011). As Bourdin in 1980s speculated, there is a danger of over protection, suggesting that the historical is "transforming" the city "into a vast museum" (as cited in Lowe, 1996). However, the exhibition warns that preservation has transformed into a new global phenomenon in which historic areas are losing their sense of identity by preventing decay and polishing façades (Ouroussoff, 2011). Observing that in new buildings often "designs are watered-down period styles, further eroding the distinction between what's real and what's fake", which produces what Rem Koolhaas describes as a "low-grade, unintended timelessness" (Ouroussoff, 2011). Lynch suggests that a city should be "something like a collage of time" with different layers added (as cited in Loew, 1996). Some planners and researchers emphasize that "unfitting" architecture that is poorly inserted can lead to poor integration with its historic area. (Hu et al., 2017) This viewpoint is usually about how to infill vacant site considering the morphological features and shapes that the new façade forms to fit in the context shown in Fig. 1a and b about the scenario.

In Sebastian Loew's book "Modern Architecture in Historic Cities: Policy, Planning and Building in Contemporary France", he highlights the conflicting public viewpoints, observing that attitudes in France and UK are quite different (1996). The French attitude "encourages and permits the presence of contemporary designs in protected historic areas" (Loew, 1996). While the British context the insertion of new buildings in conservation areas is a subject of debate in the popular press as well as in professional circles, Leow observed this in writings by Larkham 1996 and in Tiesdell (as cited in Loew, 1996). The discussions of historic preservation involve an array of stakeholders ranging from members of the public, professionals who design and build, public bodies, organizations, and even monarchy have expressed concern in the discourse of new buildings in historic areas. This is seen within newspaper articles and titles that indicate this for example: "Developers appeal after council refuses plans to redevelop historic site in Sheffield" (Ashton, 2020), "Factory roof to be assessed in campaign to save historic building" (Waller, 2020), "Prince Charles's 10 principles for architecture—and 10 much better ones" (Murphy, 2014), "Villagers lose fight to save historic cottage from demolition" (Lazzeri, 2020), "3000 sign 'Save our historic Newton Abbot theatre' petition" (Bryant, 2020), "Genuine fears Liverpool 'may lose' Unesco World Heritage status" (Houghton, 2019), "Facing up to the future: Prince Charles on 21st century architecture" (HRH Prince Charles, 2014). With media being the physical representation of the different opinions in expressing Heritage news-worthy stories, public opinion plays a significant role as seen in campaigns to "save the nation's heritage" usually driven by "strong emotional attachment", which Mark Thatcher believes is directly linked to historic buildings' relationship with nationalism (2017), suggesting that historic buildings reinforce national identity (Thatcher, 2017). While within the built environment professional circles, Tang observes that there is a "dysfunctional communication" among the different professions, which he believes stems from "subjective" opinions where "understanding of the traditional style" is formed from "experiences" (Wang & Shi, 2018). While Soutoudeh, Wen and Abdullah speculate that "architectural design" reflects the architects "personal

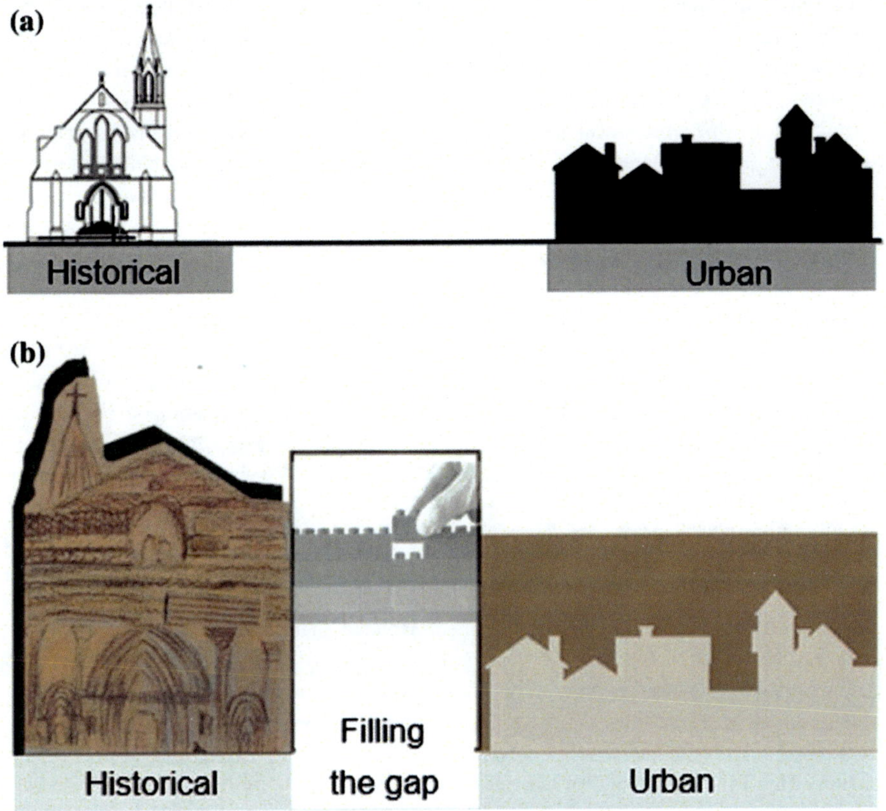

Fig. 1 **a** Vacant site and **b** aim trying to find the shape

preferences", which they argue is that the reality showcases that it is "governed by design guidelines laid out by relevant commissions, organizations and/or local councils" (Hu et al., 2017), Tugnutt and Robinson believe new buildings require increased regulation (as cited in Leow, 1996). Sebastian Leow observes that the discourse in new and old tries to suggest "objective elements that need to be regulated" elements of which include "height of buildings, roof line, street alignment and projections on the façade" (1996). Leow therefore suggests that when this occurs is when the literature on architectural design shifts and rejoins the concerns of that of the planning system (1996).

3 Planning in the UK

Planning is described as a technocratic approach Moroni (as cited in Iossifova & Moroni, 2018). The Ministry of housing communities and local government MHCLG in the UK suggests that planning aims to ensure that "the right development happens in the right place benefitting communities and the economy, through assessing if the new development is suitable" (2015), which suggests that the drivers are to create an equitable, fair society in which goods and resources are allocated in the most appropriate designation. Moroni alludes "the expression urban planning, refers to the processes and patterns of action and regulation through which public actors control the use of land and buildings" (as cited in Iossifova & Moroni, 2018). In the UK, the planning system is designed "to be applied by local government" (Department for Communities and Local Government, 2015). Therefore, local planning authorities decide whether a proposed development will be granted planning permission or not. From a legal perspective "it is an offence to carry out any work on a listed building 'that affects its character as a building of special architectural or historic interest' without listed building consent from the local planning authority" (While, 2007). Local planners are also responsible to make a local plan setting out planning policies in their area (Local Plans, 2020), which also includes those covering historic buildings and areas (Local Plans, 2020). These local plans must be consistent with planning law and national policy and guidance (Local Plans, 2020). This is very similar in Scotland however instead of non-departmental public body Historic England advising the government, it is managed by Historic Environment Scotland which is also a non-departmental public body. They usually provide a registry of details of special historic buildings and their classification. In the UK, historic sites often have a building classification hierarchy of "listing" "Grade I (buildings of exceptional importance Grade II* (particularly important buildings of more than special interest), or Grade II (buildings of special interest)" (While, 2007). Other special designations include conservation areas which were first introduced in 1967 where local authorities have been able to protect the historic character of streets and buildings by designating "Conservation Areas" (While, 2007). In Scotland "The Planning (Listed Buildings and Conservation Areas) (Scotland) Act 1997 states that conservation areas 'are areas of special architectural or historic interest, the character or appearance of which it is desirable to preserve or enhance' Conservation Area Management—Scottish Civic Trust, 2020). Local authorities have a statutory duty to identify and designate such areas" (Conservation Area Management—Scottish Civic Trust, 2020).

4 Traditional Process of Architectural Design in Historic Areas

Choay observes that buildings are no longer isolated that they are considered within their surrounding area (as cited in Leow, 1996) Suggesting that this "new phenomenon requires a new set of rules" (Leow, 1996). Currently, the building's relation to the context becomes a design implication. This is seen within "Britain where there is a continuing preoccupation with the subject of integration of new and old and a search for ways to produce and judge satisfactory examples" Larkham in 1996 (as cited in Loew, 1998). However, the judgement criteria and "gatekeepers" are still planners within the UK (Alabi & Bako, 2018). There are currently two clear schools of thought regarding the morphological façade interaction of new buildings, ones that promote contrast and ones that prefer replication (Hu et al., 2017). International organizations such as the United Nations Educational, Scientific, and Cultural Organization (UNESCO) and the International Council on Monuments and Sites (ICOMOS) "encourage new architectural interventions to be distinguishable from its settings so as to protect the integrity of the historical fabric and yet still aesthetically fit within" its context (Hu et al., 2017).

Karakul suggests that when architects design new buildings, it is expected that they integrate with their surrounding form, so that a "continuity of traditions" is seen (2007). Karakul believes this is to be achieved by expecting architects to come up with a "creative process" to do so (2007). Evaluating this further Karakul describes that such a process should stem from the architect conducting a comprehensive study of the nature of the historical buildings' design, as well as the new building's relationship with its surrounding (2007). This belief is underpinned by the view that only through understanding the architectural language that represents the traditions in the area, can Architect's design successfully (Karakul, 2007). This methodology of designing in historical areas stems from the traditional view that architectural practice or design is conducted through research, Patrick Geddes coined the term as "design as research" to be the method (as cited in Iossifova, 2007). This involves architects to have an in-understanding of the "historical, physical, social, and cultural context of a project" Gill's as cited in (Iossifova, 2007). Gill's also elaborates that this would involve surveying then designing (Iossifova, 2007). However, this traditional method is a very time-consuming process, and Gill's outlines that this is due to a "lack of time for survey and analysis" not to mention a "lack of analytical techniques" to interpret the survey information to be used (Iossifova & Moroni, 2018). However, Iossifova argues that a "search for simplicity … [viewing a project as] an artistic challenge of visual design, when it is really a problem of organized complexity obeying its own rules of evolutionary intelligence" Mehaffy and Salingaros (Doll and Gasparatos, 2018).

5 Other Methodologies—Related to the Research

A study aimed at creating a quantitative method to evaluate infill buildings in historic areas. Creating an evaluation tool uses standard deviation to judge architectural features such as "size, proportion and colour" of façades, subdivided into smaller measures within the three themes (Hu et al., 2017). Another study aims at creating a quantitative method to help standardize the process of restoration for damaged façades and infill sites in historic areas (Tang et al., 2018). This was undertaken by evaluating the historic area's traditional architectural components focusing on the historical composition of the façades to input into new designs. The study can be summarized into three workflows, the first was the data acquisition and processing technique to gather the façade's information (Tang et al., 2018). This included a questionnaire to determine what façades are considered to have historic precedent, then gathering the façade size in Gunanjie street in China using photo slicing and a DiJi drone (Tang et al., 2018). The second phase was to label and code the façade, through abstracting the façade's architectural composition. This relied on the use of monolithic structures inputted on a plane-coordinate system. This representation would then act as visual façade description of the components and the different combinations. Further, inputted into a coding matrix that can be learned by computer, using rough set theory and knowledge discovery tools (Tang et al., 2018). The research predominately used visual coding software such as Processing and Architectural software, Rhino and AutoCAD (Tang et al., 2018). Finally, the result was to generate new façade rules and types that showcase the different combinations that can be appropriate to infill gap in sites and damaged façades. These findings were then used to write a software in Visual Basic 2015, that can be used by ordinary residents (Tang et al., 2018). The software displays the historic composition and has a generative design tool which was integrated using Java. The research concludes that this can be a planning tool that can aid residential decision and help establish better conservation plans and guidelines to be used in historic areas (Tang et al., 2018).

Another research argues that the façade composition and interiors are interrelated. Alves claims that the historical character's significance is not solitary but due to these "different interactions" of the façade the levels the composition led to the historical significance to incur (2017). Another study is the historical infill problem in the historic Gaziantep Turkey. Using three-dimensional abstract forms to compose new façades in gap sites. The method relies on generative design principles, which is underpinned as a "subset of shape grammars" known as "Fractals" (Gürbüz et al., 2010). The research mentioned prior that took place in Gunjiee street was developed further, in the second paper utilizing the change in the project. Moving from a 2D façade solution to a 3D perspective using CGA Computer Generated Architecture grammar (Tang, 2019) This was due to the limitation of 2D to tackle more complex historic characters, such as the "gable roofs which depend on the depth rather than façade width" (Tang et al., 2019). Computer vision research tries to solve the incomplete façade problem, in which an image contains a façade of a building that is partially seen, but the rest covered by noise trees or distortion. One study attempted to solve this by coding the façade through grammars then learning completions through a database, evaluating the patterns through statistical modelling (Fan et al., 2014). Within computer vision, the use of "neural style transfer" can be described for example as using the style of an artist and then transferring it onto a photograph, thus making it appear that the photograph was painted in the style of a famous artist's painting (Gatys et al., 2017). This method appears to be a form of "style transfer", and it relies on decomposing the style into a perceptual factor of space, colour and scale (Gatys et al., 2017). Another method generates novel 3D objects using a 3D Generative Adversarial Networks or 3D-GAN, and the method relies on reconstructing a 3D objects from an image, then using a discriminator which is able to learn without supervision to generate novel 3D objects (Jiajun et al., 2016). For example, a chair 1 with arms and chair 2 without arms create a new chair that has arms, but with the chair backing of chair 2. This is just a basic example to showcase the shape arithmetic generation of new objects (Jiajun et al., 2016).

Semi-automated design tools intended to operationalize forms of artificial intelligence (AI) are being applied to design to generate architectural building layouts, façades, envelopes and massing (Sönmez, 2018). Sönmez further explains that "Machine Vision, Semantic Modelling, Machine Learning and Classification" are used (2018). Through techniques based on "traditional, recent Shape Grammar and Procedural Modelling studies, Case-Based Design, Similarity-Based Evaluation and Design" (Sönmez, 2018), some are integrating raw data directly from point clouds obtained through LIDAR scans. Sönmez explains one method which uses images to obtain third point clouds through Structure from Motion (SfM), classifying each point into semantic structures (window, wall, balcony, door, roof, sky, shop). Afterwards, using a trained classifier to separate the 3D façade features (Sönmez, 2018), Sönmez examines the use of machine learning approach to learn context-free grammar rules of building compositions, one method further this by applying statistical relational learning to be used to better identify the layout of the façade structure, by applying a probabilistic inference for decisions on the geometry of the façade using Markov logic networks (MLN) for the weights of the components (2018). This was then relayed to another classifier that uses a supervised method to train the classifier using large point cloud data sets to cover a variety of buildings as to account for their respective styles (2018). This is to ensure the classifier can help predict the façade structure for a variety of building types which will aid in reconstructing a range of buildings (Sönmez, 2018). The Gunanjie street research in their conclusion suggested that "Apart from the digital design tools by programming. Some new survey and analysis technologies, such as 3D scanning, image recognition, and even deep learning, were expected to be introduced into the frame of automatic workflow in the future" (Tang et al., 2013). Most papers divide up building components, abstracting the façade features then coding or labelling them, then use this to analyse and evaluate the façade.

6 Proposed Reserch methodology

Research consisted of a desktop study that followed collection of data from the Beith site tested in:

(1) Data collection using site surveying techniques to create a digital library (current stage);
(2) Data evaluation through extraction, pattern recognition, identification and labelling of characteristics by coding them, implementing knowledge discovery tools using automated or semi-automated methodologies;
(3) Design application and site resolution through iterative testing. Evaluation techniques using generative and evolutionary design principles.

Table 1 Abstract data collected into architectural characteristics

Abstract data collection indicators	Knowledge and application use
Colour	Material selection
Abstract façade shapes composition	Patterns of the different historic façade characteristics
Height	Height and floor amounts

The first steps require data collection, criteria for which is defined in Table 1. The data collection was conducted through a site survey using a desktop study and a 3D laser scanner. Some initial manual measurements will be conducted, and some knowledge is acquired about the parameters of the site to evaluate the current planning regulations.

The labelling and identification of the patterns will be trialed further in supervised and unsupervised methods to compare the time taken for the different methods to come up with a design resolution. Further work needed to undertake abstraction of the new design architectural components into shapes through their location on the façade, and this representation will help label the design and inform the options, which will allow for further evaluation of the different combinations and patterns that are present in the data set/façades. Figure 2 showcases the methodology and the different parts mentioned.

This methodology is still in its investigative early development but aims to help produce a kit of parts or system that would aid designers in preliminary means of data acquisition, which can further inform the design in historical areas. The two-way system as in following this method and attaining a design can be reversed, and traced back, thereby the method and data site acting as evidence to the planning authority. Thereby the design process and decisions justify the new building, which would confirm to have considered the site's historical character. Therefore, the design results can then be traced to the data set of the site acting as evidence for the design decisions attaining planning permission.

7 Case Study Site

The site is in a town of Beith, in the North Ayrshire council in Scotland. The area is within the town center which has fallen into decline. There used to be a building on the site but currently it is demolished. The site is located on a vacant gap site that sits on the ancillary at the start of the designated conservation area. As seen in Fig. 3a and b where the corner plot marks the site and the larger line over the street is the conservation area. As the site's surrounding is the main area of investigation, for this study only some of the conservation area in front of the site which consists of 22 buildings were used. The site also is adjacent to a listed church see Fig. 3a. The topography of the site is a slightly sloped hill with the site being at a higher point than the conservation area. The conservation area's characteristics can be described as a "small-scale rural Scottish character" (TCDG, 2016). There are 11 listed buildings, 7 of which have a grade C (s) listing and 4 with a grade B classification. The adjacent church has a grade C listing. The surrounding urban area also has no buildings on three sides of the crossing but have a new community garden in the conservation area gap site.

7.1 Planning Beith, North Ayershire, Scotland

The North Ayrshire council would like their towns to be representative of their architectural heritage having a "small-scale rural Scottish character" (2016). They outline

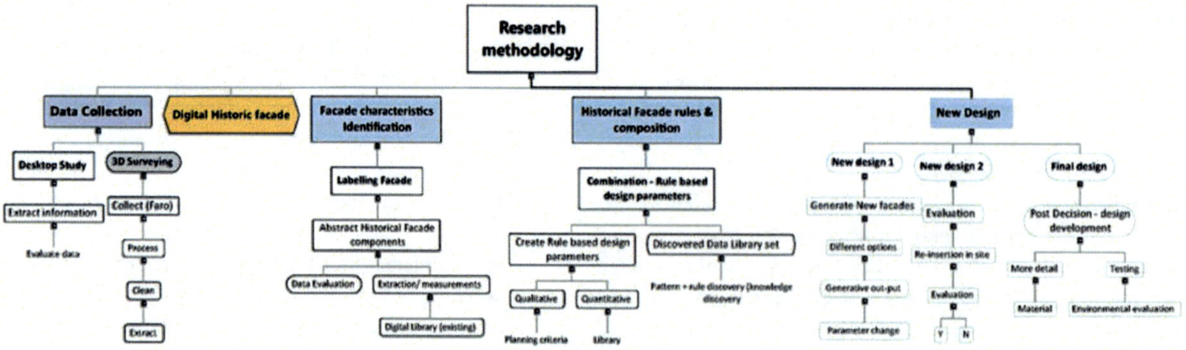

Fig. 2 Research methodology broken down in steps

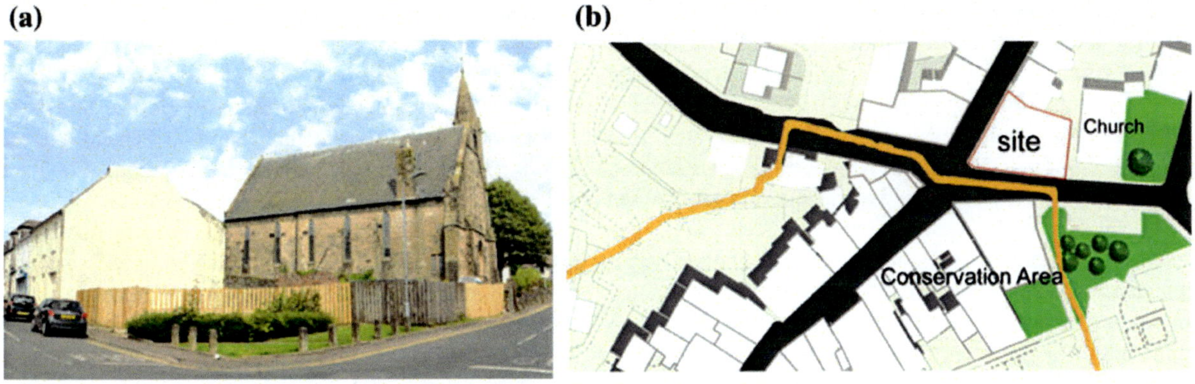

Fig. 3 a Empty site surrounded by fence and adjacent church and b site and urban surrounding

the criteria of "good design" within the planning guideline documents such as the 2010s the neighbourhood design guidance NDG and the 2016 the Town Centre design guidance TCDG, which both apply on the current gap site. Important to note that the planning guidance forms part of the judgment criteria imposed on the design, as they also contain recommendations that propose how to best achieve a "North Ayrshire-specific" design (TCDG, 2016). The documents when describing details for new designs range from prescriptiveness. Sometimes, using schematic diagrams to explain what is considered not applicable design. Some statements are up to some interpretations with suggesting to "complete" streets sympathetically. (TCDG, 2016). While some recommendations are more akin to statements for instance "Buildings in North Ayrshire-are small scale with a maximum of two or three stories" (NDG, 2010). This type of statement has implications on the way architects design as it enforces the height and scope of the building ahead of time. The façade's pattern and composition are also predetermined with new buildings requiring using local elevational "rhythms", proportions and volumes to base their new elevational designs (TCDG, 2016). See Fig. 4a for examples of Beith façades in the conservation area.

Fig. 4 a Façade photographs of the conservation area and b photographs of historic window style

The smaller-scale characteristics such as historic windows in the area are expected to be designed in a specific way. As stated in the guidelines "windows should retain their vernacular vertical proportions" (NDG, 2010). Even new buildings, whose elevations face the conservation area as in fronting a road or path that is adjacent to it, are advised to have windows that replicate the design and opening of the "original windows" (North Ayrshire Council, n.d.). The traditional window style can be seen in Fig. 4b often with a wooden frame and an inward opening (NDG, 2010). Some of the statements directly assume that new buildings are undesirable, suggesting that "New buildings will have massing and proportions like their attractive older counterparts" (TCDG, 2016). Therefore, it brings forth the design implication for architects to replicate historical buildings to ensure new buildings are deemed attractive, which does not leave much room for the architect to make any design decisions. Even the process of design has implication on attaining "good design" being achieved through "analysis of the place where they are located" (TCDG, 2016).

7.2 Initial Experimentation-Testing

To ensure a design output can be generated from the data, preliminary experiments were conducted to evaluate the best generative design strategy to be adopted. The idea of finding new novel in-between forms is present in the research of 3D GAN, and 2D painting style transfer shows a creative promise of further study. However, prior to undergoing those methods, an initial exploration of a simpler tool like tweening/shape imposition was undertaken which is a method used in computer animation to find the in-between shape of two images when animating a movement (Apress, 2006). This was tested through Flash (Phillips, 2012), in order to find the in-between form of two historic building façade images. The result although not to scale showcased potential in abstract shape or potential form see Figs. 5 and 6. This can be further developed in the last step of the methodology which aims to generate contextualized new design codes.

Pattern books in architecture have been used as a guide to attain or replicate the styles for example this can be seen taking place in the renaissance period with Vitruvius's book and Palladio (Dore & Murphy, 2013). Therefore, the development of this research's method stems from the existing ideology of pattern books. To undertake this in an existing contextual relation by trying to break down the existing patterns within the historic façade of the current site. The idea however must go further into an application method that can translate into design. This is to test the idea of an existing façade with historical precedence's pattern continued into the gap site. This ideology of the historical pattern of the surrounding site's impacting the design of the gap building is referenced in the planning guideline. Therefore, continuing this pattern to the empty site is seen in Fig. 7a and b, where transposing components of the historical façade of the church, i.e., the window and roof are

Fig. 5 Experimentation of tweening showing the shape change

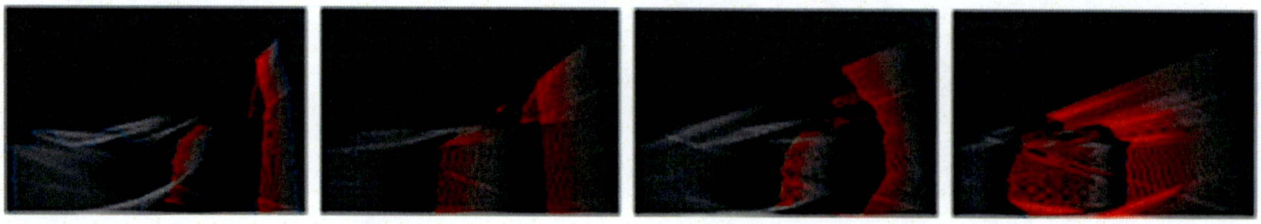

Fig. 6 Experimentation, showing the in-between forms

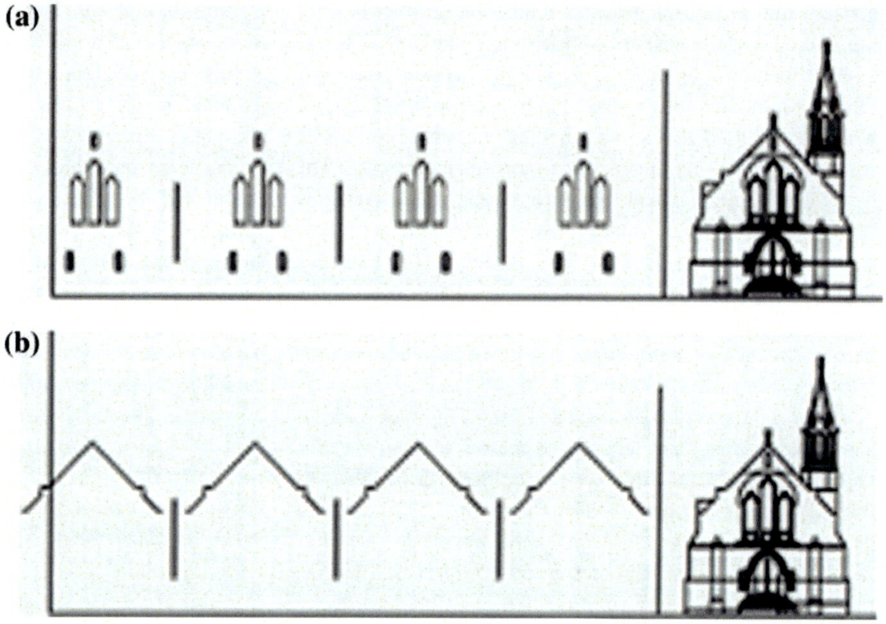

Fig. 7 **a** Window pattern applied to the site and **b** roof pattern applied to the site

replicated on the empty site to create a rhythm. The intent to compose a new digital pattern book or library of the architectural pattern is in this specific context. This can allow further guidance of the historical area's façade description, which can aid in deciding the new façade's composition. This therefore can aid in understanding the relational visual characteristics between new and old. This can be an interaction of one single composition in elevation view of the new site and the existing historic surrounding.

However, this initial experiment was not done with accurate sizes of the historical area, but proved to be successful indicating potential façade compositions for the new buildings. This experiment has demonstrated design application showcasing potential qualitative and quantitative outputs directly on the site. Figure 7a and b therefore can be developed further with a more comprehensive accurate measurements of the historic façades. The historical characteristics can be captured digitally and act as a database, which can further act as a knowledge transfer tool to facilitate in the processing of big data to identify urban patterns, (Tang et al., 2018). This can consist of different architectural elements that follow a combination of rule sets that can be identified even using semi-automated methods like the other studies discussed above. The classification strategy will further the design process accelerating it to complete an appropriate evaluation for the future developments of historic sites.

8 Site Survey-Desktop Study

A desktop study is usually used in architecture as a site analysis tool, and that allows the designer to retrieve as much information as they can without travelling to the site. The aim of the desktop study was to try and gather adequate data that can be used to evaluate the historic characteristics of the area, as well as enough information about the site to produce a 3D digital model to test and evaluate new designs. 3D models of the site are usually always a prerequisite in architectural drawings, as to attain planning permission requires drawings with their context indicating any changes made, especially for buildings in historic areas. Many planning authorities might impose the requirement of submitting a historic study of the area. This is used to showcase understanding of the historical characteristics and significance in the current location. Thus, proving that the architect has understood and evaluated the historic heritage in the context, using Digimap an ordance survey website. The downloaded data site contained informaiton on the plot size. The file type was in pdf and dwg a cad format, which enabled measurements of the width of a façade based on its size, as well as further findings of the area in order to understand the size of the forms. The results showed the range of the plot between 2.56 and 19.48 m wide. The average size of the width of the plot is 9.32 m in the conservation area. This gives an indication on the width of the future façade to consider. The average plot size area was 134 m^2 including the site and 127 m^2 without the site. Google Maps Street View was used to confirm the plot label, which showcased that some labelling was inaccurate with some façades taking the form of more than one plot. The issue with Google Street View to create an accurate 3D model is that size and measurements are not accurate and cannot be measured directly. This is due to the camera's fisheye distortion when capturing the data/images. The information is also not up to date, with one building that was demolished last year still appearing on Google Maps Street View.

LiDar data was also downloaded from Digimap. Lidar is a survey tool that uses laser pulse returns to calculate distances, and the data captured can be used to make a 3D model (Shan & Sampath, 2008). The output of the data is a dense point cloud with each point having a x- ,y- , z-coordinate that can be georefrenced. The research tried to read the data using open source software "Cloud compare" and "MeshLab" to do this. LiDAR data is collected through airplane or satelling data using infrared and green laser to collect plane data (2008). Shan explains that the conversion of lidar data to this surface model is called a digital elevation model (DEM). The surface can appear as a form of polygon mesh. However, the resutlant model showcases what can be described or classified as "noisy and ambiguous data", this "noise" is known to be caused by "atmoshpheric conditions or multiple reflections" (Shan, 2008). Trees, cars also belong as noise classificattions as they do not belong to the data points that are saught in this study being buildings (Shan & Sampath, 2008). As seen in Fig. 8a and b, Fig. 8a attempts to simplify the form and transformation inta a form of NURBS. However, the data resolution was becoming worse and as a result was more difficult to measure especially with the roof. The problem of ambiguity with roof edges or roof ridges is a common problem in Lidar as it is due to two interesecting roof planes becoming unclear is because the data belongs to more than one cluster (Shan & Sampath, 2008). This limitation of Lidar data resolution, with features like vector or roof components becoming not reliably estimated, therefore showcases the need for further cleaning (2008). This further cleaning can be conducted using trained algorithms that can help fix the resolution or remove the fuzzy noise like trees (Shan, 2008). However, for the scope of this research, this was not utilized as the main historical features on the façade were not visible. Another issue is although a surface was added on the points the data required a significant amount of cleaning and the data captures of this model can be classfied as "noisy and ambiguous data", which suggests that data resolution has to be something that has to be considered further, suggesting that architects may need to consider the quality of data more carefully when undertaking desktop site surveys. With regard to this study it has demonstrated that there may be a lack of useable data, available online to preform adequate desktop studies for historic architectural characteristics.

8.1 Data collection: Site Survey, Digital Beith, (Digital Data Acquisition)

The aim of the site survey is to digitally capture the historic characteristics of the site. This was further undertaken through a 3D laser scanner to undertake this. 3D laser scanners can capture a high degree of accuracy "Ranging error: > 2 mm at 10 m and 25 m, each at 90% and 10% reflectivity" (Faro, 2012), which increases the possibility to capture more intricate historical characteristics of façades. Additionally, the ability to take multiple scans and stitch them together to create a single 3D model gives way to capture larger urban areas, such as this one in Beith Scotland.

The model FARO FOCUS 3D-S 120 (landscape type) and a tripod was used. The FARO scanner works "by sending an infrared laser beam into the centre of its rotating mirror, The mirror deflects the laser beam on a vertical rotation around the environment being scanned, scattered light from surrounding objects is then reflected back into the scanner"(Faro, 2012). "The distance from the scanner to the object is accurately determined by measuring the phase shifts in the waves of the infrared light" (Faro, 2012). "The scanner covers a 360° × 300° field of view" (Faro, 2012), which was useful in tighter spaces. This site survey was carried out on 25/07/2019 between 12 and 4 pm BST in daylight conditions. The scanner was moved to 10 locations

Fig. 8 a Church Lidar data model then extrapolated and b model composition became more distorted

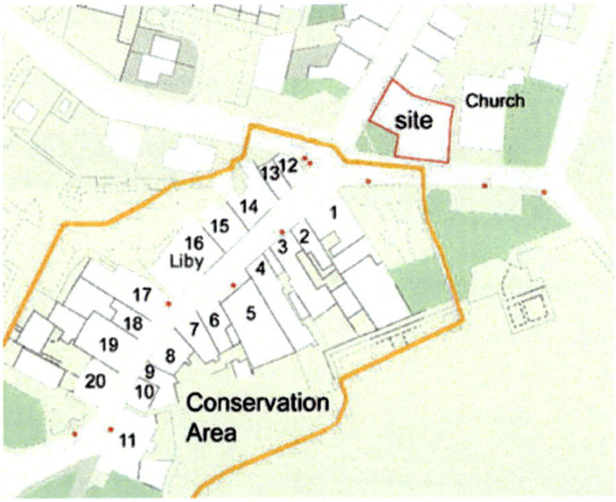

Fig. 9 Map showcasing the labels of the site and the location of the scanner

around and within the conservation area, as seen in Fig. 9 showing the location of the scanner.

The scan produces "a Point Cloud, three-dimensional data set" (Faro, 2012). The point cloud data was then processed using the device's software called "SCENE". The software is designed for laser scanners and is used for processing and managing scanned data. It can perform scan registration, which allows for "stitching" or combining the different scans using target-based designation of similar features and points that overlap in the scan. This was a largely manual selection in the target setting in the first instance. Then the calibration is automated through the software. The software does this calibration by "transforming one scan into another then tying all the scans together into a single co-ordinate system of a chosen reference scan" (Menegatti et al., 2016). As the site is quite large re-stitching the site relied on relaying information from one scan to another, see Fig. 10a and b for the point cloud model site. The scan also shows the elevations of the site with Fig. 11 showing the location of the site and church behind it, in relation with the historic conservation area buildings, the plots 1–11 in Fig. 9 showing the map. Figure 12 showing the elevation of the conservation area from plot number 12–20 as seen in the map Fig. 9. After processing the scan data, the software has inbuilt tools, that allow the scan to be directly be measured. The measuring process allows further control by assigning the coordinate from one point to another to ensure the measurement is on the same level. The façade size width and height, as well as the historic features such as windows and non-historic window openings were measured. The accumulated data was inputted in an excel table (Table 2). The tolerances and accuracy of the measurements were up to 5 decimal points in mm. Therefore, rounding of the data seemed appropriate with the final measurements recorded with 2 or 1 decimal space. Some plot sizes that were retrieved by the desktop study ordnance data of Digimap were inaccurate and updated accordingly. The window sizes included the frame due to as mentioned earlier in planning guidelines, and the frame is considered

(a) (b)

Fig. 10 **a** Point cloud model of the site and **b** point cloud model of the conservation area

Fig. 11 Point cloud elevation of the street façade of conservation site including the church and site

Fig. 12 Point cloud elevation of the street façade of the conservation area

Table 2 Measured data from point cloud model, height width of façade and window

Plot (east)	height (m)	Width (m)	Window	Plot (west)	height (m)	Width (m)	window
Church + Spire	22.7	14.38		10	11.3	19.1	
Church	16.7	14.38		11	8.8	8.8	
1	9.2	12.0	1.5 × 2	12	8	5	
2	7.6	5.28	0.8 × 1.6	13	8	5.7	
3	8.3	7.39	1.6 × 2	14	8.9	12.4	1 × 1.6
4	7.8	9.8	1 × 1.7	15	9.8	9.3	1 × 1.8
5	6.4	11.2		Liby-16	9.2	19.1	1 × 1.7
6	9.4	7	1.2 × 1.9	17	9.5	9.15	1.3 × 2.4
7	9.3	8	1.4 × 2.4	18	9.5	5.5	0.7 × 2.3
8	9.7	12	0.8 × 1.5	19	9	10.5	
9	9.2	4.95		20	8.8	11.3	1.4 × 2.6

part of the historic character. The windows in the historic area the minimum width is 0.7 m and maximum is 1.6 m, while the vertical height min. is 1.5 m and height is 2.6 m. However, it is important to consider windows with the width and height together, which would suggest the most common was 1 m × 1.7 m see Table 2.

9 Results Contradict Planning

The height range in Table 2 of the façades is 6.4 and 22.7 m, as for the width between 19.1 and 4.9 m. The nature of the site brings out the question of hierarchy. There are two elevations in the conservation area that are in front of the site as seen in Figs. 11 and 12. The conservation area on the same road as the site is seen in Fig. 11. If considering the conservation area on the east side plot 1–11, on the same road contains 3 buildings with 9.2 m. But the range of heights on that street is 6.4–11.3 m, while the adjacent road on the west site is between 8 and 9.5 m see Fig. 12. The main reason these heights are relevant is that they showcase the possibilities of what height is considered part of the context-driven design approach as it follows the surrounding building's historic heights. However, planning guidelines state that new designs can only be between 2 and 3 storeys, which would suggest that if a level is considered 3 m than a maximum of 9 m, which upon closer inspection of the site survey proved to be a limited figure. As the heights are between 6.4 and 22.7 m, if the height of the church with the spire is considered see Fig. 13a the point of reference and the number (22.7 m) translated into floor levels would be of the equivalent of 7 floors. Additionally, suggesting that seven floors on the site would be a plausible amount to be considered responding to the context. This figure is double the recommended storey. However, if taking the second highest point being the church height 16.7 m, then that would allude to five floors, while the highest point of the conservation area would suggest three floors which is the same as the guideline recommendations.

But with the church as an adjacent site that changes the elevation design context, making the church's height as a form of integer, the context becomes based on which reference height is more relevant to the site. The height and story relation is shown in Table 3. Even in-between heights can be reached taking the church height of 16.7 m and the height of the conservation area 11.3 m, finding the in-between is 14 m, equivalent to 4.6 storeys which is 4 or if rounded than five storeys. There are however several implications to undertaking this study is that it allows further accuracy to the architect, allowing them further opportunities to design, as well as an improved degree of certainty on the design decisions being taken with the consideration of the

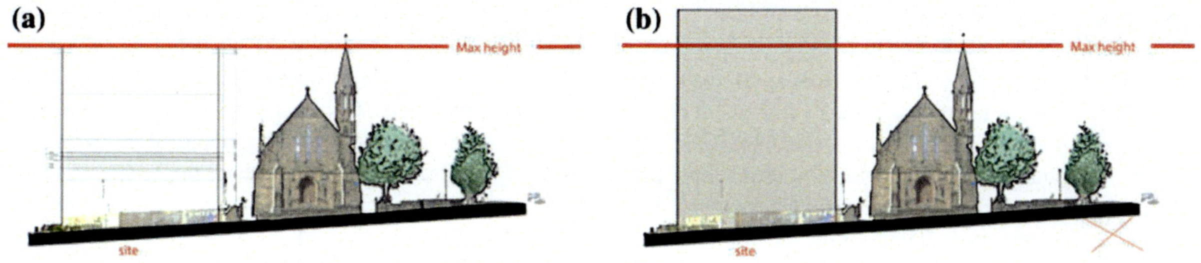

Fig. 13 **a** Diagram showing the site heights and **b** showing height that is disproportionate to the surrounding

Table 3 Height data and relations to floor amount

Name	Height (m)	Floor amounts
Planning guideline 2 storey	6	2
Planning guideline 3 storey	9	3
Church + Spire	22.7	7
Church	16.7	5
Conservation area	11.3	3

site. Therefore, when inputting all the heights gathered from Table 2 into the site elevation with the church as seen in Fig. 13a and b, it becomes clear what height is appropriate. As well as an added visual indicator. Figure 13a and b therefore acts as a more accurate planning schematic drawing that can be reused to test new designs. There are also property and financial benefits to the client as this can aid in suggesting that the façade size can increase in size, and thus increasing the floor amount. As prior design is based on just planning guidelines, while now the design can be contested and has an evidence-based justification that is rooted within the site and process of site analysis.

10 Results and Limitations

The unique context, the elevation and surrounding area of the conservation area as well as the listed church leads to a lack of clarity which takes precedent. The impact of the church's height has displaced the hierarchy of which elevation is more prevalent as well. Therefore, this unique vacant site with its unique surrounding has not been highlighted or addressed by the council in the planning documents. Hence, council planning guidelines need to be considered but also challenged as there is clearly some statements that can conflict one another. With regard to the site survey, there is missing information in the scans, where there are empty spots or cracks. The roofs of the buildings which also contribute to the façade were difficult to capture due to the street being quite narrow. Other methods were considered like drone flight; however, in Scotland this would require a specialized drone license or hiring a professional drone pilot to conduct the flight, which were not in the scope of this study. The tolerances and accuracy when rounding the data to 2 or 1 decimal spaces can result in a loss of information. However, considering the height the level of accuracy is negligible. The terrain of the landscape as it is a hill with a slight slope may need to be considered. As this would add additional height, which can provide different results of the scale of the new building.

The quality of the historical characteristics in the conservation area was not of the same quality (see Fig. 4a). There were some buildings that had a lot of changes and new signage. The changing historical character can deplete the historical significance of the area. Quality seems to be a concern that is a theme throughout, with the data captured from the 3D scan, or the inaccurate data from the desktop study. Architect's need to carefully consider the quality and speed in their workflows. The automation of cleaning data sets and ensuring more usable data is also developing which can speed up the process of data acquisition. There is a risk of quality being lost if more than one application is used as changing the file type can impact the data (Sönmez, 2018). Lastly, the information in the point cloud model can be referred to and trace the data set of the site, acting as a design development at any point in time, as well as a physical evidence of the process of design considering the historic site. The process can then be showcased to the planning officers as evidence in future consultation meetings.

11 Conclusion

The overall results indicate that 3D scanning can be a useful site survey element that can generate point cloud models, providing further information on the size of historic architectural components of historic façades. The point cloud model can also be a useful knowledge extractor tool, that if integrated through other data processing workflows can provide further knowledge discovery opportunities to understand the unique pattern and composition of these historic areas. The growing interest in automated design processes in other fields is providing new ways of working that is yet to be harnessed within the architectural design process. However, with these developing workflows, there is still more work to be done to understand how they can best be integrated. The next steps for the research are to label the architectural components and identify the historic patterns through the composition of the façade elements. As seen in Fig. 14, the method can be done manually by measuring the point cloud model, extracting the information and reproducing an abstract composition of the façades and the historic composition. This can also be further inputted to create a digital library of characteristics as seen in Fig. 15 showcasing all the plot façade sizes in the conservation area.

However, there is also more semi-automated approaches that can speed up the workflow further. This paper proposes a quantitative method to analyse and assess building façades in historical areas, allowing the surrounding site to act as a point of reference for designers alongside planning guidelines. However, this method cannot be the only approach, as architects may need to study the intangible historical heritage of the place, to help understand the historical significance and both can help enable better design decisions. As a result, this design methodology can thus act to evaluate the characteristics of the historical site, assisting architects, through both acting as a design tool in designing new proposals, as well as evaluating the design as it progresses. Through integrating new methods of working, this methodology can aid in assessing the appropriateness of new proposals at an earlier stage of the design workflow, i.e., before applying for planning permission, which can allow architects the opportunity to test their proposals ahead of time. This process of design can act as an evidence-based methodology that can demonstrate if a design is considering the surrounding site's historic character. As the exploration has suggested that in more complex sites like this one has several factors in which planning guidelines can be more difficult to integrate, the evaluation of historical sites allows for new ways for architects to design in historical areas, setting more accurate benchmarks for future designs to follow. This method has the opportunity for reproducibility within areas where the urban fabric is depleted, and the site

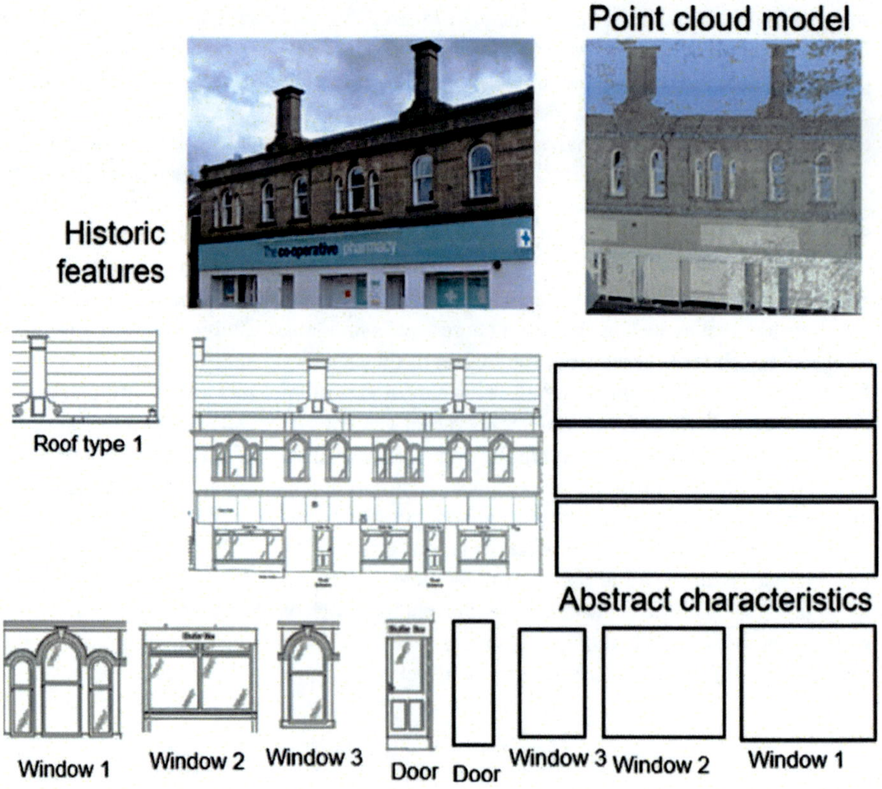

Fig. 14 Next step showing the abstraction of the architectural historic façade components

Fig. 15 Façade abstract library of each plot

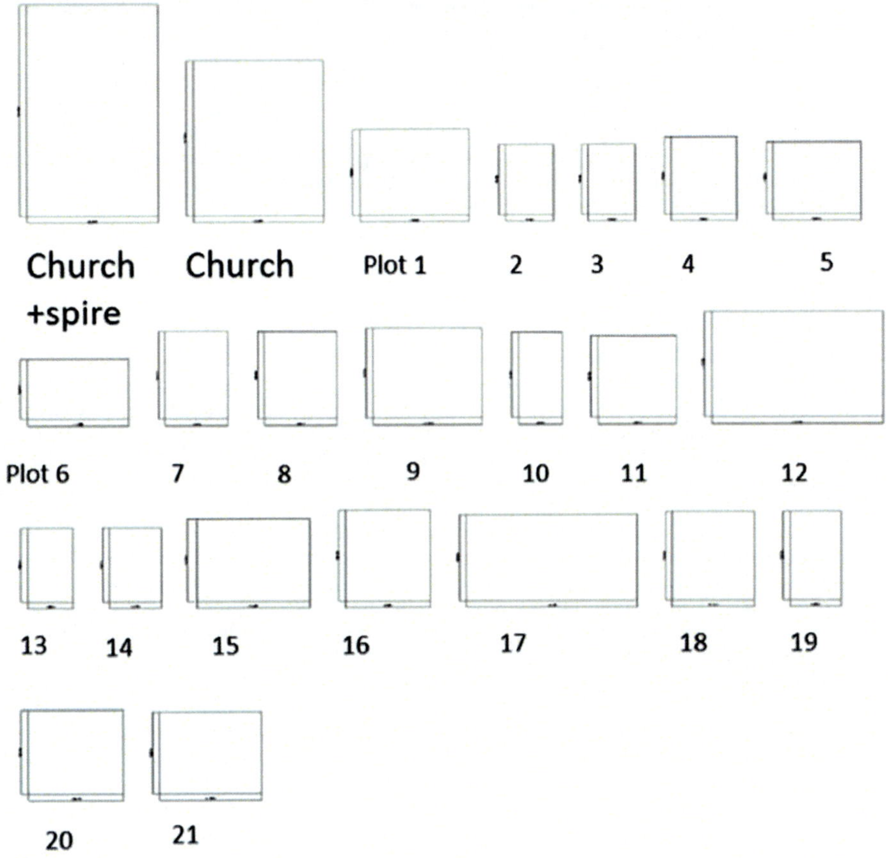

requires the historical surrounding to be considered. The potential of architects that use a more data-centric approach to inform their future designs in historical areas will be an exciting development! As then this will allow architects to generate more variety, as well as a higher prospect of data rich shared improvements to previously subjective personal design decisions. There is growing possibilities of generating previously unattainable designs and at a faster rate with more shared ownership and collective making as knowledge mobilization (Giaccardi & Palen, 2008). This can allow architects more time to focus their attention on other stages of a project. Given architecture's unique position, architects can be at the intersection of both designing and directly testing new workflows in both the digital and physical realm, therefore utilizing new ways to think and practice architecture.

References

Alabi, M., & Bako, A. (2018). Planning institutions as gatekeepers in housing production in Lagos, Nigeria. *Development in Practice, 28* (5), 681–690. https://doi.org/10.1080/09614524.2018.1468416

Alves, S. (2017). The sustainable heritage of vernacular architecture: The historic Center of Oporto. *Procedia Environmental Sciences, 38*, 187–195. https://doi.org/10.1016/j.proenv.2017.03.105

Apress, A. (2006). Motion Tweening. *Foundation Flash, 8*, 175–205. https://doi.org/10.1007/978-1-4302-0086-4_6.

Ashton, L. (2020). Developers appeal after council refuses plans to redevelop historic site in Sheffield. *The Star*.

Bryant, R. (2020). Save our historic Newton Abbot theatre' petition. *Teignmouth Post*.

Desktop study. (n.d.). Retrieved June 27, 2020, from https://www.designingbuildings.co.uk/wiki/Desktop_study.

Digimap. (n.d.). Retrieved June 27, 2020, from https://digimap.edina.ac.uk/.

Dore, C., & Murphy, M. (2013). Semi-automatic modelling of building façades with shape grammars using historic building information modelling. *ISPRS - International Archives of the Photogrammetry, Remote Sensing and Spatial Information Sciences, XL-5/W1*, 57–64. https://doi.org/10.5194/isprsarchives-xl-5-w1-57-2013.

English Heritage, & CABE. (2001). *Building in context: New development in historic areas* (vol. 1, pp. 3–35, Publication). CABE/English Heritage. Retrieved July 6, 2020, from https://www.designcouncil.org.uk/sites/default/files/asset/document/building-in-context-new-development-in-historic-areas.pdf.

Fan, L., Musialski, P., Liu, L., & Wonka, P. (2014). Structure completion for façade layouts. *ACM Transactions on Graphics, 33* (6), 1–11. https://doi.org/10.1145/2661229.2661265

Gatys, L. A., Ecker, A. S., Bethge, M., Hertzmann, A., & Shechtman, E. (2017). Controlling perceptual factors in neural style transfer. In

2017 IEEE Conference on Computer Vision and Pattern Recognition (CVPR). https://doi.org/10.1109/cvpr.2017.397

Giaccardi, E., & Palen, L. (2008). The social production of heritage through cross-media interaction: Making place for place-making. *International Journal of Heritage Studies, 14*(3), 281–297. https://doi.org/10.1080/13527250801953827

Gürbüz, E., Çagdas, G., & Alaçam, S. (2010). A generative design model for Gaziantep's traditional pattern. In *Future cities: ECAADE 2010: Proceedings of the 28th Conference on Education in Computer Aided Architectural Design in Europe, September 15–18, 2010, Zurich, Switzerland, ETH Zurich* (pp. 841–849). ECAADE.

Houghton, T. (2019). Genuine fears Liverpool 'may lose' Unesco World Heritage status. *Liverpool Echo*.

HRH The Prince of Wales, P. (2014). *Facing up to the future: Prince Charles on 21st century architecture*. Retrieved June 27, 2020, from https://www.architectural-review.com/essays/facing-up-to-the-future-prince-charles-on-21st-century-architecture/8674119.article.

Hu, Y., Heath, T., Yue, T., & Zhang, Q. (n.d.). Using quantitative analysis to assess the appropriateness of infill buildings in historic settings. *Journal of Architectural and Planning Research,* 34.

Iossifova, D., & Moroni, S. (2018). Architecture and urban design, Urban Planning. In D. Iossifova, C. Doll & A. Gasparatos (Eds.), *Defining the urban interdisciplinary and professional perspectives* (pp. 141, 109–152, 127). Routledge.

Karakul, O. (2007). Life culture sustainable principles for infilling in historic environments. In *LIVENARCH III Livable Environments and Architecture 3rd International* (pp. 239–249). VIZYON Printing Center.

Lazzeri, A. (2020). Villagers lose fight to save historic cottage from demolition. *Advertiser and Times*.

Local Government UK Website. (2020). *Local plans*. Retrieved June 27, 2020, from https://www.gov.uk/guidance/local-plans.

Loew, S. (1998). *Modern architecture in historic cities: Policy, planning and building in contemporary France*. Routledge.

Menegatti, E., Michael, N., Berns, K., & Yamaguchi, H. (2016). *Intelligent autonomous systems 13*. Springer.

Murphy, D. (2014). Prince Charles's 10 principles for architecture – and 10 much better ones. *The Guardian*.

Ouroussoff, N. (2011). An architect's fear that preservation distorts. *New York Times*.

Phillips, R. (2012). *Create a shape tween in flash*. Retrieved 2020, from https://www.youtube.com/watch?v=b7N1fAdk0Jc.

Scotland, North Ayrshire Planning Council, North Ayrshire Council. (n.d.). *Advice note window design for conservation areas & listed buildings*. North Ayrshire Council.

Scottish Civic Trust, S. (2020). *Conservation area management*. Retrieved June 27, 2020, from https://www.scottishcivictrust.org.uk/conservation-area-management/

Shan, J., & Sampath, A. (2008). Building extraction from LiDAR point clouds based on clustering techniques. In J. Shan & C. K. Toth (Eds.), *Topographic laser ranging and scanning: Principles and processing* (pp. 421–444). CRC Press/Taylor & Francis Group.

Sönmez, N. O. (2018). A review of the use of examples for automating architectural design tasks. *Computer-Aided Design, 96*, 13–30. https://doi.org/10.1016/j.cad.2017.10.005

Tang, P., Wang, X., & Shi, X. (2018). Generative design method of the façade of traditional architecture and settlement based on knowledge discovery and digital generation: a case study of Gunanjie Street in China. *International Journal of Architectural Heritage, 13* (5), 679–690. https://doi.org/10.1080/15583058.2018.1463415

Tang, P., Wang, X., & Shi, X. (2019). Generative design method of the façade of traditional architecture and settlement based on knowledge discovery and digital generation: A case study of Gunanjie Street in China. In *Proceedings of the 24th CAADRIA Conference* (vol. 1, pp. 757–766). The Association for Computer-Aided Architectural Design Research in Asia (CAADRIA) Hong Kong.

Thatcher, M. (2017). Introduction: The state and historic buildings: Preserving 'the national past.' *Nations and Nationalism, 24*(1), 22–42. https://doi.org/10.1111/nana.12372

UK, Department for Communities and Local Government. (2015). *Plain English Guide To The Planning System* (p. 4). Crown Publishing.

Waller, J. (2020). Ice factory roof to be assessed in campaign to save historic building. *Grimsbytelegraph*.

While, A. (2007). The state and the controversial demands of cultural built heritage: Modernism, dirty concrete, and postwar listing in England. *Environment and Planning B: Planning and Design, 34* (4), 645–663. https://doi.org/10.1068/b32085

Zhang, H., Xu, K., Jiang, W., Lin, J., Cohen-Or, D., & Chen, B. (2013). Layered analysis of irregular façades via symmetry maximization. *ACM Transactions on Graphics, 32*(4), 1. https://doi.org/10.1145/2461912.2461923

Importance of Digital Techniques of Documentation for the Conservation of Cultural Heritage

Mohammad Moein Danesh and Ahmad Rajabi

Abstract

Cultural heritage is an invaluable resource for societies. Inherited from past generations, it must be preserved and safeguarded for posterity. However, it is threatened by several factors, including natural disasters and those caused by human actions. In this regard, the need for conservation of cultural heritage is an indisputable reality. Digital documentation is considered an important tool, providing precision, in the recording of physical features and peculiarities of heritage. On the other hand, when acquired documents are digitally archived, they can be used for numerous purposes, such as conservation and management of heritage. In the instance of minor or major damage to built heritage, these archives can be highly useful in the restoration process. In recent years, due to considerable developments in technology and digital tools, the techniques used for documentation of historical buildings have been also significantly improved, leading to a better standard of monument conservation. Accordingly, recognition and exploitation of the most recent technologies and techniques in the field of cultural heritage are of primary importance. Deploying new methods of documentation significantly reduces costs, expedites the process of surveying, and also ensures an accurate output. This paper investigates the application of digital techniques of documentation in cultural heritage conservation. Additionally, it offers an overview of the advantages and limitations of the most widely used techniques, including terrestrial laser scanning, low-cost photogrammetry methods, and the application of unmanned aerial vehicle (UAV) platforms.

Keywords

Cultural heritage • Documentation techniques • Photogrammetry • Laser scanning • Unmanned aerial vehicles • 3D modeling

1 Introduction

Over the last few years, public awareness has increased that cultural heritage constitutes a considerable part of our history (Salazar & Marques 2005). Cultural heritage is the evidence of human achievement over the ages, representing the history and memory of the communities that constructed them and has a crucial role in characterizing the identity and culture of each society. In this regard, the conservation and protection of cultural heritage have become an important challenge attracting great attention at a regional, national, and global scale. The concept of cultural heritage is evolving considerably in multiple contexts, one of which is sustainable development that has recently emerged as a new global development model. According to the UNESCO World Heritage and Sustainable Development program, appropriate cultural heritage conservation and management has potential to contribute to sustainable development. In this framework, a well-preserved heritage site can have a great contribution to the quality of life for community residents. Heritage is regarded as an important factor for attracting investments and providing jobs, of which some can be related to tourism. Additionally, due to containing symbolic and religious values, heritage is fundamental to the spiritual well-being of society. Further, activities related to heritage enable society to make sense of dignity, continuity, and empowerment (Xiao et al. 2018).

Tangible cultural heritage of all kinds (natural, cultural, or mixed) around the world suffers from several types of threats (Pierrot-Deseilligny et al. 2011). These threats include natural disasters such as earthquakes, fires, flooding, tsunamis,

M. M. Danesh (✉) · A. Rajabi
National Research Institute of Cultural Heritage, Tehran, Iran
e-mail: mohammadmoein.danesh@mail.polimi.it

A. Rajabi
e-mail: ahmad.rajabi@mail.polimi.it

A. Versaci et al. (eds.), *Conservation of Architectural Heritage*, Advances in Science, Technology & Innovation,
https://doi.org/10.1007/978-3-030-74482-3_32

land and mud slides, winds and tropical storms (Taboroff 2000). Apart from this category, hazards including climate change, war, vandalism, urban expansion, and erosion with time can be considered among the main causes of damage. UNESCO and the Council of Europe have established specialized organizations for conservation and restoration of cultural heritage. ICOMOS is regarded as the most important one; however, CIPA Heritage Documentation, ISPRS, ICOM, and ICCROM are all active in this field (Georgopoulos 2018). Developments in the principles of conservation of cultural heritages from the mid-twenties century onward are considered by many as the most important achievements of conservation activities. Charters, declarations, and conventions were adopted by international organizations such as UNESCO and ICOMOS, with the principal purpose of safeguarding cultural sites. Of these, the most important is the International Charter for the Conservation and Restoration of Monuments and Sites known as the Venice Charter 1964. It serves as a reference point for standards related to restoration and conservation of architectural heritage while pointing out the importance of modern technology applications in conservation works (Ahmad 2006). As stated in the second article of the Venice Charter "all the sciences and techniques which can contribute to the study and safeguarding of the architectural heritage" must be used where necessary. Moreover, in the sixteenth article of the charter, the precision in the task of documentation and the necessity for dissemination of the recordings are emphasized (ICOMOS 1964).

The following paper overviews the significance of digital techniques of documentation in monuments and built heritage conservation. Furthermore, two commonly used techniques of laser scanning and photogrammetry, in addition to their advantages and limitations, are investigated. This paper also discusses the necessity of combining different acquisition methods while undertaking documentation processes. Indubitably, each documentation technique has some limitations and combining different methods is a practical approach to reduce the overall error of results and to obtain a continuous recording of objects.

2 Importance of Documentation for Cultural Heritage Conservation

The starting point of every action related to built heritage conservation is a meticulous study; therefore, in order to enhance the conservation status of historical monuments, it is essential to acquire a thorough understanding of buildings through historical research and interpretative analysis of their materials. In this context, accurate recording, documentation, and analysis of cultural heritage are regarded as crucial steps for every programmed activity, concerning their safeguard and conservation (Quintero 2017). Over the last few decades, thanks to the constant advances in technology, the metric survey has provided exceptional opportunities for studying cultural heritage in depth (Bertellini et al. 2020). Surveying as a part of the documentation process is regarded as an essential tool for studying built cultural heritage (Tucci et al. 2016). According to Salonia et al. (2007): "Survey is a fundamental step in the process of analysis and diagnosis of architectural monument state of conservation and allows to define a geometric basis to which relate every other further information, deriving from diversified disciplinary field, so that to achieve an historical-architectural integrated knowledge of the monument." In comparison to traditional documentation techniques which suffer from inaccuracies, are labor and time intensive, advanced documentation methods are inexpensive, time efficient, and also ensure accurate results and output (De Vos 2017).

In addition to the contribution that documentation makes in gaining in-depth knowledge of cultural heritage, it also provides data availability for various interested parties to disseminate knowledge for the purpose of cultural heritage valorization. There are different reasons that can be found as requisite for the transmission of such information (Böhler 2006): when different typologies of users cannot access the object, when the object has large dimensions and a complicated shape, affecting the onlooker's observation, when the object is exhibited to the public for a limited time period, when the interested parties are living far from the object and visiting it is not possible for them, and when the danger of decay or destruction threatens the object. On the other hand, contemporary technological developments in documentation have provided the outstanding potential for 3D digitalization of archaeological and cultural heritage monuments. 3D models of historical buildings, which are used as an essential tool for documenting heritage sites, can be utilized for numerous purposes such as tourism and virtual museums, educational resources for researchers and students, interaction without risk of damage, and for reconstruction or restoration, if the historic buildings are damaged (Remondino & El-Hakim 2006).

3 Techniques of Cultural Heritage Documentation

Documentation of cultural heritage normally involves professionals from different fields of expertise and is a multidimensional process. The implemented method for documentation substantially depends on the nature of the target of recording in addition to the purpose for the documentation task (Pavlidis et al. 2007). Various methods are available for the surveying of cultural heritage. Employing suitable techniques such as hardware, software, and sensor, applying the appropriate workflow and to make sure that the ultimate outcome conforms to the applied technical specifications, is dependent on the physical features and peculiarities of the object or building. The main effective factors to be considered are the object dimensions and the required amount of precision. Besides these parameters, other secondary factors may have an impact on choosing the appropriate technique. These include speed and time limitations and available budget (Patias 2006); the object accessibility and availability of ideal locations without vibration for observation; access to instruments and power supply; and the possibility of physical contact with the object and permission to utilize the chosen technique (Grussenmeyer et al. 2008b).

According to the aforementioned, precise recording of each object in view of its characteristics and the aim of documentation requires a particular technique. There are various methods available and therefore various categorizations are offered by different research studies. The classification of these techniques can enhance our understanding and helps to deal with their features. In a general categorization, the most widely applied methods for buildings documentation are classified into (Scherer 2002): topographic methods, traditional manual methods, scanning methods, and photogrammetric methods. In a research by Hassani (2015), the available techniques are categorized according to the metric data and coordinates of the points obtained from target object with or without taking images. Accordingly, documentation methods are classified into three categories: Image-based, Non-image-based, and combinative methods. However, in terms of 3D acquisition techniques using non-contact systems, the methods are divided into two categories: passive sensors which utilize natural light and active sensors that project a pattern of light in a systematic way (Guidi et al. 2004). Nowadays, the most commonly used documentation techniques for cultural heritage are (Núñez Andrés et al. 2012): digital photogrammetry (image-based method, passive sensor) and laser scanning (range-based method, active sensor). The combination of these two methods is also very common, since it allows the generation of complete and refined surface models. Moreover, unmanned aerial vehicle (UAV) platform is also being widely deployed for photogrammetric surveys (Remondino 2011), and its application for laser scanning (LiDAR) is still under significant progress (Colomina & Molina 2014).

There are numerous common characteristics between image-based and laser scanning modeling techniques which make them universally popular in the cultural heritage field. Both these methods can be implemented by easy-to-use software and hardware, and they provide time-saving benefits to the survey procedure and the elaboration phase. By using these techniques, depending on the geometric complexity of the surveyed object, a detailed 3D model can be obtained in a short time. They provide the possibility to have the measurements of the object and also to extract required 2D drawings and ortho-images. The final output of these methods which is a 3D model can be used for documentation, dissemination, and visual purposes (Fassi et al. 2013).

3.1 Photogrammetry

3.1.1 Brief History and Background

The application of photogrammetry in architecture is nearly as old as the invention of photography. Aimé Laussedat, a French military officer, in 1849 commenced his experiments on images obtained from a façade of the Hotel des Invalides. Laussedat is usually regarded as the world's first photogrammetrist. Some years later, photogrammetrical techniques were developed by the German architect, Albrecht Meydenbauer, when documenting a cathedral in the city of Wetzlar. Photography was deployed by Meydenbauer in order to avoid the traditional, often dangerous, manual methods of measuring façades. He was the person who coined the word "photogrammetry" and also the founder of the very first photogrammetric institution worldwide in 1885. It was also in that year that the potential of photogrammetry was proved and the method was applied for surveying of the most important architectural monuments (Luhmann et al. 2006). In 1860, Meydenbauer wrote a memorandum in which he described that photographic images can be used for storing the information of objects at high levels of detail and accuracy. Moreover, he was also cognisant of the fact that cultural heritage is endangered by various threats. For this reason, he worked on the idea of a "Denkmälerarchiv" (Cultural Heritage Archive) in which remarkable cultural heritage is documented in a manner that in cases of serious damage could be reconstructed according to the recorded images (Albertz 2001).

3.1.2 Defining Photogrammetry

The surveying technique of photogrammetry extracts accurate measurements and 3D information from images taken from different angles and usually, at least two pictures are required for the data processing. This technique, according to McCarthy (2014): "is the process of making measurements of features through analysis of overlapping photographs and is fundamentally based on trigonometry." Photogrammetry is the most efficient method for image data processing that is able to obtain precise metric 3D information. Images for this aim can be obtained using terrestrial, aerial, or satellite sensors and then processed following the usual photogrammetric procedure (Remondino 2011). The accuracy of output in photogrammetry depends on the quality of the used camera and its acquired photographs. In addition, the capabilities of the software used for the processing of the photographs are another important factor (Dai & Lu 2010). It has been a long time that photogrammetry is being deployed as a technique for obtaining three-dimensional data of objects as well as their textural information. In the classical method, in accordance with the stereographic principle, single metric or stereo metric cameras were used to record the images with overlapping. Subsequent to the orientation of the stereo model by a stereo plotter, the 3D measurement of the object points in the stereo model was carried out manually by an operator (Yastikli 2007). However, thanks to the introduction of new software and considerable enhancement in algorithms, traditional methods of photogrammetry are replaced with automatic methods. Large datasets of overlapping digital images of an object can be uploaded in a single batch into software with minimal manual input (McCarthy 2014). The entire process, from image orientation to the extraction of textured model and exporting orthophotograph, is fully automated. These software are cheap or free, user friendly, and they offer accurate results, using normal (non-metric) cameras (Fassi et al. 2013). The capabilities of these software are investigated in various literatures in the cultural heritage field.

Image-based modeling and photogrammetric systems have numerous advantages. Their applicability to multiscale projects is regarded by many experts to be their main advantage (Skarlatos & Kiparissi 2012). Moreover, compared to scanning techniques, their main superiority is that image-based modeling sensors are generally low cost and also portable (Remondino & El-Hakim 2006). Photogrammetry is proved to be a flexible and relatively easy technique for surveying, while it is capable of generating reliable 3D point clouds by utilizing a dense matching algorithm. A further advantage of photogrammetry is that it can provide detailed textural data, which allows the generation of photorealistic 3D models (Murtiyos et al. 2019). Another benefit of photogrammetry is its time efficiency, and it is usually quick when this technique is employed to record on site, as opposed to the processing of data in an office, which takes longer (Böhler & Marbs 2004). Based on the required accuracy for the documentation, three types of methods can be deployed in photogrammetry (Arias et al. 2007):

- Very accurate photogrammetry: for works requiring 1 mm or higher accuracy. This method is appropriate for surveying decorative elements and statues. It can also provide precise information on the conservation status of objects. For this category, frequently used scales are 1:2, 1:5, and 1:10.
- Accurate photogrammetry: this method can be utilized where the accuracy of about 1 or 2 mm is demanded. The main applications of this category are for the analysis of monuments, examination of their conservation status, and also for the recording of decorative elements. The common scales for presentation in this classification are 1:10, 1:20, and 1:50.
- Fast and simple photogrammetry: This type of method is distinguished for its simplicity and the inexpensive equipment required. In this category, the maximum error of below 5 cm is usually acceptable. The most commonly used scales for this case range from 1:100 to 1:200.

Modern photogrammetric systems utilize structure from motion (SfM), multiview stereo (MVS), and scale invariant feature transform (SIFT) methods to obtain 3D geometric information from sets of photographs. Precise 3D models of an archeological site or a monument can be generated with these algorithms in a relatively short time and at a much lower cost than laser scanner tools. Photogrammetric techniques nowadays have the potential to revolutionize how experts study, document, and preserve cultural heritage (Sapirstein 2016). These techniques have the capability of generating highly accurate datasets, while eliminating gross errors. This allows visually impressive 3D models to be obtained easily, compared to traditional stereo-based DEM generation methods (Micheletti et al. 2015). Through the utilization of these new software and algorithms, even non-experts and tourists, using their mobile phones equipped with consumer-grade digital cameras, can contribute to heritage preservation and documentation by acquiring valuable geomorphic data. The ubiquity of mobile phones is a great contribution to crowd-sourcing, which is an important method for collecting data that can also be used for the generation of three-dimensional models (Dhonju et al. 2017).

3.1.3 Theory and Fundamentals of Photogrammetry Technique

Following the acquisition of images on the field, the acquired data has to be processed through laboratory work (Arias et al. 2007). The related stages of data processing are described in this section. The location of a point in photogrammetry is specified by a three-dimensional Cartesian system, of which its orientation, origin and scale, is determined arbitrarily. The central perspective projection is the initial stage for constructing a functional model for close range photogrammetry (Cooper & Robson 1996; Yilmaz et al. 2008); which is illustrated in Fig. 32.1. As can be observed, point A by the straight line AOa passing through the perspective center denoted by O is projected on a plane

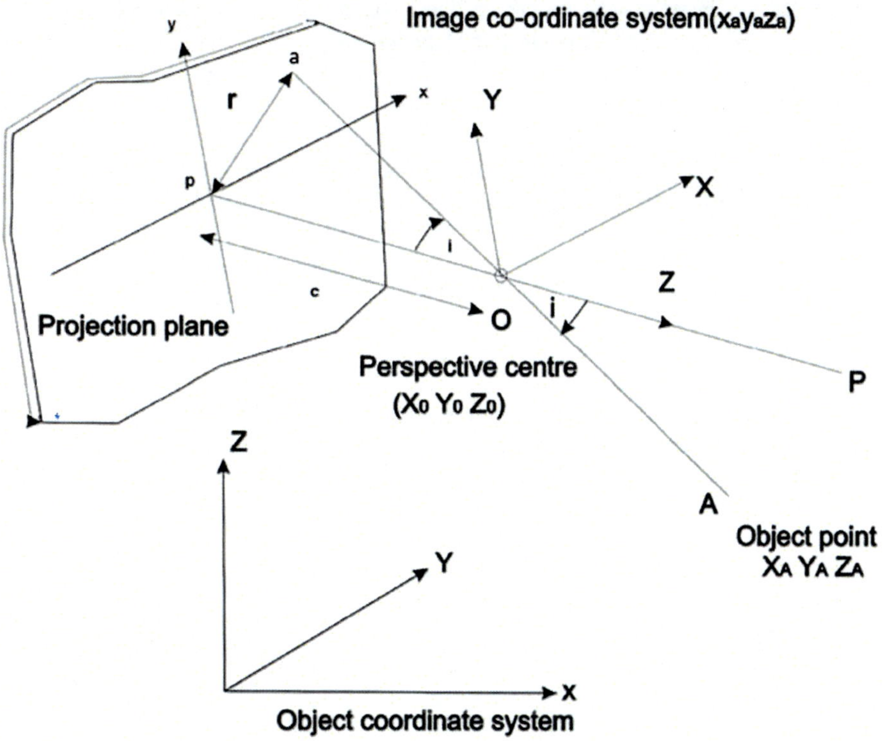

Fig. 32.1 Central perspective projection (Cooper & Robson 1996)

(image). The perspective axis *POp* is orthogonal to the projection plane and the distance of point *O* to the plane is denoted by c, which is the camera focal length. Points *A* and *a* are homologous points and in order to derive functional relationships between these two points, two three-dimensional Cartesian coordinate systems are introduced. The primary coordinate system (XYZ) is located arbitrarily in the object space in which the perspective center coordinates are (X_0, Y_0, Z_0) and for point *A* are (X_A, Y_A, Z_A). The secondary Cartesian system (*xyz*) is originated at point *O*, of which the z-axis corresponds to the principal axis (*POp*) and is directed away from the projection plane. Accordingly, the coordinates of the projected point *a* are (x_a, y_a,-c). Points *A*, *O*, and *a* are collinear and if the vectors are written in relation to the primary coordinate system, vectors (X_A- X_O) and x_a are collinear but of opposite sense. Therefore, we have $X_A = X_O - \mu R^t x_a$ and accordingly $x_a = \mu^{-1} R (X_O - X_A)$. Here, the parameter μ, is a positive scalar factor and R being the rotation matrix. The matrix notation of the second equation is (1):

$$\begin{bmatrix} x_a \\ y_a \\ -c \end{bmatrix} = \mu^{-1} \begin{bmatrix} \gamma_{11} & \gamma_{21} & \gamma_{31} \\ \gamma_{12} & \gamma_{22} & \gamma_{32} \\ \gamma_{13} & \gamma_{23} & \gamma_{33} \end{bmatrix} \begin{bmatrix} X_A - X_0 \\ Y_A - Y_0 \\ Z_A - Z_0 \end{bmatrix} \quad (1)$$

The third equation of the above matrix can be written explicitly in μ^{-1} and substituted in the other two equations, to give formulas (2) and (3). These are known as *collinearity equations* which are basic to photogrammetry techniques (Dai & Lu 2010).

$$x_a = \frac{-c[\gamma_{11}(X_0 - X_A) + \gamma_{12}(Y_0 - Y_A) + \gamma_{13}(Z_0 - Z_A)]}{[\gamma_{31}(X_0 - X_A) + \gamma_{32}(Y_0 - Y_A) + \gamma_{33}(Z_0 - Z_A)]} \quad (2)$$

And

$$y_a = \frac{-c[\gamma_{21}(X_0 - X_A) + \gamma_{22}(Y_0 - Y_A) + \gamma_{23}(Z_0 - Z_A)]}{[\gamma_{31}(X_0 - X_A) + \gamma_{32}(Y_0 - Y_A) + \gamma_{33}(Z_0 - Z_A)]} \quad (3)$$

The central perspective projection is just an idealized geometric model of the actual image formation in a camera. In fact, the operation known as *camera calibration* identifies to what extent the image formation geometry in a real camera differs from the geometry of a central perspective projection. The parameters obtained through this operation are the *interior orientation* elements of the camera. The reason for the discrepancy between the abstract model and the real is the lens. Cameras originally designed for photogrammetry are metric cameras that possess particular features to ensure close conformance to the perspective projection model (Cooper & Robson 1996). However, by means of camera calibration, the errors which arise from using non-metric cameras can also be avoided (Arias et al. 2007).

Provided that the three interior orientation parameters are acquired by camera calibration and the coordinates of *A* in the object coordinate system are known, there will be six unknown elements of exterior orientation including rotation parameters ω,φ, and κ, plus the perspective center coordinates (X_O, Y_O, Z_O). The evaluation of these parameters is known as *resection*. A minimum of three non-collinear targets or control points are necessary for resection of a camera. The method for evaluating the exterior orientation elements is dependent on the purpose of the photogrammetry, whether speed or a statistically rigorous estimation is more important. If a more statistically rigorous resection is intended, Eqs. (2) and (3) can be linearized and a least squares estimation process deployed for evaluating the six elements of exterior orientation (Cooper & Robson 1996).

The object space coordinates of target point *A*, (X_A, Y_A, Z_A) can be calculated from measurements of the two photocoordinates (x_1, y_1) and (x_2, y_2), provided that the exterior orientation elements of two cameras with perspective centers at O_1 and O_2 are known (Fig. 32.2). This process is known as *intersection* and Eqs. (2) and (3) are the basics of the method. Due to having three unknowns and four equations, a least squares estimation can be used. *Relative orientation* is known as the process of measuring the exterior orientation parameters of one camera in line with the photocoordinate system of another camera. As illustrated in Fig. 32.3, object *A* is imaged at a_1 and a_2; vectors a_1 a_2 and b are coplanar and lie in the epipolar plane of target *A* and the two perspective centers. Relative to the primary axes, these vectors are:

$$a_1 = (-\lambda)[x_1y_1(-c_1)]^t;\ a_2 = (-\mu R^t)[x_2y_2(-c_2)]^t;\ b = [b_xb_yb_z]^t$$

Here b represents the camera base and R^t being the rotation matrix. The triple scalar product of three coplanar vectors is equal to zero, therefore: $b.a_1 \times a_2 = 0$, or as it is denoted in Eq. (4). Considering the base vector b as a nonzero quantity and using $R^t\ a_1 = a_2 = [x_2'\ y_2'\ z_2']^t$ the *coplanarity Eq.* (5) for target *A* is obtained.

$$\det\begin{bmatrix} b_x & x_1 & r_{11}x_2 + r_{21}y_2 - r_{31}c_2 \\ b_y & y_1 & r_{12}x_2 + r_{22}y_2 - r_{32}c_2 \\ b_z & -c_1 & r_{13}x_2 + r_{23}y_2 - r_{33}c_2 \end{bmatrix} = 0 \quad (4)$$

$$\det\begin{bmatrix} 1 & x_1 & x_2' \\ \frac{b_y}{b_x} & y_1 & y_2' \\ \frac{b_z}{b_x} & -c_1 & z_2' \end{bmatrix} = 0 \quad (5)$$

(b_y/b_x), (b_z/b_x), ω, φ, and κ are the five elements of relative orientation. Measured photocoordinates of at least three targets, well distributed in the objects space, are necessary to evaluate the parameters of relative orientation. However, usually more than three targets are measured and the five elements are evaluated through a linear squares estimation. When relative orientation is carried out, the evaluated photocoordinates of a target can be used in order to measure, by intersection, its coordinates relative to the (x_1,y_1,z_1) axes. These coordinates are usually known as model coordinates (Cooper & Robson 1996; Yilmaz et al. 2008).

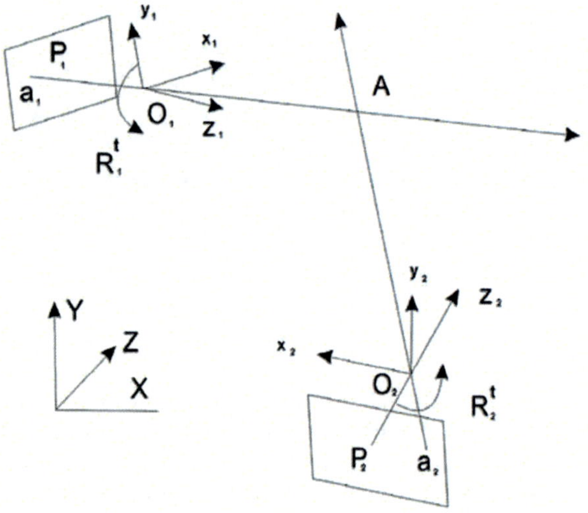

Fig. 32.2 Intersection (Cooper & Robson1996)

3.2 Laser Scanning

3.2.1 Light Detection and Ranging (LiDAR) Technology

In the last three decades, this technology has been developed significantly and the LiDAR sensor currently can be mounted on different types of platforms. In earth surface research, according to the mounted platform, LiDAR systems are divided into four groups: terrestrial laser scanning, mobile laser scanning, airborne laser scanning, and space-based laser scanning, each of which possesses different advantages and applications. The generated point cloud and data collected by these platforms are different in three main aspects: perspectives, resolutions, and content of focus (Cheng et al. 2018). A LiDAR system usually includes a laser scanning unit and a navigation system, of which the latter consists of an integrated differential global positioning system (DGPS) and an inertial navigation system (INS). By having these components on a platform, a geo-referenced point cloud can be directly obtained. The laser ranging unit can measure the distances between the sensor and objects by using the time delay method which determines the time between emission of a laser pulse and its reflection. Meanwhile, the precise location and orientation of the platform are

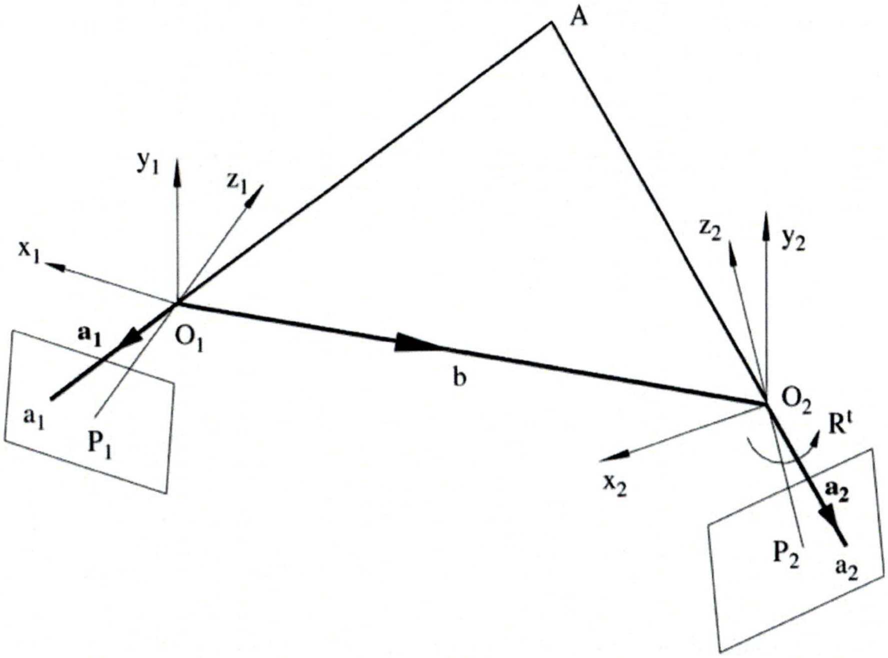

Fig. 32.3 Coplanarity (Cooper & Robson 1996)

measured by the onboard GPS/INS component (Bang et al. 2009).

3.2.2 Terrestrial Laser Scanning- TLS

At present, terrestrial laser scanning technology is being widely used in various fields. Undeniably, the technological advancements of this acquisition method have made considerable developments in the 3D modeling area. The most important features of TLS, which makes it a versatile technology for different fields, are its fast and easy 3D capture of the surroundings and the high degree of accuracy in its measurements. The TLS produces a 3D representation of a field of view, by determining distances to surfaces of objects in a spherical coordinate space. The system records a horizontal direction, an angle of elevation, and measurement of range from the object to the sensor of scanner for every point (Soudarissanane et al. 2011). The TLS mechanism works with the active emission of a pulse wave in the visible or infrared domain, which is transmitted through the air until the signal is reflected by the target surface. A photodiode is employed in the instrument that detects the reflected pulse, to measure the distance between the point of emission and object (Armesto-González et al. 2010). In laser scanning, each manufacturer provides specific precision and accuracy, which should be noted before embarking the scanning process (Boehler et al. 2003). In contrast to photogrammetry, 3D scanner is able to directly produce a robust point cloud, and the resulting color point cloud is mainly utilized to record the information of surface to produce plan and profile, 2D sections, and 3D models (Haddad 2011). When laser scanning, four major factors influence the quality of a point cloud, and these include object properties, scanning geometry, instrument calibration, and atmospheric conditions (Soudarissanane et al. 2011).

In laser scanners, RGB values are generated by external or internal digital cameras. Some types of terrestrial scanners are equipped with a digital camera set with reference to the point cloud, whereby the image and point cloud are integrated automatically. Nevertheless, the major disadvantage of scanners with integrated cameras is their low-resolution output that in most cases do not provide appropriate quality of color or texture. Therefore, to generate a point cloud with sufficient color and texture information, high-resolution images are necessary to be acquired independently of the station (Markiewicz et al. 2015). By attaching a camera on the laser scanner, the point cloud and images are obtained at the same time. As a result, the geometry of an object aquired by the laser scanner is complemented by the object's color texture captured by the camera, while dense point clouds are generated which are much more representative of the scanned objects (Xiao et al. 2007).

There are two primary active methods for optically measuring a 3D surface (Beraldin et al. 2010; El-Hakim et al.2004): The first is light transit time estimation that as illustrated in Fig. 32.4, laser light waves emitted from the sensor travel with a known speed in a particular medium. Thus, an effective method of measuring the distance is offered by evaluation of the time delay created by light travelling from a source to the object surface and back to a light sensor. The accuracy in this method does not degrade

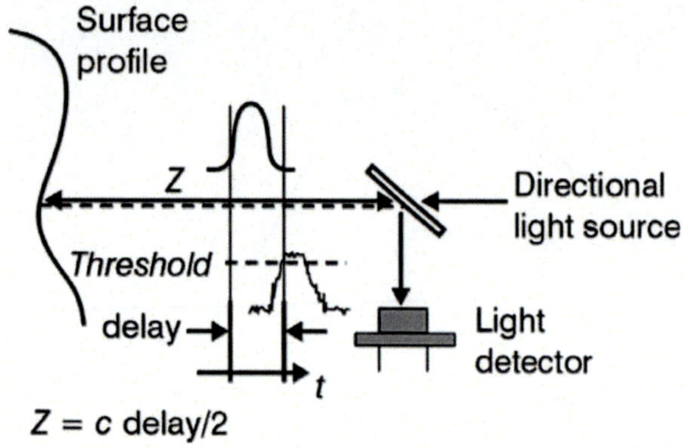

Fig. 32.4 Light transit time (Beraldin et al. 2010)

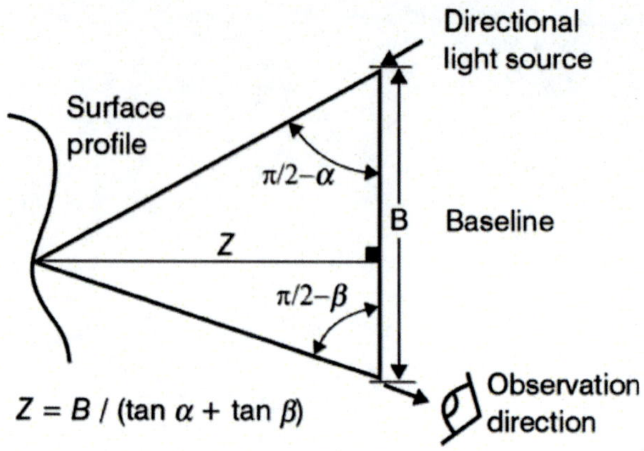

Fig. 32.5 Trangulation (Beraldin et al. 2010)

instantly as the distance increases. These sensors are able to be used for distances in the kilometer range. Such systems are also known as time-of-flight or light detection and ranging (LiDAR) systems. The second sensor type is based on the triangulation principle, which estimates the distance in an indirect way by a triangle, constructed employing an illumination angle directed at a surface and an observation direction with a known distance (base distance) from the source of illumination (Fig. 32.5). In contrast to range sensors, the accuracy of measurements in triangulation scanners diminishes by increasing the distance between the object and the sensor. For practical reasons, the base of the triangle is relatively short, and triangulation-based scanners have a short range of a few meters.

Active sensor systems like laser scanners, similar to any other surveying technology, have limitations and are not suitable for all recording tasks. Since the scanning process produces large amounts of data as point clouds, the stages of data processing for obtaining the final product can be time consuming and require adequate experience. Particularly in the case of large sites or buildings, of which the collected data is large and consists of a high number of polygons, whereby making it almost impossible to open and visualize with a typical laptop, rather requiring a high-performance desktop computer (Barsanti et al. 2014). Another limitation of this technology is that it is not efficient for recording linear elements and provides data with insufficient resolution, when edges have to be collected. The reason is that sharp corners and edges are often smoothed in the triangulated surface (Alshawabkeh and Haala 2004). These sensors are also affected by the reflective characteristics of the surface in some materials (Boehler et al. 2003). However, compared to digital photogrammetry, the high cost of 3D sensors and of the software required for processing, has been a limiting factor for extensive use of this technology in the field of cultural heritage (Guidi et al. 2007). What is more, laser scanners measure the objects from earth-bound stations and in order to scan large target objects, multiple scans will be frequently required. In this regard, the fusion of all the scans into a single entity with the same coordinate system is required (Altuntas et al. 2016).

3.3 UAV Photogrammetry

According to Eisenbeiss (2009), Unmanned Aerial Vehicle (UAV) photogrammetry is described as "a photogrammetric measurement platform, which operates remotely controlled, semiautonomously, or autonomously, without a pilot sitting in the vehicle." UAVs can be used for low-altitude aerial photography, and for gathering geospatial information through remote sensing. Over the last years, UAVs are being deployed for the aim of cultural heritage surveying due to their reliability and ease of use (Themistocleous et al. 2015). The main advantages of UAVs in the field of cultural heritage are their capability for fast data acquisition, and their ability to reach inaccessible parts of the monuments, such as rooftops for photographing. Moreover, the low-cost of UAV systems is another advantage that makes this platform popular in the field. Different phases are required for programming an image-based field surveying with UAV systems. These phases include planning of flight path, measuring the ground control points (if required for georeferencing), capturing the images, calibrating the camera, orientating and processing of images in order to extract 3D information (Remondino et al. 2012). In recent years, modern photogrammetric approaches such as SfM are being widely employed in UAV systems. Therefore, digital elevation models with high resolution can be generated from photographs obtained by consumer grade digital cameras (Fernández-Lozano & Gutiérrez-Alonso 2016; Westoby et al. 2012).

However, UAVs, especially the low-cost ones, have some limitations. These platforms have a limited capacity for weight and dimension, hence mostly sensors with low weight, like amateur cameras are selected for data acquisition. Thus, compared to large format cameras, a higher number of images for acquiring the comparable image resolution and the same image coverage are required. The limitations in payload also demand navigation systems with low weight, which results in a less accurate orientation of sensors. In addition, low-cost UAVs typically have less powerful engines which limits their access to high altitude areas (Eisenbeiss 2009).

3.4 Combination of Different Methods

Each method of documentation has advantages and limitations. Given this, and considering the geometric properties of each target object and also the aim of documentation, it is relatively difficult to obtain all the required spatial information by only a single sensor (Koch & Kaehler 2009). As an example, TLS is an optimal technique to provide an accurate geometric basis for 3D models. However, it is only used on earth-bound platforms and is not suitable to be employed for documenting building elements in height or inaccessible areas. In contrast, photogrammetry is a flexible method. A camera can be easily installed and mounted on a mast or a drone for recording the places which are not easy to be accessed (Mateus et al. 2019). Moreover, in cases where complex geometric compositions are found, the combination of different techniques is a practical solution to achieve a satisfactory and precise final outcome (Grussenmeyer et al. 2008a; Guarnieri et al. 2006; Murtiyos et al.2019). Numerous benefits can be achieved by combining data from multiple sensors such as extended spatial coverage, enhanced resolution, reduced ambiguity, robust operational performance, and increased dimensionality (Beraldin 2004).

4 Summary and Conclusion

Cultural heritage is an invaluable resource for human beings. Therefore, its conservation and protection remains a key challenge in the contemporary era. Monuments and built cultural heritage are the recordings of our achievements over the ages and their precise documentation is crucial to preserve our history. In this frame, documentation as an activity for recording the features of the objects of our cultural heritage is a fundamental step in their conservation and maintenance. Thanks to the contemporary advancements in technology and their impact on documentation methods, the conservation state of cultural heritage has significantly improved. The new techniques of documentation expedite the process of recording, reduce the costs, and also offer an accurate and precise output. Moreover, based on the data obtained by these techniques, photorealistic and detailed three-dimensional digital models can be generated which have numerous applications in the field of cultural heritage. In this research, the importance of digital techniques of documentation in the conservation of cultural heritage is overviewed. This paper offers an overview of the most commonly used documentation methods, including terrestrial laser scanning, unmanned aerial vehicles (UAV), and the low cost and widely used techniques of digital photogrammetry. The benefits of combining different acquisition methods, while undertaking documentation processes are discussed. It is found that each documentation technique has some limitations, whereby combining different methods, these disadvantages can be complemented and objects can be recorded with high accuracy.

References

Ahmad, Y. (2006). The scope and definitions of heritage: from tangible to intangible. *International Journal of Heritage Studies*, *12*, 292–300. https://doi.org/10.1080/13527250600604639.

Albertz, J. (2001). Albrecht Meydenbauer—Pioneer of photogrammetric documentation of the cultural heritage. In *Proceedings of the 18th CIPA symposium* (pp. 19–25).

Alshawabkeh, Y., & Haala, N. (2004). Integration of digital photogrammetry and laser scanning for heritage documentation. *ISPRS The International Archives of the Photogrammetry, Remote Sensing and Spatial Information Sciences*, *XXXV-B5*, 424–429

Altuntas, C., Yildiz, F., & Scaioni, M. (2016). Laser scanning and data integration for three-dimensional digital recording of complex historical structures: The case of Mevlana museum. *ISPRS Int J Geo-Information*, *5*, 18. https://doi.org/10.3390/ijgi5020018

Arias, P., Ordóñez, C., & Lorenzo, H., et al. (2007). Low-cost documentation of traditional agro-industrial buildings by close-range photogrammetry. *Building and Environment*, *42*, 1817–1827. https://doi.org/10.1016/j.buildenv.2006.02.002

Armesto-González, J., Riveiro-Rodríguez, B., González-Aguilera, D., & Rivas-Brea, M. T. (2010). Terrestrial laser scanning intensity data applied to damage detection for historical buildings. *Journal of Archaeological Science*, *37*, 3037–3047. https://doi.org/10.1016/j.jas.2010.06.031

Bang, K., Habib, A., & Müller, M. (2009). LIDAR systems calibration using overlapping strips. *Bol Ciências Geodésicas*, *15*, 725–742.

Barsanti, S. G., Remondino, F., Fenández-Palacios, B. J., & Visintini, D. (2014). Critical factors and guidelines for 3D surveying and modelling in cultural heritage. *International Journal of Heritage in the Digital Era*, *3*, 141–158. https://doi.org/10.1260/2047-4970.3.1.141

Beraldin, A., Blais, F., & Lohr, U. (2010). Laser scanning technology. In G. Vosselman & H. Maas (Eds.), *Airborne and terrestrial laser scanning* (pp. 1–42). Whittles Publishing.

Beraldin, J. (2004). Integration of laser scanning and close-range photogrammetry-The last decade and beyond. *ISPRS The International Archives of the Photogrammetry, Remote Sensing and Spatial Information Sciences*, *XXXV-B5*, 1031–1042

Bertellini, B., Gottardi, C., & Vernier, P. (2020). 3D survey techniques for the conservation and the enhancement of a Venetian historical architecture. *Applied Geomatics*, *12*, 53–68. https://doi.org/10.1007/s12518-019-00267-6.

Boehler W, Vicent M, Marbs M (2003) Investigating laser scanner accuracy. In *Proceedings of 19th CIPA symposium*. Antalya (Turkey)

Böhler, W. (2006). Comparison of 3D laser scanning and other 3D measurement techniques. In E. Baltsavias, A. Gruen, L. Gool, & M. Pateraki (Eds.), *Recording, modeling and visualization of cultural heritage* (pp. 89–99). Taylor & Francis Group.

Böhler W, Marbs A (2004) 3D scanning and photogrammetry for heritage recording: a comparison. In *Proceedings of the 12th international conference on geoinformatics* (pp. 291–298).

Colomina, I., & Molina, P. (2014). Unmanned aerial systems for photogrammetry and remote sensing: A review. *ISPRS Journal of Photogrammetry and Remote Sensing, 92*, 79–97. https://doi.org/10.1016/j.isprsjprs.2014.02.013

Cooper, M. A. R., & Robson, S. (1996) Theory of close range photogrammetry. In Atkinson K (ed) *Close range photogrammetry and machine vision*. Whittles Publishing.

Cheng, L., Chen, S., Liu, X., et al. (2018). Registration of laser scanning point clouds: A review. *Sensors, 18*, 1641. https://doi.org/10.3390/s18051641

De Vos, P. J. (2017) Documenting for posterity: Advocating the use of advanced recording techniques for documentation on in the field of building archeology. *ISPRS The International Archives of the Photogrammetry, Remote Sensing and Spatial Information Sciences, IV-2/W2*, 59–65. https://doi.org/10.5194/isprs-annals-IV-2-W2-59-2017

Dai, F., & Lu, M. (2010). Assessing the accuracy of applying photogrammetry to take geometric measurements on building products. *Journal of Construction Engineering and Management, 136*, 242–250. https://doi.org/10.1061/(ASCE)CO.1943-7862.0000114

Dhonju, H. K., Xiao, W., Sarhosis, V., et al. (2017). Feasibility study of low-cost image-based heritage documentation in Nepal. *ISPRS - The International Archives of the Photogrammetry, Remote Sensing and Spatial Information Sciences, XLII-2/W3*, 237–242. https://doi.org/10.5194/isprs-archives-XLII-2-W3-237-2017.

Eisenbeiss, H. (2009). UAV Photogrammetry. Dissertation, Institut für Geodesie und Photogrammetrie, ETH-Zürich.

El-Hakim, S. F., Beraldin, J.-A., Picard, M., & Godin, G. (2004). Detailed 3D reconstruction of large-scale heritage sites with integrated techniques. *IEEE Computer Graphics and Applications, 24*, 21–29. https://doi.org/10.1109/MCG.2004.1318815

Fernández-Lozano, J., & Gutiérrez-Alonso, G. (2016). Improving archaeological prospection using localized UAVs assisted photogrammetry: An example from the Roman Gold District of the Eria River Valley (NW Spain). *Journal of Archaeological Science: Reports, 5*, 509–520. https://doi.org/10.1016/j.jasrep.2016.01.007

Fassi F, Fregonese L, Ackermann S, De Troia V (2013) Comparison between laser scanning and automated 3D modelling techniques to reconstruct complex and extensive cultural heritage areas. *ISPRS - The International Archives of the Photogrammetry, Remote Sensing and Spatial Information Sciences, XL-5/W1*, 73–80. https://doi.org/10.5194/isprsarchives-XL-5-W1-73-2013

Georgopoulos, A. (2018). Contemporary digital technologies at the service of cultural heritage. In Chanda B, Chaudhuri S, Chaudhury S (eds) *Heritage preservation* (pp. 1–20).

Guidi, G., Beraldin, J.-A., & Atzeni, C. (2004). High-accuracy 3-D modeling of cultural heritage: The digitizing of Donatello's "Maddalena." *IEEE Transactions on Image Processing, 13*, 370–380. https://doi.org/10.1109/TIP.2003.822592

Guarnieri, A., Remondino, F., Vettore, A. (2006). Digital photogrammetry and TLS data fusion applied to cltural heritage 3D modeling. *ISPRS The International Archives of the Photogrammetry, Remote Sensing and Spatial Information Sciences,XXXVI*, Part 5.

Guidi, G., Remondino, F., Morlando, G., et al. (2007). Performances evaluation of a low cost active sensor for cultural heritage documentation. In *Proceedings of the 8th conference on optical 3D measurement techniques* (pp. 59–69)

Grussenmeyer, P., Landes, T., Voegtle, T., & Ringle, K. (2008a) Comparison methods of terrestrial laser scanning, photogrammetry and tacheometry data for recording of cultural heritage buildings. *ISPRS The International Archives of the Photogrammetry, Remote Sensing and Spatial Information Sciences, XXXVII-B5*, 213–218

Grussenmeyer, P., Patias, P., & Hanke, K. (2008b). Applications in cultural heritage documentation. In Z. Li, J. Chen, & E. Baltsavias (Eds.), *Advances in photogrammetry* (pp. 363–383). CRC Press.

Haddad, N. A. (2011). From ground surveying to 3D laser scanner: A review of techniques used for spatial documentation of historic sites. *Journal of King Saud University - Engineering Sciences, 23*, 109–118. https://doi.org/10.1016/j.jksues.2011.03.001.

Hassani, F. (2015). Documentation of cultural heritage; techniques, potentials, and constraints. *ISPRS - The International Archives of the Photogrammetry, Remote Sensing and Spatial Information Sciences, XL-5/W7*, 207–214. https://doi.org/10.5194/isprsarchives-XL-5-W7-207-2015.

ICOMOS. (1964). International charter for the conservation and restoration of monuments and sites. In *2nd international congress of architects and technicians of historic monuments*.

Koch, M., Kaehler, M. (2009). Combining 3D laser-Scanning and close-range Photogrammetry- an approach to exploit the Strength of Both methods. In *Proceedings of the computer applications to archaeology* (pp. 1–7). Williamsburg

Luhmann, T., Robson, S., Kyle, S., Hartley, I. (2006). *Close range photogrammetry: principles, techniques and applications*. Whittles, Dunbeath

McCarthy, J. (2014). Multi-image photogrammetry as a practical tool for cultural heritage survey and community engagement. *Journal of Archaeological Science, 43*, 175–185. https://doi.org/10.1016/j.jas.2014.01.010.

Markiewicz, J., Podlasiak, P., & Zawieska, D. (2015). A New Approach to the Generation of Orthoimages of Cultural Heritage Objects—Integrating TLS and Image Data. *Remote Sensor, 7*, 16963–16985. https://doi.org/10.3390/rs71215869

Micheletti, N., Chandler, J. H., & Lane, S. N. (2015). Structure from motion (SFM) photogrammetry. In L. E. Clarke & J. M. Nield (Eds.), *Geomorphological Techniques* (Online, pp. 1–12). British Society for Geomorphology.

Mateus L, Fernández J, & Ferreira V, et al. (2019). Terrestrial Laser Scanning and Digital Photogrammetry for Heritage Conservation: Case Study of the Historical Walls of Lagos, Portugal. *ISPRS - The International Archives of the Photogrammetry, Remote Sensing and Spatial Information Sciences, XLII-2/W11*, 843–847. https://doi.org/10.5194/isprs-archives-XLII-2-W11-843-2019.

Murtiyos. A., Grussenmeyer, P., & Suwardhi, D., et al. (2019). Multi-sensor 3D Recording Pipeline for the Documentation of Javanese Temples. *ISPRS - The International Archives of the Photogrammetry, Remote Sensing and Spatial Information Sciences, XLII-2/W15*, 829–834. https://doi.org/10.5194/isprs-archives-XLII-2-W15-829-2019.

Núñez Andrés, A., Buill Pozuelo, F., Regot Marimón, J., & de Mesa, G. A. (2012). Generation of virtual models of cultural heritage. *Journal of Cultural Heritage, 13*, 103–106. https://doi.org/10.1016/j.culher.2011.06.004

Patias, P. (2006). Cultural heritage documentation. In *International summer school digital recording and 3D modelling*. Aghios Nikolaos, Crete, Greece.

Pavlidis, G., Koutsoudis, A., Arnaoutoglou, F., et al. (2007). Methods for 3D digitization of Cultural Heritage. *Journal of Cultural Heritage, 8*, 93–98. https://doi.org/10.1016/j.culher.2006.10.007

Pierrot-Deseilligny M, De Luca L, Remondino F (2011) Automated Image-Based Procedures for Accurate Artifacts 3D Modeling and Orthoimage Generation. *Geoinformatics FCE CTU, 6*, 291–299. https://doi.org/10.14311/gi.6.36

Quintero, M. S. (2017. Harnessing Digital Workflows for Conserving Historic Places. *ISPRS - The International Archives of the Photogrammetry, Remote Sensing and Spatial Information Sciences, XLII-5/W1*, 9–14. https://doi.org/10.5194/isprs-archives-XLII-5-W1-9-2017

Remondino, F. (2011). Heritage Recording and 3D Modeling with Photogrammetry and 3D Scanning. *Remote Sensor, 3*, 1104–1138. https://doi.org/10.3390/rs3061104.

Remondino, F., & El-Hakim, S. (2006). Image-based 3D Modelling: A Review. *The Photogrammetric Record, 21*, 269–291. https://doi.org/10.1111/j.1477-9730.2006.00383.x

Remondino F, Barazzetti L, Nex F, et al (2012) UAV photogrammetry for mapping and 3D modeling - Current status and future perspectives. *ISPRS - The International Archives of the Photogrammetry, Remote Sensing and Spatial Information Sciences, XXXVIII-1/*, 25–31. https://doi.org/10.5194/isprsarchives-XXXVIII-1-C22-25-2011

Scherer, M. (2002). About the synthesis of different methods in surveying. In *Proceedings of 18th CIPA international symposium* (pp. 423–429).

Sapirstein, P. (2016). Accurate measurement with photogrammetry at large sites. *Journal of Archaeological Science, 66*, 137–145. https://doi.org/10.1016/j.jas.2016.01.002

Skarlatos, D., Kiparissi, S. (2012). Comparison of laser scanning, photogrammetry and SFM-MVS pipeline applied in structures and artificial surfaces. *ISPRS Annals of the Photogrammetry, Remote Sensing and Spatial Information Sciences, I-3*, 299–304. https://doi.org/10.5194/isprsannals-I-3-299-2012

Salazar, S. D. S., & Marques, J. M. (2005). Valuing cultural heritage: The social benefits of restoring and old Arab tower. *Journal of Cultural Heritage, 6*, 69–77. https://doi.org/10.1016/j.culher.2004.09.001

Salonia, P., Bellucci, V., Scolastico, S., et al. (2007). 3D survey technologies for reconstruction, analysis and diagnosis in the conservation process of cultural heritage. In *Proceedings of the 21st CIPA Symposium*. Athens (Greece)

Soudarissanane, S., Lindenbergh, R., Menenti, M., & Teunissen, P. (2011). Scanning geometry: Influencing factor on the quality of terrestrial laser scanning points. *ISPRS Journal of Photogrammetry and Remote Sensing, 66*, 389–399. https://doi.org/10.1016/j.isprsjprs.2011.01.005

Taboroff, J. (2000). Cultural Heritage and Natural Disasters: Incentives for Risk Management and Mitigation. In A. Kreimar & M. Arnold (Eds.), *Managing disaster risk in emerging economies, disaster risk management* (pp. 71–79). The World Bank.

Themistocleous, K., Ioannides, M., Agapiou, A., Hadjimitsis, D. G. (2015). The methodology of documenting cultural heritage sites using photogrammetry, UAV, and 3D printing techniques: the case study of Asinou Church in Cyprus. In *Proceedings of the third international conference on remote sensing and geoinformation of environment* (pp. 16–19)

Tucci, G., Bonora, V., Fiorini, L., Conti, A. (2016). The Florence Baptistery: 3-D Survey as a Knowledge Tool for Historical and Structural Investigations. *ISPRS - The International Archives of the Photogrammetry, Remote Sensing and Spatial Information Sciences, XLI-B5*, 977–984. https://doi.org/10.5194/isprs-archives-XLI-B5-977-2016.

Westoby, M. J., Brasington, J., Glasser, N. F., et al. (2012). 'Structure-from-Motion' photogrammetry: A low-cost, effective tool for geoscience applications. *Geomorphology, 179*, 300–314. https://doi.org/10.1016/j.geomorph.2012.08.021

Xiao Y, Zhan Q, Pang Q (2007) 3D data acquisition by terrestrial laser scanning for protection of historical buildings. In *International conference on wireless communications, networking and mobile computing* (pp. 5966–5969). IEEE.

Xiao, W., Mills, J., Guidi, G., et al. (2018). Geoinformatics for the conservation and promotion of cultural heritage in support of the UN Sustainable Development Goals. *ISPRS Journal of Photogrammetry and Remote Sensing, 142*, 389–406. https://doi.org/10.1016/j.isprsjprs.2018.01.001

Yastikli, N. (2007). Documentation of cultural heritage using digital photogrammetry and laser scanning. *Journal of Cultural Heritage, 8*, 423–427. https://doi.org/10.1016/j.culher.2007.06.003..

Yilmaz, H. M., Yakar, M., & Yildiz, F. (2008). Documentation of historical caravansaries by digital close range photogrammetry. *Automation in Construction, 17*, 489–498. https://doi.org/10.1016/j.autcon.2007.09.003

Strategic Architectural Interior Design Methodology for Recruitment of the Hieroglyphs Line in Developing Ancient Areas byHi-Tech Technology to Conserve the Egyptian Heritage

An Innovation Vision in the Development of Pyramids Area "Open Bazaars"

Hala Hassanein

Abstract

The concept of developing archaeological areas by using modern contemporary technologies is a creative vision and a new design method to improve and increase the promotional activities for the ancient cultural tourism. Employing hieroglyphic signs of the ancient Egyptian civilization in structural architecture and elements of internal architecture in an unconventional manner will convey a cultural message to visitors that expresses place identity and its heritage. We present a contemporary creative vision in design and open-commercial center including a group of bazaars as proposed addition to the existing project for development of pyramids area in Egypt that coinciding with the construction of "Egyptian Grand Museum" project currently is under-construction to be officially opened by end 2020. The philosophical concept of design of the proposed area stems from the idea of employing commercial activity as "an educational method" that expresses the concept of "Art Education." Such concept transmits different messages to the visitors about meanings of heritage of the ancient Egyptian civilization by using hieroglyphic lines and its signs and related artistic meanings in a contemporary environment through the application of technologies that enable language translation from "Ancient Egyptian Language" to other well-known languages. That will be achieved through an interactive smart design of the show-windows and displays in the commercial units. Additionally, this proposed building contains a number of commercial units, craft workshops, sales kiosks, seating areas, and entertainment places within the framework of planning and design of horizontal projections inspired by the shape of the home sign in the ancient hieroglyphic language. Eventually, the aim is to reach into a conclusion that proven preserving the heritage environment of pyramids area will be giving the modern character of new innovated approach that use ancient Egyptian language as expressed by the hieroglyphs lines and its marks into the design process. Therefore, these marks integrated into internal architecture elements in designing process of bazaars and commercial kiosks units in the proposed project building around the heritage area of pyramids by using modern interactive techniques in displaying windows of the open bazaars.

Keywords

Hieroglyphs lines • Symbols • Innovative design • Interactive techniques • Conserving heritage

Alphabetical Abbreviations

ECOROM	Indicators System Proposal of qualifying certification in ecotourism
GRC	Glass Reinforced Concrete
ICOMS	International Council of Monuments and Sites
IUCN	International Union for Conservation and Conservation of Nature and Natural Resources
LED	Light Emitting Diode
MRYT	Bazaar or market place in the Hieroglyphs language
PCAP	Projected capacitive touch screen technology
TIES	The International Ecotourism Society

H. Hassanein (✉)
Interior Architecture Design, Faculty of Fine Arts, Helwan University, Helwan, Egypt
e-mail: hala_mohamed@f-arts.helwan.edu.eg

A. Versaci et al. (eds.), *Conservation of Architectural Heritage*, Advances in Science, Technology & Innovation,
https://doi.org/10.1007/978-3-030-74482-3_33

UNESCO	United Nations Education, Scientific and Cultural Organization
UN-HABITAT	The United Nations Human Settlements Program is the United Nations program for human settlements andsustainable urban development

1 Introduction

The strategic plan for the pyramids development project as shown below in Fig. 1 is one of the biggest national projects in Egypt and one of the most important archaeological sites of international heritage, according to World Heritage Centre—UNESCO and its program of "The United Nations program for human settlements and sustainable urban development" (UN-HABITAT 2015. *"Strategic Vision for Urban Development in Greater Cairo—Pyramids area"*).

We have noted urban growth around the area that is randomly lacking good planning, that causes problems in all related services and internal movement, which accordingly wastes the region's value as a global shrine.

It is deeming necessary to highlight features of urban heritage in light of surrounding heritage values accompanying with current construction of the "Grand Egyptian Museum," which expected to be open soon to accommodate the largest number of monuments.

That is stressing the importance for restoring the previous status of Cairo and enhancing its cultural weight on the map of world tourism to restore its pioneers of art and civilization to the balance between the development of economic and human resources and the development of the cultural climate of the archaeological areas. Along with restoring the cultural and civilization spark of Egyptian civilization as one unit as a close link expresses the identity of Egyptian civilization, our emphasis will be placed on above defined dotted area in blue as the "Pyramids Development Zone" which is our target area beside the existing buildings in the transitional zone.

Classification of Heritage Areas:

Heritage should enjoy life and activity within its classification into three areas according to the required level-protection priorities. In the Egyptian National Vision as set by Ministry of Culture National Coordinating in (2010) formulating the overall guidelines for the coordination and standards of urban coordination of buildings and heritage areas for the planning pyramids area adopted by dividing it into three geographically areas as shown in Fig. 2. That done for aiming to achieve the purpose of preservation nature and related circumstances of each area and its activities as follows:

Level A—Maximum Protection Area: *is "Heritage Area"*the boundaries of archaeological area include all visible monuments, excavation areas, and surrounding environmental areas.

Level B—Medium Protection Area: *is "Scope of Protection,"* a buffer zone that does not adversely affect the area, where some requirements of service and tourism activities may adversely affect, which requires the necessary protection.

Level C—Transitional Area: is the *"Transitional Range,"* an external and transitional area within the general boundaries of the proposed area of the research project.

UNESCO International Standards and Considerations:

United Nations Educational, Scientific and Cultural Organization (UNESCO) mission is to contribute to the building

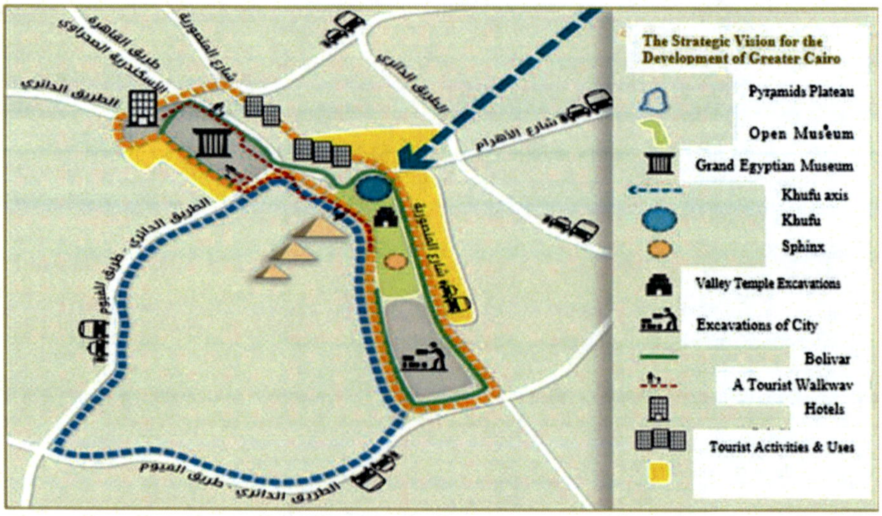

Fig. 1 Strategic vision for urban development in Greater Cairo—pyramids area

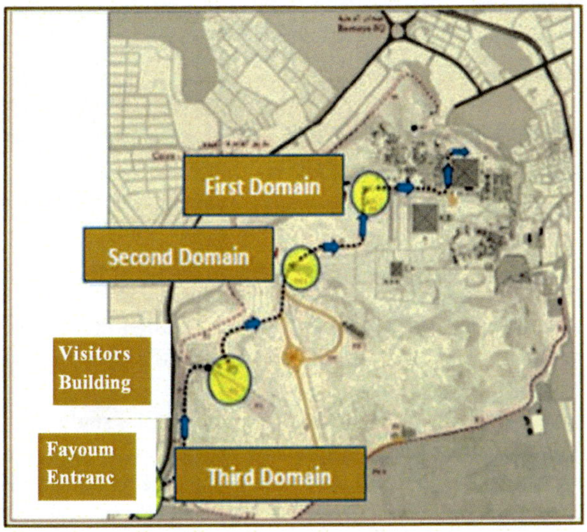

Fig. 2 Three geographical areas of the national vision

of peace, the eradication of poverty, sustainable development and intercultural dialogue through education, the sciences, culture, communication, and information. The basic program that UNESCO is undertaking at Egypt is "World Heritage Earthen Architecture Program (WHEAP)."

It has three heritage conservation organizations:

- International Council of Monuments and Sites (ICOMS) role is evaluation of all nominations of cultural and mixed properties made to the world heritage list by states parties against the criteria laid down by the World Heritage Committee.
- International Union for Conservation of Nature and Natural Resources (IUCN) is responsible for cultural heritage sites that helped to bring the natural and cultural elements of the convention closer together. While cultural landscapes are considered under the cultural rather than natural criteria, IUCN nonetheless played an important role in introducing this new concept to the convention and welcomed this development.
- ECOROM assesses the importance and safety of preservation of property and its indicators system proposal of qualifying certification in ecotourism. Ecotourism is a sector of tourism, based on the nature travel including the principles of sustainability. The International Ecotourism Society (TIES) defines ecotourism as: "responsible travel to natural areas that conserves the environment and improves the welfare of local people." Whatever definition is used, ecotourism should have a positive impact on both natural areas and the local community.

Egypt coordinates with UNESCO as a consultative body when planning its projects in the scope of standards that impose a number of international determinants of heritage protection. Some of these UNESCO standards are as follows:

- Avoid any damage to excavation and archaeology in the stages of exploration and distance from those places.
- Do not allow any heights that hinder the horizontal visibility of the archaeological site.
- No water bodies or plants in the ancient area to protect against leakage.
- Use the same kind of ores and materials that are environmentally friendly.
- Apply same color combinations of the place to achieve visual balance in the surrounding environment areas.

Research Problem:

The problem of research focuses on three axes as follows:

The First Axis:

Currently there is a huge lack of proper identification by our ancient Egyptian language: Ancient language is a language that expresses the Egyptian identity and it is far from our knowledge of it and its related alphabet. The Egyptian hieroglyphs language does not have enough public attention or continuous focus to consolidate our values and cultural language as well as ancient heritage as main theme in the recipient's minds. The ancient Egyptian language is written in four different forms: Hieroglyphs, Hieratic, Demotic and Coptic line, since the hieroglyphs language coming in the first category as the most common in use. Here, we will focus on the oldest of forms that is the hieroglyphs line with the figurative symbols of the nature and environment. The existence of a vertical line down the mark is to confirm expressing itself indicating the content of perception of it without having a certain voice value.

The Second Axis:

Existing built bazaars design is lacking many considerations in terms of.

- Not expressing surrounding nature of the place, and
- Design of commercial bazaars was not properly identified by its functions nor basic elements of architecture interior design in a manner to fit its purposes.
- As well as not properly using visual promotion and the foundations of internal architecture and its elements of façades, and interior spaces.

- Along with the most noted important element which is lack of arriving natural lighting into large number of these built bazaars.
- Eventually also not taking into account, the appropriate architectural proportions and functionally that were not enough to have the proper capacity of expected traffic activity of visitors and does not reflecting the spatial and its heritage of the area where expected to receive around 5000 visitors a day.

The Third Axis:

Lack of interest in the use of modern technological techniques to comply with national plan of 2030 by the government. That is, in regards of interior design and presentation of methods and interactive visual promotion of bazaar interfaces as well as not applying proper application of audio and visual control systems through digital technologies enhancement to convey an educational message to the recipients and visitors.

The Emphasizing Purpose:

The basic idea is incorporating modern technologies into displaying shopping windows methods as a virtual reality technology approach for commercial bazaars. It aims to enhance shopping experiences and to promote marketing its products along with educating recipients and visitors by the basics of the ancient Egyptian language via the implementation of modern and hi-tech ores and materials in the design of the proposed project as addition to the most important national project in Egypt that is "The Developing Project of Pyramids." During the design phase we have properly cope with and taken into consideration all related standards and rules set by the UNESCO that comply with its mission of supervising all Egyptian civilization heritage and the serving of cultural heritage.

The analytical study depends on selecting some ancient geometrical symbols related to design elements reached 13 symbols that expressing the ancient Egyptian language in contemporary architectural meanings properly applied in the design of the walls, main entrances, and external foyer and pharaonic cartridges.

These symbols are the core design idea for designing architecture analysis of the display windows and functional requirements for facades to suit the proposed studying area, while retaining the distinctive features and proportions of ancient Egyptian architecture and the distinctive general character of the surrounding area with using same construction materials used in the past in constructing these monuments. These natural materials are for example red granite and limestone with the integration of modern materials with advanced construction methods, lighting techniques, and acoustics in an integrated manner.

We aim to develop a methodology for designing interior architecture for these types of places as well as a strategy to achieve complementarity and the harmony in a balanced manner to achieve requirements of the modern era through various technologies while at the same time, to preserving ancient heritage using visual promotion and human interaction with computer as innovative modern methods for educating recipients and visitors.

Accordingly, aiming to achieve the concept of art education and self-education through coordination between global considerations and standards organized to protect archaeological areas and stages of the design process and requirements of the design idea and its integration with the elements of architecture.

The interior of the building proposed in the study intended to be in a manner of harmony and formative balance of the elements of internal architecture using the foundations of artistic design such as expression in line, shape, composition, and color.

Therefore, our objectives will focus on below purposes:

- Reviving the ancient heritage and Egyptian identity in a contemporary way using visual promotion methods, and to give the recipient a cultural message to concentrate knowledge of the cultural heritage.
- Use of the hieroglyphs alphabet in creative design of bazaars and kiosks in a manner that is inspired by the vocabulary of the Egyptian art.
- Increasing the value of the pharaonic civilization, emphasizing the cultural and historical identity.
- Interactive technology and its role as a link between old and modern by transmitting messages to the recipients.

Eventually, concentrating on the research to achieve its goal that is preserving heritage environment of the pyramids area does not negate the existence of contemporary tools. These tools add to and support the place identity, legacy, and its archaeological nature to attract and promote contemporary cultural tourism, through transferring educational messages merging ancient and contemporary arts, as new approach uses hieroglyphs to establish the value of the ancient Egyptian language in the minds of the recipients and the visitors.

Strategic Design Methodology:

The strategic approach to the design process aims to develop a curriculum or guide for various development and planning organizing bodies as well as for interior architecture designers for designing building near to these archaeological areas.

Where this curriculum was prepared in phased stages according to its importance that linked to each other along with the rules, regulations and laws regulating the preservation concept of these archaeological areas and their identity. Expected gain from this strategy is to create an innovative new trend combining the heritage and contemporary art to creating indicative model in a quality manner of special nature for commercial buildings located next to the archaeological areas, and as expressive and cultural tool serving the areas visitors.

As shown above in Fig. 3, the proposed methodology starts with number of linked stages in a sequence manner:

- Basic analytical studies of the civilization, environment, and geographical aspects of the place and ways of communication between them.
- Then an analytical study of the hieroglyphic signs in the ancient Egyptian language and their meanings and analysis of their distinctive graphics through calligraphy, shape, and color.
- Then comes the study of the humanitarian requirements, as well as the economic and tourist requirements of the state.
- Then the stage of design process for a number of sub-analytical studies in the fields of designing ancient Egyptian architecture, elements, and its features in terms of raw materials, determinants of the architectural form, proportions, and heights. Eventuality, designing internal architecture from the spatial analysis of internal spaces, the quality of the targeted activities, and their distribution to those spheres, then the analysis of internal movement paths and their distribution to activities.
- Last, studies on modern technological applications used in all elements of interior architecture, including audio-visual equipment, devices, modern materials, and advance lighting methods.
- Then to conclude and employ all those analytical studies and steps as mentioned above, to build and complete the creative idea and the design process to ending with the completion of proposed project as planned.

A Basic Informative Studies:

A-1: Humanitarian Needs, Civilizational, Environmental, and Economic Factors:

Humanitarian Needs:

It is the desires and requirements that people need inside the place first from practical needs point of view and to psychological needs that are together providing assistance while performing the shopping trip by pleasure, comfort, and then access to the cultural information that they are seeking for. The design aims to create a cultural and promotional environment for the shopper and visitor to get souvenirs in the shortest possible time and the fastest way while taking advantage of cultural information about the ancient Egyptian language through interactive display windows as well as each element of interior design that is embed in the project.

Civilizational Factors:

The design of touristic bazaars in the archaeological areas is influenced by several civilizational factors, namely the study of the site and its relationship with the historical nature of the area and the urban component of the structural elements as part of the overall design of the site as well as the design of car parks, layout of entrances and exits, and the relationship between the site and traffic of the surrounding streets.

Environmental Factors:

Identifying the surrounding environment is deemed necessary to serve the activities operations within the touristic bazaars because it is affected by all of, the climatic, audio, and visual effects in the surrounding environment.

Economic Factors:

It is the economic and investment return for the region by attracting the largest number of tourists and pioneers to achieve the goal of profitability and create a development area for the community surrounding those areas. The economic factor is an important element in measuring the success of any project and determining its feasibility and expected investment return. Therefore, financial and economic feasibility studies are important and essential as well, because they include in details: market analysis, marketing studies, financial ratios to measure profitability and feasibility of investing in such kind of activities within the archaeological areas.

A-2: Geographical Studies of Development Project & Existing Built Bazaars:

A-2–1:Egyptian National Plan Briefing and its Location:

The national plan is located on an area of 20 square kilometers distinguished by the uniformity of its design forms from the surroundings environment includes integrated areas including the main entrance to the development project, and the areas of the pyramids and sphinx. The environmental

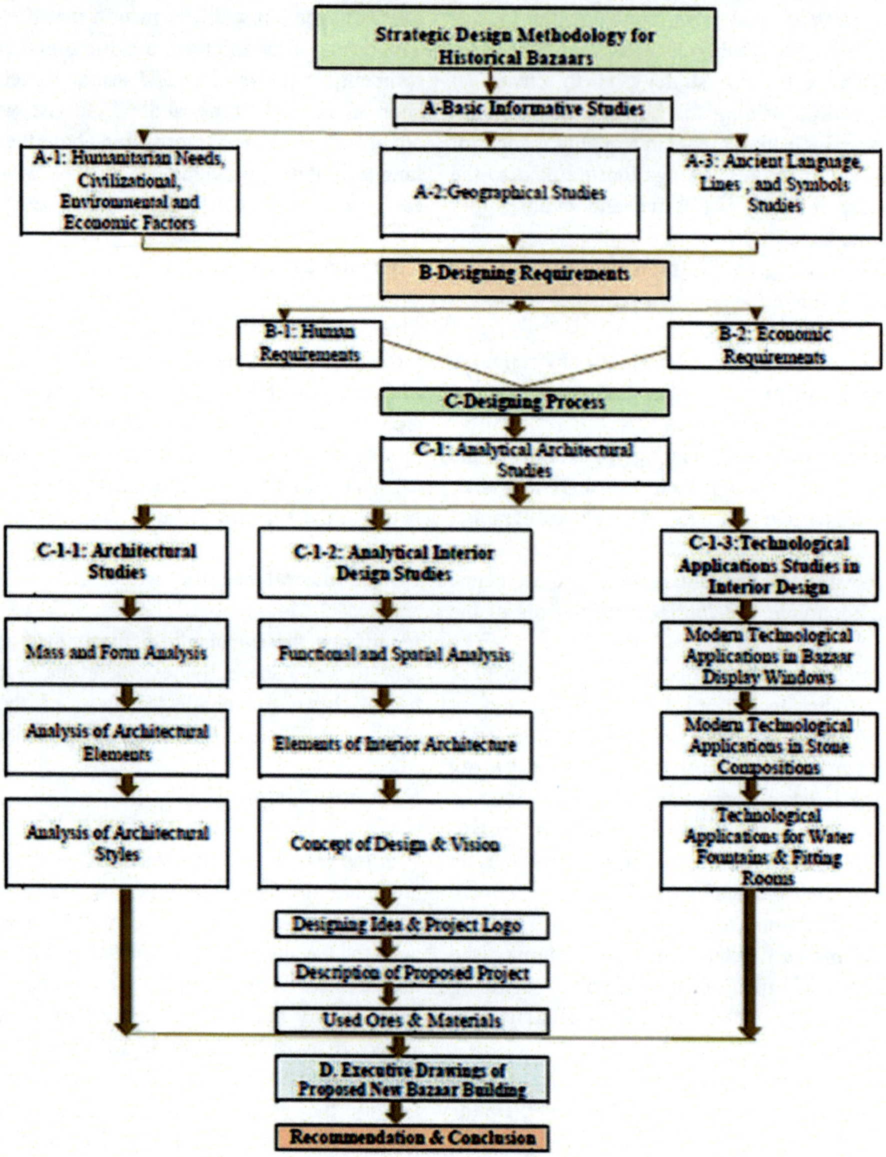

Fig. 3 Historical bazaars design methodology

impact was taken care by adding the electric cars for visitors to move to the visitors' building in the transitional area that built on 1250 square meters for each role.

It consists of a reception area, lecture halls, cinemas, panorama terrace on pyramids, administrative offices and services; it also contains an area of 500 square meters dedicated to a limited number of bazaars in small areas. As shown in Fig. 4, the plan included the identification of seven tracks of the visit starting from the main entrance on Fayoum Road to visitor center, the pyramids, and the western cemetery then to sun boats and to Sphinx then, finally returning back to the reception area.

In order to complete the development plan, the traffic routes re-planned and linked to the Grand Museum area with

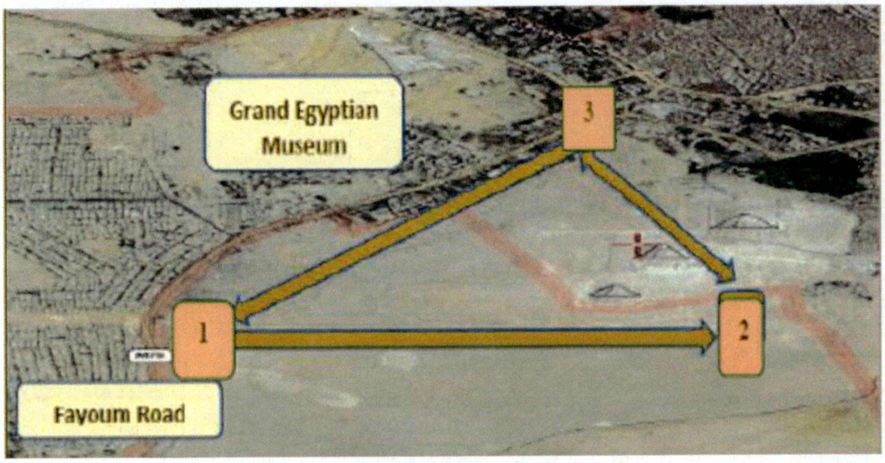

Fig. 4 Triangle of reception area, western cemetery & sun boats, and the three pyramids and sphinx **1** reception area **2** western cemetery & sun boats **3** three pyramids and sphinx

a small helipad for the visitors. The pyramids area became an integral part of the development project. It is the main objective of the development project in a modern and distinguished way.

The above plans shown in Figs. 5 and 6 are for the development of the pyramids plateau designed and submitted by CUBE Consultants Company, that is responsible for the design of "New Administrative Capital." The said grand national planning includes various projects for hotels, green garden for "Grand Egyptian Museum," "Pyramids Plaza," bolivar, tourist walker area, and pyramid plateau (scenic captures).

A-2–2: Status of Existing Bazaars:

Within the strategic plan of developing pyramids area, the existing bazaars located in the visitors' building as shown and marked in yellow color of 18 bazaars on two groups, that each one comes with nine bazaars in each side using light materials (gypsum boards) with an approximate area of 9–10 square meters. It has 2.5 square meters and its rear facades, which are not well in light and are not suitable for natural lighting resulting inefficient lighting in the place.

The shaded area in Fig. 7 is for the new proposed project marked in dark brown color:

The visitors' center designed by Tarek Waly Centre for architecture and heritage includes the entrance (1) and ticket area as Fig. 8a–d, celestial courtyard (2) Fig. 10, main hall with a general layout (3) Fig. 9, cinema hall (4), bazaars inside the building (5) as Fig. 11 a, b, then exit and parking area (6). The location of suggested project as having no (7) is shaded in dark brown in the north of above Fig. 7, some

Fig. 5 Overview of the national project upon completion

other kiosks were placed as outdoor in front of the visitors building for rending services of security and banking in Figs. 11 and 12 as follows:

A-3: Ancient Hieroglyphs Language Lines, Symbols, and Related Meanings Studies:

A-3–1: Graphic Shapes of Hieroglyphs Marks:

Hieroglyphs consist of three kinds of glyphs (Howard 2012): phonetic glyphs, including single-consonant characters that function like an alphabet; logographs, representing morphemes; and determinatives, which narrow down the

Fig. 6 Geographic layout in designing phase

meaning of logographic or phonetic words. The hieroglyphs line used in its figurative meanings and the historical value of the ancient Egyptian language, inspired by the design process of the proposed project in the stereoscopic design of various elements of interior architecture inspired by some hieroglyphs symbols:

- Ideograms means the objects drawn with vertical line below it to indicate its meaning.
- Phonograms indicate pronunciation of these phonetic symbols, as the one-sounding labels known as "alphabet."

As shown in Fig. 14 example of selected, 12 hieroglyphs symbols used in our design phase. We illustrated selective 6 symbols, figuration meanings and its related pronunciation from "Ancient Egyptian Mythology" and "Ancient Egyptian Language book" by Allen (2013) as follows:

A-3–2: Digital Hieroglyphs:

The ancient Egyptian was concerned with the numbers where the numbers are divided into two types, numerical (1, 2, 3, etc.) and alphabetical order (first, second, third…etc.), (O'Connor and Robertson, n.d.). Here is an example of the numbers on one of the walls of El-Karnack Temple as shown in Figs. 15 and 16.

Where these numbers used in the numbering of display facades in bazaars through the design of the doors to be within the general fabric of the place. That done in light of ancient Egyptian identity and civilization, so that the recipient can enjoy and attract his attention to the cultural and heritage aspects of the place expressed in all details of the design aspects of interior architecture.

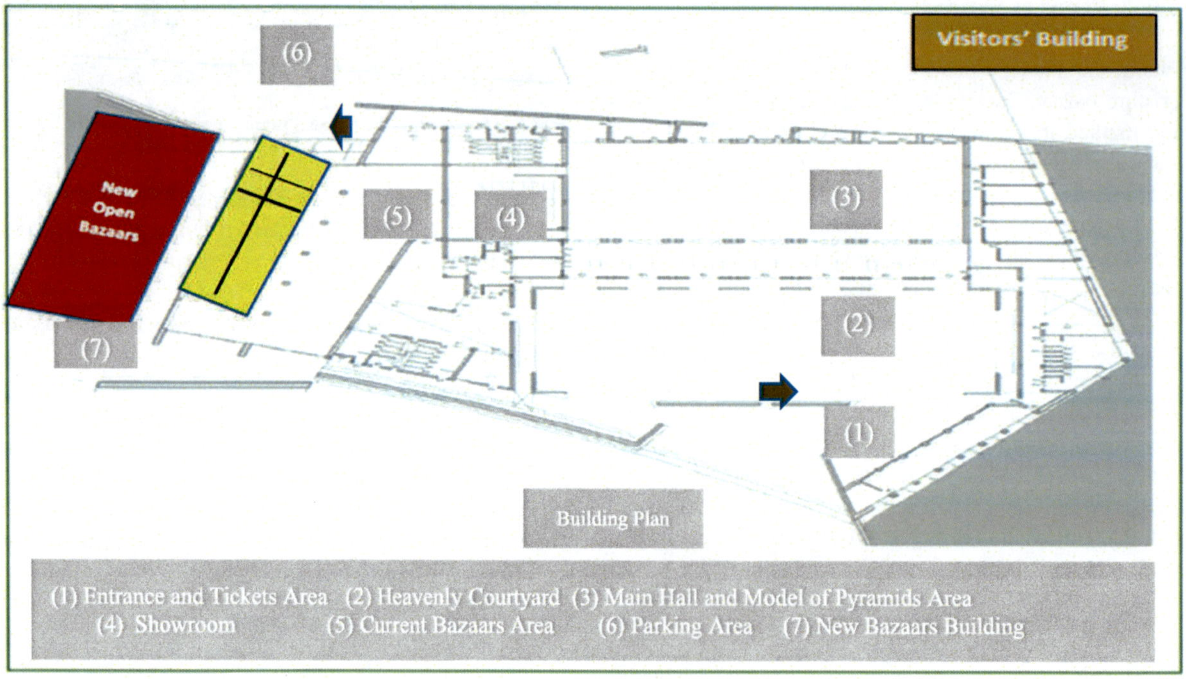

Fig. 7 Detail components of existent visitors' building—by Tarek Waly Centre

Fig. 8 Side main entrance of the visitors building. **b**. The main entrance of the visitors building. **c**. The ticketing area. **d**. The entrance

Fig. 9 Main hall with general layout

Fig. 10 Celestial courtyard

A-3–3: Using Hieroglyphs Colors:

We selected Egyptian hieroglyphs symbols from KV17, Seti I, thirteenth century BC (Dunn 2014, Wikipedia) in our designing process to achieve two goals, first is to ensure the unity in design and to provide assurance that correctly picking and selecting the same natural colors groups and its various degrees as used in pharaonic inscriptions in ancient temples and monuments. We have used the set colors group from green to white as shown its variety degrees in Fig. 17:

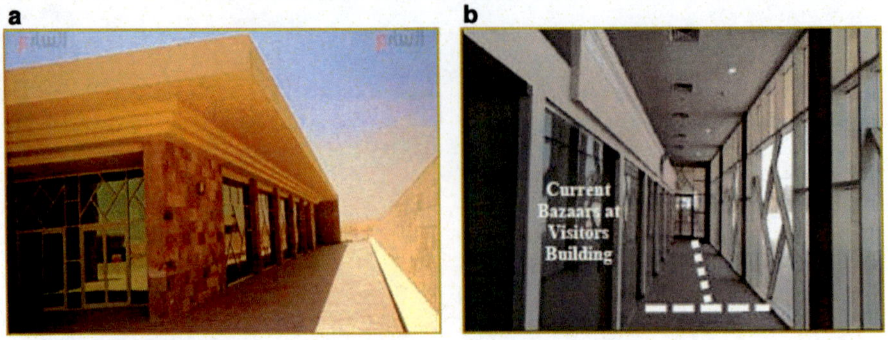

Fig. 11 Front and side walls of current visitors center. **b**. Existent bazaars on left side in small spaces

Fig. 12 Model of outdoor kiosk with seating area

Fig. 13 Outdoor banking kiosk—ATM services

B. Designing Requirements:

It is divided into two main categories, first is the human requirements and then the economic requirements as follows:

B-1: Human Requirements: The importance of design arises from the necessary elements of humanity in meeting the needs of human and has been designed in a human aspect as one of the artistic foundations of our contemporary life extending to include all the aspects of life. The goal of teaching design is to have the ability to observe using all different senses from imagination, organization and link all these information and forms together within the surrounding environment to discover the relationships and regulations in it, accompanying with practice experiments in solving the technical problems to achieve overall purpose of the design.

B-2: Economic Requirements: Economy, it is a prerequisite in architecture, which limits to some extent the freedom of the architect; however, it teaches the architect to rely on pure art made up of elements of the building itself and not from the beautiful artistic additions only.

The economy is not intended to be the economy in its basic needs, but is intended to achieve a balance between the main aspects wisely and objectively, as well as is intended to reduce wasteful and extravagance, but provided that this does not lead to a breach in the function of construction, beauty, and durability, this is achieved through:

- Appropriate selection of shapes and sizes for building elements.
- Appropriate selection of construction materials.
- Accuracy in implementation.
- Proper follow-up of the design and implementation stages.

A. Designing Process:

Concept of Creative Design and Vision:

The philosophical concept of designing the proposed area is to present the idea of employing commercial activity as an indirect education method that conveying a message to the

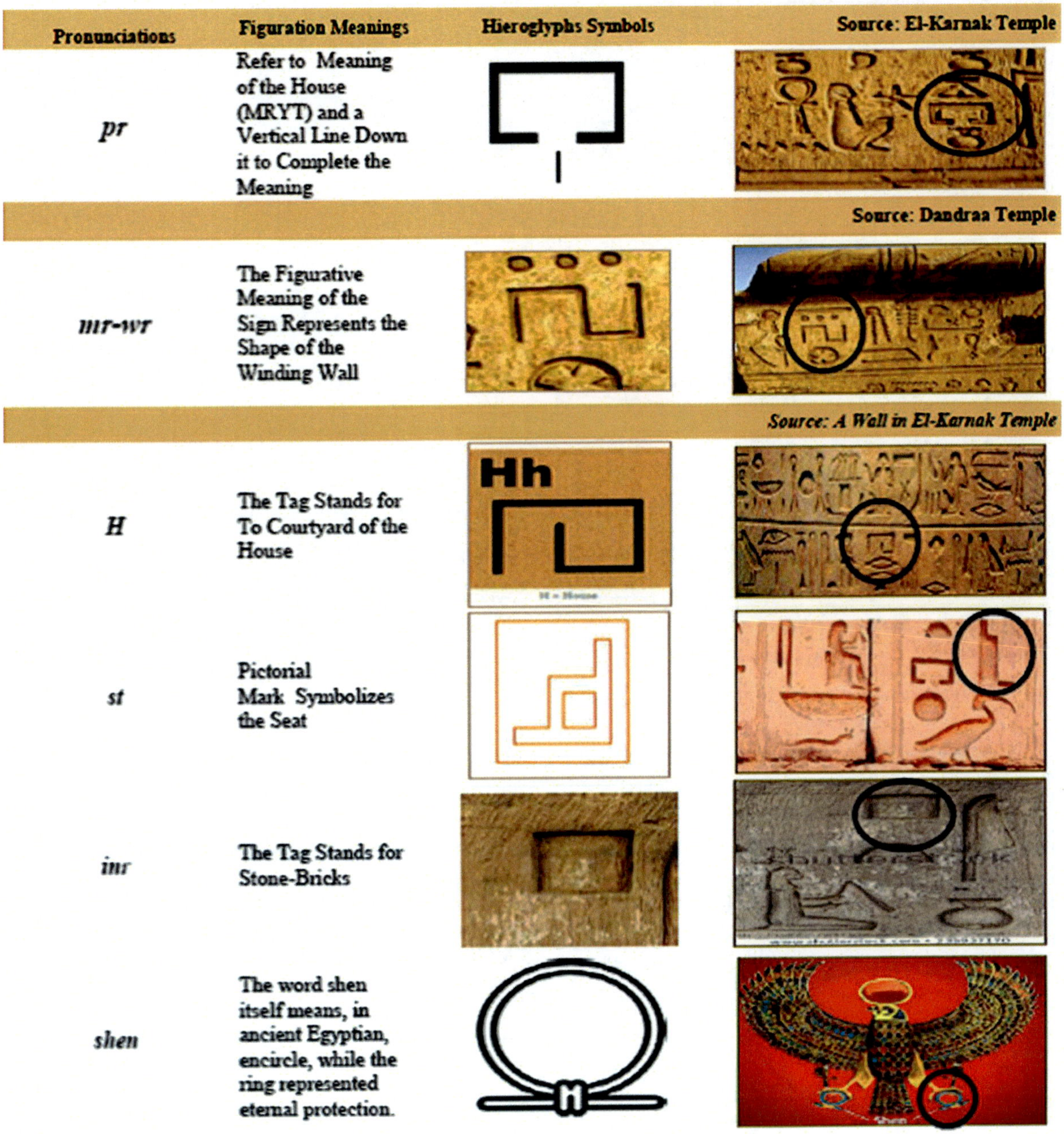

Fig. 14 Selected number of hieroglyphs mythology and symbols

recipients about the meaning and civilization concept by using the hieroglyph line and its imaginative meanings in a contemporary environment. That is doable via using the technological techniques of cultural definition and translating the ancient Egyptian language into other known languages through the interactive design of windows display interfaces. Usage of interactive screens is defining role in transmitting cultural-inspired educational messages as a key feature of establishing the concepts of ancient language and the hieroglyphs line in the minds of visitors as well as to raise and improve the marketing experience of visitors and their interaction with intellectuality letters conveyed to them. Therefore, the philosophy of designing open bazaars building inspired by the graphic representation of the hieroglyphs symbols expressed in structural parts in a creative design with the introduction of the role of technology in the delivery of information and data by using interactive techniques in windows of the commercial establishments (i.e.,

Fig. 15 Figures engraved on one of stone blocks

Fig. 16 A panel expresses numerical numbers meaning

open bazaars). The design of the new zone is in line with the strategy of preserving ancient heritage and natural civilization taking into account the global determinants, considerations set and established by UNESCO and its existing organizations to oversee the Egyptian monuments and to preserve the ancient heritage and Egyptian civilization.

C-1: Analytical Architectural Studies:

Analytical studies is core part of the paper; it consists of three sub-studies as follows:

C-1-1: Architectural Studies.
C-1–2: Analytical Interior Design Studies.
C-1–3: Technological Applications Studies in Interior Design.

C-1–1 Architectural Studies:

Architecture is the generation of social and cultural heritage in the form of built spaces. It is evolving to include other important forms of design, including space planning, building materials, and the integration of the built and ecological environments. These studies include mass and form analysis, analysis of architectural elements, and analysis of architectural styles as follows:

I. ***Mass and Form Analysis***:

There are two main aspects of the general design that must be taken into consideration, namely the architectural mass and the external facades and form of the building, where the mass can be in cubic or cylindrical or in the form of a pyramid or otherwise. The second aspect is the design form that could be in open or closed approach. The open design is in open air with large open space accompanied by as much natural lighting and few walls as possible, while closed design is in closed space depending heavily on industrial lighting rather than the nature one. Because of the nature of the area, we will implement the second aspect of the design that is the closed design approach.

II. ***Analysis of Architectural Elements***:

The idea of designing the mural shape for the proposed new project by the researcher is illustrated from Egyptian life and death, taken from the tomb chapel of Nebamun on the west bank of the Nile at Thebes in a form of a pharaonic pond garden in the tomb (Parkinson 2008). From that idea, the researcher visualizes the concept of layout for the proposed

Fig. 17 Samples of natural colors group

Fig. 18 Pond in a garden from the tomb of Nebamun, Thebes

open bazaar building in the transitional area in a form of placing the bazaars surrounding the lake in the manner as used in the below mural. That helped to coordinate the interior of the different activities of display facades, traffic paths, kiosks, the seating areas and the open courtyard, the shape of the water ripples surrounded by a group of bazaars, pergolas and seating area in a design that achieves the concept of the mural of the pharaonic garden shown in Fig. 18.

That above mural of the "garden" of the Nebamun cemetery on the west bank of the Nile at Thebes in Luxor exhibited in the British Museum painted on the plaster 72 × 62 cm.

One of the architectural influences the design of main facade of the new commercial building and inspired by it is the facade of the "Sakkara Museum" as shown in Fig. 19 and 20 known as "Amenhotep Museum" the ancient Egyptian architect. It is located to the right of the Sakkara area and contains more than 250 different antiquities, the use of the main line in the design of general shape and height ratios of the openings in main facade of the new building through the development of the design spirit of the above facade of the museum.

III. ***Analysis of Architectural Styles***:

The architectural styles in the ancient Egyptian architecture have artistic features with a beautiful expressive vision when used in the formation of different surfaces and areas. That these styles are much concern by:

- The use of geometrical design in line and composition.
- Avatar in expression.
- Using specific natural colors in the design as white, turquoise, brown, red, green, and black.
- Applying simple shapes to plan the horizontal plans such as square and rectangle shapes.

Fig. 19 Façade of Sakkara Museum

Fig. 20 Sakkara Pyramid

- Pivot and symmetry between parts of the building or the temple.
- Vastness vertical surfaces such as in walls, windows, and entrances.
- Heavenly open sky courtyard surrounded by corridors covered with columns to receive delegations and visitors.
- Last but not least, the main entrance is one between two pylons with beautiful pharaonic inscriptions bearing a number of lines, letters, and its related meanings expressed in the ancient hieroglyphic language.

C-1–2: Analytical Interior Design Studies:

The analytical study based on selection of some geometrical symbols related to building design elements reached 13 symbols such as the sign of house containment, walls, and entrances. Its applications through the geometrical analysis of the windows of the display windows and their basic elements (display windows, entrance—advertising banners

… etc.), and functional requirements of the facades to fit the proposed space while retaining the features and meanings of the hieroglyphs line. We have used major ores and raw materials, such as red granite that used in many statues, mosaics, paintings, etc., also limestone that dates back to the Neolithic era and used until the middle of the 18th Dynasty.

C-1–2-1: Functional and Spatial Analysis:

The external and internal spaces classified according to their functional needs and type of activity used, aiming to become important attractions domains for visitors and shoppers. These spaces meet three main different basic needs, the social relationships area, service exchange area, and the area for information exchange. At the start of design process, it has been planned to have specific areas for parking the electric cars and other areas for social purposes, such as a reception area includes the entrance, reception area, and the movement paths.

The spatial analysis relates to composition of architectural form of the building and to spatial components that include open areas such as the public square, corridors and the reception area that determine extent of direct or indirect contact between all domains from a functional point of view.

As shown above in Fig. 21, the planning of commercial and administrative areas in the proposed new project starts at parking area upon visitors' arrival, going to entrance and its reception heading to the area of new bazaars with its interactive and hi-tech windows ending with craft artisan workshops. Then moving forward to the exceptional beautiful view of public square area, the middle lake and sculptural formation and after finalizing the tour to the bazaars and pergolas, they moving around arriving at last to the management and control areas ending by offices of maintenance and working staff.

C-1–2-2: Elements of Interior Architecture:

The proposed design of kiosks in the proposed project building is simulated in shape of the royal palace called "The Throne." A triumphant Hunefer (Wikimedia), having passed the weighing of the heart, is presented by falcon-headed Horus to the shrine of the green-skinned Osiris, God of the underworld and the dead, accompanied by Isis and Nephthys.

It is made of wood decorated with pharaonic colors motifs. The design methodology is inspired by the beautiful decoration of the facades and floors in the same style as the pharaonic drawings. As shown in Fig. 22 presenting the papyrus of Hunefer that dating back to the Nineteenth Dynasty (1317–1285 BC), and nowadays is exhibited at the British Museum (London) representing "God of The Dead" Osiris sitting on the throne (Wikimedia Commons).

C-1–2-3: Concept of Creative Design and Vision:

The philosophical concept of designing the proposed area is to present the idea of employing commercial activity as an indirect education method that conveying a message to the recipients about the meaning and civilization concept. That done in a formal manner using the hieroglyphs line and its imaginative meanings in a contemporary environment by using the technological techniques of cultural definition and translating the ancient Egyptian language into other common languages through the interactive design of windows display interfaces.

Usage of interactive screens is playing a defining role in transmitting cultural inspired educational messages, as a key feature of establishing the concepts of ancient language and the hieroglyphs line and language in the minds of visitors and recipients. That is to add innovative value to raise and

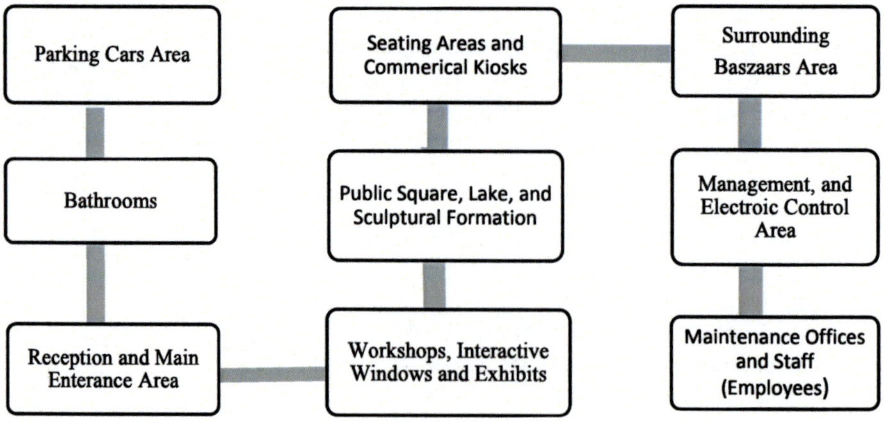

Fig. 21 Visitors circulation movement paths at the new proposed building

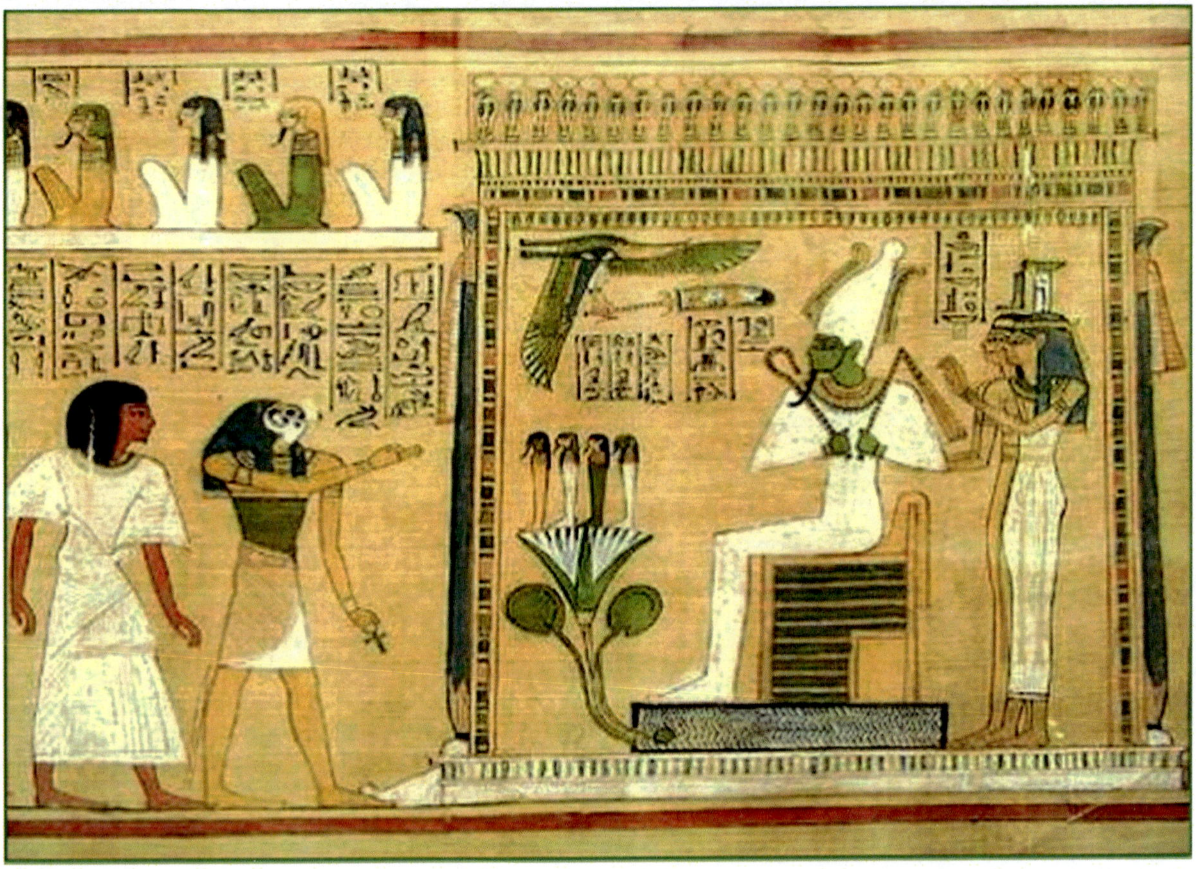

Fig. 22 Papyrus of Hunefer

improve of the marketing experience of visitors and their interaction with intellectuality letters and ancient knowledge conveyed to them. Therefore, the philosophy of design of the open bazaar building inspired by the graphic representation of the hieroglyphs symbols expressed in structural parts in a creative design with the introduction of the role of technology in the delivery of information and data in an interesting manner using modern interactive techniques in windows of the commercial establishments (open bazaars).

The design of the new zone is in line with the strategy of preserving ancient heritage and natural civilization taking into account the global determinants, considerations set and established by UNESCO and its existing organizations to oversee the Egyptian monuments and to preserve the ancient heritage and Egyptian civilization.

C-1-2-4: Designing Idea & Project Logo:

Designing idea of the project as Fig. 23 is to conceptualize the design of a business that combines a number of bazaars and commercial activities in an architectural form. Inspired

Fig. 23 Project logo

by the hieroglyphs symbol (Manley 2012), of the symbol of the house  known as "MRYT" word that means bazaar or market place in the hieroglyphs language referring to the meaning of containment of new proposed 16 bazaars and artisan workshops in our project. Such design is in around form as exposed square with an area to simulate a lake symbolized by motifs and mosaics to the water, and also includes a flat layer of transparent concrete cubes illuminating in formation refers to symbol of the house sign pointing to the natural building stones for construction of the pyramids.

Then drilling some symbols in an abstract way, in the middle in the form of a lake inside a pharaonic garden. The hieroglyphs line contains many symbols that symbolize the sound and visual expression the surrounds of the human being. To give a creative dimension using the latest technology in the interactive interface screens as of Projected Capacitive Touch Screen Technology named "PCAP touch film" and visual and audio visual systems in the framework of contemporary design to gain the greatest cognitive ability and to establish meaning and concepts of the old language in the minds of the recipient.

C-1–2-5: Description of the Proposed Project:

The project is located in the transitional area range (marked in dark brown) overlooking Fayoum road of which is the main entrance to pyramids development area, and that area where it is allowable to build buildings and its related rendering services and facilities in it.

The project area is designed in a geometrical form (oblique) so that the building open bazaars proposed on the form of a rectangle and is located between visitors' building and the educational students' building. The facade of the new project is in front of the side of the visitors' building directly. We planned to establish 16 open bazaars with a horizontal level from one floor to the existing bazaars. Movement paths from the reception to the outside waiting areas designed for and properly accounted for.

C-1–2-6: Used Ores and Materials:

We reviewed "Building Materials of the Pyramids Builders" (Dunn 2015). When approaching to select ores and materials to be used in our designing process of the new proposed building in the transitional area as shaded in Fig. 24, we have selected the most common materials that where mostly taken from "Building Materials of the Pyramids Builders." The building project is in dark brown that is based on all types of surrounding buildings to maintain and conserve the architectural style and in same the height and quality of raw materials used. Ores and materials used are units of limestone "cream stone" and the red stone of 80 cm to cover the external facades transparent in gold color. The floors in the surrounding buildings have been used raw materials tiles limestone slab materials called the head of steel sizes of 30 × 60 cm and thickness 4 cm & 5 cm to bear the density of the weight of the audience and visitors. Also in addition, other materials such as pink granite, basalt, and alabaster that were used much more sparingly. Pink granite probably most often came from the quarries around Aswan.

C-1–3: Technological Applications Studies in Interior Design:

C-1–3-1: Modern Technological Applications in the Bazaars and its Display Windows:

It is a technique of projection transforming any glass surface or acrylic into an interactive touch screen; it is one of the best modern concepts of displays in commercial units and bazaars. The interactive touch chip is manufactured from the electrostatic free projection lining through advanced transparent hi-tech technology named *Rextouch* that is ultrawide stretched LCD signage and its related sensitive sensors, (Rextouch) as below image, are glued to the back of the glass or acrylic to turn those surfaces into interactive touch screens, Figs. 25 and 26:

There are two types:

I .Dedicated to external environment that called "Sun Screen Technology" which enables the visitor to read applications displayed in the sunlight.

II .High definition and contrast "Pro-Diffusion Clear View" which uses the projection feature of this screen to the optical filter that distort and reflect the light to allow reading in light of the sun.

The sizes of electronic foils *(Tree-foil, Interactive Touch, Touch Foil)* comes in different sizes ranging from 15" to 86" with the use of projector connected to a computer loaded with information to display messages such as the vocabulary of hieroglyph dictionary translated into any language such as English, French *(projetrosette.info),* and German (Erman & Grapow 1921). That done by converting the glass front window of the commercial bazaar into a large touch screen to allow conveying information on them to the visitors as Fig. 26a–d:

In addition, the nano touch material can be used to create two nanoparticles that intersect with the nanotube layers. The microcontroller is sent to the computer by positioning the touch on the screen to provide the required information to activate the free interaction between the human and the computer—Fig. 26d.

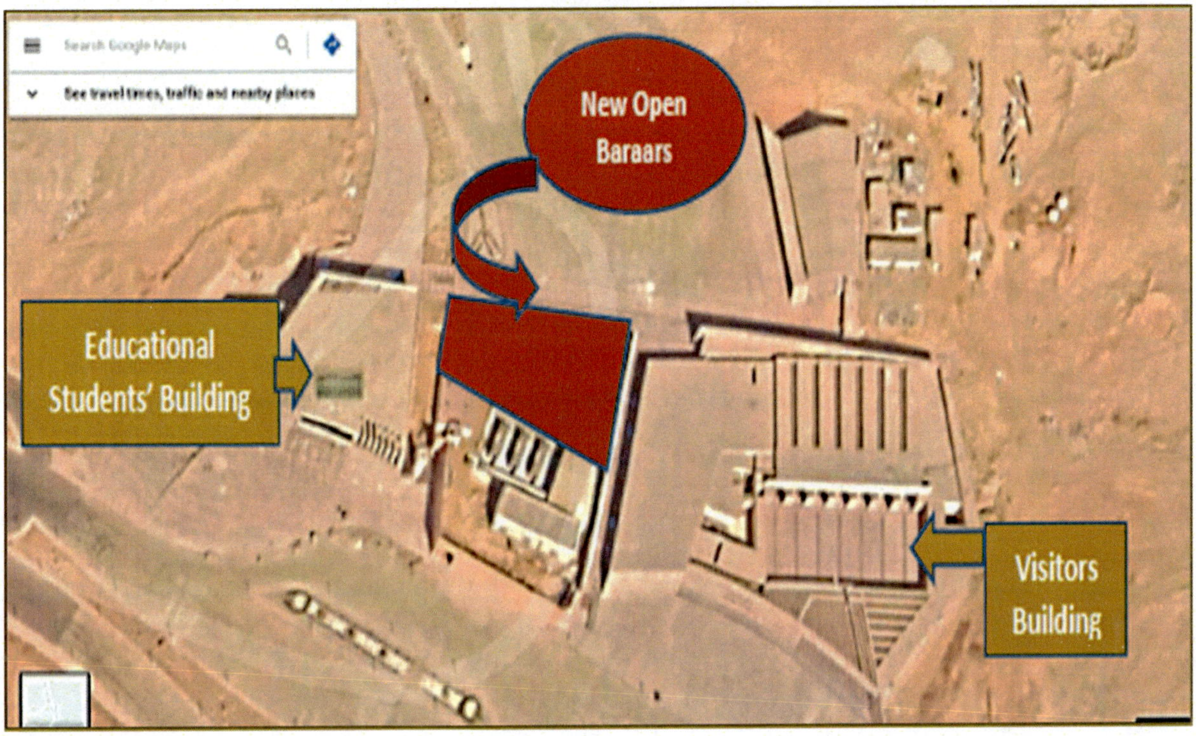

Fig. 24 New Proposed Project Location in the Transitional Area

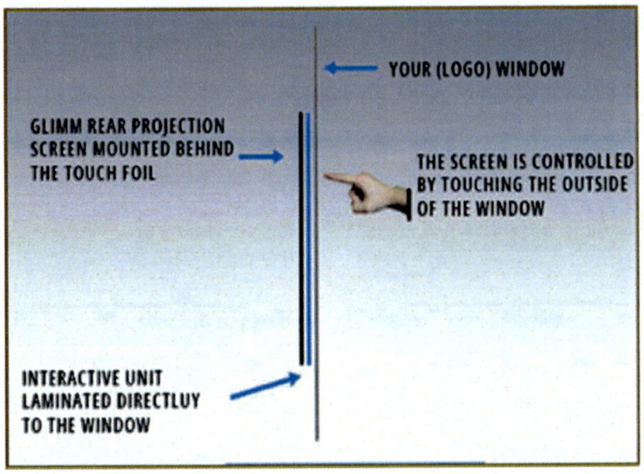

Fig. 25 Components of screen for interactive windows

C-1-3-2: Modern Technological Applications in Stone Compositions:

Transparent concrete also called as translucent concrete *(pinterest.com)* or light transmitting concrete is achieved by replacing aggregates with transparent alternate materials. The bonding material in transparent concrete may be able to transmit light by using clear resins the concrete mix. Use of optical fibers and fine concrete also used as transparent concrete. The stone composition that symbolizes the pyramid building stones engraved with a number of hieroglyphic symbols used of open courtyard overlooking the bazaar group. A modern innovation was developed to light the pharaonic stones as shown in Fig. 27 with their hieroglyphs inscriptions using the raw material technology of the concrete with transparent properties known as "Translucent Concrete" that discovered and patented in Canada in 1935 and internationally applied by Civil Engineering Home for Civil Engineers.

It is the common material of concrete but with the property of light-transferring out due to embedded light optical elements via optical fibers. As shown below in Fig. 28a, b, c, d, it has a special characteristic to activate the light and move it from one side to another by adding the fiber optic material that is not more than thickness 0.5 mm in diameter to shine in natural light of sunlight or artificial lighting sources aesthetically as follows:

This suggested hi-tech material employed in a technical work embodies the stone blocks that refer to surrounding environment of the pyramids area using the composition and sculptural composition of the symbol of the house sign as form:

These stones were also engraved with hieroglyphs symbols in design of the bazaar façades and were lit using LED lamps. Figure 29a, b explain how light appears in the light transparent concrete:

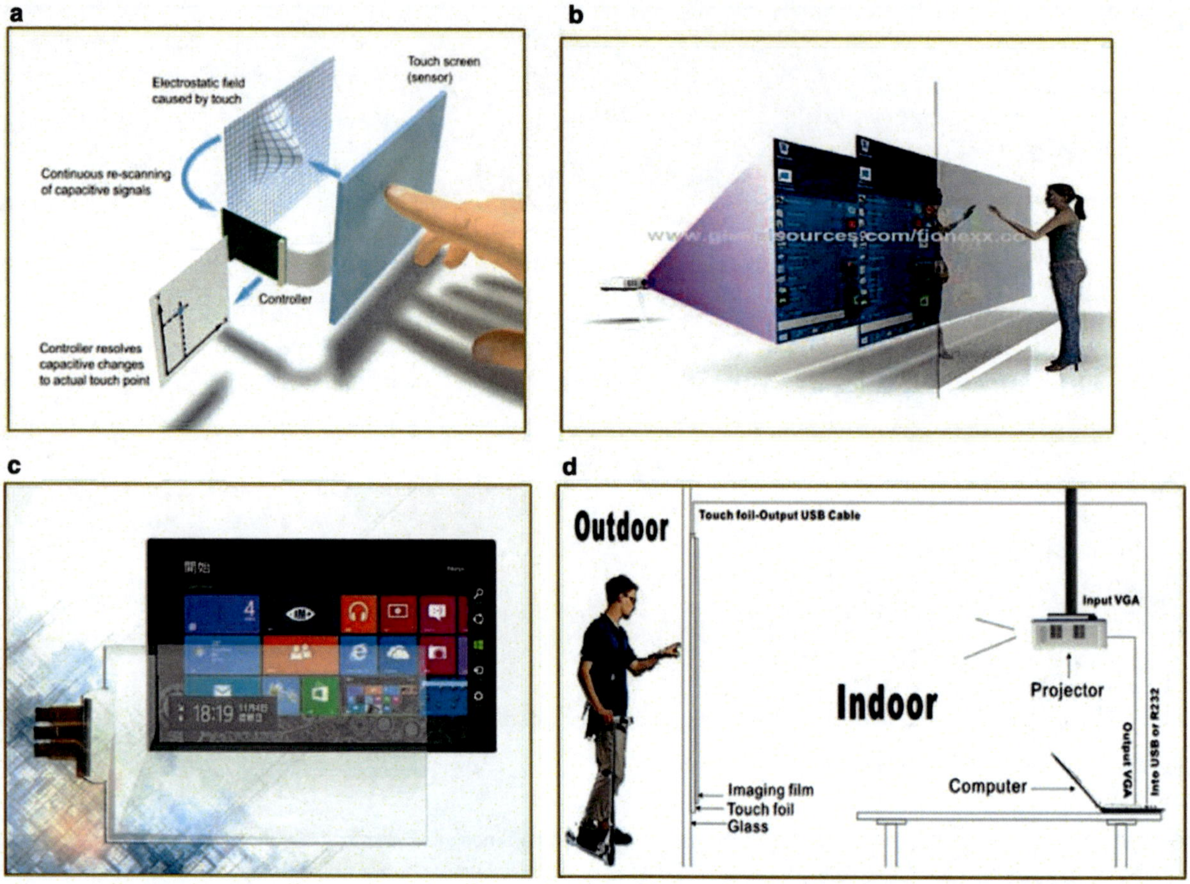

Fig. 26 Sensitive chips sensors. **b**. Human–computer interaction. **c**. Interactive screen for displaying windows. **d**. Components and detailed structure of interactive screens

Fig. 27 Examples of previous applications worldwide

Fig. 28 a Model of a sample on stone of light concrete. **b** Solid stone of light transparent concrete. **c** Lighting fibers in the concrete at daylight. **d** Lighting fibers in the concrete at night

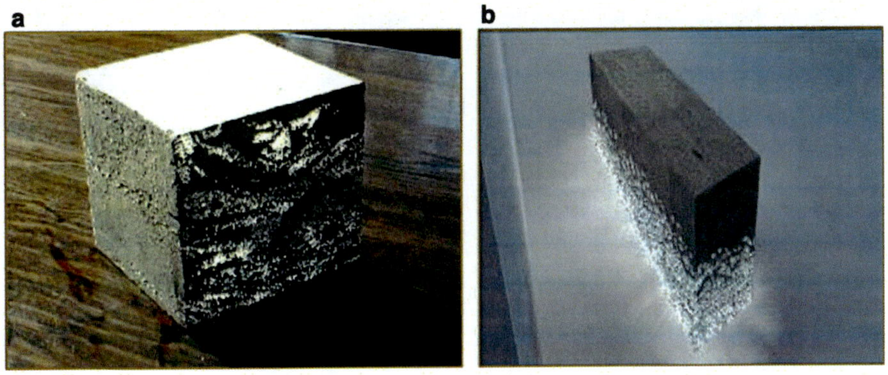

Fig. 29 a Stone of transparent concrete. **b** Showing light in the concrete

Because of its distinctive properties of new hi-tech material "The Transparent Concrete" is used on a large scale for ease of formation and elasticity of its fibers in the transmission of light in the various areas of architecture. That is place as the thickness of external finishes, as well as for walls and interior partitions, stairs, floors, and in the design of elements of furnishing. As well as it is a good energy saving component and environmentally friendly because of its light transmission properties by reducing energy consumption.

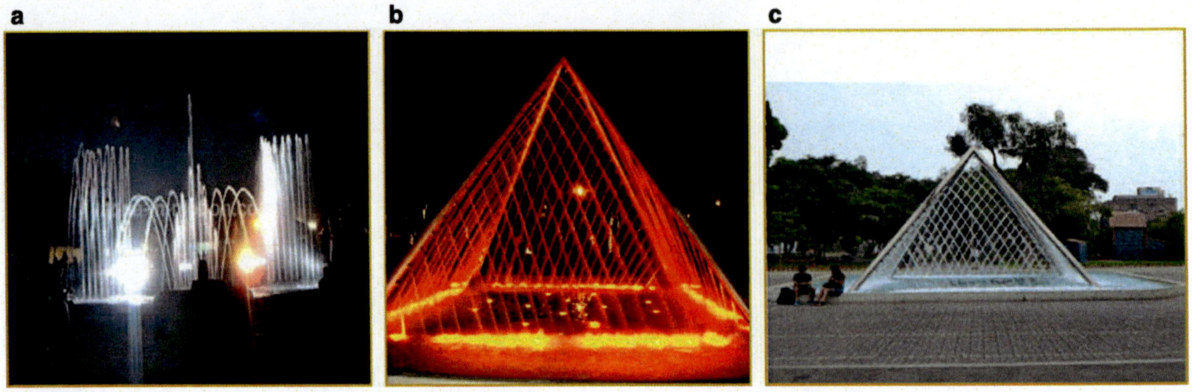

Fig. 30 a Water flows as the pyramid. b Water fountain at night. c Water fountain at daylight

C-1-3-3: Modern Technological Applications in Water Fountains and Fitting Rooms:

a. **Water Fountains**:
   The concept of technological design of JET water fountain using LED light applications (*AIM Motion Company —aqua in motion*), with different water flow rates from the nozzles of the square spray rings or the pumping points from the point in the assemblage of the pipes formed as a pyramid associated with the sound tones (musical fountains). The constant change of colors is through the laser program with the music shines in synchronize with water flow, Fig. 30a, b, c:
   One of the foam jet cells is an oxygenated compressed water air pump in a variety of plastic creations using the product of Aqua in Motion Company—AIM spray line, a set of pipes in a cross-shaped shape with 9 crankshafts. Each outlet has its own adjustable height adjustment system that can easily adjust the height to create multiple water forms in a distinctive pyramid formation as shown above in Fig. 31a, b.
b. **Fitting Rooms for Bazaars with 3D Simulation**:
   This technology of computer-generated 3D simulation images for fitting rooms began to emerge in 2005 and spread over from 2010 as a fast and alternative technology for changing rooms in regular shops (Rodriguez 2016). It is suitable for commercial shops and factories with limited space. It saves a lot of time and effort in changing clothing. It also enables the end consumer to experience a large number of goods in a short time.
   These spaces on the bazaar display windows use computer-generated 3D images to create a similar virtual experience that generates many models of virtual costumes (avatars) using multiple measurements of the final consumer's body measurements and sizes. As shown in Fig. 32, it starts by obtaining a real picture from the installed camera that sends it directly to computer showing on interactive screen in front of the consumer. Therefore, it displays many choices on it via swipe feature to turn the screen left or right between the different models in great efficiency leading to a quick decision to buy from the consumer.

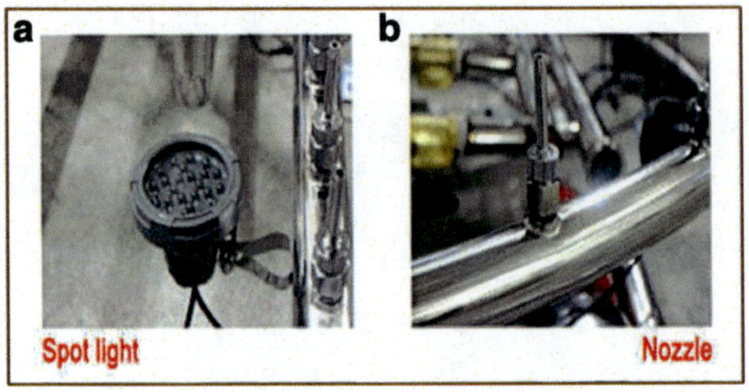

Fig. 31 a Model of AIM spray line unit. b Model of spray nozzle pipes

Fig. 32 Operating steps for 3D fitting rooms

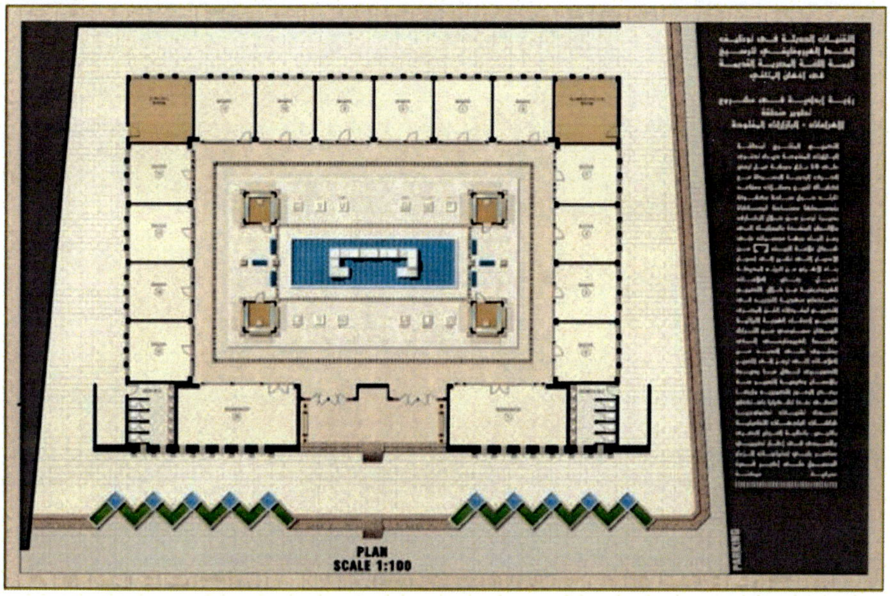

Fig. 33 Overall horizontal projection of new bazaar building—scale of 1/100

B. Executive Drawings of Proposed New Bazaar Building Project:

I. **Planning Phase**:
In Figs. 33 and 34, total building area is 1900 m^2 in a rectangular shape that symbolizing the hieroglyphic symbol of the house . It contains 16 commercial units (i.e., 14 bazaar and 2 artisan workshops units) along with 4 commercial kiosks, wooden benches and pergolas for visitors, sculptural formations and two electronic management and control units, bathrooms and digital programmable laser fountains inside and outside the building.

II. **Project Activities Measurements**:
The new proposed project includes variety of activities that differing in number and its related spaces as indicated in Table 1:

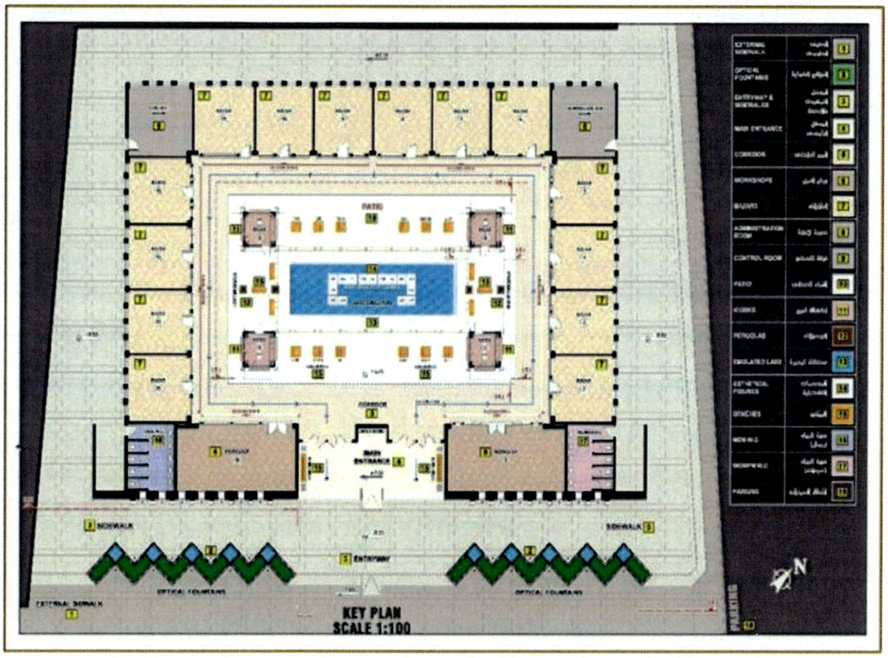

Fig. 34 Classifications table showing the spaces and type of activity in the building (horizontal key with code ke)

Table 1 Measurement Activities Table

Code	Type of Space	Quantity	Measurements
1	Commercial bazaars units	16	37 square meter
2	Artisan workshops	2	75 square meter
3	Administrative units—electronic control room	2	38.5 square meter
4	Bathrooms	2	29 square meter
5	Kiosks	4	16 square meter
6	Inner square	1	864 square meter
7	Inner lake	1	96 square meter
8	Wooden pergolas	2	32 square meter
9	Main entrance	1	105 square meter
10	Water fountains	2	76 square meter

III. **Designing the Main Facades of the New Building:**
The architectural design of the facades was inspired by main pylons facades of the "Amenhotep Museum." Top of the main wall is designed with a number of hollow hieroglyphs symbols selected from 13 symbols that employed in the interior design displaying windows, kiosks, and interior furnishing elements to emphasize the design identity of the idea using limestone cladding and formation of symbols with raw material.
The material of glass reinforced concrete—GRC is consistent with surrounding facades. The use of lighting focused on the selling windows of the facade at the bottom, to give overhead rays by laser lighting program with colorfully diversity in fountains to increase visual dazzle of the visitors as shown in Fig. 35a, b presenting the final design pylons facades at day and night.

IV. **Outdoor Furniture Design at the Reception Area**:
External reception area is the first visual contact point and first impression area for visitors as well, using a water composition representing the water symbol of the hieroglyphs line with a group of fountains and outdoors seating on both sides of entrance starting to receive them at the exit point of the visitors building. Then to the waiting area of electric buses to return back to complete the journey, and then to the Egyptian museum.

a. **Designing Water Fountains**:
As shown in Fig. 36a, b, the fountains are designed with JIT water fountain and colored waterproof light emitting diode—LED light stainless steel system using different water flow rates from the nozzles with square spray rings or pumping points from the assembly LED point in a structure of pyramid-shaped pipes. Accompanied by sound tones (musical fountains), the continuous change of colors through a laser program with music shines in concurrent with the flow of water, as shown below of two units of water formations (in day and in night) on the sides of the outer entrance of the proposed building:

b. **Designing Wooden Seatings:**
Figure 37 shows the design of 2 pharaonic sofas for visitors' comfort on both sides of the main entrance in the main reception area of the new building proposed in the research project; each of them has 2 units of flower boxes of copper material  inlaid with enamel colored hieroglyphic mark.
The lighting used in the side cushions  was through a pyramid form that emits LED lighting in order to give the use of these hieroglyphic symbols in the design rich and wonderful conceptual beauty reflecting the place and the environment surrounding the building.

V. **Interior Design of Interior Yard Area**:
Figures 38 and 39 indicate the inner courtyard as second area designed in form of a pharaonic garden characterized by the presence of a lake on its sides four kiosks as below Fig. 40 for selling antique products. That were design by sign of the house in one and courtyard sign for the other, integrated design for the facades identification of the brand identity kiosks under wooden pergolas under which seating and pharaonic lake without water in accordance with directives and considerations of the UNESCO. The floor was designed using mosaic material in blue and turquoise colors in decoration of the ancient civilization that indicates the symbol of water. In the center of the floor is a sculpture in form of a distinctive sign of the house in the architectural design of the new bazaar building. It is in the form of stone blocks, pointing to the same stone pyramids, engraved with

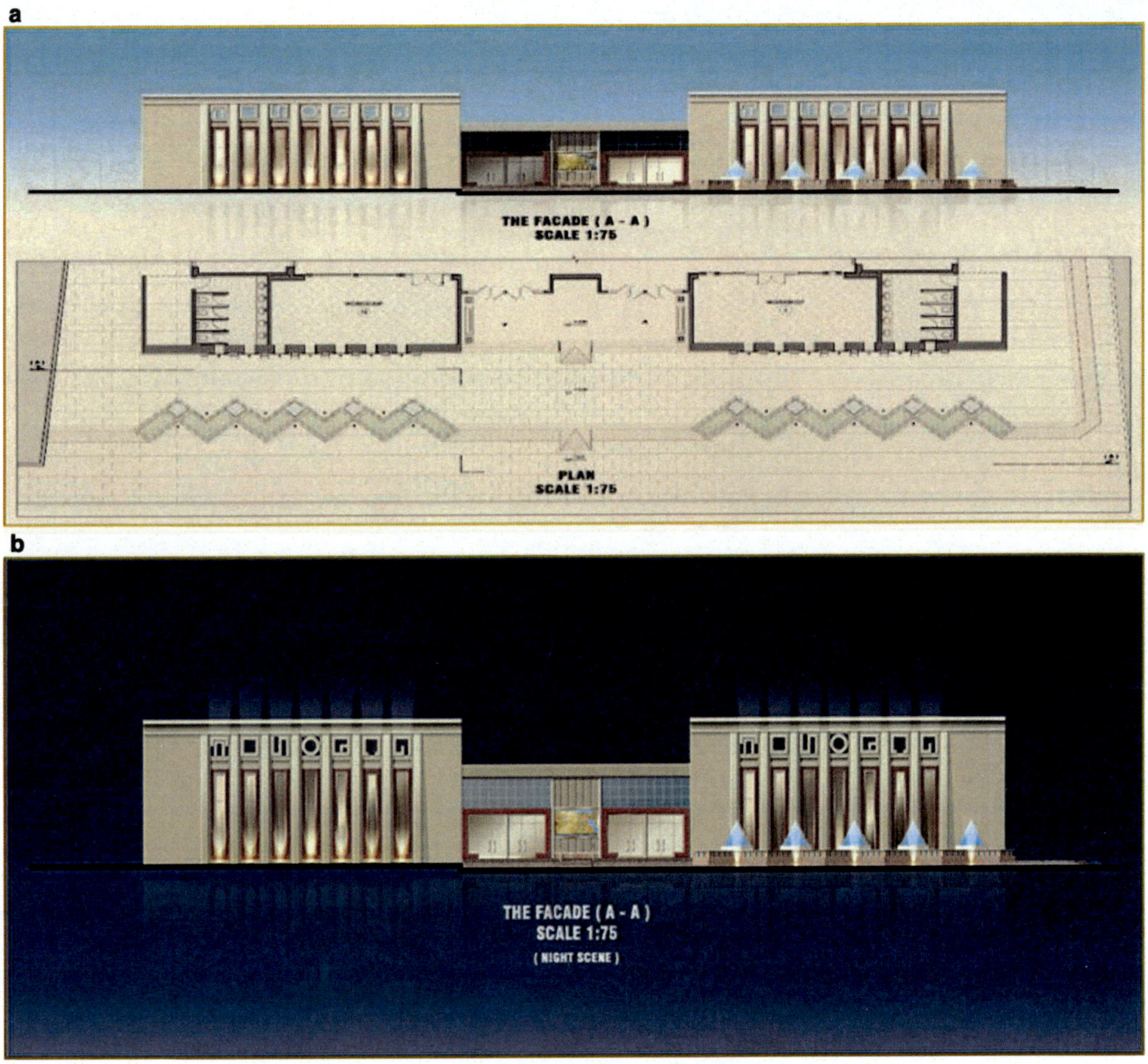

Fig. 35 **a** Day view of the main façade. **b** Night View of the Main Façade

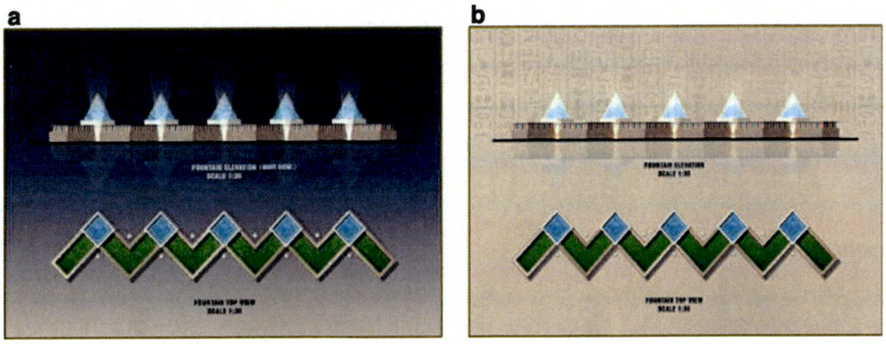

Fig. 36 Evening scene to form external fountains. **b** Day scene to form external fountains

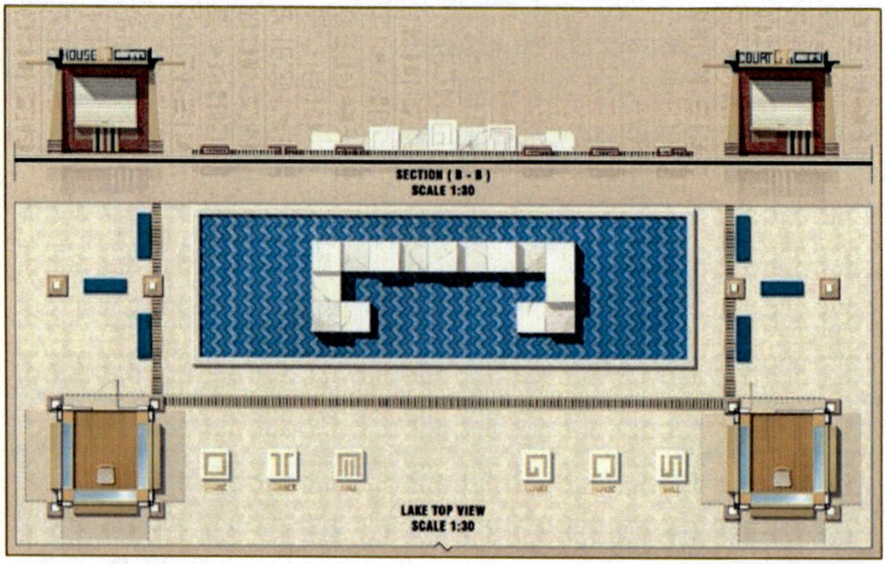

Fig. 38 Section (B-B) lake and sculptural of pyramid stones, design of kiosks and benches

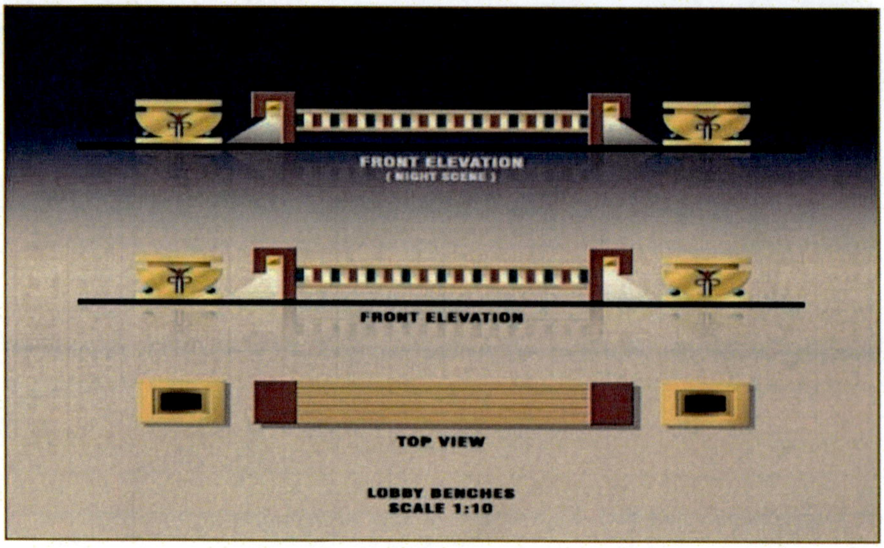

Fig. 37 Wooden Seats in a pharaonic style bearing symbols with two flower boxes

some of the hieroglyphs symbols. Optical fiber with sunlight and natural lighting in place.

Design of Marble Seats:

Seats designed from marble and granite material in Fig. 40, bearing most of the 12 hieroglyphs symbols used in the general design of bazaars display windows in the proposed new building as well as in the artisan workshops where the meaning of each sign translated separately under each seat using brass material. These seats designed in harmony shapes around the garden lake.

VI **Bazaar Interfaces Display Window and Artisan Workshops:**

The complex contains 16 commercial units and 2 artisan workshops. Two units on both sides of the new bazaar building, where the shopper/visitor starts the purchasing journey from the right direction that is counterclockwise according to the psychological studies of the human movement trends and wanders in the corridor of movement around the lake in front of other bazaars facades that shaded with a concrete roof. The visitors complete trip to the last workshop place,

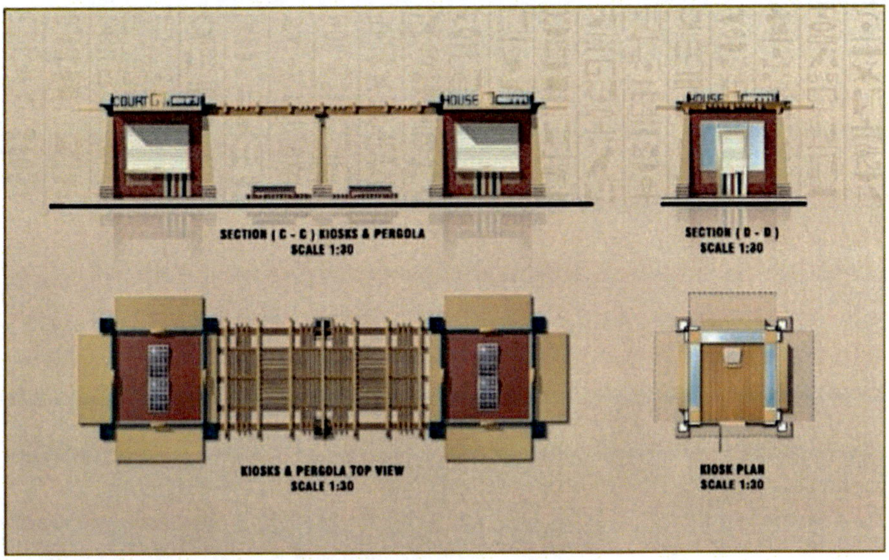

Fig. 39 Section (c- c) Horizontal and Vertical Design of Pergola, Kiosks Operates by Solar Energy

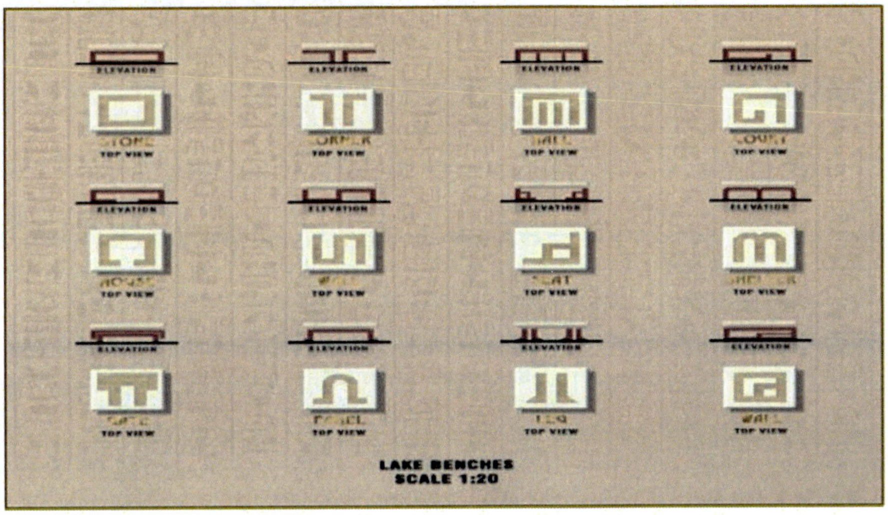

Fig. 40 Horizontal design of 12 marble seats bearing hieroglyphs symbols made from copper ore

then exit from the main entrance. The display windows contain interactive screens in Arabic, English, French, and German to translate each letter of hieroglyphs characters and their meaning to help by using embedded software's of dictionaries Egyptian-English, Egyptian-French (Rosette), and Egyptian-German (Erman & Grapow 1921) via a computer displays sound and related meaning on the interactive screen. That assists shopper or the visitor to buy. The visitor can also make some souvenirs inside the artisan workshops in the same place by writing any sentence, name, or sign of the line hieroglyphs on some products that Egypt is famous. The design revealed below Figs. 41a, b and 42.

These facades of the bazaars used the red granite material from Aswan to form the hieroglyphic symbols with side shoulders separating each window from the other bearing the mark of the distinctive cartouche of the ancient Egyptians and Egyptian civilization, which called "Ring Shen" in form of circular cartouche *(Shen, Ancient Egyptian Mythology, egyptianmythis. net)*. The center of front is a copper banner with the name of each bazaar meaning. The hieroglyphic symbols are a house, a courtyard, circular cartouche, etc. The door handle represents the line-mark below the shape of the copper material under each symbol to represent its meaning. As illustrated in the analytical studies and figures, Figs. 43, 44, and 45 show the

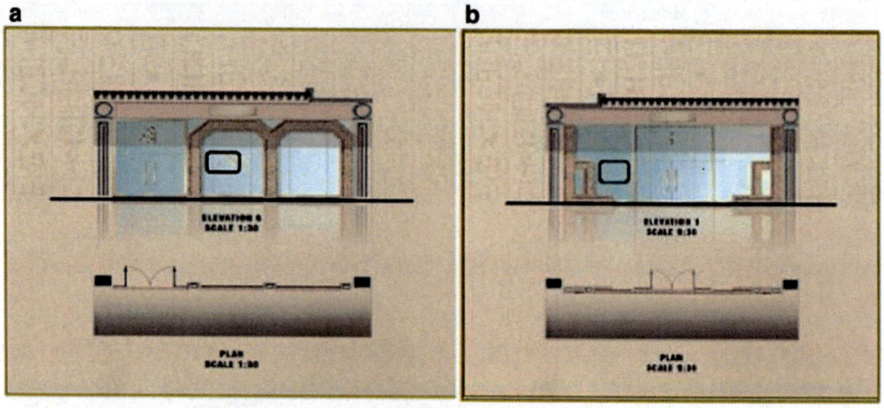

Fig. 41 Left workshop with interactive screen. **b.** Right workshop with interactive screen

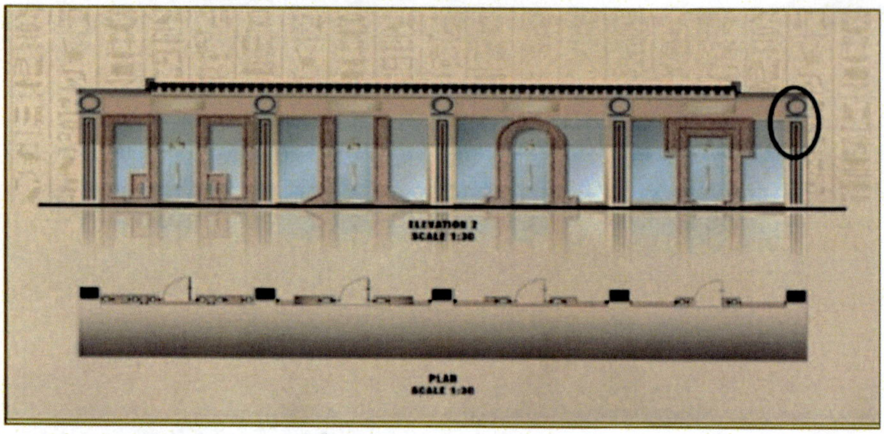

Fig. 42 Right side 4 bazaars with symbols (entrance, writing plate, human feet represents movement, and the wall)

number of display windows of the bazaars in the front facade bearing hieroglyphic symbols and are from the right side of the drawing (winding constructive wall—house—courtyard—column hall—construction wall of the corner or corner—the sign indicating the stone).

Interior Design as a Guiding Model for a Workshop:

At the end of the shopping trip, the visitor passes on the two workshop units in the building. They are an area of 75 square meters and the interior design for the workshop unit as the follows:

The above plan as set in Fig. 44 indicates two spaces; the first one is entrance space, a seating area for rest, interactive mirrors, and a panel bearing the design of hieroglyphs symbols and alphabets in Arabic and English as well as wall display units.

The second is the small unit with a laser-printing machine for fabrics, ceramics, and storage areas. Vertical projection as Fig. 45 is a transverse section showing the façade wall, exterior windows at a height of 9.50 square meters that illustrates the design of a luminous translucent concrete with copper letters representing the alphabets in hieroglyphs in Arabic and English and then in front of it a sofa length 1.60 square meters:

This section (A-A) Fig. 46 above illustrates the interior design of the sale counter of wooden display units within the bazaar as well as interactive mirrors as an alternative to the changing unit to reduce the time for the visitor.

Samples of Special Scenes of New Bazaar Building and its Exterior DesignFigs. 47, 48, and 49

Summary of Conclusions and Recommendations:

The study research contributed to the preserving of heritage environment without distorting the Egyptian heritage and surrounding of distinctive cultural character of the archaeological area of pyramids area. That done by reaching to relevant design strategy that combines all the considerations and standards of international protection rules and regulations set by UNESCO among the creative design vision aiming to translate and transform the inscriptions hieroglyphics on the walls of ancient temples into structural and architectural elements, sculptural figures, and tatted elements bearing the hieroglyphic signs.

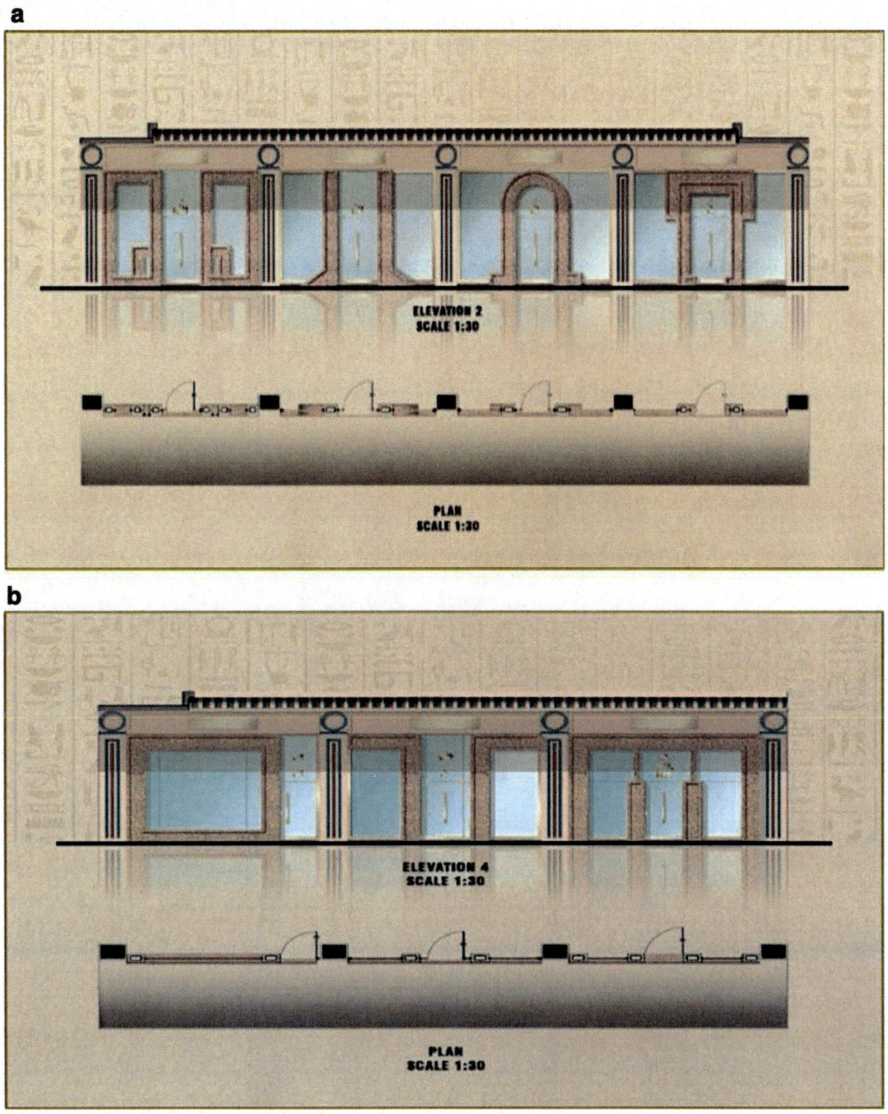

Fig. 43 For elevation 3 & 4 above showing two facades that they are on one level together

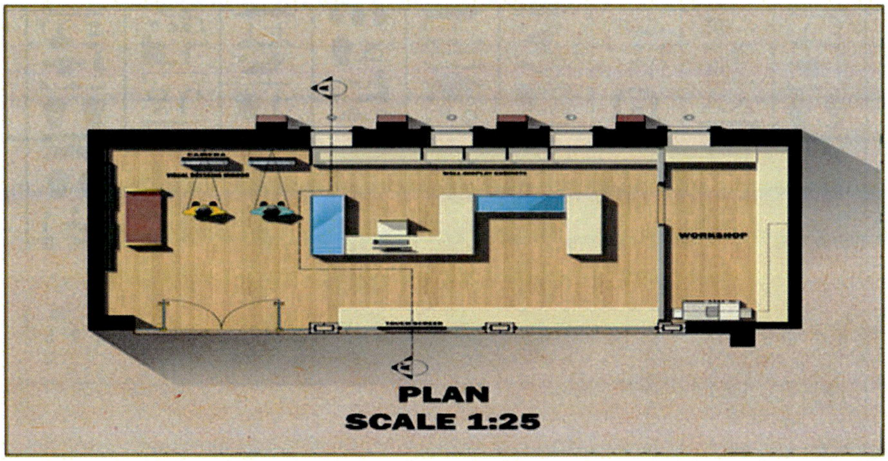

Fig. 44 Interior design for one workshop unit

Fig. 45 Side section of workshop imbedded within overall design of the building

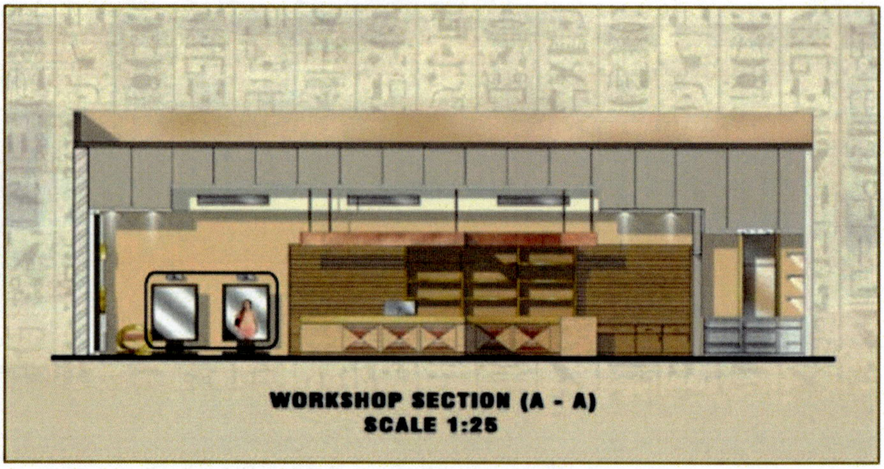

Fig. 46 Longitudinal section of the workshop showing interactive mirrors and sales counter

Fig. 47 Front view—front end

Fig. 48 Lateral perspective—side façade

Fig. 49 Main entrance of the building, the logo, entrance and exit doors

- Achieved a new method of education art in form of transferring an educational message to the visitors that accordingly affect and emphasize the importance of that language through the design form of each sign and its expression in the ancient Egyptian language.
- Integration achieved between contemporary art and preserving of the ancient Egyptian heritage through using number of interactive visual and audio technologies and applying modern raw materials. Those environmental raw materials applied in the design expressed the ancient Egyptian civilization such as red granite material that used in forming sculptural figures of hieroglyphic signs in a creative experiment and manner that supports the process of message delivering instructively to the recipient and visitors admirably through the use and enjoyment of visual and audio aids.
- The meanings of the ancient Egyptian language and its hieroglyphic signs have reached to the recipient and visitors through interactive translation integrated in the interactive display windows of commercial bazaars screens into Arabic, English, French and German languages. As a new innovative approach in designing elements of internal architecture in a simple-way that indirectly contributes to integrating the ancient hieroglyphs into visitors' minds during their purchasing journey in the bazaars area.
- The goal of communicating via art education by transferring the cultural message to the visitors has achieved indirectly in shortest possible time by using ancient symbols in achieving the elements of architecture and interior architecture in designing the generic planning of the commercial complex represented by the symbol of

"the home." The concept for these applied symbols were used as elements of the building from reception area, and the formation of water for fountains in an exciting visual journey, that attracts the visitors to enter and to complete the trip to the end. Such experience is giving him an unforgettable impression and a cultural message about the importance of the ancient Egyptian language and beauty of its paintings distinctive.

The proposed project achieved its objectives by clarifying the possibility of using ancient hieroglyphic symbols in architectural design of the building's structure. Such design inspired by the space coordination of ancient Egyptian installations similar to the design of the home, the garden, the façades of bazaars, and the craft workshops with applicable stereotypes bearing, the ancient Egyptian architectural features from the huge-mass and signs into number of symbols. That implemented through variety of interior architecture elements used in the design such as display windows, kiosks and water formations as well as the interior design of the bazaars taken as a whole.

Eventually, the researcher recommends the importance of showing the Egyptian identity in the architectural description of all building designs and construction that are located in a range close to the tourist archaeological areas or in their surroundings by expressing them clearly in its various forms including them in the components of architectural and interior design. Therefore, designing architecture based on the place identity by entrenching its historical value in the minds of the visitors considered adding value and as innovative new vision to express Egypt and its rare identity, that is a distinctive civilizational position among all other known civilizations.

Alphabetical References for Information Sources

A. Books and Documents

Allen, J. P. (2013). *The ancient Egyptian language*. Cambridge: Cambridge University Press. ISBN 9781139506090.https://doi.org/10.1017/CBO9781139506090.

Howard, M. C., (2012). *Transnationalism in ancient and medieval societies* (P. 23). ISBN 9780786468034.

Manley, B. (2012). *Egyptian hieroglyphs for complete beginners—the revolutionary new approach to reading themonuments*. London: Thames & Hudson Ltd.

Ministry of Culture National Coordinating Body for Civilization—Egypt, First Edition (2010). Guidelines for theCoordination and Standards of Urban Coordination of Buildings and Heritage Areas.

UN-HABITAT (2015). Strategic Vision for Urban Development in Greater Cairo—Pyramids area pdf, Egypt.

B. Websites

AIM, Aqua in Motion, Water Fountains, aqua-in-motion. https://www.aqua-in-motion.com/en/aim-galafountain/.

Ancient Egyptian Mythology, Shen Ring, (shen), egyptianmythis.net. http://www.egyptianmyths.net/shen.htm.

Dunn, J. (2015). Building Materials of the Pyramids Builders. http://www.touregypt.net/featurestories/material.htm.

Egyptian-French Dictionary, projectrosette.info. http://projetrosette.info/page.php?Id=599.

Erman, A. & Grapow, H. (1921). Hieroglyphic Egyptian-German dictionary, I & II - III - IV - V - VI - VII, Wörterbuch derÄgyptischen Sprache.

O'Connor, J. & Robertson, E. F. Egyptian Numbers, Andrew.ac. https://www-history.mcs.st-andrews.ac.uk/HistTopics/Egyptian_numerals.html.

Parkinson, R. B., (2008). *The painted tomb chapel of Nebamun British Museum Press* (p. 39), ISBN 9780714119793.

Rextouch, Projected Capacitive Touch Screen, Touch Foil, Android Digital Signage, Multi-touch. https://www.rextouch.com/.

Rodriguez, C. (2016). Virtual Fitting Rooms, Oa.upm.es. http://oa.upm.es/42311/1/TFM_BECERRA_RODRIGUEZ_CARLOS_a.pdf

Tarek Waly Centre for Architecture and Heritage, cuipcairo.org. https://www.cuipcairo.org/en/directory/tarek-waly-center.

The role of ICOMOS in the world heritage committee, icomos.org. https://www.icomos.org/en/what-we-do/81-english-categories/what-we-do/268-he-role-of-icomos-in-the-world-heritage-convention?showall=1.

The Tomb of Seti I, Wikipedia). https://en.wikipedia.org/wiki/KV17.

Translucent Concrete, pinterest.com. https://www.pinterest.com/pin/407505466255501232/.

Tree-foil, Interactive Touch, Touch Foil. https://treefoil.net/?gclid=Cj0KCQiAnb79BRDgARIsAOVbhRpg9pB2jS8U8stkKSTp5juyTPrrKW7x8qTxTlhDFI0o6eFunWiyhNUaAoeKEALw_wcB.

Wikimedia, "The God of The Dead", Osiris sitting on the throne, Wikimedia.https://commons.wikimedia.org/wiki/File:BD_Hunefer_cropped_2.jpg.

Preserving the Italian Heritage in Alexandria: The Don Bosco School as an Example

Heba Mahmoud Saad Abdelnaby

Abstract

The Italian community was the second largest foreign community in Egypt during the nineteenth century and the first half of the twentieth century. The Italians paid attention to the importance of Education and the need for Italian schools to serve the Italian community in Egypt. Therefore, they built many schools in Cairo, Alexandria, and the canal area. Since the largest community was in Alexandria, the Italian schools there were of special importance. The paper will focus on the Don Bosco School of Alexandria discussing the history of its building, its significance and its educational and cultural role in Alexandria. The paper will also shed light on the church, theatre and social club incorporated in the school building and their role. The aim of the study is to show the wealth of cultural heritage related to the school and the need to preserve that heritage in a suitable way. It will also discuss the potential of its use in tourism.

Keywords

Italin community • Italian Schools • Architectural Heritage • Intangible Heritage • Alexandria

1 Introduction

The Italian cities had strong relations with Egypt since medieval times because of the common interests in trade. The Mamluk sultans managed to have stable diplomatic relations with Italian cities and republics such as Venice, Genoa, Florence and Napoli. The sultans sent missions to establish trade treaties, import weapons, negotiate about a sudden problem or present gifts to enforce friendly relations (Alkhadem 1989). On the other hand, the Italian cities and republics were keen to have diplomatic representation in Egypt and the Venetians were the first to have a consul for their merchants and nationals in Egypt since 1346 AD (Mohammed 2000). Italian merchants enjoyed various privileges in Egypt during the Mamluk era and gained wealth through trade; therefore, they kept regular diplomatic missions to ensure their privileges and exemptions and to protect their interests. The Ottomans maintained the same type of relation with the Italians; whose trade flourished and varied under the Ottoman rule (Haridi 2004). The Venetians were the first commercial community in Alexandria throughout the fifteenth century and first half of the sixteenth century. Later during the seventeenth and eighteenth centuries, their numbers decreased and were replaced by the Jews; who controlled most of the trade in Alexandria; nevertheless, the Venetians retained their hotels in Alexandria and consulate in Cairo (Ibrahim 2013).

Mohammed Ali believed that the renaissance and the development he dreamt for Egypt couldn't be achieved without the aid of the European countries. Therefore, he opened the doors for foreigners to come and settle in Egypt to work in trade, invest, teach in schools, work in factories or even in the army and navy. Italy was one of the earliest countries he seeked for its assist because it had no political ambitions in Egypt. He sent his first educational missions to Italy in 1809 and then in 1813. He also encouraged the Italians to come to Egypt and work in his service. As a result, architects, artists and technicians of middle class Italians who worked in crafts and industry flowed to Egypt and worked for Mohammed Ali (Soliman 1996; Mohammed 2000). Abbas I was not in favor of foreigners, thus the flood of foreigners stopped temporarily in his regin. That changed during Saeed's and Ismael's reigns; as they both followed Mohammed Ali's policy; thus the number of foreigners in Egypt increased gradually again (Soliman 1996).

Emigration from Italy was a common practice during the second half of the nineteenth century and first half of

H. M. S. Abdelnaby (✉)
Faculty of Tourism and Hotels, Alexandria University, Alexandria, Egypt
e-mail: heba.saad@alexu.edu.eg

A. Versaci et al. (eds.), *Conservation of Architectural Heritage*, Advances in Science, Technology & Innovation,
https://doi.org/10.1007/978-3-030-74482-3_34

Table 1 The numbers of Europeans compared to numbers of Italians in Egypt

Year	Total number of Europeans in Egypt	Total number of Italians in Egypt
1843	6150	2000
1871	79,696	13,906
1907	151,414	34,926
1927	207,542	52,462
1937	179,095	47,706
1947	131,520	27,958

twentieth century and Egypt was a promising destination for Italian emigrants because of the job opportunities available there; due to the efforts of the reformer rulers of Egypt and the protection of Britain later (Lazarev 1997; Mabro 2002). The Italians were generally the second largest community of Europeans in Egypt –after the Greeks- and Table 1 shows the total numbers of Europeans and the numbers of Italians in certain years (Soliman 1996).

The foreign communities in general settled in Cairo, Alexandria and the canal cities and Alexandria was the most favored destination; especially for the Greeks and Italians.

But why Alexandria in particular was a magnate for the Italian immigrants? In addition to the obvious reasons such as close location and historic relations, the economic reasons were on top of reasons attracting the Italians to Alexandria. As the first port of Egypt since the medieval times, the city had its hotels and wikalas for merchants. The city also housed the arsenal; established by Mohammad Ali, which provided job opportunities for craftsmen. Add to this, the introduction of cotton in Egypt in 1820 and the establishment of Alexandria Stock Exchange in 1883 also provided opportunities for Italians to work on cotton trade; which was very profitable. Moreover, foreign export merchants formed and organization in Alexandria to manage their interests, such organization became Alexandria Municipality and the city was the first fully constituted in Egypt in 1890. That meant an autonomous power for the foreigners was formed; and that encouraged immigrants to come to Alexandria. The social reasons were also considered; since immigrants didn't have to search for accommodation when they come to Alexandria. Settlement offices existed for each community and nationality; welcoming immigrants and helping them to find a proper place. The mixed court which handled their interests and matters was also a social advantage in Alexandria (El-Semellawy 2011).

2 Italian Schools in Alexandria: The Don Bosco School as an Example

As the Italians flooded to Egypt seeking for job opportunities, they formed large communities mainly in Alexandria, Cairo and the Suez Canal Zone (Petricioli 1997; Turiano & Viscomi 2017). In Alexandria, the Italians were the second largest foreign community after the Greeks and their numbers increased from 7,539 in 1871 to 24,280 in 1927 (El-Semellawy 2011). The presence of such a large community required building schools that can maintain their link with homeland and language. The Italian schools in Alexandria could be divided into three types of schools: (Petricioli 1997).

- *the religious schools* established by the Salesians and the Franciscan nuns. The Salesians built the Don Bosco School for boys and Maria Ausiliatrice School for girls, while the Franciscan nuns had many schools at Marina, Nabi Daniel, Ibrahimia and Zaharia.
- *the state schools* such as the elementary schools of: Vittorio Emanuele III School, Francesco Crispi School, and Regina Elena in addition to middle schools such as Scuole medie and secondary schools such as Scuole Littorie.
- *the private schools or institutions* such as Istituto Principessa Mafalda which was set by a private citizen and and Liceo Musicale Giuseppe Verdi. Such institutions were of less importance.

The buildings of many of such schools still exist; witnessing part of the tangible Italian heritage in the city, moreover, the educational and cultural role of these schools highlights the intangible Italian heritage that was profound in Alexandria.

2.1 A Dream Come True

The schools founded by religious missions in Alexandria were varied and had a deep impact on the majority of the Italian community in the city. The Salesian mission was the earliest to settle in Alexandria as early as 1882. The Salesian missionaries aimed to provide assistance and care to the growing community of Catholic workers, the majority of which were Italians (Turiano & Viscomi 2017). Their first project was the Don Bosco school established in 1895. It was the dream of Don Bosco, an Italian Salesian father, to come to Egypt to establish a school for poor boys, but he died in 1888 before fulfilling his dream. The successor of Don Bosco; Don Rua, tried to redeem the dream by sending Don Belloni to establish a school in Alexandria; which was intended to help children of poor families by providing them good education and the chance to learn a profession that can ensure them a good chance in life.

2.2 The School Buildings: History and Architectural Features

Don Belloni chose and bought a piece of land in Bab Sidra to be the location of the school (Fig. 1). An adjacent piece of land; that was previously a fort and prison, was also bought in 1897 and the building of classrooms, residential rooms and workshops was completed by the fund sent from Don Rua in Italy. The school was named after Don Bosco and housed an elementary school and vocational institute. An additional wing was built in 1902; containing a small church and rooms for the students. In the same year an old building was bought. It was overlooking Amoud al-Sawari Street and was previously used as a prison by the British army. In 1911 a playground and a roofed hall for Gym were added. Fifteen years later, in 1926, a new building was added containing a theatre in the ground floor and classrooms and residential rooms in the upper floors. Then in 1934 the Salesians decided to add a church in their building; benefiting from the generous donations of the lawyer Alberto Lamanna and his wife Fanny (Fig. 2). The architect Loris Pagano put the design and Ditta De Farro was assigned for the construction. The church was opened in 1937 and named after Don Bosco (Figs. 3 and 4). The church was famous for its bell which was set in its tower in February 1937 and the organ which was made in Italy by Ditta Balbiani-Vigezzi di Milano and they are both still working well. The furnishing of the church was a donation from Contessa Carola De Tomich (Instituto Don Bosco 1996).

The buildings of the madrasa were generally modest since it was a religious school and was privately funded by the Salesians. The facades of elementary school for example; which is the oldest building, were simply divided into vertical recesses that contain rectangular windows. More attention was paid to the church since it was the first church to be dedicated to Don Bosco in the world; therefore, its building was the only building that was decorated in the school. The church was located on the corner of the madrasa overlooking the main street. It consisted of a rectangular building with a projection of a five-sided area to house the Alter, and the building is roofed with gable roof. The main façade was marked with two rectangular corner towers that are higher than the rest of the building, and that façade used to contain the main entrance but it is now blocked. The

Fig. 1 The search for the school location (school's photo archive)

lower part of the main façade consisted of a central entrance with three engaged columns on each side carrying three consecutive round arches. It is flanked with two side entrances, similar in shape but smaller in size. The main entrance is surmounted with a round-arched recess that contains three arched window (tracery) with stained glass, then a triangular top that has two decorated medallions (each contains a cross) flanking a decorated marble support that carries a bust of Don Bosco (Figs. 4 and 5).

The upper part is a beautiful loggia with columns and arches; consisting of: two engaged columns connected to the sides of the towers, two central columns attached to each other and six other columns all carry eight round arches. The summit is triangular in shape, decorated with a unique molding of interlacing pointed arches, and surmounted with a large cross. Each of the two rectangular side towers is decorated with a round-arched recess that contains a twin arched window, an arched window, a twin arched window, a medallion and the interlacing arches molding arranged sequentially from bottom to top. The medallion on the right contains a clock. The upper part of each tower; which is a

Fig. 2 Statue of Alberto Lamanna in the school in recognition of his efforts and support (photographed by the author)

Fig. 3 Building the church in 1935 (school's photo archive)

Fig. 4 An old photo of the façade of the church (school's photo archive)

square area, contains an opening of three round arches carried on two columns on each side and this square area caries a large pyramidal top carrying a cross and surrounded with 4 pyramidal small corners. The towers contain the church bells that are still functioning (Fig. 5). Each of the other façades was divided into five rectangular areas, each of which contains an arched recess with three arched elongated windows surmounted with a circular small window (Fig. 6).

The church is entered nowadays from a side entrance located within the school open court. The interior of the church was bright because of the high ceiling, large windows and white paint of the walls. Only the five-sided projecting area was paneled with marble and its upper part was decorated with a large oil painting of Don Bosco flanked with two elongated rectangular windows with stained glass (Fig. 16). The opposite wall has a terrace with a large round arch, flanked by two smaller terraces and this is the place that hosts the choir and organ (Fig. 17).

It is worth mentioning that the façade of the church represented some of the recurring characteristics of the Italian architecture in Alexandria during the first half of the

Fig. 5 The upper part of the façade

Fig. 6 The side façade that has the current entrance of the church (photographed by the author)

twentieth century such as the lofty overhanging towers set at corners to frame an elevation, the top-floor loggia that crowns a building, the tripartite vertical division created by windows and the decorated molding (Turchiarulo, 2012). Such features could be seen in Elie Betesh Apartments and Maurice Apartments in 1926 by Loria, Alexandria Railway Station in 1929 by Riccardo Smith and Ciceil Hotel in 1929 by Loria..etc.

2.3 The Educational System and Number of Students

Since its establishment, the Don Bosco included an elementary school and a vocational school. Like the history of the school, the educatioal system followed and its regulation were not documented. Only the school records that listed the enrolled students each year and their results could be a source of information about the educational system. Such records proved that the elementary school was teaching: Italian composition, calligraphy, drawing, Italian dictation, math, French, English and Arabic and these subjects were tested with written exams. Moreover, grammar, math, history and geography, rights and duties, gym, French, English and Arabic were also taught and were tested with oral exams (Fig. 7). The vocational school taught all subjects in Italian and had workshops specialized in carpentry, shoe-making, bookbinding, sewing and mechanics (Figs. 8, 9 and 10). The school added an institute for cinematography in 1911 and additional workshops were added in 1940 such as electricity and mechanics (Instituto Don Bosco 1996). The photo archive in the school proves that the workshops were always renovated and upgraded.

The school also taught music and had its own musical band which was established as early as the school establishment, in addition to a choir for religious hymns (Fig. 11). In 1948 the school introduced scouting to students and established the first group of Don Bosco scouts (Fig. 12).

The school records give us information about the ups and downs in students' numbers and these numbers could be understood in light of the social and political circumstances. Located in a poor Arab quarter where working-class Italians and Egyptians lived, the school attracted children of working and poor families (Petricioli 1997; Turiano & Viscomi 2017). Therefore; in the first school year, 65 students were enrolled. The number of students increased gradually after the opening of the tram line from Mohammad Ali Square to Karmouz in 1898 and also after the extensions of the school buildings in 1902, 1906 and 1926. As a result, the number of the students increased to 200 in 1902 and 341 in 1907; 220 of them in the boarding school (School records).

When Italy participated in World War II, many decisions were taken against the Italians including closing religious and state schools. Using a legal loophole to claim Vatican nationality, the Salesians managed to reopen their school which became the only available school for Italian students (Turiano & Viscomi 2017). As a result, the Don Bosco School received increasing numbers of students: 600 in 1940, 637 in 1941,700 in 1942. By the end of war, poverty and unemployment prevailed in the Italian community. One of the main causes was *the Company Law* enacted in 1947 by the Egyptian government aiming to Egyptianise the workforce; thus establishing strict quotas for foreign workers in different industries (Karanasou 1992). As a result, many Italian families departed Egypt between 1946 and 1951; some returning to Italy and others heading to Australia, South America and elsewhere in Europe (Turiano & Viscomi 2017). This is witnessed in the school records; as the number of the students decreased to 555 in 1946 and a similar average in the following years.

On the eve of 1956 war, a large number of Italians left Egypt and the school witnessed another decrease in students' numbers. When Abd el-Nasser came to power, *Egyptianisation* and *industrialization* became the keywords of his era. Therefore, the Salesians adopted a new policy trying to present the school as the place that can provide technical education to prepare highly qualified technical professionals of Egyptians on whom the national industry could depend. During the late 50 s and 60 s the number of students ranged between 650 and 450 including elementary students, secondary students and vocational students (School records). In 1970 an agreement was signed between Egypt and Italy to establish a technical industrial institute and it started

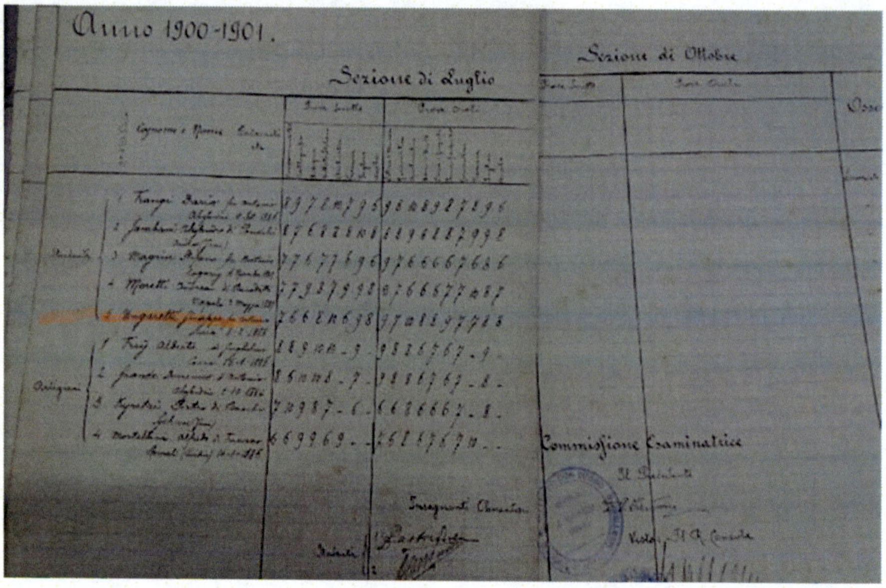

Fig. 7 Example of the school record (photographed by the author)

Fig. 8 An old photo of the carpentry workshop (school's photo archive)

Fig. 9 An old photo of the mechanics workshop (school's photo archive)

working the same year. The printing workshop was closed in 1975 but other workshops were introduced. In the following years the school gained a good reputation as a specialized vocational institution, especially that its administration is following the Italian curriculum of similar institutions, the workshops were always updated and Italian teachers were constantly available.

2.4 The Social Role of the School

The Don Bosco School was actively participating in the social life of Alexandria, not only with the Italians but with all the foreign and local communities in the city. The photo archive and the school records proved that the school had a musical band which was established as early as the school

Fig. 10 An old photo of the tailoring workshop (school's photo archive)

Fig. 11 Don Bosco Musical Band in 1897

establishment. Rapidly, the band gained fame and was continuously invited to participate in important events. The musical band celebrated the visit of the khedive Abbas Helmi II to Alexandria in 1898; to inaugurate the tram line from Mohammad Ali Square to Karmouz, and played the National anthem to him. After the khedive's return from pilgrimage in 1910, the band was also participating in the reception ceremony held in al-Baladiya square and in Ras el-Tin, then the band walked back to school through Mohammad Ali square and Sherif Street, playing the music and cheered by the Alexandrians. Moreover, the band participated in the ceremony held for the inauguration of the Italian Hospital in Alexandria in 1923 (The school photo archive). The band also had a role in uplifting the spirits in times of crisis. When Alexandria was hit by an earthquake in December 1908, an initiative was set to collect donations and the musical band accompanied the car that wandered the city; to encourage people to donate.

After building a theatre for the school, the band participated in the annual ceremony at the end of the school year till 1944 when the band's activities were halted. In April 1947 the band resumed its activities and during. During the following years the band had the chance to held separate ceremonies such as the famous concert that sold 2000 tickets and was conducted by the Maestro Costamagna in May 1950.

Sports team was not of less importance as it was well-trained and organized. It participated in the opening of

Fig. 12 Don Bosco Scouts in 1948

Alexandria Stadium in November 1929 (Al-Ahram newspaper 1929) and in the sports performance held in front of King Vittorio Emanuele III at Alexandria Stadium in March 1933, when he came to open Scuole Littorie (Al-Ahram newspaper 1933). The team also participated in the local sports competitions held in Alexandria.

Moreover, the school had an important social role during times of war; for example: during World War I the school hosted the families of those who were called for military service and the families of the detained Germans and Austrians. In recognition of such role, the Red Cross gave the school a certificate of appreciation. During World War II, the school was among few Italian schools that continued working providing education for both Italian and Egyptian children regardless the bombardment and the attacks to the school.

The church of Don Bosco was also active in its religious role for the Catholics in Alexandria in general and in the district in particular. Its social role in helping the poor, its regular religious ceremonies and sermons of the Salesian fathers in the church and in other churches were all undeniable. The school also opened a social club in 1952, which was supervised by the church.

2.5 *Important Figures in Relation to the School:*

Important names were associated with the school; that either studied or visited the school. **Giuseppe Ungaretti**; the famous Italian poet, journalist and critic was the first name because he studied in the school as witnessed from the School Records. His name was listed in the records of the school year 1900–1901 which means that he was about 12 years old and probably in the elementary level (Fig. 7). The school owns a special record that was named *the Golden Book* or *the Book of Gold* in which the visits of important figures were recorded mentioning the date, name of visitor and the words the visitors wrote to commemorate their visits in the book. From this record we recognized the visit of **Prince Ahmad Foad I**; who visited the school in 22 November 1922 and was respectably received and warmly welcomed. He visited the workshops and attended a musical ceremony followed by sport performance in the court of the school then he wrote some words in the Golden Book, praising the school and promising his support. The Golden Book also recorded the visit of **Luigi Federzoni**; president of the Senate and Royal Academy in Italy, in 30 November 1938 in addition to the visit of **Pietro Canonica** the famous sculpture the following day. The school was also visited by the **King Umberto di Savoia**; the last Italian king, in 15 May 1947 and he signed in the Golden Book (Fig. 13).

2.6 The School Nowadays

The school buildings are generally well preserved; the oldest is the building of the elementary school which also hosts the administration offices (Fig. 14). This building has a hall dedicated to Alberto Lamanna and his wife which was used as a church and it has a cupboard for storing the old musical instruments. The crypt of this building is used for storage and it contains many elements recording the history of the school such as marble capitals, marble fragments used for decoration and a marble slab with the school name from 1952.

The other buildings host the classes, the workshops and the theatre. The capacity of the theatre is over 500 persons

Fig. 13 Umberto di Savoia at Don Bosco in 1947

and its chairs are all made of wood (Fig. 15) and it is functioning on a limited scale as it is used for school events only. The church is the best preserved part of the building as its interior was recently renovated. It is richly decorated with colored glass windows, wall paintings and oil painting and portraits (Figs. 16, 17 and 18). It also owns the sacred relics of some saints (Fig. 19) in addition to the old organ which dates to 1937. Next to the entrance of the church, there is a red granite column with a statue of Virgin Mary on its top (Fig. 20) and it is thought that the column was found during the digging for the construction of the building.

The workshops are very well-maintained and updated with the latest technology (Figs. 21 and 22). In these workshops short and long training courses are offered in addition to the regular training for the enrolled students. Such trainings include Digital Electronics, Analog Electronics, Hydraulics and Electro-Hydraulics, Electrical Motor Winding, Solar Photovoltaic Energy, Refrigeration and Air Conditioning, Industrial Control and Winding…etc.

3 Preserving and Presenting the Heritage of the Don Bosco School

It is clear that the Don Bosco School had a profound role in Alexandria affecting both the Italian and the Egyptian communities. Its heritage is very rich; varying between tangible and intangible heritage but unfortunately, that heritage is threatened to be lost. Although the school is doing efforts to take good care of the school records and photo

Fig. 14 The interior of the elementary school (photographed by the author)

Fig. 15 The theatre (photographed by the author)

Fig. 16 The alter of the church (photographed by the author)

archive, but due to natural reasons, the paper is deteriorating and could be permanently lost in few years. Therefore, digital preservation is a necessity to preserve that rich heritage and to provide a better chance for researchers to further investigate these records in search for more data. The records and archives can provide us with names of notable figures that studied and graduated from the school, details about the history of the school and its teachers, details about the Italian community in the city and their role and even details about the fashions of clothes and hairstyles. So; many insights about the Italian community, other foreign communities and the Egyptians could be concluded to complete the image we have about the Italian heritage in Alexandria.

Then we need to discuss why and how that heritage could be presented. That heritage should be presented for many reasons:

It will tell the story of an important era of the history of Alexandria, thus giving the Alexandrians the complete image about their city.

It will encourage the Italians to visit Egypt to learn more about their ancestors who lived in Alexandria and experience the traces of their heritage.

It can promote both the domestic and international heritage tourism in Alexandria.

The existing Italian schools could be best used as locations to present the tangible and intangible Italian heritage. The presentation of the heritage could focus on two aspects: places and people. Starting with places; The Don Bosco School is near the catacomb and Bombay column. That means; it could be included in the programs of tourists visiting the catacomb. The visit could focus on the history of

Fig. 17 The interior of the church (photographed by the author)

Fig. 18 One of the oil paintings in the church (photographed by the author)

Fig. 19 Sacred relics of saints in the church (photographed by the author)

the site, its buildings and their architectural and artistic features. The church at Don Bosco School could attract visits since its architecture is worth-seeing; moreover, it contains oil paintings and sacred remains of saints that could be of interest to many visitors.

Focusing on people; many events could be organized about all the Italian figures who were related somehow or another with the schools. Such figures could be the architects and designers who participated in the building process, the directors of the school who had important influence and imprint, the benefactors who supported the school and donated for it, the famous people who visited the school and the well-known figures who were graduated in the school. Such events could be organized in a calendar and announced in cultural centers and on social networks to be marketed.

Fig. 20 The granite column next to the church's entrance (photographed by the author)

The events themselves could be in the form of lectures, seminars, photo or painting exhibitions, story-telling nights or even theatre performances. The School theatre is an asset that can provide the venue for many activities and events while the playground is another venue that can hot different events such as exhibitions.

It is also worth mentioning that a museum could be established to preserve and present various aspects of the Italian heritage in Alexandria. The current director of Don Bosco School; Don Jesudoss Arokiam, mentioned that his school is currently considering this idea and Don Renzo Leonarduzzi; director of the vocational school is collecting machines, artifacts and all he can use for the intended museum. The intended museum will be focusing only on the heritage of Don Bosco School, but with cooperation with other Italian institutions in the city and with consultation and collaboration with heritage experts, such museum could be a special unique site to present the Italian heritage in Alexandria. A special exhibition for industrial heritage could be held in the school and it will be the earliest of its kind in Egypt. It can present examples of the old workshops such as sewing, shoemaking and binding typography with demonstration of the skills of professionals in each workshop, the materials used and the development of the machines. It can also present the old machines and tools used to be utilized in the existing workshops and how they developed to reach digital machines. All the safety clothes and equipments could also be exhibited with interesting activities and tutorials for visitors.

Moreover, walking tours; which gained much attention in Alexandria recently, could be another method to present the Italian heritage in Alexandria and the Italian schools could be included in various walking tours. Some tours could be thematically organized focusing on certain topics that present the heritage of the city such as: secular architecture of foreign communities in Alexandria, notable figures of the twentieth century in Alexandria, The Italian heritage in Alexandria...etc. While other tours could be geographically planned; listing the entire heritage sites in a certain district or neighborhood. A guide should accompany the tour in order to reveal all the aspects of heritage in each site.

Fig. 21 The solar energy workshop (photographed by the author)

Fig. 22 The electricity workshop (photographed by the author)

4 Conclusion

The Italian community was well-established in Alexandria especially during the nineteenth century and the first half of the twentieth century. The tangible and intangible heritage the Italians left in the city is untapped wealth. The study focused on the Don Bosco School as an example of Italian schools in Alexandria as it is still functioning and its role in the society is witnessed.

The study of the history of the school, its buildings and the its role in the society depended on few studies but mainly depended on the school records, photo archive, interviews with members of the administration of the school and the only publication issued by the scool in its centennial. The study revealed that the school buildings have a significant architectural importance and the activities of the school and its social role in Alexndria narrates a lot about the inter-cultural relations between the Italians and the Egyptians. The study also concluded that school workshops with their development through ages could present the industrial heritage of the nineteenth and twentieth centuries and this aspect of heritage is not highlighted in Egypt. Moreover, the study revealed that the school records and photo archive are preserved in paper form which means that they are very likley threatened to be lost. Therefore, digital preservation is a necessity to preserve that indispensable source of information.

The study sugessted that the school, together with other Italian institutions, could be invested for the benefit of heritage tourism. Many activities could be organized to present the school's heritage in particular and the Italian heritage in general and innovative tools of inerpretation should be adopted to attracht various types of tourists and locals.

Acknowledgements is due to Father Jesudoss Arokiam; director of Don Bosco School, and Father Don Renzo Leonarduzzi, director of vocatioal school at Don Bosco, who paid all efforts to help me in my research and provided me access to the school buildings, records and photo archive and provided oral data about the history of the school.

References

Al-Ahram Newspaper. (1929) Issue 16146. p. 1, 4, 8.

Al-Ahram newspaper. (1933) Issue 17331. p. 8.

Alkhadem, S. A. (1989). *Islamic East and christian west through relations between Italian cities and the eastern mediterranean* (In Arabic). Al-Rayhani Publishing.

Don Bosco School Photo Archive (1896–2014).

Don Bosco School Records (1896–2014).

El-Semellawy, A. N. (2011). *The European imprint of the architecture of alexandria*. [MA diss, Ain Shams University: Faculty of Engineering].

Haridi, S. (2004). *The Communities in alexandria during the ottoman era (In Arabic)*. Dar Ein.

Ibrahim, N. (2013). *Alexandria during the Ottoman Era: Economic and Social Life* (In Arabic). General Egyptian Association for Book.

Instituto Don Bosco. (1996). *Cento ani per i giovanu e con i giovani*. Alexandria.

Karanasou, F. (1992). *Egyptianisation: the 1947 Company Law and the Foreign Communities in Egypt*. [PhD diss., St. Antony's College. Oxford].

Lazarev, A. (1997). Italians, italianity and fascism. In R. Ilbert, I. Yannakakis, & J. Hassoun (Eds.), *Alexandria 1860–1960: The brief life of a cosmopolitan community* (pp. 71–92). Harpocrates Publishing.

Mohammad, M. A. (2000). *The role of the foreign and arabic communities in the artistic life in Egypt during the 18th and 19th centuries* (In Arabic). [PhD thesis, Faculty of Arabic Language: Al-Azhar University]

Mabro, R. (2002). Nostalgic Literature on Alexandria. In J. Edwards (Ed.), *Historians in cairo: Essays in honor of George Scanlon* (pp. 237–265). AUC press.

Petricioli, M. (1997). Italian Schools in Egypt. *British Journal of Middle Eastern Studies.*, *24*(2), 179–191.

Soliman, M. M. (1996). *Foreigners in Egypt 1922–1952* (In Arabic). Dar Ein Cairo.

Turchiarulo, M. (2012). *Building "in a style": Italian Architecture in Alexandria, Egypt. The Work of Mario Rossi.* GanGemi Editore.

Turiano, A., & Viscomi, J. J. (2017). From immigrants to emigrants: Salesian education and the failed integration of Italians in Egypt, 1937–1960. Modern Italy, *23*(1), 1–17.

Architectural Innovative Technologies for Improving the Built Environment: A Technical Framework

Magdy M. El-Bastawisy and Mohamed M. Shawky Abou Leila

Abstract

This paper demonstrates the grave consequences and impacts upon the environment by the action of energy building adaptation practices. The term built environment encapsulates the entire sum of the persistent heritage areas, culturally symbolic sites along with the traditional districts and their contemporary counterparts. The study investigates three main aspects that contribute to these environmental issues; environmental impact of energy buildings thermal comfort requisites and architecturally innovative technologies, to improve the built environment. These architectural innovative technologies could play a significant role in reducing the despised energy building impact on the built environment. Mainly, this could be achieved by the holistic adaptation of these technologies at all architectural design stages, construction details and specifications. Tackling the built environment and thermal comfort environmental issues in terms of their improvement urgency and requisites, is to assert the worth of the architectural innovative technologies. The development of the mentioned technologies is the only means, to accomplish these requisites. Henceforth, this paper debates the capability of the alternative architectural innovative technologies, including building technology and green architecture systems, to sort out the current conflict between the built and natural environment. The findings of this research emphasize the necessity of subduing the environmental impact of energy buildings, and opting for architectural innovative technologies. These technologies are prominent for their incorporation of several methods that are capable of reducing the built environment traumatization. Consequently, the research proposes architectural design criteria and guidance, to apply, appropriate architectural innovative technical methods. This is to achieve architectural designs that could contribute to decrease energy demand of buildings, and by turn, minimize the negative impact on the built environment.

Keywords

Built environment • Architectural innovative technologies • Thermal comfort • Energy buildings • Environmental impact • Building technology and green architecture

M. M. El-Bastawisy (✉)
Urban Design and Development, Department of Islamic Architecture, Um Al-Qura University, Mecca, KSA, Saudi Arabia

M. M. El-Bastawisy
Department of Architecture & Urban Planning, Port-Said University, Port Fuad, Egypt

M. M. S. A. Leila
Architecture, Rehabilitation and Conservation, Department of Islamic Architecture, Um Al-Qura University, Mecca, KSA, Saudi Arabia

M. M. S. A. Leila
Department of Architecture, Faculty of Engineering, Mansoura University, Mansoura, Egypt

1 Introduction

Environmental issues are highly controversial, due to their instant distortion, by virtue of the buildings impact. The utilization of these buildings is held responsible for the external propagation of air pollution, compared to the other ongoing activities. In 2019, the Buildings International Energy Agency, has disclosed that the buildings sector consumes 30 percent of the total energy consumption, which contributes to the escalation of the global CO_2 emissions. However, this declaration is oblivious of the occurrence of some efficiency improvements in energy consumption in the same year. In 2018, the demand of energy in buildings, e.g., lighting, air conditioning, residential appliances, etc.; has surpassed the total energy demand (Buildings International Energy Agency 2019). In 2017, the World Green Building Council (WorldGBC), has stated that less than one percent

A. Versaci et al. (eds.), *Conservation of Architectural Heritage*, Advances in Science, Technology & Innovation,
https://doi.org/10.1007/978-3-030-74482-3_35

of all buildings worldwide are currently net zero carbon. At the same time, the annual production of Carbon dioxide (CO_2) mounts up to high consumption rates; as a result of energy consumption mainly by buildings. CO_2 makes up most of greenhouse gas emissions, which is harmful not only to the built environment but also to human health. This is because greenhouse gases induce the properties that absorb and emit infrared radiation from Earth surface and radiate it back to the Earth surface. The findings also proved that the current energy consumption and its growth rate are quite significant, particularly in residential areas. Residential areas in the UK, e.g., are responsible for 10.0% of the annual total green gas emission in 1990, which has increased to 14.1% in 2017. This occurred although the annual total green gas emission has decreased from 794.2 MtCO2e in 1990 to 455.9 MtCO2e in 2107 (WorldGBC 2017/2018).

Consequently, from 1992 onwards Earth Summits have witnessed the public concern and decrees binding the international commitments of the participant governments. Therefore, these efforts have fruited in enforcing new legislation to reduce energy consumption and the subsequent buildings environmental impact. In addition, planning legislation and new buildings regulations of European countries require that the development of buildings must cater for minimizing their environmental impact and energy consumption. Future controls of energy consumption are likely to be more demanding and wide-ranging. Agenda 21, therefore, defined that *"Many governments are committed to developing instruments which make markets work for the environment and channel development into sustainable paths. Each local authority should adopt a 'local Agenda 21' for the community"* (UN 1992).

Moreover, there are various built environments located in old and contemporary cities, contribute to the built heritage, because their buildings go back to hundred years ago. Although most of these buildings have valuable architectural character and feature, they are still occupied. Thus, they are considered that they have share in producing CO2. Therefore, the United Nations Educational, Scientific and Cultural Organization (UNESCO) and the International Council on Monuments and Sites (ICOMOS), in 2013, has vowed, to improve the environmental quality in the built heritage by reducing CO_2. To accomplish this purpose, they have agreed to provide applicable management and preservation strategies to the built heritage, particularly in Europe. This is because the built heritage can also play a major role in decreasing environmental issues (UNESCO and ICOMOS 2013).

The Agenda 21, as a result, should far-reach implications for everyone and particularly for building designers. This is because it is important to consider aspects of design which, e.g., provide flexibility for changes in use; or make use of recyclable materials as they are environmentally beneficial. Moreover, good building design, which highly considers environmental aspects at design stages and within construction details and specifications, can considerably contribute to reducing pollution and improving the environmental quality (UN 1992).

Accordingly, many trends and architectural philosophies adopted by contemporary architects have emerged with concentration on the built environment conservation. The architectural pioneers have contributed to the interpretation of these philosophical theories in architecture since the 1990s. The conservative ecological approach, for instance, believes that the use of modern technology, renewable energies and recyclable building materials will reduce the current conflict between the built environment and the natural environment and lead to the continuity of development in an ecological balance (Foster 2001).

Within this process, green architecture, as an architectural design technique, creates friendly physical development to the environment via designing energy efficient buildings and managing natural resources effectively. It depends mainly on natural resources and recycling principle, which produce no waste. It also employs natural renewable energy resources and applies recyclable materials that have least possible impact on the environment, particularly during processing, manufacturing, application and disposal. Other architects, furthermore, concentrate on environmental controls to ensure the best possible indoor thermal comfort by adapting building technology techniques, which depends on passive design for cooling and heating. It reduces the need for any mechanical controls via different methods, including natural ventilation, shading control and passive capture systems. It also maximizes the use of natural sources of heating, cooling and ventilation, to create comfortable atmosphere inside buildings and, by turn, decreases the energy need in buildings (Jona Jone, Energy & Environment 2015). Figures 1, 2 and 3 show different buildings, which was mainly designed based on different architectural technolgies.

These architectural design trends, unfortunately, are criticized because they are incomplete. They applied few methods of these technologies, such as air ventillation, shadding, photovoltic, some wall insulation, etc. In fact, this is not adequate to significantly decline energy demand in buildings. Building technology or green architecture afford different kind of methods that are compatible in decreasing the energy demand if they integrated. In other words, each of these technical methods has its own capabilities for decreasing the demand of energy comparing to the rest, which, at the same time, are compatible and overlap. Therefore, they are capable enough to considerably decrease the energy demand in buildings if they are applied holistically during the buildings architectural design different stages (Brian Edwards 1996).

Thus, the aim of the research is threefold: first, addressing the impact of energy buildings on the built environment; second, considering environmental aspects at the design stage can contribute significantly to reduce the buildings

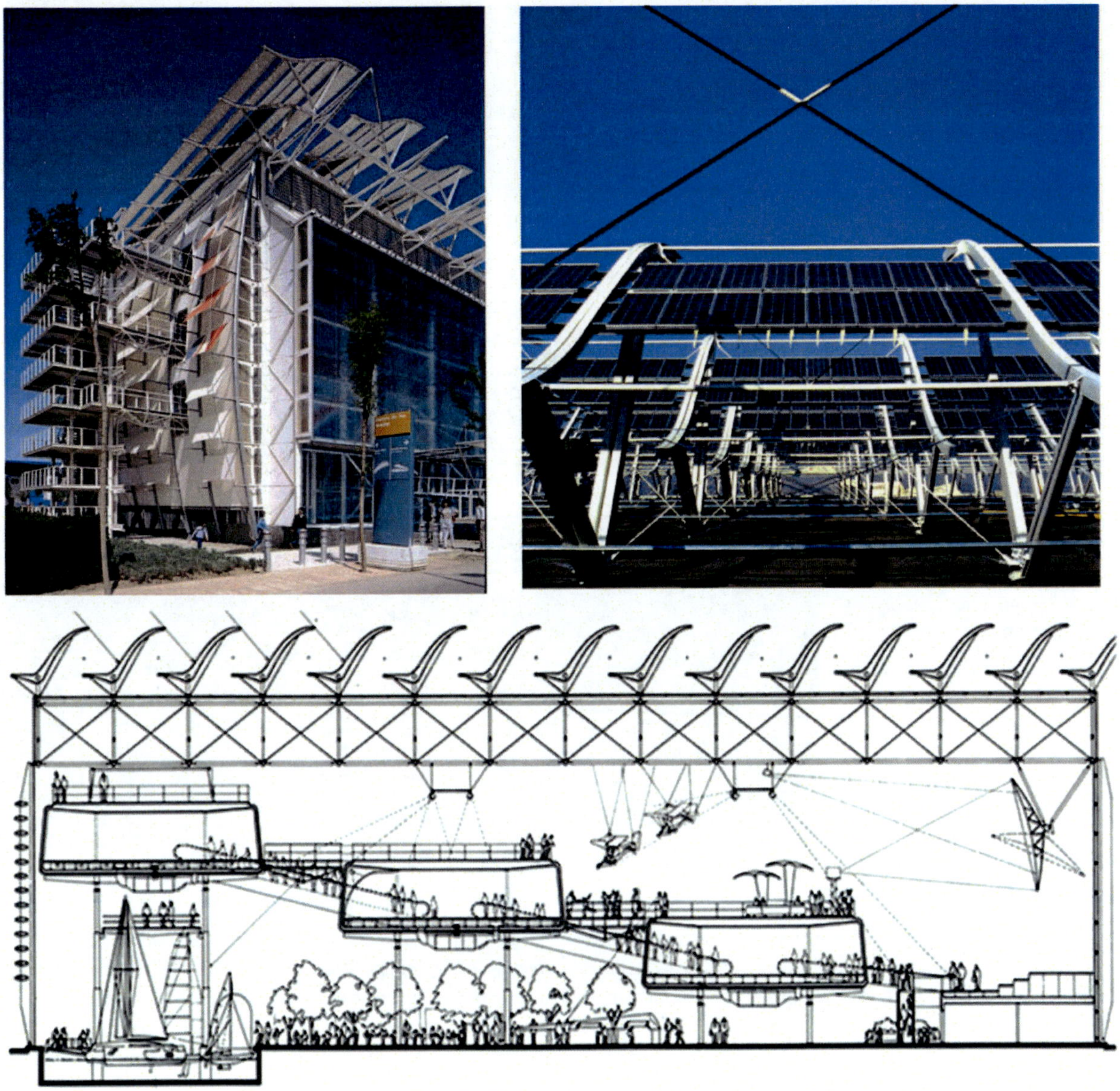

Fig. 1 British Pavilion at the Seville, Spain, Expo 92- Designed by Grimshaw. *Source* https://grimshaw.global/projects/british-pavilion-expo/

impact. In addition, the innovative technologies should be expressed holistically in the building forms, particularly within the architectural elements, features and details. It is important to consider aspects of design which provide flexibility for change in use according to requirements of different built environments. It is also essential to adapt architectural innovative technologies in buildings architectural details, which facilitate building compatibility with the built environment; and, third, giving professional design team and building users background information, design guidance and practical assessment of adapting the natural ventilation by using natural means in buildings. These aims are also appropriate to all built environment types, including the built heritage, cultural, and traditional and contemporary districts.

Therefore, the methodology of the paper is to firstly document global and indoor environmental issues occurred by the impact of energy buildings. Second, it explores alternative architectural approaches that adapt innovative technologies to decrease energy use in modern buildings. Architectural innovative technologies, in this research, cover green architecture and building technology. Third, it investigates the necessity of creating energy efficient buildings through adapting architectural innovative technologies

Fig. 2 Bahrain World Trade Center, twin tower complex, Manama, Bahrain-Designed by the Atkins, 2008. *Source* https://www.atkinsglobal.com/en-gb/projects/bahrain-world-trade-center

holistically. This is because the application methods of these technologies, in fact, are compatible for initiating efficient energy buildings. Fourth, it investigates the importance of developing architectural design to reduce buildings impact on the built environment. The efficient tools to adapt innovative technologies are during the building different design stages, including architectural design and construction details and specifications. Decisions at building different design stages could be expressed into sustainable routes by integrating the proposed architectural innovative technologies, which are expected to have long-term gainful environmental consequences. Decisions at building different design stages, in addition, are flexible to be revised and enhanced for the favorite of achieving efficient energy buildings as follows: 1. at the architectural design stages, particularly via architectural form, character, pattern, elements and features; and 2. during construction design stage through execution details and specifications. Moreover, this methodology is applicaple to new and exisitng buidlings. The difference between applying this methodology to each of these buildings could be defined mainly during the first stage, which is mainly concerned with technical data collection of buildings. Thus, the results of these data determine the building appropriate design method. In other words, the fisrt design stage is the most suitable for enhancing architectural innovative technical methods (British Standards Institution 1991).

2 Rationale Behind Developing the Architectural Design

The rationale behind developing the architectural design is to decrease buildings environmental impact; improve the quality of the built environment and increase thermal comfort of building occupants and performance. The use of buildings, as mentioned above, is responsible for external air pollution. For instance, energy buildings accounts for half of the UK's annual production of greenhouse gases. Buildings, in addition, are responsible for a large proportion of the emission of Sulphur dioxide and oxides of nitrogen, together responsible for acid rain; and the use of chlorofluorocarbons,

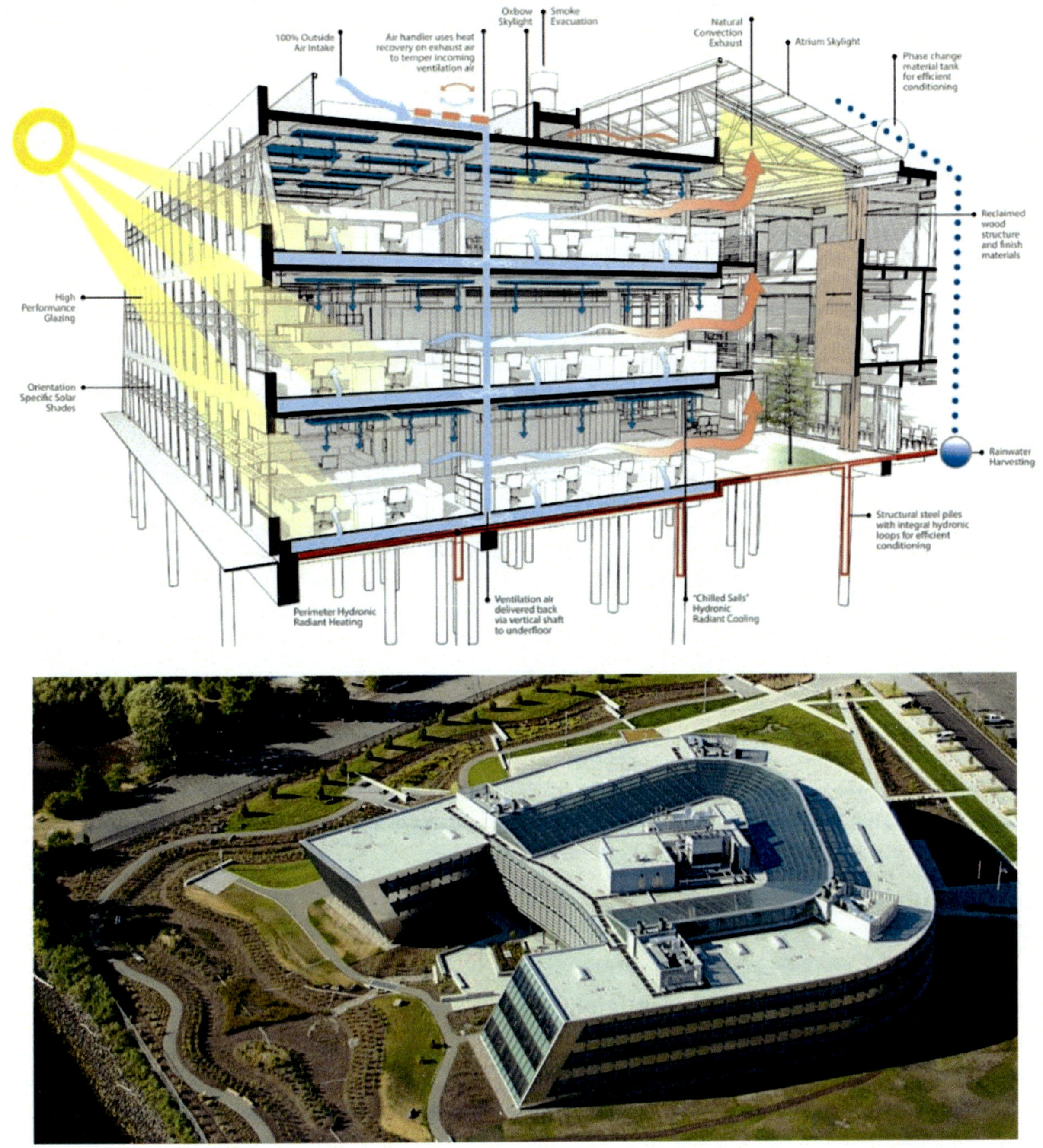

Fig. 3 Federal Center South Building, Seattle, Washington—Designed by ZGF Architects 2013. *Source* https://www.archdaily.com/447019/federal-center-south-building-2013-zgf-architects

CFCs, used in air-conditioning equipment and certain insulants, responsible for the depletion of the ozone layer.

The greenhouse effect is the name given to the phenomenon whereby trace gases in the atmosphere absorb infrared radiation emitted by the Earth's surface, causing a warming of the atmosphere. The atmospheric concentration of carbon dioxide, moreover, has grown from about 310 parts per million ppm in the 1960s to about 350 ppm during the 1990s, to about 411.2 ppm in 2018, which considered the second-highest annual increase on record. The annual growth rate was about 0.5 percent per year during 1990s, which increased to 0.9 during 2018–2019. Moreover, urban environment could be unhealthy because energy use continues to be acknowledged as a key issue of the built environment. Many problems classed as sick building syndrome have their origins in poorly designed, badly commissioned or irregularly maintained heating, ventilating and air-conditioned equipment. Buildings are complex systems and issues of the built environment impact as diverse as global warming, thermal comfort, aesthetic perceptions, health risks, energy systems engineering, ozone depletion, materials used and recycling (WorldGBC 2017/2018; Buildings International Energy Agency 2019).

The ample share of the world population spend most of their lives in buildings. That is why, health wellbeing is a matter of grand debate nowadays. Recently, the accelerated numeration increase of buildings has shown on the health condition deterioration of their occupants. Thence, the poor air quality within is in charge of the mentioned misfortune, due to the inherent atmospheric. This leads to the potential damage of their health condition. These pollutants include formaldehyde, asbestos, led-based materials and paints. No question, people are righteous of leading a healthy and comfortable life in their indoor spaces. It is their legitimate physiological right. According to the divine law of nature, people act as a source of heat emissions, due to their metabolism processes; exerting effort; or digestion; eating. This appears in energy transfer from potential to kinetic, where they can move and lift objects (Department for Business, Energy and Industrial Strategy 2017).

But if this rate of heat loss goes out of control, then a sort of distraction will result. Summing it all up, adequate ventilation is essential for the well-being and health of building occupants. Traditionally, the provision of fresh air was met by natural means. Fresh air has partly given way to air conditioning in response to the realized need to cool contemporary buildings. These buildings tended to suffer from high solar heat gains, poor natural day lighting and use of energy intensive appliances. The Building Research Establishment Environmental Assessment Method (BREEAM), concludes that if environmental concerns are recognized at the early design stage, a great impact decrease will be marked, saving the environmental and health downfall (Building Research Establishment 1990).

Within the complexity of buildings, Wozniak concluded that a logical approach must be extracted for encouraging better environmental performance by enabling sensitive design of developments and appropriate building design and operation to be recognized. In temperate climates, e.g., the Chartered Institution of Building Services Engineers (CIBSE) found that some means of warming buildings are essential in winter and it is a necessity in summer to add an element of cooling. The main object is to control the rate and modes of the cooling of the occupants, controlling the rate at which a person perspires. Consequently, increased concern over the adverse environmental impact of buildings has encouraged the design and construction of efficient buildings, in terms of adapted buildings to natural ventilation. In different living environments, the efficient buildings can provide a year-round comfort, with good controlled use, at minimum capital cost and with negligible maintenance. Obviously, many aspects of buildings design may need to be judged against best practice to avoid or minimize primary environmental damage. Environmental concerns need guidelines that derive logically from a desire to protect the environment, and these may be different from the use of other purposes such as maximizing human comfort. Wozniak added that it needs to be understood that any formula of judgment should start with the environment as the principle concern and with energy use, defining its area within a scope dedicated to environment (Chartered Institution of Building Services Engineers 2015; Wozniak 1993).

Effective methods for assessment of environmental impact of buildings are required, to enable designers to comply with the current and future legislation; to make international commitments meeting public concern expectations. As a result, an environmental impact assessment methodology is developed, for global, neighborhood and indoor issues. The main objectives of these methods are to: limit the amount of primary environmental damage caused by construction, use of buildings and development of land; provide a common set of targets and standards; encourage designers to achieve environmentally sensitive buildings, produce buildings that do not affect the health and general wellbeing of their occupants and those in the neighborhood; ensure that buildings maintain or improve their environmental performance year upon year, and provide recognition for buildings which are friendlier to the environment. Therefore, building is adopted with the environment should be tested. In other words, the extent of the environmental consideration at the design stage could decrease the impact of buildings on the environment. In fact, the objective behind this assessment is to provide evidence that considering environmental issues at the design stage could lead to a healthy building which is beneficial to the built environment (Building Research Establishment 1990).

3 Definitions

3.1 The Built Environment

Linguistic and literary speaking, a built environment, is different areas of buildings. Other linguistic definitions compare the built environment to nature. Yet, in fact, the built environment is not restricted on both perspectives. It includes all physical elements such as buildings, open spaces and infrastructures that support human activities such utilities and networks.

In 1997, Health Canada defined that the built environment "encompasses all of the buildings, spaces and products that are created or significantly modified by humans and in this context is understood to comprise: buildings and structures, facilities and networks for communication, facilities and networks for transport, plants, facilities, and networks for water supply and wastewater treatment, plants, facilities, and networks for solid waste treatment, and plants, facilities, and networks for and energy production and distribution". CMA (2013) In 2016, The Construction Industry Council (CIC), in addition, recommends that the built environment, "...encompasses all forms of building (housing, industrial, commercial, hospitals, schools, etc.), and civil engineering infrastructure, both above and below ground and includes the managed landscapes between and around buildings." CIC (2016) The Centre for Digital Built Britain also defines the built environment as; "All forms of buildings (residential, industrial, commercial, hospitals, schools), all economic infrastructure (above and below ground) and the urban space and landscape between and around buildings and infrastructure" (The Centre for Digital Built Britain and Designing Buildings Wiki 2019). Last, Tom J. Bartuska defines the built environment according to five interrelated characteristics as follows;

1. The entire sum of the manmade, modified and maintained creations.
2. The mechanisms of the man mind and the results of its purpose.
3. The innate aptitude, to serve their needs, wants, and values;
4. The manipulation instinct towards the environment for our comfort and well-being;
5. The inner drive to cooperate with or against the quality of environments both built and natural and to human–environment relationships.

He has organized the variety and scope of the built environment, into seven interrelated components, including: products, interiors, structures, landscapes, cities, regions, and Earth. These seven components outline the scope of the total built environment (Tom and Tom 2010).

On the other hand, as world population increases the requirements of the built environment are becoming more energy demanding. Over half the Earth's population now lives in cities and this percentage will rise to more than 70% by the second half of the first century. The environmental quality is facing serious concerns regarding its ability to last, and serve future generations. This is because the environmental quality has been suffering from major issues such as; Ozone Depletion, Global warming, Deforestation; air, land, and water contamination and run out of non-renewable energy resources. In the Global Report on Human Settlements 2009, Planning Sustainable Cities, Ban Ki-Moon, Secretary General of the United Nations, indicated: *"The majors challenges of the twenty first century include the rapid growth of many cities and the decline of others, the expansion of the informal sector, and the role of cities in causing or mitigating climate change. Evidence from around the world suggests that contemporary urban planning has largely failed to address these challenges"* (UN 2009). The built environment, accordingly, suppose to promote the quality of natural system. It is also regularly the need for compatibility with the nature. Therefore, the pressure for sustainable development in the built environment is dramatically increasing.

Regarding the responsibility of the built environment, different ideas and definition are emerged. The built environment, in fact, is an interdisciplinary field. Diverse specialists and experts could be involved such as architects, engineers, town planners, landscape and urban designers, central and local policy makers, etc. Their scopes are quite wide. They might comprise: 1. design, management, maintenance and monitoring of efficient and layout images of the built environments; and 2. recognizing and developing applicable methods regarding the quality and practice of the built environment. The planners and architects as ones of the main parties that have major direct influences on the environment; unfortunately, have been criticized for their role in accelerating the world's environmental crises (UNEP 1991).

3.2 Human Comfort Requirements

Human comfort inside buildings is one of the most important environmental concerns which is always affected by the architectural design. Human comfort, in fact, is a subjective matter and varies with individuals. It involves several environmental variables. These variables include: effects of temperature; air movement and humidity.

Common experience shows that writing or any light work involving manual skills is not helped by cold hands. The sensations of fatigue are also aggravated if access heat can not be got rid of adequately. Occupants clothing habits to a large extent influence the air temperature levels found comfortable for most people, but there is a wide variation between individuals in their objective sensations of thermal comfort in the same environment. Room temperature, therefore, is an important physical factor influencing the performance of such. In casual summer clothing, the optimum temperature for sedentary work at 50 percent relative humidity is about 25–26 °C. For more formal and winter clothing, the optimum temperature is 20–21 °C. Studies, in addition, have shown that comfort levels do not just relate to air temperature, but also to mean radiant temperature.

Standing beside large areas of glass can either make an occupant feel hot with shining sun, or cold at night. Radiators will also produce high gains to the side of an occupant body facing it.

The main control of internal air temperature is ventilation, which introduces fresh cooler air and removes unwanted warm air. Air movement, its effect upon human comfort, is also associated with ventilation. It is equally as important as temperature. Increased levels of air movements across the human body will aid perspiration, and hence increase heat loss through forced convection. Through various studies, the use of natural ventilation in offices has shown that positively better results have been found with regards to the sense of users' wellbeing. Freshness was also found to be associated with local stimulation of the skin by air movements. Comparing between the artificial and natural ventilation, artificial air movement can be controlled easily to give the required internal conditions. Too much air movement, e.g. anything above the recommended velocity of 0.5 m/s tend to cause disruption in moving papers and dust. Artificial ventilation can be programmed to give an output that can give any air velocity required, but these outputs tend to be of a constant value with no variation. Within this context, different architectural elements and features such as windows and stacks can play a significant role in facilitating natural ventilation requirements and, in turn, thermal comfort. Windows different types and designs, e.g., can control natural ventilation to give the desired air velocity, and then both the benefits of air movement and its variable characteristics can be enjoyed. Finally, gradual acclimatization that most experienced through annual temperature change requires different internal design temperatures to exist between the winter and the summer months.

Humidity influences the rate at which moisture can evaporate from the skin. If the relative humidity values are kept within a certain range, then this evaporation process can take place readily at rates governed by air temperature and perspiration rates. This range is generally considered to be between 40 and 70%. Air movement is again important because, in still conditions, the layer of air close to the skin rises in moisture content and the evaporation process is hindered. For relative humidity values higher than 70 percent discomfort can begin to occur, especially in higher temperature conditions when heat loss is very dependent on evaporation (World Health Organization 1987).

3.3 Architectural Innovative Technologies

3.3.1 Green Architecture Approach

Green architecture is a movement in the contemporary architecture inspired by the ecological characters of the plants and their harmony with nature. The green architecture approaches the design process; as a living plant. It has a cycle of life. It is born, grows and dies without having any negative impacts on its surrounding environment. The aim of green architecture is creating a friendly physical development to the environment. This is realized through designing energy-efficient buildings and managing natural resources effectively. It also produces green buildings that resolve many conflicting issues and satisfies the needed requirements. Each design decision has environmental repercussions and effects Thus, all resources go into buildings need to consider whether a sustainable architecture is to be produced including materials, fuels of the contribution of the users or not. This involves passively and actively harnessing natural renewable energy resources and using materials, which have least possible impact on the environment, particularly during processing, manufacturing, application and disposal of these materials. Green architecture depends mainly on natural resources and recycling principle, which produce no waste. It is also used to describe economic, energy saving, environmentally friendly architecture, which, in turn, explore the relationship between architecture and ecology (Torsten 2018).

Green architecture could also be applied for the urban development projects, including site, form and energy sustainability; material efficiency and recycling cost affords. They show the percentages of adopting renewable resources and total effects of the green elements on the projects' elements: landscape, function, air movement, solar panels, photovoltaic cells, wind energy, biomass energy, thermal energy, relation environment, recyclable materials, rain water, CO_2 emission saved, etc.

Green architecture, indeed, embraces several principles during preparation of different design phases and aspects, including: site studies, building form, energy types, recycling materials and methods. To employ direct relation with environment, and produce coherent framework for implementing a sustainable design, these design phases and aspects should highly explore green architecture principles. Many buildings, nowadays, express at least one of the various green characteristics. These buildings cannot be called green. A green building must embrace all principles of green architecture. This is because green architecture principles overlap and must applied holistically. Thus, the main principles guiding our green architecture design are: harmony with nature, site interpretation, working with climate, rationalizing new resources, creating healthy indoor environment, conserving energy, fulfilling users' needs, and rationalizing cost (Valeria & Emilia 2015).

3.3.2 Building Technology Approach

Building technology is mainly concerned with planning and design methods, which are made to achieve bioclimatic comfortable indoor spaces with minimum use of energy. The emphasis is on design of the building shell. Every component of the building shell is used to accomplish indoor thermal comfort. The passive cooling, shading methods and thermal capture, accordingly, become very important systems of climate balanced design. To minimize energy consumption. In addition, interacting design stages with technical systems is the precise scheme, within which natural cooling and heating are targeted.

There are many levels for climatic control such as microclimate, structure and mechanical control systems. This research is intended only to use building technology with passive design controls, which is the first known method for achieving thermal comfort, and best illustrated in ancient vernacular buildings, as a direct expression of adaptation to climate and resource constraints. Since the twentieth century, man has started to think of comfort as a product of energy consumption whereas building components, walls, windows, shutters, courtyards were used effectively as means of achieving comfort. Building technology, which rely on passive design controls for cooling and heating, may decrease the need for any mechanical controls. Its major application methods are mainly natural ventilation, shading control and passive capture systems (Building Research Establishment 1988).

During the architectural design process, the different passive design methods could contribute to reduce energy demand in buildings by adapting them in building orientation and form, improving building envelope performance; walls, roof, and windows, openings, architectural design elements and features to subdue the climate heating such as courts, cooling stacks, wind catchers, hydraulic cooling system, cooling structure and landscape. The principle is to provide adequate comfort standards for indoor occupants, and reduce energy consumption. Therefore, design concept and building technology details can provide significant support during planning and design process. Moreover, the purpose of environmental controls is to ensure the best possible indoor thermal conditions by relying on passive controls. This may eliminate the need for any mechanical controls, even if mechanical controls have to be resorted, their task will thereby be reduced to the minimum. The mechanical control should be used at necessary conditions. Last, after the architects begin by microclimate and passive controls to reduce the varying climatic factors (Peter 2010).

4 Architectural Design Guidance and Criteria: Technical Framework

To achieve architectural design that could contribute to decrease energy demand, and minimize buildings impact on the built environment, accordingly, the research defined architectural design guidance and criteria, technical framework applying appropriate architectural innovative technical methods. It is also appropriate to all built environmental types. The design guidance and criteria emphasize the necessity of environmental awareness for expressing architectural innovative technical methods at all building design stages, including building technology and green architecture. Application methods of both technologies are compatible and overlap because they are mainly concerned with the local nature of buildings, to utilize appropriate design concept and details. The design guidance and criteria also include case studies, conceptual architectural details that could contribute to perform energy buildings. These general design guidance and criteria, the technical framework, for architectural design could be defined as of the following. (Peter 2010; Peter et al. 2012) Fig. 4. it explains how the different application methods of building technology and green architecture could be holistically applied during architectural design process.

The design guidance and criteria, could be applied to all buildings of the built environment types, including the built heritage, cultural, and traditional and contemporary districts. The primary stage of new and/or existing buildings mainly include environmental data collection and site analysis. The data required for the existing buildings will survey all the existing buildings technical information such as: age, structure systems, materials, construction methods and materials for internal and external walls and roofs, window types and materials, floor heights, HVAC systems, building impact on the environment, etc. Accordingly, the analysis results of the data, will help architects and technical teams, to proceed within the same technical framework procedure. This would define the appropriate architectural innovative technical methods as well the new buildings. The suitable alternative detailed methods could be widely differed according to the data collection and analysis of each building (Peter et al. 2012; Zeiber 1999).

First, applying basic methods of building technology, resulted from passive design, to design concept and details can provide significant support to adapting buildings with the environment. The principle is to provide adequate comfort measures for indoor occupants and to reduce energy consumption.

1. Adapt building orientation and form to avoid climate impact on building surfaces, particularly windows and openings, Fig. 5.
2. Employ passive cooling systems, including:

- Natural ventilation systems: They are directly related to thermal comfort, e.g., by increasing air velocity and air movement in warm humid climate. It is a primary method for cooling buildings and air refreshment by using architectural elements and features such as window

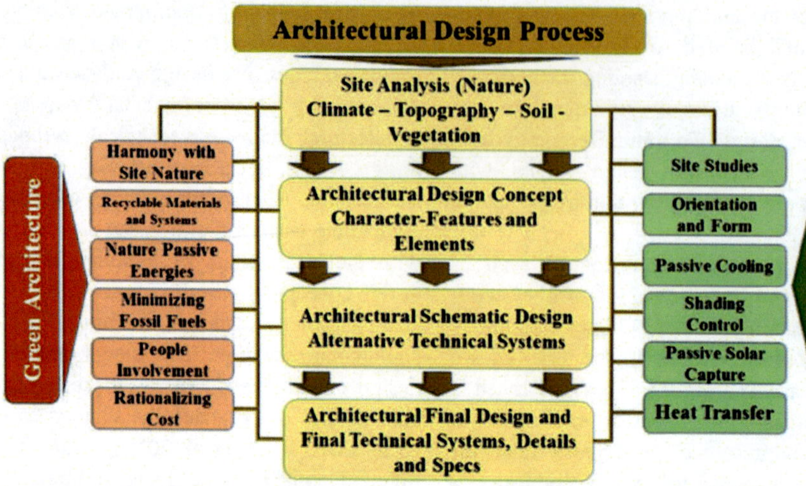

Fig. 4 Applling Different Application Methods of Building Technology and Green Architecture during Architectural Design Process, Technical Framework

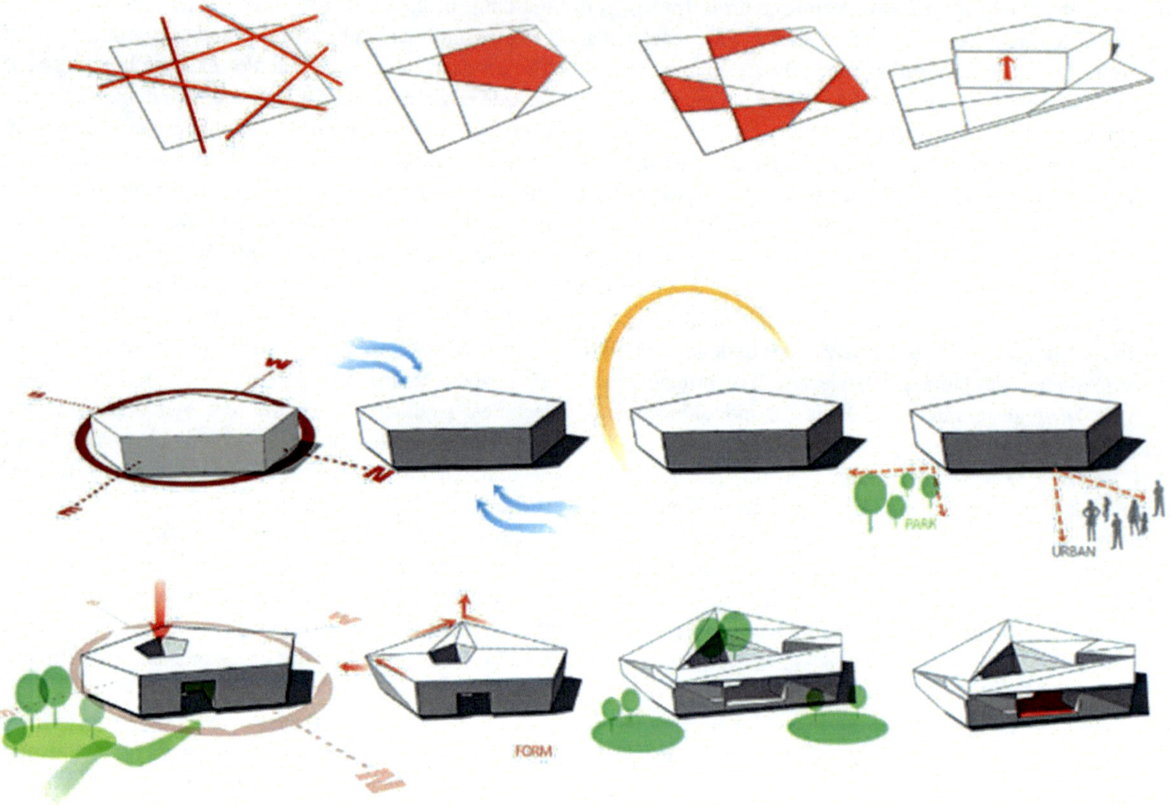

Fig. 5 Adapting Building Orientation, Form and Architectural Elements, Courts and Openings *Source* Concept Diagram, Architecture Board. https://www.pinterest.com/pin/671388256920760979/

different types and pattern, courts, cooling stacks, wind catchers, etc. It does not mean just using windows and doors for getting the natural air; and

- Evaporation cooling systems as a natural method to decrease dehydration risks.

1. Apply shading control systems that decrease sunlight penetration into buildings by using architectural elements and features such as shading different patterns and types, courts, etc., particularly during summer time.

Figure 6 shows examples of natural ventilation, sun shading, low tech and dynamic façades.

2. Utilize passive solar capture that cools and heats buildings through selecting construction materials with appropriate characteristics suitable for different location environmental conditions by using architectural elements and features such as windows, glazing type, thermal insulation, thermal mass and shading; and
3. Apply heat transfer in external walls, roof, windows and openings by using appropriate building materials for thermal storage and compatible technologies for heating and cooling. (Peter, 2010; Peter et al., 2012)

Second, green architecture can also afford considerable support during employing design concept and process via employing the fundamental active design methods. Active design methods depend mainly on natural resources and recycling principle, which produce no waste, to facilitate appropriate measures for decreasing building energy consumption.

1. Achieving harmony with on-site natural resources that produce no waste for buildings energy and materials requirements by utilizing solar energy via photovoltaic, wind energy and recycling of construction materials, and gray and waste water. Within this context, passive solar heating systems absorb, store, and distribute the sun's energy without relying on mechanical devices like pumps and fans, which require additional energy. Harmony with on-site natural resources also integrates vegetation and trees as part of buildings.
2. Achieving site interpretation by making best utilization of natural resources and interacting buildings with their sites, in terms of topography, soil types, surface and underground water, geological pattern, etc., Fig. 7.
3. Working with climate by combining architectural concept and design with the climate and striving to harness natural forces at a time, while deflecting them at another in order to create a livable indoor climate all year round.
4. Rationalizing new resources via considering traditional aesthetics of massing, proportion, scale, texture, shadow and light.
5. Creating healthy indoor environment as most of the world's population spends about 85–90 percent of their time in indoor environment. This could be achieved by utilizing materials that have least impact on the environment, during processing, manufacturing, application and disposal.
6. Conserving Energy by minimizing the need for fossil fuels in our buildings, as a non-renewable resource, and using nature passive energies to directly power all our basic life. Figure 8: explains buildings integrated photovoltaics. Figure 9: shows integrating of photovoltaic (PV) Systems into window design and shading system.
7. Fulfilling, users' needs by recognizing importance of people involvement and respecting all resources that contribute to initiating a building without excluding human beings.
8. Rationalizing cost by implementing Life Cycle Costing method (LCC), which considers the true building cost during its lifetime. LCC defines the cost of a building in

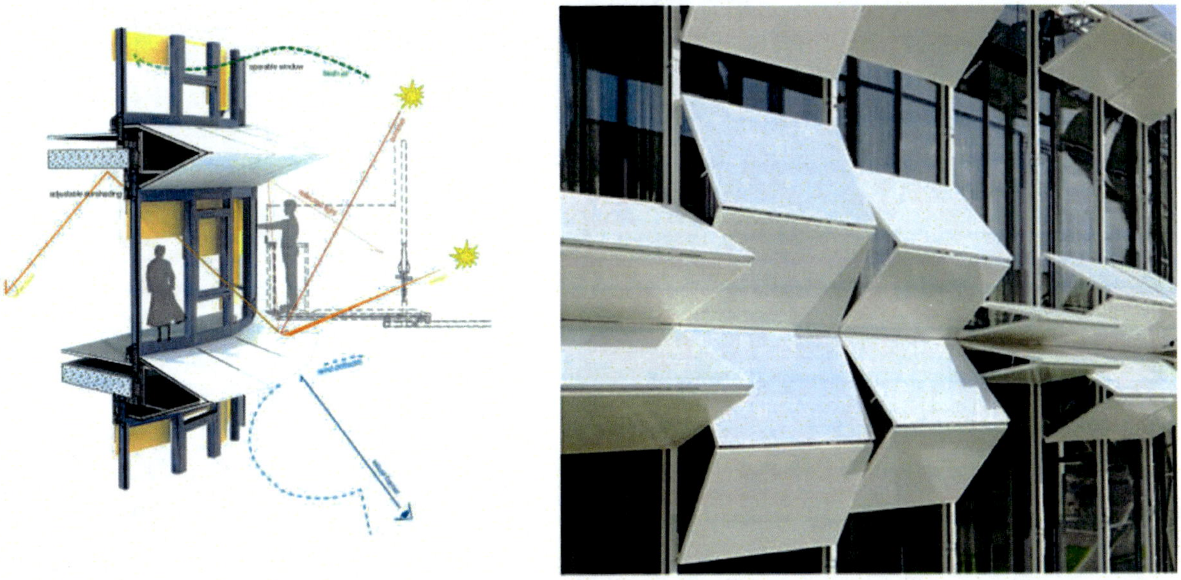

Fig. 6 Natural Ventilation, Sun Shading, Low Teck and Dynamic Façade. Low Tech Archtiecture- Facade Architecture, Building Design-Auditorium Natural Ventilation. *Source* https://www.bdonline.co.uk. Dynamic Facades: The Story. *Source* https://www.arch2o.com

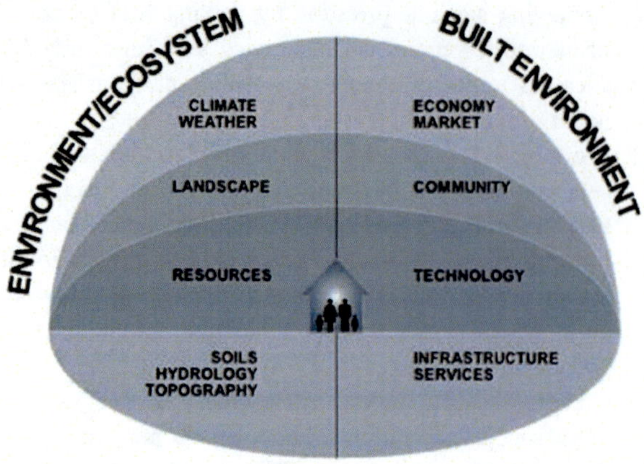

Fig. 7 Utilization of Natural Resources and Interacting Buildings with their Sites. *Source* Whole Building Design Guide (WBDG). https://www.wbdg.org/resources/building-science-concepts

use via, e.g., calculating how much it costs to heat, light, ventilate, maintain, insure and dispose of. If this considered, the low energy building performance will be far out any other building performance (Peter 2010; Peter et al. 2012).

The general design guidance for using natural ventilation include, where possible, first, occupants should be given control over their local environment. This can contribute significantly to the accepts of their environment. Second, a brief simple occupants guide using the building should be considered.

Moreover, there are issues that influence design alternatives. These issues include: site location and characteristics, heating seasons and summer overheating. Design alternatives could be defined as follows: First, green field sites pose fewest problems for natural ventilation. In urban sites, noise and external air pollution can be reduced by appropriate window design and location of spaces. Buildings in area of severe exposure, for example on hills and coastal sites, generally need smaller of fewer ventilation openings than those are sheltered. Shelter belts, plantings of trees and shrubs to act as wind break, are usually placed perpendicular to the direction of the prevailing wind. The resistance depends on the density of the foliage and is the greatest immediately behind the shelter belt. Artificial wind breaks can be provided as shelter. Ventilation design for low-rise buildings near tall buildings should consider the effect or higher wind speed at ground level as well as changes in wind induced pressure on the building faced.

In heating seasons, second, heat losses and draughts, due to uncontrolled air infiltration, should be minimized. This is by ensuring that all external openings are weather-stripped and, if possible, by making the building envelope airtight. Draught lobbies, in addition, should be provided to frequently used external doors and they should be sheltered from prevailing wind. Local radiant heat loss and down draughts should be minimized from large area of glass by using double or triple glazing, insulated blinds and low-level heating (World Health Organization 1987).

Fig. 8 Buildings Integrated Photovoltaics. York University Building in Toronto—By Perkins + Will's—March 2018. *Source* https://www.archdaily.com/tag/photovoltaic-panels. EWE and Bursagaz Headquarters, By: TAGO Architects- Bursa, TURKEY—May 2016. *Source* https://www.architectmagazine.com/project-gallery/ewe-bursagaz-headquarters

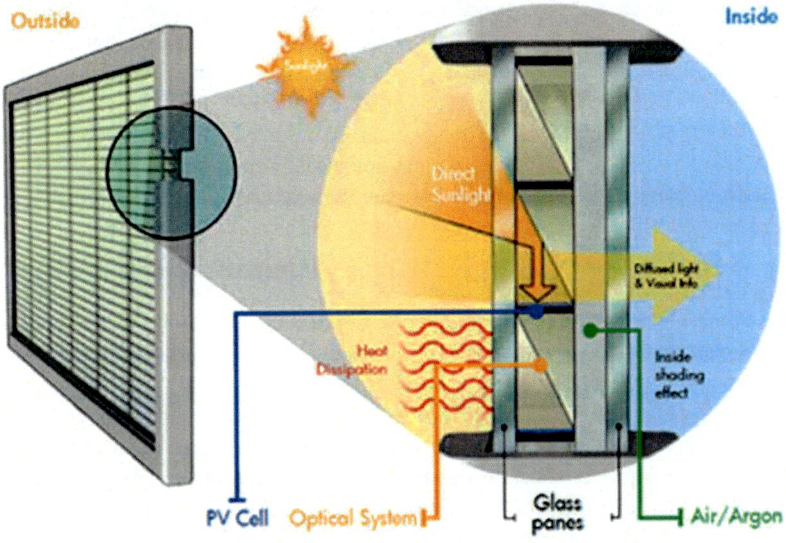

Fig. 9 Integrating of Photovoltaic (PV) Systems into Window Design and Shading System. *Source* https://www.google.com/search?safe=strict&sxsrf=ACYBGNQ6LSdJ2WLxXxeeXcl6ppuFHhtm7w:1572702628394&q=Integration+of+photovoltaic+(PV)+systems+into+window+. A Music Complex Near Paris, A Wall of Solar Panels that Move to Follow the Path of the Sun. By: Architect Shigeru- October 2017. *Source* https://www.dezeen.com/2017/10/04/shigeru-ban-la-seine-musicale-music-complex-moving-solar-panel-wall-paris-france/

Windows, third, are the most obvious controllable openings for natural ventilation, especially in summer. Windows, therefore, should: ventilate effectively, but not cause draughts; provide enough glare-free daylight and adequate view out of the building; keep out excessive solar gain; provide good insulation and avoid condensation; allow occupants to adjust the open abled area; and be simple to operate. A conflict can be occurred between security and natural ventilation. Large open window can present a security risk, especially on the ground floor. By adopting precise window opening designs or by separating the ventilation elements from the window, however, natural ventilation can be provided without compressing security. This is practically important for night ventilation (World Health Organization 1987).

In summer overheating, fourth, it should be accepted that internal conditions affecting comfort vary with time in naturally ventilation building. Thus, it should be possible to maintain a comfortable range of conditions without resorting to the provision of artificial cooling by using such measures shade, thermal mass air movement, lighting controls and low-energy lighting. Cross ventilation through operable window, in addition, should generally provide these flow rates in conventional shallow-plan buildings. However, there will not be human comfort satisfactory. Ventilation, moreover, can be used during the night to cool the building structure and, as a result, limit the temperature rise during the following day. The effectiveness of night time ventilation for cooling depends on the difference between the material the ambient temperature and the internal temperature as well as on the thermal characteristics of the building (Valeria & Emilia 2015; World Health Organization 1987).

The design of naturally ventilated building should reflect different requirements for winter and summer occupancy. In winter, minimal excess ventilation, controlled background and trickle ventilations, is required to meet occupants' needs for comfort. In summer, ventilation may be needed to exceed what is required solely to satisfy occupants' needs to avoid overheating. The distribution of fresh air is important within the space and can enhance comfort conditions and freshness. Wind pressure or thermal buoyancy, stack effect can generate natural ventilation. On a building, both function in varying parts according to the prevailing wind strength and temperature degrees. Wind pressure on a building depends on wind direction and speed; and shape of building. Temperature difference between inside and outside doors causes density differences in the air which, in turn, causes pressure differences. Airflow rate through openings is not linearly related to pressure difference. Thus, for any wind-induced and stack-induced flows cannot readily be added together. The pressure generated, wind and/ or temperature, is used to balance the resistance to air flow of all the openings on the air route through the building. The design alternatives for achieving ventilation include: Stack effect and cross ventilations, and tower head design (Valeria & Emilia 2015).

Stack Effect Ventilation will occur where there is a difference in height between two openings because the density of air decreases with increasing temperature. If a vertical column of air at one temperature is connected with similar column of air at a different temperature, the greatest pressure at the bottom of the cooler column will cause air flow into the warmer column. Taller columns and greater differences in temperature produce greater pressures. The combination of supply ducts and extract towers gives the opportunity to isolate the interior from external noise, and to some extent dust. The flow resistance should be regulated to compensate for wide variations in the driving forces, thermal buoyancy and wind pressure. It is best to provide each space with a specific ventilation tower and/or duct. Shared towers lead to balancing problems, fire risk and noise problems. Ventilation towers are appropriate for larger deep plan buildings and/or where there are areas of high density of pollutant (or heat) production. Passive stack can also be promoted through an atrium which will also act as a buffer to reduce fabric heat losses (Valeria & Emilia 2015).

Cross Ventilation occurs when air movement of air impeded by an obstruction, differences in pressure are caused around its outsides. If the obstruction is a building, these pressure differences can be used to push air into the building and to suck air out when holes, ventilators or operable windows, are placed at positions of increased or decreased pressure.

Tower Head Design will occur when air movement arises from differences in air pressure. If these differences can be controlled by innovative chimney head design, then the air can be directed and enhanced to improve a tower's performance. There are numerous ways of achieving this purpose. The simplest sample is the ventilating tower. The tower extends above the building roof to brack the wind and generate a law-pressure area, regardless the prevailing winds direction. The low pressure over this venting tower pulls air into the building from higher pressure below. This system requires producing openings on all sides of the towers head. This method is supplemented by the natural phenomena of hot air rising, e.g., the stack effect (Valeria & Emilia 2015).

5 Conclusion

The paper illustrates the impact of energy buildings on the environment. The outcomes of the investigation of the global and indoor issues provide strong evidence that buildings have adverse impact on the environment. Buildings are one of the factors that increase the warm of the Earth's surface. They are to some extent responsible for the depletion of the ozone layer. In the built areas, unlimited use of energy in buildings increases the rate of pollution which adversely affect the population health. Indoor air quality of buildings is somehow responsible for poor health. Inadequate ventilation and some materials being used in buildings are causes of pollutants, which have potential damages to human health.

It is not only the investigation of environmental building impact, but also finding a formula for decreasing this impact. Compatibility of architectural design with environmental concerns could play a great role in this process. The rationale is to influence architectural design at an early design stage,to raise the profile of environmental concerns throughout design process. Architectural design, in fact, can contribute considerably to reducing building impact and, in turn, improving the environment by enabling appropriate architectural concept, design details, and operation to be recognized.

The outcome focuses on the requirements of occupants' thermal comfort and architectural innovative technologies as factors that enhance building development. The requirements and approaches provide flexibility for design change according to different environments. They provide basis, background information, design guidance and practical assessment of adapting the natural ventilation by using architectural elements and feature in the form of buildings. They are a good practice to be expressed.

The paper provides strong evidence that architectural innovative technologies, if considered at all design stages, could contribute significantly to reduce energy building demand and their environmental impact. It is also emphasis that environmental impact assessment methods should be developed to enable designers to test their architectural design at an early stage. The result is the assurance of the compatibility of the new buildings with the environment. In addition, the existing buildings could be assessed in the favor of the environment.

Consequently, for future directions of further research on reducing the environmental building impact. First, all building designs should be examined and evaluated before proceeding on building construction, in terms of capabilities of these designs to decrease energy demand in buildings. It is also essential to enable architects and designers assessing their designs prior to establishing these buildings. Thus, future research should focus on adapting appropriate environmental impact assessment methods that assess building designs. Second, future research should also define the interactions of building technology and green architecture with advanced new technologies such as nanotechnology. In other words, to explore how application methods of these technologies could be merged, to produce efficient energy buildings, and reduce their impact on the built environment.

References

British Standards Institution. (1991). *Code of practice for ventilation principles and designing for natural ventilation*, British Standard BS 5925, London, BSI. https://www.aivc.org/sites/default/files/airbase_13035.pdf.

Building Research Establishment. (1988). *Environmental design manual: summer conditions in naturally ventilated offices,* BRE Report BR 86. Garston, BRE. https://www.brebookshop.com/details.jsp.

Building Research Establishment. (1990). *an environmental assessment for new office designs,* Building Research Establishment Environmental Assessment Method (BREEAM): Version 1/90. BRE Report. Garston, BRE.

Buildings International Energy Agency. (2019*), Energy Efficiency: Buildings*, https://www.iea.org›topics›energyefficiency›buildings.

Brian Edwards. (1996), *Towards Sustainable Architecture*, Butterworth Architecture, P.39–40,79–81,.86,87, 207.

Chartered Institution of Building Services Engineers 2015 Guide A: Environmental Design CIBSE. www.cibse.org/GuideA.

Department for Business, Energy and Industrial Strategy. (2017), *UK Greenhouse GAS Emissions, Provisional Figures*, Statistical Release: National Statistics, 29 March 2018 https://assets.publishing.service.gov.uk/government/uploads/system/uploads/attachment_data/file/695930/2017_Provisional_Emissions_statistics_2.pdf

Foster, N. (2001), *Norman Foster: Architecture & Sustainability*, London: Prestel Publication, Rebecca Lindsey, Climate Change: Atmospheric Carbon Dioxide, NOAA 2019. https://www.climate.gov›news-features›understanding-climate›climate-c.

Jona Jone, Energy & Environment. (2015). *7 Green Building Trends to Watch in 2015*, https://www.triplepundit.com/story//7-green-building-trends-watch-2015/35286.

Peter, O., A., Ezekiel, A., C., and Paul, O., O. (2012). *Design of a Sustainable Building: A Conceptual Framework for Implementing Sustainability in the Building Sector*, Buildings 2012, 2, 126–152; doi: https://doi.org/10.3390/buildings2020126. ISSN 2075–5309.

Peter A. (2010). *Passive & Active Design*, CIBSE Building Simulations Group, https://docplayer.net/20896503-Passive-active-design.html.

The Canadian Medical Association. (CMA). (2013). *Policy on the Built Environment and Health*, https://policybase.cma.ca/documents/policypdf/PD14-05.pdf.

The Centre for Digital Built Britain and Designing Buildings Wiki. (2019), *Built Environment*, https://www.designingbuildings.co.uk, wiki, Built environment. E:\IEREK\09-CAH2020\Author Files \Magdy M. El-Bastawisy\Designing Buildings Wiki, http://www.designingbuildings.co.uk›wiki›Built_environment.

The Construction Industry Council (CIC). (2016), *Essential principles for achieving an accessible and inclusive environment*, file:///C:/Users/sky/Downloads/essential-principals-guide-creating-an-accessible-and-inclusive-environment-1.pdf.

The United Nations. (2009). *Planning Sustainable Cities Global Report on Human Settlements*, United Nations Human Settlements Programme (UN-Habitat), ISBN 978–1–84407–898–1, www.unhabitat.org.

The United Nations 1992 The Rio declaration on environment and development the Earth Summit '92: The United Nations Conference on Environment and Development Rio de Janeiro 1992 The United Nations 11 15

The United Nations Educational, Scientific and Cultural Organization (UNESCO) and the International Council on Monuments and Sites (ICOMOS). (2013). *Managing Cultural World Heritage*, ISBN:978–92–3–001223–6, 978–92–3–500006–1 (chi), 978–92–3–600047–3 (ara).

Tom, J., B., and Tom, J. B. (2010), *The Built Environment: Definition, Design and Development*, https://pdfs.semanticscholar.org/ebad/e2fc6bae8ca5c8e825fbbd34e0fef7f27192.pdf.

Torsten, S. (2018). *Giving Meaning to the Concept of Sustainability in Architectural Design Practices: Setting Out the Analytical Framework of Translation, Sustainability, 10*, 1710. https://doi.org/10.3390/su10061710.

UNEP. (1991). *Caring for the Earth- A strategy for sustainable living*, Published in partnership by IUCN- The World Conservation Union, UNEP- United Nations Environment Program and WWF-World Wide Fund for Nature, Gland, Switzerland, P.4. https://portals.iucn.org/library/efiles/documents/cfe-003.pdf.

Valeria M. & Emilia C. (2015). Sustainability in the Built Environment: Integrating Scales of Action and Evaluation. *European Journal of Sustainable Development, 4, 2*, 51–60 ISSN: 2239–5938 Doi: https://doi.org/10.14207/ejsd.2015.v4n2p51.

World Green Building Council (WorldGBC). (2017/18), *Annual Report 2017/18 World*, www.worldgbc.org›news-media›worldgbc-annual-report.

World Health Organization 1987 *Air quality guidelines for Europe*, European Series No 23 Copenhagen, WHO, Regional Office for Europe 91–104 182 199

Wozniak, S. J. (1993), *Environmental assessment of buildings and building development: A logical methodology for the world*, The Jays, Bell Lane, Bedmond, Abbots Langley, Herts WD5 0QT: 1–27.

Zeiber, L. (1999). The Ecology of Architecture: A complete Guide to Creating the Environmentally Conscious Building Weston Guptill Publications

Building Virtual Environment for Safeguarding Archeological Ruins Case Study—El-Sheikh Ibada (Antinoopolis)

Noha Abbas Hamed, Mayada Belal, and Mona Raafat

Abstract

Many archaeological sites around the world have suffered over the years from serious issues such as natural disasters, looting, expanding of cities, farming, and deficiency of interest that contributed to the loss of their features. This paper looks at safeguarding these sites through interpreting their archaeological ruins. Recently, modern technology has been able to greatly help with this issue, as it can provide a detailed visual representation of the cultural heritage site in order to build a virtual environment that simulates the real world. A virtual environment can be produced by using digital documentation techniques and immersive reality applications, which can work effectively to produce a model of the missing archeological objects through integrating a detailed historical and archaeological information with an accurate metrical survey (photographs, site plans). Thus, this information can support the building of the virtual environment to visualize the lost world and interpret the cultural heritage site. Such techniques can be similarly adopted in other areas such as animation, movies and video games to promote these archaeological sites. It is therefore essential to connect our cultural heritage with such modern technology to ensure adequate technical support, particularly when dealing with archeological ruins. Moreover, providing means for simulation and interpretation can help in safeguarding these sites with its valuable resources and prevent them from neglecting. The methodology of this research starts with identifying the concepts of safeguarding cultural heritage. Then, after determining what technical requirements are involved in the field of cultural heritage, the paper identifies the most appropriate technology to the case study that can help in achieving the various purposes, such as the reconstruction of the non-existed objects, enhancing exhibition, and learning purposes. The paper concludes by proposing guidelines for producing a rich virtual environment that enables the interpretation and promotion of the case study El-Sheikh Ibada (Antinoopolis) archaeological site as well as similar cases.

Keywords

Archaeological sites • Digital cultural heritage • Archaeological ruins • Safeguarding heritage • Heritage documentation techniques • Virtual environment

1 Introduction

El-Sheikh Ibada "Antinoopolis" is an ancient city located in middle Egypt on the east bank of the Nile (Figs. 1, 2 and 3). It was established by the Roman Emperor Hadrian in AD 130. The city poses high historical, architectural, and artistic value since it integrated different cultures that represent Ancient Egypt, Greek, Roman, Byzantine, and Islamic civilizations. This mixing between these different civilizations created the unique character of this site. Unfortunately, the site has suffered from natural disasters and human neglection, which led to the destruction of most of the building, monuments, and archaeological objects of the site. Almost nothing remains from this site except archaeological ruins which hardly can be interpreted, as the absence of these physical resources, is considered being an obstacle in promoting this site. Subsequently, the site nowadays is forgotten and abandoned. Furthermore, the site is threatened by looting, city expansion, and the effects of climate that may lead it to vanish. However, it is crucial to safeguard archaeological sites such as these and guarantee its long

N. A. Hamed (✉) · M. Raafat
Helwan University, Helwan, Egypt

N. A. Hamed
BTU University, Cottbus, Germany

M. Belal
University International Relations Office (IRO), Helwan, Egypt
e-mail: mayada_belal@hq.helwan.edu.eg

A. Versaci et al. (eds.), *Conservation of Architectural Heritage*, Advances in Science, Technology & Innovation,
https://doi.org/10.1007/978-3-030-74482-3_36

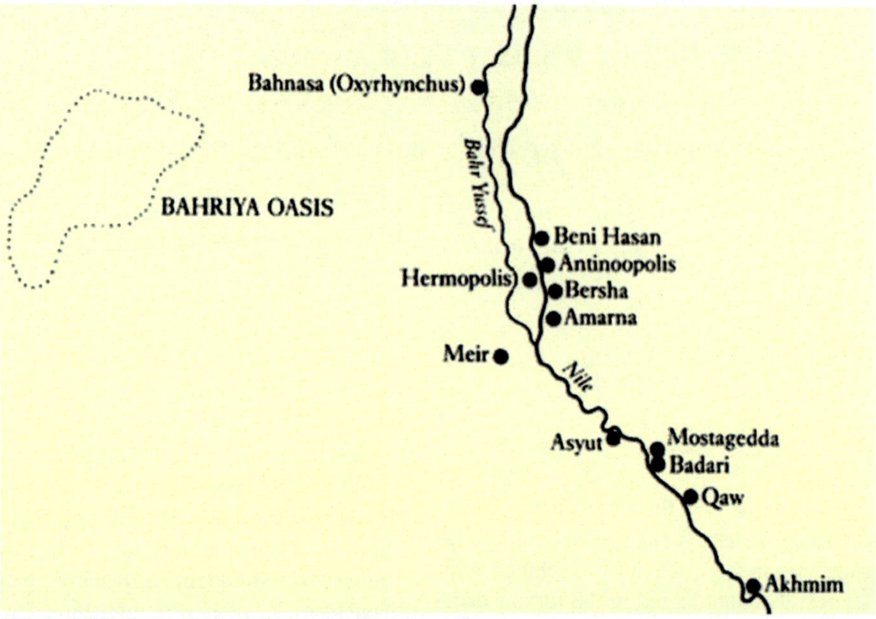

Fig. 1 Location of El-Sheikh Ibada (Antinoopolis) between the ancient cities

Fig. 2 El-Sheikh Ibada (Antinoopolis), triumphal arch in 1799

Fig. 3 El-Sheikh Ibada (Antinoopolis) in 2019

existence as well as providing open access to future generations. Accordingly, the research aims to exploit the recent technology that is currently used in cultural heritage to interpret the archaeological site of Antinoopolis through a virtual environment.

The research will focus on the reconstruction of the missing objects that can help in conceiving the lost world and understanding its ruins. In addition, provides a visualization of the historical data to reveal daily life in the past as well as building methods that can enrich the virtual environment. Technical issues that enable simulation and interaction for entertainment will also be considered. Afterwards, presenting the heritage site virtually through the appropriate virtual reality applications for the promotion of cultural heritage tourism.

2 The Safeguarding of the Archaeological Site El-Sheikh Ibada (Antinoopolis)

Safeguarding cultural heritage is the main principle of heritage conservation. As it emphasizes the protection of the cultural heritage with securing long-term viability. Accordingly, all the means of preservation, including the recent technologies, are desirable with considering the instruction of the international guidelines. Regarding to the UNESCO 2003 Convention, the safeguarding means "measures aimed at ensuring the long-term viability of the tangible/intangible cultural heritage, including the identification, documentation, research, preservation, protection, promotion, enhancement, transmission, particularly through formal and non-formal education and training, as well as the revitalization of the various aspects of such heritage" (ICOMOS, 2013).

The main objectives for the research

- Determining the contextual and technical requirements that stated by the international guidelines and agreements for safeguarding the cultural heritage sites, as well as the most suitable technology for building a virtual environment to interpret archaeological ruins.
- Examining virtual heritage applications through previous experiences to choose the most appropriate one.
- Providing workflow for the case study "Antinoopolis" considering the conceptual and technical requirements for interpreting cultural heritage.

2.1 Virtual Heritage Environment

The virtual environment can be established via the integration of 3D digital documentation techniques and immersive reality technologies (Artopoulos & Charalambous, 2019). Therefore, it is important to consider the technical component that these applications are based on, for instance, (tracking, modelling, and methods for interaction) to achieve the best experience (Bekele & Champion, 2019).

2.2 Digital Documentation Techniques

Many cultural heritage sites suffer from several risks around the world that increase the necessity of documenting these threatened sites. Recently, international organizations have a great concern for documenting cultural heritage sites with emphasizing it in conservation practices. The documentation of CH in digital form is increasingly becoming important when recording and assessing the condition of the cultural heritage sites. Accordingly, these sites can be well preserved as well as being available for the next generation. The International organization ICOMOS, UNESCO, including the famous international Charters provides the suite of documentation requirements, including the technical constraints and command specifications that must be considered while documenting the cultural heritage sites (Bekele et al., 2018).

The summary of the above-mentioned documents is:-

- To record large amounts of 4D multi-source, format, and content of information in detail with high accuracy.
- Include dated historical photographs and 3D digital records.
- Secure the best management of the 4D information by enabling them available for distribution and sharing.
- Visualization the integrated information in a suitable application for best presentation (Patias, 2006).

2.3 Modelling Techniques

Over the last few years, modelling techniques are commonly used in cultural heritage, such as 3D techniques, 3D laser scanning, CT scanning, and PTM polynomial texture mapping (Payne, 2013). These techniques can aid in the purpose of reconstruction to efficiently produce models of the missing archaeological objects with providing some conclusions about the lost world. (Rodrigues, et al., 2014).

However, the integration of laser scanning, Photogrammetry, and 3D Photo modelling techniques, can assist in sitting a developed information platform. These techniques are able to present the cultural heritage sites with high quality and accuracy of different scales as well as including the detailed objects and the monuments arrangements (Gabellone, 2009). Because of the vulnerability of the heritage sites, it is important to follow the technical consideration for 3D scans of objects that published by the international guidelines and agreements.

2.4 The Technical Requirements for Cultural Heritage Documentation

There are many Architectural applications such Photogrammetry and imaging techniques that are being used in cultural heritage. To decide the most appropriate application for the case study, the English Heritage guideline for technical specifications will be followed, particularly in documenting archaeological objects. The guideline has classified the applications in various ways and according to different parameters by the purpose of the documentation as clarified in Fig. 4.

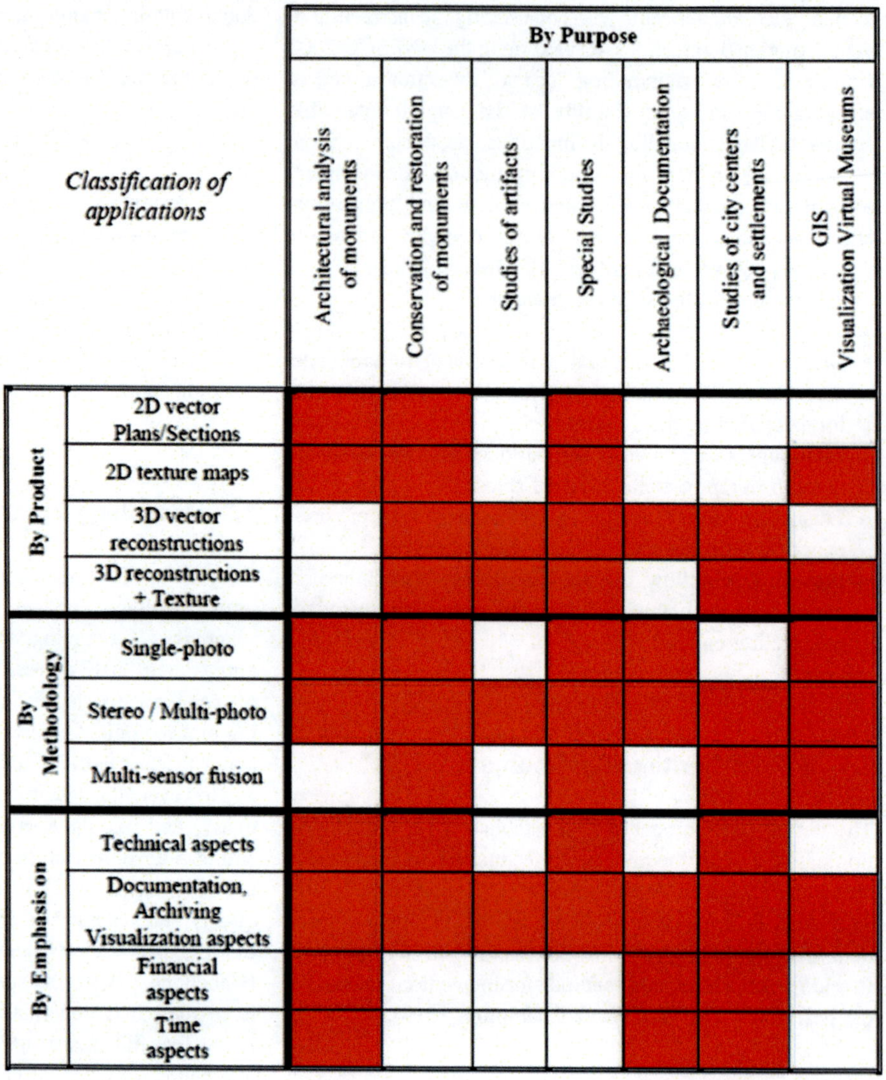

Fig. 4 A technical requirement's reference, especially for documentation of monuments, the English Heritage Guidelines

2.5 The Immersive Reality in the Applications of Virtual Heritage

The Immersive reality applications are set to become a vital tool in the presentation of cultural heritage and make it digitally accessible, as the forms of accessibility can be recognized in virtual reconstruction and virtual exhibitions. Such advantage is helpful, particularly in the case of limitation of physical access to the heritage sites. Moreover, these technologies can enrich cultural learning with enabling the interpretation of the cultural heritage perfectly to the visitors in virtual environments (Bekele & Champion, 2019).

Accordingly, the user requirements and site needs should be considered in the digital reconstruction and the best presentation (Vlahakis, et al., 2003). To decide the most reasonable application to the case study, previous studies have been analysed, particularly a detailed comparison study and recent survey of the immersive reality applications with its enabling technologies (Fig. 5).

3 Previous Experience in Building a Virtual Environment in Cultural Heritage Sites

Many projects have benefited from the recent technology with ensuring its positive impact in cultural heritage area, especially in achieving the purposes of simulation, interpretation, and promotion. Some of these projects will be overviewed as follows: -

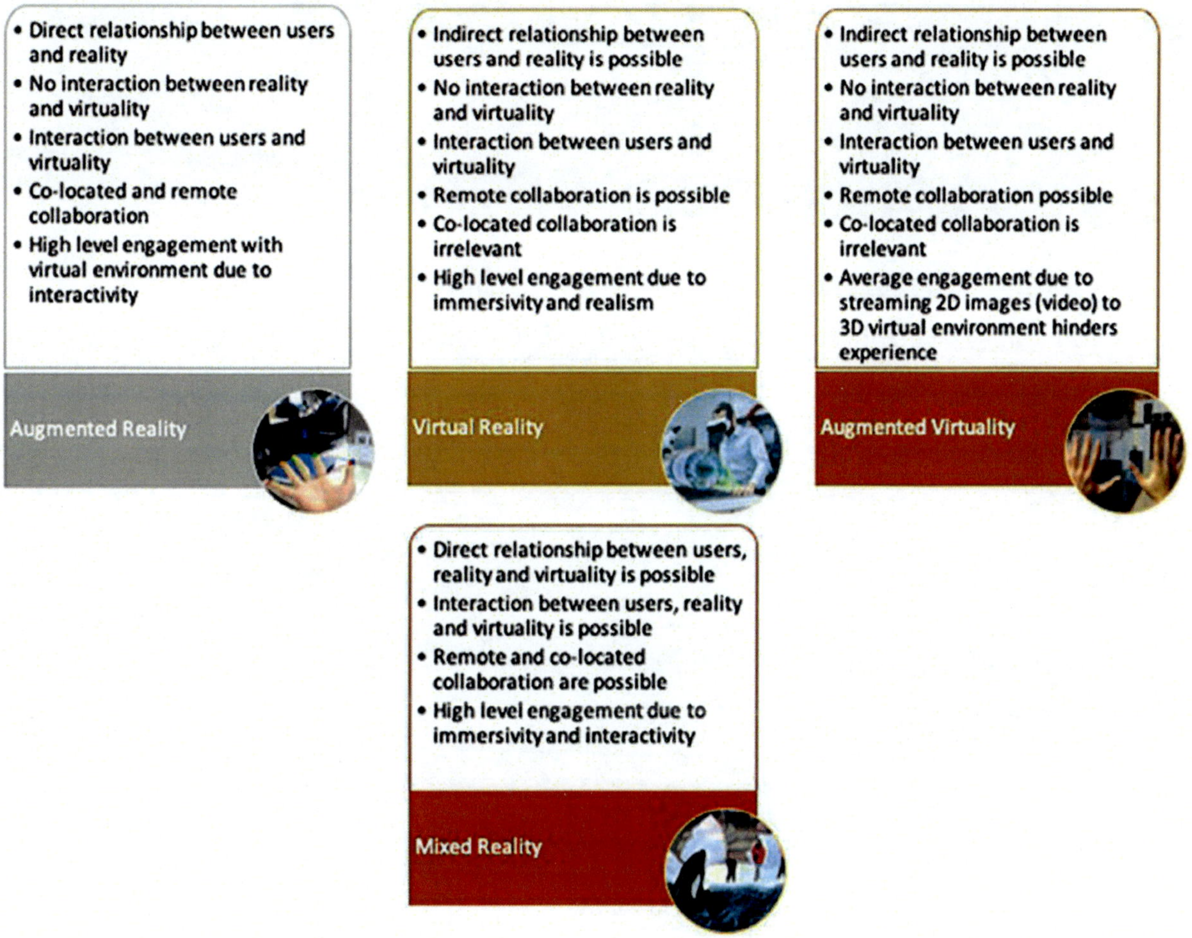

Fig. 5 An assessment of the applications of virtual reality with its enabling technologies

3.1 "Virtual Environments as a Technological Interface Between Built Heritage and the Sustainable Development of the Citthe" (2018)

The project aimed to develop a virtual environment to the historic town in Cyprus "Nicosia". As it gives the opportunity to the user to explore the city virtually in order to understand the changes around the years since the construction till the present day (Fig. 6).

The workflow of this research companies the recent immersive reality applications with digital documentation techniques "aerial photogrammetric and terrestrial tools" with designing practices for the site of study (Artopoulos & Charalambous, 2019).

3.2 "Virtual Platforms for Heritage Preservation in the Middle East: The Case of Medieval Cairo" (2017)

This study focused on developing a virtual heritage platform to historic areas through the internet (Fig. 7). For this purpose, the research referred to the recent projects in this field to be applied in the study area "Islamic Cairo" (Abdelmonem, et al., 2017).

3.3 "Archeoguide"

In this project, a platform has been developed to provide information access at the cultural heritage site by using

Fig. 6 The panoramic view of the site of study presented the gate of the archaeological site

augmented reality, mobile devices, 3D visualization, and multi-modal techniques, see Fig. 11 (Vlahakis, et al., 2001).

3.4 "I-TREASURES Project in Intangible Cultural Heritage Digitalization (2016)"

The purpose of this project was to provide an extendable system with open access ICH resources. Accordingly, the project had proposed new methods and novel paradigm technology to serve in the analysis and modelling of intangible cultural heritage (Dimitropoulos, et al., 2014).

4 Proposed Workflow for the Archaeological Site El-Sheikh Ibada (Antinoopolis)

The workflow of this research integrates the most appropriate applications of digital documentation and immersive technology that are suitable for the characteristic of the site

Fig. 7 This virtual model is representing rug shop in "Souk Al-Khayamiyyah", Medieval Cairo

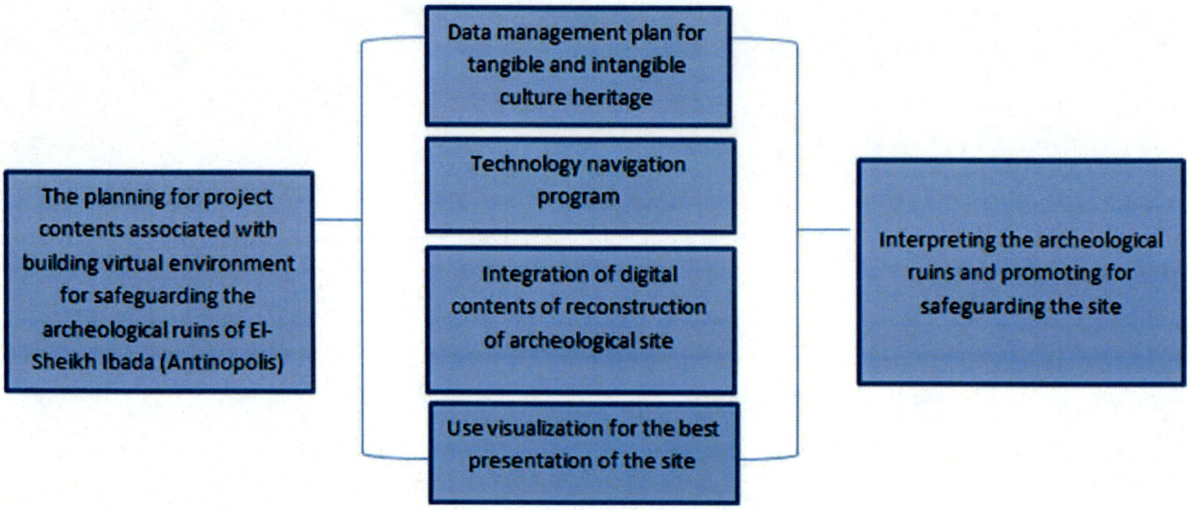

Fig. 8 The Illustration of the workflow of the research

and can safeguard its archaeological ruins. Accordingly, the workflow includes the following steps that illustrated in Fig. 8.

4.1 Data Management Plan for Tangible and Intangible Cultural Heritage

- Collecting all relevant documents and memories of the site that are related to identity and current condition.
- Include a high level of accurate historical and archaeological information.
- Provide descriptive documentary details about the archaeological site that can help in the visualization process in an interactive application.
- Recovering the eligible information about the missing elements for helping in the reconstruction process (Fig. 9).

4.2 Technology Used in Cultural Heritage

- Use 3D precise model that enables the reconstruction of the archaeological site. The model should be developed based on the available historical resources for recovering

Fig. 9 Virtual environment proposed for El-Sheikh Ibada (Antinoopolis)

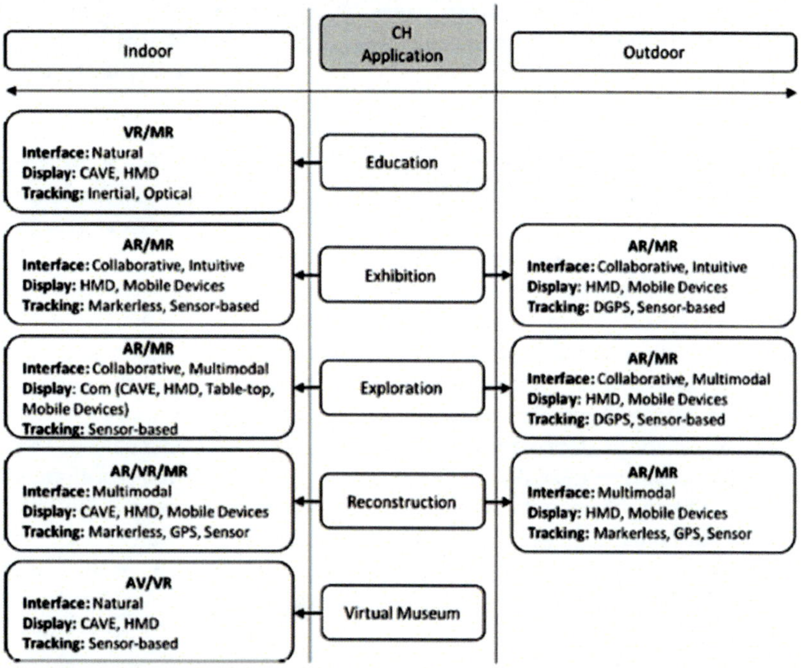

Fig. 10 Survey for suitable indoor and outdoor applications for different cultural heritage purposes

Fig. 11 ARCHEOGUIDE project, which was installed at the Olympic archaeological site in Greece

eligible missing elements. Particularly, that was elaborated in "Description of Egypt" publication, as it had recorded the building elements and monuments of the ancient Greek city, including city walls, gates, temples, and theatre. Then join it with a site survey.

- Visualize the historic context of the archaeological site by immersive technology to provide a rich virtual environment that can help user engagement within the virtual platform.
- Aerial photogrammetric and terrestrial tools are recommended as the structure of the site can be revealed better by aerial view. Regarding to immersive reality applications, Mixed reality is recommended, as this application can work effectively outdoors with achieving the best experience in relationship, collaboration, and engagement (Fig. 10).

Enable users to explore the archaeological site while they are walking in a virtual environment of the real area of the site (Fig. 11). Therefore, virtual routs can be proposed and generated as 3D objects around the site that are mapped automatically in the virtual environment. As it can be used through interface involves mobile app. Moreover, Storytelling, Assembly instruction, and Gamification can be displayed while walking in these routes by Augmented reality application for interpreting the Intangible Cultural heritage.

4.3 Establish a Museum for Intangible Cultural Heritage

All museum objects and artifacts are an expression of knowledge, beliefs, and practices which can be called "semantic baggage". These expressions have a great significance in referring to the intangible background at museums that can place the items in their cultural context. The absence of this context will make it hard or impossible to interpret these objects or produce appreciation. Moreover, they will become incomprehensible (Pearle, 2018).

To build a platform that provides access to cultural heritage at museums, digitizing and modelling are essential to visualize artifacts and objects to be presented in virtual reality applications (Vlahakis, 2003). The most useful element in a database system is metadata, which can describe CH objects and their digital representations specially for building virtual exhibitions (Fig. 12).

4.4 Representation of Intangible Culture Through a Tangible Collection at the Museum

AR application that dealing with ICH at the museum should receive the complete digital information. Hence, the information that represents the ICH (knowledge, skills, crafts, tradition, and poetics) should be coupled with physical heritage resources (heritage artifacts, heritage instruments, and objects) in digitized form (Fig. 13). Then include them in the platform that developed by the museum based on the international standards. So that it will guarantee the conceptual and technical integration. Afterwards, the application can provide information access to the museum collection (Mourkoussis, et al., 2003).

Fig. 12 3D Virtual gallery is visualized by metadata

Fig. 13 Illustration of how tangible heritage objects can refer to the intangible heritage component

Intangible culture heritage is apart from tangible cultural heritage obvious physical, or material value as an object, it also refers to knowledge, skills of using it, certain traditions, and an exclusive craft of producing it. Consequently, an equal amount of resources should be allocated to preserve its physical condition, as to safeguarding knowledge, skills, crafts, tradition, and poetics evolved around it.

Regarding the technical requirements of AR in intangible cultural heritage at a museum, the Augmented Reality application should include interaction interfaces to deal with this kind of cultural heritage. Nevertheless, ICH is a part of the fact that having no physical existence thus is not a haptic. However, encompassing the real and virtual cultural objects for safeguarding and spreading knowledge becomes possible with the help of new audio-visual means. As, the physical existence is required for visitors, particularly in local museums (Dimitropoulos, et al., 2014).

Storytelling, Assembly instruction, and Gamification in AR for Intangible Cultural Exhibitions.

Storytelling and guided instruction in AR technology are the way that Augmented reality tackles the difficult task of interpreting ICH "knowledge, skills, crafts, tradition, and poetics" (Figs. 14 and 15). In addition to Gamification to transmitting ICH knowledge concerning traditional dance, songs, and expressions. Furthermore, these techniques can help in simulation and interactions as it helps in conveying intangible culture Heritage (Papagiannakis, et al., 2018).

5 Discussion

This research emphasis on the importance of safeguarding cultural heritage sites using the recent technology. Therefore, it focuses on the following points that can contribute to the conservation practices of cultural heritage sites.

Fig. 14 An Assembly instruction-based virtual tutor for making pottery

Fig. 15 Storytelling model by using mixed reality application, ancient Pompei

To shed light on the necessity of exploiting the recent technology for the documentation of the archaeological sites digitally. Consequently, its archaeological ruins can be interpreted and presented on the site and on the website for the next generations.

To promote the neglected archaeological sites by making it more attractive. As a result, it will help in gaining public interest as well as putting them on the tourism map.

To increase the income for these sites through the tourism encouragement. For this purpose, applying tourists' needs can help in securing the financial resources for better preservation and management.

To emphasis safeguarding the tangible and intangible cultural heritage at archaeological sites.

6 Conclusions

This paper presented proposed workflow for building a virtual environment to the archaeological site El-Sheikh Ibada (Antinoopolis). For this purpose, the concepts of safeguarding cultural heritage and the technical requirements in the recent technology have been considered to provide the appropriate strategy for interpreting and promoting the case study. In addition to highlight, the necessity for safeguarding the tangible and intangible culture of the archaeological sites that are in need to our efforts to be survived.

Acknowledgements I would like to thank the IEREK team for having organized a very interesting conference on February of 2020 and their contribution in concerning our cultural heritage. Special thanks go to Prof. Mayada and Prof. Mona for their continued involvement and support for Heritage Management field. Special thanks also go to Eng. Shehab Elgabry for his support and encouragement. And lastly, I would like to thank all my family for their support.

References

Abdelmonem, M. G., Selim, G., Mushatat, S., & Almogren, A. (2017). Virtual platforms for heritage preservation in the Middle East: The case of medieval Cairo. *International Journal of Architectural Research: Archnet-IJAR, 11*(3), 28–41.

Artopoulos, G., & Charalambous, P. (2019). Virtual environments as a technological interface between built heritage and the sustainable development of the city. *International Journal of E-Planning Research (IJEPR), 8*(1).

Bekele, M. K., & Champion, E. (2019). A comparison of immersive realities and interaction methods: Cultural learning in virtual heritage. *Frontiers in Robotics and AI, 6*, 91.

Bekele, M. K., Pierdicca, R., Frontoni, E., Malinverni, E. S., & Gain, J. (2018). A survey of augmented, virtual, and mixed reality for cultural heritage. *Journal on Computing and Cultural Heritage (JOCCH), 11*(2), 1–36.

Dimitropoulos, K., Manitsaris, S., Tsalakanidou, F., Nikolopoulos, S., Denby, B., Al Kork, S., ... & Tilmanne, J. (2014, January). Capturing the intangible an introduction to the i-Treasures project. In *2014 international conference on computer vision theory and applications (VISAPP)* (Vol. 2, pp. 773–781). IEEE.

Gabellone, F. (2009). Ancient contexts and Virtual Reality: From reconstructive study to the construction of knowledge models. *Journal of Cultural Heritage, 10*, e112–e117.

ICOMOS, C. (2013). Principles for capacity building through education and training in safeguarding and integrated conservation of cultural heritage. *Submitted to CIF Members*.

Mourkoussis, N., White, M., Patel, M., Chmielewski, J., & Walczak, K. (2003, September). AMS-Metadata for cultural exhibitions using virtual reality. In *International Conference on Dublin Core and Metadata Applications* (pp. 193–202).

Papagiannakis, G., Partarakis, N., Stephanidis, C., Vassiliadi, M., Huebner, N., Grammalidis, N., ... & Sylaiou, S. (2018). Mixed reality gamified presence and storytelling for virtual museums (No. IKEEBOOKCH-2019-233). Springer.

Patias, P. (2006). Cultural heritage documentation. International Summer School Digital Recording and 3D Modeling, Aghios Nikolaos, Creta, Grecia, Abril, 24–29.

Payne, E. M. (2013). Imaging techniques in conservation. *Journal of conservation and museum studies, 10*(2).

Pearle. (2018). Performing arts employers, associations league Europe, cultural heritage in the live performance sector, Milan.

Rodrigues, N., Magalhães, L., Moura, J., & Chalmers, A. (2014). Reconstruction and generation of virtual heritage sites. *Digital Applications in Archaeology and Cultural Heritage, 1*(3–4), 92–102.

Vlahakis, V., Karigiannis, J., Tsotros, M., Gounaris, M., Almeida, L., Stricker, D., ... & Ioannidis, N. (2001). Archeoguide: first results of an augmented reality, mobile computing system in cultural heritage sites. *Virtual Reality, Archeology, and Cultural Heritage, 9* (10.1145), 584993–585015.

Vlahakis, V., Karigiannis, J., & Ioannidis, N. (2003). Augmented reality touring of archaeological sites with the ARCHEOGUIDE system. na.

Check for updates

Decision-Making Processes in Landscaping the Garden of the Museum at the Lowest Place on Earth, Dead Sea, Safi, Jordan

Alexandra-Aikaterini Andrianou, Georgios Papaioannou, and Konstantinos D. Politis

Abstract

Landscaping museum external spaces in a museum meaningful as well as environmentally sustainable way has been central in modern museums' missions and philosophies. Relevant processes involve decision making based upon cultural, environmental and managerial factors. Realizing the specific needs of an area based upon environmental aspect, especially in a semi-arid to arid environment; and at the same time respecting the historical context becomes a necessity for museum's development and prosperity. Addressing the museum's mission and supplementing its exhibitions and activities (tours, educational programmes, other) is key to a successful landscaping of the museum's exterior spaces. This paper addresses the above in the context of the recently inaugurated (May 2012) Museum at the Lowest Place on Earth (MuLPE), in the southern shores of the Dead Sea, Safi, Jordan. We explore the development of landscaping processes for MuLPE's garden design, taking into account an approach which focuses on environmental and cultural assets. We highlight the decision making process in an semi-arid to arid area presenting the first steps that will be followed in order to propose a sustainable design while respecting the historical value of the area, and we conclude by proposing future steps which will assist to the identity of the museum and the local community environmental awareness.

Keywords

Museum's external area • Semi-arid area to arid area • Landscaping processes • Museum at the lowest place on earth

A.-A. Andrianou (✉) · G. Papaioannou
Ionian University, Corfu, Greece
e-mail: gpapaioa@ionio.gr

K. D. Politis
Hellenic Society for Near Eastern Studies, Chalcis, Greece

1 Introduction

During the past decade, museum exterior spaces have become an increasingly accepted ground serving museums' missions. The spaces of a museum, internal and external, have been analysed from different theoretical perspectives (Griswold et al., 2013). Moreover, a growing demand has come into the light in which combing museological and environmental assets can provide a new perspective in landscaping designing processes when it comes into an external space of a museum (Farahat & Osman, 2018).

The external area of a museum is a landscape area in which ecological processes are implemented, *"through the understanding of what kind of landscape this area is"* (Haila, 2007). The aspects which constitute a landscape as a spatial entity provide a starting point for observation. The observation and evaluation of these aspects provide inspirations and limitations in design procedures and decisions. Moreover, the specific needs of an external area of a museum are closely related to the experiences that will take place over time (von Haaren et al., 2014). When these needs are identified and in relation to cultural contexts, the first steps of a sustainable decision-making prospect can be developed. These decisions, when it comes into a semi—arid to arid area, become crucial for the resilience of the project (Stokman & von Haaren, 2011). On the other hand, taking into account different prospects by developing the proposed area can add and deliver positive outcomes in biodiversity scale and also improving the quality of life for the local community (MacDonnell and Hahs, 2013; Wang et al., 2016).

In order to deal with the above matters, the paper proposes a decision-making processes concerning the development of the external area of the Museum at the Lowest Place on Earth (MuLPE) based upon environmental and cultural factors. In doing so, the work attempts to address three basic questions: In what way is decision-making processes in a semi-arid, arid area of a museum should be taken? How does the observation of the elements which constitute this specific

A. Versaci et al. (eds.), *Conservation of Architectural Heritage*, Advances in Science, Technology & Innovation,
https://doi.org/10.1007/978-3-030-74482-3_37

landscape shapes the area; which limitations concerning planting decisions can occur? What steps should be taken into account towards the future development and resilience of the project?

2 Literature Review

In this chapter, we address issues of landscape decision-making processes with an emphasis of semi-arid to arid areas and specific climate conditions prevailing in the area under study. As the climate change is considered a complex phenomenon with an impact on the planet's landscapes and climates, landscapes have become vulnerable and face multiple changes by this phenomenon. Landscaping decision-making processes need to tackle both landscape realities and challenges. They have to adapt to the new challenges in order to provide sustainable environments, improving well being in relation with socio-economic, natural and cultural elements (Festus, 2014).

2.1 Landscaping Decision Making Process

Landscape design addresses the aesthetic aspect of a landscape, but it cannot stop there. Other aspects, including the functional, the pragmatic, the affordable, the sustainable, even the spiritual and the mental, must be included in the equation, as they are closely related to the perception and the interaction with every specific environment. In modern landscape design, the above is achieved by finding and stressing certain elements with the priority in the harmonic coexistence between landscape and human requirements (Xu, 2019). During the design decision making process, the understanding of the natural area as well as the observation of the local environment and the cultural background, have become necessities in order to proceed towards a sustainable design proposal. To address this process and to recognize the spatial, environment, and cultural value of the landscape, the literature suggests to take into account certain overall elements as tools in the design procedures and decisions. These are the environmental/ecological element, the tangible cultural element and the intangible cultural element. The environmental one deals with the degree of correlation between the natural areas and the human perception (Ives & Kendal, 2013; Kümmerling & Müller, 2012; Miklós et al., 2019), considering the basic natural resources needs (Xu, 2019). The tangible cultural element relates to the presence of historical traces and archaeological features into the landscape. Incorporating them into the design is considered significant in arising the areas' profile, as it connects the area's past with the present (Gordon, 2012). In terms of the intangible cultural element, this relates to including the integible historical and cultural context of an area, offering spaces and/or inspiration for developing art, language, customs, religion, etc. (Pungetti, 2013; Xu, 2019). By addressing these three elements, the landscape design process bridges the past with the present with respect to the local environment, as it incorporates current needs, environmental values and cultural context.

2.2 Landscaping in Semi-Arid to Arid Areas

Many areas around the world are characterized as arid and semi-arid areas (ASARs) facing water scarcity affecting the agriculture sector (Wu et al., 2018). According to Wu the "*spatio-temporal rainfall distribution and the unpredictable rainfall patterns leads to climatic uncertainty and aridity situations*" (Wu et al., 2018). According to the Food and Agriculture Organization of the United Nations (FAO) the semi-arid zone-arid-index 0.20–0.50—has the potential to maintain planting and agricultural activities with a flexible range of water harvesting. Moreover, the planting material growing under these harsh conditions is one of the most important elements of this environment. The species that survive in these conditions are a variety of species such as "*grasses and grass-like plants, fortes and half-shrubs, shrubs and trees*" (FAO, 1989). On the other hand, the arid zone – arid index 0.03–0.20—can be cultivated with irrigation system. The species that survive in these conditions are a variety of "*annual and perennial grasses and other herbaceous vegetation, and shrubs and small trees*" (FAO, 1989).

In decision-making processes, the planting design plays a significant role in four different levels (Ghazal, 2019). From the architectural point of view, the planting design frames the space; creation of its spatial entity. From the environmental point of view, the planting design contributes to the environment and the climate monitoring (air pollution, temperatures, shade and compatibility with native wildlife) (Ghazal, 2019; Landis, 2017). Moreover, the aesthetic and experiencing prospect must be also included, as it is closely related with the perception, environmental knowledge and interaction with the specific environment (Russel et al. 2013; Hausmann et al., 2016; Ghazal, 2019).

Landscaping processes in arid regions are closely related to water scarcity in these areas. Landscape architects try to cooperate with a very low quality of water in order to irrigate the planting material and it is more likely to use also greywater for efficient irrigation in specific areas (Amiri et al., 2019). Most of the times, a different irrigation plan is needed in order to satisfy the needs of the planting material into a certain degree (Albalawneh et al., 2016; Ammar et al., 2016; Bortolini et al., 2018).

According to all the above landscape practitioners are focussing on proposing native plants ensuring their adaptation process and reducing planting stress. In this way, the field assessment and even more the soil enrichment is minimized, reducing cost (Abdullah et al., 2016; Bainbridge, 2002).

2.3 Climate Conditions in the Case Study Area (Jordan Valley)

As stated in USAID report, impacts of an expanding economy in Jordan as also as the inefficient agriculture use of water will have a huge impact in country's availability in water recourses. It is worth mentioning that the water availability has changed drastically per capita from 3,600 m^3 in 1946 to 147 m^3 today (USAID, 2013). Despite the fact that Jordan's water supplies are supported by rainfalls, the climate change phenomenon in relation to the annual rainfalls has affected the aquifers, leading them to a significant water decline (Mustafa & Rahman, 2018).

According to Al-Bakri the country has three bioclimatic zones—Jordan Valley, western highlands and Badia—arid and semi-arid regions in the eastern parts of the country (Al-Bakri et al., 2013). Also, due to the landscape diversity, the climate conditions differ from one bioclimatic zone to another. In our case, (Jordan Valley) the high temperatures and the location of the museum, southern-east end of the Dead Sea, 400 m below the sea level create semi-arid to arid conditions with warm winter and hot summer (Al-Bakri et al., 2013). Irrigation methods are used in agriculture sector across the region (Al-Bilbisi, 2013; Kool, 2016; Al Quina and Salahat, 2017).

3 Methodology

Methodological approaches play an important role in a sustainable landscape planning. "*They are focusing on presenting typologies of landscape planning methods implementing to landscape frameworks and methods*" (Ahern, 2006). On the other hand, the term of sustainability has been recognized internationally, offering policies at local and international level (IUCN n.d). The creation of a landscape design is based on the uses of a selected area, balancing between use and sustainability prospect (Ahern, 2006).

In our case study, we had to think differently in order to present a landscape design achieving our vision. Our landscaping decision process is based on adaptive planning decisions which at this point are contributing towards a sustainable design proposal (Dale et al., 2014). In terms of typology, our case study is classified as an interdisciplinary method (Arts et al., 2017). By using an interdisciplinary approach, we are taking the first steps taking into account the specific needs of the external area of a museum. Direct observation, fieldwork has been made in the summer of 2015 in order to observe the external area. During this procedure, several aspects according to the uses of the external area were discussed related to the environmental needs, users' expected experience, planting material discussions. The main goal was the identification of the basic environmental elements that needed to be observed and recorded leading to the landscape process in a semi-arid to arid area. The climate conditions in the case study area as also as the specific cultural environment proposes a framework, serving a prospect in two different levels: the ecological aspect closely connected to cultural assets. After this procedure, secondary data were reviewed in order to gather background information for the historical context of the area, the climate conditions (arid and semi-arid areas) and the appropriate planting material.

4 Towards an Ecological Planning and Design for the museum's External Area

The garden has to serve a prospect in two different levels. In terms of an ecological level, the selection of drought tolerant species that will be adapted to the local conditions will contribute towards the sustainability and resilience of the project. The environmental limitations must be taken into consideration, especially for semi-arid region planting decisions (Asgarzadeh et al., 2014). Additionally, the design of basic routes, creating a visual continuity with the internal space of the museum will lead to an external area with botanical and archaeological features. In terms of user experience, the area will create a pleasant surrounding with high aesthetic quality standards (Papafotiou et al., 2016). The basic design idea is presented below (Table 1):

In order to stress our above-mentioned environmental needs, we decided to define the conditions and objectives that will use as a tool in order propose the design concept. These are presented below:

As regards the planting material (Andrianou & Papaioannou, 2019):

- Selection of drought tolerant plant species
- Providing plant species information/informative labels
- Selection of species which have a historical importance and promotion of the historical context.

In terms of environmentally sustainable actions (Andrianou & Papaioannou, 2019):

- Developing in an environmentally friendly way the garden waste (compost bins-compost fertilizer)
- Installation of an effective rainwater harvesting system

Table 1 Basic environmental needs

Basic environmental needs
Creating designated rest areas/social relationships
Choosing the plant material
Enhancement of outdoor archaeological findings
Defining the basic routes
Ensuring safety during use

- Fostering the environmental education (ESD model) having a effective impact in everyday life.

The second aspect of this ecological planning is the decision concerning the planting material. This was based on drought tolerant species which will adapt well under these conditions (variety of evergreen and deciduous shrubs and trees) (Filippi, 2019; Taifour et al., 2017). Moreover, the suggestion of planting specific trees (Phoenix dactylifera), respects the historical use of these plants species in the area.

The selected plant species are presented below (Table 2).

5 The Case Study

5.1 Brief Description of the Landscape in Ghor Es-Safi

Located at the southern-east end of the Dead Sea, Ghor es-Safi area, at 400 m below the sea level is recognized as the lowest place on the earth's surface (HSNES n.d; Politis, 2010). By the name of "Zoar is indicated in the Old Testament which is considered to be one of the cities of the plain which was not destroyed by fire and brimstone" (HSNES n. d; Politis, 2010). Archaeological record indicates the prominence of this specific area in "early Bronze Age, in Middle Bronze Age sites and in Iron Age occupation"(HSNES n.d; Politis, 2010). Moreover, the city had an agricultural interest and this fact it is also established by the Byzantine mosaic map at Madaba, in which the city named Zoora is surrounded by date palms (Fig. 1). The historical interest of the city has been raised through "Crusader and mediaeval Islamic periods and historical sources as Yakut, al-Al-Maqdisi, and Foulcher of Chartres named the city as *Segor* and/or *Zughar*" (HSNES n.d; Politis, 2010). The city's economic status was boosted by sugar and indigo production establishing Zoora as a major market centre of these products. Furthermore, important archaeological findings such as the "Sanctuary of Lot at Deir 'Ain 'Abata and the Nabataean cemetery at Khirbet Qazone" conducted in late 1980s, 1990s gave prominence to Ghor es-Safi area (HSNES n.d; Politis, 2010).

From 1997 to 2009 the archaeological fieldwork processes were conducted and continued, bringing into the light various archaeological sites in the particular region. The excavations on this area revealed (HSNES n.d; Politis, 2010):

- "The early Byzantine–mediaeval Islamic urban centre of Khirbet esh-Sheikh 'Īsā"
- "The industrial complex of Masna' es-Sukkar (commonly known as Tawāhīnes-Sukkar)"
- "The Early Bronze Age and Byzantine cemeteries at An Naq".

Additionally, in 2004 the excavations processes were continued bringing into the light two excavations trenches "the one in Masna 'es- Sukkar (in the sugar factory complex) and at Khirbet esh-Sheikh 'Īsā (the city center)" (HSNES n. d; Politis, 2010). This discovery is considered to be one of the most importantly rich in archaeological evidences (HSNES n.d; Politis, 2010). Some of these important sites are presented below (Fig. 2).

5.2 The Museum at the Lowest Place on Earth (MuLPE) Today

The museum opened to the public on 18 May 2012 supported by the Minister of Tourism and Antiquities of Jordan (Hellenic Society for Near Eastern Studies n.d). The exhibition was designed by HNES-Hellenic Society for Near Eastern Studies—in collaboration with the British Museum (Hellenic Society for Near Eastern Studies n.d). At the museum, 22 exhibition cases and information panels are displayed, presenting material from the prehistory to the present (Politis, 2010). These presentations are dealing firstly with human activities through the exhibition and archaeological work in the region (HSNES n.d). More specifically, the thematic presentation on 'First People' emphasizes on prehistory period presenting stone tools closely related to agriculture activities (Hellenic Society for Near Eastern Studies n.d; Politis, 2010). The thematic presentation "First Cities" presents the oldest wheel-made

Table 2 Selected plant species

Trees
Phoenix dactylifera, Date palm
Cercis siliquastrum, Judas tree
Albizzia julbrissin, Silk tree
Ficus carica, Fig tree
Ceratonica siliqua, Carob
Tamarix parviflora, Early tamarisk
Prunus cerasifera, Piisard plum
Olea europea, Olea
Arbutus unedo, Strawberry tree
Elaeagnus angustifolia, Silver berry
Shrubs
Pistacia letiscus, Pistache
Limoniastrum monopetalum, Mediterranean statice
Murtus communis, Myrtle
Thymys vulgaris, Thymus
Rosmarinus officinalis, Rosemary
Laurus nobilis, Laurus
Salvia officinalis, Sage
Rosemarinus officinalis "Prostratus", Rosemary
Atriplex alimus, Mediterranean saltbush
Perennial plants
Santolina chamaecyparissus, Cotton levander
Levanda angustifolia, Levander
Euphorbia characias, Mediterranean sprunge
Crithmum maritimum, Rock samphire

pottery, stone vessels early metal work and jewellery findings dated from Early and Middle Bronze Age tombs at Bab edh-Dhra, An Naq' and Deir 'Ain 'Abata (Hellenic Society for Near Eastern Studies n.d; Politis, 2010). Moreover, the every-day life of monks in an early Christian-Byzantine monastery is presented in "Monks Life" thematic presentation (Hellenic Society for Near Eastern Studies n.d; Politis, 2010). Artefacts from Khibert Quazone are presented in "Nabateans of the Dead Sea shores" as also as their physical appearance and their clothing preferences (Hellenic Society for Near Eastern Studies n.d; Politis, 2010). The Hellenic culture in relation with one thousand years of its progression is presented in "Hellenism and Islam" exhibition (Hellenic Society for Near Eastern Studies n.d; Politis, 2010). The origins of sugar industry (the most important industry in Ghor es-Safi region) have their own thematic presentation designated as "The story of Sugar" (HSNES n.d; Politis, 2010). Other thematic presentations develop findings from recent excavations processes, mosaic art making and ancient technical knowledge artefacts (Hellenic Society for Near Eastern Studies n.d; Politis, 2010). The museum thematic presentations are concluded with an ethnographic collection display related to modern-day residents of southern-eastern Dead Sea shores (Hellenic Society for Near Eastern Studies n.d; Politis, 2010).

Fig. 1 "Depiction of Monastery of Saint Lot on sixth century AD mosaic floor map at Madaba", Jordan (K.D. Politis). *Source* Politis (2012)

5.3 The Existing Situation of the External Area

In order to propose a design for the external area of the museum, it is important take into account the existing situation of the external area. In our case, only the basic paths are designed and visitors have the opportunity to take a stroll around. They are two main entrances on the west side of the museum's plot as well as an entrance on the northeast side which serves cars and buses.

In the centre of the area, a patio area is designed which is surrounded by flower boxes without the necessary planting. Another patio area is located on the southeast area which is designed as an original demonstration space specifically for the ancient sugar compression model.

The north and the northwest side of the plot, which is considered to be the largest part of the external area, have a number of basic planting. The lack of a central path and resting areas are visible and the site at this point is considered to be quite dangerous, especially if it is going to be used in future by children. The existing situation design and some of these areas are presented below (Figs. 3, 4, 5, 6, 7, 8 and 9).

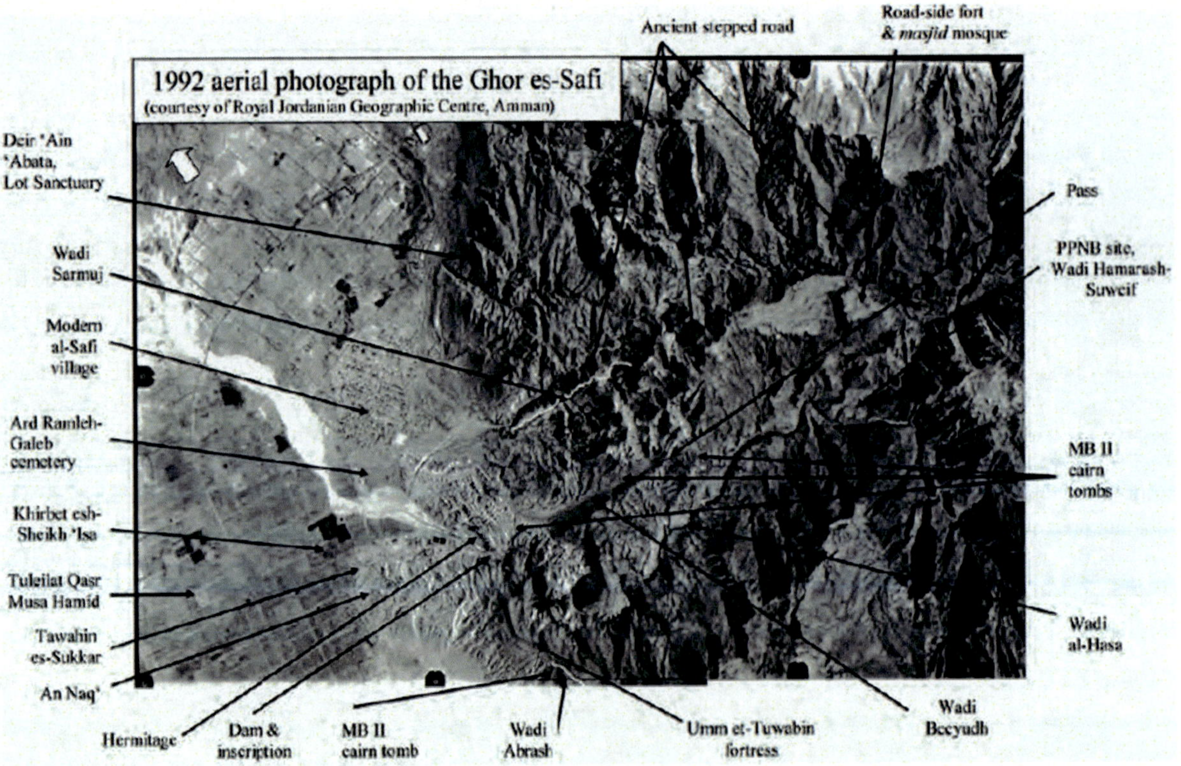

Fig. 2 "Aerial view of Ghor es-Safi, sites and wadi locations are illustrated". *Source* Politis (2010)

Fig. 3 Museum at the Lowest Place on Earth—external area, google earth pro

Fig. 4 Aerial photo of the Museum at the Lowest Place on Earth. *Source* https://www.hsnes.org/news_item002.htm

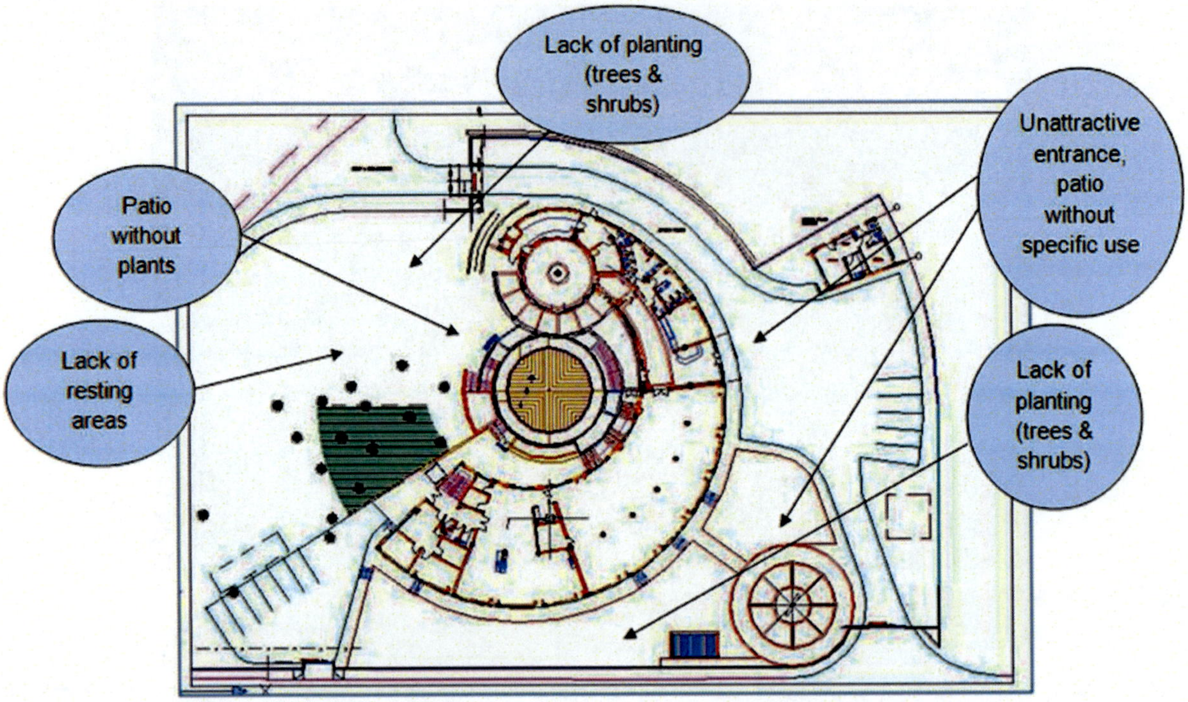

Fig. 5 Existing situation of the external area

Fig. 6 Exterior views of the Museum at the Lowest Place on Earth, entrance. *Source* Personal archive, K.D Politis

Fig. 7 Northwest views of the Museum at the Lowest Place on Earth. *Source* Personal archive, K.D Politis

Fig. 8 Northwest views of the Museum at the Lowest Place on Earth. *Source* Personal Archive, K. D Politis

5.4 The Design Proposal

The external area of the museum is divided in two sectors. The west area of the garden which will focus on the social image and the eastern area which will focus on demonstrative features. The proposal intends to satisfy a balance between conservation, archaeological attention and visitor comfort (Fig. 10). For that reason, the development of basic paths and resting areas was a high priority from the beginning. For this reason, four resting, displaying and activity areas were created. More specifically:

Area 1 Resting area in the centre (Fig. 11), where visitors will have the opportunity to participate in various activities.
Area 2 Resting area at the northwest side of the plot (Fig. 12), where will work as a second spot of interest strengthening social relations between visitors.
Area 3 Demonstration area for the ancient sugar compression model at the southeast area of the plot (Fig. 13).
Area 4 Two main areas between basic path where will be used for the exhibition of tombs/grave types and various archaeological stones (Fig. 14).

Furthermore, as concerns any additional equipment it is proposed the use of removable benches in resting areas so in case of an event they can be easily moved around. Additionally, the provision of a compost bin and the installation of a rainwater harvesting system (rain tank 3000 l) will contribute and promote the education for sustainable

Fig. 9 Northwest view of the Museum at the Lowest Place on Earth, photo: G.A Papaioannou, https://www.hsnes.org/news_item002.htm

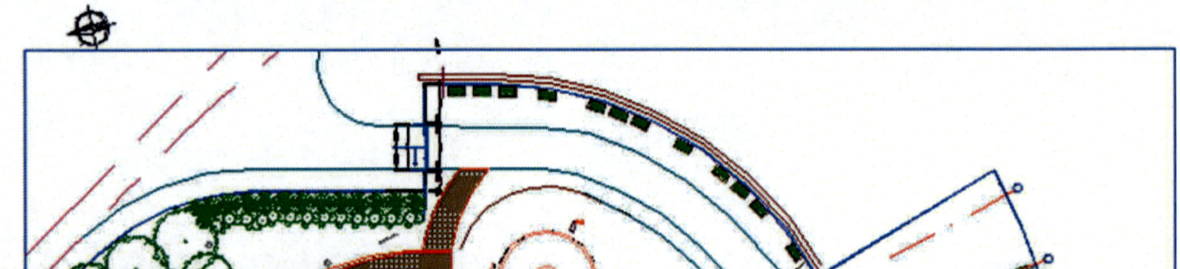

Fig. 10 External area—landscape design

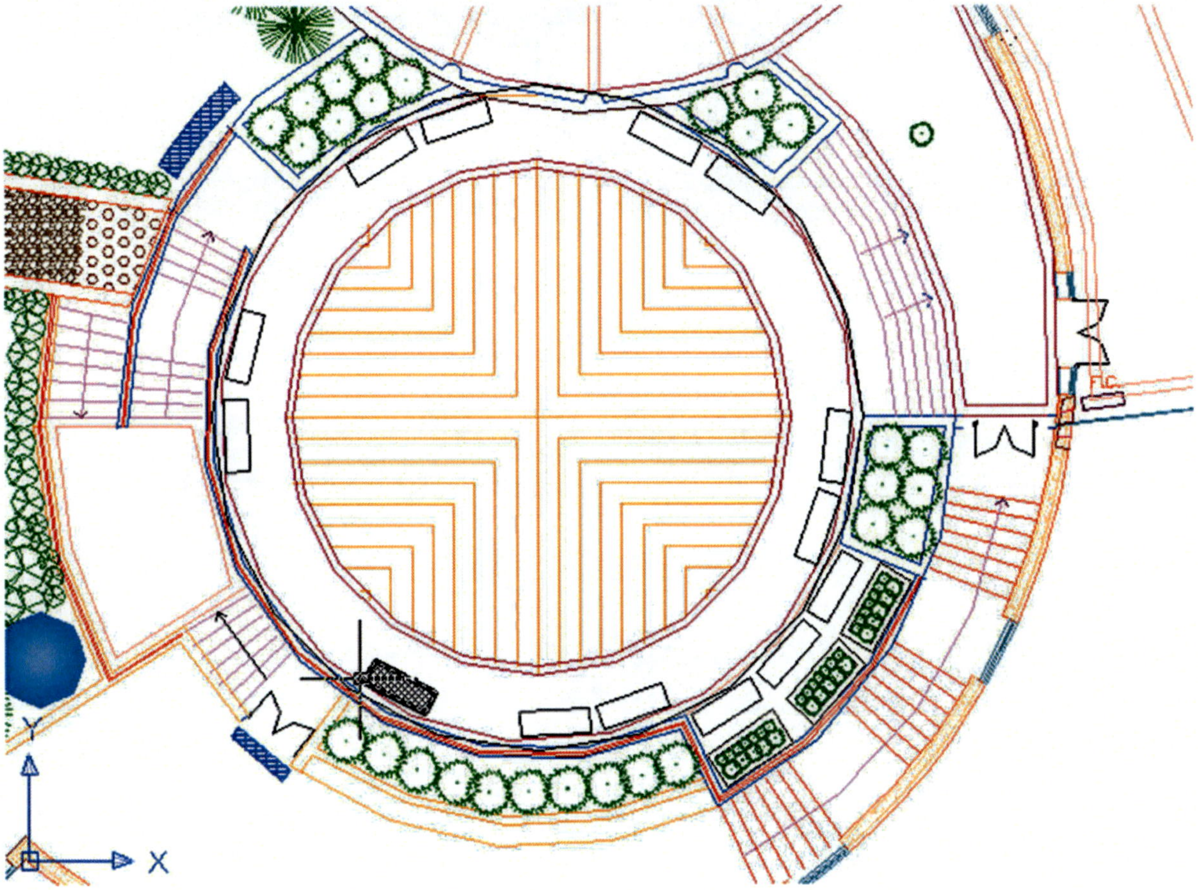

Fig. 11 Area 1 Central resting area

development model through good agricultural practices in everyday life.

Finally, informative panels will be used in order to inform the visitors concerning the garden design, the outdoor archaeological exhibits and the history of sugar industry in Ghor es-Safi region.

6 Towards the Development and Sustainability of the Garden: Future Steps

The creation and development of our proposal as also as the implementation processes are in a very early stage. The aforementioned proposal aims to satisfy a balance between the historical context of the museum and visitor comfort. In this process, we take into account a feasible way, observing the design parameters, especially the planting design process in a semi-arid to arid area.

The main aim of the design proposal was to link the existing museum area with the exterior space providing a place with various uses that will become a landmark, serving the needs and activities of locals and visitors. For this reason, the steps that will be followed will have as a result the enhancement of the local landscape character but will also respect and reflect local history.

Furthermore, the inclusion of the proposed garden design in museum's webpage will assist in this initiative. The development of a digital stroll and the presentation of the design concept will disseminate the idea to a broader audience increasing visitors and interest in the region. The knowledge that will be experienced through the abovementioned tool will benefit the visitors and potential visitors, giving enough motivation to visit, concern and participate in a landscape design concept in a more interesting and adaptive way. (Andrianou & Papaioannou, 2019; Lewi et al., 2014; Watrall, 2018).

To elaborate this further, one of our main goals in future is the development of questionnaires for the local community. This step will provide us all the information we need in order to adjust and integrate our proposal into local community needs.

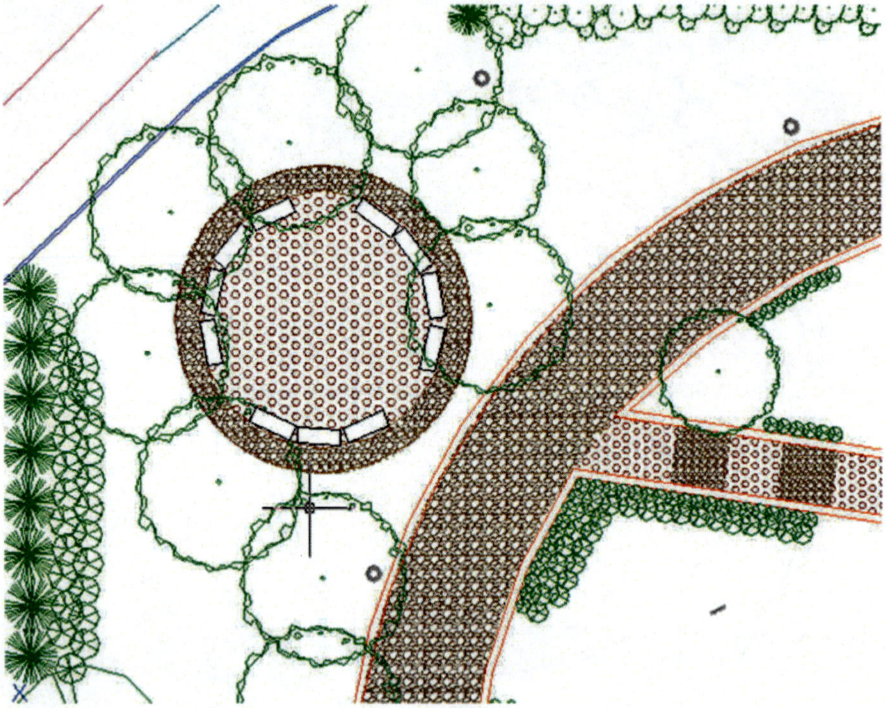

Fig. 12 Area 2 Northwest resting area

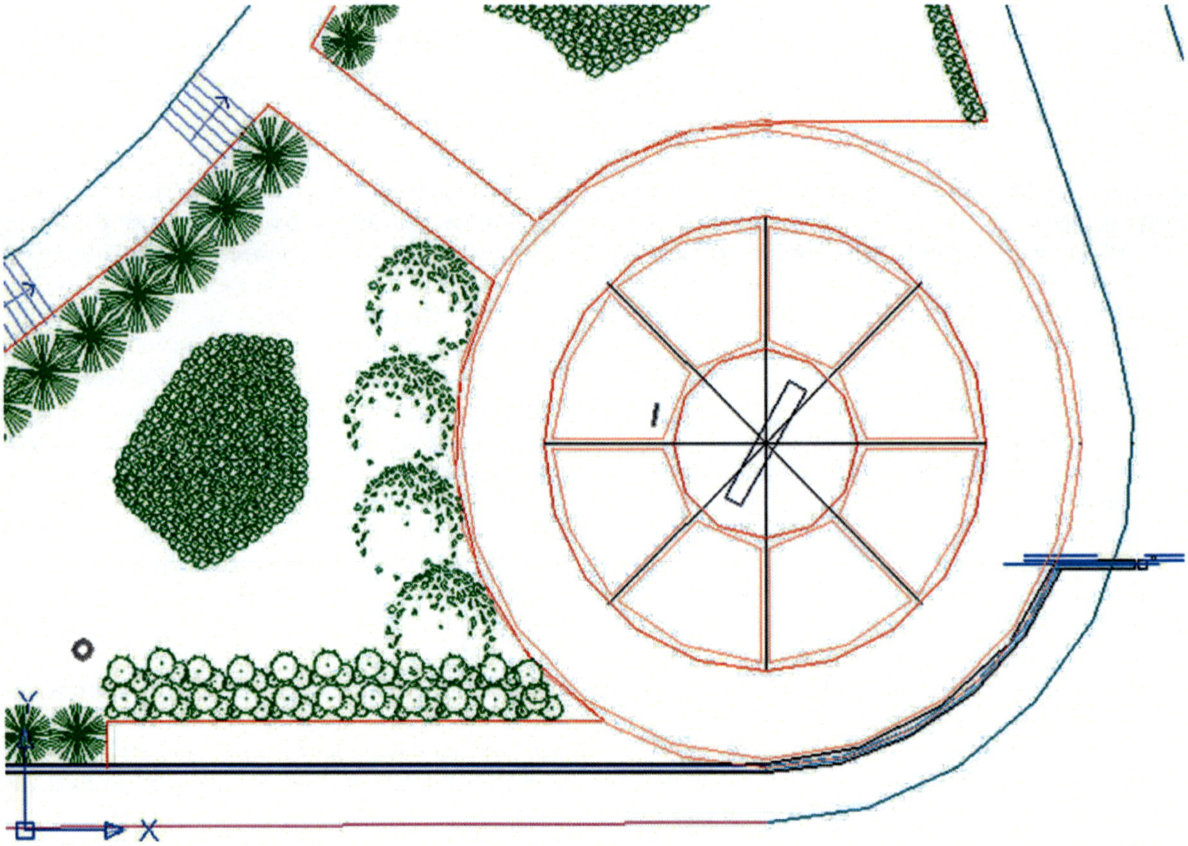

Fig. 13 Area 3 Demonstration area for the ancient sugar compression model—southeast area

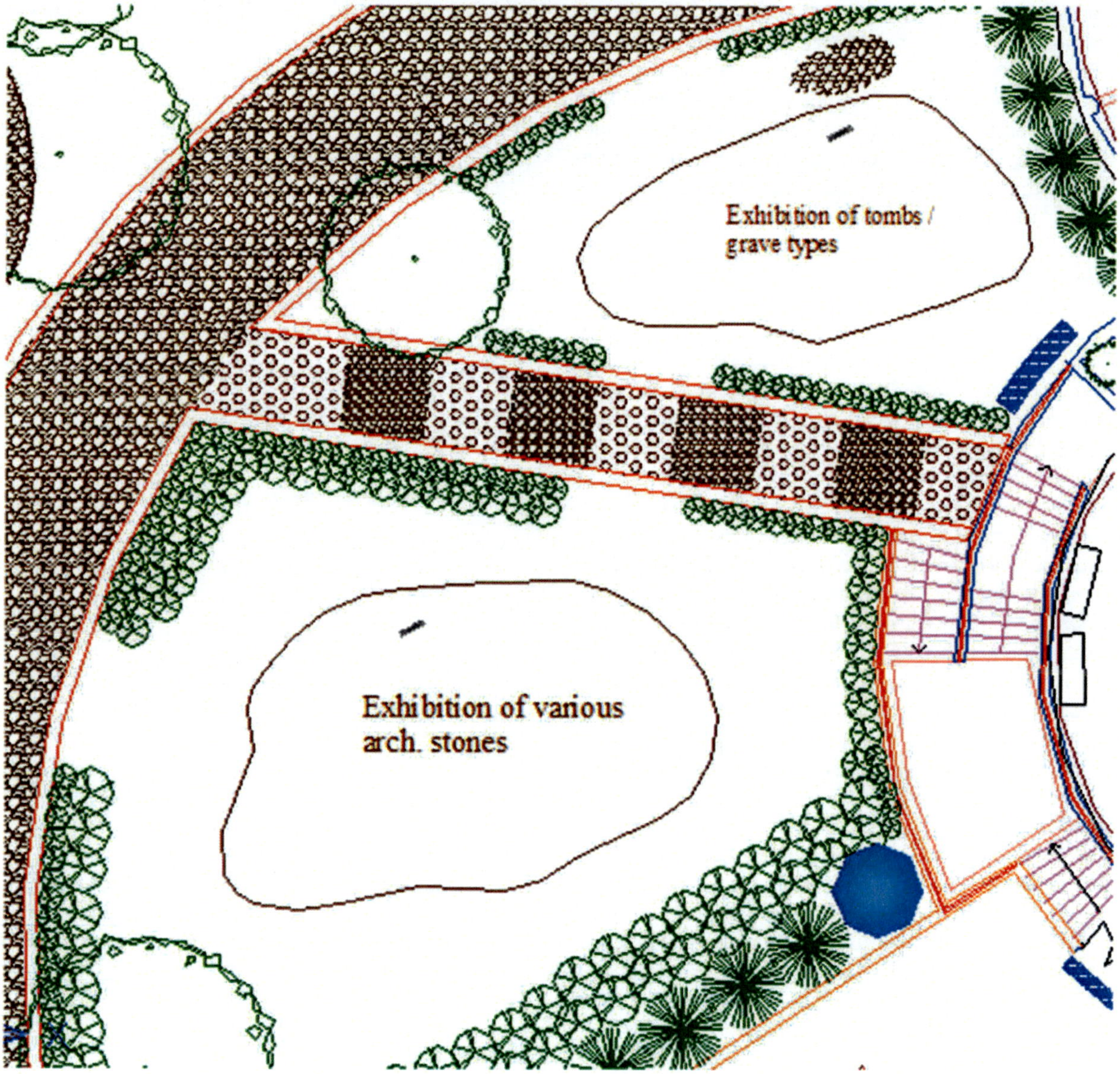

Fig. 14 Area 4, Areas for the exhibition of tombs/grave types and various archaeological stones

For this reason at this early stage, an adaptive planning decision concept was used, combining environmental and cultural assets. The decisions that they were proposed will trigger local awareness and will have a positive impact towards environmental education (Williams et al., 2015). Taking these steps will positively affect and encourage a direct environmental awareness reinforcing personal values and actions towards sustainability prospect (Sutton, 2015; Wassenberg et al. 2015).

The external area of the museum will become a pull factor, especially for local community. Local community contribution in a dynamic way assists the feeling of belonging and strengthening the social relationships. From this perspective and by giving the chance in future to the local community to participate in decisions as also as in planting procedures will attract society's curiosity, bridging the gap in favour of environmental sustainable actions (Andrianou & Papaioannou, 2019). The productive exchange of opinions, especially in the beginning of the implementation process giving the right encouragement, will emphasize on contributory approaches ensuring the future and resilience of the project.

Furthermore, developing educational programmes would have a positive impact towards society inclusion (Veghes 2019). This action will develop and promote the external area of the museum as a centre for environmental education, especially for schools. Creating curriculum-based programmes in collaboration with regional education authorities will establish strong environmental awareness playing a crucial role in children's education (Otto & Pensini, 2017).

Based on the above, the realization of MuLPE's garden landscaping process focuses on new challenges that can be addressed, proposing a realistic design alongside to environmental and cultural sustainability prospect.

7 Conclusions

This paper highlighted the decision making process in landscaping the external area of the museum at the lowest place on earth. The external area of the museum has the potential to become an important landmark balancing between conservation, archaeological attention and visitor comfort. The design approach and the steps that are proposed will enhance environmental awareness having as a result resilience and further development. In this perspective, the paper describes the development of landscaping processes for MuLPE's garden design, based on an environmental aspect, especially in a semi-arid to arid environment.

Using this approach, we recognize and emphasize the need for an ecological planning and design based on the environmental needs of this specific area. The climatic conditions are arid to semi-arid and the use of the appropriate plants will contribute to enhancing the visitor comfort and experience. Moreover, the environmental limitations concerning the proposed planting material are used in our case as a positive asset and will be applicable in other cases and planning activities which are focussing on the sustainability prospect. The shaded areas proposed as part of the design of the museum exterior will contribute to a place that various uses will take place enhancing environmental knowledge, serving the needs of the local community.

This process does not reach an end, providing new challenges in decision planning processes. The paper concludes with a number of ideas and future steps in order to ensure the sustainability of the project. As a key factor for future steps, we emphasize on the active participation of the local community in order to establish and promote good ecological outcomes. Consequently, the dissemination of our vision into a broader audience will have in future a positive impact towards visitors' attraction in regional and international level.

References

Abdullah, M. M., Feagin, R. A., Musawi, L., Whisenant, S., & Popescu, S. (2016). The use of remote sensing to develop a site history for restoration planning in an arid landscape. *Restoration Ecology, 24*(1), 91–99.

Ahern, J. (2006). Theories, methods and strategies for sustainable landscape planning. In *From Landscape Research to Landscape Planning. Aspects of Integration, Education and Application.* In B. Tress, G. Tress, G. Fry, P. Opdam (Eds.) (pp.119–131). The Netherlands: Springer.

Al-Bakri, J. T., Salahat, M., Suleiman, A., Suifan, M., Hamdan, M. R., Khresat, S., & Kandakji, T. (2013). Impact of climate and land use changes on water and food security in jordan: Implications for transcending "the tragedy of the commons." *Sustainability, 5*(2), 724–748.

Albalawneh, A., Chang, T. K., & Chou, C. S. (2016). Impacts on soil quality from long-term irrigation with treated greywater. *Paddy and Water Environment, 14*(2), 289–297.

Al-Bilbisi, H. (2013). Topography and morphology. *Atlas of Jordan: History, territories and society* (pp. 42–46). Presses de l'Ifpo, Beyrouth.

Al-Qinna, M. I., & Salahat, M. A. (2017). Investigating agricultural management practices and environmental sustainability in Jordan. *Journal of Environmental Sustainability, 5*(1), 1.

Amiri, M. J., Bahrami, M., Badkouby, M., & Kalavrouziotis, I. K. (2019). Greywater treatment using single and combined adsorbents for landscape irrigation. *Environmental Processes, 6*(1), 43–63.

Ammar, A., Riksen, M., Ouessar, M., & Ritsema, C. (2016). Identification of suitable sites for rainwater harvesting structures in arid and semi-arid regions: A review. *International Soil and Water Conservation Research, 4*(2), 108–120.

Andrianou, A. A., & Papaioannou, G. (2019). Cultural Landscapes and Botanic Gardens: The Case of Mon-Repos Garden in Corfu Island, Greece. In *Cultural Sustainable Tourism*, (pp. 99–108). Springer, Cham.

Arts, B., Buizer, M., Horlings, L., Ingram, V., Van Oosten, C., & Opdam, P. (2017). Landscape approaches: A state-of-the-art review. *Annual Review of Environment and Resources, 42*, 439–463.

Asgarzadeh, M., Vahdati, K., Lotfi, M., Arab, M., Babaei, A., Naderi, F., ... & Rouhani, G. (2014). Plant selection method for urban landscapes of semi-arid cities (a case study of Tehran). *Urban Forestry & Urban Greening,13*(3), 450–458.

Bainbridge, D. A. (2002). Alternative irrigation systems for arid land restoration. *Ecological Restoration, 20*(1), 23–30.

Bortolini, L., Maucieri, C., & Borin, M. (2018). A tool for the evaluation of irrigation water quality in the arid and semi-arid regions. *Agronomy, 8*(2), 23.

Dale, V. H., Brown, S., Haeuber, R. A., Hobbs, N. T., Huntly, N., Naiman, R. J., ... & Valone, T. J. (2014). Ecological principles and guidelines for managing the use of the land.. In *The ecological design and planning reader*, (pp. 279–298). Washington, DC: Island Press.

FAO. (1989). The arid environment. In *Arid zone forestry: A guide for field technicians*. http://www.fao.org/3/t0122e/t0122e03.htm.

Farahat, B. I., & Osman, K. A. (2018). Toward a new vision to design a museum in historical places. *HBRC Journal, 14*(1), 66–78.

Festus, I. A. (2014). Key issues on landscape planning in the context of environmental sustainability. *European Scientific Journal,10*(2).

Filippi, O. (2019). *The dry gardening handbook: plants and practices for a changing climate*. Filbert Press.

Ghazal, A. M. (2019). Criteria for sustainable planting design applications in landscape architecture projects under arid conditions. *Engineering and Technology Journal, 37* (1 Part (c) special), 10–18.

Gordon, J. E. (2012). Rediscovering a sense of wonder: Geoheritage, geotourism and cultural landscape experiences. *Geoheritage, 4*(1–2), 65–77.

Griswold, W., Mangione, G., & McDonnell, T. E. (2013). Objects, words, and bodies in space: Bringing materiality into cultural analysis. *Qualitative Sociology, 36*(4), 343–364.

Haila, Y. (2007). Enacting landscape design: From specific cases to general principles. *Managing and Designing Landscapes for Conservation: Moving from Perspectives to Principles, 22*.

Hausmann, A., Slotow, R. O. B., Burns, J. K., & Di Minin, E. (2016). The ecosystem service of sense of place: Benefits for human well-being and biodiversity conservation. *Environmental Conservation, 43*(2), 117–127.

Hellenic Society for Near Eastern Studies—HSNES- (n.d.). The museum at the Lowest Place on Earth at Safi in Jordan officially opened its doors on 18 May 2012. http://www.hsnes.org/news_item002.htm.

IUCN, (n.d). Transforming our world: The 2030 Agenda for sustainable development A/RES/70/1. https://sustainabledevelopment.un.org/content/documents/21252030%20Agenda%20for%20Sustainable%20Development%20web.pdf.

Ives, C. D., & Kendal, D. (2013). Values and attitudes of the urban public towards peri-urban agricultural land. *Land Use Policy, 34*, 80–90.

Kool, J. (2016). The Jordan Valley. In *Sustainable development in the Jordan Valley* (pp. 5–60). Cham: Springer.

Kümmerling, M., & Müller, N. (2012). The relationship between landscape design style and the conservation value of parks: A case study of a historical park in Weimar Germany. *Landscape and Urban Planning, 107*(2), 111–117.

Landis, D. A. (2017). Designing agricultural landscapes for biodiversity-based ecosystem services. *Basic and Applied Ecology, 18*, 1–12.

Lewi, H., Saniga, A., & Smith, W. (2014). Immersive and temporal experiences in historic landscapes: Designing a mobile digital guide for the Royal Botanic Gardens Melbourne. *Landscape Review, 15* (1), 4–23.

McDonnell, M. J., & Hahs, A. K. (2013). The future of urban biodiversity research: Moving beyond the 'low-hanging fruit.' *Urban Ecosystems, 16*(3), 397–409.

Miklós, L., Kočická, E., Izakovičová, Z., Kočický, D., Špinerová, A., Diviaková, A., & Miklósová, V. (2019). Landscape as a geosystem. In *Landscape as a geosystem* (pp. 11–42). Cham: Springer.

Mustafa, A., & Rahman, G. (2018). Assessing the spatio-temporal variability of meteorological drought in Jordan. *Earth Systems and Environment, 2*(2), 247–264.

Otto, S., & Pensini, P. (2017). Nature-based environmental education of children: Environmental knowledge and connectedness to nature, together, are related to ecological behavior. *Global Environmental Change, 47*, 88–94.

Papafotiou, M., Marco Martinez, G., Petrocheilou, A., & Kanellou, E. (2016). Design proposal involving native plants for the historical site of Early Christian Amphipolis, Greece. In *VI International Conference on Landscape and Urban Horticulture 1189*, (pp. 97–102).

Politis, K. D. (2010). Ancient landscapes of the Ghor es-Safi: Surveys and Excavations 1997–2009. *Acor Newsletter, 22*(2), 1–5. http://www.hsnes.org/pdf/GS_in_ACOR_22.2_low%20res.pdf.

Politis, K. D. (2012). Sanctuary of Lot at Deir 'Ain 'Abata in Jordan: Excavations 1988–2003. Amman: Jordan Distribution Agency in association with the British Museum, 355–379.

Pungetti, G. (2013). Biocultural diversity for sustainable ecological, cultural and sacred landscapes: The biocultural landscape approach. In *Landscape ecology for sustainable environment and culture* (pp. 55–76). Dordrecht: Springer.

Russell, R., Guerry, A. D., Balvanera, P., Gould, R. K., Basurto, X., Chan, K. M., ... & Tam, J. (2013). Humans and nature: How knowing and experiencing nature affect well-being. *Annual Review of Environment and Resources, 38*, 473–502.

Stokman, A., & von Haaren, C. (2011). Integrating science and creativity for landscape planning and design of urban areas. *Applied urban ecology: A global framework*, 170–185.

Sutton, S. (2015). *Environmental sustainability at historic sites and museums*. Rowman & Littlefield.

Taifour, H., El-Oqlah, A., & Ghazanfar, S. A. (2017). *The plants of Jordan: an annotated checklist*. Kew Publishing.

USAID Jordan. (2013). Country Development Cooperation Strategy (CDCS). https://www.usaid.gov/sites/default/files/documents/1883/CDCS_Jordan_Dec_2020.pdf.

Vegheş, C. (2019). Cultural heritage: The forgotten resource for marketing and sustainable development of the local communities. *European Journal of Sustainable Development, 8*(3), 41–51.

von Haaren, C., Warren-Kretzschmar, B., Milos, C., & Werthmann, C. (2014). Opportunities for design approaches in landscape planning. *Landscape and Urban Planning, 130*, 159–170.

Wang, X., Palazzo, D., & Carper, M. (2016). Ecological wisdom as an emerging field of scholarly inquiry in urban planning and design. *Landscape and Urban Planning, 155*, 100–107.

Wassenberg, C. L., Goldenberg, M. A., & Soule, K. E. (2015). Benefits of botanical garden visitation: A means-end study. *Urban Forestry & Urban Greening, 14*(1), 148–155.

Watrall, E. (2018). Public heritage at scale: Building tools for authoring mobile digital heritage and archaeology experiences. *Journal of Community Archaeology & Heritage, 5*(2), 114–127.

Williams, S. J., Jones, J. P., Gibbons, J. M., & Clubbe, C. (2015). Botanic gardens can positively influence visitors' environmental attitudes. *Biodiversity and Conservation, 24*(7), 1609–1620.

Wu, R. S., Molina, G. L. L., & Hussain, F. (2018). Optimal sites identification for rainwater harvesting in northeastern guatemala by analytical hierarchy process. *Water Resources Management, 32*(12), 4139–4153.

Xu, Y. (2019). Ecological research on the cognitive and visual features of regional culture in landscape design. *Ekoloji, 28*(107), 1487–1493.

Heritage and Sustainability: Motives Behind the Use of Colours Derived from Natural Materials in Al-Qatt

Arwa A. Alghamdi and Abeer A. Alawad

Abstract

Al-Qatt is the traditional art of interior wall decoration native to the Asir region of Saudi Arabia. Conventionally, the colours used in Al-Qatt were derived from natural materials such as turmeric and coal. But nowadays, factory-made synthetic paints are largely preferred over home-made natural colours. However, there are some Al-Qatt specialists who choose to produce their colours the traditional way. These artists are not only still using naturally made colours but also teaching the laborious process to young people. This study aims to identify the motives behind such artists' deliberate preference for traditional natural materials and understand the extent to which it may contribute to sustainable heritage preservation. The sample of this study consisted of five Al-Qatt specialists from Abha city. A descriptive analytical approach was used as the method of investigation. The results show that there are seven factors driving the deliberate use of natural materials: identity preservation, global recognition, distinction and exclusivity, aesthetic motives, psychological motives, economic motives, and environmental and health concerns. All these factors are linked to sustainable heritage preservation. The researchers recommend restoring the Al-Qatt art in heritage buildings using traditionally prepared colours, and identifying the shades used by the original artists. They also recommend that paint companies provide customized colour palettes for Al-Qatt artists to support the continuity of this art, and upgrade and maintain its authentic application. The findings of this study may benefit Al-Qatt as an industry, highlight the importance of using traditional techniques in Al-Qatt restoration, and motivate new generations to maintain the traditional features and characteristics of Al-Qatt in terms of design, materials used, and performance.

Keyword

Asir region • Sustainability • Al-Qatt natural colours • Heritage preservation

1 Introduction

Al-Qatt is an interior design element that involves wall decoration, space planning, and creation of harmony (Aref, 2016). This art, in terms of details, varies across different parts of the Asir region, such as the administrative capital Abha city, Bilad Qahtan, and Rijal Almaa. The differences are discernible to the Al-Qatt specialists and the locals. Al-Qatt is not just a decoration of vivid colours, but rather an integrated system that reflects the personality and outlook of the Asiri woman who creates it. Of the past. Many researchers have studied Al-Qatt in different areas, but they are yet to determine the approximate date of its emergence and the meaning of its symbolism.

Al-Qatt has been an original source of inspiration for many visual art creations. Asir artists such as Ibrahim Al-Almaai, Afaf Da'jam, and Fatima Faye'a are well known for Al-Qatt. The modern applications of Al-Qatt such as murals, statues, and government office decor are diverse and innovative while simultaneously representing Asir culture.

Modern applications of Al-Qatt feature a greater number of colour shades, due to the availability of synthetic colours. The art form basically consists of geometric and organic shapes, and their presentation depends on each individual artist's taste.

Recently, it has been observed that Al-Qatt specialists are choosing colours derived from natural materials (hereafter called 'natural colours'), over or in addition to synthetic

A. A. Alghamdi (✉)
King Abdulaziz University, Jeddah, Saudi Arabia

A. A. Alawad
Department of Interior Design and Furniture, King Abdulaziz University, Jeddah, Saudi Arabia
e-mail: aalawad@kau.edu.sa

A. Versaci et al. (eds.), *Conservation of Architectural Heritage*, Advances in Science, Technology & Innovation,
https://doi.org/10.1007/978-3-030-74482-3_38

colours, despite the difficulty of extracting and preparing them. This study aims to identify the motives behind this choice and determine the extent to which it helps sustainable heritage preservation. These aims are based on the researchers' belief that any cultural heritage is an integrated system, and when its core elements are altered, it may lead to an imbalance that distorts the heritage.

2 Literature Review

The Asir region is located to the southwest of the Kingdom of Saudi Arabia, which is in the middle of the historic trade route between Hejaz and Yemen. The region's total area is estimated at 80,000 km^2. Asir was named after its hard, mountainous topography, which includes numerous valleys and trails. The climate is diverse across its various sub-regions. The weather is cold throughout the year but becomes mild in the summer. Asir consists of ten administrative departments: Abha, Khamis Mushayt, Mahayel, Ahad Rufaida, Al Namas, Bisha, Tathleeth, Sarat Abidah, Sabt Al Alaya, and Dhahran Al Janoub. The population of the Asir region is estimated at nearly two million, which constitutes 7.5% of the total population of the Kingdom of Saudi Arabia (History of Aseer, 2018).

Abha is the capital of Asir region and it is located to the southwest of Asir. The city is located at an estimated altitude of 2,200 m above sea level, with an average annual temperature of 16 oC. Therefore, Abha is considered one of the most important tourist destinations in the Kingdom of Saudi Arabia. It is made up of eight regions: AlQura, Manazur, Al-Qabil, Numan, Al-Rabu', Al-Nusb, Al-Khasha, and Al-Muftahah (Marzouk, 2017). It is home to many museums and historical houses such as the historic Abu Malha Palace and the Asir Regional Museum. Some individual initiatives such as Fatima Museum, Hisn Abha, and Tahlal Museum are also located in Abha.

2.1 Al-Qatt and Heritage

Marzouk (2017) mentioned that 300 years ago the people of Asir decorated their buildings inside and outside. Mauger stated that the division of work was determined by competence and gender (Mauger, 1996). After men finished building houses and installing doors and windows, the women began their role, which included all the remaining work, such as plastering, smoothing walls and floors, and decorating and colouring interior walls with Al-Qatt (Al-Hababi, 2012).

Marzouk mentioned that, for aesthetic considerations, the facades of the stone buildings used white quartz stones to create a contrast with the dark stones used in the rest of the construction. The white quartz stone was not only aesthetically pleasing but also kept the interiors cool by reflecting sunlight. They used motifs to decorate the reception areas, bedrooms, staircases, as well as the ceilings. They revived and restored the decorations every year to preserve it for as long as possible (Marzouk, 2017).

Al-Qatt consists of certain symbols, but their composition varies according to the individual style and preference of the artist. Al Hababi states that the person who draws Al-Qatt in the house is the housewife, or a group of women who work together as a team in a room under the supervision of a 'Qattata', who is generally an expert and well-known female painter (Al Hababi, 2012). The Qattata's supervision ensures consistency, as the team follows her instructions. Moghawi argues that the Qattata is not present only for supervision; instead, she takes the lead on the more 'sensitive and delicate' parts of Al-Qatt, while the other women do larger, less detailed parts (Moghawi, 2010). In addition, Mauger explains that the value of this cooperative work lies in sharing knowledge and creating opportunities of apprenticeships to prepare younger women for the future (Mauger, 1996).

Al Hababi also explained that in the case of individual work, Al-Qatt is a self-directed art where no person asks the woman what to do or tells her what she should draw (Al-Hababi, 2012). It is not preceded by planning, sketching, or creating blueprints. Al-Qatt patterns are chosen from geometric and organic forms, and each shape has its own name that may differ across regions. In Bilad Shahran, Al-Qatt is characterized by large shapes and symbols without black lines, while in Rijal Almaa it has smaller symbols and is outlined in black, which helps each detail stand out. In Bilad Qahtan and Sarat Abidah, the symbols are simpler in compositions inside the houses. Al-Qatt is usually recoloured on special occasions and celebrations (Al-Daajam, 2018) for a fresher and brighter look.

Al Hababi stated that women choose to make their motifs as vibrant as possible. This is one of the ways a woman expresses her personality and shows her competitiveness. This indicates the importance of traditional home decoration, which has now been blurred with modern paint (Al Hababi, 2012). Al-Qatt is known by different names within the Asir region. The Sarat Mountains name its Al-Qatt; in the Tihama it is called Al-Zayan or Al-Naqsh (Aref, 2016); and in the Bilad Qahtan it is called Qatta or Kitba. Aref mentioned that the rich flora and fauna in Tihama and Rijal Almaa' influenced the structural architectural patterns and Al-Qatt designs used in the regions. That is why houses in the two regions are adorned by unique wallcoverings consisting of organic and geometric motifs (Aref, 2016).

The inclusion of Al-Qatt in the list of intangible cultural heritage of the United Nations at UNESCO in 2017 has brought much attention to the art form. UNESCO states that cultural heritage does not end at monuments and collections of objects. It also includes traditions or living expressions inherited from our ancestors and passed on to our descendants, such as oral traditions, performing arts, social practices, rituals, festive events, knowledge, and practices concerning nature and the universe or the knowledge and skills to produce traditional crafts (UNESCO, 2019). Feilden and Jokilehto have divided heritage preservation into the material framework (i.e. the architectural forms) and the intangible moral framework (all values and considerations included in the architectural work). They argue that one of the principles of preserving heritage is to preserve its design, materials, and the ingenuity with which it was made. They mentioned the importance of attention to heritage and its complementarity in characteristics such as the quality of its designs, the materials from which they were made, the craftsmanship, and the environment and its relationship with it. (Feilden & Jokilehto, 1998).

2.2 Al-Qatt's Motifs

Al-Qatt is an abstract art that includes configurations of several primary geometric elements such as point and line, and basic shapes such as triangle, square, rectangle, and circle. Aref mentioned that Al-Qatt is formed of straight lines and rarely uses curved ones. The surface of the wall is divided into three parallel horizontal sections. The upper part is Hedya, the middle part is Al-Khetam, and the lower part is Nehaya. Coloured geometric and organic motifs lie between two horizontal lines (Aref, 2016). On the other hand, Al-Sabban and Al-Daajam divided those sections differently, into four types, according to the current local vocabulary used. They are as follows:

- Al-Khetam: a designation of square shapes filled with different shapes, such as Aryash, which resembles a feather or a wheat plant because it consists of straight lines from which smaller lines are branched at 45-degree angles from the right and left; Balsanah, which is made up of diamond shapes surrounded by dots; and Al-Mathalith wa Al-Makhamis, a set of three or five horizontal lines of different colours on the lower one-third portion of the wall, below Al-Qatt motifs. Many Asiri people use only Al-Mathalith wa Al-Makhamis to decorate their houses, especially around the stairs. Al-Khetam is generally about 30–40 cm in height, often surrounded by Hedya and topped by Al-Banah and Al-Amshat, where the large triangles of Rokon are found at the corners of the walls.
- Al-Hedya: combinations of organic and geometric shapes along the wall, with a width that is much smaller than that of Al-Khetam
- Al-Batra: an area that unites all kinds of configurations, yet much smaller in size and very intricately detailed, and usually appearing at the guest reception area
- Al-Takti Al-Omary: parallel vertical lines that represent a base from which the artist's imagination can lead to Al-Khetam or Al-Hedya (Al-Daajam, 2018; Al-Sabban, 2013).

2.3 Materials and Colour

2.3.1 Natural Colours in Al-Qatt

Al-Qatt is distinguished by several properties related to the materials and colours used. Previously, Asiri Al-Qatt relied on natural materials available in the surrounding environment, such as alfalfa, barley, wheat, vegetables, and fruits; and substances derived from animals such as camels and sheep (Al-Sabban, 2013; Al-Shahrani, 2003). As Mauger stated, these materials were free; the only cost was the time taken to extract the colour (Mauger, 1996).

Colours are extracted from natural materials and mixed with gum Arabic to give them shine and firmness (Aref, 2016). Women's knowledge of certain food preparation techniques has made natural colour extracting easier for them. Some of these processes include drying, roasting, marinating, brewing, cold-pressing, and filtering (Lightner & Rand, 2014). Colours extracted from stones were ground with a mill. A mill is an old tool that consists of two tablets of large stones between which grains are crushed.

White: It is one of the basic colours used in Al-Qatt. The plaster materials ('Al-Jass' or 'Al-Qass') were used to paint the interior walls and create a blank white canvas before starting Al-Qatt work. (Al-Sabban, 2013). White colour is extracted from limestone, which contains calcium carbonate. Limestone is soaked in water until it softens; then it is mixed with gum Arabic to create a paste that can be slathered on the walls (Al-Qahtani, 2019). Women used to climb mountains and enter caves in search of limestone. At times, they had to crawl to reach it and sometimes the limestone fell on them. (Al Hababi, 2012). Limestone is characterized by a chalky texture. It is available in the surrounding environment and is easy to use. Figure 1 shows white limestone before and after grinding.

Black: Black colour is often used to mark the borders of Al-Qatt motifs using a precise brush (Al-Shahrani, 2003). There are multiple natural resources from which black colour can be extracted. Some artists use tar, known as Al-Qatran, extracted from tree trunks, while others grind coal after mixing it with gum Arabic, and some collect the soot of a burning candle or kerosene flame in a bowl (Al-Qahtani, 2019). Al Hababi refers to a woman bringing some clover sap and then pressing it and adding the soot to it

Fig. 1 White colour. *Source* The researchers

to turn it black (Al Hababi, 2012). One of the characteristics of tar is that it preserves wood and protects it from erosion. It is also used to expel insects (Al-Qahtani, 2019).

Green: In Al-Qatt, the green colour is applied on the lower one-third portion of interior walls. As the researchers mentioned, the green colour is extracted from the clover's sap and other green plants (Al-Hababi, 2012), and it is 'embellished with the finger of the hand or using a branch by making decorative effects on a layer of clay 'Labin', so it combines sunken and prominent pits' (Al-Qahtani, 2019). Because the green colour occupies a sizeable area of the artwork, it is produced in larger quantities. Its dyeing requires great effort, and it often needs periodic maintenance and repainting every year to ensure its continuity. Figure 2 shows the application of green colour.

Red and Brown: Red colour is used in Al-Qatt to inlay the motifs and fill them, as it is often one of the colours used for the lines in Al-Mathalith wa Al-Makhamis. Al-Daajam mentioned in her book that since the Al-Mashgah stone is found in abundance in the Asir region, red colour is liberally used in Al-Qatt (Al-Daajam, 2019). The participants mentioned different sources that the colour red came from, which included poppy seeds, pomegranate bark, Al-Mashgah stones, and ground fried rice. In his book, Moghawi adds that all these materials are mixed to reduce the intensity of colour and increase its volume. Figure 3 shows a picture of the red colour from the Al-Mashgah stone. Brown was common colour in Al-Qatt; it was made by soaking onion skin in hot water. The soaking reduced the intensity of smell and facilitated the extraction (Moghawi, 2010). An alternative method was to mix the colour red with black to obtain brown.

Yellow and Orange: Like the colour red, yellow and orange colours in Al-Qatt were used to inlay the motifs and fill them, fourth participant mentioned that the sources of yellow colour are turmeric, pomegranate peel, lemon peel, or the bark of a local tree called Al-Tho'ab. She also mentioned that she prefers to create orange colour by mixing yellow and red. Al-Qahtani noted that in some cases, orange was imported from Yemen (Al-Qahtani, 2019). Figure 4 shows the natural sources of yellow colour and the product achieved after grinding.

Blue: In Al-Qatt, blue colour was used to fill the motifs, and it is often one of the lines of the colours in Al-Mathalith wa Al-Makhamis. There are many claims about the sources of blue colour used in the past. Al-Qahtani mentioned that

Fig. 2 Abu Malha Palace. *Source* The researchers

Fig. 3 Red colour. *Source* The researchers

blue dye was derived from the Indigofera Tinctoria plant known as True Indigo or 'Neela'. It was sold in local markets of Asir, and either soaked in water and used as is, or used after adding some glue (Al Hababi 2012). The use of True Indigo has long been known to give white clothes a bright, bluish white. The blue arsenic, a substance that is toxic and deadly for the human body if ingested, is also known to be one of the sources used to extract blue colour (George et al., 2014). Figure 5 shows a collection of powder colours produced by grinding various natural resources. The picture was taken by the researchers at the studio of the artist Fatima Fayea during a visit to 'Fatima' museum, where Al-Qatt artists and enthusiasts are trained to use natural colours.

Other powder colours were brought from Aden in southern Arabia. Various types of red and black clay were also used for colouring. All colours were used independently, without mixing them together. This imposed a limitation on colour choice. Artists' colour choice was dependent on the locally available dyes. The use of these natural colours continued until the 1970s when women began discovering the ease of using synthetic paint. Artists pointed out that replacing the natural colours with synthetic wall paints eliminated the long and difficult process of obtaining natural materials and preparing the colours before drawing Al-Qatt (Al Hababi, 2012).

Gum Arabic: It is used as a medium for converting ground powder colours into a liquid that can be painted on walls. One of its properties is that it is colourless. It does not affect the colour that it is mixed with, but makes it durable and adds shine to it. The gum Arabic chunks extracted from trees in the Asir region are dissolved in hot water and converted into a viscous liquid, as shown in Figure 6. 'The key functional characteristics of the gum are that it forms viscous solutions only at high concentrations and that it is able to stabilise oil-in-water emulsions' (Williams & Phillips, 2009).

2.3.2 Synthetic Paint in Al-Qatt

Paint as a mixture of insoluble and suspended materials in an oily liquid medium, or suspended in an aqueous emulsion. The liquid medium consists of a group of organic or inorganic substances, and by exposing it to air and through polymerization reactions, oxidation, or drying, it turns into a coherent film that adheres to the painted surface. The paint industry has seen many scientific advancements beginning from the industrial revolution in the eighteenth century to the present day (Lambourne and Strivens, 1999). Modern paint-making allows for a lot of control over the quality and features of the paint because the manufacturer can adjust factors such as density, transparency, gloss, drying-time, ease of application and polishing, and ease of correction. Also, a multitude of hues can be produced.

Most paints are classified into two types: water-based and oil-based paints. Water-based paints include gouache, watercolours, and acrylic. Oil-based paints contain oil as a binder. They dry slowly compared to water-based paints. Many Al-Qatt artists have said that they rely primarily on

Fig. 4 Yellow colour. *Source* The researchers

Fig. 5 Collection of natural colours. *Source* The researchers

Fig. 6 Gum Arabic. *Source* The researchers

synthetic paints, especially acrylic. When they work on large areas, they use paint from well-known companies to guarantee quality and economy.

2.3.3 The Physiological and Psychological Effect of Colours Used in Interior Spaces

Colours affect human behaviour and the effect varies according to age, gender, and culture. While looking at colours, there are involuntary reactions that affect the viewer's psychological state. Colours can create a pleasant environment in the home to provide a degree of comfort and happiness. The white colour evokes purity, cleanliness, radiance, while green calms the nerves and promotes relaxation and rejuvenation (Al-Sayed & Mahmoud, 2017). Therefore, the colours white and green occupy large areas on the interior walls decorated with Al-Qatt (white as the base canvas and green in the bottom quadrant). The rest of the colours such as red, blue, yellow, and black, are found in smaller proportions compared to white and green. Birren explained the meaning of colours, stating that red has been known to be a strong colour that carries more than one meaning to the viewers, depending on their state of mind. It may stimulate the viewer to feel energetic and happy, it may aggravate a bad mood. Yellow and orange are known to leave a happy, vibrant, and healthy impression. Blue evokes a sense of stability and serenity (Birren, 1950).

3 Methodology

This study followed a descriptive and analytical method to gain information about the materials used in the making of Al-Qatt, whether natural or synthetic. A field trip to Asir was undertaken, specifically to Abha city. Researchers visited a number of villages, museums, and homes belonging to the Abha governorate. Thirty-five persons with interest in Al-Qatt were interviewed. The chosen sample was based on criteria that included experience in using both, natural and synthetic colours in Al-Qatt, and the duration of Al-Qatt practice to be no less than five years. Five of those who met these criteria were selected, and interviews were conducted in December 2019 at the participants' homes or studios. The purpose of the interviews was to identify the reasons for continuing to use natural colours in addition to synthetic materials, despite the difficulties involved in producing natural colours.

Interview questions focussed on how to obtain the materials, how to prepare the natural colours, the effort and time required to prepare them, the reasons for using natural colours, the extent of preference, and differences in application. The following questions were asked:

- How do you obtain natural materials for use in Al-Qatt? What do you do to get the right shade of colour?
- How much effort and time does it take you to prepare the colours before painting?
- Why do you continue to use natural colours even though an alternative is available?
- What is your preference—natural or synthetic colours?
- Do the shades of colours differ between natural and synthetic colours?

4 Results and Discussion

After interviewing the five professional artists, it became clear that they are proud practitioners of Al-Qatt. They are keen to preserve the artform and showcase it in an authentic manner, without corrupting it with exotic shapes or colours. The interviews were conducted to answer the two main research questions: what is the motive behind the deliberate use of using natural colours, and to what extent does it support sustainable heritage preservation. As per the findings, there are seven factors motivating the use of natural colours.

4.1 Identity

Al-Qatt professionals desire to preserve Al-Qatt's identity by preserving its vocabulary and characteristics. They want to protect its authenticity and preserve the knowledge about use of basic raw materials. The artists stated the importance of protecting heritage handicrafts from changes brought by technological progress and modern industries. One of the interviewees (the second participant) revealed that she used natural materials to paint a long mural in her house owing to her keen desire to preserve Al-Qatt's original identity. The mural ran over four metres, along the length of a corridor, as shown in Figure 7.

The fourth participant was another expert who was also keen on using natural colours along with synthetic colours. She trained other students in using natural colours and searching for raw materials from their primary source from the surrounding nature. The purpose of the training is to teach the young generation traditional techniques in extracting colours for Al-Qatt; help them preserve Al-Qatt's basic identity; and preserve the old ways of mixing different materials to produce distinct colours such as the orange and brown shown in Fig. 8.

One of the biggest challenges in using natural materials is the difficulty in obtaining them. The second and fourth participants affirmed that they are willing to endure difficulties and continue collecting rocks and plants by themselves from the mountains because doing so is necessary to preserve the original identity of Al-Qatt.

The third participant explained that 'the beauty of the Al-Qatt is distinguished by its spontaneity and randomness, and it should not be based on exact measurements'. Figures 7 and 8 serve as evidence to show that there are no exact patterns or rules that should be followed, yet both show the same unique identity of Al-Qatt.

4.2 Global Recognition

The main reason why some Al-Qatt professionals returned to using natural colours was to reach a broader audience. Al-Qatt was added to the UNESCO list of intangible cultural heritage. As explained, in the Saudi news channel showed a collaboration of more than twelve Al-Qatt artists working together to paint a mural along an eight-metre-long wall. The precision, harmony, and colour choice seen in the mural left the viewers impressed. This exposure attracted the world's attention to this valuable heritage and sparked an awareness about its characteristics and vocabulary.

In addition, the majority of participants mentioned that they were inspired by the agreement between the Kingdom of Saudi Arabia and the British TURQUOISE MOUNTAIN, which aims to preserve and renew areas and societies with a rich cultural heritage, revive traditional crafts, and create job opportunities. The agreement shows that governmental and organizational support to heritage protection is an important component of heritage sustainability and preservation.

4.3 Distinction and Exclusivity

The preparation of natural colours requires much time and effort. That is why many Al-Qatt artists use synthetic colours for the ease of access and ease of use. As a result, the distinction and uniqueness of the artworks of those who use natural colours have increased. Figure 9 shows materials personally sourced and prepared for use by the second participant. Al-Qatt artworks made with natural colours remain the focus of attention due to the complicated process of sourcing and colour preparation that goes into them.

Fig. 7 Corridor of one of the participants' homes. *Source* The researchers

Fig. 8 Mixing different materials. *Source* The researchers

4.4 Aesthetics

Colours in interior spaces, affected by lighting and surface material, play a major role in influencing feelings and reflecting a certain lifestyle. All participants mentioned that natural colours used in Al-Qatt have a 'positive effect' on the viewer. There is a difference between natural and synthetic colours in terms of the effect on viewer. Natural colours create a more comfortable and relaxing environment because they come from the natural surrounding environment and present a harmonious palette of colours that are connected and subtle. The third participant mentioned that synthetic colours 'do not reflect the true colour tone found in old Al-Qatt' since natural colours adapt to the surfaces they are applied to and blend with them. Figures 10 and 12 show traditional Al-Qatt made in the past with natural colours. Figure 11 shows modern Al-Qatt made with synthetic colours. The pictures show a clear difference between the shades of colours.

Fig. 9 Extracting colours from rocks. *Source* The researchers

4.5 Psychological Motives

As established earlier, colours have a psychological impact on the viewer. Natural colours are known to help people feel relaxed no matter how long they stay in a room. All participants agreed that many parts of Al-Qatt are very detailed and intricate and can irritate the eye if not for the calm and soothing shades of natural colours. The second participant stated that 'visitors appreciate the natural colours paintings more than the synthetic ones' This appreciation has a positive impact on the artists' motivation to use natural colours.

4.6 Economic Motives

Al-Qatt is one of the distinguishing features of urban architectural heritage of Asir. One of the principles of sustainable development is the preservation of the basic features that make up such architecture The second participant stated that artworks made with natural colours fetch higher prices and the demand for such artworks is increasing. Buyers who are interested in Asir heritage are interested in the acquisition of paintings made with natural colours. This economic opportunity was one of the reasons the artists began using natural colours.

The other factor is related to value. Natural colours are low in cost compared to high-quality synthetic colours. Although they require an effort to prepare them, the results are better. The use of natural materials requires little treatment or transportation, the environmental and economic costs are low, and they are considered recyclable. This helps the artists and professionals have a sustainable income to carry on the preservation of this valuable heritage.

Fig. 10 Original piece from artist Ibrahim Al-Almaai. *Source* The researchers

Fig. 11 Abu Malha Palace. *Source* The researchers

Fig. 12 From Thierry Mauger book p (8)

4.7 Environmental and Health Concerns

The synthetic paint industry has a proven harmful impact on the environment. The factories release volatile organic compounds (VOCs), titanium dioxide (TiO2), and other damaging components, which leads to the release of Ozone (Porwal, 2015). Other health concerns were expressed by the fourth participant when she mentioned that 'some artificial colours have odours that irritate' while working and have a negative effect on health. Porwal explains that synthetic paint can have a serious impact on the human body. It can trigger headaches, allergies, asthmatic reactions, skin and eye irritation, and threaten the health of other bodily organs (Porwal, 2015). The use of natural materials to extract colours does not release pollutants and allergens into the air. Therefore, it protects the environment and safeguards health.

5 Conclusion

Heritage reveals the history of a community and the stages of its development. Heritage can refer to things such as activities, customs, and traditions. Architectural heritage is a valuable asset. It is protected through sustainable preservation methods which seek to preserve the history and knowledge of historic civilizations or nations. Preserving heritage is of economic importance for countries and it provides a connection for people with their past. The deliberate use of natural colours has a positive impact on the sustainability of Al-Qatt and also on the lives of the practitioners. The use of natural colours has many benefits, which include preserving Al-Qatt's original identity, giving it wider recognition, creating authentic traditional art, challenging the aesthetic, creating a positive psychological impact, boosting the economy, protecting the environment, and preventing health complications linked with synthetic paint. It is clear from the study that the use of natural colours plays an important role in sustainable heritage preservation of Asir, despite the practical difficulties it entails.

In the wake of Al-Qatt professionals participating in initiatives to preserve old buildings and open them for public, the researchers recommend using natural colours in the restoration instead of industrial ones to ensure visual harmony and reflect originality, as these restored artworks would become the source that others refer to. The researchers also recommend that paint manufacturers offer natural colours and make them accessible for use in historic building renovation, and offer customized colour palettes for each of the local traditional arts.

Acknowledgements We would like to express our deepest appreciation to everyone that helped and contributed to the expansion of our knowledge of Al-Qatt. A special thanks to the artist Ibrahim Al-Almaai, the historian and writer Ali Moghawi, Mr. Ibrahim Bashasha, Mrs. Haleema Abdullah, the artist Fatima Fayea, and the artist Afaf Al-Daajam.

References

Al-Daajam, A. (2018). *Motifs of Al-qatt Al-asiri.*

Al-Hababi, H. (2012). The art of women in 'Asir (Saudi Arabia). *DOI.* https://doi.org/10.1553/wpsa25s1,10.1553/wpsa25s1

Al-Qahtani, Y. (2019). Aesthetics of the Asiri Al-Qatt and its role in enriching contemporary artistic production. *Journal of Humanities and Social Sciences, 4*(3), 77–95.

Al-Sabban, R. (2013). Designing traditional dwellings in the Asir region, south of the Kingdom of Saudi Arabia. *Journal of Educational Sciences, 116*, 1–39.

Al-Sayed, A., & Mahmoud, S. (2017). A study of the psychological dimension of color in the design of interior spaces. *Architecture, Arts and Humanistic Science Magazine, 29*(5461), 1–14.

Al-Shahrani, A. (2003). Artistic and aesthetic values in the popular decorations in Asir region. *King Khalid University—Teachers College—Educational Research Center, A2*, 95–106.

Aref, A. (2016). An analytical study on wall painting adaptation to physical and architectural environment of Asir province at the end of ottoman empire. *International Design Journal, 6*(4), 113–122.

Birren, F. (1950). Color psychology and color therapy: A factual study of the influence of color on human life. *Academic Medicine, 25*(4), 303–303.

Feilden, B. M., & Jokilehto, J. (1998). *Management guidelines for world cultural heritage sites*. ICCROM.

George, C. M., Sima, L., Arias, M., Mihalic, J., Cabrera, L. Z., Danz, D., ... & Gilman, R. H. (2014). Arsenic exposure in drinking water: an unrecognized health threat in Peru. *Bulletin of the World Health Organization, 92*, 565–572.

History of Aseer. (2018). *Aseer municipality*. https://ars.gov.sa/Ar/AsirRegion/AsirHistory/Pages/default.aspx.

Lambourne, R., & Strivens, T. A. (Eds.). (1999). *Paint and surface coatings: Theory and practice*. Elsevier.

Lightner, M., & Rand, S. (2014). The enhancement of natural colors to provoke seasonality. *International Journal of Gastronomy and Food Science, 2*(1), 55–59.

Marzouk, A. (2017). *Mural painting art in Abha South Western of KSA*. Helwan University.

Mauger, T. (1996). Impressions of Arabia: Architecture and frescoes of the Asir region. Flammarion-Pere Castor.

Moghawi, A. (2010). *Rojal, memory of Arab village*. Jeddah: A.I. Moghawi.

Porwal, T. (2015). Paint pollution harmful effects on environment. *International Journal of Environmental Problems, 3*.

Williams, P. A., & Phillips, G. O. (2009). Gum Arabic. In *Handbook of hydrocolloids* (pp. 252–273). Woodhead Publishing.